Statistical Methods in the Atmospheric Sciences

Second Edition

This is Volume 91 in the
INTERNATIONAL GEOPHYSICS SERIES

A series of monographs and textbooks
Edited by RENATA DMOWSKA, DENNIS HARTMANN, and H. THOMAS ROSSBY

A complete list of books in this series appears at the end of this volume.

STATISTICAL METHODS IN THE ATMOSPHERIC SCIENCES

Second Edition

D.S. Wilks

Department of Earth and Atmospheric Sciences
Cornell University

AMSTERDAM • BOSTON • HEIDELBERG • LONDON
NEW YORK • OXFORD • PARIS • SAN DIEGO
SAN FRANCISCO • SINGAPORE • SYDNEY • TOKYO

Academic Press is an imprint of Elsevier

Acquisitions Editor	Jennifer Helé
Publishing Services Manager	Simon Crump
Marketing Manager	Linda Beattie
Marketing Coordinator	Francine Ribeau
Cover Design	Dutton and Sherman Design
Composition	Integra Software Services
Cover Printer	Phoenix Color
Interior Printer	Maple Vail Book Manufacturing Group

Academic Press is an imprint of Elsevier
30 Corporate Drive, Suite 400, Burlington, MA 01803, USA
525 B Street, Suite 1900, San Diego, California 92101–4495, USA
84 Theobald's Road, London WC1X 8RR, UK

This book is printed on acid-free paper. ∞

Library of Congress Cataloging-in-Publication Data
Application submitted

British Library Cataloguing in Publication Data
A catalogue record for this book is available from the British Library

ISBN 13: 978-0-12-751966-1
ISBN 10: 0-12-751966-1

For information on all Elsevier Academic Press Publications
visit our Web site at www.books.elsevier.com

Printed in the United States of America
08   09   10   9   8   7   6   5   4   3   2

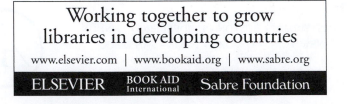

Contents

Preface to the First Edition

This text is intended as an introduction to the application of statistical methods to atmospheric data. The structure of the book is based on a course that I teach at Cornell University. The course primarily serves upper-division undergraduates and beginning graduate students, and the level of the presentation here is targeted to that audience. It is an introduction in the sense that many topics relevant to the use of statistical methods with atmospheric data are presented, but nearly all of them could have been treated at greater length and in more detail. The text will provide a working knowledge of some basic statistical tools sufficient to make accessible the more complete and advanced treatments available elsewhere.

It has been assumed that the reader has completed a first course in statistics, but basic statistical concepts are reviewed before being used. The book might be regarded as a second course in statistics for those interested in atmospheric or other geophysical data. For the most part, a mathematical background beyond first-year calculus is not required. A background in atmospheric science is also not necessary, but it will help you appreciate the flavor of the presentation. Many of the approaches and methods are applicable to other geophysical disciplines as well.

In addition to serving as a textbook, I hope this will be a useful reference both for researchers and for more operationally oriented practitioners. Much has changed in this field since the 1958 publication of the classic *Some Applications of Statistics to Meteorology*, by Hans A. Panofsky and Glenn W. Brier, and no really suitable replacement has since appeared. For this audience, my explanations of statistical tools that are commonly used in atmospheric research will increase the accessibility of the literature, and will improve readers' understanding of what their data sets mean.

Finally, I acknowledge the help I received from Rick Katz, Allan Murphy, Art DeGaetano, Richard Cember, Martin Ehrendorfer, Tom Hamill, Matt Briggs, and Pao-Shin Chu. Their thoughtful comments on earlier drafts have added substantially to the clarity and completeness of the presentation.

Preface to the Second Edition

I have been very gratified by the positive responses to the first edition of this book since it appeared about 10 years ago. Although its original conception was primarily as a textbook, it has come to be used more widely as a reference than I had initially anticipated. The entire book has been updated for this second edition, but much of the new material is oriented toward its use as a reference work. Most prominently, the single chapter on multivariate statistics in the first edition has been expanded to the final six chapters of the current edition. It is still very suitable as a textbook also, but course instructors may wish to be more selective about which sections to assign. In my own teaching, I use most of Chapters 1 through 7 as the basis for an undergraduate course on the statistics of weather and climate data; Chapters 9 through 14 are taught in a graduate-level multivariate statistics course.

I have not included large digital data sets for use with particular statistical or other mathematical software, and for the most part I have avoided references to specific URLs (Web addresses). Even though larger data sets would allow examination of more realistic examples, especially for the multivariate statistical methods, inevitable software changes would eventually render these obsolete to a degree. Similarly, Web sites can be ephemeral, although a wealth of additional information complementing the material in this book can be found on the Web through simple searches. In addition, working small examples by hand, even if they are artificial, carries the advantage of requiring that the mechanics of a procedure must be learned firsthand, so that subsequent analysis of a real data set using software is not a black-box exercise.

Many, many people have contributed to the revisions in this edition, by generously pointing out errors and suggesting additional topics for inclusion. I would like to thank particularly Matt Briggs, Tom Hamill, Ian Jolliffe, Rick Katz, Bob Livezey, and Jery Stedinger, for providing detailed comments on the first edition and for reviewing earlier drafts of new material for the second edition. This book has been materially improved by all these contributions.

PART · I

Preliminaries

CHAPTER • 1

Introduction

1.1 What Is Statistics?

This text is concerned with the use of statistical methods in the atmospheric sciences, specifically in the various specialties within meteorology and climatology. Students (and others) often resist statistics, and the subject is perceived by many to be the epitome of dullness. Before the advent of cheap and widely available computers, this negative view had some basis, at least with respect to applications of statistics involving the analysis of data. Performing hand calculations, even with the aid of a scientific pocket calculator, was indeed tedious, mind-numbing, and time-consuming. The capacity of ordinary personal computers on today's desktops is well above the fastest mainframe computers of 40 years ago, but some people seem not to have noticed that the age of computational drudgery in statistics has long passed. In fact, some important and powerful statistical techniques were not even practical before the abundant availability of fast computing. Even when liberated from hand calculations, statistics is sometimes seen as dull by people who do not appreciate its relevance to scientific problems. Hopefully, this text will help provide that appreciation, at least with respect to the atmospheric sciences.

Fundamentally, statistics is concerned with uncertainty. Evaluating and quantifying uncertainty, as well as making inferences and forecasts in the face of uncertainty, are all parts of statistics. It should not be surprising, then, that statistics has many roles to play in the atmospheric sciences, since it is the uncertainty in atmospheric behavior that makes the atmosphere interesting. For example, many people are fascinated by weather forecasting, which remains interesting precisely because of the uncertainty that is intrinsic to the problem. If it were possible to make perfect forecasts even one day into the future (i.e., if there were no uncertainty involved), the practice of meteorology would be very dull, and similar in many ways to the calculation of tide tables.

1.2 Descriptive and Inferential Statistics

It is convenient, although somewhat arbitrary, to divide statistics into two broad areas: descriptive statistics and inferential statistics. Both are relevant to the atmospheric sciences.

Descriptive statistics relates to the organization and summarization of data. The atmospheric sciences are awash with data. Worldwide, operational surface and upper-air observations are routinely taken at thousands of locations in support of weather fore-casting activities. These are supplemented with aircraft, radar, profiler, and satellite data. Observations of the atmosphere specifically for research purposes are less widespread, but often involve very dense sampling in time and space. In addition, models of the atmosphere consisting of numerical integration of the equations describing atmospheric dynamics produce yet more numerical output for both operational and research purposes.

As a consequence of these activities, we are often confronted with extremely large batches of numbers that, we hope, contain information about natural phenomena of interest. It can be a nontrivial task just to make some preliminary sense of such data sets. It is typically necessary to organize the raw data, and to choose and implement appropriate summary representations. When the individual data values are too numerous to be grasped individually, a summary that nevertheless portrays important aspects of their variations—a statistical model—can be invaluable in understanding the data. It is worth emphasizing that it is not the purpose of descriptive data analyses to play with numbers. Rather, these analyses are undertaken because it is known, suspected, or hoped that the data contain information about a natural phenomenon of interest, which can be exposed or better understood through the statistical analysis.

Inferential statistics is traditionally understood as consisting of methods and procedures used to draw conclusions regarding underlying processes that generate the data. Thiébaux and Pedder (1987) express this point somewhat poetically when they state that statistics is "the art of persuading the world to yield information about itself." There is a kernel of truth here: Our physical understanding of atmospheric phenomena comes in part through statistical manipulation and analysis of data. In the context of the atmospheric sciences it is probably sensible to interpret inferential statistics a bit more broadly as well, and include statistical weather forecasting. By now this important field has a long tradition, and is an integral part of operational weather forecasting at meteorological centers throughout the world.

1.3 Uncertainty about the Atmosphere

Underlying both descriptive and inferential statistics is the notion of uncertainty. If atmospheric processes were constant, or strictly periodic, describing them mathematically would be easy. Weather forecasting would also be easy, and meteorology would be boring. Of course, the atmosphere exhibits variations and fluctuations that are irregular. This uncertainty is the driving force behind the collection and analysis of the large data sets referred to in the previous section. It also implies that weather forecasts are inescapably uncertain. The weather forecaster predicting a particular temperature on the following day is not at all surprised (and perhaps is even pleased) if the subsequently observed temperature is different by a degree or two. In order to deal quantitatively with uncertainty it is necessary to employ the tools of probability, which is the mathematical language of uncertainty.

Before reviewing the basics of probability, it is worthwhile to examine why there is uncertainty about the atmosphere. After all, we have large, sophisticated computer models that represent the physics of the atmosphere, and such models are used routinely for forecasting its future evolution. In their usual forms these models are deterministic: they do not represent uncertainty. Once supplied with a particular initial atmospheric state (winds, temperatures, humidities, etc., comprehensively through the depth of the atmosphere and around the planet) and boundary forcings (notably solar radiation, and sea-surface and land conditions) each will produce a single particular result. Rerunning the model with the same inputs will not change that result.

In principle these atmospheric models could provide forecasts with no uncertainty, but do not, for two reasons. First, even though the models can be very impressive and give quite good approximations to atmospheric behavior, they are not complete and true representations of the governing physics. An important and essentially unavoidable cause of this problem is that some relevant physical processes operate on scales too small to be represented explicitly by these models, and their effects on the larger scales must be approximated in some way using only the large-scale information.

Even if all the relevant physics could somehow be included in atmospheric models, however, we still could not escape the uncertainty because of what has come to be known as *dynamical chaos*. This phenomenon was discovered by an atmospheric scientist (Lorenz, 1963), who also has provided a very readable introduction to the subject (Lorenz, 1993). Simply and roughly put, the time evolution of a nonlinear, deterministic dynamical system (e.g., the equations of atmospheric motion, or the atmosphere itself) depends very sensitively on the initial conditions of the system. If two realizations of such a system are started from two only very slightly different initial conditions, the two solutions will eventually diverge markedly. For the case of atmospheric simulation, imagine that one of these systems is the real atmosphere, and the other is a perfect mathematical model of the physics governing the atmosphere. Since the atmosphere is always incompletely observed, it will never be possible to start the mathematical model in exactly the same state as the real system. So even if the model is perfect, it will still be impossible to calculate what the atmosphere will do indefinitely far into the future. Therefore, deterministic forecasts of future atmospheric behavior will always be uncertain, and probabilistic methods will always be needed to describe adequately that behavior.

Whether or not the atmosphere is fundamentally a random system, for many practical purposes it might as well be. The realization that the atmosphere exhibits chaotic dynamics has ended the dream of perfect (uncertainty-free) weather forecasts that formed the philosophical basis for much of twentieth-century meteorology (an account of this history and scientific culture is provided by Friedman, 1989). "Just as relativity eliminated the Newtonian illusion of absolute space and time, and as quantum theory eliminated the Newtonian and Einsteinian dream of a controllable measurement process, chaos eliminates the Laplacian fantasy of long-term deterministic predictability" (Zeng *et al.*, 1993). Jointly, chaotic dynamics and the unavoidable errors in mathematical representations of the atmosphere imply that "all meteorological prediction problems, from weather forecasting to climate-change projection, are essentially probabilistic" (Palmer, 2001).

Finally, it is worth noting that randomness is not a state of "unpredictability," or "no information," as is sometimes thought. Rather, random means "not precisely predictable or determinable." For example, the amount of precipitation that will occur tomorrow where you live is a random quantity, not known to you today. However, a simple statistical analysis of climatological precipitation records at your location would yield relative frequencies of precipitation amounts that would provide substantially more information about tomorrow's precipitation at your location than I have as I sit writing this sentence. A still less uncertain idea of tomorrow's rain might be available to you in the form of a weather forecast. Reducing uncertainty about random meteorological events is the purpose of weather forecasts. Furthermore, statistical methods allow estimation of the precision of predictions, which can itself be valuable information.

CHAPTER • 2

Review of Probability

2.1 Background

The material in this chapter is a brief review of the basic elements of probability. More complete treatments of the basics of probability can be found in any good introductory statistics text, such as Dixon and Massey's (1983) *Introduction to Statistical Analysis* , or Winkler's (1972) *Introduction to Bayesian Inference and Decision* , among many others.

Our uncertainty about the atmosphere, or about any other system for that matter, is of different degrees in different instances. For example, you cannot be completely certain whether rain will occur or not at your home tomorrow, or whether the average temperature next month will be greater or less than the average temperature last month. But you may be more sure about one or the other of these questions.

It is not sufficient, or even particularly informative, to say that an event is uncertain. Rather, we are faced with the problem of expressing or characterizing degrees of uncertainty. A possible approach is to use qualitative descriptors such as likely, unlikely, possible, or chance of. Conveying uncertainty through such phrases, however, is ambiguous and open to varying interpretations (Beyth-Maron, 1982; Murphy and Brown, 1983). For example, it is not clear which of "rain likely" or "rain probable" indicates less uncertainty about the prospects for rain.

It is generally preferable to express uncertainty quantitatively, and this is done using numbers called *probabilities* . In a limited sense, probability is no more than an abstract mathematical system that can be developed logically from three premises called the Axioms of Probability. This system would be uninteresting to many people, including perhaps yourself, except that the resulting abstract concepts are relevant to real-world systems involving uncertainty. Before presenting the axioms of probability and a few of their more important implications, it is necessary to define some terminology.

2.2 The Elements of Probability

2.2.1 Events

An *event* is a set, or class, or group of possible uncertain outcomes. Events can be of two kinds: A *compound event* can be decomposed into two or more (sub) events, whereas an *elementary event* cannot. As a simple example, think about rolling an ordinary six-sided die. The event "an even number of spots comes up" is a compound event, since it will occur if either two, four, or six spots appear. The event "six spots come up" is an elementary event.

In simple situations like rolling dice it is usually obvious which events are simple and which are compound. But more generally, just what is defined to be elementary or compound often depends on the problem at hand and the purposes for which an analysis is being conducted. For example, the event "precipitation occurs tomorrow" could be an elementary event to be distinguished from the elementary event "precipitation does not occur tomorrow." But if it is important to distinguish further between forms of precipitation, "precipitation occurs" would be regarded as a compound event, possibly composed of the three elementary events "liquid precipitation," "frozen precipitation," and "both liquid and frozen precipitation." If we were interested further in how much precipitation will occur, these three events would themselves be regarded as compound, each composed of at least two elementary events. In this case, for example, the compound event "frozen precipitation" would occur if either of the elementary events "frozen precipitation containing at least 0.01-in. water equivalent" or "frozen precipitation containing less than 0.01-in. water equivalent" were to occur.

2.2.2 The Sample Space

The *sample space* or *event space* is the set of all possible elementary events. Thus the sample space represents the universe of all possible outcomes or events. Equivalently, it is the largest possible compound event.

The relationships among events in a sample space can be represented geometrically, using what is called a Venn Diagram. Often the sample space is drawn as a rectangle and the events within it are drawn as circles, as in Figure 2.1a. Here the sample space is the rectangle labelled **S**, which might contain the set of possible precipitation outcomes for tomorrow. Four elementary events are depicted within the boundaries of the three circles. The "No precipitation" circle is drawn not overlapping the others because neither liquid nor frozen precipitation can occur if no precipitation occurs (i.e., in the absence of precipitation). The hatched area common to both "Liquid precipitation" and "Frozen precipitation" represents the event "both liquid and frozen precipitation." That part of **S**

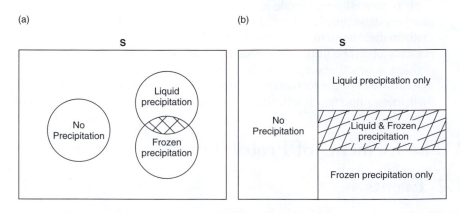

FIGURE 2.1   Venn diagrams representing the relationships of selected precipitation events. The hatched region represents the event "both liquid and frozen precipitation." (a) Events portrayed as circles in the sample space. (b) The same events portrayed as space-filling rectangles.

in Figure 2.1a not surrounded by circles is interpreted as representing the null event, which cannot occur.

It is not necessary to draw or think of Venn diagrams using circles to represent events. Figure 2.1b is an equivalent Venn diagram drawn using rectangles filling the entire sample space **S**. Drawn in this way, it is clear that **S** is composed of exactly four elementary events that represent the full range of outcomes that may occur. Such a collection of all possible elementary (according to whatever working definition is current) events is called mutually exclusive and collectively exhaustive (MECE). Mutually exclusive means that no more than one of the events can occur. Collectively exhaustive means that at least one of the events will occur. A set of MECE events completely fills a sample space.

Note that Figure 2.1b could be modified to distinguish among precipitation amounts by adding a vertical line somewhere in the right-hand side of the rectangle. If the new rectangles on one side of this line were to represent precipitation of 0.01 in. or more, the rectangles on the other side would represent precipitation less than 0.01 in. The modified Venn diagram would then depict seven MECE events.

2.2.3 The Axioms of Probability

Having carefully defined the sample space and its constituent events, the next step is to associate probabilities with each of the events. The rules for doing so all flow logically from the three Axioms of Probability. Formal mathematical definitions of the axioms exist, but they can be stated somewhat loosely as:

1) The probability of any event is nonnegative.
2) The probability of the compound event **S** is 1.
3) The probability that one or the other of two mutually exclusive events occurs is the sum of their two individual probabilities.

2.3 The Meaning of Probability

The axioms are the essential logical basis for the mathematics of probability. That is, the mathematical properties of probability can all be deduced from the axioms. A number of these properties are listed later in this chapter.

However, the axioms are not very informative about what probability actually means. There are two dominant views of the meaning of probability—the Frequency view and the Bayesian view—and other interpretations exist as well (Gillies, 2000). Perhaps surprisingly, there has been no small controversy in the world of statistics as to which is correct. Passions have actually run so high on this issue that adherents to one interpretation or the other have been known to launch personal (verbal) attacks on those supporting a different view!

It is worth emphasizing that the mathematics are the same in any case, because both Frequentist and Bayesian probability follow logically from the same axioms. The differences are entirely in interpretation. Both of these dominant interpretations of probability have been accepted and useful in the atmospheric sciences, in much the same way that the particle/wave duality of the nature of electromagnetic radiation is accepted and useful in the field of physics.

2.3.1 Frequency Interpretation

The frequency interpretation is the mainstream view of probability. Its development in the eighteenth century was motivated by the desire to understand games of chance, and to optimize the associated betting. In this view, the probability of an event is exactly its long-run relative frequency. This definition is formalized in the Law of Large Numbers, which states that the ratio of the number of occurrences of event $\{E\}$ to the number of opportunities for $\{E\}$ to have occurred converges to the probability of $\{E\}$, denoted $\text{Pr}\{E\}$, as the number of opportunities increases. This idea can be written formally as

$$\text{Pr}\left\{\left|\frac{a}{n} - \text{Pr}\{E\}\right| \geq \varepsilon\right\} \to 0 \quad \text{as } n \to \infty, \tag{2.1}$$

where a is the number of occurrences, n is the number of opportunities (thus a/n is the relative frequency), and ε is an arbitrarily small number.

The frequency interpretation is intuitively reasonable and empirically sound. It is useful in such applications as estimating climatological probabilities by computing historical relative frequencies. For example, in the last 50 years there have been $31 \times 50 = 1550$ August days. If rain has occurred at a location of interest on 487 of those days, a natural estimate for the climatological probability of precipitation at that location on an August day would be $487/1550 = 0.314$.

2.3.2 Bayesian (Subjective) Interpretation

Strictly speaking, employing the frequency view of probability requires a long series of identical trials. For estimating climatological probabilities from historical weather data this requirement presents essentially no problem. However, thinking about probabilities for events like {the football team at your college or alma mater will win at least half of their games next season} presents some difficulty in the relative frequency framework. Although abstractly we can imagine a hypothetical series of football seasons identical to the upcoming one, this series of fictitious football seasons is of no help in actually estimating a probability for the event.

The subjective interpretation is that probability represents the degree of belief, or quantified judgement, of a particular individual about the occurrence of an uncertain event. For example, there is now a long history of weather forecasters routinely (and very skillfully) assessing probabilities for events like precipitation occurrence on days in the near future. If your college or alma mater is a large enough school that professional gamblers take an interest in the outcomes of its football games, probabilities regarding those outcomes are also regularly assessed—subjectively.

Two individuals can have different subjective probabilities for an event without either necessarily being wrong, and often such differences in judgement are attributable to differences in information and/or experience. However, that different individuals may have different subjective probabilities for the same event does not mean that an individual is free to choose any numbers and call them probabilities. The quantified judgement must be a consistent judgement in order to be a legitimate subjective probability. This consistency means, among other things, that subjective probabilities must be consistent with the axioms of probability, and thus with the mathematical properties of probability implied by the axioms. The monograph by Epstein (1985) provides a good introduction to Bayesian methods in the context of atmospheric problems.

2.4 Some Properties of Probability

One reason Venn diagrams can be so useful is that they allow probabilities to be visualized geometrically as areas. Familiarity with geometric relationships in the physical world can then be used to better grasp the more ethereal world of probability. Imagine that the area of the rectangle in Figure 2.1b is 1, according to the second axiom. The first axiom says that no areas can be negative. The third axiom says that the total area of nonoverlapping parts is the sum of the areas of those parts.

A number of mathematical properties of probability that follow logically from the axioms are listed in this section. The geometric analog for probability provided by a Venn diagram can be used to help visualize them.

2.4.1 Domain, Subsets, Complements, and Unions

Together, the first and second axioms imply that the probability of any event will be between zero and one, inclusive. This limitation on the domain of probability can be expressed mathematically as

$$0 \leq \Pr\{E\} \leq 1. \tag{2.2}$$

If $\Pr\{E\} = 0$ the event will not occur. If $\Pr\{E\} = 1$ the event is absolutely sure to occur.

If event $\{E_2\}$ necessarily occurs whenever event $\{E_1\}$ occurs, $\{E_1\}$ is said to be a subset of $\{E_2\}$. For example, $\{E_1\}$ and $\{E_2\}$ might denote occurrence of frozen precipitation, and occurrence of precipitation of any form, respectively. In this case the third axiom implies

$$\Pr\{E_1\} \leq \Pr\{E_2\}. \tag{2.3}$$

The complement of event $\{E\}$ is the (generally compound) event that $\{E\}$ does not occur. In Figure 2.1b, for example, the complement of the event "liquid and frozen precipitation" is the compound event "either no precipitation, or liquid precipitation only, or frozen precipitation only." Together the second and third axioms imply

$$\Pr\{E\}^C = 1 - \Pr\{E\}, \tag{2.4}$$

where $\{E\}^C$ denotes the complement of $\{E\}$. (Many authors use an overbar as an alternative notation to represent complements. This use of the overbar is very different from its most common statistical meaning, which is to denote an arithmetic average.)

The union of two events is the compound event that one or the other, or both, of the events occur. In set notation, unions are denoted by the symbol $\cup$. As a consequence of the third axiom, probabilities for unions can be computed using

$$\Pr\{E_1 \cup E_2\} = \Pr\{E_1 \text{ or } E_2 \text{ or both}\}$$
$$= \Pr\{E_1\} + \Pr\{E_2\} - \Pr\{E_1 \cap E_2\}. \tag{2.5}$$

The symbol ∩ is called the intersection operator, and

$$\Pr\{E_1 \cap E_2\} = \Pr\{E_1, E_2\} = \Pr\{E_1 \text{ and } E_2\} \tag{2.6}$$

is the event that both$\{E_1\}$ and $\{E_2\}$ occur. The notation$\{E_1, E_2\}$ is equivalent to $\{E_1 \cap E_2\}$. Another name for $\Pr\{E_1, E_2\}$ is the joint probability of $\{E_1\}$ and $\{E_2\}$. Equation 2.5 is sometimes called the Additive Law of Probability. It holds whether or not$\{E_1\}$ and $\{E_2\}$ are mutually exclusive. However, if the two events are mutually exclusive the probability of their intersection is zero, since mutually exclusive events cannot both occur.

The probability for the joint event, $\Pr\{E_1, E_2\}$ is subtracted in Equation 2.5 to compensate for its having been counted twice when the probabilities for events$\{E_1\}$ and $\{E_2\}$ are added. This can be seen most easily by thinking about how to find the total geometric area surrounded by the two overlapping circles in Figure 2.1a. The hatched region in Figure 2.1a represents the intersection event {liquid precipitation and frozen precipitation}, and it is contained within both of the two circles labelled "Liquid precipitation" and "Frozen precipitation."

The additive law, Equation 2.5, can be extended to the union of three or more events by thinking of $\{E_1\}$ or $\{E_2\}$ as a compound event (i.e., a union of other events), and recursively applying Equation 2.5. For example, if $\{E_2\} = \{E_3 \cup E_4\}$, substituting into Equation 2.5 yields, after some rearrangement,

$$\begin{aligned}
\Pr\{E_1 \cup E_3 \cup E_4\} = {} & \Pr\{E_1\} + \Pr\{E_3\} + \Pr\{E_4\} \\
& - \Pr\{E_1 \cap E_3\} - \Pr\{E_1 \cap E_4\} - \Pr\{E_3 \cap E_4\} \\
& + \Pr\{E_1 \cap E_3 \cap E_4\}.
\end{aligned} \tag{2.7}$$

This result may be difficult to grasp algebraically, but is fairly easy to visualize geometrically. Figure 2.2 illustrates the situation. Adding together the areas of the three circles individually (first line in Equation 2.7) results in double-counting of the areas with two overlapping hatch patterns, and triple-counting of the central area contained in all three circles. The second line of Equation 2.7 corrects the double-counting, but subtracts the area of the central region three times. This area is added back a final time in the third line of Equation 2.7.

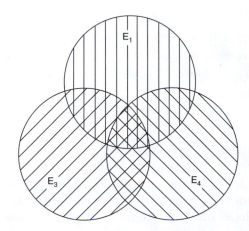

FIGURE 2.2   Venn diagram illustrating computation of probability of the union of three intersecting events in Equation 2.7. The regions with two overlapping hatch patterns have been double-counted, and their areas must be subtracted to compensate. The central region with three overlapping hatch patterns has been triple-counted, but then subtracted three times when the double-counting is corrected. Its area must be added back again.

2.4.2 DeMorgan's Laws

Manipulating probability statements involving complements of unions or intersections, or statements involving intersections of unions or complements, is facilitated by the two relationships known as DeMorgan's Laws,

$$\Pr\{A \cup B)^C\} = \Pr\{A^C \cap B^C\} \tag{2.8a}$$

and

$$\Pr\{(A \cap B)^C\} = \Pr\{A^C \cup B^C\}. \tag{2.8b}$$

The first of these laws, Equation 2.8a, expresses the fact that the complement of a union of two events is the intersection of the complements of the two events. In the geometric terms of the Venn diagram, the events outside the union of $\{A\}$ and $\{B\}$ (left-hand side) are simultaneously outside of both $\{A\}$ and $\{B\}$ (right-hand side). The second of DeMorgan's laws, Equation 2.8b, says that the complement of an intersection of two events is the union of the complements of the two individual events. Here, in geometric terms, the events not in the overlap between $\{A\}$ and $\{B\}$ (left-hand side) are those either outside of $\{A\}$ or outside of $\{B\}$ or both (right-hand side).

2.4.3 Conditional Probability

It is often the case that we are interested in the probability of an event, given that some other event has occurred or will occur. For example, the probability of freezing rain, given that precipitation occurs, may be of interest; or perhaps we need to know the probability of coastal windspeeds above some threshold, given that a hurricane makes landfall nearby. These are examples of conditional probabilities. The event that must be "given" is called the conditioning event. The conventional notation for conditional probability is a vertical line, so denoting $\{E_1\}$ as the event of interest and $\{E_2\}$ as the conditioning event, we write

$$\Pr\{E_1 | E_2\} = \Pr\{E_1 \text{ given that } E_2 \text{ has occurred or will occur}\}. \tag{2.9}$$

If the event $\{E_2\}$ has occurred or will occur, the probability of $\{E_1\}$ is the conditional probability $\Pr\{E_1 | E_2\}$. If the conditioning event has not occurred or will not occur, this in itself gives no information on the probability of $\{E_1\}$.

More formally, conditional probability is defined in terms of the intersection of the event of interest and the conditioning event, according to

$$\Pr\{E_1 | E_2\} = \frac{\Pr\{E_1 \cap E_2\}}{\Pr\{E_2\}}, \tag{2.10}$$

provided the probability of the conditioning event is not zero. Intuitively, it makes sense that conditional probabilities are related to the joint probability of the two events in question, $\Pr\{E_1 \cap E_2\}$. Again, this is easiest to understand through the analogy to areas in a Venn diagram, as shown in Figure 2.3. We understand the unconditional probability of $\{E_1\}$ to be represented by that proportion of the sample space $\mathbf{S}$ occupied by the rectangle labelled E_1. Conditioning on $\{E_2\}$ means that we are interested only in those outcomes containing $\{E_2\}$. We are, in effect, throwing away any part of $\mathbf{S}$ not contained in $\{E_2\}$. This amounts to considering a new sample space, $\mathbf{S}'$, that is coincident

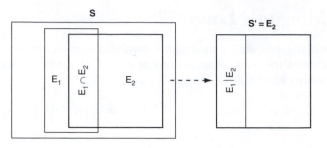

FIGURE 2.3   Illustration of the definition of conditional probability. The unconditional probability of $\{E_1\}$ is that fraction of the area of **S** occupied by $\{E_1\}$ on the left side of the figure. Conditioning on $\{E_2\}$ amounts to considering a new sample space, **S′** composed only of $\{E_2\}$, since this means we are concerned only with occasions when $\{E_2\}$ occurs. Therefore the conditional probability $\Pr\{E_1|E_2\}$ is given by that proportion of the area of the new sample space **S′** occupied by both $\{E_1\}$ and $\{E_2\}$. This proportion is computed in Equation 2.10.

with $\{E_2\}$. The conditional probability $\Pr\{E_1|E_2\}$ therefore is represented geometrically as that proportion of the new sample space area occupied by both $\{E_1\}$ and $\{E_2\}$. If the conditioning event and the event of interest are mutually exclusive, the conditional probability clearly must be zero, since their joint probability will be zero.

2.4.4 Independence

Rearranging the definition of conditional probability, Equation 2.10 yields the form of this expression called the Multiplicative Law of Probability:

$$\Pr\{E_1 \cap E_2\} = \Pr\{E_1|E_2\}\Pr\{E_2\} = \Pr\{E_2|E_1\}\Pr\{E_1\}. \qquad (2.11)$$

Two events are said to be independent if the occurrence or nonoccurrence of one does not affect the probability of the other. For example, if we roll a red die and a white die, the probability of an outcome on the red die does not depend on the outcome of the white die, and vice versa. The outcomes for the two dice are independent. Independence between $\{E_1\}$ and $\{E_2\}$ implies $\Pr\{E_1|E_2\} = \Pr\{E_1\}$ and $\Pr\{E_2|E_1\} = \Pr\{E_2\}$. Independence of events makes the calculation of joint probabilities particularly easy, since the multiplicative law then reduces to

$$\Pr\{E_1 \cap E_2\} = \Pr\{E_1\}\Pr\{E_2\}, \qquad \text{for } \{E_1\} \text{ and } \{E_2\} \text{ independent.} \qquad (2.12)$$

Equation 2.12 is extended easily to the computation of joint probabilities for more than two independent events, by simply multiplying all the probabilities of the independent unconditional events.

EXAMPLE 2.1   Conditional Relative Frequency

Consider estimating climatological (i.e., long-run, or relative frequency) estimates of probabilities using the data given in Table A.1 of Appendix A. Climatological probabilities conditional on other events can be computed. Such probabilities are sometimes referred to as conditional climatological probabilities, or *conditional climatologies* .

Suppose it is of interest to estimate the probability of at least 0.01 in. of liquid equivalent precipitation at Ithaca in January, given that the minimum temperature is at least 0°F. Physically, these two events would be expected to be related since very cold temperatures typically occur on clear nights, and precipitation occurrence requires clouds. This physical relationship would lead us to expect that these two events would be statistically related (i.e., not independent), and that the conditional probabilities of precipitation given different temperature conditions will be different from each other, and from the unconditional probability. In particular, on the basis of our understanding of the underlying physical processes, we expect the probability of precipitation given minimum temperature of 0°F or higher will be larger than the conditional probability given the complementary event of minimum temperature colder than 0°F.

To estimate this probability using the conditional relative frequency, we are interested only in those data records for which the Ithaca minimum temperature was at least 0°F. There are 24 such days in Table A.1. Of these 24 days, 14 show measurable precipitation (ppt), yielding the estimate $\Pr\{\text{ppt} \geq 0.01 \text{ in.}|T_{min} \geq 0°F\} = 14/24 \approx 0.58$. The precipitation data for the seven days on which the minimum temperature was colder than 0°F has been ignored. Since measurable precipitation was recorded on only one of these seven days, we could estimate the conditional probability of precipitation given the complementary conditioning event of minimum temperature colder than 0°F as $\Pr\{\text{ppt} \geq 0.01 \text{ in.}|T_{min} < 0°F\} = 1/7 \approx 0.14$. The corresponding estimate of the unconditional probability of precipitation would be $\Pr\{\text{ppt} \geq 0.01 \text{ in.}\} = 15/31 \approx 0.48$. ◊

The differences in the conditional probability estimates calculated in Example 2.1 reflect statistical dependence. Since the underlying physical processes are well understood, we would not be tempted to speculate that relatively warmer minimum temperatures somehow cause precipitation. Rather, the temperature and precipitation events show a statistical relationship because of their (different) physical relationships to clouds. When dealing with statistically dependent variables whose physical relationships may not be known, it is well to remember that statistical dependence does not necessarily imply a physical cause-and-effect relationship.

EXAMPLE 2.2   Persistence as Conditional Probability

Atmospheric variables often exhibit statistical dependence with their own past or future values. In the terminology of the atmospheric sciences, this dependence through time is usually known as *persistence*. Persistence can be defined as the existence of (positive) statistical dependence among successive values of the same variable, or among successive occurrences of a given event. Positive dependence means that large values of the variable tend to be followed by relatively large values, and small values of the variable tend to be followed by relatively small values. It is usually the case that statistical dependence of meteorological variables in time is positive. For example, the probability of an above-average temperature tomorrow is higher if today's temperature was above average. Thus, another name for persistence is *positive serial dependence*. When present, this frequently occurring characteristic has important implications for statistical inferences drawn from atmospheric data, as will be seen in Chapter 5.

Consider characterizing the persistence of the event {precipitation occurrence} at Ithaca, again with the small data set of daily values in Table A.1 of Appendix A. Physically, serial dependence would be expected in these data because the typical time scale for the midlatitude synoptic waves associated with most winter precipitation at this location is several days, which is longer than the daily observation interval. The statistical consequence should be that days for which measurable precipitation is reported should tend to occur in runs, as should days without measurable precipitation.

To evaluate serial dependence for precipitation events it is necessary to estimate conditional probabilities of the type Pr{ppt. today|ppt. yesterday}. Since data set A.1 contains no records for either 31 December 1986 or 1 February 1987, there are 30 yesterday/today data pairs to work with. To estimate Pr{ppt. today|ppt. yesterday} we need only count the number of days reporting precipitation (as the conditioning, or "yesterday" event) that are followed by the subsequent day reporting precipitation (as the event of interest, or "today"). When estimating this conditional probability, we are not interested in what happens following days on which no precipitation is reported. Excluding 31 January, there are 14 days on which precipitation is reported. Of these, 10 are followed by another day with precipitation reported, and four are followed by dry days. The conditional relative frequency estimate therefore would be Pr{ppt. today|ppt. yesterday} $= 10/14 \approx 0.71$. Similarly, conditioning on the complementary event (no precipitation "yesterday") yields Pr{ppt. today|no ppt. yesterday} $= 5/16 \approx 0.31$. The difference between these conditional probability estimates confirms the serial dependence in this data, and quantifies the tendency of the wet and dry days to occur in runs. These two conditional probabilities also constitute a "conditional climatology." ◊

2.4.5 Law of Total Probability

Sometimes probabilities must be computed indirectly because of limited information. One relationship that can be useful in such situations is the Law of Total Probability. Consider a set of MECE events, $\{E_i\}, i = 1, \ldots, I$; on the sample space of interest. Figure 2.4 illustrates this situation for $I = 5$ events. If there is an event $\{A\}$, also defined on this sample space, its probability can be computed by summing the joint probabilities

$$Pr\{A\} = \sum_{i=1}^{I} Pr\{A \cap E_i\}. \tag{2.13}$$

The notation on the right-hand side of this equation indicates summation of terms defined by the mathematical template to the right of the uppercase sigma, for all integer values of the index i between 1 and I, inclusive. Substituting the multiplicative law of probability yields

$$Pr\{A\} = \sum_{i=1}^{I} Pr\{A|E_i\}Pr\{E_i\}. \tag{2.14}$$

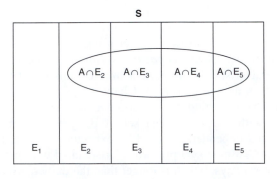

FIGURE 2.4   Illustration of the Law of Total Probability. The sample space **S** contains the event $\{A\}$, represented by the oval, and five MECE events, $\{E_i\}$.

If the unconditional probabilities $\Pr\{E_i\}$ and the conditional probabilities of $\{A\}$ given the MECE events $\{E_i\}$ are known, the unconditional probability of $\{A\}$ can be computed. It is important to note that Equation 2.14 is correct only if the events $\{E_i\}$ constitute a MECE partition of the sample space.

EXAMPLE 2.3   Combining Conditional Probabilities Using the Law of Total Probability

Example 2.2 can also be viewed in terms of the Law of Total Probability. Consider that there are only $I = 2$ MECE events partitioning the sample space: $\{E_1\}$ denotes precipitation yesterday and $\{E_2\} = \{E_1\}^c$ denotes no precipitation yesterday. Let the event $\{A\}$ be the occurrence of precipitation today. If the data were not available, we could compute $\Pr\{A\}$ using the conditional probabilities through the Law of Total Probability. That is, $\Pr\{A\} = \Pr\{A|E_1\}\Pr\{E_1\} + \Pr\{A|E_2\}\Pr\{E_2\} = (10/14)(14/30) + (5/16)(16/30) = 0.50$. Since the data are available in Appendix A, the correctness of this result can be confirmed simply by counting. $\Diamond$

2.4.6 Bayes' Theorem

Bayes' Theorem is an interesting combination of the Multiplicative Law and the Law of Total Probability. In a relative frequency setting, Bayes' Theorem is used to "invert" conditional probabilities. That is, if $\Pr\{E_1|E_2\}$ is known, Bayes' Theorem may be used to compute $\Pr\{E_2|E_1\}$. In the Bayesian framework it is used to revise or update subjective probabilities consistent with new information.

Consider again a situation such as that shown in Figure 2.4, in which there is a defined set of MECE events $\{E_i\}$, and another event $\{A\}$. The Multiplicative Law (Equation 2.11) can be used to find two expressions for the joint probability of $\{A\}$ and any of the events $\{E_i\}$. Since

$$\Pr\{A, E_i\} = \Pr\{A|E_i\}\Pr\{E_i\}$$
$$= \Pr\{E_i|A\}\Pr\{A\}, \tag{2.15}$$

combining the two right-hand sides and rearranging yields

$$\Pr\{E_i|A\} = \frac{\Pr\{A|E_i\}\Pr\{E_i\}}{\Pr\{A\}} = \frac{\Pr\{A|E_i\}\Pr\{E_i\}}{\sum_{j=1}^{J}\Pr\{A|E_j\}\Pr\{E_j\}}. \tag{2.16}$$

The Law of Total Probability has been used to rewrite the denominator. Equation (2.16) is the expression for Bayes' Theorem. It is applicable separately for each of the MECE events $\{E_i\}$. Note, however, that the denominator is the same for each E_i, since $\Pr\{A\}$ is obtained each time by summing over all the events, indexed in the denominator by the subscript j.

EXAMPLE 2.4   Bayes' Theorem from a Relative Frequency Standpoint

Conditional probabilities for precipitation occurrence given minimum temperatures above or below 0°F were estimated in Example 2.1. Bayes' Theorem can be used to compute the converse conditional probabilities, concerning temperature events given that precipitation did or did not occur. Let $\{E_1\}$ represent minimum temperature of 0°F or above, and $\{E_2\} = \{E_1\}^C$ be the complementary event that minimum temperature is colder than 0°F. Clearly the two events constitute a MECE partition of the sample space. Recall that minimum temperatures of at least 0°F were reported on 24 of the 31 days, so that the unconditional climatological estimates of the probabilities for the temperature events would be $\Pr\{E_1\} = 24/31$ and $\Pr\{E_2\} = 7/31$. Recall also that $\Pr\{A|E_1\} = 14/24$ and $\Pr\{A|E_2\} = 1/7$.

Equation 2.16 can be applied separately for each of the two events $\{E_i\}$. In each case the denominator is $\Pr\{A\} = (14/24)(24/31) + (1/7)(7/31) = 15/31$. (This differs slightly from the estimate for the probability of precipitation obtained in Example 2.2, since there the data for 31 December could not be included.) Using Bayes' Theorem, the conditional probability for minimum temperature at least 0°F given precipitation occurrence is $(14/24)(24/31)/(15/31) = 14/15$. Similarly, the conditional probability for minimum temperature below 0°F given nonzero precipitation is $(1/7)(7/31)/(15/31) = 1/15$. Since all the data are available in Appendix A, these calculations can be verified directly. ◊

EXAMPLE 2.5   Bayes' Theorem from a Subjective Probability Standpoint

A subjective (Bayesian) probability interpretation of Example 2.4 can also be made. Suppose a weather forecast specifying the probability of the minimum temperature being at least 0°F is desired. If no more sophisticated information were available, it would be natural to use the unconditional climatological probability for the event, $\Pr\{E_1\} = 24/31$, as representing the forecaster's uncertainty or degree of belief in the outcome. In the Bayesian framework this baseline state of information is known as the *prior probability*. Suppose, however, that the forecaster could know whether or not precipitation will occur on that day. That information would affect the forecaster's degree of certainty in the temperature outcome. Just how much more certain the forecaster can become depends on the strength of the relationship between temperature and precipitation, expressed in the conditional probabilities for precipitation occurrence given the two minimum temperature outcomes. These conditional probabilities, $\Pr\{A|E_i\}$ in the notation of this example, are known as the *likelihoods*. If precipitation occurs the forecaster is more certain that the minimum temperature will be at least 0°F, with the revised probability given by Equation 2.16 as $(14/24)(24/31)/(15/31) = 14/15$. This modified, or updated (in light of the additional information regarding precipitation occurrence), judgement regarding the probability of a very cold minimum temperature not occurring is called the posterior probability. Here the posterior probability is larger than the prior probability of 24/31. Similarly, if precipitation does not occur, the forecaster is more confident that the minimum temperature will not be 0°F or warmer. Note that the differences between this example and Example 2.4 are entirely in the interpretation, and that the computations and numerical results are identical. ◊

2.5 Exercises

2.1. In the climatic record for 60 winters at a given location, single-storm snowfalls greater than 35 cm occurred in nine of those winters (define such snowfalls as event "A"), and the lowest temperature was below -25°C in 36 of the winters (define this as event "B"). Both events "A" and "B" occurred in three of the winters.

  a. Sketch a Venn diagram for a sample space appropriate to this data.

  b. Write an expression using set notation for the occurrence of 35-cm snowfalls, -25°C temperatures, or both. Estimate the climatological probability for this compound event.

c. Write an expression using set notation for the occurrence of winters with 35-cm snowfalls in which the temperature does not fall below $-25°C$. Estimate the climatological probability for this compound event.

d. Write an expression using set notation for the occurrence of winters having neither 25°C temperatures nor 35-cm snowfalls. Again, estimate the climatological probability.

2.2. Using the January 1987 data set in Table A.1, define event "A" as Ithaca $T_{max} > 32°F$, and event "B" as Canandaigua $T_{max} > 32°F$.

a. Explain the meanings of $\Pr(A), \Pr(B), \Pr(A, B), \Pr(A \cup B), \Pr(A|B)$, and $\Pr(B|A)$.

b. Estimate, using relative frequencies in the data, $\Pr(A), \Pr(B)$, and $\Pr(A, B)$.

c. Using the results from part (b), calculate $\Pr(A|B)$.

d. Are events "A" and "B" independent? How do you know?

2.3. Again using the data in Table A.1, estimate probabilities of the Ithaca maximum temperature being at or below freezing (32°F), given that the previous day's maximum temperature was at or below freezing,

a. Accounting for the persistence in the temperature data.

b. Assuming (incorrectly) that sequences of daily temperatures are independent.

2.4. Three radar sets, operating independently, are searching for "hook" echoes (a radar signature associated with tornados). Suppose that each radar has a probability of 0.05 of failing to detect this signature when a tornado is present.

a. Sketch a Venn diagram for a sample space appropriate to this problem.

b. What is the probability that a tornado will escape detection by all three radars?

c. What is the probability that a tornado will be detected by all three radars?

2.5. The effect of cloud seeding on suppression of damaging hail is being studied in your area, by randomly seeding or not seeding equal numbers of candidate storms. Suppose the probability of damaging hail from a seeded storm is 0.10, and the probability of damaging hail from an unseeded storm is 0.40. If one of the candidate storms has just produced damaging hail, what is the probability that it was seeded?

PART · II

Univariate Statistics

CHAPTER • 3

Empirical Distributions and Exploratory Data Analysis

3.1 Background

One very important application of statistical ideas in meteorology and climatology is in making sense of a new set of data. As mentioned in Chapter 1, meteorological observing systems and numerical models, supporting both operational and research efforts, produce torrents of numerical data. It can be a significant task just to get a feel for a new batch of numbers, and to begin to make some sense of them. The goal is to extract insight about the processes underlying the generation of the numbers.

Broadly speaking, this activity is known as Exploratory Data Analysis, or EDA. Its systematic use increased substantially following Tukey's (1977) pathbreaking and very readable book of the same name. The methods of EDA draw heavily on a variety of graphical methods to aid in the comprehension of the sea of numbers confronting an analyst. Graphics are a very effective means of compressing and summarizing data, portraying much in little space, and exposing unusual features of a data set. The unusual features can be especially important. Sometimes unusual data points result from errors in recording or transcription, and it is well to know about these as early as possible in an analysis. Sometimes the unusual data are valid, and may turn out be the most interesting and informative parts of the data set.

Many EDA methods originally were designed to be applied by hand, with pencil and paper, to small (up to perhaps 200-point) data sets. More recently, graphically oriented computer packages have come into being that allow fast and easy use of these methods on desktop computers (e.g., Velleman, 1988). The methods can also be implemented on larger computers with a modest amount of programming.

3.1.1 Robustness and Resistance

Many of the classical techniques of statistics work best when fairly stringent assumptions about the nature of the data are met. For example, it is often assumed that data will follow the familiar bell-shaped curve of the Gaussian distribution. Classical procedures can behave very badly (i.e., produce quite misleading results) if their assumptions are not satisfied by the data to which they are applied.

The assumptions of classical statistics were not made out of ignorance, but rather out of necessity. Invocation of simplifying assumptions in statistics, as in other fields, has allowed progress to be made through the derivation of elegant analytic results, which are relatively simple but powerful mathematical formulas. As has been the case in many quantitative fields, the relatively recent advent of cheap computing power has freed the data analyst from sole dependence on such results, by allowing alternatives requiring less stringent assumptions to become practical. This does not mean that the classical methods are no longer useful. However, it is much easier to check that a given set of data satisfies particular assumptions before a classical procedure is used, and good alternatives are computationally feasible in cases where the classical methods may not be appropriate.

Two important properties of EDA methods are that they are robust and resistant. Robustness and resistance are two aspects of insensitivity to assumptions about the nature of a set of data. A robust method is not necessarily optimal in any particular circumstance, but performs reasonably well in most circumstances. For example, the sample average is the best characterization of the center of a set of data if it is known that those data follow a Gaussian distribution. However, if those data are decidedly non-Gaussian (e.g., if they are a record of extreme rainfall events), the sample average will yield a misleading characterization of their center. In contrast, robust methods generally are not sensitive to particular assumptions about the overall nature of the data.

A resistant method is not unduly influenced by a small number of outliers, or "wild data." As indicated previously, such points often show up in a batch of data through errors of one kind or another. The results of a resistant method change very little if a small fraction of the data values are changed, even if they are changed drastically. In addition to not being robust, the sample average is not a resistant characterization of the center of a data set, either. Consider the small set {11, 12, 13, 14, 15, 16, 17, 18, 19}. Its average is 15. However, if the set {11, 12, 13, 14, 15, 16, 17, 18, 91} had resulted from a transcription error, the "center" of the data (erroneously) characterized using the sample average instead would be 23. Resistant measures of the center of a batch of data, such as those to be presented later, would be changed little or not at all by the substitution of "91" for "19" in this simple example.

3.1.2 Quantiles

Many common summary measures rely on the use of selected sample quantiles (also known as fractiles). Quantiles and fractiles are essentially equivalent to the more familiar term, percentile. A sample quantile, q_p, is a number having the same units as the data, which exceeds that proportion of the data given by the subscript p, with $0 \leq p \leq 1$. The sample quantile q_p can be interpreted approximately as that value expected to exceed a randomly chosen member of the data set, with probability p. Equivalently, the sample quantile q_p would be regarded as the $[p \times 100]^{th}$ percentile of the data set.

The determination of sample quantiles requires that a batch of data first be arranged in order. Sorting small sets of data by hand presents little problem. Sorting larger sets of data is best accomplished by computer. Historically, the sorting step constituted a major bottleneck in the application of robust and resistant procedures to large data sets. Today the sorting can be done easily using either a spreadsheet or data analysis program on a desktop computer, or one of many sorting algorithms available in collections of general-purpose computing routines (e.g., Press *et al.*, 1986).

The sorted, or ranked, data values from a particular sample are called the order statistics of that sample. Given a set of data $\{x_1, x_2, x_3, x_4, x_5, \ldots, x_n\}$, the order statistics for this sample would be the same numbers, sorted in ascending order. These sorted values are conventionally denoted using parenthetical subscripts, that is, by the set $\{x_{(1)}, x_{(2)}, x_{(3)}, x_{(4)}, x_{(5)}, \ldots, x_{(n)}\}$. Here the ith smallest of the n data values is denoted $x_{(i)}$.

Certain sample quantiles are used particularly often in the exploratory summarization of data. Most commonly used is the median, or $q_{0.5}$, or 50th percentile. This is the value at the center of the data set, in the sense that equal proportions of the data fall above and below it. If the data set at hand contains an odd number of values, the median is simply the middle order statistic. If there are an even number, however, the data set has two middle values. In this case the median is conventionally taken to be the average of these two middle values. Formally,

$$
q_{0.5} = \begin{cases} x_{([n+1]/2)}, & n \text{ odd} \\ \dfrac{x_{(n/2)} + x_{([n/2]+1)}}{2}, & n \text{ even.} \end{cases} \tag{3.1}
$$

Almost as commonly used as the median are the quartiles, $q_{0.25}$ and $q_{0.75}$. Usually these are called the lower and upper quartiles, respectively. They are located half-way between the median, $q_{0.5}$, and the extremes, $x_{(1)}$ and $x_{(n)}$. In typically colorful terminology, Tukey (1977) calls $q_{0.25}$ and $q_{0.75}$ the "hinges," apparently imagining that the data set has been folded first at the median, and then at the quartiles. The quartiles are thus the two medians of the half-data sets between $q_{0.5}$ and the extremes. If n is odd, these half-data sets each consist of $(n+1)/2$ points, and both include the median. If n is even these half-data sets each contain $n/2$ points, and do not overlap. Other quantiles that also are used frequently enough to be named are the two terciles, $q_{0.333}$ and $q_{0.667}$; the four quintiles, $q_{0.2}, q_{0.4}, q_{0.6}$, and $q_{0.8}$; the eighths, $q_{0.125}, q_{0.375}, q_{0.625}$, and $q_{0.875}$ (in addition to the quartiles and median); and the deciles, $q_{0.1}, q_{0.2}, \ldots q_{0.9}$.

EXAMPLE 3.1 Computation of Common Quantiles

If there are $n = 9$ data values in a batch of data, the median is $q_{0.5} = x_{(5)}$, or the fifth largest of the nine. The lower quartile is $q_{0.25} = x_{(3)}$, and the upper quartile is $q_{0.75} = x_{(7)}$.

If $n = 10$, the median is the average of the two middle values, and the quartiles are the single middle values of the upper and lower halves of the data. That is, $q_{0.25}, q_{0.5}$, and $q_{0.75}$ are $x_{(3)}, [x_{(5)} + x_{(6)}]/2$, and $x_{(8)}$, respectively.

If $n = 11$ then there is a unique middle value, but the quartiles are determined by averaging the two middle values of the upper and lower halves of the data. That is, $q_{0.25}, q_{0.5}$, and $q_{0.75}$ are $[x_{(3)} + x_{(4)}]/2, x_{(6)}$, and $[x_{(8)} + x_{(9)}]/2$, respectively.

For $n = 12$ both quartiles and the median are determined by averaging pairs of middle values; $q_{0.25}, q_{0.5}$, and $q_{0.75}$ are $[x_{(3)} + x_{(4)}]/2, [x_{(6)} + x_{(7)}]/2$, and $[x_{(9)} + x_{(10)}]/2$, respectively. ◊

3.2 Numerical Summary Measures

Some simple robust and resistant summary measures are available that can be used without hand plotting or computer graphic capabilities. Often these will be the first quantities to be computed from a new and unfamiliar set of data. The numerical summaries listed in this section can be subdivided into measures of location, spread, and symmetry. Location refers to the central tendency, or general magnitude of the data values. Spread denotes the degree of variation or dispersion around the central value. Symmetry describes the balance with which the data values are distributed about their center. Asymmetric data tend to spread more either on the high side (have a long right tail), or the low side (have a long left tail). These three types of numerical summary measures correspond to the first three statistical moments of a data sample, but the classical measures of these moments (i.e., the sample mean, sample variance, and sample coefficient of skewness, respectively) are neither robust nor resistant.

3.2.1 Location

The most common robust and resistant measure of central tendency is the median, $q_{0.5}$. Consider again the data set $\{11, 12, 13, 14, 15, 16, 17, 18, 19\}$. The median and mean are both 15. If, as noted before, the "19" is replaced erroneously by "91," the mean

$$\bar{x} = \frac{1}{n} \sum_{i=1}^{n} x_i, \tag{3.2}$$

($=23$) is very strongly affected, illustrating its lack of resistance to outliers. The median is unchanged by this common type of data error.

A slightly more complicated measure of location that takes into account more information about the magnitudes of the data is the *trimean*. The trimean is a weighted average of the median and the quartiles, with the median receiving twice the weight of each of the quartiles:

$$\text{Trimean} = \frac{q_{0.25} + 2\, q_{0.5} + q_{0.75}}{4}. \tag{3.3}$$

The trimmed mean is another resistant measure of location, whose sensitivity to outliers is reduced by removing a specified proportion of the largest and smallest observations. If the proportion of observations omitted at each end is α, then the α-trimmed mean is

$$\bar{x}_\alpha = \frac{1}{n - 2k} \sum_{i=k+1}^{n-k} x_{(i)}, \tag{3.4}$$

where k is an integer rounding of the product αn, the number of data values "trimmed" from each tail. The trimmed mean reduces to the ordinary mean (Equation 3.2) for $\alpha = 0$.

Other methods of characterizing location can be found in Andrews *et al.* (1972), Goodall (1983), Rosenberger and Gasko (1983), and Tukey (1977).

3.2.2 Spread

The most common, and simplest, robust and resistant measure of spread, also known as dispersion or scale, is the Interquartile Range (IQR). The Interquartile Range is simply the difference between the upper and lower quartiles:

$$\text{IQR} = q_{0.75} - q_{0.25}. \tag{3.5}$$

The IQR is a good index of the spread in the central part of a data set, since it simply specifies the range of the central 50% of the data. The fact that it ignores the upper and lower 25% of the data makes it quite resistant to outliers. This quantity is sometimes called the fourth-spread.

It is worthwhile to compare the IQR with the conventional measure of scale of a data set, the sample standard deviation

$$s = \sqrt{\frac{1}{n-1} \sum_{i=1}^{n} (x_i - \bar{x})^2}. \tag{3.6}$$

The square of the sample standard deviation, s^2, is known as the sample variance. The standard deviation is neither robust nor resistant. It is very nearly just the square root of the average squared difference between the data points and their sample mean. (The division by $n-1$ rather than n often is done in order to compensate for the fact that the x_i are closer, on average, to their sample mean than to the true population mean: dividing by $n-1$ exactly counters the resulting tendency for the sample standard deviation to be too small, on average.) Because of the square root in Equation 3.6, the standard deviation has the same physical dimensions as the underlying data. Even one very large data value will be felt very strongly because it will be especially far away from the mean, and that difference will be magnified by the squaring process. Consider again the set $\{11, 12, 13, 14, 15, 16, 17, 18, 19\}$. The sample standard deviation is 2.74, but it is greatly inflated to 25.6 if "91" erroneously replaces "19". It is easy to see that in either case IQR = 4.

The IQR is very easy to compute, but it does have the disadvantage of not making much use of a substantial fraction of the data. A more complete, yet reasonably simple alternative is the median absolute deviation (MAD). The MAD is easiest to understand by imagining the transformation $y_i = |x_i - q_{0.5}|$. Each transformed value y_i is the absolute value of the difference between the corresponding original data value and the median. The MAD is then just the median of the transformed (y_i) values:

$$MAD = \text{median}|x_i - q_{0.5}|. \tag{3.7}$$

Although this process may seem a bit elaborate at first, a little thought illustrates that it is analogous to computation of the standard deviation, but using operations that do not emphasize outlying data. The median (rather than the mean) is subtracted from each data value, any negative signs are removed by the absolute value (rather than squaring) operation, and the center of these absolute differences is located by their median (rather than their mean).

A still more elaborate measure of spread is the trimmed variance. The idea, as for the trimmed mean (Equation 3.4), is to omit a proportion of the largest and smallest values and compute the analogue of the sample variance (the square of Equation 3.6)

$$s_\alpha^2 = \frac{1}{n-2k} \sum_{i=k+1}^{n-k} (x_{(i)} - \bar{x}_\alpha)^2. \tag{3.8}$$

Again, k is the nearest integer to αn, and squared deviations from the consistent trimmed mean (Equation 3.4) are averaged. The trimmed variance is sometimes multiplied by an adjustment factor to make it more consistent with the ordinary sample variance, s^2 (Graedel and Kleiner, 1985).

Other measures of spread can be found in Hosking (1990) and Iglewicz (1983).

3.2.3 Symmetry

The conventional moments-based measure of symmetry in a batch of data is the sample skewness coefficient,

$$\gamma = \frac{\dfrac{1}{n-1}\sum_{i=1}^{n}(x_i - \bar{x})^3}{s^3}. \tag{3.9}$$

This measure is neither robust nor resistant. The numerator is similar to the sample variance, except that the average is over cubed deviations from the mean. Thus the sample skewness coefficient is even more sensitive to outliers than is the standard deviation. The average cubed deviation in the numerator is divided by the cube of the sample standard deviation in order to standardize the skewness coefficient and make comparisons of skewness between different data sets more meaningful. The standardization also serves to make the sample skewness a dimensionless quantity.

Notice that cubing differences between the data values and their mean preserves the signs of these differences. Since the differences are cubed, the data values farthest from the mean will dominate the sum in the numerator of Equation 3.9. If there are a few very large data values, the sample skewness will tend to be positive. Therefore batches of data with long right tails are referred to both as right-skewed and positively skewed. Data that are physically constrained to lie above a minimum value (such as precipitation or wind speed, both of which must be nonnegative) are often positively skewed. Conversely, if there are a few very small (or large negative) data values, these will fall far below the mean. The sum in the numerator of Equation 3.9 will then be dominated by a few large negative terms, so that the skewness coefficient will be negative. Data with long left tails are referred to as left-skewed, or negatively skewed. For essentially symmetric data, the skewness coefficient will be near zero.

A robust and resistant alternative to the sample skewness is the Yule-Kendall index,

$$\gamma_{YK} = \frac{(q_{0.75} - q_{0.5}) - (q_{0.5} - q_{0.25})}{IQ} = \frac{q_{0.25} - 2\,q_{0.5} + q_{0.75}}{IQ}, \tag{3.10}$$

which is computed by comparing the distance between the median and each of the two quartiles. If the data are right-skewed, at least in the central 50% of the data, the distance to the median will be greater from the upper quartile than from the lower quartile. In this case the Yule-Kendall index will be greater than zero, consistent with the usual convention of right-skewness being positive. Conversely, left-skewed data will be characterized by a negative Yule-Kendall index. Analogously to Equation 3.9, division by the interquartile range nondimensionalizes γ_{YK} (i.e., scales it in a way that the physical dimensions, such as meters or millibars, cancel) and thus improves its comparability between data sets.

Alternative measures of skewness can be found in Brooks and Carruthers (1953) and Hosking (1990).

3.3 Graphical Summary Techniques

Numerical summary measures are quick and easy to compute and display, but they can express only a small amount of detail. In addition, their visual impact is limited. A number of graphical displays for exploratory data analysis have been devised that require only slightly more effort to produce.

3.3.1 *Stem-and-Leaf Display*

The stem-and-leaf display is a very simple but effective tool for producing an overall view of a new set of data. At the same time it provides the analyst with an initial exposure to the individual data values. In its simplest form, the stem-and-leaf display groups the data values according to their all-but-least significant digits. These values are written in either ascending or descending order to the left of a vertical bar, constituting the "stems." The least significant digit for each data value is then written to the right of the vertical bar, on the same line as the more significant digits with which it belongs. These least significant values constitute the "leaves."

Figure 3.1a shows a stem-and-leaf display for the January 1987 Ithaca maximum temperatures in Table A.1. The data values are reported to whole degrees, and range from 9°F to 53°F. The all-but-least significant digits are thus the tens of degrees, which are written to the left of the bar. The display is built up by proceeding through the data values one by one, and writing its least significant digit on the appropriate line. For example, the temperature for 1 January is 33°F, so the first "leaf" to be plotted is the first "3" on the stem of temperatures in the 30s. The temperature for 2 January is 32°F, so a "2" is written to the right of the "3" just plotted for 1 January.

The initial stem-and-leaf display for this particular data set is a bit crowded, since most of the values are in the 20s and 30s. In cases like this, better resolution can be obtained by constructing a second plot, like that in Figure 3.1b, in which each stem has been split to contain only the values 0–4 or 5–9. Sometimes the opposite problem will occur, and the initial plot is too sparse. In that case (if there are at least three significant digits), replotting can be done with stem labels omitting the two least significant digits. Less stringent groupings can also be used. Regardless of whether or not it may be desirable to split or consolidate stems, it is often useful to rewrite the display with the leaf values sorted, as has also been done in Figure 3.1b.

The stem-and-leaf display is much like a quickly plotted histogram of the data, placed on its side. In Figure 3.1, for example, it is evident that these temperature data are reasonably symmetrical, with most of the values falling in the upper 20s and lower 30s. Sorting the leaf values also facilitates extraction of quantiles of interest. In this case it is easy to count inward from the extremes to find that the median is 30, and that the two quartiles are 26 and 33.

It can happen that there are one or more outlying data points that are far removed from the main body of the data set. Rather than plot many empty stems, it is usually more convenient to just list these extreme values separately at the upper and lower ends

```
(a)                                          (b)

                                             5·   3
                                             4*   5
  5 | 3                                      4·
                                             3*   6 7 7
  4 | 5                                      3·   0 0 0 0 2 2 2 3 3 4 4
                                             2*   5 5 6 6 6 7 7 8 8 9 9
  3 | 3 2 0 0 7 7 0 6 2 3 4 2 0 4            2·   2 4
                                             1*   7
  2 | 9 5 9 5 8 7 6 8 4 6 2 6 7              1·
                                             0*   9
  1 | 7

  0 | 9
```

FIGURE 3.1   Stem-and-leaf displays for the January 1987 Ithaca maximum temperatures in Table A.1. The plot in panel (a) results after the first pass through the data, using the 10's as "stem" values. A bit more resolution is obtained in panel (b) by creating separate stems for least-significant digits from 0 to 4 (●), and from 5 to 9 (*). At this stage it is also easy to sort the data values before rewriting them.

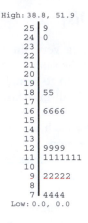

```
High: 38.8, 51.9
        25 │ 9
        24 │ 0
        23 │
        22 │
        21 │
        20 │
        19 │
        18 │ 55
        17 │
        16 │ 6666
        15 │
        14 │
        13 │
        12 │ 9999
        11 │ 1111111
        10 │
         9 │ 22222
         8 │
         7 ┃ 4444
      Low: 0.0, 0.0
```

FIGURE 3.2   Stem-and-leaf display of 1:00 A.M. wind speeds (km/hr) at Newark, New Jersey, Airport during December 1974. Very high and very low values are written outside the plot itself to avoid having many blank stems. The striking grouping of repeated leaf values suggests that a rounding process has been applied to the original observations. From Graedel and Kleiner (1985).

of the display, as in Figure 3.2. This display is of data, taken from Graedel and Kleiner (1985), of wind speeds in kilometers per hour (km/h) to the nearest tenth. Merely listing two extremely large values and two values of calm winds at the top and bottom of the plot has reduced the length of the display by more than half. It is quickly evident that the data are strongly skewed to the right, as often occurs for wind data.

The stem-and-leaf display in Figure 3.2 also reveals something that might have been missed in a tabular list of the daily data. All the leaf values on each stem are the same. Evidently a rounding process has been applied to the data, knowledge of which could be important to some subsequent analyses. In this case the rounding process consists of transforming the data from the original units (knots) to km/hr. For example, the four observations of 16.6 km/hr result from original observations of 9 knots. No observations on the 17 km/hr stem would be possible, since observations of 10 knots transform to 18.5 km/hr.

3.3.2 Boxplots

The boxplot, or box-and-whisker plot, is a very widely used graphical tool introduced by Tukey (1977). It is a simple plot of five sample quantiles: the minimum, $x_{(1)}$, the lower quartile, $q_{0.25}$, the median, $q_{0.5}$, the upper quartile, $q_{0.75}$, and the maximum, $x_{(n)}$. Using these five numbers, the boxplot essentially presents a quick sketch of the distribution of the underlying data.

Figure 3.3 shows a boxplot for the January 1987 Ithaca maximum temperature data in Table A.1. The box in the middle of the diagram is bounded by the upper and lower quartiles, and thus locates the central 50% of the data. The bar inside the box locates the median. The whiskers extend away from the box to the two extreme values.

Boxplots can convey a surprisingly large amount of information at a glance. It is clear from the small range occupied by the box in Figure 3.3, for example, that the data are concentrated quite near 30°F. Being based only on the median and the quartiles, this portion of the boxplot is highly resistant to any outliers that might be present. The full range of the data is also apparent at a glance. Finally, we can see easily that these data

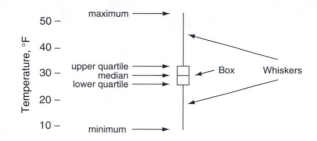

FIGURE 3.3   A simple boxplot, or box-and-whiskers plot, for the January 1987 Ithaca maximum temperature data. The upper and lower ends of the box are drawn at the quartiles, and the bar through the box is drawn at the median. The whiskers extend from the quartiles to the maximum and minimum data values.

are nearly symmetrical, since the median is near the center of the box, and the whiskers are of comparable length.

3.3.3 Schematic Plots

A shortcoming of the boxplot is that information about the tails of the data is highly generalized. The whiskers extend to the highest and lowest values, but there is no information about the distribution of data points within the upper and lower quarters of the data. For example, although Figure 3.3 shows that the highest maximum temperature is 53°F, it gives no information as to whether this is an isolated point (with the remaining warm temperatures cooler than, say, 40°F) or whether the warmer temperatures are more or less evenly distributed between the upper quartile and the maximum.

It is often useful to have some idea of the degree of unusualness of the extreme values. A refinement of the boxplot that presents more detail in the tails is the schematic plot, which was also originated by Tukey (1977). The schematic plot is identical to the boxplot, except that extreme points deemed to be sufficiently unusual are plotted individually. Just how extreme is sufficiently unusual depends on the variability of the data in the central part of the sample, as reflected by the IQR. A given extreme value is regarded as being less unusual if the two quartiles are far apart (i.e., if the IQR is large), and more unusual if the two quartiles are very near each other (the IQR is small).

The dividing lines between less- and more-unusual points are known in Tukey's idiosyncratic terminology as the "fences." Four fences are defined: inner and outer fences, above and below the data, according to

$$\text{Upper outer fence} = q_{0.75} + 3 \text{ IQR}$$

$$\text{Upper inner fence} = q_{0.75} + \frac{3 \text{ IQR}}{2}$$

$$\text{Lower inner fence} = q_{0.25} - \frac{3 \text{ IQR}}{2}$$

$$\text{Lower outer fence} = q_{0.25} - 3 \text{ IQR}. \tag{3.11}$$

Thus the two outer fences are located three times the distance of the interquartile range above and below the two quartiles. The inner fences are mid-way between the outer fences and the quartiles, being one and one-half times the distance of the interquartile range away from the quartiles.

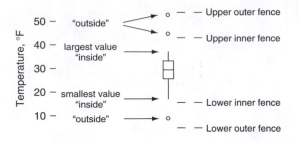

FIGURE 3.4   A schematic plot for the January 1987 Ithaca maximum temperature data. The central box portion of the figure is identical to the boxplot of the same data in Figure 3.3. The three values outside the inner fences are plotted separately. None of the values are beyond the outer fences, or far out. Notice that the whiskers extend to the most extreme inside data values, and not to the fences.

In the schematic plot, points within the inner fences are called "inside." The range of the inside points is shown by the extent of the whiskers. Data points between the inner and outer fences are referred to as being "outside," and are plotted individually in the schematic plot. Points above the upper outer fence or below the lower outer fence are called "far out," and are plotted individually with a different symbol. These differences are illustrated in Figure 3.4. In common with the boxplot, the box in a schematic plot shows the locations of the quartiles and the median.

EXAMPLE 3.2   Construction of a Schematic Plot

Figure 3.4 is a schematic plot for the January 1987 Ithaca maximum temperature data. As can be determined from Figure 3.1, the quartiles for this data are at 33°F and 26°F, and the IQR = $33 - 26 = 7$°F. From this information it is easy to compute the locations of the inner fences at $33 + (3/2)(7) = 43.5$°F and $26 - (3/2)(7) = 15.5$°F. Similarly, the outer fences are at $33 + (3)(7) = 54$°F and $26 - (3)(7) = 5$°F.

The two warmest temperatures, 53°F and 45°F, are greater than the upper inner fence, and are shown individually by circles. The coldest temperature, 9°F, is less than the lower inner fence, and is also plotted individually. The whiskers are drawn to the most extreme temperatures inside the fences, 37°F and 17°F. If the warmest temperature had been 55°F rather than 53°F, it would have fallen outside the outer fence (far out), and would have been plotted individually with a different symbol. This separate symbol for the far out points is often an asterisk. ◊

One important use of schematic plots or box plots is simultaneous graphical comparison of several batches of data. This use of schematic plots is illustrated in Figure 3.5, which shows side-by-side plots for all four of the batches of temperatures data in Table A.1. Of course it is known in advance that the maximum temperatures are warmer than the minimum temperatures, and comparing their schematic plots brings out this difference quite strongly. Apparently, Canandaigua was slightly warmer than Ithaca during this month, and more strongly so for the minimum temperatures. The Ithaca minimum temperatures were evidently more variable than the Canandaigua minimum temperatures. For both locations, the minimum temperatures are more variable than the maximum temperatures, especially in the central parts of the distributions represented by the boxes. The location of the median in the upper end of the boxes of the minimum temperature schematic plots suggests a tendency toward negative skewness, as does the inequality of the whisker lengths for the Ithaca minimum temperature data. The maximum temperatures appear to

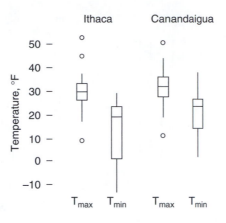

FIGURE 3.5   Side-by-side schematic plots for the January 1987 temperatures in Table A.1. The minimum temperature data for both locations are all inside, so the schematic plots are identical to ordinary boxplots.

be reasonably symmetrical for both locations. Note that none of the minimum temperature data are outside the inner fences, so that boxplots of the same data would be identical.

3.3.4 Other Boxplot Variants

Two variations on boxplots or schematic plots suggested by McGill *et al.* (1978) are sometimes used, particularly when comparing side-by-side plots. The first is to plot each box width proportional to $\sqrt{n}$. This simple variation allows plots of data having larger sample sizes to stand out and give a stronger visual impact.

The second variant is the notched boxplot or schematic plot. The boxes in these plots resemble hourglasses, with the constriction, or waist, located at the median. The lengths of the notched portions of the box differ from plot to plot, reflecting estimates of preselected confidence limits (see Chapter 5) for the median. The details of constructing these intervals are given in Velleman and Hoaglin (1981). Combining both of these techniques, that is, constructing notched, variable-width plots, is straightforward. If the notched portion needs to extend beyond the quartiles, however, the overall appearance of the plot can begin to look a bit strange (an example can be seen in Graedel and Kleiner, 1985). A nice alternative to notching is to add shading or stippling in the box to span the computed interval, rather than deforming its outline with notches (e.g., Velleman 1988).

3.3.5 Histograms

The histogram is a very familiar graphical display device for a single batch of data. The range of the data is divided into class intervals or *bins*, and the number of values falling into each interval is counted. The histogram then consists of a series of rectangles whose widths are defined by the class limits implied by the bin width, and whose heights depend on the number of values in each bin. Example histograms are shown in Figure 3.6. Histograms quickly reveal such attributes of the data as location, spread, and symmetry. If the data are multimodal (i.e., more than one "hump" in the distribution of the data), this is quickly evident as well.

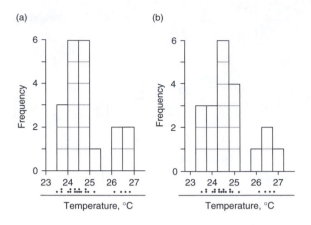

FIGURE 3.6   Histograms of the June Guayaquil temperature data in Table A.3, illustrating differences that can arise due to arbitrary shifts in the horizontal placement of the bins. This figure also illustrates that each histogram bar can be viewed as being composed of stacked "building blocks" (grey) equal in number to the number of data values in the bin. Dotplots below each histogram locate the original data.

Usually the widths of the bins are chosen to be equal. In this case the heights of the histogram bars are then simply proportional to the numbers of counts. The vertical axis can be labelled to give either the number of counts represented by each bar (the absolute frequency), or the proportion of the entire sample represented by each bar (the relative frequency). More properly, however, it is the areas of the histogram bars (rather than their heights) that are proportional to probabilities. This point becomes important if the histogram bins are chosen to have unequal widths, or when a parametric probability function (see Chapter 4) is to be superimposed on the histogram. Accordingly, it may be desirable to scale the histogram so the total area contained in the histogram bars sums to 1.

The main issue to be confronted when constructing a histogram is choice of the bin width. Intervals that are too wide will result in important details of the data being masked (the histogram is too smooth). Intervals that are too narrow will result in a plot that is irregular and difficult to interpret (the histogram is too rough). In general, narrower histogram bins are justified by larger data samples, but the nature of the data also influences the choice. One approach to selecting the binwidth, h, is to begin by computing

$$h \cong \frac{c \; IQR}{n^{1/3}}, \tag{3.12}$$

where c is a constant in the range of perhaps 2.0 to 2.6. Results given in Scott (1992) indicate that $c = 2.6$ is optimal for Gaussian (bell-shaped) data, and that smaller values are more appropriate for skewed and/or multimodal data.

The initial binwidth computed using Equation 3.12, or arrived at according to any other rule, should be regarded as just a guideline, or rule of thumb. Other considerations also will enter into the choice of the bin width, such as the practical desirability of having the class boundaries fall on values that are natural with respect to the data at hand. (Computer programs that plot histograms must use rules such as that in Equation 3.12, and one indication of the care with which the software has been written is whether the resulting histograms have natural or arbitrary bin boundaries.) For example, the January 1987 Ithaca maximum temperature data has IQR = 7°F, and $n = 31$. A bin width of 5.7°F would be suggested initially by Equation 3.12, using $c = 2.6$ since the data look at least approximately Gaussian. A natural choice in this case might be to choose 10 bins of width 5°F, yielding a histogram looking much like the stem-and-leaf display in Figure 3.1b.

3.3.6 Kernel Density Smoothing

One interpretation of the histogram is as a nonparametric estimator of the underlying probability distribution from which the data have been drawn. That is, fixed mathematical forms of the kind presented in Chapter 4 are not assumed. However, the alignment of the histogram bins on the real line is an arbitrary choice, and construction of a histogram requires essentially that each data value is rounded to the center of the bin into which it falls. For example, in Figure 3.6a the bins have been aligned so that they are centered at integer temperature values ±0.25°C, whereas the equally valid histogram in Figure 3.6b has shifted these by 0.25°C. The two histograms in Figure 3.6 present somewhat different impressions of the data, although both indicate bimodality in the data that can be traced (through the asterisks in Table A.3) to the occurrence of El Niño. Another, possibly less severe, difficulty with the histogram is that the rectangular nature of the histogram bars presents a rough appearance, and appears to imply that any value within a given bin is equally likely.

An alternative to the histogram that does not require arbitrary rounding to bin centers, and which presents a smooth result, is kernel density smoothing. The application of kernel smoothing to the frequency distribution of a data set produces the kernel density estimate, which is a nonparametric alternative to the fitting of a parametric probability density function (see Chapter 4). It is easiest to understand kernel density smoothing as an extension to the histogram. As illustrated in Figure 3.6, after rounding each data value to its bin center the histogram can be viewed as having been constructed by stacking rectangular building blocks above each bin center, with the number of the blocks equal to the number of data points in each bin. In Figure 3.6 the distribution of the data are indicated below each histogram in the form of dotplots, which locate each data value with a dot, and indicate instances of repeated data with stacks of dots.

The rectangular building blocks in Figure 3.6 each have area equal to the bin width (0.5°F), because the vertical axis is just the raw number of counts in each bin. If instead the vertical axis had been chosen so the area of each building block was $1/n$ ($= 1/20$ for these data), the resulting histograms would be quantitative estimators of the underlying probability distribution, since the total histogram area would be 1 in each case, and total probability must sum to 1.

Kernel density smoothing proceeds in an analogous way, using characteristic shapes called kernels, that are generally smoother than rectangles. Table 3.1 lists four commonly used smoothing kernels, and Figure 3.7 shows their shapes graphically. These are all nonnegative functions with unit area, that is, $\int K(t)\,dt = 1$ in each case, so each is a proper probability density function (discussed in more detail in Chapter 4). In addition, all are centered at zero. The support (value of the argument t for which $K(t) > 0$) is $-1 < t < 1$ for the triangle, quadratic and quartic kernels; and covers the entire real line for the Gaussian kernel. The kernels listed in Table 3.1 are appropriate for use with continuous data (taking on values over all or some portion of the real line). Some kernels appropriate to discrete data (able to take on only a finite number of values) are presented in Rajagopalan *et al.* (1997).

Instead of stacking rectangular kernels centered on bin midpoints (which is one way of looking at histogram construction), kernel density smoothing is achieved by stacking

TABLE 3.1   Some commonly used smoothing kernels.

Name	K(t)	Support [t for which K(t) > 0]	$1/\sigma_k$
Triangular	$1 - |t|$	$-1 < t < 1$	$\sqrt{6}$
Quadratic (Epanechnikov)	$(3/4)(1 - t^2)$	$-1 < t < 1$	$\sqrt{5}$
Quartic (Biweight)	$(15/16)(1 - t^2)^2$	$-1 < t < 1$	$\sqrt{7}$
Gaussian	$(2\pi)^{-1/2}\exp[-t^2/2]$	$-\infty < t < \infty$	1

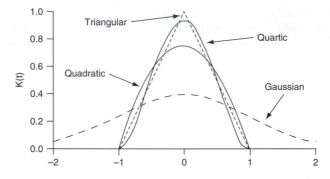

FIGURE 3.7   Four commonly used smoothing kernels defined in Table 3.1.

kernel shapes, equal in number to the number of data values, with each stacked element being centered at the data value it represents. Of course in general kernel shapes do not fit together like building blocks, but kernel density smoothing is achieved through the mathematical equivalent of stacking, by adding the heights of all the kernel functions contributing to the smoothed estimate at a given value, x_0,

$$\hat{f}(x_0) = \frac{1}{n\,h} \sum_{i=1}^{n} K\left(\frac{x_0 - x_i}{h}\right). \tag{3.13}$$

The argument within the kernel function indicates that each of the kernels employed in the smoothing (corresponding to the data values x_i close enough to the point x_0 that the kernel height is not zero) is centered at its respective data value x_i; and is scaled in width relative to the shapes as plotted in Figure 3.7 by the smoothing parameter h. Consider, for example, the triangular kernel in Table 3.1, with $t = (x_0 - x_i)/h$. The function $K[(x_0 - x_i)/h] = 1 - |(x_0 - x_i)/h|$ is an isosceles triangle with support (i.e., nonzero height) for $x_i - h < x_0 < x_i + h$; and the area within this triangle is h, because the area within $1 - |t|$ is 1 and its base has been expanded (or contracted) by a factor of h. Therefore, in Equation 3.13 the kernel heights stacked at the value x_0 will be those corresponding to any of the x_i at distances closer to x_0 than h. In order for the area under the entire function in Equation 3.13 to integrate to 1, which is desirable if the result is meant to estimate a probability density function, each of the n kernels to be superimposed should have area $1/n$. This is achieved by dividing each $K[(x_0 - x_i)/h]$, or equivalently dividing their sum, by the product nh.

   The choice of kernel type seems to be less important than choice of the smoothing parameter. The Gaussian kernel is intuitively appealing, but it is computationally slower both because of the exponential function calls, and because its infinite support leads to all data values contributing to the smoothed estimate at any x_0 (none of the n terms in Equation 3.13 are ever zero). On the other hand, all the derivatives of the resulting function will exist, and nonzero probability is estimated everywhere on the real line, whereas these are not characteristics of the other kernels listed in Table 3.1.

EXAMPLE 3.3   Kernel Density Estimates for the Guayaquil Temperature Data

Figure 3.8 shows kernel density estimates for the June Guayaquil temperature data in Table A.3, corresponding to the histograms in Figure 3.6. The four density estimates have been constructed using the quartic kernel and four choices for the smoothing parameter h, which increase from panels (a) through (d). The role of the smoothing parameter is analogous to that of the histogram bin width, also called h, in that larger values result in smoother shapes that progressively suppress details. Smaller values result in more irregular shapes that reveal more details, including the sampling variability. Figure 3.8b, plotted using $h = 0.6$, also shows the individual kernels that have been summed to produce the smoothed density estimate. Since $h = 0.6$ and the support of the quartic kernel is $-1 < t < 1$ (see Table 3.1) the widths of each of the individual kernels in Figure 3.8b is 1.2. The five repeated data values 23.7, 24.1, 24.3, 24.5 and 24.8 (cf. dotplots at the

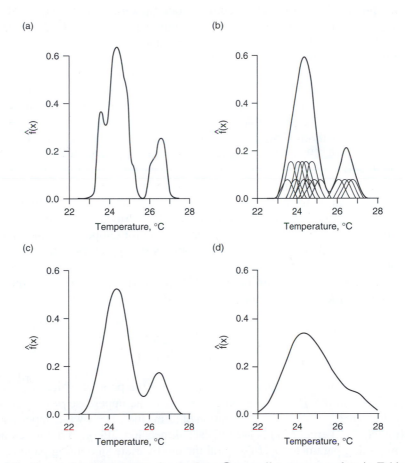

FIGURE 3.8   Kernel density estimates for the June Guayaquil temperature data in Table A.3, constructed using the quartic kernel and (a) $h = 0.3$, (b) $h = 0.6$, (c) $h = 0.92$, and (d) $h = 2.0$. Also shown in panel (b) are the individual kernels that have been added together to construct the estimate. These same data are shown as histograms in Figure 3.6.

bottom of Figure 3.6) are represented by the five taller kernels, the areas of which are each $2/n$. The remaining 10 data values are unique, and their kernels each have area $1/n$. ◊

Comparing the panels in Figure 3.8 emphasizes that a good choice for the smoothing parameter h is critical. Silverman (1986) suggests that a reasonable initial choice for use with the Gaussian kernel could be

$$h = \frac{\min\left\{0.9\,s, \frac{2}{3}\,\text{IQR}\right\}}{n^{1/5}}, \tag{3.14}$$

which indicates that less smoothing (smaller h) is justified for larger sample sizes n, although h should not decrease with sample size as quickly as the histogram bin width (Equation 3.12). Since the Gaussian kernel is intrinsically broader than the others listed in Table 3.1 (cf. Figure 3.7), smaller smoothing parameters are appropriate for these, in proportion to the reciprocals of the kernel standard deviations (Scott, 1992), which are listed in the last column in Table 3.1. For the Guayaquil temperature data, $s = 0.98$ and IQR $= 0.95$, so 2/3 IQR is smaller than $0.9s$, and Equation 3.14 yields $h = (2/3)(0.95)/20^{1/5} = 0.35$ for smoothing these data with a Gaussian kernel. But Figure 3.8 was prepared using the more compact quartic kernel, whose standard deviation is $1/\sqrt{7}$, yielding an initial choice for the smoothing parameter $h = (\sqrt{7})(0.35) = 0.92$.

When kernel smoothing is used for exploratory analyses or construction of an aesthetically pleasing data display, a recommended smoothing parameter computed in this way will often be the starting point for a subjective choice following some exploration through trial and error, and this process may even enhance the exploratory data analysis. In instances where the kernel density estimate will be used in subsequent quantitative analyses it may be preferable to estimate the smoothing parameter objectively using cross-validation methods similar to those presented in Chapter 6 (Scott, 1992; Silverman, 1986; Sharma et al., 1998). Adopting the exploratory approach, both $h = 0.92$ (see Figure 3.8c) and $h = 0.6$ (see Figure 3.8b) appear to produce reasonable balances between display of the main data features (here, the bimodality related to El Niño) and suppression of irregular sampling variability. Figure 3.8a, with $h = 0.3$, is too rough for most purposes, as it retains irregularities that can probably be ascribed to sampling variations, and (almost certainly spuriously) indicates zero probability for temperatures near 25.5°C. On the other hand, Figure 3.8d is clearly too smooth, as it suppresses entirely the bimodality in the data.

Kernel smoothing can be extended to bivariate, and higher-dimensional, data using the product-kernel estimator

$$\hat{f}(\mathbf{x}_0) = \frac{1}{n\,h_1 h_2 \cdots h_k} \sum_{i=1}^{n} \left[\prod_{j=1}^{k} K\left(\frac{x_{0j} - x_{ij}}{h_j} \right) \right]. \tag{3.15}$$

Here there are k data dimensions, x_{0j} denotes the point at which the smoothed estimate is produced in the jth of these dimensions, and the uppercase pi indicates multiplication of factors analogously to the summation of terms indicated by the uppercase sigma. The same (univariate) kernel $K(\cdot)$ is used in each dimension, although not necessarily with the same smoothing parameter h_j. In general the multivariate smoothing parameters h_j will need to be larger than for the same data smoothed alone (that is, for a univariate smoothing of the corresponding j^{th} variable in $\mathbf{x}$), and should decrease with sample size in proportion to $n^{-1/(k+4)}$. Equation 3.15 can be extended to include also nonindependence of the kernels among the k dimensions by using a multivariate probability density (for example, the

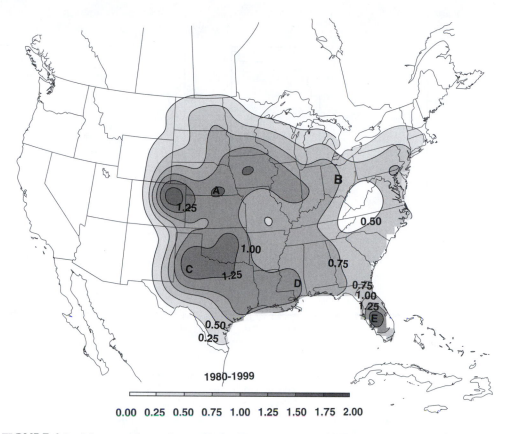

FIGURE 3.9   Mean numbers of tornado days per year in the United States, as estimated using a three-dimensional (time, latitude, longitude) kernel smoothing of daily, 80×80 km gridded tornado occurrence counts. From Brooks *et al.* (2003).

multivariate normal distribution described in Chapter 10) for the kernel (Scott, 1992; Silverman, 1986; Sharma *et al.*, 1998).

Finally, note that kernel smoothing can be applied in settings other than estimation of probability distribution functions. For example, Figure 3.9, from Brooks *et al.* (2003), shows mean numbers of tornado days per year, based on daily tornado occurrence counts in 80×80 km grid squares, for the period 1980–1999. The figure was produced after a three-dimensional smoothing using a Gaussian kernel, smoothing in time with $h = 15$ days, and smoothing in latitude and longitude with $h = 120$ km. The figure allows a smooth interpretation of the underlying data, which in raw form are very erratic in both space and time.

3.3.7 Cumulative Frequency Distributions

The cumulative frequency distribution is a display related to the histogram. It is also known as the empirical cumulative distribution function. The cumulative frequency distribution is a two-dimensional plot in which the vertical axis shows cumulative probability estimates associated with data values on the horizontal axis. That is, the plot represents relative frequency estimates for the probability that an arbitrary or random future datum will not

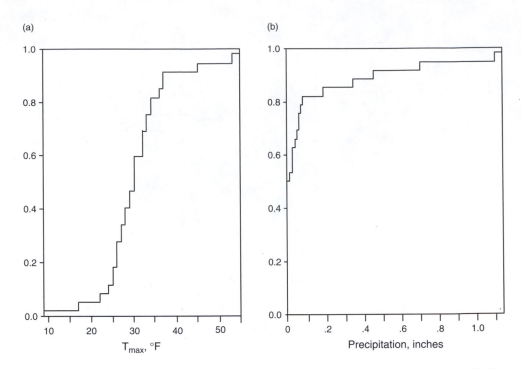

FIGURE 3.10   Empirical cumulative frequency distribution functions for the January 1987 Ithaca maximum temperature data (a), and precipitation data (b). The S shape exhibited by the temperature data is characteristic of reasonably symmetrical data, and the concave downward look exhibited by the precipitation data is characteristic of data that are skewed to the right.

exceed the corresponding value on the horizontal axis. Thus, the cumulative frequency distribution is like the integral of a histogram with arbitrarily narrow bin width. Figure 3.10 shows two empirical cumulative distribution functions, illustrating that they are step functions with probability jumps occurring at the data values. Just as histograms can be smoothed using kernel density estimators, smoothed versions of empirical cumulative distribution functions can be obtained by integrating the result of a kernel smoothing.

The vertical axes in Figure 3.10 show the empirical cumulative distribution function, $p(x)$, which can be expressed as

$$p(x) \approx \mathrm{Pr}\ \{X \leq x\}. \tag{3.16}$$

The notation on the right side of this equation can be somewhat confusing at first, but is standard in statistical work. The uppercase letter X represents the generic random variable, or the "arbitrary or random future" value referred to in the previous paragraph. The lowercase x, on both sides of Equation 3.16, represents a specific value of the random quantity. In the cumulative frequency distribution, these specific values are plotted on the horizontal axis.

In order to construct a cumulative frequency distribution, it is necessary to estimate $p(x)$ using the ranks, i, of the order statistics, $x_{(i)}$. In the literature of hydrology these estimates are known as plotting positions (e.g., Harter, 1984), reflecting their historical use in graphically comparing the empirical distribution of a batch of data with candidate parametric functions (Chapter 4) that might be used to represent them. There is substantial

TABLE 3.2 Some common plotting position estimators for cumulative probabilities corresponding to the ith order statistic, $x_{(i)}$, and the corresponding values of the parameter a in Equation 3.17.

Name	Formula	a	Interpretation
Weibull	$i/(n+1)$	0	mean of sampling distribution
Benard & Bos-Levenbach	$(i-0.3)/(n+0.4)$	0.3	approximate median of sampling distribution
Tukey	$(i-1/3)/(n+1/3)$	1/3	approximate median of sampling distribution
Gumbel	$(i-1)/(n-1)$	1	mode of sampling distribution
Hazen	$(i-1/2)/n$	1/2	midpoints of n equal intervals on $[0, 1]$
Cunnane	$(i-2/5)/(n+1/5)$	2/5	subjective choice, commonly used in hydrology

literature devoted to equations that can be used to calculate plotting positions, and thus to estimate cumulative probabilities. Most are particular cases of the formula

$$p(x_{(i)}) = \frac{i-a}{n+1-2a}, \quad 0 \le a \le 1, \tag{3.17}$$

in which different values for the constant a result in different plotting position estimators, some of which are shown in Table 3.2. The names in this table relate to authors who proposed the various estimators, and not to particular probability distributions that may be named for the same authors. The first four plotting positions in Table 3.2 are motivated by characteristics of the sampling distributions of the cumulative probabilities associated with the order statistics. The notion of a sampling distribution is considered in more detail in Chapter 5, but briefly, think about hypothetically obtaining a large number of data samples of size n from some unknown distribution. The ith order statistics from these samples will differ somewhat from each other, but each will correspond to some cumulative probability in the distribution from which the data were drawn. In aggregate over the large number of hypothetical samples there will be a distribution—the sampling distribution—of cumulative probabilities corresponding to the ith order statistic. One way to imagine this sampling distribution is as a histogram of cumulative probabilities for, say, the smallest (or any of the other order statistics) of the n values in each of the batches. This notion of the sampling distribution for cumulative probabilities is expanded upon more fully in a climatological context by Folland and Anderson (2002).

The mathematical form of the sampling distribution of cumulative probabilities corresponding to the ith order statistic is known to be a Beta distribution (see Section 4.4.4), with parameters $p = i$ and $q = n - i + 1$, regardless of the distribution from which the x's have been independently drawn. Thus the Weibull ($a = 0$) plotting position estimator is the mean of the cumulative probabilities corresponding to a particular $x_{(i)}$, averaged over many hypothetical samples of size n. Similarly, the Benard & Bos-Levenbach ($a = 0.3$) and Tukey ($a = 1/3$) estimators approximate the medians of these distributions. The Gumbel ($a = 1$) plotting position locates the modal (single most frequent) cumulative probability, although it ascribes zero and unit cumulative probability to $x_{(1)}$ and $x_{(n)}$, respectively, leading to the unwarranted implication that the probabilities of observing data more extreme than these are zero. It is possible also to derive plotting position formulas using the reverse perspective, thinking about the sampling distributions of data quantiles x_i corresponding to particular, fixed cumulative probabilities (e.g., Cunnane,

1978; Stedinger *et al.*, 1993). Unlike the first four plotting positions in Table 3.2, the plotting positions resulting from this approach depend on the distribution from which the data have been drawn, although the Cunnane ($a = 2/5$) plotting position is a compromise approximation to many of them. In practice most of the various plotting position formulas produce quite similar results, especially when judged in relation to the intrinsic variability (Equation 4.50b) of the sampling distribution of the cumulative probabilities, which is much larger than the differences among the various plotting positions in Table 3.2. Generally very reasonable results are obtained using moderate (in terms of the parameter a) plotting positions such as Tukey or Cunnane.

Figure 3.10a shows the cumulative frequency distribution for the January 1987 Ithaca maximum temperature data, using the Tukey ($a = 1/3$) plotting position to estimate the cumulative probabilities. Figure 3.10b shows the Ithaca precipitation data displayed in the same way. For example, for the coldest of the 31 temperatures in Figure 3.10a is $x_{(1)} = 9°F$, and $p(x_{(1)})$ is plotted at $(1 - .333)/(31 + .333) = 0.0213$. The steepness in the center of the plot reflects the concentration of data values in the center of the distribution, and the flatness at high and low temperatures results from their being more rare. The S-shaped character of this cumulative distribution is indicative of a reasonably symmetric distribution, with comparable numbers of observations on either side of the median at a given distance from the median. The cumulative distribution function for the precipitation data (see Figure 3.10b) rises quickly on the left because of the high concentration of data values there, and then rises more slowly in the center and right of the figure because of the relatively fewer large observations. This concave downward look to the cumulative distribution function is indicative of positively skewed data. A plot for a batch of negatively skewed data would show just the reverse characteristics: a very shallow slope in the left and center of the diagram, rising steeply toward the right, yielding a function that would be concave upward.

3.4 Reexpression

It is possible that the original scale of measurement may obscure important features in a set of data. If so, an analysis can be facilitated, or may yield more revealing results if the data are first subjected to a mathematical transformation. Such transformations can also be very useful for helping atmospheric data conform to the assumptions of regression analysis (see Section 6.2), or allowing application of multivariate statistical methods that may assume Gaussian distributions (see Chapter 10). In the terminology of exploratory data analysis, such data transformations are known as *reexpression* of the data.

3.4.1 *Power Transformations*

Often data transformations are undertaken in order to make the distribution of values more nearly symmetric, and the resulting symmetry may allow use of more familiar and traditional statistical techniques. Sometimes a symmetry-producing transformation can make exploratory analyses, such as those described in this chapter, more revealing. These transformations can also aid in comparing different batches of data, for example by straightening the relationship between two variables. Another important use of transformations is to make the variations or dispersion (i.e., the spread) of one variable less dependent on the value of another variable, in which case the transformation is called *variance stabilizing*.

Undoubtedly the most commonly used (although not the only possible—see, for example, Equation 10.9) symmetry-producing transformations are the power transformations, defined by the two closely related functions

$$T_1(x) = \begin{cases} x^\lambda, & \lambda > 0 \\ \ln(x), & \lambda = 0, \\ -(x^\lambda), & \lambda < 0 \end{cases} \tag{3.18a}$$

and

$$T_2(x) = \begin{cases} \dfrac{x^\lambda - 1}{\lambda}, & \lambda \neq 0 \\ \ln(x), & \lambda = 0. \end{cases} \tag{3.18b}$$

These transformations are useful when dealing with unimodal (single-humped) distributions of positive data variables. Each of these functions defines a family of transformations indexed by the single parameter λ. The name power transformation derives from the fact that the important work of these transformations—changing the shape of the data distribution—is accomplished by the exponentiation, or raising the data values to the power λ. Thus the sets of transformations in Equations 3.18a and 3.18b are actually quite comparable, and a particular value of λ produces the same effect on the overall shape of the data in either case. The transformations in Equation 3.18a are of a slightly simpler form, and are often employed because of the greater ease. The transformations in Equation 3.18b, also known as the Box-Cox transformations, are simply shifted and scaled versions of Equation 3.18a, and are sometimes more useful when comparing among different transformations. Also Equation 3.18b is mathematically "nicer" since the limit of the transformation as $\lambda \rightarrow 0$ is actually the function $\ln(x)$.

For transformation of data that include some zero or negative values, the original recommendation by Box and Cox (1964) was to modify the transformation by adding a positive constant to each data value, with the magnitude of the constant being large enough for all the data to be shifted onto the positive half of the real line. This easy approach is often adequate, but it is somewhat arbitrary. Yeo and Johnson (2000) have proposed a unified extension of the Box-Cox transformations that accommodate data anywhere on the real line.

In both Equations 3.18a and 3.18b, adjusting the value of the parameter λ yields specific members of an essentially continuously varying set of smooth transformations. These transformations are sometimes referred to as the "ladder of powers." A few of these transformation functions are plotted in Figure 3.11. The curves in this figure are functions specified by Equation 3.18b, although the corresponding curves from Equation 3.18a have the same shapes. Figure 3.11 makes it clear that use of the logarithmic transformation for $\lambda = 0$ fits neatly into the spectrum of the power transformations. This figure also illustrates another property of the power transformations, which is that they are all increasing functions of the original variable, x. This property is achieved in Equation 3.18a by the negative sign in the transformations with $\lambda < 1$. For the transformations in Equation 3.18b this sign reversal is achieved by dividing by λ. This strictly increasing property of the power transformations implies that they are order preserving, so that the smallest value in the original data set will correspond to the smallest value in the transformed data set, and likewise for the largest values. In fact, there will be a one-to-one correspondence

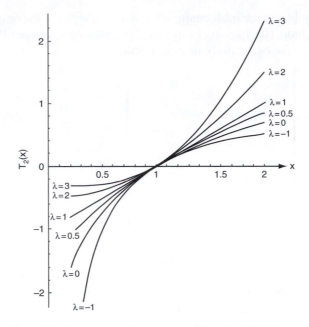

FIGURE 3.11   Graphs of the power transformations in Equation 3.18b for selected values of the transformation parameter λ. For $\lambda = 1$ the transformation is linear, and produces no change in the shape of the data. For $\lambda < 1$ the transformation reduces all data values, with larger values more strongly affected. The reverse effect is produced by transformations with $\lambda > 1$.

between all the order statistics of the original and transformed distributions. Thus the median, quartiles, and so on, of the original data will become the corresponding quantiles of the transformed data.

Clearly for $\lambda = 1$ the data remain essentially unchanged. For $\lambda > 1$ the data values are increased (except for the subtraction of $1/\lambda$ and division by λ, if Equation 3.18b is used), with the larger values being increased more than the smaller ones. Therefore power transformations with $\lambda > 1$ will help produce symmetry when applied to negatively skewed data. The reverse is true for $\lambda < 1$, where larger data values are decreased more than smaller values. Power transformations with $\lambda < 1$ are therefore generally applied to data that are originally positively skewed, in order to produce more nearly symmetric data. Figure 3.12 illustrates the mechanics of this process for an originally positively

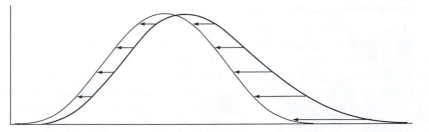

FIGURE 3.12   Effect of a power transformation with $\lambda < 1$ on a batch of data with positive skew (heavy curve). Arrows indicate that the transformation moves all the points to the left, with the larger values being moved much more. The resulting distribution (light curve) is reasonably symmetric.

skewed distribution (heavy curve). Applying a power transformation with $\lambda < 1$ reduces all the data values, but affects the larger values more strongly. An appropriate choice of λ can often produce at least approximate symmetry through this process (light curve). Choosing an excessively small or negative value for λ would produce an overcorrection, resulting in the transformed distribution being negatively skewed.

Initial inspection of an exploratory data plot such as a schematic diagram can indicate quickly the direction and approximate magnitude of the skew in a batch of data. It is thus usually clear whether a power transformation with $\lambda > 1$ or $\lambda < 1$ is appropriate, but a specific value for the exponent will not be so obvious. A number of approaches to choosing an appropriate transformation parameter have been suggested. The simplest of these is the d_λ statistic (Hinkley 1977),

$$d_\lambda = \frac{|\text{mean}(\lambda) - \text{median}(\lambda)|}{\text{spread}(\lambda)}. \tag{3.19}$$

Here, spread is some resistant measure of dispersion, such as the IQR or MAD. Each value of λ will produce a different mean, median, and spread in a particular set of data, and these dependencies on λ are indicated in the equation. The Hinkley d_λ is used to decide among power transformations essentially by trial and error, by computing its value for each of a number of different choices for λ. Usually these trial values of λ are spaced at intervals of $1/2$ or $1/4$. That choice of λ producing the smallest d_λ is then adopted to transform the data. One very easy way to do the computations is with a spreadsheet program on a desk computer.

The basis of the d_λ statistic is that, for symmetrically distributed data, the mean and median will be very close. Therefore, as successively stronger power transformations (values of λ increasingly far from 1) move the data toward symmetry, the numerator in Equation 3.19 will move toward zero. As the transformations become too strong, the numerator will begin to increase relative to the spread measure, resulting in the d_λ increasing again.

Equation 3.19 is a simple and direct approach to finding a power transformation that produces symmetry or near-symmetry in the transformed data. A more sophisticated approach was suggested in the original Box and Cox (1964) paper, which is particularly appropriate when the transformed data should have a distribution as close as possible to the bell-shaped Gaussian, for example when the results of multiple transformations will be summarized simultaneously through the multivariate Gaussian, or multivariate normal distribution (see Chapter 10). In particular, Box and Cox suggested choosing the power transformation exponent to maximize the log-likelihood function (see Section 4.6) for the Gaussian distribution

$$L(\lambda) = -\frac{n}{2} \ln[s^2(\lambda)] + (\lambda - 1) \sum_{i=1}^{n} \ln[x_i]. \tag{3.20}$$

Here n is the sample size, and $s^2(\lambda)$ is the sample variance (computed with a divisor of n rather than $n - 1$, see Equation 4.70b) of the data after transformation with the exponent λ. As was the case for using the Hinkley statistic (Equation 3.19), different values of λ are tried, and the one yielding the largest value of $L(\lambda)$ is chosen as most appropriate. It is possible that the two criteria will yield different choices for λ since Equation 3.19 addresses only symmetry of the transformed data, whereas Equation 3.20 tries to accommodate all aspects of the Gaussian distribution, including but not limited to its symmetry. Note, however, that choosing λ by maximizing Equation 3.20 does not necessarily produce transformed data that are close to Gaussian if the original data are not well suited to the transformations in Equation 3.18.

EXAMPLE 3.4   Choosing an Appropriate Power Transformation

Table 3.3 shows the 1933–1982 January Ithaca precipitation data from Table A.2 in Appendix A, sorted in ascending order and subjected to the power transformations in Equation 3.18b, for $\lambda = 1$, $\lambda = 0.5$, $\lambda = 0$ and $\lambda = -0.5$. For $\lambda = 1$ this transformation amounts only to subtracting 1 from each data value. Note that even for the negative exponent $\lambda = -0.5$ the ordering of the original data is preserved in all the transformations, so that it is easy to determine the medians and the quartiles of the original and transformed data.

Figure 3.13 shows schematic plots for the data in Table 3.3. The untransformed data (leftmost plot) are clearly positively skewed, which is usual for distributions of

TABLE 3.3   Ithaca January precipitation 1933–1982, from Table A.2 ($\lambda = 1$). The data have been sorted, with the power transformations in Equation 3.18b applied for $\lambda = 1$, $\lambda = 0.5$, $\lambda = 0$, and $\lambda = -0.5$. For $\lambda = 1$ the transformation subtracts 1 from each data value. Schematic plots of these data are shown in Figure 3.13.

Year	$\lambda = 1$	$\lambda = 0.5$	$\lambda = 0$	$\lambda = -0.5$	Year	$\lambda = 1$	$\lambda = 0.5$	$\lambda = 0$	$\lambda = -0.5$
1933	−0.56	−0.67	−0.82	−1.02	1948	0.72	0.62	0.54	0.48
1980	−0.48	−0.56	−0.65	−0.77	1960	0.75	0.65	0.56	0.49
1944	−0.46	−0.53	−0.62	−0.72	1964	0.76	0.65	0.57	0.49
1940	−0.28	−0.30	−0.33	−0.36	1974	0.84	0.71	0.61	0.53
1981	−0.13	−0.13	−0.14	−0.14	1962	0.88	0.74	0.63	0.54
1970	0.03	0.03	0.03	0.03	1951	0.98	0.81	0.68	0.58
1971	0.11	0.11	0.10	0.10	1954	1.00	0.83	0.69	0.59
1955	0.12	0.12	0.11	0.11	1936	1.08	0.88	0.73	0.61
1946	0.13	0.13	0.12	0.12	1956	1.13	0.92	0.76	0.63
1967	0.16	0.15	0.15	0.14	1965	1.17	0.95	0.77	0.64
1934	0.18	0.17	0.17	0.16	1949	1.27	1.01	0.82	0.67
1942	0.30	0.28	0.26	0.25	1966	1.38	1.09	0.87	0.70
1963	0.31	0.29	0.27	0.25	1952	1.44	1.12	0.89	0.72
1943	0.35	0.32	0.30	0.28	1947	1.50	1.16	0.92	0.74
1972	0.35	0.32	0.30	0.28	1953	1.53	1.18	0.93	0.74
1957	0.36	0.33	0.31	0.29	1935	1.69	1.28	0.99	0.78
1969	0.36	0.33	0.31	0.29	1945	1.74	1.31	1.01	0.79
1977	0.36	0.33	0.31	0.29	1939	1.82	1.36	1.04	0.81
1968	0.39	0.36	0.33	0.30	1950	1.82	1.36	1.04	0.81
1973	0.44	0.40	0.36	0.33	1959	1.94	1.43	1.08	0.83
1941	0.46	0.42	0.38	0.34	1976	2.00	1.46	1.10	0.85
1982	0.51	0.46	0.41	0.37	1937	2.66	1.83	1.30	0.95
1961	0.69	0.60	0.52	0.46	1979	3.55	2.27	1.52	1.06
1975	0.69	0.60	0.52	0.46	1958	3.90	2.43	1.59	1.10
1938	0.72	0.62	0.54	0.48	1978	5.37	3.05	1.85	1.21

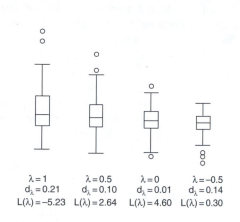

λ=1   λ=0.5   λ=0   λ=-0.5
d_λ=0.21 d_λ=0.10 d_λ=0.01 d_λ=0.14
L(λ)=-5.23   L(λ)=2.64   L(λ)=4.60   L(λ)=0.30

FIGURE 3.13   The effect of the power transformations in Equation 3.18b on the January total precipitation data for Ithaca, 1933–1982 (Table A.2). The original data (λ = 1) are strongly skewed to the right, with the largest value being far out. The square root transformation (λ = 0.5) improves the symmetry somewhat. The logarithmic transformation (λ = 0) produces a reasonably symmetric distribution. When subjected to the more extreme inverse square root transformation (λ = −0.5) the data begins to exhibit negative skewness. The logarithmic transformation would be chosen as best by both the Hinkley d_λ statistic (Equation 3.19), and the Gaussian log-likelihood (Equation 3.20).

precipitation amounts. All three of the values outside the fences are large amounts, with the largest being far out. The three other schematic plots show the results of progressively stronger power transformations with λ < 1. The logarithmic transformation (λ = 0) both minimizes the Hinkley d_λ statistic (Equation 3.19) with IQR as the measure of spread, and maximizes the Gaussian log-likelihood (Equation 3.20). The near symmetry exhibited by the schematic plot for the logarithmically transformed data supports the conclusion that it is best among the possibilities considered according to both criteria. The more extreme inverse square-root transformation (λ = −0.5) has evidently overcorrected for the positive skewness, as the three smallest amounts are now outside the lower fence. ◊

3.4.2 Standardized Anomalies

Transformations can also be useful when we are interested in working simultaneously with batches of data that are related, but not strictly comparable. One instance of this situation occurs when the data are subject to seasonal variations. Direct comparison of raw monthly temperatures, for example, will usually show little more than the dominating influence of the seasonal cycle. A record warm January will still be much colder than a record cool July. In situations of this sort, reexpression of the data in terms of standardized anomalies can be very helpful.

The standardized anomaly, z, is computed simply by subtracting the sample mean of the raw data x, and dividing by the corresponding sample standard deviation:

$$z = \frac{x - \bar{x}}{s_x} = \frac{x'}{s_x}. \tag{3.21}$$

In the jargon of the atmospheric sciences, an anomaly x' is understood to be the subtraction from a data value of a relevant average, as in the numerator of Equation 3.21. The term anomaly does not connote a data value or event that is abnormal or necessarily even unusual. The standardized anomaly in Equation 3.21 is produced by dividing the anomaly in the numerator by the corresponding standard deviation. This transformation is sometimes also referred to as a normalization. It would also be possible to construct standardized anomalies using resistant measures of location and spread, for example, subtracting the median and dividing by IQR, but this is rarely done. Use of standardized anomalies is motivated by ideas deriving from the bell-shaped Gaussian distribution, which are explained in Section 4.4.2. However, it is not necessary to assume that a batch of data follows any particular distribution in order to reexpress them in terms of standardized anomalies, and transforming non-Gaussian data according to Equation 3.21 will not make them any more Gaussian.

The idea behind the standardized anomaly is to try to remove the influences of location and spread from a batch of data. The physical units of the original data cancel, so standardized anomalies are always dimensionless quantities. Subtracting the mean produces a series of anomalies, x', located somewhere near zero. Division by the standard deviation puts excursions from the mean in different batches of data on equal footings. Collectively, a batch of data that has been transformed to a set of standardized anomalies will exhibit a mean of zero and a standard deviation of 1.

For example, we often find that summer temperatures are less variable than winter temperatures. We might find that the standard deviation for average January temperature at some location is around 3°C, but that the standard deviation for average July temperature at the same location is close to 1°C. An average July temperature 3°C colder than the long-term mean for July would then be quite unusual, corresponding to a standardized anomaly of −3. An average January temperature 3°C warmer than the long-term mean January temperature at the same location would be a fairly ordinary occurrence, corresponding to a standardized anomaly of only +1. Another way to look at the standardized anomaly is as a measure of distance, in standard deviation units, between a data value and its mean.

EXAMPLE 3.5   Expressing Climatic Data in Terms of Standardized Anomalies

Figure 3.14 illustrates the use of standardized anomalies in an operational context. The plotted points are values of an index of the Southern Oscillation Index, which

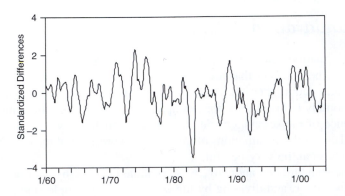

FIGURE 3.14   Standardized differences between the standardized monthly sea level pressure anomalies at Tahiti and Darwin (Southern Oscillation Index), 1960–2002. Individual monthly values have been smoothed in time.

is an index of the El-Niño-Southern Oscillation (ENSO) phenomenon that is used by the Climate Prediction Center of the U.S. National Centers for Environmental Prediction (Ropelewski and Jones, 1987). The values of this index in the figure are derived from month-by-month differences in the standardized anomalies of sea-level pressure at two tropical locations: Tahiti, in the central Pacific Ocean; and Darwin, in northern Australia. In terms of Equation 3.21 the first step toward generating Figure 3.14 is to calculate the difference $\Delta z = z_{\text{Tahiti}} - z_{\text{Darwin}}$ for each month during the years plotted. The standardized anomaly z_{Tahiti} for January 1960, for example, is computed by subtracting the average pressure for all Januaries at Tahiti from the observed monthly pressure for January 1960. This difference is then divided by the standard deviation characterizing the year-to-year variations of January atmospheric pressure at Tahiti.

Actually, the curve in Figure 3.14 is based on monthly values that are themselves standardized anomalies of this difference of standardized anomalies Δz, so that Equation 3.21 has been applied twice to the original data. The first of the two standardizations is undertaken to minimize the influences of seasonal changes in the average monthly pressures and the year-to-year variability of the monthly pressures. The second standardization, calculating the standardized anomaly of the difference Δz, ensures that the resulting index will have unit standard deviation. For reasons that will be made clear in the discussion of the Gaussian distribution in Section 4.4.2, this attribute aids qualitative judgements about the unusualness of a particular index value.

Physically, during El Niño events the center of tropical Pacific precipitation activity shifts eastward from the western Pacific (near Darwin) to the central Pacific (near Tahiti). This shift is associated with higher than average surface pressures at Darwin and lower than average surface pressures at Tahiti, which together produce a negative value for the index plotted in Figure 3.14. The exceptionally strong El-Niño event of 1982–1983 is especially prominent in this figure. ◊

3.5 Exploratory Techniques for Paired Data

The techniques presented so far in this chapter have pertained mainly to the manipulation and investigation of single batches of data. Some comparisons have been made, such as the side-by-side schematic plots in Figure 3.5. There, several distributions of data from Appendix A were plotted, but potentially important aspects of the structure of that data were not shown. In particular, the relationships between variables observed on a given day were masked when the data from each batch were separately ranked prior to construction of schematic plots. However, for each observation in one batch there is a corresponding observation from the same date in any one of the others. In this sense, the observations are paired. Elucidating relationships among sets of data pairs often yields important insights.

3.5.1 Scatterplots

The nearly universal format for graphically displaying paired data is the familiar scatterplot, or *x-y* plot. Geometrically, a scatterplot is simply a collection of points in the plane whose two Cartesian coordinates are the values of each member of the data pair. Scatterplots allow easy examination of such features in the data as trends, curvature in

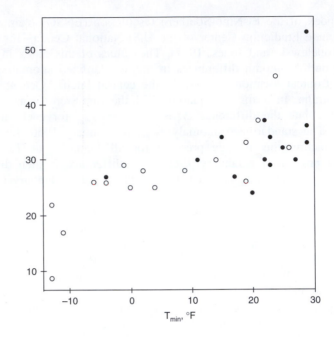

FIGURE 3.15   Scatterplot for daily maximum and minimum temperatures during January 1987 at Ithaca, New York. Closed circles represent days with at least 0.01 in. of precipitation (liquid equivalent).

the relationship, clustering of one or both variables, changes of spread of one variable as a function of the other, and extraordinary points or outliers.

Figure 3.15 is a scatterplot of the maximum and minimum temperatures for Ithaca during January 1987. It is immediately apparent that very cold maxima are associated with very cold minima, and there is a tendency for the warmer maxima to be associated with the warmer minima. This scatterplot also shows that the central range of maximum temperatures is not strongly associated with minimum temperature, since maxima near $30°$ F occur with minima anywhere in the range of $-5°$ to $20°$ F, or warmer.

Also illustrated in Figure 3.15 is a useful embellishment on the scatterplot, namely the use of more than one type of plotting symbol. Here points representing days on which at least 0.01 in. (liquid equivalent) of precipitation were recorded are plotted using the filled circles. As was evident in Example 2.1 concerning conditional probability, precipitation days tend to be associated with warmer minimum temperatures. The scatterplot indicates that the maximum temperatures tend to be warmer as well, but that the effect is not as pronounced.

3.5.2 Pearson (Ordinary) Correlation

Often an abbreviated, single-valued measure of association between two variables, say x and y, is needed. In such situations, data analysts almost automatically (and sometimes fairly uncritically) calculate a correlation coefficient. Usually, the term correlation coefficient is used to mean the "Pearson product-moment coefficient of linear correlation" between two variables x and y.

One way to view the Pearson correlation is as the ratio of the sample covariance of the two variables to the product of the two standard deviations,

$$r_{xy} = \frac{Cov\ (x, y)}{s_x s_y} = \frac{\dfrac{1}{n-1} \sum_{i=1}^{n} [(x_i - \bar{x})(y_i - \bar{y})]}{\left[\dfrac{1}{n-1} \sum_{i=1}^{n} (x_i - \bar{x})^2\right]^{1/2} \left[\dfrac{1}{n-1} \sum_{i=1}^{n} (y_i - \bar{y})^2\right]^{1/2}}$$

$$= \frac{\sum_{i=1}^{n} (x_i' y_i')}{\left[\sum_{i=1}^{n} (x_i')^2\right]^{1/2} \left[\sum_{i=1}^{n} (y_i')^2\right]^{1/2}}, \tag{3.22}$$

where the primes denote anomalies, or subtraction of mean values, as before. Note that the sample variance is a special case of the covariance (numerator in Equation 3.22), with $x = y$. One application of the covariance is in the mathematics used to describe turbulence, where the average product of, for example, the horizontal velocity anomalies u' and v' is called the *eddy covariance*, and is used in the framework of Reynolds averaging (e.g., Stull, 1988).

The Pearson product-moment correlation coefficient is neither robust nor resistant. It is not robust because strong but nonlinear relationships between the two variables x and y may not be recognized. It is not resistant since it can be extremely sensitive to one or a few outlying point pairs. Nevertheless it is often used, both because its form is well suited to mathematical manipulation, and it is closely associated with regression analysis (see Section 6.2), and the bivariate (Equation 4.33) and multivariate (see Chapter 10) Gaussian distributions.

The Pearson correlation has two important properties. First, it is bounded by -1 and 1; that is, $-1 \le r_{xy} \le 1$. If $r_{xy} = -1$ there is a perfect, negative linear association between x and y. That is, the scatterplot of y versus x consists of points all falling along one line, and that line has negative slope. Similarly if $r_{xy} = 1$ there is a perfect positive linear association. (But note that $|r_{xy}| = 1$ says nothing about the slope of the perfect linear relationship between x and y, except that it is not zero.) The second important property is that the square of the Pearson correlation, r_{xy}^2, specifies the proportion of the variability of one of the two variables that is linearly accounted for, or described, by the other. It is sometimes said that r_{xy}^2 is the proportion of the variance of one variable "explained" by the other, but this interpretation is imprecise at best and is sometimes misleading. The correlation coefficient provides no explanation at all about the relationship between the variables x and y, at least not in any physical or causative sense. It may be that x physically causes y or vice versa, but often both result physically from some other or many other quantities or processes.

The heart of the Pearson correlation coefficient is the covariance between x and y in the numerator of Equation 3.22. The denominator is in effect just a scaling constant, and is always positive. Thus, the Pearson correlation is essentially a nondimensional-ized covariance. Consider the hypothetical cloud of (x, y) data points in Figure 3.16, recognizable immediately as exhibiting positive correlation. The two perpendicular lines passing through the two sample means define four quadrants, labelled conventionally using Roman numerals. For each of the n points, the quantity within the summation in the numerator of Equation 3.22 is calculated. For points in quadrant I, both the x and y values are larger than their respective means ($x' > 0$ and $y' > 0$), so that both factors being multiplied will be positive. Therefore points in quadrant I contribute positive terms to the

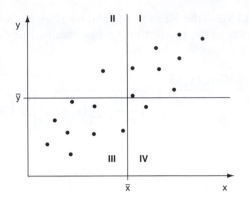

FIGURE 3.16   Hypothetical cloud of points in two dimensions, illustrating the mechanics of the Pearson correlation coefficient (Equation 3.22). The two sample means divide the plane into four quadrants, numbered I–IV.

sum in the numerator of Equation 3.22. Similarly, for points in quadrant III, both x and y are smaller than their respective means ($x' < 0$ and $y' < 0$), and again the product of their anomolies will be positive. Thus points in quadrant III will also contribute positive terms to the sum in the numerator. For points in quadrants II and IV one of the two variables x and y is above its mean and the other is below. Therefore the product in the numerator of Equation 3.22 will be negative for points in quadrants II and IV, and these points will contribute negative terms to the sum.

In Figure 3.16 most of the points are in either quadrants I or III, and therefore most of the terms in the numerator of Equation 3.22 are positive. Only the two points in quadrants II and IV contribute negative terms, and these are small in absolute value since the x and y values are relatively close to their respective means. The result is a positive sum in the numerator and therefore a positive covariance. The two standard deviations in the denominator of Equation 3.22 must always be positive, which yields a positive correlation coefficient overall for the points in Figure 3.16. If most of the points had been in quadrants II and IV, the point cloud would slope downward rather than upward, and the correlation coefficient would be negative. If the point cloud were more or less evenly distributed among the four quadrants, the correlation coefficient would be near zero, since the positive and negative terms in the sum in the numerator of Equation 3.20 would tend to cancel.

Another way of looking at the Pearson correlation coefficient is produced by moving the scaling constants in the denominator (the standard deviations), inside the summation of the numerator. This operation yields

$$r_{xy} = \frac{1}{n-1} \sum_{i=1}^{n} \left[\frac{(x_i - \bar{x})}{s_x} \frac{(y_i - \bar{y})}{s_y} \right] = \frac{1}{n-1} \sum_{i=1}^{n} z_{x_i} z_{y_i}, \qquad (3.23)$$

so that another way of looking at the Pearson correlation is as (nearly) the average product of the variables after conversion to standardized anomalies.

From the standpoint of computational economy, the formulas presented so far for the Pearson correlation are awkward. This is true whether or not the computation is to be done by hand or by a computer program. In particular, they all require two passes through a data set before the result is achieved: the first to compute the sample means, and the second to accumulate the terms involving deviations of the data values from their sample means (the anomalies). Passing twice through a data set requires twice the effort and provides double the opportunity for keying errors when using a hand calculator, and can amount to substantial increases in computer time if working with large data sets. Therefore, it is often useful to know the computational form of the Pearson correlation, which allows its computation with only one pass through a data set.

The computational form arises through an easy algebraic manipulation of the summations in the correlation coefficient. Consider the numerator in Equation 3.22. Carrying out the indicated multiplication yields

$$\sum_{i=1}^{n}[(x_i - \bar{x})(y_i - \bar{y})] = \sum_{i=1}^{n}[x_iy_i - x_i\bar{y} - y_i\bar{x} + \bar{x}\bar{y}]$$

$$= \sum_{i=1}^{n}(x_iy_i) - \bar{y}\sum_{i=1}^{n}x_i - \bar{x}\sum_{i=1}^{n}y_i + \bar{x}\bar{y}\sum_{i=1}^{n}(1)$$

$$= \sum_{i=1}^{n}(x_iy_i) - n\bar{x}\bar{y} - n\bar{x}\bar{y} + n\bar{x}\bar{y}$$

$$= \sum_{i=1}^{n}(x_iy_i) - \frac{1}{n}\left[\sum_{i=1}^{n}x_i\right]\left[\sum_{i=1}^{n}y_i\right]. \tag{3.24}$$

The second line in Equation 3.24 is arrived at through the realization that the sample means are constant, once the individual data values are determined, and therefore can be moved (factored) outside the summations. In the last term on this line there is nothing left inside the summation but the number 1, and the sum of n of these is simply n. The third step recognizes that the sample size multiplied by the sample mean yields the sum of the data values, which follows directly from the definition of the sample mean (Equation 3.2). The fourth step simply substitutes again the definition of the sample mean, to emphasize that all the quantities necessary for computing the numerator of the Pearson correlation can be known after one pass through the data. These are the sum of the x's, the sum of the y's, and the sum of their products.

It should be apparent from the similarity in form of the summations in the denominator of the Pearson correlation that analogous formulas can be derived for them or, equivalently, for the sample standard deviation. The mechanics of the derivation are exactly as followed in Equation 3.24, with the result being

$$s_x = \left[\frac{\sum x_i^2 - n\bar{x}^2}{n-1}\right]^{1/2} = \left[\frac{\sum x_i^2 - \frac{1}{n}(\sum x_i)^2}{n-1}\right]^{1/2}. \tag{3.25}$$

A similar result, of course, is obtained for y. Mathematically, Equation 3.25 is exactly equivalent to the formula for the sample standard deviation in Equation 3.6. Thus Equations 3.24 and 3.25 can be substituted into the form of the Pearson correlation given in Equations 3.22 or 3.23, to yield the computational form for the correlation coefficient

$$r_{xy} = \frac{\sum_{i=1}^{n}x_iy_i - \frac{1}{n}\left(\sum_{i=1}^{n}x_i\right)\left(\sum_{i=1}^{n}y_i\right)}{\left[\sum_{i=1}^{n}x_i^2 - \frac{1}{n}\left(\sum_{i=1}^{n}x_i\right)^2\right]^{1/2}\left[\sum_{i=1}^{n}y_i^2 - \frac{1}{n}\left(\sum_{i=1}^{n}y_i\right)^2\right]^{1/2}}. \tag{3.26}$$

Analogously, a computational form for the sample skewness coefficient (Equation 3.9) is

$$\gamma = \frac{\frac{1}{n-1}\left[\Sigma x_i^3 - \frac{3}{n}(\Sigma x_i)(\Sigma x_i^2) + \frac{2}{n^2}(\Sigma x_i)^3\right]}{s^3}.$$ (3.27)

It is important to mention a cautionary note regarding the computational forms just derived. There is a potential problem inherent in their use, which stems from the fact that they are very sensitive to round-off errors. The problem arises because each of these formulas involve the difference of two numbers that may be of comparable magnitude. To illustrate, suppose that the two terms on the last line of Equation 3.24 have each been saved to five significant digits. If the first three of these digits are the same, their difference will then be known only to two significant digits rather than five. The remedy to this potential problem is to retain as many as possible (preferably all) of the significant digits in each calculation, for example by using the double-precision representation when programming floating-point calculations on a computer.

EXAMPLE 3.6   Some Limitations of Linear Correlation

Consider the two artificial data sets in Table 3.4. The data values are few and small enough that the computational form of the Pearson correlation can be used without discarding any significant digits. For Set I, the Pearson correlation is $r_{xy} = +0.88$, and for Set II the Pearson correlation is $r_{xy} = +0.61$. Thus moderately strong linear relationships appear to be indicated for both sets of paired data.

The Pearson correlation is neither robust nor resistant, and these two small data sets have been constructed to illustrate these deficiencies. Figure 3.17 shows scatterplots of the two data sets, with Set I in panel (a) and Set II in panel (b). For Set I the relationship between x and y is actually stronger than indicated by the linear correlation of 0.88. The data points all fall very nearly on a smooth curve, but since that curve is not a straight line the Pearson coefficient underestimates the strength of the relationship. It is not robust to deviations from linearity in a relationship.

Figure 3.17b illustrates that the Pearson correlation coefficient is not resistant to outlying data. Except for the single outlying point, the data in Set II exhibit very little

TABLE 3.4   Artificial paired data sets for correlation examples.

Set I		Set II	
x	y	x	y
0	0	2	8
1	3	3	4
2	6	4	9
3	8	5	2
5	11	6	5
7	13	7	6
9	14	8	3
12	15	9	1
16	16	10	7
20	16	20	17

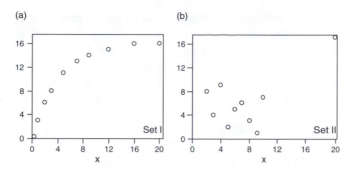

FIGURE 3.17   Scatterplots of the two artificial sets of paired data in Table 3.4. Pearson correlation for the data in panel (a) (Set I in Table 3.4) of only 0.88 underrepresents the strength of the relationship, illustrating that this measure of correlation is not robust. The Pearson correlation for the data in panel (b) (Set II) is 0.61, reflecting the overwhelming influence of the single outlying point, and illustrating lack of resistance.

structure. If anything these remaining nine points are weakly negatively correlated. However, the values $x = 20$ and $y = 17$ are so far from their respective sample means that the product of the resulting two large positive differences in the numerator of Equation 3.22 or Equation 3.23 dominates the entire sum, and erroneously indicates a moderately strong positive relationship among the ten data pairs overall. ◊

3.5.3 *Spearman Rank Correlation and Kendall's* τ

Robust and resistant alternatives to the Pearson product-moment correlation coefficient are available. The first of these is known as the Spearman rank correlation coefficient. The Spearman correlation is simply the Pearson correlation coefficient computed using the ranks of the data. Conceptually, either Equation 3.22 or Equation 3.23 is applied, but to the ranks of the data rather than to the data values themselves. For example, consider the first data pair, (2, 8), in Set II of Table 3.4. Here $x = 2$ is the smallest of the 10 values of x and therefore has rank 1. Being the eighth smallest of the 10, $y = 8$ has rank 8. Thus this first data pair would be transformed to (1,8) before computation of the correlation. Similarly, both x and y values in the outlying pair (20,17) are the largest of their respective batches of 10, and would be transformed to (10,10) before computation of the Spearman correlation coefficient.

In practice it is not necessary to use Equation 3.22, 3.23, or 3.26 to compute the Spearman rank correlation. Rather, the computations are simplified because we know in advance what the transformed values will be. Because the data are ranks, they consist simply of all the integers from 1 through the sample size n. For example, the average of the ranks of any of the four data batches in Table 3.4 is $(1+2+3+4+5+6+7+8+9+10)/10 = 5.5$. Similarly, the standard deviation (Equation 3.25) of these first ten positive integers is about 3.028. More generally, the average of the integers from 1 to n is $(n+1)/2$, and their variance is $n(n^2 - 1)/[12(n-1)]$. Taking advantage of this information, computation of the Spearman rank correlation can be simplified to

$$r_{rank} = 1 - \frac{6 \sum\limits_{i=1}^{n} D_i^2}{n(n^2 - 1)}, \tag{3.28}$$

where D_i is the difference in ranks between the ith pair of data values. In cases of ties, where a particular data value appears more than once, all of these equal values are assigned their average rank before computing the D_i's.

Kendall's τ is a second robust and resistant alternative to the conventional Pearson correlation. Kendall's τ is calculated by considering the relationships among all possible matchings of the data pairs (x_i, y_i), of which there are $n(n-1)/2$ in a sample of size n. Any such matching in which both members of one pair are larger than their counterparts in the other pair are called concordant. For example, the pairs (3, 8) and (7, 83) are concordant because both numbers in the latter pair are larger than their counterparts in the former. Match-ups in which each pair has one of the larger values, for example (3, 83) and (7, 8), are called discordant. Kendall's τ is calculated by subtracting the number of discordant pairs, N_D, from the number of concordant pairs, N_C, and dividing by the number of possible match-ups among the n observations,

$$\tau = \frac{N_C - N_D}{n(n-1)/2}. \tag{3.29}$$

Identical pairs contribute 1/2 to both N_C and N_D.

EXAMPLE 3.7   Comparison of Spearman and Kendall Correlations for the Table 3.4 Data

In Set I of Table 3.4, there is a monotonic relationship between x and y, so that each of the two batches of data are already arranged in ascending order. Therefore both members of each of the n pairs has the same rank within its own batch, and the differences D_i are all zero. Actually, the two largest y values are equal, and each would be assigned the rank 9.5. Other than this tie, the sum in the numerator of the second term in Equation 3.28 is zero, and the Spearman rank correlation is essentially 1. This result better reflects the strength of the relationship between x and y than does the Pearson correlation of 0.88. Thus, the Pearson correlation coefficient reflects the strength of linear relationships, but the Spearman rank correlation reflects the strength of monotone relationships.

Because the data in Set I exhibit an essentially perfect positive monotone relationship, all of the $10(10-1)/2 = 45$ possible match-ups between data pairs yield concordant relationships. For data sets with perfect negative monotone relationships (one of the variables is strictly decreasing as a function of the other), all comparisons among data pairs yield discordant relationships. Except for one tie, all comparisons for Set I are concordant relationships. $N_C = 45$, so that Equation 3.29 would produce $\tau = (45 - 0)/45 = 1$.

For the data in Set II, the x values are presented in ascending order, but the y values with which they are paired are jumbled. The difference of ranks for the first record is $D_1 = 1 - 8 = -7$. There are only three data pairs in Set II for which the ranks match (the fifth, sixth, and the outliers of the tenth pair). The remaining seven pairs will contribute nonzero terms to the sum in Equation 3.28, yielding $r_{rank} = 0.018$ for Set II. This result reflects much better the very weak overall relationship between x and y in Set II than does the Pearson correlation of 0.61.

Calculation of Kendall's τ for Set II is facilitated by their being sorted according to increasing values of the x variable. Given this arrangement, the number of concordant combinations can be determined by counting the number of subsequent y variables that are larger than each of the first through $(n-1)^{st}$ listings in the table. Specifically, there are two y variables larger than 8 in (2, 8) among the nine values below it, five y variables larger than 4 in (3, 4) among the eight values below it, one y variable larger than 9 in (4, 9) among the seven values below it, ..., and one y variable larger than 7 in (10, 7) in the single value below it. Together there are $2+5+1+5+3+2+2+2+1 = 23$ concordant combinations, and $45 - 23 = 22$ discordant combinations, yielding $\tau = (23 - 22)/45 = 0.022$. ◊

3.5.4 Serial Correlation

In Chapter 2 meteorological *persistence*, or the tendency for weather in successive time periods to be similar, was illustrated in terms of conditional probabilities for the two discrete events "precipitation" and "no precipitation." For continuous variables (e.g., temperature), persistence typically is characterized in terms of serial correlation, or temporal autocorrelation. The prefix "auto" in autocorrelation denotes the correlation of a variable with itself, so that temporal autocorrelation indicates the correlation of a variable with its own future and past values. Sometimes such correlations are referred to as lagged correlations. Almost always, autocorrelations are computed as Pearson product-moment correlation coefficients, although there is no reason why other forms of lagged correlation cannot be computed as well.

The process of computing autocorrelations can be visualized by imagining two copies of a sequence of observations being written, with one of the series shifted by one unit of time. This shifting is illustrated in Figure 3.18, using the January 1987 Ithaca maximum temperature data from Table A.1. This data series has been rewritten, with the middle part of the month represented by ellipses, on the first line. The same record has been recopied on the second line, but shifted to the right by one day. This process results in 30 pairs of temperatures within the box, which are available for the computation of a correlation coefficient.

Autocorrelations are computed by substituting the lagged data pairs into the formula for the Pearson correlation (Equation 3.22). For the lag-1 autocorrelation there are $n - 1$ such pairs. The only real confusion arises because the mean values for the two series will in general be slightly different. In Figure 3.18, for example, the mean of the 30 boxed values in the upper series is 29.77°F, and the mean for the boxed values in the lower series is 29.73°F. This difference arises because the upper series does not include the temperature for 1 January, and the lower series does not include the temperature for 31 January. Denoting the sample mean of the first $n - 1$ values with the subscript "−" and that of the last $n - 1$ values with the subscript "+," the lag-1 autocorrelation is

$$r_1 = \frac{\sum_{i=1}^{n-1}[(x_i - \bar{x}_-)(x_{i+1} - \bar{x}_+)]}{\left[\sum_{i=1}^{n-1}(x_i - \bar{x}_-)^2 \sum_{i=2}^{n}(x_i - \bar{x}_+)^2\right]^{1/2}}. \tag{3.30}$$

For the January 1987 Ithaca maximum temperature data, for example, $r_1 = 0.52$.

$$33 \;|\; 32 \quad 30 \quad 29 \quad 25 \quad 30 \quad 53 \;\bullet\bullet\bullet\; 17 \quad 26 \quad 27 \quad 30 \quad 34 \;|$$
$$|\; 33 \quad 32 \quad 30 \quad 29 \quad 25 \quad 30 \quad 53 \;\bullet\bullet\bullet\; 17 \quad 26 \quad 27 \quad 30 \;|\; 34$$

FIGURE 3.18 Construction of a shifted time series of January 1987 Ithaca maximum temperature data. Shifting the data by one day leaves 30 data pairs (enclosed in the box) with which to calculate the lag-1 autocorrelation coefficient.

The lag-1 autocorrelation is the most commonly computed measure of persistence, but it is also sometimes of interest to compute autocorrelations at longer lags. Conceptually, this is no more difficult than the procedure for the lag-1 autocorrelation, and computationally the only difference is that the two series are shifted by more than one time unit. Of course, as a time series is shifted increasingly relative to itself there is progressively less overlapping data to work with. Equation 3.30 can be generalized to the lag-k autocorrelation coefficient using

$$r_k = \frac{\sum_{i=1}^{n-k} [(x_i - \bar{x}_-)(x_{i+k} - \bar{x}_+)]}{\left[\sum_{i=1}^{n-k} (x_i - \bar{x}_-)^2 \sum_{i=k+1}^{n} (x_i - \bar{x}_+)^2\right]^{1/2}}. \tag{3.31}$$

Here the subscripts "$-$" and "$+$" indicate sample means over the first and last $n - k$ data values, respectively. Equation 3.31 is valid for $0 \leq k < n - 1$, although it is usually only the lowest few values of k that will be of interest. So much data is lost at large lags that correlations for roughly $k > n/2$ or $k > n/3$ rarely are computed.

In situations where a long data record is available it is sometimes acceptable to use an approximation to Equation 3.31, which simplifies the calculations and allows use of a computational form. In particular, if the data series is sufficiently long, the overall sample mean will be very close to the subset averages of the first and last $n - k$ values. The overall sample standard deviation will be close to the two subset standard deviations for the first and last $n - k$ values as well. Invoking these assumptions leads to the very commonly used approximation

$$r_k \approx \frac{\sum_{i=1}^{n-k} [(x_i - \bar{x})(x_{i+k} - \bar{x})]}{\sum_{i=1}^{n} (x_i - \bar{x})^2} \approx \frac{\sum_{i=1}^{n-k} (x_i x_{i+k}) - \frac{n-k}{n^2}\left(\sum_{i=1}^{n} x_i\right)^2}{\sum_{i=1}^{n} x_i^2 - \frac{1}{n}\left(\sum_{i=1}^{n} x_i\right)^2}. \tag{3.32}$$

3.5.5 Autocorrelation Function

Together, the collection of autocorrelations computed for various lags are called the *autocorrelation function*. Often autocorrelation functions are displayed graphically with the autocorrelations plotted as a function of lag. Figure 3.19 shows the first seven values of the autocorrelation function for the January 1987 Ithaca maximum temperature data. An autocorrelation function always begins with $r_0 = 1$, since any unshifted series of data will exhibit perfect correlation with itself. It is typical for an autocorrelation function to exhibit a more or less gradual decay toward zero as the lag k increases, reflecting the generally weaker statistical relationships between data points further removed from each other in time. It is instructive to relate this observation to the context of weather forecasting. If the autocorrelation function did not decay toward zero after a few days, making reasonably accurate forecasts at that range would be very easy: simply forecasting today's observation (the persistence forecast) or some modification of today's observation would give good results.

Sometimes it is convenient to rescale the autocorrelation function, by multiplying all the autocorrelations by the variance of the data. The result, which is proportional to the numerators of Equations 3.31 and 3.32, is called the *autocovariance* function,

$$\gamma_k = \sigma^2 r_k, \quad k = 0, 1, 2, \ldots. \tag{3.33}$$

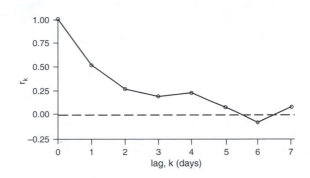

FIGURE 3.19   Sample autocorrelation function for the January 1987 Ithaca maximum temperature data. The correlation is 1 for $k = 0$, since the unlagged data are perfectly correlated with themselves. The autocorrelation function decays to essentially zero for $k \geq 5$.

The existence of autocorrelation in meteorological and climatological data has important implications regarding the applicability of some standard statistical methods to atmospheric data. In particular, uncritical application of classical methods requiring independence of data within a sample will often give badly misleading results when applied to strongly persistent series. In some cases it is possible to successfully modify these techniques, by accounting for the temporal dependence using sample autocorrelations. This topic will be discussed in Chapter 5.

3.6 Exploratory Techniques for Higher-Dimensional Data

When exploration, analysis, or comparison of matched data consisting of more than two variables are required, the methods presented so far can be applied only to pairwise subsets of the variables. Simultaneous display of three or more variables is intrinsically difficult due to a combination of geometric and cognitive problems. The geometric problem is that most available display media (i.e., paper and computer screens) are two-dimensional, so that directly plotting higher-dimensional data requires a geometric projection onto the plane, during which process information is inevitably lost. The cognitive problem derives from the fact that our brains have evolved to deal with life in a three-dimensional world, and visualizing four or more dimensions simultaneously is difficult or impossible. Nevertheless clever graphical tools have been devised for multivariate (three or more variables simultaneously) EDA. In addition to the ideas presented in this section, some multivariate graphical EDA devices designed particularly for ensemble forecasts are shown in Section 6.6.6, and a high-dimensional EDA approach based on principal component analysis is described in Section 11.7.3.

3.6.1 The Star Plot

If the number of variables, K, is not too large, each of a set of n K-dimensional observations can be displayed graphically as a star plot. The star plot is based on K coordinate axes with the same origin, spaced $360°/K$ apart on the plane. For each of the n observations, the value of the kth of the K variables is proportional to the radial plotting distance on the corresponding axis. The "star" consists of line segments connecting these points to their counterparts on adjacent radial axes.

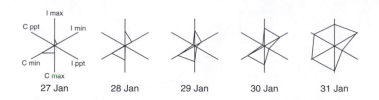

FIGURE 3.20   Star plots for the last five days in the January 1987 data in Table A.1, with axes labelled for the 27 January star only. Approximate radial symmetry in these plots reflects correlation between like variables at the two locations, and expansion of the stars through the time period indicates warmer and wetter days at the end of the month.

For example, Figure 3.20 shows star plots for the last 5 (of $n = 31$) days of the January 1987 data in Table A.1. Since there are $K = 6$ variables, the six axes are separated by angles of $360°/6 = 60°$, and each is identified with one of the variables as indicated in the panel for 27 January. In general the scales of proportionality on star plots are different for different variables, and are designed so the smallest value (or some value near but below it) corresponds to the origin, and the largest value (or some value near and above it) corresponds to the full length of the axis. Because the variables in Figure 3.20 are matched in type, the scales for the three types of variables have been chosen identically in order to better compare them. For example, the origin for both the Ithaca and Canandaigua maximum temperature axes corresponds to 10°F, and the ends of these axes correspond to 40°F. The precipitation axes have zero at the origin and 0.15 in. at the ends, so that the double-triangle shapes for 27 and 28 January indicate zero precipitation at both locations for those days. The near-symmetry of the stars suggests strong correlations for the pairs of like variables (since their axes have been plotted 180° apart), and the tendency for the stars to get larger through time indicates warmer and wetter days at the end of the month.

3.6.2 The Glyph Scatterplot

The glyph scatterplot is an extension of the ordinary scatterplot, in which the simple dots locating points on the two-dimensional plane defined by two variables are replaced by "glyphs," or more elaborate symbols that encode the values of additional variables in their sizes and/or shapes. Figure 3.15 is a primitive glyph scatterplot, with the open/closed circular glyphs indicating the binary precipitation/no-precipitation variable.

Figure 3.21 is a simple glyph scatterplot displaying three variables relating to evaluation of a small set of winter maximum temperature forecasts. The two scatterplot axes are the forecast and observed temperatures, rounded to 5°F bins, and the circular glyphs are drawn so that their areas are proportional to the numbers of forecast-observation pairs in a given 5°F × 5°F square bin. Choosing area to be proportional to the third variable (here, data counts in each bin) is preferable to radius or diameter because the glyph areas correspond better to the visual impression of size.

Essentially Figure 3.21 is a two-dimensional histogram for this bivariate set of temperature data, but is more effective than a direct generalization to three dimensions of a conventional two-dimensional histogram for a single variable. Figure 3.22 shows such a perspective-view bivariate histogram, which is fairly ineffective because projection of the three dimensions onto the two-dimensional page has introduced ambiguities about the locations of individual points. This is so, even though each point in Figure 3.22 is tied to its location on the forecast-observed plane at the apparent base of the plot through

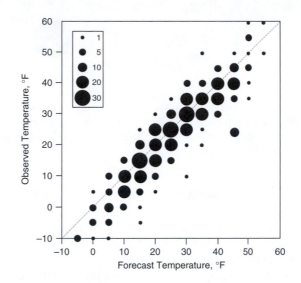

FIGURE 3.21   Glyph scatterplot of the bivariate frequency distribution of forecast and observed winter daily maximum temperatures for Minneapolis, 1980–1981 through 1985–1986. Temperatures have been rounded to 5°F intervals, and the circular glyphs have been scaled to have areas proportional to the counts (inset).

(a)

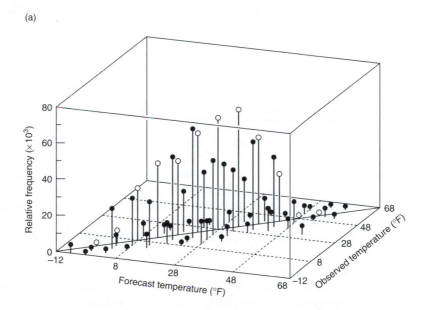

FIGURE 3.22   Bivariate histogram rendered in perspective view, of the same data plotted as a glyph scatterplot in Figure 3.21. Even though data points are located on the forecast-observation plane by the vertical tails, and points on the 1:1 diagonal are further distinguished by open circles, the projection from three dimensions to two makes the figure difficult to interpret. From Murphy *et al.* (1989).

the vertical tails, and the points falling exactly on the diagonals are indicated by open plotting symbols. Figure 3.21 speaks more clearly than Figure 3.22 about the data, for example showing immediately that there is an overforecasting bias (forecast temperatures systematically warmer than the corresponding observed temperatures, on average). An effective alternative to the glyph scatterplot in Figure 3.21 for displaying the bivariate

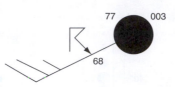

FIGURE 3.23   An elaborate glyph, known as a meteorological station model, simultaneously depicting seven quantities. When plotted on a map, two location variables (latitude and longitude) are added as well, increasing the dimensionality of the depiction to nine quantities, in what amounts to a glyph scatterplot of the weather data.

frequency distribution might be a contour plot of the bivariate kernel density estimate (see Section 3.3.6) for these data.

More elaborate glyphs than the circles in Figure 3.21 can be used to display data with more than three variables simultaneously. For example, star glyphs as described in Section 3.6.1 could be used as the plotting symbols in a glyph scatter plot. Virtually any shape that might be suggested by the data or the scientific context can be used in this way as a glyph. For example, Figure 3.23 shows a glyph that simultaneously displays seven meteorological quantities: wind direction, wind speed, sky cover, temperature, dew-point temperature, pressure, and current weather condition. When these glyphs are plotted as a scatterplot defined by longitude (horizontal axis) and latitude (vertical axis), the result is a raw weather map, which is, in effect, a graphical EDA depiction of a nine-dimensional data set describing the spatial distribution of weather at a particular time.

3.6.3 The Rotating Scatterplot

Figure 3.22 illustrates that it is generally unsatisfactory to attempt to extend the two-dimensional scatterplot to three dimensions by rendering it as a perspective view. The problem occurs because the three-dimensional counterpart of the scatterplot consists of a point cloud located in a volume rather than on the plane, and geometrically projecting this volume onto any one plane results in ambiguities about distances perpendicular to that plane. One solution to this problem is to draw larger and smaller symbols, respectively, that are closer to and further from the front of the direction of the projection, in a way that mimics the change in apparent size of physical objects with distance.

More effective, however, is to view the three-dimensional data in a computer animation known as a rotating scatterplot. At any instant the rotating scatterplot is a projection of the three-dimensional point cloud, together with its three coordinate axes for reference, onto the two-dimensional surface of the computer screen. But the plane onto which the data are projected can be changed smoothly in time, typically using the computer mouse, in a way that produces the illusion that we are viewing the points and their axes rotating around the three-dimensional coordinate origin, "inside" the computer monitor. The apparent motion can be rendered quite smoothly, and it is this continuity in time that allows a subjective sense of the shape of the data in three dimensions to be developed as we watch the changing display. In effect, the animation substitutes time for the missing third dimension.

It is not really possible to convey the power of this approach in the static form of a book page. However, an idea of how this works can be had from Figure 3.24, which shows four snapshots from a rotating scatterplot sequence, using the June Guayaquil data for temperature, pressure, and precipitation in Table A.3, with the five El Niño years

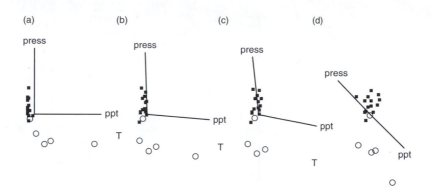

FIGURE 3.24 Four snapshots of the evolution of a three-dimensional rotating plot of the June Guayaquil data in Table A.3, in which the five El Niño years are shown as circles. The temperature axis is perpendicular to, and extends out of the page in panel (a), and the three subsequent panels show the changing perspectives as the temperature axis is rotated into the plane of the page, in a direction down and to the left. The visual illusion of a point cloud suspended in a three-dimensional space is much greater in a live rendition with continuous motion.

indicated with the open circles. Initially (see Figure 3.24a) the temperature axis is oriented out of the plane of the page, so what appears is a simple two-dimensional scatterplot of precipitation versus pressure. In Figure 3.24 (b)–(d), the temperature axis is rotated into the plane of the page, which allows a gradually changing perspective on the arrangement of the points relative to each other and relative to the projections of the coordinate axes. Figure 3.24 shows only about 90° of rotation. A "live" examination of these data with a rotating plot usually would consist of choosing an initial direction of rotation (here, down, and to the left), allowing several full rotations in that direction, and then possibly repeating the process for other directions of rotation until an appreciation of the three-dimensional shape of the point cloud has developed.

3.6.4 The Correlation Matrix

The correlation matrix is a very useful device for simultaneously displaying correlations among more than two batches of matched data. For example, the data set in Table A.1 contains matched data for six variables. Correlation coefficients can be computed for each of the 15 distinct pairings of these six variables. In general, for K variables, there are $(K)(K-1)/2$ distinct pairings, and the correlations between them can be arranged systematically in a square array, with as many rows and columns as there are matched data variables whose relationships are to be summarized. Each entry in the array, $r_{i,j}$, is indexed by the two subscripts, i and j, that point to the identity of the two variables whose correlation is represented. For example, $r_{2,3}$ would denote the correlation between the second and third variables in a list. The rows and columns in the correlation matrix are numbered correspondingly, so that the individual correlations are arranged as shown in Figure 3.25.

The correlation matrix was not designed for exploratory data analysis, but rather as a notational shorthand that allows mathematical manipulation of the correlations in the framework of linear algebra (see Chapter 9). As a format for an organized exploratory arrangement of correlations, parts of the correlation matrix are redundant, and some are

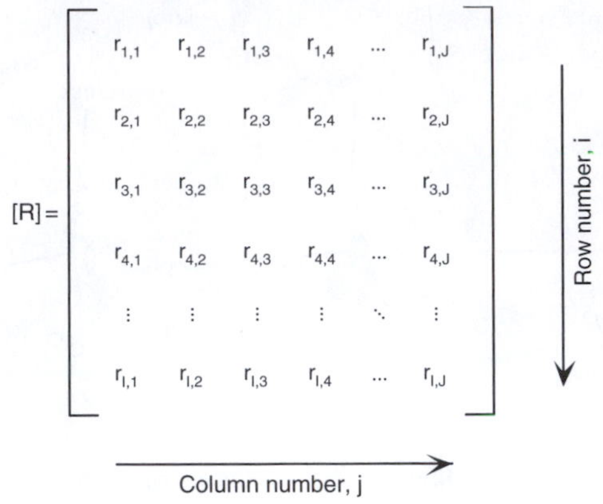

$$[R] = \begin{bmatrix} r_{1,1} & r_{1,2} & r_{1,3} & r_{1,4} & \cdots & r_{1,J} \\ r_{2,1} & r_{2,2} & r_{2,3} & r_{2,4} & \cdots & r_{2,J} \\ r_{3,1} & r_{3,2} & r_{3,3} & r_{3,4} & \cdots & r_{3,J} \\ r_{4,1} & r_{4,2} & r_{4,3} & r_{4,4} & \cdots & r_{4,J} \\ \vdots & \vdots & \vdots & \vdots & \ddots & \vdots \\ r_{I,1} & r_{I,2} & r_{I,3} & r_{I,4} & \cdots & r_{I,J} \end{bmatrix} \quad \Big\downarrow \text{Row number, } i$$

Column number, j $\longrightarrow$

FIGURE 3.25 The layout of a correlation matrix, [R]. Correlations $r_{i,j}$ between all possible pairs of variables are arranged so that the first subscript, i, indexes the row number, and the second subscript, j, indexes the column number.

simply uninformative. Consider first the diagonal elements of the matrix, arranged from the upper left to the lower right corners; that is, $r_{1,1}, r_{2,2}, r_{3,3}, \ldots, r_{K,K}$. These are the correlations of each of the variables with themselves, and are always equal to 1. Realize also that the correlation matrix is symmetric. That is, the correlation $r_{i,j}$ between variables i and j is exactly the same number as the correlation $r_{j,i}$, between the same pair of variables, so that the correlation values above and below the diagonal of 1's are mirror images of each other. Therefore, as noted earlier, only $(K)(K-1)/2$ of the K^2 entries in the correlation matrix provide distinct information.

Table 3.5 shows correlation matrices for the data in Table A.1. The matrix on the left contains Pearson product-moment correlation coefficients, and the matrix on the right contains Spearman rank correlation coefficients. As is consistent with usual practice when using correlation matrices for display rather than computational purposes, only the lower triangles of each matrix actually are printed. Omitted are the uninformative diagonal elements and the redundant upper triangular elements. Only the $(6)(5)/2 = 15$ distinct correlation values are presented.

Important features in the underlying data can be discerned by studying and comparing these two correlation matrices. First, notice that the six correlations involving only temperature variables have comparable values in both matrices. The strongest Spearman

TABLE 3.5 Correlation matrices for the data in Table A.1. Only the lower triangle of the matrices are shown, to omit redundancies and the uninformative diagonal values. The left matrix contains Pearson product-moment correlations, and the right matrix contains Spearman rank correlations.

	Ith. Ppt	Ith. Max	Ith. Min	Can. Ppt	Can. Max	Ith. Ppt	Ith. Max	Ith. Min	Can. Ppt	Can. Max
Ith. Max	−.024					.319				
Ith. Min	.287	.718				.597	.761			
Can. Ppt	.965	.018	.267			.750	.281	.546		
Can. Max	−.039	.957	.762	−.015		.267	.944	.749	.187	
Can. Min	.218	.761	.924	.188	.810	.514	.790	.916	.352	.776

correlations are between like temperature variables at the two locations. Correlations between maximum and minimum temperatures at the same location are moderately large, but weaker. The correlations involving one or both of the precipitation variables differ substantially between the two correlation matrices. There are only a few very large precipitation amounts for each of the two locations, and these tend to dominate the Pearson correlations, as explained previously. On the basis of this comparison between the correlation matrices, we therefore would suspect that the precipitation data contained some outliers, even without the benefit of knowing the type of data, or of having seen the individual numbers. The rank correlations would be expected to better reflect the degree of association for data pairs involving one or both of the precipitation variables. Subjecting the precipitation variables to a monotonic transformation appropriate to reducing the skewness would produce no changes in the matrix of Spearman correlations, but would be expected to improve the agreement between the Pearson and Spearman correlations.

Where there are a large number of variables being related through their correlations, the very large number of pairwise comparisons can be overwhelming, in which case this arrangement of the numerical values is not particularly effective as an EDA device. However, different colors or shading levels can be assigned to particular ranges of correlation, and then plotted in the same two-dimensional arrangement as the numerical correlations on which they are based, in order to more directly gain a visual appreciation of the patterns of relationship.

3.6.5 The Scatterplot Matrix

The scatterplot matrix is a graphical extension of the correlation matrix. The physical arrangement of the correlation coefficients in a correlation matrix is convenient for quick comparisons of relationships between pairs of variables, but distilling these relationships down to a single number such as a correlation coefficient inevitably hides important details. A scatterplot matrix is an arrangement of individual scatterplots according to the same logic governing the placement of individual correlation coefficients in a correlation matrix.

Figure 3.26 is a scatterplot matrix for the January 1987 data, with the scatterplots arranged in the same pattern as the correlation matrices in Table 3.5. The complexity of a scatterplot matrix can be bewildering at first, but a large amount of information about the joint behavior of the data is displayed very compactly. For example, quickly evident from a scan of the precipitation rows and columns is the fact that there are just a few large precipitation amounts at each of the two locations. Looking vertically along the column for Ithaca precipitation, or horizontally along the row for Canandaigua precipitation, the eye is drawn to the largest few data values, which appear to line up. Most of the precipitation points correspond to small amounts and therefore hug the opposite axes. Focusing on the plot of Canandaigua versus Ithaca precipitation, it is apparent that the two locations received most of their precipitation for the month on the same few days. Also evident is the association of precipitation with milder minimum temperatures that was seen in previous looks at this same data. The closer relationships between maximum and maximum, or minimum and minimum temperature variables at the two locations—as compared to the maximum versus minimum temperature relationships at one location—can also be seen clearly.

The scatterplot matrix in Figure 3.26 has been drawn without the diagonal elements in the positions that correspond to the unit correlation of a variable with itself in a correlation

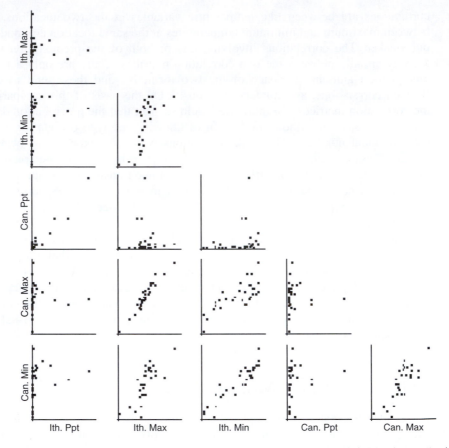

FIGURE 3.26   Scatterplot matrix for the January 1987 data in Table A.1 of Appendix A.

matrix. A scatterplot of any variable with itself would be equally dull, consisting only of a straight-line collection of points at a 45° angle. However, it is possible to use the diagonal positions in a scatterplot matrix to portray useful univariate information about the variable corresponding to that matrix position. One simple choice would be schematic plots of each of the variables in the diagonal positions. Another potentially useful choice is the Q-Q plot (Section 4.5.2) for each variable, which graphically compares the data with a reference distribution; for example, the bell-shaped Gaussian distribution.

The scatterplot matrix can be even more revealing if constructed using software allowing "brushing" of data points in related plots. When brushing, the analyst can select a point or set of points in one plot, and the corresponding points in the same data record then also light up or are otherwise differentiated in all the other plots then visible. For example, when preparing Figure 3.15, the differentiation of Ithaca temperatures occurring on days with measurable precipitation was achieved by brushing another plot (that plot was not reproduced in Figure 3.15) involving the Ithaca precipitation values. The solid circles in Figure 3.15 thus constitute a temperature scatterplot conditional on nonzero precipitation. Brushing can also sometimes reveal surprising relationships in the data by keeping the brushing action of the mouse in motion. The resulting "movie" of brushed points in the other simultaneously visible plots essentially allows the additional dimension of time to be used in differentiating relationships in the data.

3.6.6 Correlation Maps

Correlation matrices such as those in Table 3.5 are understandable and informative, so long as the number of quantities represented (six, in the case of Table 3.5) remains reasonably small. When the number of variables becomes large it may not be possible to easily make sense of the individual values, or even to fit their correlation matrix on a single page. A frequent cause of atmospheric data being excessively numerous for effective display in a correlation or scatterplot matrix is the necessity of working with data from a large number of locations. In this case the geographical arrangement of the locations can be used to organize the correlation information in map form.

Consider, for example, summarization of the correlations among surface pressure at perhaps 200 locations around the world. By the standards of the discipline, this would be only a modestly large set of data. However, this many batches of pressure data would lead to $(200)(199)/2 = 19,100$ distinct station pairs, and as many correlation coefficients. A technique that has been used successfully in such situations is construction of a series of one-point correlation maps.

Figure 3.27, taken from Bjerknes (1969), is a one-point correlation map for annual surface pressure data. Displayed on this map are contours of Pearson correlations between the pressure data at roughly 200 locations with that at Djakarta, Indonesia. Djakarta is thus the "one point" in this one-point correlation map. Essentially, the quantities being contoured are the values in the row (or column) corresponding to Djakarta in the very large correlation matrix containing all the 19,100 or so correlation values. A complete representation of that large correlation matrix in terms of one-point correlation maps would require as many maps as stations, or in this case about 200. However, not all the maps would be as interesting as Figure 3.27, although the maps for nearby stations (for example, Darwin, Australia) would look very similar.

Clearly Djakarta is located under the +1.0 on the map, since the pressure data there are perfectly correlated with themselves. Not surprisingly, pressure correlations for locations

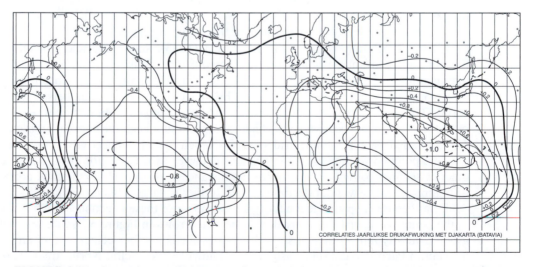

FIGURE 3.27   One-point correlation map of annual surface pressures at locations around the globe with those at Djakarta, Indonesia. The strong negative correlation of -0.8 at Easter Island is related to the El Niño-Southern Oscillation phenomenon. From Bjerknes (1969).

near Djakarta are quite high, with gradual declines toward zero at locations somewhat further away. This pattern is the spatial analog of the tailing off of the (temporal) autocorrelation function indicated in Figure 3.19. The surprising feature in Figure 3.27 is the region in the eastern tropical Pacific, centered on Easter Island, for which the correlations with Djakarta pressure are strongly negative. This negative correlation implies that in years when average pressure at Djakarta (and nearby locations, such as Darwin) are high, pressures in the eastern Pacific are low, and vice versa. This correlation pattern is an expression in the surface pressure data of the ENSO phenomenon, sketched earlier in this chapter, and is an example of what has come to be known as a *teleconnection* pattern. In the ENSO warm phase, the center of tropical Pacific convection moves eastward, producing lower than average pressures near Easter Island and higher than average pressures at Djakarta. When the precipitation shifts westward during the cold phase, pressures are low at Djakarta and high at Easter Island.

Not all geographically distributed correlation data exhibit teleconnection patterns such as the one shown in Figure 3.27. However, many large-scale fields, especially pressure (or geopotential height) fields, show one or more teleconnection patterns. A device used to simultaneously display these aspects of the large underlying correlation matrix is the teleconnectivity map. To construct a teleconnectivity map, the row (or column) for each station or gridpoint in the correlation matrix is searched for the largest negative value. The teleconnectivity value for location i, T_i, is the absolute value of that most negative correlation,

$$T_i = |\min_j r_{i,j}|. \tag{3.34}$$

Here the minimization over j (the column index for [R]) implies that all correlations $r_{i,j}$ in the ith row of [R] are searched for the smallest (most negative) value. For example, in Figure 3.27 the largest negative correlation with Djakarta pressures is with Easter Island, is -0.80. The teleconnectivity for Djakarta surface pressure would therefore be 0.80, and this value would be plotted on a teleconnectivity map at the location of Djakarta. To construct the full teleconnectivity map for surface pressure, the other 199 or so rows of the correlation matrix, each corresponding to another station, would be examined for the largest negative correlation (or, if none were negative, then the smallest positive one), and its absolute value would be plotted at the map position of that station.

Figure 3.28, from Wallace and Blackmon (1983), shows the teleconnectivity map for northern hemisphere winter 500 mb heights. The density of the shading indicates the magnitude of the individual gridpoint teleconnectivity values. The locations of local maxima of teleconnectivity are indicated by the positions of the numbers, expressed as ×100. The arrows in Figure 3.28 point from the teleconnection centers (i.e., the local maxima in T_i) to the location with which each maximum negative correlation is exhibited. The unshaded regions include gridpoints for which the teleconnectivity is relatively low. The one-point correlation maps for locations in these unshaded regions would tend to show gradual declines toward zero at increasing distances, analogously to the time correlations in Figure 3.19, but without declining much further to large negative values.

It has become apparent that a fairly large number of these teleconnection patterns exist in the atmosphere, and the many double-headed arrows in Figure 3.28 indicate that these group naturally into patterns. Especially impressive is the four-center pattern arcing from the central Pacific to the southeastern U.S., known as the Pacific-North America, or PNA pattern. Notice, however, that these patterns emerged here from a statistical, exploratory analysis of a large mass of atmospheric data. This type of work actually had its roots in the early part of the twentieth century (see Brown and Katz, 1991), and is a good example

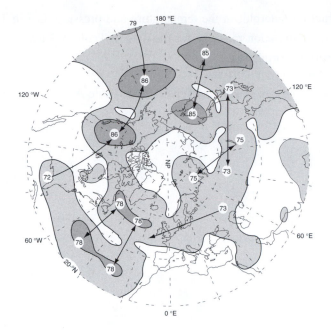

FIGURE 3.28   Teleconnectivity, or absolute value of the strongest negative correlation from each of many one-point correlation maps plotted at the base grid point, for winter 500 mb heights. From Wallace and Blackmon (1983).

of exploratory data analysis in the atmospheric sciences turning up interesting patterns in very large data sets.

3.7 Exercises

3.1. Compare the median, trimean, and the mean of the precipitation data in Table A.3.

3.2. Compute the MAD, the IQR, and the standard deviation of the pressure data in Table A.3.

3.3. Draw a stem-and-leaf display for the temperature data in Table A.3.

3.4. Compute the Yule-Kendall Index and the skewness coefficient using the temperature data in Table A.3.

3.5. Draw the empirical cumulative frequency distribution for the pressure data in Table A.3. Compare it with a histogram of the same data.

3.6. Compare the boxplot and the schematic plot representing the precipitation data in Table A.3.

3.7. Use Hinkley's d_λ to find an appropriate power transformation for the precipitation data in Table A.2 using Equation 3.18a, rather than Equation 3.18b as was done in Example 3.4. Use IQR in the denominator of Equation 3.19.

3.8. Construct side-by-side schematic plots for the candidate, and final, transformed distributions derived in Exercise 3.7. Compare the result to Figure 3.13.

3.9. Express the June 1951 temperature in Table A.3 as a standardized anomaly.

3.10. Plot the autocorrelation function up to lag 3, for the Ithaca minimum temperature data in Table A.1.

3.11. Construct a scatterplot of the temperature and pressure data in Table A.3.

3.12. Construct correlation matrices for the data in Table A.3 using
  a. The Pearson correlation.
  b. The Spearman rank correlation.

3.13. Draw and compare star plots of the data in Table A.3 for each of the years 1965 through 1969.

CHAPTER ◆ 4

Parametric Probability Distributions

4.1 Background

4.1.1 Parametric vs. Empirical Distributions

In Chapter 3, methods for exploring and displaying variations in data sets were presented. These methods had at their heart the expression of how, empirically, a particular set of data are distributed through their range. This chapter presents an approach to the summarization of data that involves imposition of particular mathematical forms, called parametric distributions, to represent variations in the underlying data. These mathematical forms amount to idealizations of real data, and are theoretical constructs.

It is worth taking a moment to understand why we would commit the violence of forcing real data to fit an abstract mold. The question is worth considering because parametric distributions *are* abstractions. They will represent real data only approximately, although in many cases the approximation can be very good indeed. Basically, there are three ways in which employing parametric probability distributions may be useful.

- **Compactness**. Particularly when dealing with large data sets, repeatedly manipulating the raw data can be cumbersome, or even severely limiting. A well-fitting parametric distribution reduces the number of quantities required for characterizing properties of the data from the full n order statistics $(x_{(1)}, x_{(2)}, x_{(3)}, \ldots, x_{(n)})$ to a few distribution parameters.
- **Smoothing and interpolation**. Real data are subject to sampling variations that lead to gaps or rough spots in the empirical distributions. For example, in Figures 3.1 and 3.10a there are no maximum temperature values between 10°F and 16°F, although certainly maximum temperatures in this range can and do occur during January at Ithaca. Imposition of a parametric distribution on this data would represent the possibility of these temperatures occurring, as well as allowing estimation of their probabilities of occurrence.
- **Extrapolation**. Estimating probabilities for events outside the range of a particular data set requires assumptions about as-yet unobserved behavior. Again referring to Figure 3.10a, the empirical cumulative probability associated with the coldest temperature, 9°F, was estimated as 0.0213 using the Tukey plotting position. The probability of a maximum temperature this cold or colder could be estimated as 0.0213,

but nothing can be said quantitatively about the probability of January maximum temperatures colder than 5°F or 0°F without the imposition of a probability model such as a parametric distribution.

The distinction has been drawn between empirical and theoretical data representations, but it should be emphasized that use of parametric probability distributions is not independent of empirical considerations. In particular, before embarking on the representation of data using theoretical functions, we must decide among candidate distribution forms, fit parameters of the chosen distribution, and check that the resulting function does, indeed, provide a reasonable fit. All three of these steps require use of real data.

4.1.2 What Is a Parametric Distribution?

A parametric distribution is an abstract mathematical form, or characteristic shape. Some of these mathematical forms arise naturally as a consequence of certain kinds of data-generating processes, and when applicable these are especially plausible candidates for concisely representing variations in a set of data. Even when there is not a strong natural basis behind the choice of a particular parametric distribution, it may be found empirically that the distribution represents a set of data very well.

The specific nature of a parametric distribution is determined by particular values for entities called parameters of that distribution. For example, the Gaussian (or "normal") distribution has as its characteristic shape the familiar symmetric bell. However, merely asserting that a particular batch of data, say average September temperatures at a location of interest, is well-represented by the Gaussian distribution is not very informative about the nature of the data, without specifying *which* Gaussian distribution represents the data. There are, in fact, infinitely many particular examples of the Gaussian distribution, corresponding to all possible values of the two distribution parameters μ and σ. But knowing, for example, that the monthly temperature for September is well-represented by the Gaussian distribution with $\mu = 60°F$ and $\sigma = 2.5°F$ conveys a large amount of information about the nature and magnitudes of the variations of September temperatures at that location.

4.1.3 Parameters vs. Statistics

There is a potential for confusion between the distribution parameters and sample statistics. Distribution parameters are abstract characteristics of a particular distribution. They succinctly represent underlying population properties. By contrast, a statistic is any quantity computed from a sample of data. Usually, the notation for sample statistics involves Roman (i.e., ordinary) letters, and that for parameters involves Greek letters.

The confusion between parameters and statistics arises because, for some common parametric distributions, certain sample statistics are good estimators for the distribution parameters. For example, the sample standard deviation, s (Equation 3.6), a statistic, can be confused with the parameter σ of the Gaussian distribution because the two often are equated when finding a particular Gaussian distribution to best match a data sample. Distribution parameters are found (fitted) using sample statistics. However, it is not always the case that the fitting process is as simple as that for the Gaussian distribution, where the sample mean is equated to the parameter μ and the sample standard deviation is equated to the parameter σ.

4.1.4 Discrete vs. Continuous Distributions

There are two distinct types of parametric distributions, corresponding to different types of data, or random variables. Discrete distributions describe random quantities (i.e., the data of interest) that can take on only particular values. That is, the allowable values are finite, or at least countably infinite. For example, a discrete random variable might take on only the values 0 or 1, or any of the nonnegative integers, or one of the colors red, yellow, or blue. A continuous random variable typically can take on any value within a specified range of the real numbers. For example, a continuous random variable might be defined on the real numbers between 0 and 1, or the nonnegative real numbers, or, for some distributions, any real number.

Strictly speaking, using a continuous distribution to represent observable data implies that the underlying observations are known to an arbitrarily large number of significant figures. Of course this is never true, but it is convenient and not too inaccurate to represent as continuous those variables that are continuous conceptually but reported discretely. Temperature and precipitation are two obvious examples that really range over some portion of the real number line, but which are usually reported to discrete multiples of 1°F and 0.01 in. in the United States. Little is lost when treating these discrete observations as samples from continuous distributions.

4.2 Discrete Distributions

There are a large number of parametric distributions appropriate for discrete random variables. Many of these are listed in the encyclopedic volume by Johnson *et al.* (1992), together with results concerning their properties. Only four of these, the binomial distribution, the geometric distribution, the negative binomial distribution, and the Poisson distribution, are presented here.

4.2.1 Binomial Distribution

The binomial distribution is one of the simplest parametric distributions, and therefore is employed often in textbooks to illustrate the use and properties of parametric distributions more generally. This distribution pertains to outcomes of situations where, on some number of occasions (sometimes called trials), one or the other of two MECE events will occur. Classically the two events have been called success and failure, but these are arbitrary labels. More generally, one of the events (say, the success) is assigned the number 1, and the other (the failure) is assigned the number zero.

The random variable of interest, X, is the number of event occurrences (given by the sum of 1's and 0's) in some number of trials. The number of trials, N, can be any positive integer, and the variable X can take on any of the nonnegative integer values from 0 (if the event of interest does not occur at all in the N trials) to N (if the event occurs on each occasion). The binomial distribution can be used to calculate probabilities for each of these $N+1$ possible values of X if two conditions are met: (1) the probability of the event occurring does not change from trial to trial (i.e., the occurrence probability is stationary), and (2) the outcomes on each of the N trials are mutually independent. These conditions are rarely strictly met, but real situations can be close enough to this ideal that the binomial distribution provides sufficiently accurate representations.

One implication of the first restriction, relating to constant occurrence probability, is that events whose probabilities exhibit regular cycles must be treated carefully. For example, the event of interest might be thunderstorm or dangerous lightning occurrence, at a location where there is a diurnal or annual variation in the probability of the event. In cases like these, subperiods (e.g., hours or months, respectively) with approximately constant occurrence probabilities usually would be analyzed separately.

The second necessary condition for applicability of the binomial distribution, relating to event independence, is usually more troublesome for atmospheric data. For example, the binomial distribution usually would not be directly applicable to daily precipitation occurrence or nonoccurrence. As illustrated by Example 2.2, there is often substantial day-to-day dependence between such events. For situations like this the binomial distribution can be generalized to a theoretical stochastic process called a Markov chain, discussed in Section 8.2. On the other hand, the year-to-year statistical dependence in the atmosphere is usually weak enough that occurrences or nonoccurrences of an event in consecutive annual periods can be considered to be effectively independent (12-month climate forecasts would be much easier if they were not!). An example of this kind will be presented later.

The usual first illustration of the binomial distribution is in relation to coin flipping. If the coin is fair, the probability of either heads or tails is 0.5, and does not change from one coin-flipping occasion (or, equivalently, from one coin) to the next. If $N>1$ coins are flipped simultaneously, the outcome on one of the coins does not affect the other outcomes. The coin-flipping situation thus satisfies all the requirements for description by the binomial distribution: dichotomous, independent events with constant probability.

Consider a game where $N = 3$ fair coins are flipped simultaneously, and we are interested in the number, X, of heads that result. The possible values of X are 0, 1, 2, and 3. These four values are a MECE partition of the sample space for X, and their probabilities must therefore sum to 1. In this simple example, you may not need to think explicitly in terms of the binomial distribution to realize that the probabilities for these four events are 1/8, 3/8, 3/8, and 1/8, respectively.

In the general case, probabilities for each of the $N+1$ values of X are given by the probability distribution function for the binomial distribution,

$$\Pr\{X = x\} = \binom{N}{x} p^x (1-p)^{N-x}, \quad x = 0, 1, \ldots, N. \qquad (4.1)$$

Here, consistent with the usage in Equation 3.16, the uppercase X indicates the random variable whose precise value is unknown, or has yet to be observed. The lowercase x denotes a specific, particular value that the random variable can take on. The binomial distribution has two parameters, N and p. The parameter p is the probability of occurrence of the event of interest (the success) on any one of the N independent trials. For a given pair of the parameters N and p, Equation 4.1 is a function associating a probability with each of the discrete values $x = 0, 1, 2, \ldots, N$, such that $\Sigma_x \Pr\{X = x\} = 1$. That is, the probability distribution function distributes probability over all events in the sample space. Note that the binomial distribution is unusual in that both of its parameters are conventionally represented by Roman letters.

The right-hand side of Equation 4.1 consists of two parts: a combinatorial part and a probability part. The combinatorial part specifies the number of distinct ways of realizing

x success outcomes from a collection of N trials. It is pronounced "N choose x," and is computed according to

$$\binom{N}{x} = \frac{N!}{x!(N-x)!}. \tag{4.2}$$

By convention, $0! = 1$. For example, when tossing $N = 3$ coins, there is only one way that $x = 3$ heads can be achieved: all three coins must come up heads. Using Equation 4.2, "three choose three" is given by $3!/(3!0!) = (1 \cdot 2 \cdot 3)/(1 \cdot 2 \cdot 3 \cdot 1) = 1$. There are three ways in which $x = 1$ can be achieved: either the first, the second, or the third coin can come up heads, with the remaining two coins coming up tails; using Equation 4.2 we obtain $3!/(1!2!) = (1 \cdot 2 \cdot 3)/(1 \cdot 1 \cdot 2) = 3$.

The probability part of Equation 4.1 follows from the multiplicative law of probability for independent events (Equation 2.12). The probability of a particular sequence of exactly x event occurrences and $N - x$ nonoccurrences is simply p multiplied by itself x times, and then multiplied by $1 - p$ (the probability of nonoccurrence) $N - x$ times. The number of these particular sequences of exactly x event occurrences and $N - x$ nonoccurrences is given by the combinatorial part, for each x, so that the product of the combinatorial and probability parts in Equation 4.1 yields the probability for x event occurrences, regardless of their locations in the sequence of N trials.

EXAMPLE 4.1 Binomial Distribution and the Freezing of Cayuga Lake, I

Consider the data in Table 4.1, which lists years during which Cayuga Lake, in central New York state, was observed to have frozen. Cayuga Lake is rather deep, and will freeze only after a long period of exceptionally cold and cloudy weather. In any given winter, the lake surface either freezes or it does not. Whether or not the lake freezes in a given winter is essentially independent of whether or not it froze in recent years. Unless there has been appreciable climate change in the region over the past two hundred years, the probability that the lake will freeze in a given year is effectively constant through the period of the data in Table 4.1. Therefore, we expect the binomial distribution to provide a good statistical description of the freezing of this lake.

In order to use the binomial distribution as a representation of the statistical properties of the lake-freezing data, we need to *fit* the distribution to the data. Fitting the distribution simply means finding particular values for the distribution parameters, p and N in this case, for which Equation 4.1 will behave as much as possible like the data in Table 4.1. The binomial distribution is somewhat unique in that the parameter N depends on the question we want to ask, rather than on the data per se. If we want to compute the probability of the lake freezing next winter, or in any single winter in the future, $N = 1$. (The special case of Equation 4.1 with $N = 1$ is called the Bernoulli distribution.) If we want to compute probabilities for the lake freezing, say, at least once during some decade in the future, $N = 10$.

TABLE 4.1 Years in which Cayuga Lake has frozen, as of 2004.

1796	1904
1816	1912
1856	1934
1875	1961
1884	1979

The binomial parameter p in this application is the probability that the lake freezes in any given year. It is natural to estimate this probability using the relative frequency of the freezing events in the data. This is a straightforward task here, except for the small complication of not knowing exactly when the climatic record starts. The written record clearly starts no later than 1796, but probably began some years before that. Suppose that the data in Table 4.1 represent a 220-year record. The 10 observed freezing events then lead to the relative frequency estimate for the binomial p of $10/220 = 0.045$.

We are now in a position to use Equation 4.1 to estimate probabilities of a variety of events relating to the freezing of this lake. The simplest kinds of events to work with have to do with the lake freezing exactly a specified number of times, x, in a specified number of years, N. For example, the probability of the lake freezing exactly once in 10 years is

$$\Pr\{X = 1\} = \binom{10}{1} 0.045^1 (1 - 0.045)^{10-1} = \frac{10!}{1!9!}(0.045)(0.955^9) = 0.30. \qquad (4.3)$$

◇

EXAMPLE 4.2   Binomial Distribution and the Freezing of Cayuga Lake, II

A somewhat harder class of events to deal with is exemplified by the problem of calculating the probability that the lake freezes at least once in 10 years. It is clear from Equation 4.3 that this probability will be no smaller than 0.30, since the probability for the compound event will be given by the sum of the probabilities $\Pr\{X = 1\} + \Pr\{X = 2\} + \cdots + \Pr\{X = 10\}$. This result follows from Equation 2.5, and the fact that these events are mutually exclusive: the lake cannot freeze both exactly once and exactly twice in the same decade.

The brute-force approach to this problem is to calculate all 10 probabilities in the sum, and then add them up. This is rather tedious, however, and quite a bit of work can be saved by giving the problem a bit more thought. Consider that the sample space here is composed of 11 MECE events: that the lake freezes exactly $0, 1, 2, \ldots$, or 10 times in a decade. Since the probabilities for these 11 events must sum to 1, it is much easier to proceed using

$$\Pr\{X \geq 1\} = 1 - \Pr\{X = 0\} = 1 - \frac{10!}{0!10!}(0.045)^0(0.955)^{10} = 0.37. \qquad (4.4)$$

◇

It is worth noting that the binomial distribution can be applied to situations that are not intrinsically binary, through a suitable redefinition of events. For example, temperature is not intrinsically binary, and is not even intrinsically discrete. However, for some applications it is of interest to consider the probability of frost; that is, $\Pr\{T \leq 32°F\}$. Together with the probability of the complementary event, $\Pr\{T > 32°F\}$, the situation is one concerning dichotomous events, and therefore could be a candidate for representation using the binomial distribution.

4.2.2 Geometric Distribution

The geometric distribution is related to the binomial distribution, describing a different aspect of the same conceptual situation. Both distributions pertain to a collection of independent trials in which one or the other of a pair of dichotomous events occurs. The trials are independent in the sense that the probability of the success occurring, p, does not depend on the outcomes of previous trials, and the sequence is stationary in the sense that p does not change over the course of the sequence (as a consequence of, for example, an annual cycle). For the geometric distribution to be applicable, the collection of trials must occur in a sequence.

The binomial distribution pertains to probabilities that particular numbers of successes will be realized in a fixed number of trials. The geometric distribution specifies probabilities for the number of trials that will be required to observe the next success. For the geometric distribution, this number of trials is the random variable X, and the probabilities corresponding to its possible values are given by the geometric probability distribution function

$$\Pr\{X = x\} = p(1-p)^{x-1}, x = 1, 2, \ldots. \tag{4.5}$$

Here X can take on any positive integer value, since at least one trial will be required in order to observe a success, and it is possible (although vanishingly probable) that we would have to wait indefinitely for this outcome. Equation 4.5 can be viewed as an application of the multiplicative law of probability for independent events, as it multiplies the probability for a success by the probability of a sequence of $x-1$ consecutive failures. The function $k = 1$ in Figure 4.1a shows an example geometric probability distribution, for the Cayuga Lake freezing probability $p = 0.045$.

Usually the geometric distribution is applied to trials that occur consecutively through time, so it is sometimes called the *waiting* distribution. The distribution has been used to describe lengths of weather regimes, or spells. One application of the geometric distribution is description of sequences of dry time periods (where we are waiting for a wet event) and wet periods (during which we are waiting for a dry event), when the time dependence of events follows the first-order Markov process (Waymire and Gupta 1981; Wilks 1999a), described in Section 8.2.

4.2.3 Negative Binomial Distribution

The negative binomial distribution is closely related to the geometric distribution, although this relationship is not indicated by its name, which comes from a technical derivation with

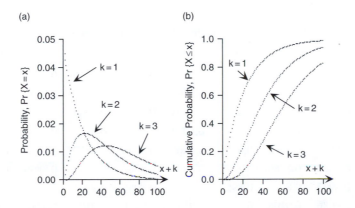

FIGURE 4.1   Probability distribution functions (a), and cumulative probability distribution functions (b), for the waiting time $x + k$ years for Cayuga Lake to freeze k times, using the Negative Binomial distribution, Equation 4.6.

parallels to a similar derivation for the binomial distribution. The probability distribution function for the negative binomial distribution is defined for nonnegative integer values of the random variable x,

$$\Pr\{X = x\} = \frac{\Gamma(k+x)}{x!\,\Gamma(k)} p^k (1-p)^x, \quad x = 0, 1, 2, \ldots \tag{4.6}$$

The distribution has two parameters, p, $0 < p < 1$ and k, $k > 0$. For integer values of k the negative binomial distribution is called the Pascal distribution, and has an interesting interpretation as an extension of the geometric distribution of waiting times for the first success in a sequence of independent Bernoulli trials with probability p. In this case, the negative binomial X pertains to the number of failures until the kth success, so that $x + k$ is the total waiting time required to observe the kth success.

The notation $\Gamma(k)$ on the left-hand side of Equation 4.6 indicates a standard mathematical function known as the gamma function, defined by the definite integral

$$\Gamma(k) = \int_0^\infty t^{k-1} e^{-t} \, dt. \tag{4.7}$$

In general, the gamma function must be evaluated numerically (e.g., Abramowitz and Stegun, 1984; Press *et al.*, 1986) or approximated using tabulated values, such as those given in Table 4.2. It satisfies the factorial recurrence relationship,

$$\Gamma(k+1) = k\Gamma(k), \tag{4.8}$$

allowing Table 4.2 to be extended indefinitely. For example, $\Gamma(3.50) = (2.50)\Gamma(2.50) = (2.50)(1.50)\Gamma(1.50) = (2.50)(1.50)(0.8862) = 3.323$. Similarly, $\Gamma(4.50) = (3.50)\Gamma(3.50) = (3.50)(3.323) = 11.631$. The gamma function is also known as the factorial function, the reason for which is especially clear when its argument is an integer (for example, in Equation 4.6 when k is an integer); that is, $\Gamma(k+1) = k!$.

With this understanding of the gamma function, it is straightforward to see the connection between the negative binomial distribution with integer k as a waiting distribution for k successes, and the geometric distribution (Equation 4.5) as a waiting distribution for the first success, in a sequence of Bernoulli trials with success probability p. Since X is the

TABLE 4.2   Values of the gamma function, $\Gamma(k)$ (Equation 4.7), for $1.00 \le k \le 1.99$.

k	0.00	0.01	0.02	0.03	0.04	0.05	0.06	0.07	0.08	0.09
1.0	1.0000	0.9943	0.9888	0.9835	0.9784	0.9735	0.9687	0.9642	0.9597	0.9555
1.1	0.9514	0.9474	0.9436	0.9399	0.9364	0.9330	0.9298	0.9267	0.9237	0.9209
1.2	0.9182	0.9156	0.9131	0.9108	0.9085	0.9064	0.9044	0.9025	0.9007	0.8990
1.3	0.8975	0.8960	0.8946	0.8934	0.8922	0.8912	0.8902	0.8893	0.8885	0.8879
1.4	0.8873	0.8868	0.8864	0.8860	0.8858	0.8857	0.8856	0.8856	0.8857	0.8859
1.5	0.8862	0.8866	0.8870	0.8876	0.8882	0.8889	0.8896	0.8905	0.8914	0.8924
1.6	0.8935	0.8947	0.8959	0.8972	0.8986	0.9001	0.9017	0.9033	0.9050	0.9068
1.7	0.9086	0.9106	0.9126	0.9147	0.9168	0.9191	0.9214	0.9238	0.9262	0.9288
1.8	0.9314	0.9341	0.9368	0.9397	0.9426	0.9456	0.9487	0.9518	0.9551	0.9584
1.9	0.9618	0.9652	0.9688	0.9724	0.9761	0.9799	0.9837	0.9877	0.9917	0.9958

number of failures before observing the kth success, the total number of trials to achieve k successes will be $x + k$, so for $k = 1$, Equations 4.5 and 4.6 pertain to the same situation. The numerator in the first factor on the right-hand side of Equation 4.6 is $\Gamma(x+1) = x!$, cancelling the $x!$ in the denominator. Realizing that $\Gamma(1) = 1$ (see Table 4.2), Equation 4.6 reduces to Equation 4.5 except that Equation 4.6 pertains to $k = 1$ additional trial since it also includes that $k = 1^{st}$ success.

EXAMPLE 4.3   Negative Binomial Distribution, and the Freezing of Cayuga Lake, III

Assuming again that the freezing of Cayuga Lake is well-represented statistically by a series of annual Bernoulli trials with $p = 0.045$, what can be said about the probability distributions for the number of years, $x + k$, required to observe k winters in which the lake freezes? As noted earlier, these probabilities will be those pertaining to X in Equation 4.6.

Figure 4.1a shows three of these negative binomial distributions, for $k = 1, 2$, and 3, shifted to the right by k years in order to show the distributions of waiting times, $x + k$. That is, the leftmost points in the three functions in Figure 4.1a all correspond to $X = 0$ in Equation 4.6. For $k = 1$ the probability distribution function is the same as for the geometric distribution (Equation 4.5), and the figure shows that the probability of freezing in the next year is simply the Bernoulli $p = 0.045$. The probabilities that year $x + 1$ will be the next freezing event decrease smoothly at a fast enough rate that probabilities for the first freeze being more than a century away are quite small. It is impossible for the lake to freeze $k = 2$ times before next year, so the first probability plotted in Figure 4.1a for $k = 2$ is at $x + k = 2$ years, and this probability is $p^2 = 0.045^2 = 0.0020$. These probabilities rise through the most likely waiting time for two freezes at $x + 2 = 23$ years before falling again, although there is a nonnegligible probability that the lake still will not have frozen twice within a century. When waiting for $k = 3$ freezes, the probability distribution of waiting times is flattened more and shifted even further into the future.

An alternative way of viewing these distributions of waiting times is through their cumulative probability distribution functions,

$$\Pr\{X \leq x\} = \sum_{X \leq x} \Pr\{X = x\}, \tag{4.9}$$

which are plotted in Figure 4.1b. Here all the probabilities for waiting times less than or equal to a waiting time of interest have been summed, analogously to Equation 3.16 for the empirical cumulative distribution function. For $k = 1$, the cumulative distribution function rises rapidly at first, indicating that the probability of the first freeze occurring within the next few decades is quite high, and that it is nearly certain that the lake will freeze next within a century (assuming that the annual freezing probability p is stationary so that, e.g., it is not decreasing through time as a consequence of a changing climate). These functions rise more slowly for the waiting times for $k = 2$ and $k = 3$ freezes; and indicate a probability of 0.94 that the lake will freeze at least twice, and a probability of 0.83 that the lake will freeze at least three times, during the next century, again assuming that the climate is stationary. ◊

 Use of the negative binomial distribution is not limited to integer values of the parameter k, and when k is allowed to take on any positive value the distribution may be appropriate for flexibly describing variations in data on counts. For example, the negative binomial distribution has been used (in slightly modified form) to represent the distributions of spells of consecutive wet and dry days (Wilks 1999a), in a way that is more flexible than Equation 4.5 because values of k different from 1 produce different shapes for the distribution, as in Figure 4.1a. In general, appropriate parameter values must be determined by the data to which the distribution will be fit. That is, specific values for the parameters p and k must be determined that will allow Equation 4.6 to look as much as possible like the empirical distribution of the data that it will be used to represent.

The simplest way to find appropriate values for the parameters, that is, to fit the distribution, is to use the method of moments. To use the method of moments we mathematically equate the sample moments and the distribution (or population) moments. Since there are two parameters, it is necessary to use two distribution moments to define them. The first moment is the mean and the second moment is the variance. In terms of the distribution parameters, the mean of the negative binomial distribution is $\mu = k(1-p)/p$, and the variance is $\sigma^2 = k(1-p)/p^2$. Estimating p and k using the method of moments involves simply setting these expressions equal to the corresponding sample moments and solving the two equations simultaneously for the parameters. That is, each data value x is an integer number of counts, and the mean and variance of these x's are calculated, and substituted into the equations

$$p = \frac{\bar{x}}{s^2},$$
(4.10a)

and

$$k = \frac{\bar{x}^2}{s^2 - \bar{x}}.$$
(4.10b)

4.2.4 Poisson Distribution

The Poisson distribution describes the numbers of discrete events occurring in a series, or a sequence, and so pertains to data on counts that can take on only nonnegative integer values. Usually the sequence is understood to be in time; for example, the occurrence of Atlantic Ocean hurricanes during a particular hurricane season. However, it is also possible to apply the Poisson distribution to counts of events occurring in one or more spatial dimensions, such as the number of gasoline stations along a particular stretch of highway, or the distribution of hailstones in a small area.

The individual events being counted are independent in the sense that they do not depend on whether or how many other events may have occurred elsewhere in the sequence. Given the rate of event occurrence, the probabilities of particular numbers of events in a given interval depends only on the size of the interval, usually the length of the time interval over which events will be counted. The numbers of occurrences do not depend on where in time the interval is located or how often events have occurred in other nonoverlapping intervals. Thus Poisson events occur randomly, but at a constant average rate. A sequence of such events is sometimes said to have been generated by a Poisson process. As was the case for the binomial distribution, strict adherence to this independence condition is often difficult to demonstrate in atmospheric data, but the Poisson distribution can still yield a useful representation if the degree of dependence is not too strong. Ideally, Poisson events should be rare enough that the probability of more than one occurring simultaneously is very small. Another way of motivating the Poisson distribution mathematically is as the limiting case of the binomial distribution, as p approaches zero and N approaches infinity.

The Poisson distribution has a single parameter, μ, that specifies the average occurrence rate. The Poisson parameter is sometimes called the *intensity*, and has physical dimensions of occurrences per unit time. The probability distribution function for the Poisson distribution is

$$\Pr\{X = x\} = \frac{\mu^x e^{-\mu}}{x!} x = 0, 1, 2, \ldots, \tag{4.11}$$

which associates probabilities with all possible numbers of occurrences, X, from zero to infinitely many. Here $e = 2.718 \ldots$ is the base of the natural logarithms. The sample space for Poisson events therefore contains (countably) infinitely many elements. Clearly the summation of Equation 4.11 for x running from zero to infinity must be convergent, and equal to 1. The probabilities associated with very large numbers of counts are vanishingly small, since the denominator in Equation 4.11 is $x!$.

To use the Poisson distribution it must be fit to a sample of data. Again, fitting the distribution means finding the a specific value for the single parameter μ that makes Equation 4.11 behave as similarly as possible to a set data at hand. For the Poisson distribution, a good way to estimate the parameter μ is by the method of moments. Fitting the Poisson distribution is thus especially easy, since its one parameter is the mean number of occurrences per unit time, which can be estimated directly as the sample average of the number occurrences per unit time.

EXAMPLE 4.4   Poisson Distribution and Annual Tornado Counts in New York State

Consider the Poisson distribution in relation to the annual tornado counts in New York state for 1959–1988, in Table 4.3. During the 30 years covered by these data, 138 tornados were reported in New York state. The average, or mean, rate of tornado occurrence is simply $138/30 = 4.6$ tornados/year, so this average is the method-of-moments estimate of the Poisson intensity for these data. Having fit the distribution by estimating a value for its parameter, the Poisson distribution can be used to compute probabilities that particular numbers of tornados will be reported in New York annually.

TABLE 4.3   Numbers of tornados reported annually in New York state, 1959–1988.

1959	3	1969	7	1979	3
1960	4	1970	4	1980	4
1961	5	1971	5	1981	3
1962	1	1972	6	1982	3
1963	3	1973	6	1983	8
1964	1	1974	6	1984	6
1965	5	1975	3	1985	7
1966	1	1976	7	1986	9
1967	2	1977	5	1987	6
1968	2	1978	8	1988	5

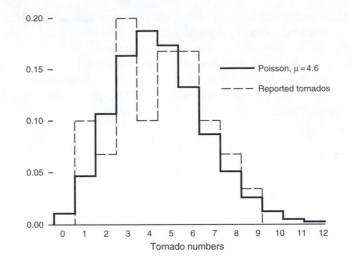

FIGURE 4.2   Histogram of number of tornados reported annually in New York state for 1959–1988 (dashed), and fitted Poisson distribution with $\mu = 4.6$ tornados/year (solid).

The first 13 of these probabilities (pertaining to zero through 12 tornados per year) are plotted in the form of a histogram in Figure 4.2, together with a histogram of the actual data.

The Poisson distribution allocates probability smoothly (given the constraint that the data are discrete) among the possible outcomes, with the most probable numbers of tornados being near the mean rate of 4.6. The distribution of the data shown by the dashed histogram resembles that of the fitted Poisson distribution, but is much more irregular, due at least in part to sampling variations. For example, there does not seem to be a physically based reason why four tornados per year should be substantially less likely than three tornados per year, or why two tornados per year should be less likely than only one tornado per year. Fitting the Poisson distribution to this data provides a sensible way to smooth out these variations, which is desirable if the irregular variations in the data histogram are not physically meaningful. Similarly, using the Poisson distribution to summarize the data allows quantitative estimation of probabilities for zero tornados per year, or for greater than nine tornados per year, even though these numbers of tornados do not happen to have been reported during 1959–1988.◊

4.3 Statistical Expectations

4.3.1 Expected Value of a Random Variable

The expected value of a random variable or function of a random variable is simply the probability-weighted average of that variable or function. This weighted average is called the expected value, although we do not necessarily expect this outcome to occur in the informal sense of an "expected" event being likely. Paradoxically, it can happen that the statistical expected value is an impossible outcome. Statistical expectations are closely tied to probability distributions, since the distributions will provide the weights or weighting function for the weighted average. The ability to work easily with statistical expectations can be a strong motivation for choosing to represent data using parametric distributions rather than empirical distribution functions.

It is easiest to see expectations as probability-weighted averages in the context of a discrete probability distribution, such as the binomial. Conventionally, the expectation operator is denoted E[ ], so that the expected value for a discrete random variable is

$$E[X] = \sum_x x \Pr\{X = x\}. \qquad (4.12)$$

The equivalent notation $<X> = E[X]$ is sometimes used for the expectation operator. The summation in Equation 4.12 is taken over all allowable values of X. For example, the expected value of X when X follows the binomial distribution is

$$E[X] = \sum_{x=0}^{N} x \binom{N}{x} p^x (1-p)^{N-x}. \qquad (4.13)$$

Here the allowable values of X are the nonnegative integers up to and including N, and each term in the summation consists of the specific value of the variable, x, multiplied by the probability of its occurrence from Equation 4.1.

The expectation $E[X]$ has a special significance, since it is the mean of the distribution of X. Distribution (or population) means are commonly denoted using the symbol μ. It is possible to analytically simplify Equation 4.12 to obtain, for the binomial distribution, the result $E[X] = Np$. Thus the mean of any binomial distribution is given by the product $\mu = Np$. Expected values for all four of the discrete probability distributions described in Section 4.2 are listed in Table 4.4, in terms of the distribution parameters. The New York tornado data in Table 4.3 constitute an example of the expected value $E[X] = 4.6$ tornados being impossible to realize in any year.

4.3.2 *Expected Value of a Function of a Random Variable*

It can be very useful to compute expectations, or probability-weighted averages, of functions of random variables, $E[g(x)]$. Since the expectation is a linear operator, expectations of functions of random variables have the following properties:

$$E[c] = c \qquad (4.14a)$$

$$E[c\, g_1(x)] = c\, E[g_1(x)] \qquad (4.14b)$$

$$E\left[\sum_{j=1}^{J} g_j(x)\right] = \sum_{j=1}^{J} E[g_j(x)], \qquad (4.14c)$$

TABLE 4.4   Expected values (means) and variances for the four discrete probability distribution functions described in Section 4.2, in terms of their distribution parameters.

Distribution	Probability Distribution Function	$\mu = E[X]$	$\sigma^2 = Var[X]$
Binomial	Equation 4.1	Np	$Np(1-p)$
Geometric	Equation 4.5	$1/p$	$(1-p)/p^2$
Negative Binomial	Equation 4.6	$k(1-p)/p$	$k(1-p)/p^2$
Poisson	Equation 4.11	μ	μ

where c is any constant, and $g_j(x)$ is any function of x. Because the constant c does not depend on x, $E[c] = \Sigma_x c\Pr\{X = x\} = c\Sigma_x \Pr\{X = x\} = c \cdot 1 = c$. Equation 4.14a and 4.14b reflect the fact that constants can be factored out when computing expectations. Equation 4.14c expresses the important property that the expectation of a sum is equal to the sum of the separate expected values.

Use of the properties expressed in Equation 4.14 can be illustrated with the expectation of the function $g(x) = (x - \mu)^2$. The expected value of this function is called the variance, and is often denoted by σ^2. Substituting into Equation 4.12, multiplying out terms, and applying the properties in Equations 4.14 yields

$$\begin{aligned}
\text{Var}[X] = E[(X - \mu)^2] &= \sum_x (x - \mu)^2 \Pr\{X = x\} \\
&= \sum_x (x^2 - 2\mu x + \mu^2) \Pr\{X = x\} \\
&= \sum_x x^2 \Pr\{X = x\} - 2\mu \sum_x x \Pr\{X = x\} + \mu^2 \sum_x \Pr\{X = x\} \\
&= E[X^2] - 2\mu E[X] + \mu^2 \cdot 1 \\
&= E[X^2] - \mu^2.
\end{aligned} \tag{4.15}$$

Notice the similarity of the first right-hand side in Equation 4.15 to the sample variance, given by the square of Equation 3.6. Similarly, the final equality in Equation 4.15 is analogous to the computational form for the sample variance, given by the square of Equation 3.25. Notice also that combining the first line of Equation 4.15 with the properties in Equation 4.14 yields

$$\text{Var}[c\, g(x)] = c^2 \text{Var}[g(x)]. \tag{4.16}$$

Variances for the four discrete distributions described in Section 4.2 are listed in Table 4.4.

EXAMPLE 4.5   Expected Value of a Function of a Binomial Random Variable

Table 4.5 illustrates the computation of statistical expectations for the binomial distribution with $N = 3$ and $p = 0.5$. These parameters correspond to the situation of simultaneously flipping three coins, and counting $X = $ the number of heads. The first column shows the possible outcomes of X, and the second column shows the probabilities for each of the outcomes, computed according to Equation 4.1.

TABLE 4.5   Binomial probabilities for $N = 3$ and $p = 0.5$, and the construction of the expectations $E[X]$ and $E[X^2]$ as probability-weighted averages.

X	$\Pr(X = x)$	$x \cdot \Pr(X = x)$	$x^2 \cdot \Pr(X = x)$
0	0.125	0.000	0.000
1	0.375	0.375	0.375
2	0.375	0.750	1.500
3	0.125	0.375	1.125
		$E[X] = 1.500$	$E[X^2] = 3.000$

The third column in Table 4.5 shows the individual terms in the probability-weighted average $E[X] = \Sigma_x[x \Pr(X = x)]$. Adding these four values yields $E[X] = 1.5$, as would be obtained by multiplying the two distribution parameters $\mu = Np$, in Table 4.4.

The fourth column in Table 4.5 similarly shows the construction of the expectation $E[X^2] = 3.0$. We might imagine this expectation in the context of a hypothetical game, in which the player receives $\$X^2$; that is, nothing if zero heads come up, \$1 if one head comes up, \$4 if two heads come up, and \$9 if three heads come up. Over the course of many rounds of this game, the long-term average payout would be $E[X^2] = \$3.00$. An individual willing to pay more than \$3 to play this game would be either foolish, or inclined toward taking risks.

Notice that the final equality in Equation 4.15 can be verified for this particular binomial distribution using Table 4.5. Here $E[X^2] - \mu^2 = 3.0 - (1.5)^2 = 0.75$, agreeing with $Var[X] = Np(1-p) = 3(0.5)(1-0.5) = 0.75$. ◊

4.4 Continuous Distributions

Most atmospheric variables can take on any of a continuum of values. Temperature, precipitation amount, geopotential height, wind speed, and other quantities are at least conceptually not restricted to integer values of the physical units in which they are measured. Even though the nature of measurement and reporting systems is such that atmospheric measurements are rounded to discrete values, the set of reportable values is large enough that most variables can still be treated as continuous quantities.

Many continuous parametric distributions exist. Those used most frequently in the atmospheric sciences are discussed later. Encyclopedic information on these and many other continuous distributions can be found in Johnson *et al.* (1994, 1995).

4.4.1 *Distribution Functions and Expected Values*

The mathematics of probability for continuous variables are somewhat different, although analogous, to those for discrete random variables. In contrast to probability calculations for discrete distributions, which involve summation over a discontinuous probability distribution function (e.g., Equation 4.1), probability calculations for continuous random variables involve integration over continuous functions called *probability density functions* (PDFs). A PDF is sometimes referred to more simply as a density.

Conventionally, the probability density function for a random variable X is denoted $f(x)$. Just as a summation of a discrete probability distribution function over all possible values of the random quantity must equal 1, the integral of any PDF over all allowable values of x must equal 1:

$$\int_x f(x)dx = 1. \tag{4.17}$$

A function cannot be a PDF unless it satisfies this equation. Furthermore, a PDF $f(x)$ must be nonnegative for all values of x. No specific limits of integration have been included in Equation 4.17, because different probability densities are defined over different ranges of the random variable (i.e., have different support).

Probability density functions are the continuous, and theoretical, analogs of the familiar histogram (see Section 3.3.5) and of the nonparametric kernel density estimate

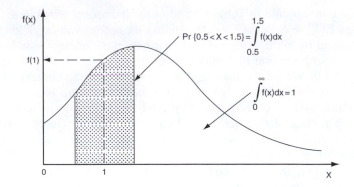

FIGURE 4.3   Hypothetical probability density function $f(x)$ for a nonnegative random variable, X. Evaluation of $f(x)$ is not, by itself, meaningful in terms of probabilities for specific values of X. Probabilities are obtained by integrating portions of $f(x)$.

(see Section 3.3.6). However, the meaning of the PDF is often initially confusing precisely because of the analogy with the histogram. In particular, the height of the density function $f(x)$, obtained when it is evaluated at a particular value of the random variable, is not in itself meaningful in the sense of defining a probability. The confusion arises because often it is not realized that probability is proportional to area, and not to height, in both the PDF and the histogram.

Figure 4.3 shows a hypothetical PDF, defined on nonnegative values of a random variable X. A probability density function can be evaluated for specific values of the random variable, say $X = 1$, but by itself $f(1)$ is not meaningful in terms of probabilities for X. In fact, since X varies continuously over some segment of the real numbers, the probability of *exactly* $X = 1$ is infinitesimally small. It is meaningful, however, to think about and compute probabilities for values of a random variable in noninfinitesimal neighborhoods around $X = 1$. Figure 4.3 shows the probability of X between 0.5 and 1.5 as the integral of the PDF between these limits.

An idea related to the PDF is that of the *cumulative distribution function* (CDF). The CDF is a function of the random variable X, given by the integral of the PDF up to a particular value of x. Thus, the CDF specifies probabilities that the random quantity X will not exceed particular values. It is therefore the continuous counterpart to the empirical CDF, Equation 3.16; and the discrete CDF, Equation 4.9. Conventionally, CDFs are denoted $F(x)$:

$$F(x) = Pr\{X \leq x\} = \int_{X \leq x} f(x)dx. \tag{4.18}$$

Again, specific integration limits have been omitted from Equation 4.18 to indicate that the integration is performed from the minimum allowable value of X to the particular value, x, that is the argument of the function. Since the values of $F(x)$ are probabilities, $0 \leq F(x) \leq 1$.

Equation 4.18 transforms a particular value of the random variable to a cumulative probability. The value of the random variable corresponding to a particular cumulative probability is given by the inverse of the cumulative distribution function,

$$F^{-1}(p) = x(F), \tag{4.19}$$

where p is the cumulative probability. That is, Equation 4.19 specifies the upper limit of the integration in Equation 4.18 that will yield a particular cumulative probability $p = F(x)$. Since this inverse of the CDF specifies the data quantile corresponding to a particular probability, Equation 4.19 is also called the *quantile function*. Depending on the parametric distribution being used, it may or may not be possible to write an explicit formula for the CDF or its inverse.

Statistical expectations also are defined for continuous random variables. As is the case for discrete variables, the expected value of a variable or a function is the probability-weighted average of that variable or function. Since probabilities for continuous random variables are computed by integrating their density functions, the expected value of a function of a random variable is given by the integral

$$E[g(x)] = \int_x g(x)f(x)dx. \tag{4.20}$$

Expectations of continuous random variables also exhibit the properties in Equations 4.14 and 4.16. For $g(x) = x$, $E[X] = \mu$ is the mean of the distribution whose PDF is $f(x)$. Similarly, the variance of a continuous variable is given by the expectation of the function $g(x) = (x - E[X])^2$,

$$Var[X] = E[(x - E[X])^2] = \int_x (x - E[X])^2 f(x)dx \tag{4.21a}$$

$$= \int_x x^2 f(x)dx - (E[X])^2 = E[X^2] - \mu^2. \tag{4.21b}$$

Note that, depending on the particular functional form of $f(x)$, some or all of the integrals in Equations 4.18, 4.20, and 4.21 may not be analytically computable, and for some distributions the integrals may not even exist.

Table 4.6 lists means and variances for the distributions to be described in this section, in terms of the distribution parameters.

TABLE 4.6 Expected values (means) and variances for continuous probability density functions described in this section, in terms of their parameters.

Distribution	PDF	E[X]	Var[X]
Gaussian	Equation 4.23	μ	σ^2
Lognormal[1]	Equation 4.30	$\exp[\mu + \sigma^2/2]$	$(\exp[\sigma^2] - 1)\exp[2\mu + \sigma^2]$
Gamma	Equation 4.38	$\alpha\beta$	$\alpha\beta^2$
Exponential	Equation 4.45	β	β^2
Chi-square	Equation 4.47	ν	2ν
Pearson III	Equation 4.48	$\zeta + \alpha\beta$	$\alpha\beta^2$
Beta	Equation 4.49	$p/(p+q)$	$(pq)/[(p+q)^2(p+q+1)]$
GEV[2]	Equation 4.54	$\zeta - \beta[1 - \Gamma(1-\kappa)]/\kappa$	$\beta^2(\Gamma[1-2\kappa] - \Gamma^2[1-\kappa])/\kappa^2$
Gumbel[3]	Equation 4.57	$\zeta + \gamma\beta$	$\beta\pi/\sqrt{6}$
Weibull	Equation 4.60	$\beta\Gamma[1 + 1/\alpha]$	$\beta^2(\Gamma[1+2/\alpha] - \Gamma^2[1+1/\alpha])$
Mixed Exponential	Equation 4.66	$w\beta_1 + (1-w)\beta_2$	$w\beta_1^2 + (1-w)\beta_2^2 + w(1-w)(\beta_1 - \beta_2)^2$

1. For the lognormal distribution, μ and σ^2 refer to the mean and variance of the log-transformed variable $y = \ln(x)$.
2. For the GEV the mean exists (is finite) only for $\kappa < 1$, and the variance exists only for $\kappa < 1/2$.
3. $\gamma = 0.57721\ldots$ is Euler's constant.

4.4.2 *Gaussian Distributions*

The Gaussian distribution plays a central role in classical statistics, and has many applications in the atmospheric sciences as well. It is sometimes also called the *normal* distribution, although this name carries the unwanted connotation that it is in some way universal, or that deviations from it are in some way unnatural. Its PDF is the bell-shaped curve, familiar even to people who have not studied statistics.

The breadth of applicability of the Gaussian distribution follows in large part from a very powerful theoretical result, known as the Central Limit Theorem. Informally, the Central Limit Theorem states that in the limit, as the sample size becomes large, the sum (or, equivalently, the arithmetic mean) of a set of independent observations will have a Gaussian distribution. This is true regardless of the distribution from which the original observations have been drawn. The observations need not even be from the same distribution! Actually, the independence of the observations is not really necessary for the shape of the resulting distribution to be Gaussian, which considerably broadens the applicability of the Central Limit Theorem for atmospheric data.

What is not clear for particular data sets is just how large the sample size must be for the Central Limit Theorem to apply. In practice this sample size depends on the distribution from which the summands are drawn. If the summed observations are themselves taken from a Gaussian distribution, the sum of any number of them (including, of course, $n = 1$) will also be Gaussian. For underlying distributions not too unlike the Gaussian (unimodal and not too asymmetrical), the sum of a modest number of observations will be nearly Gaussian. Summing daily temperatures to obtain a monthly averaged temperature is a good example of this situation. Daily temperature values can exhibit noticeable asymmetry (e.g., Figure 3.5), but are usually much more symmetrical than daily precipitation values. Conventionally, average daily temperature is approximated as the average of the daily maximum and minimum temperatures, so that the average monthly temperature is computed as

$$\bar{T} = \frac{1}{30} \sum_{i=1}^{30} \frac{T_{max}(i) + T_{min}(i)}{2}, \tag{4.22}$$

for a month with 30 days. Here the average monthly temperature derives from the sum of 60 numbers drawn from two more or less symmetrical distributions. It is not surprising, in light of the Central Limit Theorem, that monthly temperature values are often very successfully represented by Gaussian distributions.

A contrasting situation is that of the monthly total precipitation, constructed as the sum of, say, 30 daily precipitation values. There are fewer numbers going into this sum than is the case for the average monthly temperature in Equation 4.22, but the more important difference has to do with the distribution of the underlying daily precipitation amounts. Typically most daily precipitation values are zero, and most of the nonzero amounts are small. That is, the distributions of daily precipitation amounts are usually very strongly skewed to the right. Generally, the distribution of sums of 30 such values is also skewed to the right, although not so extremely. The schematic plot for $\lambda = 1$ in Figure 3.13 illustrates this asymmetry for total January precipitation at Ithaca. Note, however, that the distribution of Ithaca January precipitation totals in Figure 3.13 is much more symmetrical than the corresponding distribution for the underlying daily precipitation amounts in Table A.1. Even though the summation of 30 daily values has not produced a Gaussian distribution for the monthly totals, the shape of the distribution of monthly precipitation is much closer to the Gaussian than the very strongly skewed distribution of the daily precipitation amounts. In humid climates, the distributions of seasonal (i.e., 90-day) precipitation totals begin to approach the Gaussian, but even annual precipitation totals at arid locations can exhibit substantial positive skewness.

The PDF for the Gaussian distribution is

$$f(x) = \frac{1}{\sigma\sqrt{2\pi}} \exp\left[-\frac{(x-\mu)^2}{2\sigma^2}\right], \quad -\infty < x < \infty. \tag{4.23}$$

The two distribution parameters are the mean, μ, and the standard deviation, σ; π is the mathematical constant 3.14159 . . . . Gaussian random variables are defined on the entire real line, so Equation 4.23 is valid for $-\infty < x < \infty$. Graphing Equation 4.23 results in the familiar bell-shaped curve shown in Figure 4.4. This figure shows that the mean locates the center of this symmetrical distribution, and the standard deviation controls the degree to which the distribution spreads out. Nearly all the probability is within $\pm 3\sigma$ of the mean.

In order to use the Gaussian distribution to represent a set of data, it is necessary to fit the two distribution parameters. Good parameter estimates for this distribution are easily obtained using the method of moments. Again, the method of moments amounts to nothing more than equating the sample moments and the distribution, or population, moments. The first moment is the mean, μ, and the second moment is the variance, σ^2. Therefore, we simply estimate μ as the sample mean (Equation 3.2), and σ as the sample standard deviation (Equation 3.6).

If a data sample follows at least approximately a Gaussian distribution, these parameter estimates will make Equation 4.23 behave similarly to the data. Then, in principle, probabilities for events of interest can be obtained by integrating Equation 4.23. Practically, however, analytic integration of Equation 4.23 is impossible, so that a formula for the CDF, $F(x)$, for the Gaussian distribution does not exist. Rather, Gaussian probabilities are obtained in one of two ways. If the probabilities are needed as part of a computer program, the integral of Equation 4.23 can be economically approximated (e.g., Abramowitz and Stegun 1984) or computed by numerical integration (e.g., Press *et al.* 1986) to precision that is more than adequate. If only a few probabilities are needed, it is practical to compute them by hand using tabulated values such as those in Table B.1 in Appendix B.

In either of these two situations, a data transformation will nearly always be required. This is because Gaussian probability tables and algorithms pertain to the standard Gaussian

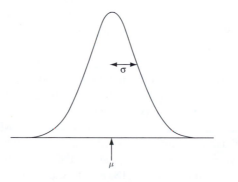

FIGURE 4.4   Probability density function for the Gaussian distribution. The mean, μ, locates the center of this symmetrical distribution, and the standard deviation, σ, controls the degree to which the distribution spreads out. Nearly all of the probability is within $\pm 3\sigma$ of the mean.

distribution; that is, the Gaussian distribution having $\mu = 0$ and $\sigma = 1$. Conventionally, the random variable described by the standard Gaussian distribution is denoted z. Its probability density simplifies to

$$\phi(z) = \frac{1}{\sqrt{2\pi}} \exp\left[-\frac{z^2}{2}\right]. \tag{4.24}$$

The symbol $\phi(z)$ is often used for the CDF of the standard Gaussian distribution, rather than $f(z)$. Any Gaussian random variable, x, can be transformed to standard form, x, simply by subtracting its mean and dividing by its standard deviation,

$$z = \frac{x - \mu}{\sigma}. \tag{4.25}$$

In practical settings, the mean and standard deviation usually need to be estimated using the corresponding sample statistics, so that we use

$$z = \frac{x - \bar{x}}{s}. \tag{4.26}$$

Note that whatever physical units characterize x cancel in this transformation, so that the standardized variable, z, is always dimensionless.

Equation 4.26 is exactly the same as the standardized anomaly of Equation 3.21. Any batch of data can be transformed by subtracting the mean and dividing by the standard deviation, and this transformation will produce transformed values having a sample mean of zero and a sample standard deviation of one. However, the transformed data will not follow a Gaussian distribution unless the untransformed variable does. The use of the standardized variable in Equation 4.26 to obtain probabilities is illustrated in the following example.

EXAMPLE 4.6 Evaluating Gaussian Probabilities

Consider a Gaussian distribution characterized by $\mu = 22.2°F$ and $\sigma = 4.4°F$. These values were fit to a set of monthly averaged January temperatures at Ithaca. Suppose we are interested in evaluating the probability that an arbitrarily selected, or future, January will have average temperature as cold as or colder than 21.4°F, the value observed in 1987 (see Table A.1). Transforming this temperature using the standardization in Equation 4.26 yields $z = (21.4°F - 22.2°F)/4.4°F = -0.18$. Thus the probability of a temperature as cold as or colder than 21.4°F is the same as the probability of a value of z as small as or smaller than $-0.18 : \Pr\{X \leq 21.4°F\} = \Pr\{Z \leq -0.18\}$.

Evaluating $\Pr\{Z \leq -0.18\}$ is easy, using Table B.1 in Appendix B, which contains cumulative probabilities for the standard Gaussian distribution. This cumulative distribution function is so commonly used that it is conventionally given its own symbol, $\Phi(z)$, rather than $F(z)$ as would be expected from Equation 4.18. Looking across the row labelled -0.1 to the column labelled 0.08 yields the desired probability, 0.4286. Evidently, there is a rather substantial probability that an average temperature this cold or colder will occur in January at Ithaca.

Notice that Table B.1 contains no rows for positive values of z. These are not necessary because the Gaussian distribution is symmetric. This means, for example, that $\Pr\{Z \geq +0.18\} = \Pr\{Z \leq -0.18\}$, since there will be equal areas under the curve in Figure 4.4 to the left of $z = -0.18$, and to the right of $z = +0.18$. Therefore, Table B.1 can be used more generally to evaluate probabilities for $z > 0$ by applying the relationship

$$\Pr\{Z \leq z\} = 1 - \Pr\{Z \leq -z\}, \tag{4.27}$$

which follows from the fact that the total area under the curve of any probability density function is 1 (Equation 4.17).

Using Equation 4.27 it is straightforward to evaluate $Pr\{Z \le +0.18\} = 1 - 0.4286 = 0.5714$. The average January temperature at Ithaca to which $z = +0.18$ corresponds is obtained by inverting Equation 4.26,

$$x = s\,z + \bar{x}. \tag{4.28}$$

The probability is 0.5714 that an average January temperature at Ithaca will be no greater than $(4.4°F)(0.18) + 22.2°F = 23.0°F$.

It is only slightly more complicated to compute probabilities for outcomes between two specific values, say Ithaca January temperatures between 20°F and 25°F. Since the event $\{T \le 20°F\}$ is a subset of the event $\{T \le 25°F\}$; the desired probability, $Pr\{20°F < T \le 25°F\}$ can be obtained by the subtraction $\Phi(z_{25}) - \Phi(z_{20})$. Here $z_{25} = (25.0°F - 22.2°F)/4.4°F = 0.64$, and $z_{20} = (20.0°F - 22.2°F)/4.4°F = -0.50$. Therefore (from Table B.1), $Pr\{20°F < T \le 25°F\} = \Phi(z_{25}) - \Phi(z_{20}) = 0.739 - 0.309 = 0.430$.

It is also occasionally required to evaluate the inverse of the standard Gaussian CDF; that is, the standard Gaussian quantile function, $\Phi^{-1}(p)$. This function specifies values of the standard Gaussian variate, z, corresponding to particular cumulative probabilities, p. Again, an explicit formula for this function cannot be written, but Φ^{-1} can be evaluated using Table B.1 in reverse. For example, to find the average January Ithaca temperature defining the lowest decile (i.e., the coldest 10% of Januaries), the body of Table B1 would be searched for $\Phi(z) = 0.10$. This cumulative probability corresponds almost exactly to $z = -1.28$. Using Equation 4.28, $z = -1.28$ corresponds to a January temperature of $(4.4°F)(-1.28) + 22.2°F = 16.6°F$. ◊

When high precision is not required for Gaussian probabilities, a "pretty good" approximation to the standard Gaussian CDF can be used,

$$\Phi(z) \approx \frac{1}{2}\left[1 \pm \sqrt{1 - \exp\left(\frac{-2\,z^2}{\pi}\right)}\,\right]. \tag{4.29}$$

The positive root is taken for $z > 0$ and the negative root is used for $z < 0$. The maximum errors produced by Equation 4.29 are about 0.003 (probability units) in magnitude, which occur at $z = \pm 1.65$. Equation 4.29 can be inverted to yield an approximation to the Gaussian quantile function, but the approximation is poor for the tail (i.e., for extreme) probabilities that are often of greatest interest.

As noted in Section 3.4.1, one approach to dealing with skewed data is to subject them to a power transformation that produces an approximately Gaussian distribution. When that power transformation is logarithmic (i.e., $\lambda = 0$ in Equation 3.18), the (original, untransformed) data are said to follow the lognormal distribution, with PDF

$$f(x) = \frac{1}{x\sigma_y\sqrt{2\pi}}\exp\left[-\frac{(\ln x - \mu_y)^2}{2\sigma_y^2}\right], \quad x > 0. \tag{4.30}$$

Here μ_y and σ_y are the mean and standard deviation, respectively, of the transformed variable, $y = \ln(x)$. Actually, the lognormal distribution is somewhat confusingly named, since the random variable x is the *anti*log of a variable y that follows a Gaussian distribution.

Parameter fitting for the lognormal is simple and straightforward: the mean and standard deviation of the log-transformed data values y—that is, μ_y and σ_y, respectively—are estimated by their sample counterparts. The relationships between these parameters, in Equation 4.30, and the mean and variance of the original variable X are

$$\mu_x = \exp\left[\mu_y + \frac{\sigma_y^2}{2}\right] \tag{4.31a}$$

and

$$\sigma_x^2 = (\exp[\sigma_y^2] - 1)\exp[2\mu_y + \sigma_y^2]. \tag{4.31b}$$

Lognormal probabilities are evaluated simply by working with the transformed variable $y = \ln(x)$, and using computational routines or probability tables for the Gaussian distribution. In this case the standard Gaussian variable

$$z = \frac{\ln(x) - \mu_y}{\sigma_y}, \tag{4.32}$$

follows a Gaussian distribution with $\mu_z = 0$ and $\sigma_z = 1$.

The lognormal distribution is sometimes somewhat arbitrarily assumed for positively skewed data. In particular, the lognormal too frequently is used without checking whether a different power transformation might produce more nearly Gaussian behavior. In general it is recommended that other candidate power transformations be investigated as explained in Section 3.4.1 before the lognormal is assumed for a particular data set.

In addition to the power of the Central Limit Theorem, another reason that the Gaussian distribution is used so frequently is that it easily generalizes to higher dimensions. That is, it is usually straightforward to represent joint variations of multiple Gaussian variables through what is called the multivariate Gaussian, or multivariate normal distribution. This distribution is discussed more extensively in Chapter 10, since in general the mathematical development for the multivariate Gaussian distribution requires use of matrix algebra.

However, the simplest case of the multivariate Gaussian distribution, describing the joint variations of two Gaussian variables, can be presented without vector notation. This two-variable distribution is known as the bivariate Gaussian, or bivariate normal distribution. It is sometimes possible to use this distribution to describe the behavior of two non-Gaussian distributions if the variables are first subjected to transformations such as those in Equations 3.18. In fact the opportunity to use the bivariate normal can be a major motivation for using such transformations.

Let the two variables considered be x and y. The bivariate normal distribution is defined by the PDF

$$f(x, y) = \frac{1}{2\pi\sigma_x\sigma_y\sqrt{1-\rho^2}}\exp\left\{-\frac{1}{2(1-\rho^2)}\left[\left(\frac{x-\mu_x}{\sigma_x}\right)^2 + \left(\frac{y-\mu_y}{\sigma_y}\right)^2\right.\right.$$
$$\left.\left. -2\,\rho\left(\frac{x-\mu_x}{\sigma_x}\right)\left(\frac{y-\mu_y}{\sigma_y}\right)\right]\right\}. \tag{4.33}$$

As a generalization of Equation 4.23 from one to two dimensions, this function defines a surface above the $x - y$ plane rather than a curve above the x-axis. For continuous bivariate distributions, including the bivariate normal, probability corresponds geometrically to the volume under the surface defined by the PDF so that, analogously to Equation 4.17, a necessary condition to be fulfilled by any bivariate PDF is

$$\int_x \int_y f(x, y) dy\, dx = 1, \ f(x, y) \geq 0. \qquad (4.34)$$

The bivariate normal distribution has five parameters: the two means and standard deviations for the variables x and y, and the correlation between them, ρ. The two marginal distributions for the variables x and y (i.e., the univariate probability density functions $f(x)$ and $f(y)$) must both be Gaussian distributions. It is usual, although not guaranteed, for the joint distribution of any two Gaussian variables to be bivariate normal. The two marginal distributions have parameters μ_x, σ_x, and μ_y, σ_y, respectively. Fitting the bivariate normal distribution is very easy. These four parameters are estimated using their sample counterparts for the x and y variables separately, and the parameter ρ is estimated as the Pearson product-moment correlation between x and y, Equation 3.22.

Figure 4.5 illustrates the general shape of the bivariate normal distribution. It is mound-shaped in three dimensions, with properties that depend on the five parameters. The function achieves its maximum height above the point (μ_x, μ_y). Increasing σ_x stretches the density in the x direction and increasing σ_y stretches it in the y direction. For $\rho = 0$ the density is axially symmetric around the point (μ_x, μ_y), and curves of constant height are concentric circles if $\sigma_x = \sigma_y$ and are ellipses otherwise. As ρ increases in absolute value the density function is stretched diagonally, with the lines of constant height becoming increasingly elongated ellipses. For negative ρ the orientation of these ellipses is as depicted in Figure 4.5: larger values of x are more likely with smaller values of y, and smaller values of x are more likely with larger values of y. The ellipses have the opposite orientation (positive slope) for positive values of ρ.

Probabilities for joint outcomes of x and y are given by the double integral of Equation 4.33 over the relevant region in the plane, for example

$$\Pr\{(y_1 < Y \leq y_2) \cap (x_1 < X \leq x_2)\} = \int_{x_1}^{x_2} \int_{y_1}^{y_2} f(x, y) dy\, dx. \qquad (4.35)$$

This integration cannot be done analytically, and in practice numerical methods usually are used. Probability tables for the bivariate normal do exist (National Bureau of Standards 1959), but they are lengthy and cumbersome. It is possible to compute probabilities

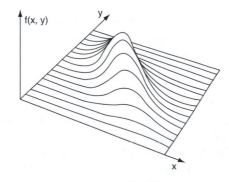

FIGURE 4.5   Perspective view of a bivariate normal distribution with $\sigma_x = \sigma_y$, and $\rho = -0.75$. The individual lines depicting the hump of the bivariate distribution have the shape of the (univariate) Gaussian distribution, illustrating that conditional distributions of x given a particular value of y are themselves Gaussian.

for elliptically shaped regions, called probability ellipses, centered on (μ_x, μ_y) using the method illustrated in Example 10.1. When computing probabilities for other regions, it can be more convenient to work with the bivariate normal distribution in standardized form. This is the extension of the standardized univariate Gaussian distribution (Equation 4.24), and is achieved by subjecting both the x and y variables to the transformation in Equation 4.25. Thus $\mu_{z_y} = \mu_{z_y} = 0$ and $\sigma_{z_y} = \sigma_{z_y} = 1$, leading to the bivariate density

$$f(z_x, z_y) = \frac{1}{2\pi\sqrt{1-\rho^2}} \exp\left[-\frac{z_x^2 - 2\rho z_x z_y + z_y^2}{2(1-\rho^2)} \right]. \tag{4.36}$$

A very useful property of the bivariate normal distribution is that the conditional distribution of one of the variables, given any particular value of the other, is Gaussian. This property is illustrated graphically in Figure 4.5, where the individual lines defining the shape of the distribution in three dimensions themselves have Gaussian shapes. Each indicates a function proportional to a conditional distribution of x given a particular value of y. The parameters for these conditional Gaussian distributions can be calculated from the five parameters of the bivariate normal. For the distribution of x given a particular value of y, the conditional Gaussian density function $f(x|Y = y)$ has parameters

$$\mu_{x|y} = \mu_x + \rho\sigma_x \frac{(y - \mu_y)}{\sigma_y} \tag{4.37a}$$

and

$$\sigma_{x|y} = \sigma_x \sqrt{1-\rho^2}. \tag{4.37b}$$

Equation 4.37a relates the mean of x to the standardized anomaly of y. It indicates that the conditional mean $\mu_{x|y}$ is larger than the unconditional mean μ_x if y is greater than its mean and ρ is positive, or if y is less than its mean and ρ is negative. If x and y are uncorrelated, knowing a value of y gives no additional information about y, and $\mu_{x|y} = \mu_x$ since $\rho = 0$. Equation 4.37b indicates that, unless the two variables are uncorrelated, $\sigma_{x|y} < \sigma_x$, regardless of the sign of ρ. Here knowing y provides some information about x, and the diminished uncertainty about x is reflected in a smaller standard deviation. In this sense, ρ^2 is often interpreted as the proportion of the variance in x that is accounted for by y.

EXAMPLE 4.7 Bivariate Normal Distribution and Conditional Probability

Consider the maximum temperature data for January 1987 at Ithaca and Canandaigua, in Table A.1. Figure 3.5 indicates that these data are fairly symmetrical, so that it may be reasonable to model their joint behavior as bivariate normal. A scatterplot of these two variables is shown in one of the panels of Figure 3.26. The average maximum temperatures are 29.87°F and 31.77°F at Ithaca and Canandaigua, respectively. The corresponding sample standard deviations are 7.71°F and 7.86°F. Table 3.5 shows their Pearson correlation to be 0.957.

With such a high correlation, knowing the temperature at one location should give very strong information about the temperature at the other. Suppose it is known that the Ithaca maximum temperature is 25°F and probability information about the Canandaigua maximum temperature is desired. Using Equation 4.37a, the conditional mean for the distribution of maximum temperature at Canandaigua, given that the Ithaca maximum temperature is 25°F, is 27.1°F—substantially lower than the unconditional mean of 31.77°F.

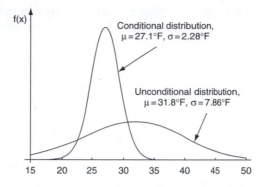

FIGURE 4.6 Gaussian distributions, representing the unconditional distribution for daily January maximum temperature at Canandaigua, and the conditional distribution given that the Ithaca maximum temperature was 25°F. The high correlation between maximum temperatures at the two locations results in the conditional distribution being much sharper, reflecting substantially diminished uncertainty.

Using Equation 4.37b, the conditional standard deviation is 2.28°F. (This would be the conditional standard deviation regardless of the particular value of the Ithaca temperature chosen, since Equation 4.37b does not depend on the value of the conditioning variable.) The conditional standard deviation is so much lower than the unconditional standard deviation because of the high correlation of maximum temperature between the two locations. As illustrated in Figure 4.6, this reduced uncertainty means that any of the conditional distributions for Canandaigua temperature given the Ithaca temperature will be much sharper than the unmodified, unconditional distribution for Canandaigua maximum temperature.

Using these parameters for the conditional distribution of maximum temperature at Canandaigua, we can compute such quantities as the probability that the Canandaigua maximum temperature is at or below freezing, given that the Ithaca maximum is 25°F. The required standardized variable is $z = (32 - 27.1)/2.28 = 2.14$, which corresponds to a probability of 0.984. By contrast, the corresponding climatological probability (without benefit of knowing the Ithaca maximum temperature) would be computed from $z = (32 - 31.8)/7.86 = 0.025$, corresponding to the much lower probability 0.510. ◊

4.4.3 Gamma Distributions

The statistical distributions of many atmospheric variables are distinctly asymmetric, and skewed to the right. Often the skewness occurs when there is a physical limit on the left that is relatively near the range of the data. Common examples are precipitation amounts or wind speeds, which are physically constrained to be nonnegative. Although it is mathematically possible to fit Gaussian distributions in such situations, the results are generally not useful. For example, the January 1933–1982 precipitation data in Table A.2 can be characterized by a sample mean of 1.96 in., and a sample standard deviation of 1.12 in. These two statistics are sufficient to fit a Gaussian distribution to this data, but applying this fitted distribution leads to nonsense. In particular, using Table B.1, we can compute the probability of negative precipitation as $\Pr\{Z \leq (0.00 - 1.96)/1.12\} = \Pr\{Z \leq -1.75\} = 0.040$. This computed probability is not especially large, but neither is it vanishingly small. The true probability is exactly zero: observing negative precipitation is impossible.

There are a variety of continuous distributions that are bounded on the left by zero and positively skewed. One common choice, used especially often for representing precipitation data, is the gamma distribution. The gamma distribution is defined by the PDF

$$f(x) = \frac{(x/\beta)^{\alpha-1} \exp(-x/\beta)}{\beta \Gamma(\alpha)}, \quad x, \alpha, \beta > 0. \tag{4.38}$$

The two parameters of the distribution are α, the shape parameter; and β, the scale parameter. The quantity $\Gamma(\alpha)$ is the gamma function, defined in Equation 4.7.

The PDF of the gamma distribution takes on a wide variety of shapes depending on the value of the shape parameter, α. As illustrated in Figure 4.7, for $\alpha < 1$ the distribution is very strongly skewed to the right, with $f(x) \rightarrow \infty$ as $x \rightarrow 0$. For $\alpha = 1$ the function intersects the vertical axis at $1/\beta$ for $x = 0$ (this special case of the gamma distribution is called the exponential distribution, which is described more fully later in this section). For $\alpha > 1$ the gamma distribution density function begins at the origin, $f(0) = 0$. Progressively larger values of α result in less skewness, and a shifting of probability density to the right. For very large values of α (larger than perhaps 50 to 100) the gamma distribution approaches the Gaussian distribution in form. The parameter α is always dimensionless.

The role of the scale parameter, β, effectively is to stretch or squeeze (i.e., to scale) the gamma density function to the right or left, depending on the overall magnitudes of the data values represented. Notice that the random quantity x in Equation 4.38 is divided by β in both places where it appears. The scale parameter β has the same physical dimensions as x. As the distribution is stretched to the right by larger values of β, its height must drop in order to satisfy Equation 4.17. Conversely, as the density is squeezed to the left its height must rise. These adjustments in height are accomplished by the β in the denominator of Equation 4.38.

The versatility in shape of the gamma distribution makes it an attractive candidate for representing precipitation data, and it is often used for this purpose. However, it is more difficult to work with than the Gaussian distribution, because obtaining good parameter estimates from particular batches of data is not as straightforward. The simplest (although certainly not best) approach to fitting a gamma distribution is to use the method of moments. Even here, however, there is a complication, because the two parameters for the gamma distribution do not correspond exactly to moments of the distribution, as was the case for the Gaussian. The mean of the gamma distribution is given by the product $\alpha\beta$, and the variance is $\alpha\beta^2$. Equating these expressions with the corresponding sample

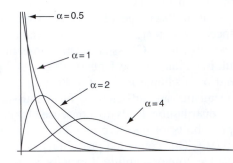

FIGURE 4.7   Gamma distribution density functions for four values of the shape parameter, α.

quantities yields a set of two equations in two unknowns, which can be solved to yield the moments estimators

$$\hat{\alpha} = \bar{x}^2/s^2 \tag{4.39a}$$

and

$$\hat{\beta} = s^2/\bar{x}. \tag{4.39b}$$

The moments estimators for the gamma distribution are not too bad for large values of the shape parameter, perhaps $\alpha > 10$, but can give very bad results for small values of α (Thom 1958; Wilks 1990). The moments estimators are said to be inefficient, in the technical sense of not making maximum use of the information in a data set. The practical consequence of this inefficiency is that particular values of the parameters in Equation 4.39 are erratic, or unnecessarily variable, from data sample to data sample.

A much better approach to parameter fitting for the gamma distribution is to use the method of maximum likelihood. For many distributions, including the gamma distribution, maximum likelihood fitting requires an iterative procedure that is really only practical using a computer. Section 4.6 presents the method of maximum likelihood for fitting parametric distributions, including the gamma distribution in Example 4.12.

There are two approximations to the maximum likelihood estimators for the gamma distribution that are simple enough to compute by hand. Both employ the sample statistic

$$D = \ln(\bar{x}) - \frac{1}{n}\sum_{i=1}^{n}\ln(x_i), \tag{4.40}$$

which is the difference between the natural log of the sample mean, and the mean of the logs of the data. Equivalently, the sample statistic D is the difference between the logs of the arithmetic and geometric means. Notice that the sample mean and standard deviation are not sufficient to compute the statistic D, since each datum must be used to compute the second term in Equation 4.40.

The first of the two maximum likelihood approximations for the gamma distribution is due to Thom (1958). The Thom estimator for the shape parameter is

$$\hat{\alpha} = \frac{1 + \sqrt{1 + 4\,D/3}}{4\,D}, \tag{4.41}$$

after which the scale parameter is obtained from

$$\hat{\beta} = \frac{\bar{x}}{\hat{\alpha}}. \tag{4.42}$$

The second approach is a polynomial approximation to the shape parameter (Greenwood and Durand 1960). One of two equations is used,

$$\hat{\alpha} = \frac{0.5000876 + 0.1648852\,D - 0.0544274\,D^2}{D}, \quad 0 \le D \le 0.5772, \tag{4.43a}$$

or

$$\hat{\alpha} = \frac{8.898919 + 9.059950\,D + 0.9775373\,D^2}{17.79728\,D + 11.968477\,D^2 + D^3}, \quad 0.5772 \le D \le 17.0, \tag{4.43b}$$

depending on the value of D. The scale parameter is again subsequently estimated using Equation 4.42.

As was the case for the Gaussian distribution, the gamma density function is not analytically integrable. Gamma distribution probabilities must therefore be obtained either by computing approximations to the CDF (i.e., to the integral of Equation 4.38), or from tabulated probabilities. Formulas and computer routines for this purpose can be found in Abramowitz and Stegun (1984), and Press et al. (1986), respectively. A table of gamma distribution probabilities is included as Table B.2 in Appendix B.

In any of these cases, gamma distribution probabilities will be available for the standard gamma distribution, with $\beta = 1$. Therefore, it is nearly always necessary to transform by rescaling the variable X of interest (characterized by a gamma distribution with arbitrary scale parameter β) to the standardized variable

$$\xi = x/\beta, \tag{4.44}$$

which follows a gamma distribution with $\beta = 1$. The standard gamma variate ξ is dimensionless. The shape parameter, α, will be the same for both x and ξ. The procedure is analogous to the transformation to the standardized Gaussian variable, z, in Equation 4.26.

Cumulative probabilities for the standard gamma distribution are given by a mathematical function known as the incomplete gamma function, $P(\alpha, \xi) = \Pr\{\Xi \le \xi\} = F(\xi)$. It is this function that was used to compute the probabilities in Table B.2. The cumulative probabilities for the standard gamma distribution in Table B.2 are arranged in an inverse sense to the Gaussian probabilities in Table B.1. That is, quantiles (transformed data values, ξ) of the distributions are presented in the body of the table, and cumulative probabilities are listed as the column headings. Different probabilities are obtained for different shape parameters, α, which appear in the first column.

EXAMPLE 4.8   Evaluating Gamma Distribution Probabilities

Consider again the data for January precipitation at Ithaca during the 50 years 1933–1982 in Table A.2. The average January precipitation for this period is 1.96 in. and the mean of the logarithms of the monthly precipitation totals is 0.5346, so Equation 4.40 yields $D = 0.139$. Both Thom's method (Equation 4.41) and the Greenwood and Durand formula (Equation 4.43a) yield $\alpha = 3.76$ and $\beta = 0.52$ in. By contrast, the moments estimators (Equation 4.39) yield $\alpha = 3.09$ and $\beta = 0.64$ in.

Adopting the approximate maximum likelihood estimators, the unusualness of the January 1987 precipitation total at Ithaca can be evaluated with the aid of Table B.2. That is, by representing the climatological variations in Ithaca January precipitation by the fitted gamma distribution with $\alpha = 3.76$ and $\beta = 0.52$ in., the cumulative probability corresponding to 3.15 in. (sum of the daily values for Ithaca in Table A.1) can be computed.

First, applying Equation 4.44, the standard gamma variate $\xi = 3.15$ in./0.52 in. $= 6.06$. Adopting $\alpha = 3.75$ as the closest tabulated value to the fitted $\alpha = 3.76$, it can be seen that $\xi = 6.06$ lies between the tabulated values for $F(5.214) = 0.80$ and $F(6.354) = 0.90$. Interpolation yields $F(6.06) = 0.874$, indicating that there is approximately one chance in eight for a January this wet or wetter to occur at Ithaca. The probability estimate could be refined slightly by interpolating between the rows for $\alpha = 3.75$ and $\alpha = 3.80$ to yield $F(6.06) = 0.873$, although this additional calculation would probably not be worth the effort.

Table B.2 can also be used to invert the gamma CDF to find precipitation values corresponding to particular cumulative probabilities, $\xi = F^{-1}(p)$, that is, to evaluate the quantile function. Dimensional precipitation values are then recovered by reversing the transformation in Equation 4.44. Consider estimation of the median January precipitation at Ithaca. This will correspond to the value of ξ satisfying $F(\xi) = 0.50$ which, in the row for $\alpha = 3.75$ in Table B.2, is 3.425. The corresponding dimensional precipitation amount is given by the product $\xi\beta = (3.425)(0.52\,\text{in.}) = 1.78\,\text{in.}$ By comparison, the sample median of the precipitation data in Table A.2 is 1.72 in. It is not surprising that the median is less than the mean of 1.96 in., since the distribution is positively skewed. The (perhaps surprising, but often unappreciated) implication of this comparison is that below normal (i.e., below average) precipitation is more likely than above normal precipitation, as a consequence of the positive skewness of the distribution of precipitation. ◊

EXAMPLE 4.9   Gamma Distribution in Operational Climatology

The gamma distribution can be used to report monthly and seasonal precipitation amounts in a way that allows comparison with locally applicable climatological distributions. Figure 4.8 shows an example of this format for United States precipitation for January 1989. The precipitation amounts for this month are not shown as accumulated depths, but rather as quantiles corresponding to local climatological gamma distributions. Five categories are mapped: less than the 10th percentile $q_{0.1}$, between the 10th and 30th percentile $q_{0.3}$, between the 30th and 70th percentile $q_{0.7}$, between the 70th and 90th percentile $q_{0.9}$, and wetter than the 90th percentile.

It is immediately clear which regions received substantially less, slightly less, about the same, slightly more, or substantially more precipitation in January 1989 as compared to the underlying climatological distribution. The shapes of these distributions vary widely, as can be seen in Figure 4.9. Comparing with Figure 4.7, it is clear that the distributions of January precipitation in much of the southwestern U.S. are very strongly skewed, and the corresponding distributions in much of the east, and in the Pacific northwest are much more symmetrical. (The corresponding scale parameters can be obtained from the mean monthly precipitation and the shape parameter using $\beta = \mu/\alpha$.) One of the advantages of expressing monthly precipitation amounts with respect to climatological gamma distributions is that these very strong differences in the shapes of the precipitation climatologies

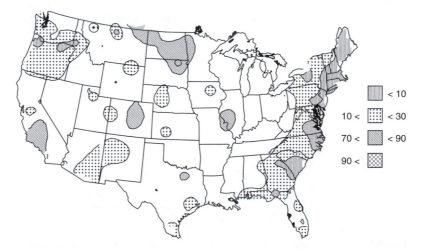

FIGURE 4.8   Precipitation totals for January 1989 over the conterminous U.S., expressed as percentile values of local gamma distributions. Portions of the east and west were drier than usual, and parts of the central portion of the country were wetter. From Arkin (1989).

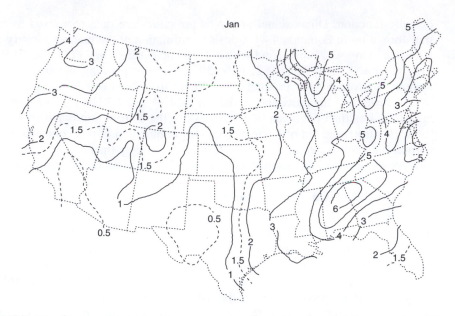

FIGURE 4.9   Gamma distribution shape parameters for January precipitation over the conterminous United States. The distributions in the southwest are strongly skewed, and those for most locations in the east are much more symmetrical. The distributions were fit using data from the 30 years 1951–1980. From Wilks and Eggleston (1992).

do not confuse comparisons between locations. Also, representing the climatological variations with parametric distributions both smooth the climatological data, and simplify the map production by summarizing each precipitation climate using only the two gamma distribution parameters for each location rather than the entire raw precipitation climatology for the United States.

Figure 4.10 illustrates the definition of the percentiles using a gamma probability density function with $\alpha = 2$. The distribution is divided into five categories corresponding to the five shading levels in Figure 4.8, with the precipitation amounts $q_{0.1}$, $q_{0.3}$, $q_{0.7}$, and $q_{0.9}$ separating regions of the distribution containing 10%, 20%, 40%, 20%, and 10% of

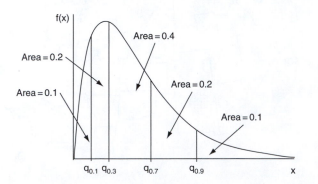

FIGURE 4.10   Illustration of the precipitation categories in Figure 4.8 in terms of a gamma distribution density function with $\alpha = 2$. Outcomes drier than the 10th percentile lie to the left of $q_{0.1}$. Areas with precipitation between the 30th and 70th percentiles (between $q_{0.3}$ and $q_{0.7}$) would be unshaded on the map. Precipitation in the wettest 10% of the climatological distribution lie to the right of $q_{0.9}$.

the probability, respectively. As can be seen in Figure 4.9, the shape of the distribution in Figure 4.10 is characteristic of January precipitation for many locations in the midwestern U.S. and southern plains. For the stations in northeastern Oklahoma reporting January 1989 precipitation above the 90th percentile in Figure 4.8, the corresponding precipitation amounts would have been larger than the locally defined $q_{0.9}$. ◊

There are two important special cases of the gamma distribution, which result from particular restrictions on the parameters α and β. For $\alpha = 1$, the gamma distribution reduces to the exponential distribution, with PDF

$$f(x) = \frac{1}{\beta} \exp\left(-\frac{x}{\beta}\right), \quad x \geq 0. \tag{4.45}$$

The shape of this density is simply an exponential decay, as indicated in Figure 4.7, for $\alpha = 1$. Equation 4.45 is analytically integrable, so the CDF for the exponential distribution exists in closed form,

$$F(x) = 1 - \exp\left(-\frac{x}{\beta}\right). \tag{4.46}$$

The quantile function is easily derived by solving Equation 4.46 for x (Equation 4.80). Since the shape of the exponential distribution is fixed by the restriction $\alpha = 1$, it is usually not suitable for representing variations is quantities like precipitation, although mixtures of two exponential distributions (see Section 4.4.6) can represent daily nonzero precipitation values quite well.

An important use of the exponential distribution in atmospheric science is in the characterization of the size distribution of raindrops, called drop-size distributions (e.g., Sauvageot 1994). When the exponential distribution is used for this purpose, it is called the Marshall-Palmer distribution, and generally denoted $N(D)$, which indicates a distribution over the numbers of droplets as a function of their diameters. Drop-size distributions are particularly important in radar applications where, for example, reflectivities are computed as expected values of a quantity called the backscattering cross-section, with respect to a drop-size distribution such as the exponential.

The second important special case of the gamma distribution is the chi-square (χ^2) distribution. Chi-square distributions are gamma distributions with scale parameter $\beta = 2$. Chi-square distributions are expressed conventionally in terms of an integer-valued parameter called the *degrees of freedom*, denoted ν. The relationship to the gamma distribution more generally is that the degrees of freedom are twice the gamma distribution shape parameter, or $\alpha = \nu/2$, yielding the Chi-square PDF

$$f(x) = \frac{x^{(\nu/2-1)} \exp\left(-\frac{x}{2}\right)}{2^{\nu/2} \Gamma\left(\frac{\nu}{2}\right)}, \quad x > 0. \tag{4.47}$$

Since it is the gamma scale parameter that is fixed at $\beta = 2$ to define the chi-square distribution, Equation 4.47 is capable of the same variety of shapes as the full gamma distribution. Because there is no explicit horizontal scale in Figure 4.7, it could be interpreted as showing chi-square densities with $\nu = 1, 2, 4$, and 8. The chi-square distribution arises as the distribution of the sum of ν squared independent standard Gaussian variates, and is used in several ways in the context of statistical testing (see Chapter 5). Table B.3 lists right-tail quantiles for chi-square distributions.

The gamma distribution is also sometimes generalized to a three-parameter distribution by moving the PDF to the left or right according to a shift parameter ζ. This three-parameter gamma distribution is also known as the Pearson Type III, or simply Pearson III distribution, and has PDF

$$f(x) = \frac{\left(\frac{x-\zeta}{\beta}\right)^{\alpha-1} \exp\left(-\frac{x-\zeta}{\beta}\right)}{|\beta|\Gamma(\alpha)}, \quad x > \zeta \text{ for } \beta > 0, \text{ or } x < \zeta \text{ for } \beta < 0. \tag{4.48}$$

Usually the scale parameter β is positive, which results in the Pearson III being a gamma distribution shifted to the right if $\zeta > 0$, with support $x > \zeta$. However, Equation 4.48 also allows $\beta < 0$, in which case the PDF is reflected (and so has a long left tail and negative skewness) and the support is $x < \zeta$. Sometimes, analogously to the lognormal distribution, the random variable x in Equation 4.48 has been log-transformed, in which case the distribution of the original variable $[= \exp(x)]$ is said to be log-Pearson III. Other transformations might also be used here, but assuming a logarithmic transformation is not as arbitrary as in the case of the lognormal. In contrast to the fixed bell shape of the Gaussian distribution, quite different distribution forms can be accommodated by Equation 4.48 in a way that is similar to adjusting the transformation exponent λ in Equation 3.18, by different values of the shape parameter α.

4.4.4 Beta Distributions

Some variables are restricted to segments of the real line that are bounded on two sides. Often such variables are restricted to the interval $0 \le x \le 1$. Examples of physically important variables subject to this restriction are cloud amount (observed as a fraction of the sky) and relative humidity. An important, more abstract, variable of this type is probability, where a parametric distribution can be useful in summarizing the frequency of use of forecasts, for example, of daily rainfall probability. The parametric distribution usually chosen to describe these types of variables is the beta distribution.

The PDF of the beta distribution is

$$f(x) = \left[\frac{\Gamma(p+q)}{\Gamma(p)\Gamma(q)}\right] x^{p-1}(1-x)^{q-1} \quad 0 \le x \le 1, \quad p, q > 0. \tag{4.49}$$

This is a very flexible function, taking on many different shapes depending on the values of its two parameters, p and q. Figure 4.11 illustrates five of these. In general, for $p \le 1$ probability is concentrated near zero (e.g., $p = .25$ and $q = 2$, or $p = 1$ and $q = 2$, in Figure 4.11), and for $q \le 1$ probability is concentrated near 1. If both parameters are less than one the distribution is U-shaped. For $p > 1$ and $q > 1$ the distribution has a single mode (hump) between 0 and 1 (e.g., $p = 2$ and $q = 4$, or $p = 10$ and $q = 2$, in Figure 4.11), with more probability shifted to the right for $p > q$, and more probability shifted to the left for $q > p$. Beta distributions with $p = q$ are symmetric. Reversing the values of p and q in Equation 4.49 results in a density function that is the mirror image (horizontally flipped) of the original.

Beta distribution parameters usually are fit using the method of moments. Using the expressions for the first two moments of the distribution,

$$\mu = p/(p+q) \tag{4.50a}$$

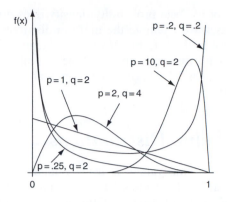

FIGURE 4.11 Five example probability density functions for beta distributions. Mirror images of these distributions are obtained by reversing the parameters p and q.

and

$$\sigma^2 = \frac{pq}{(p+q)^2(p+q+1)}, \tag{4.50b}$$

the moments estimators

$$\hat{p} = \frac{\bar{x}^2(1-\bar{x})}{s^2} - \bar{x} \tag{4.51a}$$

and

$$\hat{q} = \frac{\hat{p}(1-\bar{x})}{\bar{x}} \tag{4.51b}$$

are easily obtained.

An important special case of the beta distribution is the uniform, or rectangular distribution, with $p = q = 1$, and PDF $f(x) = 1$. The uniform distribution plays a central role in the computer generation of random numbers (see Section 4.7.1).

Use of the beta distribution is not limited only to variables having support on the unit interval [0,1]. A variable, say y, constrained to any interval $[a, b]$ can be represented by a beta distribution after subjecting it to the transformation

$$x = \frac{y-a}{b-a}. \tag{4.52}$$

In this case parameter fitting is accomplished using

$$\bar{x} = \frac{\bar{y}-a}{b-a} \tag{4.53a}$$

and

$$s_x^2 = \frac{s_y^2}{(b-a)^2}, \tag{4.53b}$$

which are then substituted into Equation 4.51.

The integral of the beta probability density does not exist in closed form except for a few special cases, for example the uniform distribution. Probabilities can be obtained through numerical methods (Abramowitz and Stegun 1984; Press *et al.* 1986), where the CDF for the beta distribution is known as the incomplete beta function, $I_x(p, q) = \Pr\{0 \le X \le x\} = F(x)$. Tables of beta distribution probabilities are given in Epstein (1985) and Winkler (1972b).

4.4.5 *Extreme-Value Distributions*

The statistics of extreme values is usually understood to relate to description of the behavior of the largest of m values. These data are extreme in the sense of being unusually large, and by definition are also rare. Often extreme-value statistics are of interest because the physical processes generating extreme events, and the societal impacts that occur because of them, are also large and unusual. A typical example of extreme-value data is the collection of annual maximum, or block maximum (largest in a block of m values), daily precipitation values. In each of n years there is a wettest day of the $m = 365$ days in each year, and the collection of these n wettest days is an extreme-value data set. Table 4.7 shows a small example annual maximum data set, for daily precipitation at Charleston, South Carolina. For each of the $n = 20$ years, the precipitation amount for the wettest of the $m = 365$ days is shown in the table.

A basic result from the theory of extreme-value statistics states (e.g., Leadbetter *et al.* 1983; Coles 2001) that the largest of m independent observations from a fixed distribution will follow a known distribution increasingly closely as m increases, regardless of the (single, fixed) distribution from which the observations have come. This result is called the Extremal Types Theorem, and is the analog within the statistics of extremes of the Central Limit Theorem for the distribution of sums converging to the Gaussian distribution. The theory and approach are equally applicable to distributions of extreme minima (smallest of m observations) by analyzing the variable $-X$.

The distribution toward which the sampling distributions of largest-of-m values converges is called the generalized extreme value, or GEV, distribution, with PDF

$$f(x) = \frac{1}{\beta} \left[1 + \frac{\kappa(x - \zeta)}{\beta} \right]^{1-1/\kappa} \exp \left\{ - \left[1 + \frac{\kappa(x - \zeta)}{\beta} \right]^{-1/\kappa} \right\}, \quad 1 + \kappa(x - \zeta)/\beta > 0.$$

(4.54)

Here there are three parameters: a location (or shift) parameter ζ, a scale parameter β, and a shape parameter κ. Equation 4.54 can be integrated analytically, yielding the CDF

$$F(x) = \exp \left\{ - \left[1 + \frac{\kappa(x - \zeta)}{\beta} \right]^{-1/\kappa} \right\},$$

(4.55)

TABLE 4.7   Annual maxima of daily precipitation amounts (inches) at Charleston, South Carolina, 1951–1970.

1951	2.01	1956	3.86	1961	3.48	1966	4.58
1952	3.52	1957	3.31	1962	4.60	1967	6.23
1953	2.61	1958	4.20	1963	5.20	1968	2.67
1954	3.89	1959	4.48	1964	4.93	1969	5.24
1955	1.82	1960	4.51	1965	3.50	1970	3.00

and this CDF can be inverted to yield an explicit formula for the quantile function,

$$F^{-1}(p) = \zeta + \frac{\beta}{\kappa}\{[-\ln(p)]^{-\kappa} - 1\}. \tag{4.56}$$

Particularly in the hydrological literature, Equations 4.54 through 4.56 are often written with the sign of the shape parameter κ reversed.

Because the moments of the GEV (see Table 4.6) involve the gamma function, estimating GEV parameters using the method of moments is no more convenient than alternative methods that yield more precise results. The distribution usually is fit using either the method of maximum likelihood (see Section 4.6), or a method known as L-moments (Hosking 1990; Stedinger *et al.* 1993) that is used frequently in hydrological applications. L-moments fitting tends to be preferred for small data samples (Hosking 1990). Maximum likelihood methods can be adapted easily to include effects of *covariates*, or additional influences; for example, the possibility that one or more of the distribution parameters may have a trend due to climate changes (Katz *et al.* 2002; Smith 1989; Zhang *et al.* 2004). For moderate and large sample sizes the results of the two parameter estimation methods are usually similar. Using the data in Table 4.7, the maximum likelihood estimates for the GEV parameters are $\zeta = 3.50$, $\beta = 1.11$, and $\kappa = -0.29$; and the corresponding L-moment estimates are $\zeta = 3.49$, $\beta = 1.18$, and $\kappa = -0.32$.

Three special cases of the GEV are recognized, depending on the value of the shape parameter κ. The limit of Equation 4.54 as κ approaches zero yields the PDF

$$f(x) = \frac{1}{\beta}\exp\left\{-\exp\left[-\frac{(x-\zeta)}{\beta}\right] - \frac{(x-\zeta)}{\beta}\right\}, \tag{4.57}$$

known as the Gumbel, or Fisher-Tippett Type I, distribution. The Gumbel distribution is the limiting form of the GEV for extreme data drawn independently from distributions with well-behaved (i.e., exponential) tails, such as the Gaussian and the gamma. The Gumbel distribution is so frequently used to represent the statistics of extremes that it is sometimes called "the" extreme value distribution. The Gumbel PDF is skewed to the right, and exhibits its maximum at $x = \zeta$. Gumbel distribution probabilities can be obtained from the cumulative distribution function

$$F(x) = \exp\left\{-\exp\left[-\frac{(x-\zeta)}{\beta}\right]\right\}. \tag{4.58}$$

Gumbel distribution parameters can be estimated through maximum likelihood or L-moments, as described earlier for the more general case of the GEV, but the simplest way to fit this distribution is to use the method of moments. The moments estimators for the Gumbel distribution parameters are computed using the sample mean and standard deviation. The estimation equations are

$$\hat{\beta} = \frac{s\sqrt{6}}{\pi} \tag{4.59a}$$

and

$$\hat{\zeta} = \bar{x} - \gamma\hat{\beta}, \tag{4.59b}$$

where $\gamma = 0.57721 \ldots$ is Euler's constant.

For $\kappa > 0$ the Equation 4.54 is called the Frechet, or Fisher-Tippett Type II distribution. These distributions exhibit what are called "heavy" tails, meaning that the PDF decreases rather slowly for large values of x. One consequence of heavy tails is that some of the moments of Frechet distributions are not finite. For example, the integral defining the variance (Equation 4.21) is infinite for $\kappa > 1/2$, and even the mean [Equation 4.20 with $g(x) = x$] is not finite for $\kappa > 1$. Another consequence of heavy tails is that quantiles associated with large cumulative probabilities (i.e., Equation 4.56 with $p \approx 1$) will be quite large.

The third special case of the GEV distribution occurs for $\kappa < 0$, and is known as the Weibull, or Fisher-Tippett Type III distribution. Usually Weibull distributions are written with the shift parameter $\zeta = 0$, and a parameter transformation, yielding the PDF

$$f(x) = \left(\frac{\alpha}{\beta}\right)\left(\frac{x}{\beta}\right)^{\alpha-1} \exp\left[-\left(\frac{x}{\beta}\right)^{\alpha}\right], \quad x, \alpha, \beta > 0. \tag{4.60}$$

As is the case for the gamma distribution, the two parameters α and β are called the shape and scale parameters, respectively. The form of the Weibull distribution also is controlled similarly by the two parameters. The response of the shape of the distribution to different values of α is shown in Figure 4.12. In common with the gamma distribution, $\alpha \leq 1$ produces reverse "J" shapes and strong positive skewness, and for $\alpha = 1$ the Weibull distribution also reduces to the exponential distribution (Equation 4.45) as a special case. Also in common with the gamma distribution, the scale parameter acts similarly to either stretch or compress the basic shape along the x axis, for a given value of α. For $\alpha = 3.6$ the Weibull is very similar to the Gaussian distribution.

The PDF for the Weibull distribution is analytically integrable, resulting in the CDF

$$F(x) = \Pr\{X \leq x\} = 1 - \exp\left[-\left(\frac{x}{\beta}\right)^{\alpha}\right]. \tag{4.61}$$

This function can easily be solved for x to yield the quantile function. As is the case for the GEV more generally, the moments of the Weibull distribution involve the gamma

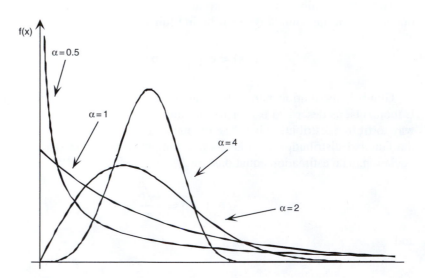

FIGURE 4.12   Weibull distribution probability density functions for four values of the shape parameter, α.

function (see Table 4.6), so there is no computational advantage to parameter fitting by the method of moments. Usually Weibull distributions are fit using either maximum likelihood (see Section 4.6) or L-moments (Stedinger *et al.* 1993).

One important motivation for studying and modeling the statistics of extremes is to estimate annual probabilities of rare and potentially damaging events, such as extremely large daily precipitation amounts that might cause flooding, or extremely large wind speeds that might cause damage to structures. In applications like these, the assumptions of classical extreme-value theory, namely that the underlying events are independent and come from the same distribution, and that the number of individual (usually daily) values m is sufficient for convergence to the GEV, may not be met. Most problematic for the application of extreme-value theory is that the underlying data often will not be drawn from the same distribution, for example because of an annual cycle in the statistics of the m ($= 365$, usually) values, and/or because the largest of the m values are generated by different processes in different blocks (years). For example, some of the largest daily precipitation values may occur because of hurricane landfalls, some may occur because of large and slowly moving thunderstorm complexes, and some may occur as a consequence of near-stationary frontal boundaries; and the statistics of (i.e., the underlying PDFs corresponding to) the different physical processes may be different (e.g., Walshaw 2000).

These considerations do not invalidate the GEV (Equation 4.54) as a candidate distribution to describe the statistics of extremes, and empirically this distribution often is found to be an excellent choice even when the assumptions of extreme-value theory are not met. However, in the many practical settings where the classical assumptions are not valid the GEV is not guaranteed to be the most appropriate distribution to represent a set of extreme-value data. The appropriateness of the GEV should be evaluated along with other candidate distributions for particular data sets (Madsen *et al.* 1997; Wilks 1993), possibly using approaches presented in Sections 4.6 or 5.2.6.

Another practical issue that arises when working with statistics of extremes is choice of the extreme data that will be used to fit a distribution. As already noted, a typical choice is to choose the largest single daily value in each of n years, known as the block maximum, or annual maximum series. Potential disadvantages of this approach are that a large fraction of the data are not used, including values that are not largest in their year of occurrence but may be larger than the maxima in other years. An alternative approach to assembling a set of extreme-value data is to choose the largest n values regardless of their year of occurrence. The result is called *partial-duration* data in hydrology. This approach is known more generally as peaks-over-threshold, or POT, since any values larger than a minimum level are chosen, and we are not restricted to choosing the same number of extreme values as there may be years in the climatological record. Because the underlying data may exhibit substantial serial correlation, some care is required to ensure that selected partial-duration data represent distinct events. In particular it is usual that only the largest of consecutive values above the selection threshold are incorporated into an extreme-value data set.

Are annual maximum or partial-duration data are more useful in particular applications? Interest usually focuses on the extreme right-hand tail of an extreme-value distribution, which corresponds to the same data regardless of whether they are chosen as annual maxima or peaks over a threshold. This is because the largest of the partial-duration data will also have been the largest single values in their years of occurrence. Usually the choice between annual and partial-duration data is best made empirically, according to which better allows the fitted extreme-value distribution to estimate the extreme tail probabilities (Madsen *et al.* 1997; Wilks 1993).

The result of an extreme-value analysis is often simply a summary of quantiles corresponding to large cumulative probabilities, for example the event with an annual probability of 0.01 of being exceeded. Unless n is rather large, direct estimation of these extreme quantiles will not be possible (cf. Equation 3.17), and a well-fitting extreme-value distribution provides a reasonable and objective way to extrapolate to probabilities that may be substantially larger than $1 - 1/n$. Often these extreme probabilities are expressed as average return periods,

$$R(x) = \frac{1}{\omega[1 - F(x)]}.$$
(4.62)

The return period $R(x)$ associated with a quantile x typically is interpreted to be the average time between occurrence of events of that magnitude or greater. The return period is a function of the CDF evaluated at x, and the average sampling frequency ω. For annual maximum data $\omega = 1\,\text{yr}^{-1}$, so the event x corresponding to a cumulative probability $F(x) = 0.99$ will have probability $1 - F(x)$ of being exceeded in any given year. This value of x would be associated with a return period of 100 years, and would be called the 100-year event. For partial-duration data, ω need not necessarily be $1\,\text{yr}^{-1}$, and $\omega = 1.65\,\text{yr}^{-1}$ has been suggested by some authors (Madsen *et al.* 1997; Stedinger *et al.* 1993). As an example, if the largest $2n$ daily values in n years are chosen regardless of their year of occurrence, then $\omega = 2.0\,\text{yr}^{-1}$. In that case the 100-year event would correspond to $F(x) = 0.995$.

EXAMPLE 4.10   Return Periods and Cumulative Probability

As noted earlier, a maximum-likelihood fit of the GEV distribution to the annual maximum precipitation data in Table 4.7 yielded the parameter estimates $\zeta = 3.50, \beta = 1.11$, and $\kappa = -0.29$. Using Equation 4.56 with cumulative probability $p = 0.5$ yields a median of 3.89 in. This is the precipitation amount that has a 50% chance of being exceeded in a given year. This amount will therefore be exceeded on average in half of the years in a hypothetical long climatological record, and so the average time separating daily precipitation events of this magnitude or greater is two years (Equation 4.62).

Because $n = 20$ years for these data, the median can be well estimated directly as the sample median. But consider estimating the 100-year 1-day precipitation event from these data. According to Equation 4.62 this corresponds to the cumulative probability $F(x) = 0.99$, whereas the empirical cumulative probability corresponding to the most extreme precipitation amount in Table 4.7 might be estimated as $p \approx 0.967$, using the Tukey plotting position (see Table 3.2). However, using the GEV quantile function Equation 4.56, together with Equation 4.62, a reasonable estimate for the 100-year amount is calculated to be 6.32 in. (The corresponding 2- and 100-year precipitation amounts derived from the L-moment parameter estimates, $\zeta = 3.49, \beta = 1.18$, and $\kappa = -0.32$, are 3.90 in. and 6.33 in., respectively.)

It is worth emphasizing that the T-year event is in no way guaranteed to occur within a particular period of T years. The probability that the T-year event occurs in any given year is $1/T$, for example $1/T = 0.01$ for the $T = 100$-year event. In any particular year, the occurrence of the T-year event is a Bernoulli trial, with $p = 1/T$. Therefore, the geometric distribution (Equation 4.5) can be used to calculate probabilities of waiting particular numbers of years for the event. Another interpretation of the return period is as the mean of the geometric distribution for the waiting time. The probability of the 100-year event occurring in an arbitrarily chosen century can be calculated as $\Pr\{X \leq 100\} = 0.634$ using Equation 4.5. That is, there is more than a 1/3 chance that the 100-year event will not occur in any particular 100 years. Similarly, the probability of the 100-year event not occurring in 200 years is approximately 0.134. ◊

4.4.6 *Mixture Distributions*

The parametric distributions presented so far in this chapter may be inadequate for data that arise from more than one generating process or physical mechanism. An example is the Guayaquil temperature data in Table A.3, for which histograms are shown in Figure 3.6. These data are clearly bimodal; with the smaller, warmer hump in the distribution associated with El Niño years, and the larger, cooler hump consisting mainly of the non-El Niño years. Although the Central Limit Theorem suggests that the Gaussian distribution should be a good model for monthly averaged temperatures, the clear differences in the Guayaquil June temperature climate associated with El Niño make the Gaussian a poor choice to represent these data overall. However, separate Gaussian distributions for El Niño years and non-El Niño years might provide a good probability model for these data.

Cases like this are natural candidates for representation with mixture distributions, or weighted averages of two or more PDFs. Any number of PDFs can be combined to form a mixture distribution (Everitt and Hand 1981; McLachlan and Peel 2000; Titterington *et al.* 1985), but by far the most commonly used mixture distributions are weighted averages of two component PDFs,

$$f(x) = w f_1(x) + (1 - w) f_2(x). \tag{4.63}$$

The component PDFs, $f_1(x)$ and $f_2(x)$ can be any distributions, although usually they are of the same parametric form. The weighting parameter $w, 0 < w < 1$, determines the contribution of each component density to the mixture PDF, $f(x)$, and can be interpreted as the probability that a realization of the random variable X will have come from $f_1(x)$.

Of course the properties of the mixture distribution depend on the component distributions and the weight parameter. The mean is simply the weighted average of the two component means,

$$\mu = w\mu_1 + (1 - w)\mu_2. \tag{4.64}$$

On the other hand, the variance

$$\sigma^2 = [w\sigma_1^2 + (1 - w)\sigma_2^2] + [w(\mu_1 - \mu)^2 + (1 - w)(\mu_2 - \mu)^2]$$
$$= w\sigma_1^2 + (1 - w)\sigma_2^2 + w(1 - w)(\mu_1 - \mu_2)^2, \tag{4.65}$$

has contributions from the weighted variances of the two distributions (first square-bracketed terms on the first line), plus the additional dispersion deriving from the difference of the two means (second square-bracketed terms). Mixture distributions are clearly capable of representing bimodality (or, when the mixture is composed of three or more component distributions, multimodality), but mixture distributions can also be unimodal if the differences between component means are small enough relative to the component standard deviations or variances.

Usually mixture distributions are fit using maximum likelihood, using the EM algorithm (see Section 4.6.3). Figure 4.13 shows the PDF for a maximum likelihood fit of a mixture of 2 Gaussian distributions to the June Guayaquil temperature data in Table A.3, with parameters $\mu_1 = 24.34°C, \sigma_1 = 0.46°C, \mu_2 = 26.48°C, \sigma_2 = 0.26°C,$ and $w = 0.80$ (see Example 4.13). Here μ_1 and σ_1 are the parameters of the first (cooler and more probable) Gaussian distribution, $f_1(x)$, and μ_2 and σ_2 are the parameters of the second (warmer and less probable) Gaussian distribution, $f_2(x)$. The mixture PDF in Figure 4.13

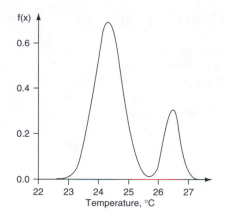

FIGURE 4.13   Probability density function for the mixture (Equation 4.63) of two Gaussian distributions fit to the June Guayaquil temperature data (Table A3). The result is very similar to the kernel density estimate derived from the same data, Figure 3.8b.

results as a simple (weighted) addition of the two component Gaussian distributions, in a way that is similar to the construction of the kernel density estimates for the same data in Figure 3.8, as a sum of scaled kernels that are themselves probability density functions. Indeed, the Gaussian mixture in Figure 4.13 resembles the kernel density estimate derived from the same data in Figure 3.8b. The means of the two component Gaussian distributions are well separated relative to the dispersion characterized by the two standard deviations, resulting in the mixture distribution being strongly bimodal.

Gaussian distributions are the most common choice for components of mixture distributions, but mixtures of exponential distributions (Equation 4.45) are also important and frequently used. In particular, the mixture distribution composed of two exponential distributions is called the mixed exponential distribution, with PDF

$$f(x) = \frac{w}{\beta_1} \exp\left(-\frac{x}{\beta_1}\right) + \frac{1-w}{\beta_2} \exp\left(-\frac{x}{\beta_2}\right).\tag{4.66}$$

The mixed exponential distribution has been found to be particularly well suited for nonzero daily precipitation data (Woolhiser and Roldan 1982; Foufoula-Georgiou and Lettenmaier 1987; Wilks 1999a), and is especially useful for simulating (see Section 4.7) spatially correlated daily precipitation amounts (Wilks 1998).

Mixture distributions are not limited to combinations of univariate continuous PDFs. The form of Equation 4.63 can as easily be used to form mixtures of discrete probability distribution functions, or mixtures of multivariate joint distributions. For example, Figure 4.14 shows the mixture of two bivariate Gaussian distributions (Equation 4.33) fit to a 51-member ensemble forecast (see Section 6.6) for temperature and wind speed. The distribution was fit using the maximum likelihood algorithm for multivariate Gaussian mixtures given in Smyth et al. (1999) and Hannachi and O'Neill (2001). Although multivariate mixture distributions are quite flexible in accommodating unusual-looking data, this flexibility comes at the price of needing to estimate a large number of parameters, so use of relatively elaborate probability models of this kind may be limited by the available sample size. The mixture distribution in Figure 4.14 requires 11 parameters to characterize it: two means, two variances, and one correlation for each of the two component bivariate distributions, plus the weight parameter w.

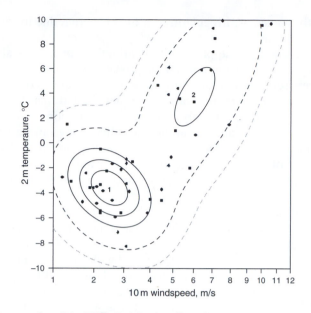

FIGURE 4.14   Contour plot of the PDF of a bivariate Gaussian mixture distribution, fit to an ensemble of 51 forecasts for 2 m temperature and 10 m windspeed, made at 180 h lead time. The windspeeds have first been square-root transformed to make their univariate distribution more Gaussian. Dots indicate individual forecasts made by the 51 ensemble members. The two constituent bivariate Gaussian densities $f_1(x)$ and $f_2(x)$ are centered at 1 and 2, respectively, and the smooth lines indicate level curves of their mixture $f(x)$, formed with $w = 0.57$. Solid contour interval is 0.05, and the heavy and light dashed lines are 0.01 and 0.001, respectively. Adapted from Wilks (2002b).

4.5 Qualitative Assessments of the Goodness of Fit

Having fit a parametric distribution to a batch of data, it is of more than passing interest to verify that the theoretical probability model provides an adequate description. Fitting an inappropriate distribution can lead to erroneous conclusions being drawn. Quantitative methods for evaluating the closeness of fitted distributions to underlying data rely on ideas from formal hypothesis testing, and a few such methods will be presented in Section 5.2.5. This section describes some qualitative, graphical methods useful for subjectively discerning the goodness of fit. These methods are instructive even if a formal goodness-of-fit test of fit is to be computed. A formal test may indicate an inadequate fit, but it may not inform the analyst as to the specific nature of the problem. Graphical comparisons of the data and the fitted distribution allow diagnosis of where and how the theoretical representation may be inadequate.

4.5.1 Superposition of a Fitted Parametric Distribution and Data Histogram

Probably the simplest and most intuitive means of comparing a fitted parametric distribution to the underlying data is superposition of the fitted distribution and a histogram. Gross departures from the data can readily be seen in this way. If the data are sufficiently numerous, irregularities in the histogram due to sampling variations will not be too distracting.

For discrete data, the probability distribution function is already very much like the histogram. Both the histogram and the probability distribution function assign probability to a discrete set of outcomes. Comparing the two requires only that the same discrete data values, or ranges of the data values, are plotted, and that the histogram and distribution function are scaled comparably. This second condition is met by plotting the histogram in terms of relative, rather than absolute, frequency on the vertical axis. Figure 4.2 is an example of the superposition of a Poisson probability distribution function on the histogram of observed annual numbers of tornados in New York state.

The procedure for superimposing a continuous PDF on a histogram is entirely analogous. The fundamental constraint is that the integral of any probability density function, over the full range of the random variable, must be one. That is, Equation 4.17 is satisfied by all probability density functions. One approach to matching the histogram and the density function is to rescale the density function. The proper scaling factor is obtained by computing the area occupied collectively by all the bars in the histogram plot. Denoting this area as A, it is easy to see that multiplying the fitted density function $f(x)$ by A produces a curve whose area is also A because, as a constant, A can be taken out of the integral: $\int A \cdot f(x) dx = A \cdot \int f(x) dx = A \cdot 1 = A$. Note that it is also possible to rescale the histogram heights so that the total area contained in the bars is 1. This latter approach is more traditional in statistics, since the histogram is regarded as an estimate of the density function.

EXAMPLE 4.11   Superposition of PDFs onto a Histogram

Figure 4.15 illustrates the procedure of superimposing fitted distributions and a histogram, for the 1933–1982 January precipitation totals at Ithaca from Table A.2. Here $n = 50$ years of data, and the bin width for the histogram (consistent with Equation 3.12) is 0.5 in., so the area occupied by the histogram rectangles is $A = (50)(0.5) = 25$. Superimposed on this histogram are PDFs for the gamma distribution fit using Equation 4.41 or 4.43a

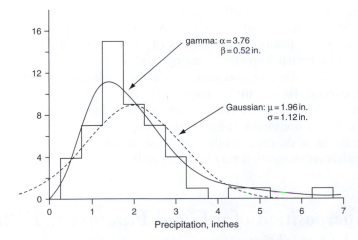

FIGURE 4.15   Histogram of the 1933–1982 Ithaca January precipitation data from Table A2, with the fitted gamma (solid) and the Gaussian (broken) PDFs. Each of the two density functions has been multiplied by $A = 25$, since the bin width is 0.5 in. and there are 50 observations. Apparently the gamma distribution provides a reasonable representation of the data. The Gaussian distribution underrepresents the right tail and implies nonzero probability for negative precipitation.

(solid curve), and the Gaussian distribution fit by matching the sample and distribution moments (dashed curve). In both cases the PDFs (Equations 4.38 and 4.23, respectively) have been multiplied by 25 so that their areas are equal to that of the histogram. It is clear that the symmetrical Gaussian distribution is a poor choice for representing these positively skewed precipitation data, since too little probability is assigned to the largest precipitation amounts and nonnegligible probability is assigned to impossible negative precipitation amounts. The gamma distribution represents these data much more closely, and provides a quite plausible summary of the year-to-year variations in the data. The fit appears to be worst for the 0.75 in. −1.25 in. and 1.25 in. −1.75 in. bins, although this easily could have resulted from sampling variations. This same data set will also be used in Section 5.2.5 to test formally the fit of these two distributions. ◊

4.5.2 Quantile-Quantile (Q–Q) Plots

Quantile-quantile (Q–Q) plots compare empirical (data) and fitted CDFs in terms of the dimensional values of the variable (the empirical quantiles). The link between observations of the random variable x and the fitted distribution is made through the quantile function, or inverse of the CDF (Equation 4.19), evaluated at estimated levels of cumulative probability.

The Q–Q plot is a scatterplot. Each coordinate pair defining the location of a point consists of a data value, and the corresponding estimate for that data value derived from the quantile function of the fitted distribution. Adopting the Tukey plotting position formula (see Table 3.2) as the estimator for empirical cumulative probability (although others could reasonably be used), each point in a Q–Q plot would have the Cartesian coordinates $(F^{-1}[(i-1/3)/(n+1/3)], x_{(i)})$. Thus the ith point on the Q–Q plot is defined by the ith smallest data value, $x_{(i)}$, and the value of the random variable corresponding to the sample cumulative probability $p = (i-1/3)/(n+1/3)$ in the fitted distribution. A Q–Q plot for a fitted distribution representing the data perfectly would have all points falling on the 1:1 diagonal line.

Figure 4.16 shows Q–Q plots comparing the fits of gamma and Gaussian distributions to the 1933–1982 Ithaca January precipitation data in Table A.2 (the parameter estimates are shown in Figure 4.15). Figure 4.16 indicates that the fitted gamma distribution corresponds well to the data through most of its range, since the quantile function evaluated at the estimated empirical cumulative probabilities is quite close to the observed data values, yielding points very close to the 1:1 line. The fitted distribution seems to underestimate the largest few points, suggesting that the tail of the fitted gamma distribution may be too thin, although at least some of these discrepancies might be attributable to sampling variations.

On the other hand, Figure 4.16 shows the Gaussian fit to these data is clearly inferior. Most prominently, the left tail of the fitted Gaussian distribution is too heavy, so that the smallest theoretical quantiles are too small, and in fact the smallest two are actually negative. Through the bulk of the distribution the Gaussian quantiles are further from the 1:1 line than the gamma quantiles, indicating a less accurate fit, and on the right tail the Gaussian distribution underestimates the largest quantiles even more than does the gamma distribution.

It is possible also to compare fitted and empirical distributions by reversing the logic of the Q–Q plot, and producing a scatterplot of the empirical cumulative probability (estimated using a plotting position, Table 3.2) as a function of the fitted CDF, $F(x)$,

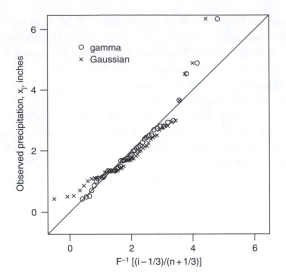

FIGURE 4.16   Quantile-quantile plots for gamma (o) and Gaussian (x) fits to the 1933–1982 Ithaca January precipitation in Table A2. Observed precipitation amounts are on the vertical, and amounts inferred from the fitted distributions using the Tukey plotting position are on the horizontal. Diagonal line indicates 1:1 correspondence.

evaluated at the corresponding data value. Plots of this kind are called probability-probability, or P–P plots. P–P plots seem to be used less frequently than Q–Q plots, perhaps because comparisons of dimensional data values can be more intuitive than comparisons of cumulative probabilities. P–P plots are also less sensitive to differences in the extreme tails of a distribution, which are often of most interest. Both Q–Q and P–P plots belong to a broader class of plots known as probability plots.

4.6 Parameter Fitting Using Maximum Likelihood

4.6.1 The Likelihood Function

For many distributions, fitting parameters using the simple method of moments produces inferior results that can lead to misleading inferences and extrapolations. The method of maximum likelihood is a versatile and important alternative. As the name suggests, the method seeks to find values of the distribution parameters that maximize the likelihood function. The procedure follows from the notion that the likelihood is a measure of the degree to which the data support particular values of the parameter(s) (e.g., Lindgren 1976). A Bayesian interpretation of the procedure (except for small sample sizes) would be that the maximum likelihood estimators are the most probable values for the parameters, given the observed data.

Notationally, the likelihood function for a single observation, x, looks identical to the probability density (or, for discrete variables, the probability distribution) function, and the difference between the two initially can be confusing. The distinction is that the PDF is a function of the data for fixed values of the parameters, whereas the likelihood function is a function of the unknown parameters for fixed values of the (already observed) data. Just as the joint PDF of n independent variables is the product of the n individual

PDFs, the likelihood function for the parameters of a distribution given a sample of n independent data values is just the product of the n individual likelihood functions. For example, the likelihood function for the Gaussian parameters μ and σ, given a sample of n observations, $x_i, i = 1, \ldots, n$, is

$$\Lambda(\mu, \sigma) = \sigma^{-n}(\sqrt{2\pi})^{-n} \prod_{i=1}^{n} \exp\left[-\frac{(x_i - \mu)^2}{2\sigma^2}\right]. \tag{4.67}$$

Here the uppercase pi indicates multiplication of terms of the form indicated to its right, analogously to the addition implied by the notation of uppercase sigma. Actually, the likelihood can be any function proportional to Equation 4.67, so the constant factor involving the square root of 2π could have been omitted because it does not depend on either of the two parameters. It has been included to emphasize the relationship between Equations 4.67 and 4.23. The right-hand side of Equation 4.67 looks exactly the same as the joint PDF for n independent Gaussian variables, except that the parameters μ and σ are the variables, and the x_i denote fixed constants. Geometrically, Equation 4.67 describes a surface above the $\mu - \sigma$ plane that takes on a maximum value above a specific pair of parameter values, depending on the particular data set given by the x_i values.

Usually it is more convenient to work with the logarithm of the likelihood function, known as the *log-likelihood*. Since the logarithm is a strictly increasing function, the same parameter values will maximize both the likelihood and log-likelihood functions. The log-likelihood function for the Gaussian parameters, corresponding to Equation 4.67 is

$$L(\mu, \sigma) = \ln[\Lambda(\mu, \sigma)] = -n\ln(\sigma) - n\ln(\sqrt{2\pi}) - \frac{1}{2\sigma^2}\sum_{i=1}^{n}(x_i - \mu)^2, \tag{4.68}$$

where, again, the term involving 2π is not strictly necessary for locating the maximum of the function because it does not depend on the parameters μ and σ.

Conceptually, at least, maximizing the log-likelihood is a straightforward exercise in calculus. For the Gaussian distribution the exercise really is simple, since the maximization can be done analytically. Taking derivatives of Equation 4.68 with respect to the parameters yields

$$\frac{\partial L(\mu, \sigma)}{\partial \mu} = \frac{1}{\sigma^2}\left[\sum_{i=1}^{n} x_i - n\mu\right] \tag{4.69a}$$

and

$$\frac{\partial L(\mu, \sigma)}{\partial \sigma} = -\frac{n}{\sigma} + \frac{1}{\sigma^3}\sum_{i=1}^{n}(x_i - \mu)^2. \tag{4.69b}$$

Setting each of these derivatives equal to zero and solving yields, respectively,

$$\hat{\mu} = \frac{1}{n}\sum_{i=1}^{n} x_i \tag{4.70a}$$

and

$$\hat{\sigma} = \sqrt{\frac{1}{n}\sum_{i=1}^{n}(x_i - \hat{\mu})^2}. \tag{4.70b}$$

These are the maximum likelihood estimators (MLEs) for the Gaussian distribution, which are readily recognized as being very similar to the moments estimators. The only difference is the divisor in Equation 4.70b, which is n rather than $n-1$. The divisor $n-1$ is often adopted when computing the sample standard deviation, because that choice yields an unbiased estimate of the population value. This difference points out the fact that the maximum likelihood estimators for a particular distribution may not be unbiased. In this case the estimated standard deviation (Equation 4.70b) will tend to be too small, on average, because the x_i are on average closer to the sample mean computed from them in Equation 4.70a than to the true mean, although these differences are small for large n.

4.6.2 The Newton-Raphson Method

The MLEs for the Gaussian distribution are somewhat unusual, in that they can be computed analytically. It is more typical for approximations to the MLEs to be calculated iteratively. One common approach is to think of the maximization of the log-likelihood as a nonlinear rootfinding problem to be solved using the multidimensional generalization of the Newton-Raphson method (e.g., Press *et al.* 1986). This approach follows from the truncated Taylor expansion of the derivative of the log-likelihood function

$$L'(\boldsymbol{\theta}^*) \approx L'(\boldsymbol{\theta}) + (\boldsymbol{\theta}^* - \boldsymbol{\theta})L''(\boldsymbol{\theta}), \tag{4.71}$$

where $\boldsymbol{\theta}$ denotes a generic vector of distribution parameters and $\boldsymbol{\theta}^*$ are the true values to be approximated. Since it is the *derivative* of the log-likelihood function, $L'(\boldsymbol{\theta}^*)$, whose roots are to be found, Equation 4.71 requires computation of the second derivatives of the log-likelihood, $L''(\boldsymbol{\theta})$. Setting Equation 4.71 equal to zero (to find a maximum in the log-likelihood, L) and rearranging yields the expression describing the algorithm for the iterative procedure,

$$\boldsymbol{\theta}^* = \boldsymbol{\theta} - \frac{L'(\boldsymbol{\theta})}{L''(\boldsymbol{\theta})}. \tag{4.72}$$

Beginning with an initial guess, $\boldsymbol{\theta}$, we compute an updated set of estimates, $\boldsymbol{\theta}^*$, which are in turn used as the guesses for the next iteration.

EXAMPLE 4.12   Algorithm for Maximum Likelihood Estimates of Gamma Distribution Parameters

In practice, use of Equation 4.72 is somewhat complicated by the fact that usually more than one parameter must be estimated, so that $L'(\boldsymbol{\theta})$ is a vector of first derivatives, and $L''(\boldsymbol{\theta})$ is a matrix of second derivatives. To illustrate, consider the gamma distribution (Equation 4.38). For this distribution, Equation 4.72 becomes

$$\begin{bmatrix} \alpha^* \\ \beta^* \end{bmatrix} = \begin{bmatrix} \alpha \\ \beta \end{bmatrix} - \begin{bmatrix} \partial^2 L/\partial\alpha^2 & \partial^2 L/\partial\alpha\partial\beta \\ \partial^2 L/\partial\beta\partial\alpha & \partial^2 L/\partial\beta^2 \end{bmatrix}^{-1} \begin{bmatrix} \partial L/\partial\alpha \\ \partial L/\partial\beta \end{bmatrix}$$

$$= \begin{bmatrix} \alpha \\ \beta \end{bmatrix} - \begin{bmatrix} -n\Gamma''(\alpha) & -n/\beta \\ -n/\beta & \dfrac{n\alpha}{\beta^2} - \dfrac{2\Sigma x}{\beta^3} \end{bmatrix}^{-1} \begin{bmatrix} \Sigma \ln(x) - n\ln(\beta) - n\Gamma'(\alpha) \\ \Sigma x/\beta^2 - n\alpha/\beta \end{bmatrix}, \tag{4.73}$$

where $\Gamma'(\alpha)$ and $\Gamma''(\alpha)$ are the first and second derivatives of the gamma function (Equation 4.7), which must be evaluated or approximated numerically (e.g., Abramowitz and Stegun 1984). The matrix-algebra notation in this equation is explained in Chapter 9. Equation 4.73 would be implemented by starting with initial guesses for the parameters α and β, perhaps using the moments estimators (Equations 4.39). Updated values, α^* and β^*, would then result from an application of Equation 4.73. The updated values would then be substituted into the right-hand side of Equation 4.73, and the process repeated until convergence of the algorithm. Convergence could be diagnosed by the parameter estimates changing sufficiently little, perhaps by a small fraction of a percent, between iterations. Note that in practice the Newton-Raphson algorithm may overshoot the likelihood maximum on a given iteration, which could result in a decline from one iteration to the next in the current approximation to the log-likelihood. Often the Newton-Raphson algorithm is programmed in a way that checks for such likelihood decreases, and tries smaller changes in the estimated parameters (although in the same direction specified by, in this case, Equation 4.73). $\lozenge$

4.6.3 *The EM Algorithm*

Maximum likelihood estimation using the Newton-Raphson method is generally fast and effective in applications where estimation of relatively few parameters is required. However, for problems involving more than perhaps three parameters, the computations required can expand dramatically. Even worse, the iterations can be quite unstable (sometimes producing "wild" updated parameters θ^* well away from the maximum likelihood values being sought) unless the initial guesses are so close to the correct values that the estimation procedure itself is almost unnecessary.

An alternative to Newton-Raphson that does not suffer these problems is the EM, or Expectation-Maximization algorithm (McLachlan and Krishnan 1997). It is actually somewhat imprecise to call the EM algorithm an "algorithm," in the sense that there is not an explicit specification (like Equation 4.72 for the Newton-Raphson method) of the steps required to implement it in a general way. Rather, it is more of a conceptual approach that needs to be tailored to particular problems.

The EM algorithm is formulated in the context of parameter estimation given "incomplete" data. Accordingly, on one level, it is especially well suited to situations where some data may be missing, or unobserved above or below known thresholds (censored data, and truncated data), or recorded imprecisely because of coarse binning. Such situations are handled easily by the EM algorithm when the estimation problem would be easy (for example, reducing to an analytic solution such as Equation 4.70) if the data were "complete." More generally, an ordinary (i.e., not intrinsically "incomplete") estimation problem can be approached with the EM algorithm if the existence of some additional unknown (and possibly hypothetical or unknowable) data would allow formulation of a straightforward (e.g., analytical) maximum likelihood estimation procedure. Like the Newton-Raphson method, the EM algorithm requires iterated calculations, and therefore an initial guess at the parameters to be estimated. When the EM algorithm can be formulated for a maximum likelihood estimation problem, the difficulties experienced by the Newton-Raphson approach do not occur, and in particular the updated log-likelihood will not decrease from iteration to iteration, regardless of how many parameters are being estimated simultaneously. For example, construction of Figure 4.14 required simultaneous estimation of 11 parameters, which would have been numerically impractical with the Newton-Raphson approach unless the correct answer had been known to good approximation initially.

Just what will constitute the sort of "complete" data allowing the machinery of the EM algorithm to be used smoothly will differ from problem to problem, and may require some creativity to define. Accordingly, it is not practical to outline the method here in enough generality to serve as stand-alone instruction in its use, although the following example illustrates the nature of the process. Further examples of its use in the atmospheric science literature include Hannachi and O'Neill (2001), Katz and Zheng (1999), Sansom and Thomson (1992) and Smyth *et al.* (1999). The original source paper is Dempster *et al.* (1977), and the authoritative book-length treatment is McLachlan and Krishnan (1997).

EXAMPLE 4.13   Fitting a Mixture of Two Gaussian Distributions with the EM Algorithm

Figure 4.13 shows a PDF fit to the Guayaquil temperature data in Table A.3, assuming a mixture distribution in the form of Equation 4.63, where both component PDFs $f_1(x)$ and $f_2(x)$ have been assumed to be Gaussian (Equation 4.23). As noted in connection with Figure 4.13, the fitting method was maximum likelihood, using the EM algorithm.

One interpretation of Equation 4.63 is that each datum x has been drawn from either $f_1(x)$ or $f_2(x)$, with overall relative frequencies w and $(1-w)$, respectively. It is not known which x's might have been drawn from which PDF, but if this more complete information were somehow to be available, then fitting the mixture of two Gaussian distributions indicated in Equation 4.63 would be straightforward: the parameters μ_1 and σ_1 defining the PDF $f_1(x)$ could be estimated using Equation 4.70 on the basis of the $f_1(x)$ data only, the parameters μ_2 and σ_2 defining the PDF $f_2(x)$ could be estimated using Equation 4.70 on the basis of the $f_2(x)$ data only, and the mixing parameter w could be estimated as the proportion of $f_1(x)$ data.

Even though the labels identifying particular x's as having been drawn from either $f_1(x)$ or $f_2(x)$ are not available (so that the data set is "incomplete"), the parameter estimation can proceed using the expected values of these hypothetical identifiers at each iteration step. If the hypothetical identifier variable would have been binary (equal to 1 for $f_1(x)$, and equal to 0 for $f_2(x)$) its expected value, given each data value x_i, would correspond to the probability that x_i was drawn from $f_1(x)$. The mixing parameter w would be equal to the average of the n hypothetical binary variables.

Equation 13.32 specifies the expected values of the hypothetical indicator variables (i.e., the n conditional probabilities) in terms of the two PDFS $f_1(x)$ and $f_2(x)$, and the mixing parameter w:

$$P(f_1|x_i) = \frac{w\, f_1(x_i)}{w\, f_1(x_i) + (1-w)f_2(x_i)}, \quad i = 1, \dots, n. \tag{4.74}$$

Having calculated these n posterior probabilities, the updated estimate for the mixing parameter is

$$w = \frac{1}{n}\sum_{i=1}^{n} P(f_1|x_i). \tag{4.75}$$

Equations 4.74 and 4.75 define the E-(or expectation-) part of this implementation of the EM algorithm, where statistical expectations have been calculated for the unknown (and hypothetical) binary group membership data. Having calculated these probabilities, the -M (or -maximization) part of the EM algorithm is ordinary maximum-likelihood estimation (Equations 4.70, for Gaussian-distribution fitting), using these expected quantities in place of their unknown "complete-data" counterparts:

$$\hat{\mu}_1 = \frac{1}{n\,w} \sum_{i=1}^{n} P(f_1|x_i)x_i, \qquad (4.76a)$$

$$\hat{\mu}_2 = \frac{1}{n(1-w)} \sum_{i=1}^{n} [1 - P(f_1|x_i)]x_i, \qquad (4.76b)$$

$$\hat{\mu}_2 = \left[\frac{1}{n\,w} \sum_{i=1}^{n} P(f_1|x_i)(x_i - \hat{\mu}_1)^2 \right]^{1/2}, \qquad (4.76c)$$

and

$$\hat{\sigma}_2 = \left[\frac{1}{n(1-w)} \sum_{i=1}^{n} [1 - P(f_1|x_i)](x_i - \hat{\mu}_2)^2 \right]^{1/2}. \qquad (4.76d)$$

That is, Equation 4.76 implements Equation 4.70 for each of the two Gaussian distributions $f_1(x)$ and $f_2(x)$, using expected values for the hypothetical indicator variables, rather than sorting the x's into two disjoint groups. If these hypothetical labels could be known, this sorting would correspond to the $P(f_1|x_i)$ values being equal to the corresponding binary indicators, so that Equation 4.75 would be the relative frequency of $f_1(x)$ observations; and each x_i would contribute to either Equations 4.76a and 4.76c, or to Equations 4.76b and 4.76d, only.

This implementation of the EM algorithm, for estimating parameters of the mixture PDF for two Gaussian distributions in Equation 4.63, begins with initial guesses for the five distribution parameters μ_1, σ_1, μ_2 and σ_2, and w. These initial guesses are used in Equations 4.74 and 4.75 to obtain the initial estimates for the posterior probabilities $P(f_1|x_i)$ and the corresponding updated mixing parameter w. Updated values for the two means and two standard deviations are then obtained using Equations 4.76, and the process is repeated until convergence. It is not necessary for the initial guesses to be particularly good ones. For example, Table 4.8 outlines the progress of the EM algorithm in fitting the mixture distribution that is plotted in Figure 4.13, beginning with the rather poor initial guesses $\mu_1 = 22°C$, $\mu_2 = 28°C$, and $\sigma_1 = \sigma_2 = 1°C$, and $w = 0.5$. Note that the initial guesses for the two means are not even within the range of the data. Nevertheless,

TABLE 4.8 Progress of the EM algorithm over the seven iterations required to fit the mixture of Gaussian PDFs shown in Figure 4.13.

Iteration	w	μ_1	μ_2	σ_1	σ_2	Log-likelihood
0	0.50	22.00	28.00	1.00	1.00	−79.73
1	0.71	24.26	25.99	0.42	0.76	−22.95
2	0.73	24.28	26.09	0.43	0.72	−22.72
3	0.75	24.30	26.19	0.44	0.65	−22.42
4	0.77	24.31	26.30	0.44	0.54	−21.92
5	0.79	24.33	26.40	0.45	0.39	−21.09
6	0.80	24.34	26.47	0.46	0.27	−20.49
7	0.80	24.34	26.48	0.46	0.26	−20.48

Table 4.8 shows that the updated means are quite near their final values after only a single iteration, and that the algorithm has converged after seven iterations. The final column in this table shows that the log-likelihood increases monotonically with each iteration. ◊

4.6.4 *Sampling Distribution of Maximum-Likelihood Estimates*

Even though maximum-likelihood estimates may require elaborate computations, they are still sample statistics that are functions of the underlying data. As such, they are subject to sampling variations for the same reasons and in the same ways as more ordinary statistics, and so have sampling distributions that characterize the precision of the estimates. For sufficiently large sample sizes, these sampling distributions are approximately Gaussian, and the joint sampling distribution of simultaneously estimated parameters is approximately multivariate Gaussian (e.g., the sampling distribution of the estimates for α and β in Equation 4.73 would be approximately bivariate normal).

Let $\boldsymbol{\theta} = [\theta_1, \theta_2, \ldots, \theta_k]$ represent a k-dimensional vector of parameters to be estimated. For example in Equation 4.73, $k = 2$, $\theta_1 = \alpha$, and $\theta_2 = \beta$. The estimated variance-covariance matrix for the multivariate Gaussian ($[\Sigma]$, in Equation 10.1) sampling distribution is given by the inverse of the information matrix, evaluated at the estimated parameter values $\hat{\boldsymbol{\theta}}$,

$$\hat{\mathrm{Var}}(\hat{\boldsymbol{\theta}}) = [\mathrm{I}(\hat{\boldsymbol{\theta}})]^{-1} \tag{4.77}$$

(the matrix algebra notation is defined in Chapter 9). The information matrix is computed in turn from the second derivatives of the log-likelihood function, with respect to the vector of parameters, and evaluated at their estimated values,

$$[\mathrm{I}(\hat{\boldsymbol{\theta}})] = -\begin{bmatrix} \partial^2 \mathrm{L}/\partial\hat{\theta}_1^2 & \partial^2 \mathrm{L}/\partial\hat{\theta}_1\partial\hat{\theta}_2 & \cdots & \partial^2 \mathrm{L}/\partial\hat{\theta}_1\partial\hat{\theta}_k \\ \partial^2 \mathrm{L}/\partial\hat{\theta}_2\partial\hat{\theta}_1 & \partial^2 \mathrm{L}/\partial\hat{\theta}_2^2 & \cdots & \partial^2 \mathrm{L}/\partial\hat{\theta}_2\partial\hat{\theta}_k \\ \vdots & \vdots & & \vdots \\ \partial^2 \mathrm{L}/\partial\hat{\theta}_k\partial\hat{\theta}_1 & \partial^2 \mathrm{L}/\partial\hat{\theta}_k\partial\hat{\theta}_2 & & \partial^2 \mathrm{L}/\partial\hat{\theta}_k^2 \end{bmatrix}. \tag{4.78}$$

Note that the inverse of the information matrix appears as part of the Newton-Raphson iteration for the estimation itself, for example for parameter estimation for the gamma distribution in Equation 4.73. One advantage of using this algorithm is that the estimated variances and covariances for the joint sampling distribution for the estimated parameters will already have been calculated at the final iteration. The EM algorithm does not automatically provide these quantities, but they can, of course, be computed from the estimated parameters; either by substitution of the parameter estimates into analytical expressions for the second derivatives of the log-likelihood function, or through a finite-difference approximation to the derivatives.

4.7 Statistical Simulation

An underlying theme of this chapter is that uncertainty in physical processes can be described by suitable probability distributions. When a component of a physical phenomenon or process of interest is uncertain, that phenomenon or process can still be studied through computer simulations, using algorithms that generate numbers that can be regarded as random samples from the relevant probability distribution(s). The generation of these apparently random numbers is called statistical simulation.

This section describes algorithms that are used in statistical simulation. These algorithms consist of deterministic recursive functions, so their output is not really random at all. In fact, their output can be duplicated exactly if desired, which can help in the debugging of code and in executing controlled replication of numerical experiments. Although these algorithms are sometimes called random-number generators, the more correct name is pseudo-random number generator, since their deterministic output only appears to be random. However, quite useful results can be obtained by regarding them as being effectively random.

Essentially all random number generation begins with simulation from the uniform distribution, with PDF $f(u) = 1, 0 \leq u \leq 1$, which is described in Section 4.7.1. Simulating values from other distributions involves transformation of one or more uniform variates. Much more on this subject than can be presented here, including code and pseudocode for many particular algorithms, can be found in such references as Boswell *et al.* (1993), Bratley *et al.* (1987), Dagpunar (1988), Press *et al.* (1986), Tezuka (1995), and the encyclopedic Devroye (1986).

The material in this section pertains to generation of scalar, independent random variates. The discussion emphasizes generation of continuous variates, but the two general methods described in Sections 4.7.2 and 4.7.3 can be used for discrete distributions as well. Extension of statistical simulation to correlated sequences is included in Sections 8.2.4 and 8.3.7 on time-domain time series models. Extensions to multivariate simulation are presented in Section 10.4.

4.7.1 *Uniform Random Number Generators*

As noted earlier, statistical simulation depends on the availability of a good algorithm for generating apparently random and uncorrelated samples from the uniform (0, 1) distribution, which can be transformed to simulate random sampling from other distributions. Arithmetically, uniform random number generators take an initial value of an integer, called the seed, operate on it to produce an updated seed value, and then rescale the updated seed to the interval (0, 1). The initial seed value is chosen by the programmer, but usually subsequent calls to the uniform generating algorithm operate on the most recently updated seed. The arithmetic operations performed by the algorithm are fully deterministic, so restarting the generator with a previously saved seed will allow exact reproduction of the resulting "random" number sequence.

The most commonly encountered algorithm for uniform random number generation is the linear congruential generator, defined by

$$S_n = a \, S_{n-1} + c, \quad \mod M \tag{4.79a}$$

and

$$u_n = S_n / M. \tag{4.79b}$$

Here S_{n-1} is the seed brought forward from the previous iteration, S_n is the updated seed, and a, c, and M are integer parameters called the multiplier, increment, and modulus, respectively. U_n in Equation 4.79b is the uniform variate produced by the iteration defined by Equation 4.79. Since the updated seed S_n is the remainder when $aS_{n-1} + c$ is divided by M, S_n is necessarily smaller than M, and the quotient in Equation 4.79b will be less than 1. For $a > 0$ and $c \geq 0$ Equation 4.79b will be greater than 0. The parameters in Equation 4.79a must be chosen carefully if a linear congruential generator is to work at all well. The sequence S_n repeats with a period of at most M, and it is common to choose the modulus as a prime number that is nearly as large as the largest integer that can be represented by the computer on which the algorithm will be run. Many computers use 32-bit (i.e., 4-byte) integers, and $M = 2^{31} - 1$ is a usual choice, often in combination with $a = 16807$ and $c = 0$.

Linear congruential generators can be adequate for some purposes, particularly in low-dimensional applications. In higher dimensions, however, their output is patterned in a way that is not space-filling. In particular, pairs of successive u's from Equation 4.79b fall on a set of parallel lines in the $u_n - u_{n+1}$ plane, triples of successive u's from Equation 4.79b fall on a set of parallel planes in the volume defined by the $u_n - u_{n+1} - u_{n+2}$ axes, and so on, with the number of these parallel features diminishing rapidly as the dimension k increases, approximately according to $(k! M)^{1/k}$. Here is another reason for choosing the modulus M to be as large as reasonably possible, since for $M = 2^{31} - 1$ and $k = 2$, $(k! M)^{1/k}$ is approximately 65,000.

Figure 4.17 shows a magnified view of a portion of the unit square, onto which 1000 nonoverlapping pairs of uniform variates generated using Equation 4.79 have been plotted. This small domain contains 17 of the parallel lines that successive pairs from this generator fall, which are spaced at an interval of 0.000059. Note that the minimum separation of the points in the vertical is much closer, indicating that the spacing of the near-vertical lines of points does not define the resolution of the generator. The relatively close horizontal spacing in Figure 4.17 suggests that simple linear congruential generators may not be too crude for some low-dimensional purposes (although see Section 4.7.4 for a pathological interaction with a common algorithm for generating Gaussian variates in two dimensions). However, in higher dimensions the number of hyperplanes onto which successive groups of values from a linear congruential generator are constrained

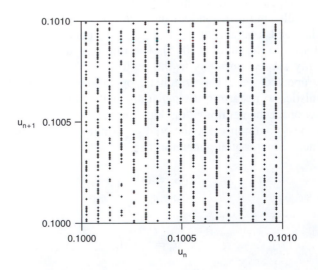

FIGURE 4.17   1000 non-overlapping pairs of uniform random variates in a small portion of the square defined by $0 < u_n < 1$ and $0 < u_{n+1} < 1$; generated using Equation 4.79, and $a = 16807$, $c = 0$, and $M = 2^{31} - 1$. This small domain contains 17 of the roughly 65,000 parallel lines onto which the successive pairs fall over the whole unit square.

decreases rapidly, so that it is impossible for algorithms of this kind to generate many of the combinations that should be possible: for $k = 3, 5, 10$, and 20 dimensions, the number of hyperplanes containing all the supposedly randomly generated points is smaller than 2350, 200, 40, and 25, respectively, even for the relatively large modulus $M = 2^{31} - 1$. Note that the situation can be very much worse than this if the parameters are chosen poorly: a notorious but formerly widely used generator known as RANDU (Equation 4.79 with $a = 65539$, $c = 0$, and $M = 2^{31}$) is limited to only 15 planes in three dimensions.

Direct use of linear congruential uniform generators cannot be recommended because of their patterned results in two or more dimensions. Better algorithms can be constructed by combining two or more independently running linear congruential generators, or by using one such generator to shuffle the output of another; examples are given in Bratley *et al.* (1987) and Press *et al.* (1986). An attractive alternative with apparently very good properties is a relatively recent algorithm called the Mersenne twister (Matsumoto and Nishimura 1998), which is freely available and easily found through a Web search on that name.

4.7.2 Nonuniform Random Number Generation by Inversion

Inversion is the easiest method of nonuniform variate generation to understand and program, when the quantile function $F^{-1}(p)$ (Equation 4.19) exists in closed form. It follows from the fact that, regardless of the functional form of the CDF $F(x)$, the distribution of the variable defined by that transformation, $u = F(x)$ is uniform on $[0, 1]$. The converse is also true, so that the CDF of the transformed variable $x(F) = F^{-1}(u)$ is $F(x)$, where the distribution of u is uniform on $[0, 1]$. Therefore, to generate a variate with CDF $F(x)$, for which the quantile function exists in closed form, we need only to generate a uniform variate as described in Section 4.7.1, and invert the CDF by substituting that value into the corresponding quantile function.

Inversion also can be used for distributions without closed-form quantile functions, by using numerical approximations, iterative evaluations, or interpolated table look-ups. Depending on the distribution, however, these workarounds might be insufficiently fast or accurate, in which case other methods would be more appropriate.

EXAMPLE 4.14   Generation of Exponential Variates Using Inversion

The exponential distribution (Equations 4.45 and 4.46) is a simple continuous distribution, for which the quantile function exists in closed form. In particular, solving Equation 4.46 for the cumulative probability p yields

$$F^{-1}(p) = -\beta \ln(1 - p). \tag{4.80}$$

Generating exponentially distributed variates requires only that a uniform variate be substituted for the cumulative probability p in Equation 4.80, so $x(F) = F^{-1}(u) = -\beta \ln(1 - u)$. Figure 4.18 illustrates the process for an arbitrarily chosen u, and the exponential distribution with mean $\beta = 2.7$. Note that the numerical values in Figure 4.18 have been rounded to a few significant figures for convenience, but in practice all the significant digits would be retained in a computation.◊

Since the uniform distribution is symmetric around its middle value 0.5, the distribution of $1 - u$ is the same uniform distribution as that of u, so that exponential variates can

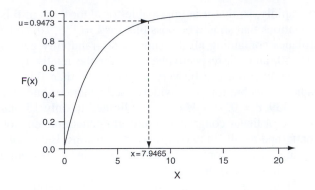

FIGURE 4.18   Illustration of the generation of an exponential variate by inversion. The smooth curve is the CDF (Equation 4.46) with mean $\beta = 2.7$. The uniform variate $u = 0.9473$ is transformed, through the inverse of the CDF, to the generated exponential variate $x = 7.9465$. This figure also illustrates that inversion produces a monotonic transformation of the underlying uniform variates.

be generated just as easily using $x(F) = F^{-1}(1 - u) = -\beta \ln(u)$. Even though this is somewhat simpler computationally, it may be worthwhile to use $-\beta \ln(1 - u)$ anyway in order to maintain the monotonicity of the inversion method; namely that the quantiles of the underlying uniform distribution correspond exactly to the quantiles of the distribution of the generated variates, so the smallest u's correspond to the smallest x's, and the largest u's correspond to the largest x's. One instance where this property can be useful is in the comparison of simulations that might depend on different parameters or different distributions. Maintaining monotonicity across such a collection of simulations (and beginning each with the same random number seed) can allow more precise comparisons among the different simulations, because a greater fraction of the variance of differences between simulations is then attributable to differences in the simulated processes, and less is due to sampling variations in the random number streams. This technique is known as variance reduction in the simulation literature.

4.7.3 Nonuniform Random Number Generation by Rejection

The inversion method is mathematically and computationally convenient when the quantile function can be evaluated simply, but it can be awkward otherwise. A more general approach is the rejection method, or acceptance-rejection method, which requires only that the PDF, $f(x)$, of the distribution to be simulated can be evaluated explicitly. However, in addition, an envelope PDF, $g(x)$, must also be found. The envelope density $g(x)$ must have the same support as $f(x)$, and should be easy to simulate from (for example, by inversion). In addition a constant c must be found such that $f(x) \leq cg(x)$, for all x having nonzero probability. That is, $f(x)$ must be dominated by the function $cg(x)$ for all relevant x. The difficult part of designing a rejection algorithm is finding an appropriate envelope PDF with a shape similar to that of the distribution to be simulated, so that the constant c can be as close to 1 as possible.

Once the envelope PDF and a constant c sufficient to ensure domination have been found, simulation by rejection proceeds in two steps, each of which requires an independent call to the uniform generator. First, a candidate variate is generated from $g(x)$ using

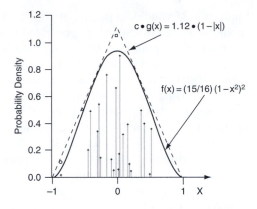

FIGURE 4.19   Illustration of simulation from the quartic (biweight) density, $f(x) = (15/16)(1-x^2)^2$ (Table 3.1), using a triangular density (Table 3.1) as the envelope, with $c = 1.12$. Twenty-five candidate x's have been simulated from the triangular density, of which 21 have been accepted ($+$) because they also fall under the distribution $f(x)$ to be simulated, and 4 have been rejected (O) because they fall outside it. Light grey lines point to the values simulated, on the horizontal axis.

the first uniform variate u_1, perhaps by inversion as $x = G^{-1}(u_1)$. Second, the candidate x is subjected to a random test using the second uniform variate: the candidate x is accepted if $u_2 \le f(x)/[cg(x)]$, otherwise the candidate x is rejected and the procedure is tried again with a new pair of uniform variates.

Figure 4.19 illustrates the rejection method, to simulate from the quartic density (see Table 3.1). The PDF for this distribution is a fourth-degree polynomial, so its CDF could be found easily by integration to be a fifth-degree polynomial. However, explicitly inverting the CDF (solving the fifth-degree polynomial) could be problematic, so rejection is a plausible method to simulate from this distribution. The triangular distribution (also given in Table 3.1) has been chosen as the envelope distribution $g(x)$; and the constant $c = 1.12$ is sufficient for $cg(x)$ to dominate $f(x)$ over $-1 \le x \le 1$. The triangular function is a reasonable choice for the envelope density because it dominates $f(x)$ with a relatively small value for the stretching constant c, so that the probability for a candidate x to be rejected is relatively small. In addition, it is simple enough that we easily can derive its quantile function, allowing simulation through inversion. In particular, integrating the triangle PDF yields the CDF

$$G(x) = \begin{cases} \dfrac{x^2}{2} + x + \dfrac{1}{2}, & -1 \le x \le 0, & \text{(4.81a)} \\[2mm] -\dfrac{x^2}{2} + x + \dfrac{1}{2}, & 0 \le x \le 1, & \text{(4.81b)} \end{cases}$$

which can be inverted to obtain the quantile function

$$x(G) = G^{-1}(p) = \begin{cases} \sqrt{2p} - 1, & 0 \le p \le \dfrac{1}{2}, & \text{(4.82a)} \\[2mm] 1 - \sqrt{2(1-p)}, & \dfrac{1}{2} \le p \le 1. & \text{(4.82b)} \end{cases}$$

Figure 4.19 indicates 25 candidate points, of which 21 have been accepted (X), with light grey lines pointing to the corresponding generated values on the horizontal axis. The horizontal coordinates of these points are $G^{-1}(u_1)$; that is, random draws from the triangular kernel $g(x)$ using the uniform variate u_1. Their vertical coordinates are $u_2 cg[G^{-1}(u_1)]$, which is a uniformly distributed distance between the horizontal axis and $cg(x)$, evaluated at the candidate x using the second uniform variate u_2. Essentially, the rejection algorithm works because the two uniform variates define points distributed uniformly (in two dimensions) under the function $cg(x)$, and a candidate x is accepted according to the conditional probability that it is also under the PDF $f(x)$. The rejection method is thus very similar to Monte-Carlo integration of $f(x)$. An illustration of simulation from this distribution by rejection is included in Example 4.15.

One drawback of the rejection method is that some pairs of uniform variates are wasted when a candidate x is rejected, and this is the reason that it is desirable for the constant c to be as small as possible: the probability that a candidate x will be rejected is $1 - 1/c$ ($= 0.107$ for the situation in Figure 4.19). Another property of the method is that an indeterminate, random number of uniform variates is required for one call to the algorithm, so that the synchronization of random number streams allowed by the inversion method is difficult, at best, to achieve when using rejection.

4.7.4 Box-Muller Method for Gaussian Random Number Generation

One of the most frequently needed distributions in simulation is the Gaussian (Equation 4.23). Since the CDF for this distribution does not exist in closed form, neither does its quantile function, so generation of Gaussian variates by inversion can be done only approximately. Alternatively, standard Gaussian (Equation 4.24) variates can be generated in pairs using a clever transformation of a pair of independent uniform variates, through an algorithm called the Box-Muller method. Corresponding dimensional (nonstandard) Gaussian variables can then be reconstituted using the distribution mean and variance, according to Equation 4.28.

The Box-Muller method generates pairs of independent standard bivariate normal variates z_1 and z_2—that is, a random sample from the bivariate PDF in Equation 4.36— with the correlation $\rho = 0$, so that the level contours of the PDF are circles. Because the level contours are circles, any direction away from the origin is equally likely, so that in polar coordinates the PDF for the angle of a random point is uniform on $[0, 2\pi]$. A uniform angle on this interval can be easily simulated from the first of the pair of independent uniform variates as $\theta = 2\pi u_1$. The CDF for the radial distance of a standard bivariate Gaussian variate is

$$F(r) = 1 - \exp\left[-\frac{r^2}{2}\right], \quad 0 \le r \le \infty, \tag{4.83}$$

which is known as the Rayleigh distribution. Equation 4.83 is easily invertable to yield the quantile function $r(F) = F^{-1}(u_2) = -2\ln(1 - u_2)$. Transforming back to Cartesian coordinates, the generated pair of independent standard Gaussian variates is

$$z_1 = \cos(2\pi u_1)\sqrt{-2\ln(u_2)} \tag{4.84a}$$

$$z_2 = \sin(2\pi u_1)\sqrt{-2\ln(u_2)}. \tag{4.84b}$$

The Box-Muller method is very common and popular, but caution must be exercised in the choice of a uniform generator with which to drive it. In particular, the lines in the $u_1 - u_2$ plane produced by simple linear congruential generators, illustrated in Figure 4.17, are operated upon by the polar transformation in Equation 4.84 to yield spirals in the $z_1 - z_2$ plane, as discussed in more detail by Bratley *et al.* (1987). This patterning is clearly undesirable, and more sophisticated uniform generators are essential when generating Box-Muller Gaussian variates.

4.7.5 Simulating from Mixture Distributions and Kernel Density Estimates

Simulation from mixture distributions (Equation 4.63) is only slightly more complicated than simulation from one of the component PDFs. It is a two-step procedure, in which a component distribution is chosen according to weights, w, which can be considered to be probabilities with which the component distributions will be chosen. Having randomly chosen a component distribution, a variate from that distribution is generated and returned as the simulated sample from the mixture.

Consider, for example, simulation from the mixed exponential distribution, Equation 4.66, which is a probability mixture of two exponential PDFs. Two independent uniform variates are required in order to produce one realization from this distribution: one uniform variate to choose one of the two exponential distributions, and the other to simulate from that distribution. Using inversion for the second step (Equation 4.80) the procedure is simply

$$x = \begin{cases} -\beta_1 \ln(1 - u_2), & u_1 \leq w, & (4.85a) \\ -\beta_2 \ln(1 - u_2), & u_1 > w. & (4.85b) \end{cases}$$

Here the exponential distribution with mean β_1 is chosen with probability w, using u_1; and the inversion of whichever of the two distributions is chosen is implemented using the second uniform variate u_2.

The kernel density estimate, described in Section 3.3.6 is an interesting instance of a mixture distribution. Here the mixture consists of n equiprobable PDFs, each of which corresponds to one of n observations of a variable x. These PDFs are often of one of the forms listed in Table 3.1. Again, the first step is to choose which of the n data values on which to center the kernel to be simulated from in the second step, which can be done according to:

$$\text{choose } x_i \text{ if } \quad \frac{i-1}{n} \leq u < \frac{i}{n}, \qquad (4.86a)$$

which yields

$$i = \text{int}[n\, u + 1]. \qquad (4.86b)$$

Here $\text{int}[\bullet]$ indicated retention of the integer part only, or truncation of fractions.

EXAMPLE 4.15   Simulation from the Kernel Density Estimate in Figure 3.8b

Figure 3.8b shows a kernel density estimate representing the Guayaquil temperature data in Table A.3; constructed using Equation 3.13, the quartic kernel (see Table 3.1), and smoothing parameter $h = 0.6$. Using rejection to simulate from the quartic kernel density, at least three independent uniform variates will be required to simulate one random sample from this distribution. Suppose these three uniform variates are generated as $u_1 = 0.257990$, $u_2 = 0.898875$, and $u_3 = 0.465617$.

The first step is to choose which of the $n = 20$ temperature values in Table A.3 will be used to center the kernel to be simulated from. Using Equation 4.86b, this will be x_i, where $i = \text{int}[20 \cdot 0.257990 + 1] = \text{int}[6.1598] = 6$, yielding $T_i = 24.3°C$, because $i = 6$ corresponds to the year 1956.

The second step is to simulate from a quartic kernel, which can be done by rejection, as illustrated in Figure 4.19. First, a candidate x is generated from the dominating triangular distribution by inversion (Equation 4.82b) using the second uniform variate, $u_2 = 0.898875$. This calculation yields $x(G) = 1 - [2(1 - 0.898875)]^{1/2} = 0.550278$. Will this value be accepted or rejected? This question is answered by comparing u_3 to the ratio $f(x)/[cg(x)]$, where $f(x)$ is the quartic PDF, $g(x)$ is the triangular PDF, and $c = 1.12$ in order for $cg(x)$ to dominate $f(x)$. We find, then, that $u_3 = 0.465617 < 0.455700/[1.12 \cdot 0.449722] = 0.904726$, so the candidate $x = 0.550278$ is accepted.

The value x just generated is a random draw from a standard quartic kernel, centered on zero and having unit smoothing parameter. Equating it with the argument of the kernel function K in Equation 3.13 yields $x = 0.550278 = (T - T_i)/h = (T - 24.3°C)/0.6$, which centers the kernel on T_i and scales it appropriately, so that the final simulated value is $T = (0.550278)(0.6) + 24.3 = 24.63°C$. ◊

4.8 Exercises

4.1. Using the binomial distribution as a model for the freezing of Cayuga Lake as presented in Examples 4.1 and 4.2, calculate the probability that the lake will freeze at least once during the four-year stay of a typical Cornell undergraduate in Ithaca.

4.2. Compute probabilities that Cayuga Lake will freeze next

a. In exactly 5 years.
b. In 25 or more years.

4.3. In an article published in the journal *Science*, Gray (1990) contrasts various aspects of Atlantic hurricanes occurring in drought vs. wet years in sub-Saharan Africa. During the 18-year drought period 1970–1987, only one strong hurricane (intensity 3 or higher) made landfall on the east coast of the United States, but 13 such storms hit the eastern United States during the 23-year wet period 1947–1969.

a. Assume that the number of hurricanes making landfall in the eastern U.S. follows a Poisson distribution whose characteristics depend on African rainfall. Fit two Poisson distributions to Gray's data (one conditional on drought, and one conditional on a wet year, in West Africa).
b. Compute the probability that at least one strong hurricane will hit the eastern United States, given a dry year in West Africa.
c. Compute the probability that at least one strong hurricane will hit the eastern United States, given a wet year in West Africa.

4.4. Assume that a strong hurricane making landfall in the eastern U.S. causes, on average, $5 billion in damage. What are the expected values of annual hurricane damage from such storms, according to each of the two conditional distributions in Exercise 4.3?

4.5. Using the June temperature data for Guayaquil, Ecuador, in Table A.3,

a. Fit a Gaussian distribution.
b. Without converting the individual data values, determine the two Gaussian parameters that would have resulted if this data had been expressed in °F.
c. Construct a histogram of this temperature data, and superimpose the density function of the fitted distribution on the histogram plot.

4.6. Using the Gaussian distribution with $\mu = 19°C$ and $\sigma = 1.7°C$:

a. Estimate the probability that January temperature (for Miami, Florida) will be colder than 15°C.
b. What temperature will be higher than all but the warmest 1% of Januaries at Miami?

4.7. For the Ithaca July rainfall data given in Table 4.9,

a. Fit a gamma distribution using Thom's approximation to the maximum likelihood estimators.
b. Without converting the individual data values, determine the values of the two parameters that would have resulted if the data had been expressed in mm.
c. Construct a histogram of this precipitation data and superimpose the fitted gamma density function.

4.8. Use the result from Exercise 4.7 to compute:

a. The 30th and 70th percentiles of July precipitation at Ithaca.
b. The difference between the sample mean and the median of the fitted distribution.
c. The probability that Ithaca precipitation during any future July will be at least 7 in.

4.9. Using the lognormal distribution to represent the data in Table 4.9, recalculate Exercise 4.8.

4.10. The average of the greatest snow depths for each winter at a location of interest is 80 cm, and the standard deviation (reflecting year-to-year differences in maximum snow depth) is 45 cm.

a. Fit a Gumbel distribution to represent this data, using the method of moments.
b. Derive the quantile function for the Gumbel distribution, and use it to estimate the snow depth that will be exceeded in only one year out of 100, on average.

4.11. Consider the bivariate normal distribution as a model for the Canandaigua maximum and Canandaigua minimum temperature data in Table A.1.

a. Fit the distribution parameters.
b. Using the fitted distribution, compute the probability that the maximum temperature will be as cold or colder than 20°F, given that the minimum temperature is 0°F.

TABLE 4.9   July precipitation at Ithaca, New York, 1951–1980 (inches).

1951	4.17	1961	4.24	1971	4.25
1952	5.61	1962	1.18	1972	3.66
1953	3.88	1963	3.17	1973	2.12
1954	1.55	1964	4.72	1974	1.24
1955	2.30	1965	2.17	1975	3.64
1956	5.58	1966	2.17	1976	8.44
1957	5.58	1967	3.94	1977	5.20
1958	5.14	1968	0.95	1978	2.33
1959	4.52	1969	1.48	1979	2.18
1960	1.53	1970	5.68	1980	3.43

4.12. Construct a Q-Q plot for the temperature data in Table A.3, assuming a Gaussian distribution.

4.13. a. Derive a formula for the maximum likelihood estimate for the exponential distribution (Equation 4.45) parameter, β.

b. Derive a formula for the standard deviation of the sampling distribution for β, assuming n is large.

4.14. Design an algorithm to simulate from the Weibull distribution by inversion.

CHAPTER ♦ 5

Hypothesis Testing

5.1 Background

Formal testing of statistical hypotheses, also known as significance testing, usually is covered extensively in introductory courses in statistics. Accordingly, this chapter will review only the basic concepts behind formal hypothesis tests, and subsequently emphasize aspects of hypothesis testing that are particularly relevant to applications in the atmospheric sciences.

5.1.1 Parametric vs. Nonparametric Tests

There are two contexts in which hypothesis tests are performed; broadly, there are two types of tests. Parametric tests are those conducted in situations where we know or assume that a particular theoretical distribution is an appropriate representation for the data and/or the test statistic. Nonparametric tests are conducted without assumptions that particular parametric forms are appropriate in a given situation.

Very often, parametric tests consist essentially of making inferences about particular distribution parameters. Chapter 4 presented a number of parametric distributions that have been found to be useful for describing atmospheric data. Fitting such a distribution amounts to distilling the information contained in a sample of data, so that the distribution parameters can be regarded as representing (at least some aspects of) the nature of the underlying physical process of interest. Thus a statistical test concerning a physical process of interest can reduce to a test pertaining to a distribution parameter, such as the Gaussian mean μ.

Nonparametric, or distribution-free tests proceed without the necessity of assumptions about what, if any, parametric distribution pertains to the data at hand. Nonparametric tests proceed along one of two basic lines. One approach is to construct the test in such a way that the distribution of the data is unimportant, so that data from any distribution can be treated in the same way. In the following, this approach is referred to as *classical* nonparametric testing, since the methods were devised before the advent of cheap computing power. In the second approach, crucial aspects of the relevant distribution are inferred directly from the data, by repeated computer manipulations of the observations. These nonparametric tests are known broadly as *resampling* procedures.

5.1.2 The Sampling Distribution

The concept of the sampling distribution is fundamental to all statistical tests. Recall that a statistic is some numerical quantity computed from a batch of data. The sampling distribution for a statistic is the probability distribution describing batch- to-batch variations of that statistic. Since the batch of data from which any sample statistic (including the test statistic for a hypothesis test) has been computed is subject to sampling variations, sample statistics are subject to sampling variations as well. The value of a statistic computed from a particular batch of data in general will be different from that for the same statistic computed using a different batch of the same kind of data. For example, average January temperature is obtained by averaging daily temperatures during that month at a particular location for a given year. This statistic is different from year to year.

The random variations of sample statistics can be described using probability distributions just as the random variations of the underlying data can be described using probability distributions. Thus, sample statistics can be viewed as having been drawn from probability distributions, and these distributions are called sampling distributions. The sampling distribution provides a probability model describing the relative frequencies of possible values of the statistic.

5.1.3 The Elements of Any Hypothesis Test

Any hypothesis test proceeds according to the following five steps:

1) Identify a *test statistic* that is appropriate to the data and question at hand. The test statistic is the quantity computed from the data values that will be the subject of the test. In parametric settings the test statistic will often be the sample estimate of a parameter of a relevant distribution. In nonparametric resampling tests there is nearly unlimited freedom in the definition of the test statistic.
2) Define a *null hypothesis*, usually denoted H_0. The null hypothesis defines a specific logical frame of reference against which to judge the observed test statistic. Often the null hypothesis will be a straw man that we hope to reject.
3) Define an *alternative hypothesis*, H_A. Many times the alternative hypothesis will be as simple as "H_0 is not true," although more complex alternative hypotheses are also possible.
4) Obtain the *null distribution*, which is simply the sampling distribution for the test statistic, if the null hypothesis is true. Depending on the situation, the null distribution may be an exactly known parametric distribution, a distribution that is well approximated by a known parametric distribution, or an empirical distribution obtained by resampling the data. Identifying the null distribution is the crucial step defining the hypothesis test.
5) Compare the observed test statistic to the null distribution. If the test statistic falls in a sufficiently improbable region of the null distribution, H_0 is rejected as too unlikely to have been true given the observed evidence. If the test statistic falls within the range of ordinary values described by the null distribution, the test statistic is seen as consistent with H_0, which is then not rejected. Note that not rejecting H_0 does not mean that the null hypothesis is necessarily true, only that there is insufficient evidence to reject this hypothesis. When H_0 is not rejected, we can really only say that it is not inconsistent with the observed data.

5.1.4 Test Levels and p Values

The sufficiently improbable region of the null distribution just referred to is defined by the rejection level, or simply the level, of the test. The null hypothesis is rejected if the probability (according to the null distribution) of the observed test statistic, *and all other results at least as unfavorable to the null hypothesis*, is less than or equal to the test level. The test level is chosen in advance of the computations, but it depends on the particular investigator's judgment and taste, so that there is usually a degree of arbitrariness about its specific value. Commonly the 5% level is chosen, although tests conducted at the 10% level or the 1% level are not unusual. In situations where penalties can be associated quantitatively with particular test errors (e.g., erroneously rejecting H_0), however, the test level can be optimized (see Winkler 1972b).

The p value is the specific probability that the observed value of the test statistic, together with all other possible values of the test statistic that are at least as unfavorable to the null hypothesis, will occur (according to the null distribution). Thus, the null hypothesis is rejected if the p value is less than or equal to the test level, and is not rejected otherwise.

5.1.5 Error Types and the Power of a Test

Another way of looking at the level of a test is as the probability of falsely rejecting the null hypothesis, given that it is true. This false rejection is called a Type I error, and its probability (the level of the test) is often denoted α. Type I errors are defined in contrast to Type II errors, which occur if H_0 is not rejected when it is in fact false. The probability of a Type II error usually is denoted β.

Figure 5.1 illustrates the relationship of Type I and Type II errors for a test conducted at the 5% level. A test statistic falling to the right of a critical value, corresponding to the test level, results in rejection of the null hypothesis. Since the area under the probability density function of the null distribution to the right of the critical value in Figure 5.1 (horizontal shading) is 0.05, this is the probability of a Type I error. The portion of the

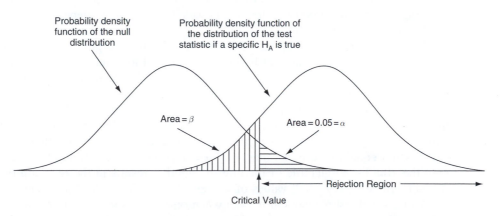

FIGURE 5.1   Illustration of the relationship of the rejection level, α, corresponding to the probability of a Type I error (horizontal hatching); and the probability of a Type II error, β (vertical hatching); for a test conducted at the 5% level. The horizontal axis represents possible values of the test statistic. Decreasing the probability of a Type I error necessarily increases the probability of a Type II error, and vice versa.

horizontal axis corresponding to H_0 rejection is sometimes called the rejection region, or the critical region. Outcomes in this range are not impossible under H_0, but rather have some small probability α of occurring. Usually the rejection region is defined by that value of the specified probability of a Type I error, α, according to the null hypothesis. It is clear from this illustration that, although we would like to minimize the probabilities of both Type I and Type II errors, this is not, in fact, possible. Their probabilities, α and β, can be adjusted by adjusting the level of the test, which corresponds to moving the critical value to the left or right; but decreasing α in this way necessarily increases β, and vice versa.

The level of the test, α, can be prescribed, but the probability of a Type II error, β, usually cannot. This is because the alternative hypothesis is defined more generally than the null hypothesis, and usually consists of the union of many specific alternative hypotheses. The probability α depends on the null distribution, which must be known in order to conduct a test, but β depends on which specific alternative hypothesis would be applicable, and this is generally not known. Figure 5.1 illustrates the relationship between α and β for only one of a potentially infinite number of possible alternative hypotheses.

It is sometimes useful, however, to examine the behavior of β over a range of the possibilities for H_A. This investigation usually is done in terms of the quantity $1 - \beta$, which is known as the power of the test against a specific alternative. Geometrically, the power of the test illustrated in Figure 5.1 is the area under the sampling distribution on the right (i.e., for a particular H_A) that does not have vertical shading. The relationship between the power of a test and a continuum of specific alternative hypotheses is called the *power function*. The power function expresses the probability of rejecting the null hypothesis, as a function of how far wrong it is. One reason why we might like to choose a less stringent test level (say, $\alpha = 0.10$) would be to better balance error probabilities for a test known to have low power.

5.1.6 One-Sided vs. Two-Sided Tests

A statistical test can be either one-sided or two-sided. This dichotomy is sometimes expressed in terms of tests being either one-tailed or two-tailed, since it is the probability in the extremes (tails) of the null distribution that governs whether a test result is interpreted as being significant. Whether a test is one-sided or two-sided depends on the nature of the hypothesis being tested.

A one-sided test is appropriate if there is a prior (e.g., a physically based) reason to expect that violations of the null hypothesis will lead to values of the test statistic on a particular side of the null distribution. This situation is illustrated in Figure 5.1, which has been drawn to imply that alternative hypotheses producing smaller values of the test statistic have been ruled out on the basis of prior information. In such cases the alternative hypothesis would be stated in terms of the true value being larger than the null hypothesis value (e.g., H_A: $\mu > \mu_0$), rather than the more vague alternative hypothesis that the true value is not equal to the null value (H_A: $\mu \neq \mu_0$). In Figure 5.1, any test statistic larger than the $100(1 - \alpha)$ quantile of the null distribution results in the rejection of H_0 at the α level, whereas very small values of the test statistic do not lead to a rejection of H_0.

A one-sided test is also appropriate when only values on one tail or the other of the null distribution are unfavorable to H_0, because of the way the test statistic has been constructed. For example, a test statistic involving a squared difference will be near zero if the difference is small, but will take on large positive values if the difference is large. In this case, results on the left tail of the null distribution could be quite supportive of H_0, in which case only right-tail probabilities would be of interest.

Two-sided tests are appropriate when either very large or very small values of the test statistic are unfavorable to the null hypothesis. Usually such tests pertain to the very general alternative hypothesis "H_0 is not true." The rejection region for two-sided tests consists of both the extreme left and extreme right tails of the null distribution. These two portions of the rejection region are delineated in such a way that the sum of their two probabilities under the null distribution yield the level of the test, α. That is, the null hypothesis is rejected at the α level if the test statistic is larger than $100(1-\alpha)/2\%$ of the null distribution on the right tail, or is smaller than the $100(\alpha/2)\%$ of this distribution on the left tail. Thus, a test statistic must be further out on the tail (i.e., more unusual with respect to H_0) to be declared significant in a two-tailed test as compared to a one-tailed test, at a specified test level. That the test statistic must be more extreme to reject the null hypothesis in a two-tailed test is appropriate, because generally one-tailed tests are used when additional (i.e., external to the test data) information exists, which then allows stronger inferences to be made.

5.1.7 Confidence Intervals: Inverting Hypothesis Tests

Hypothesis testing ideas can be used to construct confidence intervals around sample statistics. These are intervals constructed to be wide enough to contain, with a specified probability, the population quantity (often a distribution parameter) corresponding to the sample statistic. A typical use of confidence intervals is to construct error bars around plotted sample statistics in a graphical display.

In essence, a confidence interval is derived from the hypothesis test in which the value of an observed sample statistic plays the role of the population parameter value, under a hypothetical null hypothesis. The confidence interval around this sample statistic then consists of other possible values of the sample statistic for which that hypothetical H_0 would not be rejected. Hypothesis tests evaluate probabilities associated with an observed test statistic in the context of a null distribution, and conversely confidence intervals are constructed by finding the values of the test statistic that would not fall into the rejection region. In this sense, confidence interval construction is the inverse operation to hypothesis testing.

EXAMPLE 5.1   A Hypothesis Test Involving the Binomial Distribution

The testing procedure can be illustrated with a simple, although artificial, example. Suppose that advertisements for a tourist resort in the sunny desert southwest claim that, on average, six days out of seven are cloudless during winter. To verify this claim, we would need to observe the sky conditions in the area on a number of winter days, and then compare the fraction observed to be cloudless with the claimed proportion of $6/7 = 0.857$. Assume that we could arrange to take observations on 25 independent occasions. (These will not be consecutive days, because of the serial correlation of daily weather values.) If cloudless skies are observed on 15 of those 25 days, is this observation consistent with, or does it justify questioning, the claim?

This problem fits neatly into the parametric setting of the binomial distribution. A given day is either cloudless or it is not, and observations have been taken sufficiently far apart in time that they can be considered to be independent. By confining observations to only a relatively small portion of the year, we can expect that the probability, p, of a cloudless day is approximately constant from observation to observation.

The first of the five hypothesis testing steps has already been completed, since the test statistic of $X = 15$ out of $N = 25$ days has been dictated by the form of the problem.

The null hypothesis is that the resort advertisement was correct in claiming $p = 0.857$. Understanding the nature of advertising, it is reasonable to anticipate that, should the claim be false, the true probability will be lower. Thus the alternative hypothesis is that $p < 0.857$. That is, the test will be one-tailed, since results indicating $p > 0.857$ are not of interest with respect to discerning the truth of the claim. Our prior information regarding the nature of advertising claims will allow stronger inference than would have been the case if we assumed that alternatives with $p > 0.857$ were plausible.

Now the crux of the problem is to find the null distribution; that is, the sampling distribution of the test statistic X if the true probability of cloudless conditions is 0.857. This X can be thought of as the sum of 25 independent 0's and 1's, with the 1's having some constant probability of occurring on each of the 25 occasions. These are again the conditions for the binomial distribution. Thus, for this test the null distribution is also binomial, with parameters $p = 0.857$ and $N = 25$.

It remains to compute the probability that 15 or fewer cloudless days would have been observed on 25 independent occasions if the true probability p is in fact 0.857. (This probability is the p value for the test, which is a different usage for this symbol than the binomial distribution parameter, p.) The direct, but tedious, approach to this computation is summation of the terms given by

$$\Pr\{X \leq 15\} = \sum_{x=0}^{15} \binom{25}{x} 0.857^x (1 - 0.857)^{25-x}. \tag{5.1}$$

Here the terms for the outcomes for $X < 15$ must be included in addition to $\Pr\{X = 15\}$, since observing, say, only 10 cloudless days out of 25 would be even more unfavorable to H_0 than $X = 15$. The p value for this test as computed from Equation 5.1 is only 0.0015. Thus, $X = 15$ is a highly improbable result if the true probability of a cloudless day is 6/7, and this null hypothesis would be resoundingly rejected. According to this test, the observed data are very convincing evidence that the true probability is smaller than 6/7.

A much easier approach to the p-value computation is to use the Gaussian approximation to the binomial distribution. This approximation follows from the Central Limit Theorem since, as the sum of some number of 0's and 1's, the variable X will follow approximately the Gaussian distribution if N is sufficiently large. Here sufficiently large means roughly that $0 < p \pm 3[p(1-p)/N]^{1/2} < 1$, in which case the binomial X can be characterized to good approximation using a Gaussian distribution with

$$\mu \approx Np \tag{5.2a}$$

and

$$\sigma \approx \sqrt{N\,p(1-p)}. \tag{5.2b}$$

In the current example these parameters are $\mu \approx (25)(0.857) = 21.4$ and $\sigma \approx [(25)(0.857)(1 - 0.857)]^{1/2} = 1.75$. However, $p + 3[p(1-p)/N]^{1/2} = 1.07$, which suggests that use of the Gaussian approximation is questionable in this example. Figure 5.2 compares the exact binomial null distribution with its Gaussian approximation. The correspondence is close, although the Gaussian approximation ascribes more probability than we would like to the impossible outcomes $\{X > 25\}$, and correspondingly too little probability is assigned to the left tail. Nevertheless, the approximation will be carried forward here to illustrate its use.

One small technical issue that must be faced here relates to the approximation of discrete probabilities using a continuous probability density function. The p value for

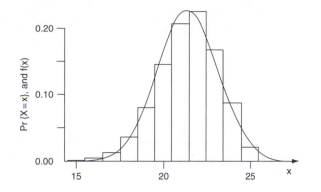

FIGURE 5.2 Relationship of the binomial null distribution (histogram bars) for Example 5.1, and its Gaussian approximation (smooth curve). The observed $X = 15$ falls on the far left tail of the null distribution. The exact p value is $\Pr\{X \leq 15\} = 0.0015$. Its approximation using the Gaussian distribution, including the continuity correction, is $\Pr\{X \leq 15.5\} = \Pr\{Z \leq -3.37\} = 0.00038$.

the exact binomial test is given by $\Pr\{X \leq 15\}$, but its Gaussian approximation is given by the integral of the Gaussian PDF over the corresponding portion of the real line. This integral should include values greater than 15 but closer to 15 than 16, since these also approximate the discrete $X = 15$. Thus the relevant Gaussian probability will be $\Pr\{X \leq 15.5\} = \Pr\{Z \leq (15.5 - 21.4)/1.75\} = \Pr\{Z \leq -3.37\} = 0.00038$, again leading to rejection but with too much confidence (too small a p value) because the Gaussian approximation puts insufficient probability on the left tail. The additional increment of 0.5 between the discrete $X = 15$ and the continuous $X = 15.5$ is called a *continuity correction*.

The Gaussian approximation to the binomial, Equations 5.2, can also be used to construct a confidence interval (error bars) around the observed estimate of the binomial $p = 15/25 = 0.6$. To do this, imagine a test whose null hypothesis is that the true binomial probability for this situation is 0.6. This test is then solved in an inverse sense to find the values of the test statistic defining the boundaries of the rejection regions. That is, how large or small a value of x/N would be tolerated before this new null hypothesis would be rejected?

If a 95% confidence region is desired, the test to be inverted will be at the 5% level. Since the true binomial p could be either larger or smaller than the observed x/N, a two-tailed test (rejection regions for both very large and very small x/N) is appropriate. Referring to Table B.1, since this null distribution is approximately Gaussian, the standardized Gaussian variable cutting off probability equal to $0.05/2 = 0.025$ at the upper and lower tails is $z = \pm1.96$. (This is the basis of the useful rule of thumb that a 95% confidence interval consists approximately of the mean value ± 2 standard deviations.) Using Equation 5.2a, the mean number of cloudless days should be $(25)(0.6) = 15$, and from Equation 5.2b the corresponding standard deviation is $[(25)(0.6)(1 - 0.6)]^{1/2} = 2.45$. Using Equation 4.28 with $z = \pm1.96$ yields $x = 10.2$ and $x = 19.8$, leading to the 95% confidence interval bounded by $p = x/N = 0.408$ and 0.792. Notice that the claimed binomial p of $6/7 = 0.857$ falls outside this interval. For the test used to construct this confidence interval, $p \pm 3[p(1 - p)/N]^{1/2}$ ranges from 0.306 to 0.894, which is comfortably within the range $[0, 1]$. The confidence interval computed exactly from the binomial probabilities is $[0.40, 0.76]$, with which the Gaussian approximation agrees very nicely.

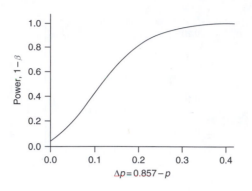

FIGURE 5.3   Power function for the test in Example 5.1. The vertical axis shows the probability of rejecting the null hypothesis; as a function of the difference between the true (and unknown) binomial p, and the binomial p for the null distribution (0.857).

Finally, what is the power of this test? That is, we might like to calculate the probability of rejecting the null hypothesis as a function of the true binomial p. As illustrated in Figure 5.1 the answer to this question will depend on the level of the test, since it is more likely (with probability $1 - \beta$) to correctly reject a false null hypothesis if α is relatively large. Assuming a test at the 5% level, and again assuming the Gaussian approximation to the binomial distribution for simplicity, the critical value will correspond to $z = -1.645$ relative to the null distribution; or $-1.645 = (Np - 21.4)/1.75$, yielding $Np = 18.5$. The power of the test for a given alternative hypothesis is the probability observing the test statistic $X = \{$number of cloudless days out of $N\}$ less than or equal to 18.5, given the true binomial p corresponding to that alternative hypothesis, and will equal the area to the left of 18.5 in the sampling distribution for X defined by that binomial p and $N = 25$. Collectively, these probabilities for a range of alternative hypotheses constitute the power function for the test.

Figure 5.3 shows the resulting power function. Here the horizontal axis indicates the difference between the true binomial p and that assumed by the null hypothesis ($= 0.857$). For $\Delta p = 0$ the null hypothesis is true, and Figure 5.3 indicates a 5% chance of rejecting it, which is consistent with the test being conducted at the 5% level. We do not know the true value of p, but Figure 5.3 shows that the probability of rejecting the null hypothesis increases as the true p is increasingly different from 0.857, until we are virtually assured of rejecting H_0 with a sample size of $N = 25$ if the true probability is smaller than about 1/2. If $N > 25$ days had been observed, the resulting power curves would be above that shown in Figure 5.3, so that probabilities of rejecting false null hypotheses would be greater (i.e., their power functions would climb more quickly toward 1), indicating more sensitive tests. Conversely, corresponding tests involving fewer samples would be less sensitive, and their power curves would lie below the one shown in Figure 5.3. ◊

5.2 Some Parametric Tests

5.2.1 One-Sample t Test

By far, most of the work on parametric tests in classical statistics has been done in relation to the Gaussian distribution. Tests based on the Gaussian are so pervasive because of the strength of the Central Limit Theorem. As a consequence of this theorem, many non-Gaussian problems can be treated at least approximately in the Gaussian framework. The example test for the binomial parameter p in Example 5.1 is one such case.

Probably the most familiar statistical test is the one-sample t test, which examines the null hypothesis that an observed sample mean has been drawn from a population centered at some previously specified mean, μ_0. If the number of data values making up the sample mean is large enough for its sampling distribution to be essentially Gaussian (by the Central Limit Theorem), then the test statistic

$$t = \frac{\bar{x} - \mu_0}{[\hat{Var}(\bar{x})]^{1/2}} \tag{5.3}$$

follows a distribution known as Student's t, or simply the t distribution. Equation 5.3 resembles the standard Gaussian variable z (Equation 4.25), except that a sample estimate of the variance of the sample mean (denoted by the "hat" accent) has been substituted in the denominator.

The t distribution is a symmetrical distribution that is very similar to the standard Gaussian distribution, although with more probability assigned to the tails. That is, the t distribution has heavier tails than the Gaussian distribution. The t distribution is controlled by a single parameter, ν, called the degrees of freedom. The parameter ν can take on any positive integer value, with the largest differences from the Gaussian being produced for small values of ν. For the test statistic in Equation 5.3, $\nu = n - 1$, where n is the number of independent observations being averaged in the sample mean in the numerator.

Tables of t distribution probabilities are available in almost any introductory statistics textbook. However, for even moderately large values of n (and therefore of ν) the variance estimate in the denominator becomes sufficiently precise that the t distribution is closely approximated by the standard Gaussian distribution (the differences in tail quantiles are about 4% and 1% for $\nu = 30$ and 100, respectively), so it is usually quite acceptable to evaluate probabilities associated with the test statistic in Equation 5.3 using standard Gaussian probabilities.

Use of the standard Gaussian PDF (Equation 4.24) as the null distribution for the test statistic in Equation 5.3 can be understood in terms of the central limit theorem, which implies that the sampling distribution of the sample mean in the numerator will be approximately Gaussian if n is sufficiently large. Subtracting the mean μ_0 in the numerator will center that Gaussian distribution on zero (if the null hypothesis, to which μ_0 pertains, is true). If n is also large enough that the standard deviation of that sampling distribution (in the denominator) can be estimated sufficiently precisely, then the resulting sampling distribution will also have unit standard deviation. A Gaussian distribution with zero mean and unit standard deviation is the standard Gaussian distribution.

The variance of the sampling distribution of a mean of n independent observations, in the denominator of Equation 5.3, is estimated according to

$$\hat{Var}[\bar{x}] = s^2/n, \tag{5.4}$$

where s^2 is the sample variance (the square of Equation 3.6) of the individual x's being averaged. Equation 5.4 is clearly true for the simple case of $n = 1$, but also makes intuitive sense for larger values of n. We expect that averaging together, say, pairs ($n = 2$) of x's will give quite irregular results from pair to pair. That is, the sampling distribution of the average of two numbers will have a high variance. On the other hand, averaging together batches of $n = 1000$ x's will give very consistent results from batch to batch, because the occasional very large x will tend to be balanced by the occasional very small x: a sample of $n = 1000$ will tend to have nearly equally many very large and very small values. The variance (i.e., the batch-to- batch variability) of the sampling distribution of the average of 1000 numbers will thus be small.

For small values of t in Equation 5.3, the difference in the numerator is small in comparison to the standard deviation of the sampling distribution of the difference, implying a quite ordinary sampling fluctuation for the sample mean, which should not trigger rejection of H_0. If the difference in the numerator is more than about twice as large as the denominator in absolute value, the null hypothesis would usually be rejected, corresponding to a two-sided test at the 5% level (cf. Table B.1).

5.2.2 Tests for Differences of Mean under Independence

Another common statistical test is that for the difference between two independent sample means. Plausible atmospheric examples of this situation might be differences of average winter 500 mb heights when one or the other of two synoptic regimes had prevailed, or perhaps differences in average July temperature at a location as represented in a climate model under a doubling vs. no doubling of atmospheric carbon dioxide (CO_2) concentrations.

In general, two sample means calculated from different batches of data, even if they are drawn from the same population or generating process, will be different. The usual test statistic in this situation is a function of the difference of the two sample means being compared, and the actual observed difference will almost always be some number other than zero. The null hypothesis is usually that the true difference is zero. The alternative hypothesis is either that the true difference is not zero (the case where no *a priori* information is available as to which underlying mean should be larger, leading to a two-tailed test), or that one of the two underlying means is larger than the other (leading to a one-tailed test). The problem is to find the sampling distribution of the difference of the two sample means, given the null hypothesis assumption that their population counterparts are the same. It is in this context that the observed difference of means can be evaluated for unusualness.

Nearly always—and sometimes quite uncritically—the assumption is tacitly made that the sampling distributions of the two sample means being differenced are Gaussian. This assumption will be true either if the data composing each of the sample means are Gaussian, or if the sample sizes are sufficiently large that the Central Limit Theorem can be invoked. If both of the two sample means have Gaussian sampling distributions their difference will be Gaussian as well, since any linear combination of Gaussian variables will itself follow a Gaussian distribution. Under these conditions the test statistic

$$z = \frac{(\bar{x}_1 - \bar{x}_2) - E[\bar{x}_1 - \bar{x}_2]}{(\hat{Var}[\bar{x}_1 - \bar{x}_2])^{1/2}} \tag{5.5}$$

will be distributed as standard Gaussian (Equation 4.24) for large samples. Note that this equation has a form similar to both Equations 5.3 and 4.26.

If the null hypothesis is equality of means of the two populations from which values of x_1 and x_2 are drawn,

$$E[\bar{x}_1 - \bar{x}_2] = E[\bar{x}_1] - E[\bar{x}_2] = \mu_1 - \mu_2 = 0. \tag{5.6}$$

Thus, a specific hypothesis about the magnitude of the two means is not required. If some other null hypothesis is appropriate to the problem at hand, that difference of underlying means would be substituted in the numerator of Equation 5.5.

The variance of a difference (or sum) of two independent random quantities is the sum of the variances of those quantities. Intuitively this makes sense since contributions to the variability of the difference are made by the variability of each the two quantities being differenced. With reference to the denominator of Equation 5.5,

$$\hat{V}ar\,[\bar{x}_1 - \bar{x}_2] = \hat{V}ar\,[\bar{x}_1] + \hat{V}ar\,[\bar{x}_2] = \frac{s_1^2}{n_1} + \frac{s_2^2}{n_2}, \tag{5.7}$$

where the last equality is achieved using Equation 5.4. Thus if the batches making up the two averages are independent, Equation 5.5 can be transformed to the standard Gaussian z by rewriting this test statistic as

$$z = \frac{\bar{x}_1 - \bar{x}_2}{\left[\dfrac{s_1^2}{n_1} + \dfrac{s_2^2}{n_2}\right]^{1/2}}, \tag{5.8}$$

when the null hypothesis is that the two underlying means μ_1 and μ_2 are equal. This expression for the test statistic is appropriate when the variances of the two distributions from which the x_1's and x_2's are drawn are not equal. For relatively small sample sizes its sampling distribution is (approximately, although not exactly) the t distribution, with $\nu = \min(n_1, n_2) - 1$. For moderately large samples the sampling distribution is close to the standard Gaussian, for the same reasons presented in relation to its one-sample counterpart, Equation 5.3.

When it can be assumed that the variances of the distributions from which the x_1's and x_2's have been drawn are equal, that information can be used to calculate a single, pooled, estimate for that variance. Under this assumption of equal population variances, Equation 5.5 becomes instead

$$z = \frac{\bar{x}_1 - \bar{x}_2}{\left[\left(\dfrac{1}{n_1} + \dfrac{1}{n_2}\right)\left\{\dfrac{(n_1 - 1)s_1^2 + (n_2 - 1)s_2^2}{n_1 + n_2 - 2}\right\}\right]^{1/2}}. \tag{5.9}$$

The quantity in curly brackets in the denominator is the pooled estimate of the population variance for the data values, which is just a weighted average of the two sample variances, and has in effect been substituted for both s_1^2 and s_2^2 in Equations 5.7 and 5.8. The sampling distribution for Equation 5.9 is the t distribution with $\nu = n_1 + n_2 - 2$. However, it is again usually quite acceptable to evaluate probabilities associated with the test statistic in Equation 5.9 using the standard Gaussian distribution.

For small values of z in either Equations 5.8 or 5.9, the difference of sample means in the numerator is small in comparison to the standard deviation of the sampling distribution of their difference, indicating a quite ordinary value in terms of the null distribution. As before, if the difference in the numerator is more than about twice as large as the denominator in absolute value, and the sample size is moderate or large, the null hypothesis would be rejected at the 5% level for a two-sided test.

5.2.3 *Tests for Differences of Mean for Paired Samples*

Equation 5.7 is appropriate when the x_1's and x_2's are observed independently. An important form of nonindependence occurs when the data values making up the two averages are paired, or observed simultaneously. In this case, necessarily, $n_1 = n_2$. For example, the daily temperature data in Table A.1 of Appendix A are of this type, since there is an observation of each variable at both locations on each day. When paired atmospheric data are used in a two-sample t test, the two averages being differenced are generally correlated. When this correlation is positive, as will often be the case, Equation 5.7 or the denominator of Equation 5.9 will overestimate the variance of the sampling distribution of the difference in the numerators of Equation 5.8 or 5.9. The result is that the test statistic will be too small (in absolute value), on average, so that null hypotheses that should be rejected will not be.

Intuitively, we expect the sampling distribution of the difference in the numerator of the test statistic to be affected if pairs of x's going into the averages are strongly correlated. For example, the appropriate panel in Figure 3.26 indicates that the maximum temperatures at Ithaca and Canandaigua are strongly correlated, so that a relatively warm average monthly maximum temperature at one location would likely be associated with a relatively warm average at the other. A portion of the variability of the monthly averages is thus common to both, and that portion cancels in the difference in the numerator of the test statistic. That cancellation must also be accounted for in the denominator if the sampling distribution of the test statistic is to be approximately standard Gaussian.

The easiest and most straightforward approach to dealing with the t test for paired data is to analyze differences between corresponding members of the $n_1 = n_2 = n$ pairs, which transforms the problem to the one-sample setting. That is, consider the sample statistic

$$\Delta = x_1 - x_2, \tag{5.10a}$$

with sample mean

$$\bar{\Delta} = \frac{1}{n} \sum_{i=1}^{n} \Delta_i = \bar{x}_1 - \bar{x}_2. \tag{5.10b}$$

The corresponding population mean will be $\mu_\Delta = \mu_1 - \mu_2$, which is often zero under H_0. The resulting test statistic is then of the same form as Equation 5.3,

$$z = \frac{\bar{\Delta} - \mu_\Delta}{(s_\Delta^2/n)^{1/2}}, \tag{5.11}$$

where s_Δ^2 is the sample variance of the n differences in Equation 5.10a. Any joint variation in the pairs making up the difference $\Delta = x_1 - x_2$ is also automatically reflected in the sample variance s_Δ^2 of those differences.

Equation 5.11 is an instance where positive correlation in the data is beneficial, in the sense that a more sensitive test can be conducted. Here a positive correlation results in a smaller standard deviation for the sampling distribution of the difference of means being tested, implying less underlying uncertainty. This sharper null distribution produces a more powerful test, and allows smaller differences in the numerator to be detected as significantly different from zero.

Intuitively this effect on the sampling distribution of the difference of sample means makes sense as well. Consider again the example of Ithaca and Canandaigua temperatures for January 1987, which will be revisited in Example 5.2. The positive correlation between daily temperatures at the two locations will result in the batch-to-batch (i.e., January-to-January, or interannual) variations in the two monthly averages moving together for the two locations: months when Ithaca is warmer than usual tend also to be months when Canandaigua is warmer than usual. The more strongly correlated are x_1 and x_2, the less likely are the pair of corresponding averages from a particular batch of data to differ because of sampling variations. To the extent that the two sample averages are different, then, the evidence against their underlying means not being the same is stronger, as compared to the situation when their correlation is near zero.

5.2.4 Test for Differences of Mean under Serial Dependence

The material in the previous sections is essentially a recapitulation of the classical tests for comparing sample means, presented in almost every elementary statistics textbook. A key assumption underlying these tests is the independence among the individual observations of comprising each of the sample means in the test statistic. That is, it is assumed that all the x_1 values are mutually independent and that the x_2 values are mutually independent, whether or not the data values are paired. This assumption of independence leads to the expression in Equation 5.4 that allows estimation of the variance of the null distribution.

Atmospheric data often do not satisfy the independence assumption. Frequently the averages to be tested are time averages, and the persistence, or time dependence, often exhibited is the cause of the violation of the assumption of independence. Lack of independence invalidates Equation 5.4. In particular, meteorological persistence implies that the variance of a time average is larger than specified by Equation 5.4. Ignoring the time dependence thus leads to underestimation of the variance of sampling distributions of the test statistics in Sections 5.2.2. and 5.2.3. This underestimation leads in turn to an inflated value of the test statistic, and consequently to overconfidence regarding the significance of the difference in the numerator. Equivalently, properly representing the effect of persistence in the data will require larger sample sizes to reject a null hypothesis for a given magnitude of the difference in the numerator.

Figure 5.4 may help to understand why serial correlation leads to a larger variance for the sampling distribution of a time average. The upper panel of this figure is an artificial time series of 100 independent Gaussian variates drawn from a generating process with $\mu = 0$, as described in Section 4.7.4. The series in the lower panel also consists of Gaussian variables having $\mu = 0$, but in addition this series has a lag-1 autocorrelation (Equation 3.23) of $\rho_1 = 0.6$. This value of the autocorrelation was chosen here because it is typical of the autocorrelation exhibited by daily temperatures (e.g., Madden 1979). Both panels have been scaled to produce unit (population) variance. The two plots look similar because the autocorrelated series was generated from the independent series according to what is called a first-order autoregressive process (Equation 8.16).

The outstanding difference between the independent and autocorrelated pseudo-data in Figure 5.4 is that the correlated series is smoother, so that adjacent and nearby values tend to be more alike than in the independent series. The autocorrelated series exhibits longer runs of points away from the (population) mean value. As a consequence, averages computed over subsets of the autocorrelated record are less likely to contain compensating points with large absolute value but of different sign, and those averages are therefore more likely to be far from zero (the true underlying average) than their counterparts computed using the independent values. That is, these averages will be less consistent from batch to batch. This is just another way of saying that the sampling distribution of an average of autocorrelated data has a higher variance than that of independent data. The gray horizontal lines in Figure 5.4 are subsample averages over consecutive sequences of

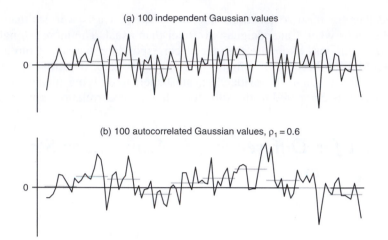

FIGURE 5.4   Comparison of artificial time series of (a) independent Gaussian variates, and (b) auto-correlated Gaussian variates having $\rho_1 = 0.6$. Both series were drawn from a generating process with $\mu = 0$, and the two panels have been scaled to have unit variances for the data points. Nearby values in the autocorrelated series tend to be more alike, with the result that averages over segments with $n = 10$ (horizontal grey bars) of the autocorrelated time series are more likely to be far from zero than are averages from the independent series. The sampling distribution of averages computed from the autocorrelated series accordingly has larger variance: the sample variances of the 10 subsample averages in panels (a) and (b) are 0.0825 and 0.2183, respectively.

$n = 10$ points, and these are visually more variable in Figure 5.4b. The sample variances of the 10 subsample means are 0.0825 and 0.2183 in panels (a) and (b), respectively.

Not surprisingly, the problem of estimating the variance of the sampling distribution of a time average has received considerable attention in the meteorological literature (e.g., Jones 1975; Katz 1982; Madden 1979; Zwiers and Thiébaux 1987; Zwiers and von Storch 1995). One convenient and practical approach to dealing with the problem is to think in terms of the effective sample size, or equivalent number of independent samples, n'. That is, imagine that there is a fictitious sample size, $n' < n$ of independent values, for which the sampling distribution of the average has the same variance as the sampling distribution of the average over the n autocorrelated values at hand. Then, n' could be substituted for n in Equation 5.4, and the classical tests described in the previous section could be carried through as before.

Estimation of the effective sample size is most easily approached if it can be assumed that the underlying data follow a first-order autoregressive process (Equation 8.16). It turns out that first-order autoregressions are often reasonable approximations for representing the persistence of daily meteorological values. This assertion can be appreciated informally by looking at Figure 5.4b. This plot consists of random numbers, but resembles statistically the day-to-day fluctuations in a meteorological variable like surface temperature.

The persistence in a first-order autoregression is completely characterized by the single parameter ρ_1, the lag-1 autocorrelation coefficient, which can be estimated from a data series using the sample estimate, r_1 (Equation 3.30). Using this correlation, the effective sample size can be estimated using the approximation

$$n' \cong n \frac{1 - \rho_1}{1 + \rho_1}. \tag{5.12}$$

When there is no time correlation, $\rho_1 = 0$ and $n' = n$. As ρ_1 increases the effective sample size becomes progressively smaller. When a more complicated time-series model is necessary to describe the persistence, appropriate but more complicated expressions for the effective sample size can be derived (see Katz 1982, 1985; and Section 8.3.5). Note that Equation 5.12 is applicable only to sampling distributions of the mean, and different expressions will be appropriate for use with different statistics (Livezey 1995a; Matalas and Langbein 1962; Thiebaux and Zwiers 1984; von Storch and Zwiers 1999; Zwiers and von Storch 1995).

Using Equation 5.12, the counterpart to Equation 5.4 for the variance of a time average over a sufficiently large sample becomes

$$\hat{Var} [\bar{x}] \cong \frac{s^2}{n'} = \frac{s^2}{n} \left(\frac{1+\rho_1}{1-\rho_1} \right). \tag{5.13}$$

The ratio $(1+\rho_1)/(1-\rho_1)$ acts as a variance inflation factor, adjusting the variance of the sampling distribution of the time average to reflect the influence of the serial correlation. Sometimes this variance inflation factor is called the time between effectively independent samples, T_0 (e.g., Leith 1973). Equation 5.4 can be seen as a special case of Equation 5.13, with $\rho_1 = 0$.

EXAMPLE 5.2   Two-Sample t Test for Correlated Data

Consider testing whether the average maximum temperatures at Ithaca and Canandaigua for January 1987 (Table A.1 in Appendix A) are significantly different. This is equivalent to testing whether the difference of the two sample means is significantly different from zero, so that Equation 5.6 will hold for the null hypothesis. It has been shown previously (see Figure 3.5) that these two batches of daily data are reasonably symmetric and well-behaved, so the sampling distribution of the monthly average should be nearly Gaussian under the Central Limit Theorem. Thus, the parametric test just described (which assumes the Gaussian form for the sampling distribution) should be appropriate.

The data for each location were observed on the same 31 days in January 1987, so the two batches are paired samples. Equation 5.11 is therefore the appropriate choice for the test statistic. Furthermore, we know that the daily data underlying the two time averages exhibit serial correlation (Figure 3.19 for the Ithaca data) so it is expected that the effective sample size corrections in Equations 5.12 and 5.13 will be necessary as well.

Table A1 shows the mean January 1987 temperatures, so the difference (Ithaca – Canandaigua) in mean maximum temperature is $29.87 - 31.77 = -1.9°F$. Computing the standard deviation of the differences between the 31 pairs of maximum temperatures yields $s_\Delta = 2.285°F$. The lag-1 autocorrelation for these differences is 0.076, yielding $n' = 31(1 - .076)/(1 + .076) = 26.6$. Since the null hypothesis is that the two population means are equal, $\mu_\Delta = 0$, and Equation 5.11 (using the effective sample size n' rather than the actual sample size n) yields $z = -1.9/(2.285^2/26.6)^{1/2} = -4.29$. This is a sufficiently extreme value not to be included in Table B.1, although Equation 4.29 estimates $\Phi(-4.29) \approx 0.000002$; so the 2-tailed p-value would be 0.000004, which is clearly significant. This extremely strong result is possible in part because much of the variability of the two temperature series is shared (the correlation between them is 0.957), and removing shared variance results in a rather small denominator for the test statistic.

Finally, notice that the lag-1 autocorrelation for the paired temperature differences is only 0.076, which is much smaller than the autocorrelations in the two individual series: 0.52 for Ithaca and 0.61 for Canandaigua. Much of the temporal dependence is also exhibited jointly by the two series, and so is removed when calculating the differences Δ.

Here is another advantage of using the series of differences to conduct this test, and another major contribution to the strong result. The relatively low autocorrelation of the difference series translates into an effective sample size of 26.6 rather than only 9.8 (Ithaca) and 7.5 (Canandaigua), which produces an even more sensitive test. ◊

5.2.5 *Goodness-of-Fit Tests*

When discussing the fitting of parametric distributions to data samples in Chapter 4, methods for visually and subjectively assessing the goodness of fit were presented. Formal, quantitative tests of the goodness of fit also exist, and these are carried out within the framework of hypothesis testing. The graphical methods can still be useful when formal tests are conducted, for example in pointing out where and how a lack of fit is manifested. Many goodness-of-fit tests have been devised, but only a few common ones are presented here.

Assessing goodness of fit presents an atypical hypothesis test setting, in that these tests usually are computed to obtain evidence in favor of H_0, that the data at hand were drawn from a hypothesized distribution. The interpretation of confirmatory evidence is then that the data are not inconsistent with the hypothesized distribution, so the power of these tests is an important consideration. Unfortunately, because there are any number of ways in which the null hypothesis can be wrong in this setting, it is usually not possible to formulate a single best (most powerful) test. This problem accounts in part for the large number of goodness-of-fit tests that have been proposed (D'Agostino and Stephens 1986), and the ambiguity about which might be most appropriate for a particular problem.

The χ^2 test is a simple and common goodness-of-fit test. It essentially compares a data histogram with the probability distribution (for discrete variables) or probability density (for continuous variables) function. The χ^2 test actually operates more naturally for discrete random variables, since to implement it the range of the data must be divided into discrete classes, or bins. When alternative tests are available for continuous data they are usually more powerful, presumably at least in part because the rounding of data into bins, which may be severe, discards information. However, the χ^2 test is easy to implement and quite flexible, being, for example, very straightforward to implement for multivariate data.

For continuous random variables, the probability density function is integrated over each of some number of MECE classes to obtain the theoretical probabilities for observations in each class. The test statistic involves the counts of data values falling into each class in relation to the computed theoretical probabilities,

$$\chi^2 = \sum_{classes} \frac{(\# \text{ Observed} - \# \text{ Expected})^2}{\# \text{ Expected}}$$

$$= \sum_{classes} \frac{(\# \text{ Observed} - n \Pr\{\text{data in class}\})^2}{n \Pr\{\text{data in class}\}}. \tag{5.14}$$

In each class, the number (#) of data values expected to occur, according to the fitted distribution, is simply the probability of occurrence in that class multiplied by the sample size, n. The number of expected occurrences need not be an integer value. If the fitted distribution is very close to the data, the expected and observed counts will be very close for each class; and the squared differences in the numerator of Equation 5.14 will all be very small, yielding a small χ^2. If the fit is not good, at least a few of the classes will exhibit large discrepancies. These will be squared in the numerator of Equation 5.14 and lead to large values of χ^2. It is not necessary for the classes to be of equal width or equal probability, but classes with small numbers of expected counts should be avoided. Sometimes a minimum of five expected events per class is imposed.

Under the null hypothesis that the data were drawn from the fitted distribution, the sampling distribution for the test statistic is the χ^2 distribution with parameter $\nu = ($ # of classes – # of parameters fit – 1 $)$ degrees of freedom. The test will be one-sided, because the test statistic is confined to positive values by the squaring process in the numerator of Equation 5.14, and small values of the test statistic support H_0. Right-tail quantiles for the χ^2 distribution are given in Table B.3.

EXAMPLE 5.3   Comparing Gaussian and Gamma Distribution Fits Using the χ^2 Test

Consider the fits of the gamma and Gaussian distributions to the 1933–1982 Ithaca January precipitation data in Table A.2. The approximate maximum likelihood estimators for the gamma distribution parameters (Equations 4.41 or 4.43a, and Equation 4.42) are $\alpha = 3.76$ and $\beta = 0.52$ in. The sample mean and standard deviation (i.e., the Gaussian parameter estimates) for these data are 1.96 in. and 1.12 in., respectively. The two fitted distributions are illustrated in relation to the data in Figure 4.15. Table 5.1 contains the information necessary to conduct the χ^2 test for these two distributions. The precipitation amounts have been divided into six classes, or bins, the limits of which are indicated in the first row of the table. The second row indicates the number of years in which the January precipitation total was within each class. Both distributions have been integrated over these classes to obtain probabilities for precipitation in each class. These probabilities were then multiplied by $n = 50$ to obtain the expected number of counts.

Applying Equation 5.14 yields $\chi^2 = 5.05$ for the gamma distribution and $\chi^2 = 14.96$ for the Gaussian distribution. As was also evident from the graphical comparison in Figure 4.15, these test statistics indicate that the Gaussian distribution fits these precipitation data substantially less well. Under the respective null hypotheses, these two test statistics are drawn from a χ^2 distribution with degrees of freedom $\nu = 6 - 2 - 1 = 3$; because Table 5.1 contains six classes, and two parameters (α and β, or μ and σ, for the gamma and Gaussian, respectively) were fit for each distribution.

Referring to the $\nu = 3$ row of Table B.3, $\chi^2 = 5.05$ is smaller than the 90th percentile value of 6.251, so the null hypothesis that the data have been drawn from the fitted gamma distribution would not be rejected even at the 10% level. For the Gaussian fit, $\chi^2 = 14.96$ is between the tabulated values of 11.345 for the 99th percentile and 16.266

TABLE 5.1   The χ^2 goodness-of-fit test applied to gamma and Gaussian distributions for the 1933–1982 Ithaca January precipitation data. Expected numbers of occurrences in each bin are obtained by multiplying the respective probabilities by $n = 50$.

Class	<1"	1 — 1.5"	1.5 — 2"	2 — 2.5"	2.5 — 3"	≥3"
Observed #	5	16	10	7	7	5
Gamma:						
Probability	0.161	0.215	0.210	0.161	0.108	0.145
Expected #	8.05	10.75	10.50	8.05	5.40	7.25
Gaussian:						
Probability	0.195	0.146	0.173	0.173	0.132	0.176
Expected #	9.75	7.30	8.65	8.90	6.60	8.80

for the 99.9th percentile, so this null hypothesis would be rejected at the 1% level, but not at the 0.1% level. ◊

A very frequently used test of the goodness of fit is the one-sample Kolmogorov-Smirnov (K-S) test. The χ^2 test essentially compares the histogram and the PDF or discrete distribution function, and the K-S test compares the empirical and theoretical CDFs. Again, the null hypothesis is that the observed data were drawn from the distribution being tested, and a sufficiently large discrepancy will result in the null hypothesis being rejected. For continuous distributions the K-S test usually will be more powerful than the χ^2 test, and so usually will be preferred.

In its original form, the K-S test is applicable to any distributional form (including but not limited to any of the distributions presented in Chapter 4), provided that the parameters have *not* been estimated from the data sample. In practice this provision can constitute a serious limitation to the use of the original K-S test, since it is often the correspondence between a fitted distribution and the particular batch of data used to fit it that is of interest. This may seem like a trivial problem, but it can have serious consequences, as has been pointed out by Crutcher (1975). Estimating the parameters from the same batch of data used to test the goodness of fit results in the fitted distribution parameters being tuned to the data sample. When erroneously using K-S critical values that assume independence between the test data and the estimated parameters, it will often be the case that the null hypothesis (that the distribution fits well) will not be rejected when in fact it should be.

With modification, the K-S framework can be used in situations where the distribution parameters have been fit to the same data used in the test. In this situation, the K-S test is often called the Lilliefors test, after the statistician who did much of the early work on the subject (Lilliefors 1967). Both the original K- S test and the Lilliefors test use the test statistic

$$D_n = \max_x |F_n(x) - F(x)|, \tag{5.15}$$

where $F_n(x)$ is the empirical cumulative probability, estimated as $F_n(x_{(i)}) = i/n$ for the ith smallest data value; and $F(x)$ is the theoretical cumulative distribution function evaluated at x (Equation 4.18). Thus the K-S test statistic D_n looks for the largest difference, in absolute value, between the empirical and fitted cumulative distribution functions. Any real and finite batch of data will exhibit sampling fluctuations resulting in a nonzero value for D_n, even if the null hypothesis is true and the theoretical distribution fits very well. If D_n is sufficiently large, the null hypothesis can be rejected. How large is large enough depends on the level of the test, of course; but also on the sample size, whether or not the distribution parameters have been fit using the test data, and if so also on the particular distribution form being fit.

When the parametric distribution to be tested has been specified completely externally to the data—the data have not been used in any way to fit the parameters—the original K-S test is appropriate. This test is distribution-free, in the sense that its critical values are applicable to any distribution. These critical values can be obtained to good approximation (Stephens 1974) using

$$C_\alpha = \frac{K_\alpha}{\sqrt{n} + 0.12 + 0.11/\sqrt{n}}, \tag{5.16}$$

where $K_\alpha = 1.224, 1.358$, and 1.628, for $\alpha = 0.10, 0.05$ and 0.01, respectively. The null hypothesis is rejected for $D_n \geq C_\alpha$.

Usually the original K-S test (and therefore Equation 5.16) is not appropriate because the parameters of the distribution being tested have been fit using the test data. But even in this case bounds on the true CDF, whatever its form, can be computed and displayed graphically using $F_n(x) \pm C_\alpha$ as limits covering the actual cumulative probabilities, with probability $1 - \alpha$. Values of C_α can also be used in an analogous way to calculate probability bounds on empirical quantiles consistent with a particular theoretical distribution (Loucks *et al.* 1981). Because the D_n statistic is a maximum over the entire data set, these bounds are valid jointly, for the entire distribution.

When the distribution parameters have been fit using the data at hand, Equation 5.16 is not sufficiently stringent, because the fitted distribution "knows" too much about the data to which it is being compared, and the Lilliefors test is appropriate. Here, however, the critical values of D_n depend on the distribution that has been fit. Table 5.2, from Crutcher (1975), lists critical values of D_n (above which the null hypothesis would be rejected) for four test levels for the gamma distribution. These critical values depend on both the sample size and the estimated shape parameter, α. Larger samples will be less subject to irregular sampling variations, so the tabulated critical values decline for larger n. That is, smaller maximum deviations from the fitted theoretical distribution (Equation 5.15) are tolerated for larger sample sizes. Critical values in the last row of the table, for $\alpha = \infty$, pertain to the Gaussian distribution, since as the gamma shape parameter becomes very large the gamma distribution converges toward the Gaussian.

It is interesting to note that critical values for Lilliefors tests are usually derived through statistical simulation (see Section 4.7). The procedure is that a large number of samples from a known distribution are generated, estimates of the distribution parameters are calculated from each of these samples, and the agreement, for each synthetic data batch, between data generated from the known distribution and the distribution fit to it is assessed using Equation 5.15. Since the null hypothesis is true in each case by construction, the α-level critical value is approximated as the $(1 - \alpha)$ quantile of that collection of synthetic D_n's. Thus, Lilliefors test critical values for any distribution that may be of interest can be computed using the methods in Section 4.7.

EXAMPLE 5.4   Comparing Gaussian and Gamma Fits Using the K-S Test

Again consider the fits of the gamma and Gaussian distributions to the 1933–1982 Ithaca January precipitation data, from Table A.2, shown in Figure 4.15. Figure 5.5 illustrates the Lilliefors test for these two fitted distributions. In each panel of Figure 5.5, the black dots are the empirical cumulative probability estimates, $F_n(x)$, and the smooth curves are the fitted theoretical CDFs, $F(x)$, both plotted as functions of the observed monthly precipitation. Coincidentally, the maximum differences between the empirical and fitted theoretical cumulative distribution functions occur at the same (highlighted) point, yielding $D_n = 0.068$ for the gamma distribution (a) and $D_n = 0.131$ for the Gaussian distribution (b).

In each of the two tests to be conducted the null hypothesis is that the precipitation data were drawn from the fitted distribution, and the alternative hypothesis is that they were not. These will necessarily be one-sided tests, because the test statistic D_n is the absolute value of the largest difference between the theoretical and empirical cumulative probabilities. Therefore values of the test statistic on the far right tail of the null distribution will indicate large discrepancies that are unfavorable to H_0, whereas values of the test statistic on the left tail of the null distribution will indicate $D_n \approx 0$, or near-perfect fits that are very supportive of the null hypothesis.

The critical values in Table 5.2 are the minimum D_n necessary to reject H_0; that is, the leftmost bounds of the relevant rejection, or critical regions. The sample size of

TABLE 5.2   Critical values for the K-S statistic D_n used in the Lilliefors test to assess goodness of fit of gamma distributions, as a function of the estimated shape parameter, α, when the distribution parameters have been fit using the data to be tested. The row labeled $\alpha = \infty$ pertains to the Gaussian distribution with parameters estimated from the data. From Crutcher (1975).

α	20% level			10% level			5% level			1% level		
	n = 25	n = 30	large n	n = 25	n = 30	large n	n = 25	n = 30	large n	n = 25	n = 30	large n
1	0.165	0.152	$0.84/\sqrt{n}$	0.185	0.169	$0.95/\sqrt{n}$	0.204	0.184	$1.05/\sqrt{n}$	0.241	0.214	$1.20/\sqrt{n}$
2	0.159	0.146	$0.81/\sqrt{n}$	0.176	0.161	$0.91/\sqrt{n}$	0.190	0.175	$0.97/\sqrt{n}$	0.222	0.203	$1.16/\sqrt{n}$
3	0.148	0.136	$0.77/\sqrt{n}$	0.166	0.151	$0.86/\sqrt{n}$	0.180	0.165	$0.94/\sqrt{n}$	0.214	0.191	$1.80/\sqrt{n}$
4	0.146	0.134	$0.75/\sqrt{n}$	0.164	0.148	$0.83/\sqrt{n}$	0.178	0.163	$0.91/\sqrt{n}$	0.209	0.191	$1.06/\sqrt{n}$
8	0.143	0.131	$0.74/\sqrt{n}$	0.159	0.146	$0.81/\sqrt{n}$	0.173	0.161	$0.89/\sqrt{n}$	0.203	0.187	$1.04/\sqrt{n}$
∞	0.142	0.131	$0.736/\sqrt{n}$	0.158	0.144	$0.805/\sqrt{n}$	0.173	0.161	$0.886/\sqrt{n}$	0.200	0.187	$1.031/\sqrt{n}$

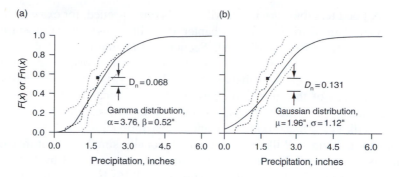

FIGURE 5.5 Illustration of the Kolmogorov-Smirnov D_n statistic as applied to the 1933–1982 Ithaca January precipitation data, fitted to a gamma distribution (a) and a Gaussian distribution (b). Solid curves indicate theoretical cumulative distribution functions, and black dots show the corresponding empirical estimates. The maximum difference between the empirical and theoretical CDFs occurs for the highlighted square point, and is substantially greater for the Gaussian distribution. Grey dots show limits of the 95% confidence interval for the true CDF from which the data were drawn (Equation 5.16).

$n = 50$ is sufficient to evaluate the tests using critical values from the large n columns. In the case of the Gaussian distribution, the relevant row of the table is for $\alpha = \infty$. Since $0.886/\sqrt{50} = 0.125$ and $1.031/\sqrt{50} = 0.146$ bound the observed $D_n = 0.131$, the null hypothesis that the precipitation data were drawn from this Gaussian distribution would be rejected at the 5% level, but not the 1% level. For the fitted gamma distribution the nearest row in Table 5.2 is for $\alpha = 4$, where even at the 20% level the critical value of $0.75/\sqrt{50} = 0.106$ is substantially larger than the observed $D_n = 0.068$. Thus the data are quite consistent with the proposition of their having been drawn from this gamma distribution.

Regardless of the distribution from which these data were drawn, it is possible to use Equation 5.16 to calculate confidence intervals on its CDF. Using $C_\alpha = 1.358$, the grey dots in Figure 5.5 show the 95% confidence intervals for $n = 50$ as $F_n(x) \pm 0.188$. The intervals defined by these points cover the true CDF with 95% probability, throughout the range of the data, because the K-S statistic pertains to the largest difference between $F_n(x)$ and $F(x)$, regardless of where in the distribution that maximum discrepancy may occur for a particular sample. ◊

A related test is the two-sample K-S test, or Smirnov test. Here the idea is to compare two batches of data to one another under the null hypothesis that they were drawn from the same (but unspecified) distribution or generating process. The Smirnov test statistic,

$$D_S = \max_x |F_n(x_1) - F_m(x_2)|, \tag{5.17}$$

looks for the largest (in absolute value) difference between the empirical cumulative distribution functions of samples of n_1 observations of x_1 and n_2 observations of x_2. Again, the test is one-sided because of the absolute values in Equation 5.17, and the null hypothesis that the two data samples were drawn from the same distribution is rejected at the $\alpha \cdot 100\%$ level if

$$D_S > \left[-\frac{1}{2} \left(\frac{1}{n_1} + \frac{1}{n_2} \right) \ln \left(\frac{\alpha}{2} \right) \right]^{1/2}. \tag{5.18}$$

A good test for Gaussian distribution is often needed, for example when the multivariate Gaussian distribution (see Chapter 10) will be used to represent the joint variations of (possibly power-transformed, Section 3.4.1) multiple variables. The Lilliefors test (Table 5.2, with $\alpha = \infty$) is an improvement in terms of power over the chi-square test for this purpose, but tests that are generally better (D'Agostino 1986) can be constructed on the basis of the correlation between the empirical quantiles (i.e., the data), and the Gaussian quantile function based on their ranks. This approach was introduced by Shapiro and Wilk (1965), and both the original test formulation and its subsequent variants are known as Shapiro-Wilk tests. A computationally simple variant that is nearly as powerful as the original Shapiro-Wilk formulation was proposed by Filliben (1975). The test statistic is simply the correlation (Equation 3.26) between the empirical quantiles $x_{(i)}$ and the Gaussian quantile function $\Phi^{-1}(p_i)$, with p_i estimated using a plotting position (see Table 3.2) approximating the median cumulative probability for the ith order statistic (e.g., the Tukey plotting position, although Filliben (1975) used Equation 3.17 with $a = 0.3175$). That is, the test statistic is simply the correlation computed from the points on a Gaussian Q-Q plot. If the data are drawn from a Gaussian distribution these points should fall on a straight line, apart from sampling variations.

Table 5.3 shows critical values for the Filliben test for Gaussian distribution. The test is one tailed, because high correlations are favorable to the null hypothesis that the data are Gaussian, so the null hypothesis is rejected if the correlation is smaller than the appropriate critical value. Because the points on a Q-Q plot are necessarily nondecreasing, the critical values in Table 5.3 are much larger than would be appropriate for testing the significance of the linear association between two independent (according to a null hypothesis) variables. Notice that, since the correlation will not change if the data are first standardized (Equation 3.21), this test does not depend in any way on the accuracy with which the distribution parameters may have been estimated. That is, the test addresses

TABLE 5.3   Critical values for the Filliben (1975) test for Gaussian distribution, based on the Q-Q plot correlation. H_0 is rejected if the correlation is smaller than the appropriate critical value.

n	0.5% level	1% level	5% level	10% level
10	.860	.876	.917	.934
20	.912	.925	.950	.960
30	.938	.947	.964	.970
40	.949	.958	.972	.977
50	.959	.965	.977	.981
60	.965	.970	.980	.983
70	.969	.974	.982	.985
80	.973	.976	.984	.987
90	.976	.978	.985	.988
100	.9787	.9812	.9870	.9893
200	.9888	.9902	.9930	.9942
300	.9924	.9935	.9952	.9960
500	.9954	.9958	.9970	.9975
1000	.9973	.9976	.9982	.9985

the question of whether the data were drawn from a Gaussian distribution, but does not address the question of what the parameters of that distribution might be.

EXAMPLE 5.5   Filliben Q-Q Correlation Test for Gaussian Distribution

The Q-Q plots in Figure 4.16 showed that the Gaussian distribution fits the 1933–1982 Ithaca January precipitation data in Table A.2 less well that the gamma distribution. That Gaussian Q-Q plot is reproduced in Figure 5.6 (X's), with the horizontal axis scaled to correspond to standard Gaussian quantiles, z, rather than to dimensional precipitation amounts. Using the Tukey plotting position (see Table 3.2), estimated cumulative probabilities corresponding to (for example) the smallest and largest of these $n = 50$ precipitation amounts are $0.67/50.33 = 0.013$ and $49.67/50.33 = 0.987$. Standard Gaussian quantiles, z corresponding to these cumulative probabilities (see Table B.1) are ± 2.22. The correlation for these $n = 50$ points is $r = 0.917$, which is smaller than all of the critical values in that row of Table 5.3. Accordingly, the Filliben test would reject the null hypothesis that these data were drawn from a Gaussian distribution, at the 0.5% level. The fact that the horizontal scale is the nondimensional z rather than dimensional precipitation (as in Figure 4.16) is immaterial, because the correlation is unaffected by linear transformations of either or both of the two variables being correlated.

Figure 3.13, in Example 3.4, indicated that a logarithmic transformation of these data was effective in producing approximate symmetry. Whether this transformation is also effective at producing a plausibly Gaussian shape for these data can be addressed with the Filliben test. Figure 5.6 also shows the standard Gaussian Q-Q plot for the log-transformed Ithaca January precipitation totals (O's). This relationship is substantially more linear than for the untransformed data, and is characterized by a correlation of $r = 0.987$. Again looking on the $n = 50$ row of Table 5.3, this correlation is larger than the 10% critical value, so the null hypothesis of Gaussian distribution would not be rejected. ◊

Using statistical simulation (see Section 4.7) tables of critical Q-Q correlations can be obtained for other distributions, by generating large numbers of batches of size n from the distribution of interest, computing Q-Q plot correlations for each of these batches,

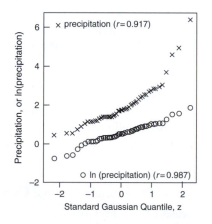

FIGURE 5.6   Standard Gaussian Q-Q plots for the 1933–1982 Ithaca January precipitation in Table A.2 (X's), and for the log-transformed data (O's). Using Table 5.3, null hypotheses that these data were drawn from Gaussian distributions would be rejected for the original data ($p < 0.005$), but not rejected for the log-transformed data ($p > 0.10$).

and defining the critical value as that delineating the extreme $\alpha \cdot 100\%$ smallest of them. Results of this approach have been tabulated for the Gumbel distribution (Vogel 1986), the uniform distribution (Vogel and Kroll 1989), the GEV distribution (Chowdhury *et al.* 1991), and the Pearson III distribution (Vogel and McMartin 1991).

5.2.6 Likelihood Ratio Test

Sometimes we need to construct a test in a parametric setting, but the hypothesis is sufficiently complex that the simple, familiar parametric tests cannot be brought to bear. A flexible alternative, known as the likelihood ratio test, can be used if two conditions are satisfied. First, we must be able to cast the problem in such a way that the null hypothesis pertains to some number, k_0 of free (i.e., fitted) parameters, and the alternative hypothesis pertains to some larger number, $k_A > k_0$, of parameters. Second, it must be possible to regard the k_0 parameters of the null hypothesis as a special case of the full parameter set of k_A parameters. Examples of this second condition on H_0 could include forcing some of the k_A parameters to have fixed values, or imposing equality between two or more of them. As the name implies, the likelihood ratio test compares the likelihoods associated with H_0 vs. H_A, when the k_0 and k_A parameters, respectively, have been fit using the method of maximum likelihood (see Section 4.6).

Even if the null hypothesis is true, the likelihood associated with H_A will always be at least as large as that for H_0. This is because the greater number of parameters $k_A > k_0$ allows the likelihood function for the former greater freedom in accommodating the observed data. The null hypothesis is therefore rejected only if the likelihood associated with the alternative is sufficiently large that the difference is unlikely to have resulted from sampling variations.

The test statistic for the likelihood ratio test is

$$\Lambda^* = 2 \ln\left[\frac{\Lambda(H_A)}{\Lambda(H_0)}\right] = 2[L(H_A) - L(H_0)]. \tag{5.19}$$

Here $\Lambda(H_0)$ and $\Lambda(H_A)$ are the likelihood functions (see Section 4.6) associated with the null and alternative hypothesis, respectively. The second equality, involving the difference of the log-likelihoods $L(H_0) = \ln[\Lambda(H_0)]$ and $L(H_A) = \ln[\Lambda(H_A)]$, is used in practice since it is generally the log-likelihoods that are maximized (and thus computed) when fitting the parameters.

Under H_0, and given a large sample size, the sampling distribution of the statistic in Equation 5.19 is χ^2, with degrees of freedom $\nu = k_A - k_0$. That is, the degrees-of-freedom parameter is given by the difference between H_A and H_0 in the number of empirically estimated parameters. Since small values of Λ^* are not unfavorable to H_0, the test is one-sided and H_0 is rejected only if the observed Λ^* is in a sufficiently improbable region on the right tail.

EXAMPLE 5.6   Testing for Climate Change Using the Likelihood Ratio Test

Suppose there is a reason to suspect that the first 25 years (1933–1957) of the Ithaca January precipitation data in Table A.2 have been drawn from a different gamma distribution than the second half (1958–1982). This question can be tested against the null hypothesis that all 50 precipitation totals were drawn from the same gamma distribution using a likelihood ratio test. To perform the test it is necessary to fit gamma distributions

TABLE 5.4 Gamma distribution parameters (MLEs) and log-likelihoods for fits to the first and second halves of the 1933–1982 Ithaca January precipitation data, and to the full data set.

	Dates	α	β	$L(\alpha, \beta)$
H_A:				
	1933–1957	4.525	0.4128	−30.2796
	1958–1982	3.271	0.6277	−35.8965
H_0:				
	1933–1982	3.764	0.5209	−66.7426

separately to the two halves of the data, and compare these two distributions with the single gamma distribution fit using the full data set.

The relevant information is presented in Table 5.4, which indicates some differences between the two 25-year periods. For example, the average January precipitation $(= \alpha\beta)$ for 1933–1957 was 1.87 in., and the corresponding average for 1958–1982 was 2.05 in. The year-to-year variability $(= \alpha\beta^2)$ of January precipitation was greater in the second half of the period as well. Whether the extra two parameters required to represent the January precipitation using two gamma distributions rather than one are justified by the data can be evaluated using the test statistic in Equation 5.19. For this specific problem the test statistic is

$$\Lambda^* = 2\left\{\left[\sum_{i=1933}^{1957} L(\alpha_1, \beta_1; x_i)\right] + \left[\sum_{i=1958}^{1982} L(\alpha_2, \beta_2; x_i)\right]\right.$$
$$\left. - \left[\sum_{i=1933}^{1982} L(\alpha_0, \beta_0; x_i)\right]\right\}, \tag{5.20}$$

where the subscripts 1, 2, and 0 on the parameters refer to the first half, the second half, and the full period (null hypothesis), respectively, and the log-likelihood for the gamma distribution given a single observation, x_i, is (compare Equation 4.38)

$$L(\alpha, \beta; x_i) = (\alpha - 1)\ln\frac{x_i}{\beta} - \frac{x_i}{\beta} - \ln\beta - \ln\Gamma(\alpha). \tag{5.21}$$

The three terms in square brackets in Equation 5.20 are given in the last column of Table 5.4.

Using the information in Table 5.4, $\Lambda^* = 2(-30.2796 - 35.8965 + 66.7426) = 1.130$. Since there are $k_A = 4$ parameters under $H_A(\alpha_1, \beta_1, \alpha_2, \beta_2)$ and $k_0 = 2$ parameters under $H_0(\alpha_0, \beta_0)$, the null distribution is the χ^2 distribution with $\nu = 2$. Looking on the $\nu = 2$ row of Table B.3, we find $\chi^2 = 1.130$ is smaller than the median value, leading to the conclusion that the observed Λ^* is quite ordinary in the context of the null hypothesis that the two data records were drawn from the same gamma distribution, which would not be rejected. More precisely, recall that the χ^2 distribution with $\nu = 2$ is a gamma distribution with $\alpha = 1$ and $\beta = 2$, which in turn is the exponential distribution with $\beta = 2$. The exponential distribution has the closed form CDF in Equation 4.46, which yields the right-tail probability (p value) $1 - F(1.130) = 0.5684$. ◊

5.3 Nonparametric Tests

Not all formal hypothesis tests rest on assumptions involving theoretical distributions for the data, or theoretical sampling distributions of the test statistics. Tests not requiring such assumptions are called *nonparametric*, or distribution-free. Nonparametric methods are appropriate if either or both of the following conditions apply:

1) We know or suspect that the parametric assumption(s) required for a particular test are not met, for example grossly non-Gaussian data in conjunction with the *t* test for the difference of means in Equation 5.5.
2) A test statistic that is suggested or dictated by the physical problem at hand is a complicated function of the data, and its sampling distribution is unknown and/or cannot be derived analytically.

The same hypothesis testing ideas apply to both parametric and nonparametric tests. In particular, the five elements of the hypothesis test presented at the beginning of this chapter apply also to nonparametric tests. The difference between parametric and nonparametric tests is in the means by which the null distribution is obtained in Step 4.

We can recognize two branches of nonparametric testing. The first, called classical nonparametric testing in the following, consists of tests based on mathematical analysis of selected hypothesis test settings. These are older methods, devised before the advent of cheap and widely available computing. They employ analytic mathematical results (formulas) that are applicable to data drawn from any distribution. Only a few classical nonparametric tests for location will be presented here, although the range of classical nonparametric methods is much more extensive (e.g., Conover 1999; Daniel 1990; Sprent and Smeeton 2001).

The second branch of nonparametric testing includes procedures collectively called resampling tests. Resampling tests build up a discrete approximation to the null distribution using a computer, by repeatedly operating on (resampling) the data set at hand. Since the null distribution is arrived at empirically, the analyst is free to use virtually any test statistic that may be relevant, regardless of how mathematically complicated it may be.

5.3.1 Classical Nonparametric Tests for Location

Two classical nonparametric tests for the differences in location between two data samples are especially common and useful. These are the Wilcoxon-Mann-Whitney rank-sum test for two independent samples (analogous to the parametric test in Equation 5.8) and the Wilcoxon signed-rank test for paired samples (corresponding to the parametric test in Equation 5.11).

The Wilcoxon-Mann-Whitney rank-sum test was devised independently in the 1940s by Wilcoxon, and by Mann and Whitney, although in different forms. The notations from both forms of the test are commonly used, and this can be the source of some confusion. However, the fundamental idea behind the test is not difficult to understand. The test is resistant, in the sense that a few wild data values that would completely invalidate the *t* test of Equation 5.8 will have little or no influence. It is robust in the sense that, even if all the assumptions required for the *t* test in Equation 5.8 are met, the rank- sum test is almost as good (i.e., nearly as powerful).

Given two samples of independent (i.e., both serially independent, and unpaired) data, the aim is to test for a possible difference in location. Here location is meant in the EDA sense of overall magnitude, or the nonparametric analog of the mean. The null hypothesis is that the two data samples have been drawn from the same distribution. Both one-sided (the center of one sample is expected in advance to be larger or smaller than the other if the null hypothesis is not true) and two-sided (no prior information on which sample should be larger) alternative hypotheses are possible. Importantly, the effect of serial correlation on the Wilcoxon-Mann-Whitney test is qualitatively similar to the effect on the t test: the variance of the sampling distribution of the test statistic is inflated, possibly leading to unwarranted rejection of H_0 if the problem is ignored (Yue and Wang 2002). The same effect occurs in other classical nonparametric tests as well (von Storch 1995).

Under the null hypothesis that the two data samples are from the same distribution, the labeling of each data value as belonging to one group or the other is entirely arbitrary. That is, if the two data samples are really drawn from the same population, each observation is as likely to have been placed in one sample as the other by the process that generated the data. Under the null hypothesis, then, there are not n_1 observations in Sample 1 and n_2 observations in Sample 2, but rather $n = n_1 + n_2$ observations making up a single empirical distribution. The notion that the data labels are arbitrary because they have all been drawn from the same distribution under H_0 is known as the principle of *exchangeability*, which also underlies permutation tests, as discussed in Section 5.3.3.

The rank-sum test statistic is a function not of the data values themselves, but of their ranks within the n observations that are pooled under the null hypothesis. It is this feature that makes the underlying distribution(s) of the data irrelevant. Define R_1 as the sum of the ranks held by the members of Sample 1 in this pooled distribution, and R_2 as the sum of the ranks held by the members of Sample 2. Since there are n members of the pooled empirical distribution implied by the null distribution, $R_1 + R_2 = 1 + 2 + 3 + 4 + \cdots + n = (n)(n+1)/2$. If the two samples really have been drawn from the same distribution (i.e., if H_0 is true), then R_1 and R_2 will be similar in magnitude if $n_1 = n_2$. Regardless of whether or not the sample sizes are equal, however, R_1/n_1 and R_2/n_2 should be similar in magnitude if the null hypothesis is true.

The null distribution for R_1 and R_2 is obtained in a way that exemplifies the approach of nonparametric tests more generally. If the null hypothesis is true, the observed partitioning of the data into two groups of size n_1 and n_2 is only one of very many equally likely ways in which the n values could have been split and labeled. Specifically, there are $(n!)/[(n_1!)(n_2!)]$ such equally likely partitions of the data under the null hypothesis. For example, if $n_1 = n_2 = 10$, this number of possible distinct pairs of samples is 184,756. Conceptually, imagine the statistics R_1 and R_2 being computed for each of these 184,756 possible arrangements of the data. It is simply this very large collection of (R_1, R_2) pairs, or, more specifically, the collection of 184,756 scalar test statistics computed from these pairs, that constitutes the null distribution. If the observed test statistic falls comfortably within this large empirical distribution, then that particular partition of the n observations is quite consistent with H_0. If, however, the observed R_1 and R_2 are more different from each other than under most of the other possible partitions of the data, H_0 would be rejected.

It is not actually necessary to compute the test statistic for all $(n!)/[(n_1!)(n_2!)]$ possible arrangements of the data. Rather, the Mann-Whitney U-statistic,

$$U_1 = R_1 - \frac{n_1}{2}(n_1 + 1) \tag{5.22a}$$

or

$$U_2 = R_2 - \frac{n_2}{2}(n_2 + 1), \tag{5.22b}$$

is computed for one or the other of the two Wilcoxon rank-sum statistics, R_1 or R_2. Both U_1 and U_2 carry the same information, since $(U_1 + U_2) = (n_1)(n_2)$, although some tables of null distribution probabilities for the rank-sum test require the smaller of U_1 and U_2. For even moderately large values of n_1 and n_2 (both larger than about 10), however, a simple method for evaluating null distribution probabilities is available. In this case, the null distribution of the Mann-Whitney U-statistic is approximately Gaussian, with

$$\mu_U = \frac{n_1 n_2}{2} \qquad (5.23a)$$

and

$$\sigma_U = \left[\frac{n_1 n_2 (n_1 + n_2 + 1)}{12} \right]^{1/2}. \qquad (5.23b)$$

For smaller samples, tables of critical values (e.g., Conover 1999) can be used.

EXAMPLE 5.7 Evaluation of a Cloud Seeding Experiment Using the Wilcoxon-Mann-Whitney Test

Table 5.5 contains data from a weather modification experiment investigating the effect of cloud seeding on lightning strikes (Baughman *et al.* 1976). It was suspected in advance that seeding the storms would reduce lightning. The experimental procedure involved randomly seeding or not seeding candidate thunderstorms, and recording a number of characteristics of the lightning, including the counts of strikes presented in Table 5.5. There were $n_1 = 12$ seeded storms, exhibiting an average of 19.25 cloud-to-ground lightning strikes; and $n_2 = 11$ unseeded storms, with an average of 69.45 strikes.

Inspecting the data in Table 5.5, it is apparent that the distribution of lightning counts for the unseeded storms is distinctly non-Gaussian. In particular, the set contains one very large outlier of 358 strikes. We suspect, therefore, that uncritical application of the t test (Equation 5.8) to test the significance of the difference in the observed mean numbers of

TABLE 5.5 Counts of cloud-to-ground lightning for experimentally seeded and nonseeded storms. From Baughman *et al.* (1976).

	Seeded		Unseeded
Date	Lightning strikes	Date	Lightning strikes
7/20/65	49	7/2/65	61
7/21/65	4	7/4/65	33
7/29/65	18	7/4/65	62
8/27/65	26	7/8/65	45
7/6/66	29	8/19/65	0
7/14/66	9	8/19/65	30
7/14/66	16	7/12/66	82
7/14/66	12	8/4/66	10
7/15/66	2	9/7/66	20
7/15/66	22	9/12/66	358
8/29/66	10	7/3/67	63
8/29/66	34		

lightning strikes could produce misleading results. This is because the single very large value of 358 strikes leads to a sample standard deviation for the unseeded storms of 98.93 strikes. This large sample standard deviation would lead us to attribute a very large spread to the assumed t-distributed sampling distribution of the difference of means, so that even rather large values of the test statistic would be judged as being fairly ordinary.

The mechanics of applying the rank-sum test to the data in Table 5.5 are shown in Table 5.6. In the left-hand portion of the table, the 23 data points are pooled and ranked, consistent with the null hypothesis that all the data came from the same population, regardless of the labels S or N. There are two observations of 10 lightning strikes, and as is conventional each has been assigned the average rank $(5+6)/2 = 5.5$. This expedient poses no real problem if there are few tied ranks, but the procedure can be modified slightly

TABLE 5.6   Illustration of the procedure of the rank-sum test using the cloud-to-ground lightning data in Table 5.5. In the left portion of this table, the $n_1 + n_2 = 23$ counts of lightning strikes are pooled and ranked. In the right portion of the table, the observations are segregated according to their labels of seeded (S) or not seeded (N) and the sums of the ranks for the two categories (R_1 and R_2) are computed.

Pooled Data			Segregated Data			
Strikes	Seeded?	Rank				
0	N	1			N	1
2	S	2	S	2		
4	S	3	S	3		
9	S	4	S	4		
10	N	5.5			N	5.5
10	S	5.5	S	5.5		
12	S	7	S	7		
16	S	8	S	8		
18	S	9	S	9		
20	N	10			N	10
22	S	11	S	11		
26	S	12	S	12		
29	S	13	S	13		
30	N	14			N	14
33	N	15			N	15
34	S	16	S	16		
45	N	17			N	17
49	S	18	S	18		
61	N	19			N	19
62	N	20			N	20
63	N	21			N	21
82	N	22			N	22
358	N	23			N	23
	Sums of Ranks:		$R_1 = 108.5$		$R_2 = 167.5$	

(Conover 1999) if there are many ties. In the right-hand portion of the table, the data are segregated according to their labels, and the sums of the ranks of the two groups are computed. It is clear from this portion of Table 5.6 that the smaller numbers of strikes tend to be associated with the seeded storms, and the larger numbers of strikes tend to be associated with the unseeded storms. These differences are reflected in the differences in the sums of the ranks: R_1 for the seeded storms is 108.5, and R_2 for the unseeded storms is 167.5. The null hypothesis that seeding does not affect the number of lightning strikes can be rejected if this difference between R_1 and R_2 is sufficiently unusual against the backdrop of all possible $(23!)/[(12!)(11!)] = 1,352,078$ distinct arrangements of these data under H_0.

The Mann-Whitney U-statistic, Equation 5.22, corresponding to the sum of the ranks of the seeded data, is $U_1 = 108.5 - (6)(12 + 1) = 30.5$. The null distribution of all 1,352,078 possible values of U_1 for this data is closely approximated by the Gaussian distribution having (Equation 5.23) $\mu_U = (12)(11)/2 = 66$ and $\sigma_U = [(12)(11)(12 + 11 + 1)/12]^{1/2} = 16.2$. Within this Gaussian distribution, the observed $U_1 = 30.5$ corresponds to a standard Gaussian $z = (30.5 - 66)/16.2 = -2.19$. Table B.1 shows the (one-tailed) p value associated with this z to be 0.014, indicating that approximately 1.4% of the 1,352,078 possible values of U_1 under H_0 are smaller than the observed U_1. Accordingly, H_0 usually would be rejected. ◊

There is also a classical nonparametric test, the Wilcoxon signed-rank test, analogous to the paired two-sample parametric test of Equation 5.11. As is the case for its parametric counterpart, the signed-rank test takes advantage of positive correlation between the pairs of data in assessing possible differences in location. In common with the unpaired rank-sum test, the signed-rank test statistic is based on ranks rather than the numerical values of the data. Therefore this test also does not depend on the distribution of the underlying data, and is resistant to outliers.

Denote the data pairs (x_i, y_i), for $i = 1, \ldots, n$. The signed-rank test is based on the set of n differences, D_i, between the n data pairs. If the null hypothesis is true, and the two data sets represent paired samples from the same population, roughly equally many of these differences will be positive and negative, and the overall magnitudes of the positive and negative differences should be comparable. The comparability of the positive and negative differences is assessed by ranking them in absolute value. That is the n values of D_i are transformed to the series of ranks,

$$T_i = \text{rank}|D_i| = \text{rank}|x_i - y_i|. \tag{5.24}$$

Data pairs for which $|D_i|$ are equal are assigned the average rank of the tied values of $|D_i|$, and pairs for which $x_i = y_i$ (implying $D_i = 0$) are not included in the subsequent calculations. Denote as n' the number of pairs for which $x_i \neq y_i$.

If the null hypothesis is true, the labeling of a given data pair as (x_i, y_i) could just as well have been reversed, so that the ith data pair was just as likely to be labeled (y_i, x_i). Changing the ordering reverses the sign of D_i, but yields the same $|D_i|$. The unique information in the pairings that actually were observed is captured by separately summing the ranks, T_i, corresponding to pairs having positive or negative values of D_i, denoting as T either the statistic

$$T^+ = \sum_{D_i > 0} T_i \tag{5.25a}$$

or

$$T^- = \sum_{D_i < 0} T_i, \tag{5.25b}$$

respectively. Tables of null distribution probabilities sometimes require choosing the smaller of Equations 5.25a and 5.25b. However, knowledge of one is sufficient for the other, since $T^+ + T^- = n'(n'+1)/2$.

The null distribution of T is arrived at conceptually by considering again that H_0 implies the labeling of one or the other of each datum in a pair as x_i or y_i is arbitrary. Therefore, under the null hypothesis there are $2^{n'}$ equally likely arrangements of the $2n'$ data values at hand, and the resulting $2^{n'}$ possible values of T constitute the relevant null distribution. As before, it is not necessary to compute all possible values of the test statistic, since for moderately large n' (greater than about 20) the null distribution is approximately Gaussian, with parameters

$$\mu_T = \frac{n'(n'+1)}{4} \tag{5.26a}$$

and

$$\sigma_T = \left[\frac{n'(n'+1)(2n'+1)}{24} \right]^{1/2}. \tag{5.26b}$$

For smaller samples, tables of critical values for T^+ (e.g., Conover 1999) can be used. Under the null hypothesis, T will be close to μ_T because the numbers and magnitudes of the ranks T_i will be comparable for the negative and positive differences D_i. If there is a substantial difference between the x and y values in location, most of the large ranks will correspond to either the negative or positive D_i's, implying that T will be either very large or very small.

EXAMPLE 5.8   Comparing Thunderstorm Frequencies Using the Signed Rank Test

The procedure for the Wilcoxon signed-rank test is illustrated in Table 5.7. Here the paired data are counts of thunderstorms reported in the northeastern United States (x) and the Great Lakes states (y) for the $n = 21$ years 1885–1905. Since the two areas are close geographically, we expect that large-scale flow conducive to thunderstorm formation in one of the regions would be generally conducive in the other region as well. It is thus not surprising that the reported thunderstorm counts in the two regions are substantially positively correlated.

For each year the difference in reported thunderstorm counts, D_i, is computed, and the absolute values of these differences are ranked. None of the $D_i = 0$, so $n' = n = 21$. Years having equal differences, in absolute value, are assigned the average rank (e.g., 1892, 1897, and 1901 have the eighth, ninth, and tenth smallest $|D_i|$, and are all assigned the rank 9). The ranks for the years with positive and negative D_i, respectively, are added in the final two columns, yielding $T^+ = 78.5$ and $T^- = 152.5$.

If the null hypothesis that the reported thunderstorm frequencies in the two regions are equal is true, then labeling of counts in a particular year as being Northeastern or Great Lakes is arbitrary and thus so is the sign of each D_i. Consider, arbitrarily, the test statistic T as the sum of the ranks for the positive differences, $T^+ = 78.5$. Its unusualness in the context of H_0 is assessed in relation to the $2^{21} = 2,097,152$ values of T^+ that could result from all the possible permutations of the data under the null hypothesis. This null distribution is closely approximated by the Gaussian distribution having $\mu_T = (21)(22)/4 = 115.5$ and $\sigma_T = [(21)(22)(42+1)/24]^{1/2} = 28.77$. The p value for this test is then obtained by computing the standard Gaussian $z = (78.5 - 115.5)/28.77 = -1.29$. If there is no reason to expect one or the other of the two regions to have had more

TABLE 5.7   Illustration of the procedure of the Wilcoxon signed-rank test using data for counts of thunderstorms reported in the northeastern United States (x) and the Great Lakes states (y) for the period 1885–1905, from Brooks and Carruthers (1953). Analogously to the procedure of the rank-sum test (see Table 5.6), the absolute values of the annual differences, $|D_i|$, are ranked and then segregated according to whether D_i is positive or negative. The sum of the ranks of the segregated data constitute the test statistic.

| Year | Paired Data | | Differences | | Segregated Ranks | |
	X	Y	D_i	Rank$|D_i|$	$D_i > 0$	$D_i < 0$
1885	53	70	−17	20		20
1886	54	66	−12	17.5		17.5
1887	48	82	−34	21		21
1888	46	58	−12	17.5		17.5
1889	67	78	−11	16		16
1890	75	78	−3	4.5		4.5
1891	66	76	−10	14.5		14.5
1892	76	70	+6	9	9	
1893	63	73	−10	14.5		14.5
1894	67	59	+8	11.5	11.5	
1895	75	77	−2	2		2
1896	62	65	−3	4.5		4.5
1897	92	86	+6	9	9	
1898	78	81	−3	4.5		4.5
1899	92	96	−4	7		7
1900	74	73	+1	1	1	
1901	91	97	−6	9		9
1902	88	75	+13	19	19	
1903	100	92	+8	11.5	11.5	
1904	99	96	+3	4.5	4.5	
1905	107	98	+9	13	13	
					Sums of Ranks: $T^+ = 78.5$	$T^- = 152.5$

reported thunderstorms, the test is two-tailed (H_A is simply "not H_0"), so the p value is $\Pr\{z \le -1.29\} + \Pr\{z > +1.29\} = 2\Pr\{z \le -1.29\} = 0.197$. The null hypothesis would not be rejected in this case. Note that the same result would be obtained if the test statistic $T^- = 152.5$ had been chosen instead. ◊

5.3.2 *Introduction to Resampling Tests*

Since the advent of inexpensive and fast computing, another approach to nonparametric testing has become practical. This approach is based on the construction of artificial data sets from a given collection of real data, by resampling the observations in a manner consistent with the null hypothesis. Sometimes such methods are also known as randomization tests, rerandomization tests, or Monte-Carlo tests. Resampling methods are highly adaptable to different testing situations, and there is considerable scope for the analyst to creatively design new tests to meet particular needs.

The basic idea behind resampling tests is to build up a collection of artificial data batches of the same size as the actual data at hand, using a procedure that is consistent with the null hypothesis, and then to compute the test statistic of interest for each artificial batch. The result is as many artificial values of the test statistic as there are artificially generated data batches. Taken together, these reference test statistics constitute an estimated null distribution against which to compare the test statistic computed from the original data.

As a practical matter, we program a computer to do the resampling. Fundamental to this process are the uniform [0,1] random number generators described in Section 4.7.1 These algorithms produce streams of numbers that resemble independent values drawn independently from the probability density function $f(u) = 1, 0 \le u \le 1$. The synthetic uniform variates are used to draw random samples from the data to be tested.

In general, resampling tests have two very appealing advantages. The first is that no assumptions regarding an underlying parametric distribution for the data or the sampling distribution for the test statistic are necessary, because the procedures consist entirely of operations on the data themselves. The second is that any statistic that may be suggested as important by the physical nature of the problem can form the basis of the test, so long as it can be computed from the data. For example, when investigating location (i.e., overall magnitudes) of a sample of data, we are not confined to the conventional tests involving the arithmetic mean or sums of ranks; because it is just as easy to use alternative measures such as the median, geometric mean, or more exotic statistics if any of these are more meaningful to the problem at hand. The data being tested can be scalar (each data point is one number) or vector-valued (data points are composed of pairs, triples, etc.), as dictated by the structure of each particular problem. Resampling procedures involving vector-valued data can be especially useful when the effects of spatial correlation must be captured by a test, in which case each element in the data vector corresponds to a different location.

Any computable statistic (i.e., any function of the data) can be used as a test statistic in a resampling test, but not all will be equally good. In particular, some choices may yield tests that are more powerful than others. Good (2000) suggests the following desirable attributes for candidate test statistics.

1) **Sufficiency**. All the information about the distribution attribute or physical phenomenon of interest contained in the data is also reflected in the chosen statistic. Given a sufficient statistic, the data have nothing additional to say about the question being addressed.
2) **Invariance**. A test statistic should be constructed in a way that the test result does not depend on arbitrary transformations of the data, for example from °F to °C.
3) **Loss**. The mathematical penalty for discrepancies that is expressed by the test statistic should be consistent with the problem at hand, and the use to which the test result will be put. Often squared-error losses are assumed in parametric tests because of mathematical tractability and connections with the Gaussian distribution, although squared-error loss is disproportionately sensitive to large differences. In a resampling test there is no reason to avoid absolute-error loss or other loss functions if these make more sense in the context of a particular problem.

In addition, Hall and Wilson (1991) point out that better results are obtained when the resampled statistic does not depend on unknown quantities; for example, unknown parameters.

5.3.3 *Permutation Tests*

Two- (or more) sample problems can often be approached using permutation tests. These have been described in the atmospheric science literature, for example, by Mielke *et al.* (1981) and Priesendorfer and Barnett (1983). The concept behind permutation tests is not new (Pittman 1937), but they did not become practical until the advent of fast and abundant computing.

Permutation tests are a natural generalization of the Wilcoxon- Mann-Whitney test described in Section 5.3.1, and also depend on the principle of exchangeability. That is, exchangeability implies that, under the null hypothesis, all the data were drawn from the same distribution. Therefore, the labels identifying particular data values as belonging to one sample or another are arbitrary. Under H_0 these data labels are exchangeable.

The key difference between permutation tests generally, and the Wilcoxon-Mann-Whitney test as a special case, is that any test statistic that may be meaningful can be employed, including but certainly not limited to the particular function of the ranks given in Equation 5.22. Among other advantages, the lifting of restrictions on the mathematical form of possible test statistics expands the range of applicability of permutation tests to vector-valued data. For example, Mielke *et al.* (1981) provide a simple illustrative example using two batches of bivariate data ($x = [x, y]$) and the Euclidian distance measure examining the tendency of the two batches to cluster in the $[x, y]$ plane. Zwiers (1987) gives an example of a permutation test that uses higher-dimensional multivariate Gaussian variates.

The exchangeability principle leads logically to the construction of the null distribution using samples drawn by computer from a pool of the combined data. As was the case for the Wilcoxon-Mann- Whitney test, if two batches of size n_1 and n_2 are to be compared, the pooled set to be resampled contains $n = n_1 + n_2$ points. However, rather than computing the test statistic using all possible $n!/(n_1!)(n_2!)$ groupings (i.e., permutations) of the pooled data, the pool is merely sampled some large number (perhaps 10000) of times. (An exception can occur when n is small enough for a full enumeration of all possible permutations to be practical.) For permutation tests the samples are drawn *without replacement*, so that each of the individual n observations is represented once and once only in one or the other of the artificial samples of size n_1 and n_2. In effect, the data labels are randomly permuted for each resample. For each of these pairs of synthetic samples the test statistic is computed, and the resulting distribution (of perhaps 10000) outcomes forms the null distribution against which the observed test statistic can be compared, in the usual way.

An efficient permutation algorithm can be implemented in the following way. Assume for convenience that $n_1 \geq n_2$. The data values (or vectors) are first arranged into a single array of size $n = n_1 + n_2$. Initialize a reference index $m = n$. The algorithm proceeds by implementing the following steps n_2 times:

- Randomly choose x_i, $i = 1, \ldots, m$; using Equation 4.86 (i.e., randomly draw from the first m array positions).
- Exchange the array positions of (or, equivalently, the indices pointing to) x_i and x_m (i.e., the chosen x's will be placed at the end of the n-dimensional array).
- Decrement the reference index by 1 (i.e., $m = m - 1$).

At the end of this process there will be a random selection of the n pooled observations in the first n_1 positions, which can be treated as Sample 1, and the remaining n_2 data values at the end of the array can be treated as Sample 2. The scrambled array can be operated upon directly for subsequent random permutations—it is not necessary first to restore the data to their original ordering.

EXAMPLE 5.9   Two-Sample Permutation Test for a Complicated Statistic

Consider again the lightning data in Table 5.5. Assume that their dispersion is best (from the standpoint of some criterion external to the hypothesis test) characterized by the L-scale statistic (Hosking 1990),

$$\lambda_2 = \frac{(n-2)!}{n!} \sum_{i=1}^{n-1} \sum_{j=i+1}^{n} |x_i - x_j|. \tag{5.27}$$

Equation 5.27 amounts to half the average difference, in absolute value, between all possible pairs of points in the sample of size n. For a tightly clustered sample of data each term in the sum will be small, and therefore λ_2 will be small. For a data sample that is highly variable, some of the terms in Equation 5.27 will be very large, and λ_2 will be correspondingly large.

To compare sample λ_2 values from the seeded and unseeded storms in Table 5.5, we probably would use either the ratio or the difference of λ_2 for the two samples. A resampling test procedure provides the freedom to choose the one (or some other) making more sense for the problem at hand. Suppose the most relevant test statistic is the ratio $[\lambda_2(\text{seeded})]/[\lambda_2(\text{unseeded})]$. Under the null hypothesis that the two samples have the same L- scale, this statistic should be near one. If the seeded storms are more variable with respect to the numbers of lightning strikes, the ratio statistic should be greater than one. If the seeded storms are less variable, the statistic should be less than one. The ratio of L-scales has been chosen for this example arbitrarily, to illustrate that any computable function of the data can be used as the basis of a permutation test, regardless of how unusual or complicated it may be.

The null distribution of the test statistic is built up by sampling some (say 10,000) of the $23!/(12!11!) = 1,352,078$ distinct partitions, or permutations, of the $n = 23$ data points into two batches of $n_1 = 12$ and $n_2 = 11$. For each partition, λ_2 is computed according to Equation 5.27 for each of the two synthetic samples, and their ratio (with the value for the $n_1 = 12$ batch in the numerator) is computed and stored. The observed value of the ratio of the L-scales, 0.188, is then evaluated with respect to this empirically generated null distribution.

Figure 5.7 shows a histogram for the null distribution of the ratios of the L-scales constructed from 10000 permutations of the original data. The observed value of 0.188 is smaller than all except 49 of these 10000 values, which would lead to the null hypothesis being soundly rejected. Depending on whether a 1-sided or 2-sided test would be appropriate on the basis of external information, the p values would be 0.0049 or 0.0098, respectively. Notice that this null distribution has the unusual feature of being bimodal, having two humps. This characteristic results from the large outlier in Table 5.5, 358 lightning strikes on 9/12/66, producing a very large L-scale in whichever partition it has been assigned. Partitions for which this observation has been assigned to the unseeded group are in the left hump, and those for which the outlier has been assigned to the seeded group are in the right hump.

The conventional test for differences in dispersion involves the ratio of variances, the null distribution for which would be the F distribution, if the two underlying data samples are both Gaussian. The F test is not robust to violations of the Gaussian assumption. The permutation distribution corresponding to Figure 5.7 for the variance ratios would

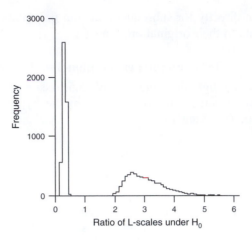

FIGURE 5.7   Histogram for the null distribution of the ratio of the L-scales for lightning counts of seeded vs. unseeded storms in Table 5.5. The observed ratio of 0.188 is smaller than all but 49 of the 10,000 permutation realizations of the ratio, which provides very strong evidence that the lightning production by seeded storms was less variable than by unseeded storms. This null distribution is bimodal because the one outlier (353 strikes on 9/12/66) produces a very large L-scale in whichever of the two partitions it has been randomly assigned.

be even more extreme, because the sample variance is less resistant to outliers than is λ_2, and the F distribution would be a very poor approximation to it. ◊

5.3.4 The Bootstrap

Permutation schemes are very useful in multiple-sample settings where the exchangeability principle applies. But in one-sample settings permutation procedures are useless because there is nothing to permute: there is only one way to resample a single data batch with replacement, and that is to replicate the original sample by choosing each of the original n data values exactly once. When the exchangeability assumption cannot be supported, the justification for pooling multiple samples before permutation disappears, because the null hypothesis no longer implies that all data, regardless of their labels, were drawn from the same population.

In either of these situations an alternative computer-intensive resampling procedure called the *bootstrap* is available. The bootstrap is a newer idea than permutation, dating from Efron (1979). The idea behind the bootstrap is known as the plug-in principle, under which we estimate any function of the underlying (population) distribution using (plugging into) the same function, but using the empirical distribution, which puts probability $1/n$ on each of the n observed data values. Put another way, the idea behind the bootstrap is to treat a finite sample at hand as similarly as possible to the unknown distribution from which it was drawn. In practice, this perspective leads to resampling *with replacement*, since an observation of a particular value from an underlying distribution does not preclude subsequent observation of an equal data value. In general the bootstrap is less accurate than the permutation approach when permutation is appropriate, but can be used in instances where permutation cannot. Fuller exposition of the bootstrap than is possible here can be found in Efron and Gong (1983), Efron and Tibshirani (1993), and Leger *et al.*

(1992), among others. Some examples of its use in climatology are given in Downton and Katz (1993) and Mason and Mimmack (1992).

Resampling with replacement is the primary distinction in terms of the mechanics between the bootstrap and the permutation approach, in which the resampling is done without replacement. Conceptually, the resampling process is equivalent to writing each of the n data values on separate slips of paper and putting all n slips of paper in a hat. To construct one bootstrap sample, n slips of paper are drawn from the hat and their data values recorded, but each slip is put back in the hat and mixed (this is the meaning of "with replacement") before the next slip is drawn. Generally some of the original data values will be drawn into a given bootstrap sample multiple times, and some will not be drawn at all. If n is small enough, all possible distinct bootstrap samples can be fully enumerated. In practice, we usually program a computer to perform the resampling, using Equation 4.86 in conjunction with a uniform random number generator (Section 4.7.1). This process is repeated a large number, perhaps $n_B = 10,000$ times, yielding n_B bootstrap samples, each of size n. The statistic of interest is computed for each of these n_B bootstrap samples. The resulting frequency distribution is then used to approximate the true sampling distribution of that statistic.

EXAMPLE 5.10   1-Sample Bootstrap: Confidence Interval for a Complicated Statistic

The bootstrap is often used in one-sample settings to estimate confidence intervals around observed values of a test statistic. Because we do not need to know the analytical form of its sampling distribution, the procedure can be applied to any test statistic, regardless of how mathematically complicated it may be. To take a hypothetical example, consider the standard deviation of the logarithms, $s_{\ln x}$, of the 1933–1982 Ithaca January precipitation data in Table A.2 of Appendix A. This statistic has been chosen for this example arbitrarily, to illustrate that any computable sample statistic can be bootstrapped. Here scalar data are used, but Efron and Gong (1983) illustrate the bootstrap using vector- valued (paired) data, for which a confidence interval of the Pearson correlation coefficient is estimated.

The value of $s_{\ln x}$ computed from the $n = 50$ data values is 0.537, but in order to make inferences about the true value, we need to know or estimate its sampling distribution. Figure 5.8 shows a histogram of the sample standard deviations computed from $n_B = 10000$ bootstrap samples of size $n = 50$ from the logarithms of this data set. The necessary calculations required less than one second of computer time. This empirical distribution approximates the sampling distribution of the $s_{\ln x}$ for these data.

Confidence regions for $s_{\ln x}$ most easily are approached using the straightforward and intuitive percentile method (Efron and Tibshirani 1993; Efron and Gong 1983). To form a $(1 - \alpha)\%$ confidence interval, we simply find the values of the parameter estimates defining largest and smallest $n_B \cdot \alpha/2$ of the n_B bootstrap estimates. These values also define the central $n_B \cdot (1 - \alpha)$ of the estimates, which is the region of interest. In Figure 5.8, for example, the estimated 95% confidence interval for $s_{\ln x}$ is between 0.41 and 0.65. Better and more sophisticated methods of bootstrap confidence interval construction, called bias-corrected or BC_a intervals, are described in Efron (1987) and Efron and Tibshirani (1993). Zwiers (1990) and Downton and Katz (1993) also sketch the mechanics of BC_a bootstrap confidence intervals. ◊

The previous example illustrates use of the bootstrap in a one-sample setting where permutations are not possible. Bootstrapping is also applicable in multiple-sample situations where the data labels are not exchangable, so that pooling and permutation of data is not consistent with the null hypothesis. Such data can still be resampled with replacement

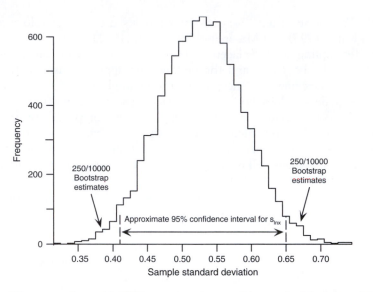

FIGURE 5.8   Histogram of $n_B = 10,000$ bootstrap estimates of the standard deviation of the logarithms of the 1933–1982 Ithaca January precipitation data. The sample standard deviation computed directly from the data is 0.537. The 95% confidence interval for the statistic, as estimated using the percentile method, is also shown.

using the bootstrap, while maintaining the separation of samples having meaningfully different labels. To illustrate, consider investigating differences of means using the test statistic in Equation 5.5. Depending on the nature of the underlying data and the available sample sizes, we might not trust the Gaussian approximation to the sampling distribution of this statistic, in which case a natural alternative would be to approximate it through resampling. If the data labels were exchangeable, it would be natural to compute a pooled estimate of the variance and use Equation 5.9 as the test statistic, estimating its sampling distribution through a permutation procedure because both the means and variances are equal under the null hypothesis. On the other hand, if the null hypothesis did not include equality of the variances, Equation 5.8 would be the correct test statistic, but it would not be appropriate to estimate its sampling distribution through permutation, because in this case the data labels would be meaningful, even under H_0. However, the two samples could be separately resampled with replacement to build up a bootstrap approximation to the sampling distribution of Equation 5.8. We would need to be careful in generating the bootstrap distribution for Equation 5.8 to construct the bootstrapped quantities consistent with the null hypothesis of equality of means. In particular, we could not bootstrap the raw data directly, because they have different means (whereas the two population means are equal according to the null hypothesis). One option would be to recenter each of the data batches to the overall mean (which would equal the estimate of the common, pooled mean, according to the plug-in principle). A more straightforward approach would be to estimate the sampling distribution of the test statistic directly, and then exploit the duality between hypothesis tests and confidence intervals to address the null hypothesis. This second approach is illustrated in the following example.

EXAMPLE 5.11   Two-Sample Bootstrap Test for a Complicated Statistic

Consider again the situation in Example 5.9, in which we are interested in the ratio of L-scales (Equation 5.27) for the numbers of lightning strikes in seeded vs. unseeded storms in Table 5.5. The permutation test in Example 5.9 was based on the assumption that, under the null hypothesis, *all* aspects of the distribution of lightning strikes were the same for the seeded and unseeded storms. But pooling and permutation would not be appropriate if we wish to allow for the possibility that, even if the L-spread does not depend on seeding, other aspects of the distributions (for example, the median numbers of lightning strikes) may be different.

That less restrictive null hypothesis can be accommodated by separately and repeatedly bootstrapping the $n_1 = 12$ seeded and $n_2 = 11$ unseeded lightning counts, and forming $n_B = 10,000$ samples of the ratio of one bootstrap realization of each, yielding bootstrap realizations of the test statistic $\lambda_2(\text{seeded})/\lambda_2(\text{unseeded})$. The result, shown in Figure 5.9 is a bootstrap estimate of the sampling distribution of this ratio for the data at hand. Its center is near the observed ratio of 0.188, which is the $q_{.4835}$ quantile of this bootstrap distribution. Even though this is not the bootstrap null distribution—which would be the sampling distribution if $\lambda_2(\text{seeded})/\lambda_2(\text{unseeded}) = 1$—it can be used to evaluate the null hypothesis by examining the unusualness of $\lambda_2(\text{seeded})/\lambda_2(\text{unseeded}) = 1$ with respect to this sampling distribution. The horizontal grey arrow indicates the estimated 95% confidence interval for the L-scale ratio, which ranges from 0.08 to 0.75. Since this interval does not include 1, H_0 would be rejected at the 5% level (two-sided). The bootstrap L-scale ratios are greater than 1 for only 33 of the $n_B = 10000$ resamples, so the actual p value would be estimated as either 0.0033 (one-sided) or 0.0066 (two-sided), and so H_0 could be rejected at the 1% level as well. ◊

It is important to note that direct use of either bootstrap or permutation methods only makes sense when the underlying data to be resampled are independent. If the data are mutually correlated (exhibiting, for example, time correlation or persistence) the results of these approaches will be misleading (Zwiers 1987, 1990), in the same way and for the same reason that autocorrelation affects parametric tests. The random sampling used in either permutation or the bootstrap shuffles the original data, destroying the ordering that produces the autocorrelation.

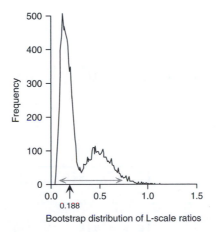

Bootstrap distribution of L-scale ratios

FIGURE 5.9   Bootstrap distribution for the ratio of L-scales for lightning strikes in seeded and unseeded storms, Table 5.5. The ratio is greater than 1 for only 33 of 10,000 bootstrap samples, indicating that a null hypothesis of equal L-scales would be rejected. Also shown (grey arrows) is the 95% confidence interval for the ratio, which ranges from 0.08−0.75.

Solow (1985) has suggested a way around this problem, which involves transformation of the data to an uncorrelated series using time-series methods, for example by fitting an autoregressive process (see Section 8.3). The bootstrapping or permutation test is then carried out on the transformed series, and synthetic samples exhibiting correlation properties similar to the original data can be obtained by applying the inverse transformation. Another approach, called nearest-neighbor bootstrapping (Lall and Sharma 1996), accommodates serial correlation by resampling according to probabilities that depend on similarity to the previous few data points, rather the unvarying $1/n$ implied by the independence assumption. Essentially, the nearest-neighbor bootstrap resamples from relatively close analogs rather than from the full data set. The closeness of the analogs can be defined for both scalar and vector (multivariate) data.

The bootstrap can be used for dependent data more directly through a modification known as the moving-block bootstrap (Efron and Tibshirani 1993; Lahiri 2003; Leger *et al.* 1992; Wilks 1997b). Instead of resampling individual data values or data vectors, contiguous sequences of length L are resampled in order to build up a synthetic sample of size n. Figure 5.10 illustrates resampling a data series of length $n = 12$ by choosing $b = 3$ contiguous blocks of length $L = 4$, with replacement. The resampling works in the same way as the ordinary bootstrap, except that instead of resampling from a collection of n individual, independent values, the objects to be resampled with replacement are all the $n - L + 1$ contiguous subseries of length L.

The idea behind the moving-blocks bootstrap is to choose the blocklength L to be large enough for data values separated by this time period or more to be essentially independent (so the blocklength should increase as the strength of the autocorrelation increases), while retaining the time correlation in the original series at lags L and shorter. The blocklength should also increase as n increases. The choice of the blocklength is important, with null hypotheses that are true being rejected too often if L is too small and too rarely if L is too large. If it can be assumed that the data follow a first-order autoregressive process (Equation 8.16), good results are achieved by choosing the blocklength according to the implicit equation (Wilks 1997b)

$$L = (n - L + 1)^{(2/3)(1 - n'/n)}, \tag{5.28}$$

where n' is defined by Equation 5.12.

5.4 Field Significance and Multiplicity

Special problems occur when statistical tests involving atmospheric (or other geophysical) fields must be performed. In this context the term atmospheric field usually connotes

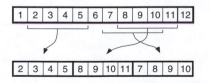

FIGURE 5.10   Schematic illustration of the moving-block bootstrap. Beginning with a time series of length $n = 12$ (above), $b = 3$ blocks of length $L = 4$ are drawn with replacement. The resulting time series (below) is one of $(n - L + 1)^b = 729$ equally likely bootstrap samples. From Wilks (1997b).

a two-dimensional (horizontal) array of geographical locations at which data are available. It may be, for example, that two atmospheric models (one, perhaps, reflecting an increase of the atmospheric carbon dioxide concentration) both produce realizations of surface temperature at each of many gridpoints, and the question is whether the average temperatures portrayed by the two models are significantly different.

In principle, multivariate methods of the kind described in Section 10.5 would be preferred for this kind of problem, but often in practice the data are insufficient to implement them effectively if at all. Accordingly, hypothesis tests for this kind of data are often approached by first conducting individual tests at each of the gridpoints using a procedure such as that in Equation 5.8. If appropriate, a correction for serial correlation of the underlying data such as that in Equation 5.13 would be part of each of these local tests. Having conducted the local tests, however, it still remains to evaluate, collectively, the overall significance of the differences between the fields, or field significance. This evaluation of overall significance is sometimes called determination of global or pattern significance. There are two major difficulties associated with this step. These derive from the problems of test multiplicity, and from spatial correlation of the underlying data.

5.4.1 The Multiplicity Problem for Independent Tests

Consider first the relatively simple problem of evaluating the collective significance of N independent local tests. In the context of atmospheric field significance testing, this situation would correspond to evaluating results from a collection of gridpoint tests when there is no spatial correlation among the underlying data at different grid points. Actually, the basic situation is not unique to geophysical settings, and arises any time the results of multiple, independent tests must be jointly evaluated. It is usually called the problem of multiplicity.

To fix ideas, imagine that there is a spatial array of $N = 20$ gridpoints at which local tests for the central tendencies of two data sources have been conducted. These tests might be t tests, Wilcoxon-Mann-Whitney tests, or any other test appropriate to the situation at hand. From these 20 tests, it is found that $x = 3$ of them declare significant differences at the 5% level. It is sometimes naively supposed that, since 5% of 20 is 1, finding that any one of the 20 tests indicated a significant difference at the 5% level would be grounds for declaring the two fields to be significantly different, and that by extension, three significant tests out of 20 would be very strong evidence. Although this reasoning sounds superficially plausible, it is really only even approximately true if there are very many, perhaps 1000, independent tests (Livezey and Chen 1983; von Storch 1982).

Recall that declaring a significant difference at the 5% level means that, if the null hypothesis is true and there are really no significant differences, there is a probability no greater than 0.05 that evidence against H_0 as strong as or stronger than observed would have appeared by chance. For a single test, the situation is analogous to rolling a 20-sided die, and observing that the side with the 1 on it has come up. However, conducting $N = 20$ tests is like rolling this die 20 times: there is a substantially higher chance than 5% that the side with 1 on it comes up at least once in 20 throws, and it is this latter situation that is analogous to the evaluation of the results from $N = 20$ independent hypothesis tests.

Thinking about this analogy between multiple tests and multiple rolls of the 20-sided die suggests that we can quantitatively analyze the multiplicity problem for independent tests in the context of the binomial distribution. In effect, we must conduct a hypothesis test on the results of the N individual independent hypothesis tests. The global null hypothesis is that all N of the local null hypotheses are correct. Recall that the binomial distribution specifies probabilities for X successes out of N independent trials if the probability of success on any one trial is p. In the testing multiplicity context, X is the number of significant individual tests out of N tests conducted, and p is the level of the test.

EXAMPLE 5.12   A Simple Illustration of the Multiplicity Problem

In the hypothetical example just discussed, $N = 20$ tests, $p = 0.05$ is the level of each of these tests, and $x = 3$ of the 20 tests yielded significant differences. The question of whether the differences are (collectively) significant at the $N = 20$ gridpoints thus reduces to evaluating $\Pr\{X \geq 3\}$, given that the null distribution for the number of significant tests is binomial with $N = 20$ and $p = 0.05$. Using the binomial probability distribution function (Equation 4.1) with these two parameters, we find $\Pr\{X = 0\} = 0.358, \Pr\{X = 1\} = 0.377$, and $\Pr\{X = 2\} = 0.189$. Thus, $\Pr\{X \geq 3\} = 1 - \Pr\{X < 3\} = 0.076$, and the null hypothesis that the two mean fields, as represented by the $N = 20$ gridpoints, are equal would not be rejected at the 5% level. Since $\Pr\{X = 3\} = 0.060$, finding four or more significant local tests would result in a declaration of field significance, at the 5% level.

Even if there are no real differences, the chances of finding at least one significant test result out of 20 are almost 2 out of 3, since $\Pr\{X = 0\} = 0.358$. Until we are aware of and accustomed to the issue of multiplicity, results such as these seem counterintuitive. Livezey and Chen (1983) have pointed out some instances in the literature of the atmospheric sciences where a lack of awareness of the multiplicity problem has lead to conclusions that were not really supported by the data. ◊

5.4.2 Field Significance Given Spatial Correlation

Dealing with the multiplicity problem in the context of the binomial distribution, as described earlier, is a straightforward and satisfactory solution when evaluating a collection of independent tests. When the tests are performed using data from spatial fields, however, the positive interlocation correlation of the underlying data produces statistical dependence among the local tests.

Informally, we can imagine that positive correlation between data at two locations would result in the probability of a Type I error (falsely rejecting H_0) at one location being larger if a Type I error had occurred at the other location. This is because a test statistic is a statistic like any other—a function of the data—and, to the extent that the underlying data are correlated, the statistics calculated from them will be also. Thus, false rejections of the null hypothesis tend to cluster in space, leading (if we are not careful) to the erroneous impression that a spatially coherent and physically meaningful difference between the fields exists. Loosely, we can imagine that there are some number $N' < N$ of effectively independent gridpoint tests. Therefore, as pointed out by Livezey and Chen (1983), the binomial test of local test results to correct for the multiplicity problem only provides a lower limit on the p value pertaining to field significance. It is useful to perform that simple test, however, because a set of local tests providing insufficient evidence to reject H_0 under the assumption independence will certainly not give a stronger result when the spatial dependence has been accounted for. In such cases there is no point in carrying out the more elaborate calculations required to account for the spatial dependence.

To date, most approaches to incorporating the effects of spatial dependence into tests of field significance have involved resampling procedures. As described previously, the idea is to generate an approximation to the sampling distribution of the test statistic (in this case the number of significant local, or gridpoint tests—called the counting norm statistic by Zwiers (1987)) by repeated resampling of the data in a way that mimics the actual data generation process if the null hypothesis is true. Usually the null hypothesis specifies that there are no real differences with respect to some attribute of interest as reflected by the field of test statistics. Different testing situations present different challenges, and considerable creativity sometimes is required for the analyst to devise a consistent resampling procedure. Further discussion on this topic can be found in Livezey and Chen (1983), Livezey (1985a, 1995), Preisendorfer and Barnett (1983), Wigley and Santer (1990), and Zwiers (1987).

EXAMPLE 5.13   The Multiplicity Problem with Spatial Correlation

An instructive example of the use of a permutation test to assess the joint results of a field of hypothesis tests is presented by Livezey and Chen (1983), using data from Chen (1982b). The basic problem is illustrated in Figure 5.11a, which shows the field of correlations between northern hemisphere winter (December–February) 700 mb heights, and values of the Southern Oscillation Index (SOI) (see Figure 3.14) for the previous summer (June–August). The areas of large positive and negative correlation suggest that the SOI might be a useful as one element of a long-range (six months ahead) forecast procedure for winter weather. First, however, a formal test that the field of correlations in Figure 5.11a is different from zero is in order.

The testing process begins with individual tests for the significance of the correlation coefficients at each gridpoint. If the underlying data (here the SOI values and 700 mb heights) approximately follow Gaussian distributions with negligible year-to-year correlation, an easy approach to this suite of tests is to use the Fisher Z transformation,

$$Z = \frac{1}{2} \ln \left[\frac{1+r}{1-r} \right], \tag{5.29}$$

where r is the Pearson correlation (Equation 3.22). Under the null hypothesis that the correlation r is zero, the distribution of Z approximates the Gaussian distribution with $\mu = 0$ and $\sigma = (n-3)^{-1/2}$. (If a different null hypothesis were appropriate, the mean of the corresponding Gaussian distribution would be the Z transform of the correlation under that null hypothesis.) Since Chen (1982b) used $n = 29$ years, values of Z larger in absolute value than $1.96/\sqrt{26} = 0.384$, corresponding to correlations larger than 0.366 in absolute value, would be declared (locally) significant at the 5% level. A sufficiently large area in Figure 5.11a exhibits correlations larger than this magnitude that the preliminary (binomial) test, accounting for the multiplicity problem, rejects the null hypothesis of zero correlation under the assumption that the local tests are mutually independent.

The need for a more sophisticated test, accounting also for the effects of spatial correlation of the 700 mb heights, is underscored by the correlation field in Figure 5.10b. This shows the correlations between the same 29-year record of northern hemisphere 700 mb heights with 29 independent Gaussian random numbers; that is, a random series similar to Figure 5.4a. Clearly the real correlations of 700 mb heights with this series of random numbers is zero, but the substantial spatial correlations among the 700 mb heights yields spatially coherent areas of chance sample correlations that are deceptively high.

(a)

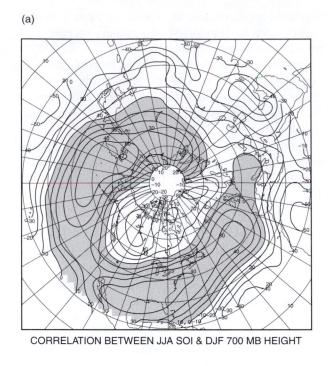

CORRELATION BETWEEN JJA SOI & DJF 700 MB HEIGHT

(b)

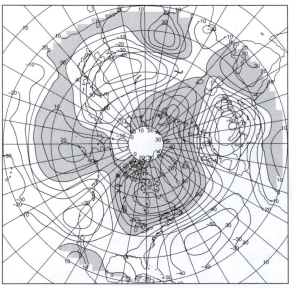

CORRELATION BETWEEN NOISE AND DJF 700 MB HEIGHT

FIGURE 5.11   Correlations of northern hemisphere winter (December–February) 700 mb heights with (a) the Southern Oscillation Index for the previous summer (June–August), and (b) a realization of independent Gaussian random numbers. Shaded areas show positive correlation, and the contour interval is 0.1. The strong spatial correlation of the 700 mb height field produces the spatially coherent correlation with the random number series as an artifact, complicating interpretation of gridpoint hypothesis tests. From Chen (1982b).

The approach to this particular problem taken by Livezey and Chen (1983) was to repeatedly generate sequences of 29 independent Gaussian random variables, as a null hypothesis stand-in for the observed series of SOI values, and tabulate frequencies of local tests erroneously rejecting H_0 for each sequence. Because the gridpoints are distributed on a regular latitude-longitude grid, points nearer the pole represent smaller areas, and accordingly the test statistic is the fraction of the hemispheric area with locally significant tests. This is an appropriate design, since under H_0 there is no real correlation between the 700 mb heights and the SOI, and it is essential for the spatial correlation of the 700 mb heights to be preserved in order to simulate the true data generating process. Maintaining each winter's 700 mb map as a discrete unit ensures automatically that the observed spatial correlations in these data are maintained, and reflected in the null distribution. A possibly better approach could have been to repeatedly use the 29 observed SOI values, but in random orders, or to block-bootstrap them, in place of the Gaussian random numbers. Alternatively, sequences of correlated values generated from a time-series model (see Section 8.3) mimicking the SOI could have been used.

Livezey and Chen chose to resample the distribution of rejected tests 200 times, with the resulting 200 hemispheric fractions with significant local tests constituting the relevant null distribution. Given modern computing capabilities they would undoubtedly have generated many more synthetic samples. A histogram of this distribution is shown in Figure 5.12, with the largest 5% of the values shaded. The fraction of hemispheric area from Figure 5.11a that was found to be locally significant is indicated by the arrow labeled LAG 2 (two seasons ahead of winter). Clearly the results for correlation with the summer SOI are located well out on the tail, and therefore are judged to be globally significant. Results for correlation with fall (September–November) are indicated by the arrow labeled LAG 1, and are evidently significantly different from zero at the 5% level, but less strongly so than for the summer SOI observations. The corresponding results for contemporaneous (winter SOI and winter 700 mb heights) correlations, by contrast, do not fall in the most extreme 5% of the null distribution values, and H_0 would not be rejected for these data. ◊

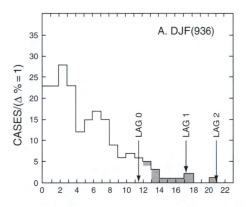

FIGURE 5.12 Null distribution for the percentage of hemispheric area exhibiting significant local tests for correlation between northern hemisphere winter 700 mb heights and series of independent Gaussian random numbers. Percentages of the hemispheric area corresponding to significant local tests for correlations between the heights and concurrent SOI values, the SOI for the previous fall, and the SOI for the previous summer are indicated as LAG 0, LAG 1, and LAG 2, respectively. The result for correlations with summer SOI are inconsistent with the null hypothesis of no relationship between the height field and the SOI, which would be rejected. Results for lag-0 (i.e., contemporaneous) correlation are not strong enough to reject this H_0. From Livezey and Chen (1983).

5.5 Exercises

5.1. For the June temperature data in Table A.3,

   a. Use a two-sample t test to investigate whether the average June temperatures in El Niño and non-El Niño years are significantly different. Assume the variances are unequal and that the Gaussian distribution is an adequate approximation to the distribution of the test statistic.

   b. Construct a 95% confidence interval for the difference in average June temperature between El Niño and non-El Niño years.

5.2. Calculate n', the equivalent number of independent samples, for the two sets of minimum air temperatures in Table A.1.

5.3. Use the data set in Table A.1 of the text to test the null hypothesis that the average minimum temperatures for Ithaca and Canandaigua in January 1987 are equal. Compute p values, assuming the Gaussian distribution is an adequate approximation to the null distribution of the test statistic, and

   a. $H_A =$ the minimum temperatures are different for the two locations.

   b. $H_A =$ the Canandaigua minimum temperatures are warmer.

5.4. Given that the correlations in Figure 5.11a were computed using 29 years of data, use the Fisher Z transformation to compute the magnitude of the correlation coefficient that was necessary for the null hypothesis to be rejected at a single gridpoint at the 5% level, versus the alternative that $r \neq 0$.

5.5. Test the fit of the Gaussian distribution to the July precipitation data in Table 4.9, using

   a. A K-S (i.e., Lilliefors) test.

   b. A Chi-square test.

   c. A Filliben Q-Q correlation test.

5.6. Test whether the 1951–1980 July precipitation data in Table 4.9 might have been drawn from the same distribution as the 1951–1980 January precipitation comprising part of Table A.2, using the likelihood ratio test, assuming gamma distributions.

5.7. Use the Wilcoxon-Mann-Whitney test to investigate whether the magnitudes of the pressure data in Table A.3 are lower in El Niño years,

   a. Using the exact one-tailed critical values 18, 14, 11, and 8 for tests at the 5%, 2.5%, 1%, and 0.5% levels, respectively, for the smaller of U_1 and U_2.

   b. Using the Gaussian approximation to the sampling distribution of U.

5.8. Discuss how the sampling distribution of the skewness coefficient (Equation 3.9) of June precipitation at Guayaquil could be estimated using the data in Table A.3, by bootstrapping. How could the resulting bootstrap distribution be used to estimate a 95% confidence interval for this statistic? If the appropriate computing resources are available, implement your algorithm.

5.9. Discuss how to construct a resampling test to investigate whether the variance of June precipitation at Guayaquil is different in El Niño versus non-El Niño years, using the data in Table A.3.

   a. Assuming that the precipitation distributions are the same under H_0.

   b. Allowing other aspects of the precipitation distributions to be different under H_0.

   If the appropriate computing resources are available, implement your algorithm.

5.10. A published article contains a statistical analysis of historical summer precipitation data in relation to summer temperatures using individual t tests for 121 locations at the 10% level. The study investigates the null hypothesis of no difference in total precipitation between the 10 warmest summers in the period 1900–1969 and the remaining 60 summers, reports that 15 of the 121 tests exhibit significant results, and claims that the overall pattern is therefore significant. Evaluate this claim.

CHAPTER • 6

Statistical Forecasting

6.1 Background

Much of operational weather and long-range (climate) forecasting has a statistical basis. As a nonlinear dynamical system, the atmosphere is not perfectly predictable in a deterministic sense. Consequently, statistical methods are useful, and indeed necessary, parts of the forecasting enterprise. This chapter provides an introduction to statistical forecasting of scalar (single-number) quantities. Some methods suited to statistical prediction of vector (multiple values simultaneously) quantities, for example spatial patterns, are presented in Sections 12.2.3 and 13.4.

Some statistical forecast methods operate without information from the fluid-dynamical Numerical Weather Prediction (NWP) models that have become the mainstay of weather forecasting for lead times ranging from one day to a week or so in advance. Such pure statistical forecast methods are sometimes referred to as Classical, reflecting their prominence in the years before NWP information was available. These methods are still viable and useful at very short lead times (hours in advance), or very long lead times (weeks or more in advance), for which NWP information is not available with either sufficient promptness or accuracy, respectively.

Another important application of statistical methods to weather forecasting is in conjunction with NWP information. Statistical forecast equations routinely are used to post-process and enhance the results of dynamical forecasts at operational weather forecasting centers throughout the world, and are essential as guidance products to aid weather forecasters. The combined statistical and dynamical approaches are especially important for providing forecasts for quantities and locations (e.g., particular cities rather than gridpoints) not represented by the NWP models.

The types of statistical forecasts mentioned so far are objective, in the sense that a given set of inputs always produces the same particular output. However, another important aspect of statistical weather forecasting is in the subjective formulation of forecasts, particularly when the forecast quantity is a probability or set of probabilities. Here the Bayesian interpretation of probability as a quantified degree of belief is fundamental. Subjective probability assessment forms the basis of many operationally important forecasts, and is a technique that could be used more broadly to enhance the information content of operational forecasts.

6.2 Linear Regression

Much of statistical weather forecasting is based on the statistical procedure known as linear, least-squares regression. In this section, the fundamentals of linear regression are reviewed. Much more complete treatments can be found in Draper and Smith (1998) and Neter *et al.* (1996).

6.2.1 Simple Linear Regression

Regression is most easily understood in the case of simple linear regression, which describes the linear relationship between two variables, say x and y. Conventionally the symbol x is used for the independent, or predictor variable, and the symbol y is used for the dependent variable, or predictand. Very often, more than one predictor variable is required in practical forecast problems, but the ideas for simple linear regression generalize easily to this more complex case of multiple linear regression. Therefore, most of the important ideas about regression can be presented in the context of simple linear regression.

Essentially, simple linear regression seeks to summarize the relationship between two variables, shown graphically in their scatterplot, by a single straight line. The regression procedure chooses that line producing the least error for predictions of y given observations of x. Exactly what constitutes least error can be open to interpretation, but the most usual error criterion is minimization of the sum (or, equivalently, the average) of the squared errors. It is the choice of the squared-error criterion that is the basis of the name least-squares regression, or ordinary least squares (OLS) regression. Other error measures are possible, for example minimizing the average (or, equivalently, the sum) of absolute errors, which is known as least absolute deviation (LAD) regression (Gray *et al.* 1992; Mielke *et al.* 1996). Choosing the squared-error criterion is conventional not because it is necessarily best, but rather because it makes the mathematics analytically tractable. Adopting the squared-error criterion results in the line-fitting procedure being fairly tolerant of small discrepancies between the line and the points. However, the fitted line will adjust substantially to avoid very large discrepancies. It is thus not resistant to outliers. Alternatively, LAD regression is resistant because the errors are not squared, but the lack of analytic results (formulas) for the regression function means that the estimation must be iterative.

Figure 6.1 illustrates the situation. Given a data set of (x, y) pairs, the problem is to find the particular straight line,

$$\hat{y} = a + bx, \tag{6.1}$$

minimizing the squared vertical distances (thin lines) between it and the data points. The circumflex ("hat") accent signifies that the equation specifies a predicted value of y. The inset in Figure 6.1 indicates that the vertical distances between the data points and the line, also called errors or residuals, are defined as

$$e_i = y_i - \hat{y}(x_i). \tag{6.2}$$

There is a separate residual e_i for each data pair (x_i, y_i). Note that the sign convention implied by Equation 6.2 is for points above the line to be regarded as positive errors, and points below the line to be negative errors. This is the usual convention in statistics,

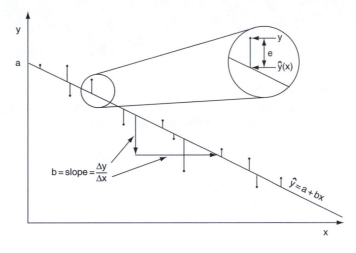

FIGURE 6.1 Schematic illustration of simple linear regression. The regression line, $\hat{y} = a + bx$, is chosen as the one minimizing some measure of the vertical differences (the residuals) between the points and the line. In least-squares regression that measure is the sum of the squared vertical distances. Inset shows the residual, e, as the difference between the data point and the regression line.

but is opposite to what often is seen in the atmospheric sciences, where forecasts smaller than the observations (the line being below the point) are regarded as having negative errors, and vice versa. However, the sign convention for the residuals is unimportant, since it is the minimization of the sum of squared residuals that defines the best-fitting line. Combining Equations 6.1 and 6.2 yields the regression equation,

$$y_i = y_i + e_i = a + bx_i + e_i, \tag{6.3}$$

which says that the true value of the predictand is the sum of the predicted value (Equation 6.1) and the residual.

Finding analytic expressions for the least-squares intercept, a, and the slope, b, is a straightforward exercise in calculus. In order to minimize the sum of squared residuals,

$$\sum_{i=1}^{n}(e_i)^2 = \sum_{i=1}^{n}(y_i - \hat{y}_i)^2 = \sum_{i=1}^{n}(y_i - [a + bx_i])^2, \tag{6.4}$$

it is only necessary to set the derivatives of Equation 6.4 with respect to the parameters a and b to zero and solve. These derivatives are

$$\frac{\partial \sum_{i=1}^{n}(e_i)^2}{\partial a} = \frac{\partial \sum_{i=1}^{n}(y_i - a - bx_i)^2}{\partial a} = -2\sum_{i=1}^{n}(y_i - a - bx_i) = 0 \tag{6.5a}$$

and

$$\frac{\partial \sum_{i=1}^{n}(e_i)^2}{\partial b} = \frac{\partial \sum_{i=1}^{n}(y_i - a - bx_i)^2}{\partial b} = -2\sum_{i=1}^{n}[x_i(y_i - a - bx_i)] = 0. \tag{6.5b}$$

Rearranging Equations 6.5 leads to the so-called normal equations,

$$\sum_{i=1}^{n}y_i = n\,a + b\sum_{i=1}^{n}x_i \tag{6.6a}$$

and

$$\sum_{i=1}^{n} x_i y_i = a \sum_{i=1}^{n} x_i + b \sum_{i=1}^{n} (x_i)^2. \tag{6.6b}$$

Dividing Equation 6.6a by n leads to the observation that the fitted regression line must pass through the point located by the two sample means of x and y. Finally, solving the normal equations for the regression parameters yields

$$b = \frac{\sum_{i=1}^{n}[(x_i - \bar{x})(y_i - \bar{y})]}{\sum_{i=1}^{n}(x_i - \bar{x})^2} = \frac{n\sum_{i=1}^{n} x_i y_i - \sum_{i=1}^{n} x_i \sum_{i=1}^{n} y_i}{n\sum_{i=1}^{n}(x_i)^2 - \left(\sum_{i=1}^{n} x_i\right)^2} \tag{6.7a}$$

and

$$a = \bar{y} - b\bar{x}. \tag{6.7b}$$

Equation 6.7a, for the slope, is similar in form to the Pearson correlation coefficient, and can be obtained with a single pass through the data using the computational form given as the second equality. Note that, as was the case for the correlation coefficient, careless use of the computational form of Equation 6.7a can lead to roundoff errors since the numerator is the generally the difference between two large numbers.

6.2.2 Distribution of the Residuals

Thus far, fitting the straight line has involved no statistical ideas at all. All that has been required was to define least error to mean minimum squared error. The rest has followed from straightforward mathematical manipulation of the data, namely the (x, y) pairs. To bring in statistical ideas, it is conventional to assume that the quantities e_i are independent random variables with zero mean and constant variance. Often, the additional assumption is made that these residuals follow a Gaussian distribution.

Assuming that the residuals have zero mean is not at all problematic. In fact, one convenient property of the least-squares fitting procedure is the guarantee that

$$\sum_{i=1}^{n} e_i = 0, \tag{6.8}$$

from which it is clear that the sample mean of the residuals (dividing this equation by n) is also zero.

Imagining that the residuals can be characterized in terms of a variance is really the point at which statistical ideas begin to come into the regression framework. Implicit in their possessing a variance is the idea that the residuals scatter randomly about some mean value (recall Equations 4.21 or 3.6). Equation 6.8 says that the mean value around which they will scatter is zero, so it is the regression line around which the data points will scatter. We then need to imagine a series of distributions of the residuals *conditional* on the x values, with each observed residual regarded as having been drawn from one of these conditional distributions. The constant variance assumption really means that the variance of the residuals is constant in x, or that all the conditional distributions of the

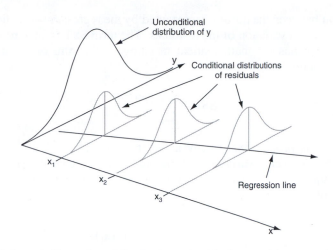

FIGURE 6.2   Schematic illustration of distributions of residuals around the regression line, conditional on values of the predictor variable, x. The actual residuals are regarded as having been drawn from these distributions.

residuals have the same variance. Therefore a given residual (positive or negative, large or small) is by assumption equally likely to occur at any part of the regression line.

Figure 6.2 is a schematic illustration of the idea of a suite of conditional distributions centered on the regression line. The three distributions are identical, except that their means are shifted higher or lower depending on the level of the regression line (predicted value of y) for each x. Extending this thinking slightly, it is not difficult to see that the regression equation can be regarded as specifying the conditional mean of the predictand, given a specific value of the predictor. Also shown in Figure 6.2 is a schematic representation of the unconditional distribution of the predictand, y. The distributions of residuals are less spread out (have smaller variance) than the unconditional distribution of y, indicating that there is less uncertainty about y if a corresponding x value is known.

Central to the making of statistical inferences in the regression setting is the estimation of this (constant) residual variance from the sample of residuals. Since the sample average of the residuals is guaranteed by Equation 6.8 to be zero, the square of Equation 3.6 becomes

$$s_e^2 = \frac{1}{n-2} \sum_{i=1}^{n} e_i^2, \tag{6.9}$$

where the sum of squared residuals is divided by $n-2$ because two parameters (a and b) have been estimated. Substituting Equation 6.2 then yields

$$s_e^2 = \frac{1}{n-2} \sum_{i=1}^{n} [y_i - \hat{y}(x_i)]^2. \tag{6.10}$$

Rather than compute the estimated residual variance using 6.10, however, it is more usual to use a computational form based on the relationship,

$$SST = SSR + SSE, \tag{6.11}$$

which proved in most regression texts. The notation in Equation 6.11 consists of acronyms describing, respectively, the variation in the predictand, y; and a partitioning of that variation between the portion represented by the regression, and the unrepresented portion ascribed to the variation of the residuals. The term SST is an acronym for sum of squares, total, which has the mathematical meaning of the sum of squared deviations of the y values around their mean,

$$SST = \sum_{i=1}^{n}(y_i - \bar{y})^2 = \sum_{i=1}^{n} y_i^2 - n\bar{y}^2. \tag{6.12}$$

This term is proportional to (by a factor of $n - 1$) the sample variance of y, and thus measures the overall variability of the predictand. The term SSR stands for the regression sum of squares, or the sum of squared differences between the regression predictions and the sample mean of y,

$$SSR = \sum_{i=1}^{n}[\hat{y}(x_i) - \bar{y}]^2, \tag{6.13a}$$

which relates to the regression equation according to

$$SSR = b^2 \sum_{i=1}^{n}(x_i - \bar{x})^2 = b^2 \left[\sum_{i=1}^{n} x_i^2 - n\bar{x}^2\right]. \tag{6.13b}$$

Equation 6.13 indicates that a regression line differing little from the sample mean of the y values will have a small slope and produce a very small SSR, whereas one with a large slope will exhibit some large differences from the sample mean of the predictand and therefore produce a large SSR. Finally, SSE refers to the sum of squared differences between the residuals and their mean, which is zero, or sum of squared errors:

$$SSE = \sum_{i=1}^{n} e_i^2. \tag{6.14}$$

Since this differs from Equation 6.9 only by a factor of $n - 2$, rearranging Equation 6.11 yields the computational form

$$s_e^2 = \frac{1}{n-2}\{SST - SSR\} = \frac{1}{n-2}\left\{\sum_{i=1}^{n} y_i^2 - n\bar{y}^2 - b^2\left[\sum_{i=1}^{n} x_i^2 - n\bar{x}^2\right]\right\}. \tag{6.15}$$

6.2.3 The Analysis of Variance Table

In practice, regression analysis is now almost universally done using computer software. A central part of the regression output of such packages is a summary of the foregoing information in an Analysis of Variance, or ANOVA table. Usually, not all the information in an ANOVA table will be of interest, but it is such a universal form of regression output that you should understand its components. Table 6.1 outlines the arrangement of an ANOVA table for simple linear regression, and indicates where the quantities described in the previous section are reported. The three rows correspond to the partition of the variation of the predictand as expressed in Equation 6.11. Accordingly, the Regression and Residual entries in the df (degrees of freedom) and SS (sum of squares) columns will sum to the corresponding entry in the Total row. Therefore, the ANOVA table contains

TABLE 6.1 Generic Analysis of Variance (ANOVA) table for simple linear regression. The column headings df, SS, and MS stand for degrees of freedom, sum of squares, and mean square, respectively. Regression df $= 1$ is particular to simple linear regression (i.e., a single predictor x). Parenthetical references are to equation numbers in the text.

Source	df	SS	MS	
Total	$n-1$	SST (6.12)		
Regression	1	SSR (6.13)	MSR $=$ SSR/1	(F $=$ MSR/MSE)
Residual	$n-2$	SSE (6.14)	MSE $= s_e^2$	

some redundant information, and as a consequence the output from some regression packages will omit the Total row entirely.

The entries in the MS (mean squared) column are given by the corresponding quotients of SS/df. For simple linear regression, the regression df $= 1$, and SSR $=$ MSR. Comparing with Equation 6.15, it can be seen that the MSE (mean squared error) is the sample variance of the residuals. The total mean square, left blank in Table 6.1 and in the output of most regression packages, would be SST $/(n-1)$, or simply the sample variance of the predictand.

6.2.4 Goodness-of-Fit Measures

The ANOVA table also presents (or provides sufficient information to compute) three related measures of the fit of a regression, or the correspondence between the regression line and a scatterplot of the data. The first of these is the MSE. From the standpoint of forecasting, the MSE is perhaps the most fundamental of the three measures, since it indicates the variability of, or the uncertainty about, the observed values (the quantities being forecast) around the forecast regression line. As such, it directly reflects the average accuracy of the resulting forecasts. Referring again to Figure 6.2, since MSE $= s_e^2$ this quantity indicates the degree to which the distributions of residuals cluster tightly (small MSE), or spread widely (large MSE) around a regression line. In the limit of a perfect linear relationship between x and y, the regression line coincides exactly with all the point pairs, the residuals are all zero, SST will equal SSR, SSE will be zero, and the variance of the residual distributions is also zero. In the opposite limit of absolutely no linear relationship between x and y, the regression slope will be zero, the SSR will be zero, SSE will equal SST, and the MSE will very nearly equal the sample variance of the predictand itself. In this unfortunate case, the three conditional distributions in Figure 6.2 would be indistinguishable from the unconditional distribution of y.

The relationship of the MSE to the strength of the regression fit is also illustrated in Figure 6.3. Panel (a) shows the case of a reasonably good regression, with the scatter of points around the regression line being fairly small. Here SSR and SST are nearly the same. Panel (b) shows an essentially useless regression, for values of the predictand spanning the same range as in panel (a). In this case the SSR is nearly zero since the regression has nearly zero slope, and the MSE is essentially the same as the sample variance of the y values themselves.

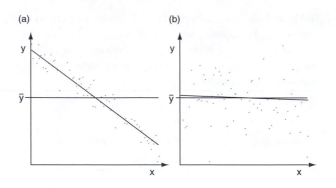

FIGURE 6.3   Illustration of the distinction between a fairly good regression relationship (a) and an essentially useless relationship (b). The points in panel (a) cluster closely around the regression line (solid), indicating small MSE, and the line deviates strongly from the average value of the predictand (dotted), producing a large SSR. In panel (b) the scatter around the regression line is large, and the regression line is almost indistinguishable from the mean of the predictand.

The second usual measure of the fit of a regression is the coefficient of determination, or R^2. This can be computed from

$$R^2 = \frac{SSR}{SST} = 1 - \frac{SSE}{SST},$$
(6.16)

and is often also displayed as part of standard regression output. The R^2 can be interpreted as the proportion of the variation of the predictand (proportional to SST) that is described or accounted for by the regression (SSR). Sometimes we see this concept expressed as the proportion of variation explained, although this claim is misleading: a regression analysis can quantify the nature and strength of a relationship between two variables, but can say nothing about which variable (if either) causes the other. This is the same caveat offered in the discussion of the correlation coefficient in Chapter 3. For the case of simple linear regression, the square root of the coefficient of determination is exactly (the absolute value of) the Pearson correlation between x and y.

For a perfect regression, SSR = SST and SSE = 0, so $R^2 = 1$. For a completely useless regression, SSR = 0 and SSE = SST, so that $R^2 = 0$. Again, Figure 6.3b shows something close to this latter case. Comparing Equation 6.13a, the least-squares regression line is almost indistinguishable from the sample mean of the predictand, so SSR is very small. In other words, little of the variation in y can be ascribed to the regression so the proportion SSR/SST is nearly zero.

The third commonly used measure of the strength of the regression is the F ratio, generally given in the last column of the ANOVA table. The ratio MSR/MSE increases with the strength of the regression, since a strong relationship between x and y will produce a large MSR and a small MSE. Assuming that the residuals are independent and follow the same Gaussian distribution, and under the null hypothesis no real linear relationship, the sampling distribution of the F ratio has a known parametric form. This distribution forms the basis of a test that is applicable in the case of simple linear regression, but in the more general case of multiple regression (more than one x variable) problems of test multiplicity, to be discussed later, invalidate it. However, even if the F ratio cannot be used for quantitative statistical inference, it is still a valid qualitative index of the strength of a regression. See, for example, Draper and Smith (1998) or Neter *et al.* (1996) for discussions of the F test for overall significance of the regression.

6.2.5 Sampling Distributions of the Regression Coefficients

Another important use of the estimated residual variance is to obtain estimates of the sampling distributions of the regression coefficients. As statistics computed from a finite set of data subject to sampling variations, the computed regression intercept and slope, a and b, also exhibit sampling variability. Estimation of their sampling distributions allows construction of confidence intervals for the true population counterparts, around the sample intercept and slope values a and b, and provides a basis for hypothesis tests about the corresponding population values.

Under the assumptions listed previously, the sampling distributions for both intercept and slope are Gaussian. On the strength of the central limit theorem, this result also holds approximately for any regression when n is large enough, because the estimated regression parameters (Equation 6.7) are obtained as the sums of large numbers of random variables. For the intercept the sampling distribution has parameters

$$\mu_a = a \tag{6.17a}$$

and

$$\sigma_a = s_e \left[\frac{\sum_{i=1}^{n} x_i^2}{n \sum_{i=1}^{n} (x_i - \bar{x})^2} \right]^{1/2}. \tag{6.17b}$$

For the slope the parameters of the sampling distribution are

$$\mu_b = b \tag{6.18a}$$

and

$$\sigma_b = \frac{s_e}{\left[\sum_{i=1}^{n} (x_i - \bar{x})^2 \right]^{1/2}}. \tag{6.18b}$$

Equations 6.17a and 6.18a indicate that the least-squares regression parameter estimates are unbiased. Equations 6.17b and 6.18b show that the precision with which the intercept and slope can be estimated from the data depend directly on the estimated standard deviation of the residuals, s_e, which is the square root of the MSE from the ANOVA table (see Table 6.1). Additionally, the estimated slope and intercept are not independent, having correlation

$$r_{a,b} = \frac{-\bar{x}}{\frac{1}{n} \left(\sum_{i=1}^{n} x_i^2 \right)^{1/2}}. \tag{6.19}$$

Taken together with the (at least approximately) Gaussian sampling distributions for a and b, Equations 6.17 through 6.19 define their joint bivariate normal (Equation 4.33) distribution. Equations 6.17b, 6.18b, and 6.19 are valid only for simple linear regression. With more than one predictor variable, analogous (vector) equations (Equation 9.40) must be used.

The output from regression packages will almost always include the standard errors (Equations 6.17b and 6.18b) in addition to the parameter estimates themselves. Some packages also include the ratios of the estimated parameters to their standard errors in a column labeled t ratio. When this is done, a one-sample t test (Equation 5.3) is implied, with the null hypothesis being that the underlying (population) mean for the parameter is zero. Sometimes a p value associated with this test is also automatically included in the regression output.

For the case of the regression slope, this implicit t test bears directly on the meaningfulness of the fitted regression. If the estimated slope is small enough that its true value could plausibly (with respect to its sampling distribution) be zero, then the regression is not informative, or useful for forecasting. If the slope is actually zero, then the value of the predictand specified by the regression equation is always the same, and equal to its sample mean (cf. Equations 6.1 and 6.7b). If the assumptions regarding the regression residuals are satisfied, we would reject this null hypothesis at the 5% level if the estimated slope is, roughly, at least twice as large (in absolute value) as its standard error.

The same hypothesis test for the regression intercept often is offered by computerized statistical packages as well. Depending on the problem at hand, however, this test for the intercept may or may not be meaningful. Again, the t ratio is just the parameter estimate divided by its standard error, so the implicit null hypothesis is that the true intercept is zero. Occasionally, this null hypothesis is physically meaningful, and if so the test statistic for the intercept is worth looking at. On the other hand, it often happens that there is no physical reason to expect that the intercept might be zero. It may even be that a zero intercept is physically impossible. In such cases this portion of the automatically generated computer output is meaningless.

EXAMPLE 6.1   A Simple Linear Regression

To concretely illustrate simple linear regression, consider the January 1987 minimum temperatures at Ithaca and Canandaigua from Table A.1 in Appendix A. Let the predictor variable, x, be the Ithaca minimum temperature, and the predictand y, be the Canandaigua minimum temperature. The scatterplot of this data is shown the middle panel of the bottom row of the scatterplot matrix in Figure 3.26, and as part of Figure 6.10. A fairly strong, positive, and reasonably linear relationship is indicated.

Table 6.2 shows what the output from a typical statistical computer package would look like for this regression. The data set is small enough that the computational formulas can be worked through to verify the results. (A little work with a hand calculator will verify that $\Sigma x = 403$, $\Sigma y = 627$, $\Sigma x^2 = 10803$, $\Sigma y^2 = 15009$, and $\Sigma xy = 11475$.) The upper

TABLE 6.2   Example output typical of that produced by computer statistical packages, for prediction of Canandaigua minimum temperature (y) using Ithaca minimum temperature (x) from the January 1987 data set in Table A.1.

Source	df	SS	MS	F
Total	30	2327.419		
Regression	1	1985.798	1985.798	168.57
Residual	29	341.622	11.780	

Variable	Coefficient	s.e.	t ratio
Constant	12.4595	0.8590	14.504
IthacaMin	0.5974	0.0460	12.987

portion of Table 6.2 corresponds to the template in Table 6.1, with the relevant numbers filled in. Of particular importance is $MSE = 11.780$, yielding as its square root the estimated sample standard deviation for the residuals, $s_e = 3.43°F$. This standard deviation addresses directly the precision of specifying the Canandaigua temperatures on the basis of the concurrent Ithaca temperatures, since we expect about 95% of the actual predictand values to be within $\pm 2s_e = \pm 6.9°F$ of the temperatures given by the regression. The coefficient of determination is easily computed as $R^2 = 1985.798/2327.419 = 85.3\%$. The Pearson correlation is $\sqrt{0.853} = 0.924$, as was given in Table 3.5. The value of the F statistic is very high, considering that the 99th percentile of its distribution under the null hypothesis of no real relationship is about 7.5. We also could compute the sample variance of the predictand, which would be the total mean square cell of the table, as $2327.419/30 = 77.58°F^2$.

The lower portion of Table 6.2 gives the regression parameters, a and b, their standard errors, and the ratios of these parameter estimates to their standard errors. The specific regression equation for this data set, corresponding to Equation 6.1, would be

$$T_{Can.} = \underset{(0.859)}{12.46} + \underset{(0.046)}{0.597} \, T_{Ith.} \tag{6.20}$$

Thus, the Canandaigua temperature would be estimated by multiplying the Ithaca temperature by 0.597 and adding $12.46°F$. The intercept $a = 12.46° F$ has no special physical significance except as the predicted Canandaigua temperature when the Ithaca temperature is $0°F$. Notice that the standard errors of the two coefficients have been written parenthetically below the coefficients themselves. Although this is not a universal practice, it is very informative to someone reading Equation 6.20 without the benefit of the information in Table 6.2. In particular, it allows the reader to get a sense for the significance of the slope (i.e., the parameter b). Since the estimated slope is about 13 times larger than its standard error it is almost certainly not really zero. This conclusion speaks directly to the question of the meaningfulness of the fitted regression. On the other hand, the corresponding implied significance test for the intercept would be much less interesting, unless the possibility of a zero intercept would in itself be meaningful. ◊

6.2.6 Examining Residuals

It is not sufficient to feed data to a computer regression package and uncritically accept the results. Some of the results can be misleading if the assumptions underlying the computations are not satisfied. Since these assumptions pertain to the residuals, it is important to examine the residuals for consistency with the assumptions made about their behavior.

One easy and fundamental check on the residuals can be made by examining a scatterplot of the residuals as a function of the predicted value $\hat{y}$. Many statistical computer packages provide this capability as a standard regression option. Figure 6.4a shows a scatterplot of a hypothetical data set, with the least-squares regression line, and Figure 6.4b shows a plot of the resulting residuals as a function of the predicted values. The residual plot presents the impression of fanning, or exhibition of increasing spread as $\hat{y}$ increases. That is, the variance of the residuals appears to increase as the predicted value increases. This condition of nonconstant residual variance is called *heteroscedasticity*. Since the computer program that fit the regression has assumed constant residual variance, the MSE given in the ANOVA table is an overestimate for smaller values of x and y (where the points cluster closer to the regression line), and an underestimate of the residual variance

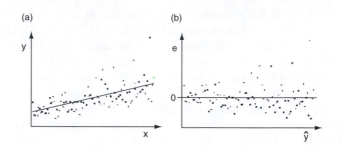

FIGURE 6.4   Hypothetical linear regression (a), and plot of the resulting residuals against the predicted values (b), for a case where the variance of the residuals is not constant. The scatter around the regression line in (a) increases for larger values of x and y, producing a visual impression of fanning in the residual plot (b). A transformation of the predictand is indicated.

for larger values of x and y (where the points tend to be far from the regression line). If the regression is used as a forecasting tool, we would be overconfident about forecasts for larger values of y, and underconfident about forecasts for smaller values of y. In addition, the sampling distributions of the regression parameters will be more variable than implied by Equations 6.17 and 6.18. That is, the parameters will not have been estimated as precisely as the standard regression output would lead us to believe.

Often nonconstancy of residual variance of the sort shown in Figure 6.4b can be remedied by transforming the predictand y, perhaps by using a power transformation (Equations 3.18). Figure 6.5 shows the regression and residual plots for the same data as in Figure 6.4 after logarithmically transforming the predictand. Recall that the logarithmic transformation reduces all the data values, but reduces the larger values more strongly than the smaller ones. Thus, the long right tail of the predictand has been pulled in relative to the shorter left tail, as in Figure 3.12. As a result, the transformed data points appear to cluster more evenly around the new regression line. Instead of fanning, the residual plot in Figure 6.5b gives the visual impression of a horizontal band, indicating appropriately constant variance of the residuals (homoscedasticity). Note that if the fanning in Figure 6.4b had been in the opposite sense, with greater residual variability for smaller values of $\hat{y}$ and lesser residual variability for larger values of $\hat{y}$, a transformation that stretches the right tail relative to the left tail (e.g., y^2) would have been appropriate.

It can also be informative to look at scatterplots of residuals vs. a predictor variable. Figure 6.6 illustrates some of the forms such plots can take, and their diagnostic

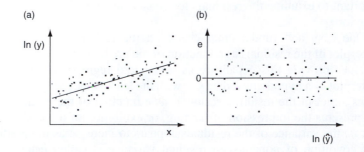

FIGURE 6.5   Scatterplot with regression (a), and resulting residual plot (b), for the same data in Figure 6.4 after logarithmically transforming the predictand. The visual impression of a horizontal band in the residual plot supports the assumption of constant variance of the residuals.

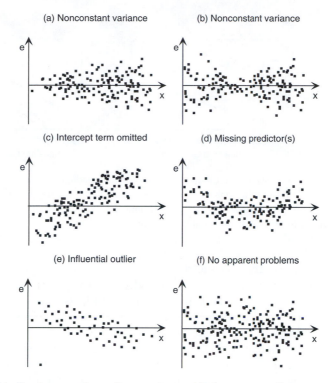

FIGURE 6.6   Idealized scatterplots of regression residuals vs. a predictor x, with corresponding diagnostic interpretations.

interpretations. Figure 6.6a is similar to Figure 6.4b, in that the fanning of the residuals indicates nonconstancy of variance. Figure 6.6b illustrates a different form of heteroscedasticity, that might be more challenging to remedy through a variable transformation. The type of residual plot in Figure 6.6c, with a linear dependence on the predictor of the linear regression, indicates that either the intercept a has been omitted, or that the calculations have been done incorrectly. Figure 6.6d shows a form for the residual plot that can occur when additional predictors would improve a regression relationship. Here the variance is reasonably constant in x, but the (conditional) average residual exhibits a dependence on x. Figure 6.6e illustrates the kind of behavior that can occur when a single outlier in the data has undue influence on the regression. Here the regression line has been pulled toward the outlying point in order to avoid the large squared error associated with it, leaving a trend in the other residuals. If the outlier were determined not to be a valid data point, it should either be corrected if possible or otherwise discarded. If it is a valid data point, a resistant approach such as LAD regression might be more appropriate. Figure 6.6f again illustrates the desirable horizontally banded pattern of residuals, similar to Figure 6.5b.

A graphical impression of whether the residuals follow a Gaussian distribution can be obtained through a Q-Q plot. The capacity to make these plots is also often a standard option in statistical computer packages. Figures 6.7a and 6.7b show Q-Q plots for the residuals in Figures 6.4b and 6.5b, respectively. The residuals are plotted on the vertical, and the standard Gaussian variables corresponding to the empirical cumulative probability of each residual are plotted on the horizontal. The curvature apparent in Figure 6.7a indicates that the residuals from the regression involving the untransformed y are positively

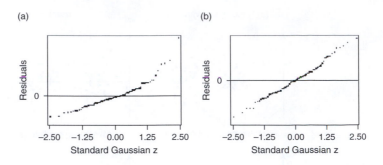

(a)   (b)

FIGURE 6.7   Gaussian quantile-quantile plots of the residuals for predictions of the untransformed y in Figure 6.4a (a), and the logarithmically transformed y in Figure 6.5b (b). In addition to producing essentially constant residual variance, logarithmic transformation of the predictand has rendered the distribution of the residuals essentially Gaussian.

skewed relative to the (symmetric) Gaussian distribution. The Q-Q plot of residuals from the regression involving the logarithmically transformed y is very nearly linear. Evidently the logarithmic transformation has produced residuals that are close to Gaussian, in addition to stabilizing the residual variances. Similar conclusions could have been reached using a goodness of fit test (see Section 5.2.5).

It is also possible and desirable to investigate the degree to which the residuals are uncorrelated. This question is of particular interest when the underlying data are serially correlated, which is a common condition for atmospheric variables. A simple graphical evaluation can be obtained by plotting the regression residuals as a function of time. If groups of positive and negative residuals tend to cluster together (qualitatively resembling Figure 5.4b) rather than occurring more irregularly (as in Figure 5.4a), then time correlation can be suspected.

A popular formal test for serial correlation of regression residuals, included in many computer regression packages, is the Durbin-Watson test. This test examines the null hypothesis that the residuals are serially independent, against the alternative that they are consistent with a first-order autoregressive process (Equation 8.16). The Durbin-Watson test statistic,

$$d = \frac{\sum_{i=2}^{n}(e_i - e_{i-1})^2}{\sum_{i=1}^{n}e_i^2},$$ (6.21)

computes the squared differences between pairs of consecutive residuals, divided by a scaling factor. If the residuals are positively correlated, adjacent residuals will tend to be similar in magnitude, so the Durbin-Watson statistic will be relatively small. If the residuals are randomly distributed in time, the sum in the numerator will tend to be larger. Therefore we reject the null hypothesis that the residuals are independent if the Durbin-Watson statistic is sufficiently small.

Figure 6.8 shows critical values for Durbin-Watson tests at the 5% level. These vary depending on the sample size, and the number of predictor(x) variables, K. For simple linear regression, $K = 1$. For each value of K, Figure 6.8 shows two curves. If the observed value of the test statistic falls below the lower curve, the null hypothesis is rejected and we conclude that the residuals exhibit significant serial correlation. If the test statistic falls above the upper curve, we do not reject the null hypothesis that the

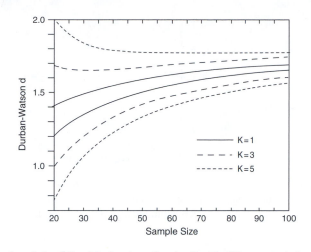

FIGURE 6.8 Graphs of the 5% critical values for the Durbin-Watson statistic as a function of the sample size, for $K = 1$, 3, and 5 predictor variables. A test statistic d below the relevant lower curve results in a rejection of the null hypothesis of zero serial correlation. If the test statistic is above the relevant upper curve the null hypothesis is not rejected. If the test statistic is between the two curves the test is indeterminate.

residuals are serially uncorrelated. If the test statistic falls between the two relevant curves, the test is indeterminate. The reason behind the existence of this unusual indeterminate condition is that the null distribution of the Durban-Watson statistic depends on the data set being considered. In cases where the test result is indeterminate according to Figure 6.8, some additional calculations (Durban and Watson 1971) can be performed to resolve the indeterminacy, that is, to find the specific location of the critical value between the appropriate pair of curves, for the particular data at hand.

EXAMPLE 6.2   Examination of the Residuals from Example 6.1

A regression equation constructed using autocorrelated variables as predictand and predictors does not necessarily exhibit strongly autocorrelated residuals. Consider again the regression between Ithaca and Canandaigua minimum temperatures for January 1987 in Example 6.1. The lag-1 autocorrelations (Equation 3.30) for the Ithaca and Canandaigua minimum temperature data are 0.651 and 0.672, respectively. The residuals for this regression are plotted as a function of time in Figure 6.9. A strong serial correlation for these residuals is not apparent, and their lag-1 autocorrelation as computed using Equation 3.30 is only 0.191.

Having computed the residuals for the Canandaigua vs. Ithaca minimum temperature regression, it is straightforward to compute the Durbin-Watson d (Equation 6.21). In fact, the denominator is simply the SSE from the ANOVA Table 6.2, which is 341.622. The numerator in Equation 6.21 must be computed from the residuals, and is 531.36. These yield $d = 1.55$. Referring to Figure 6.8, the point at $n = 31$, $d = 1.55$ is well above the upper solid (for $K = 1$, since there is a single predictor variable) line, so the null hypothesis of uncorrelated residuals would not be rejected at the 5% level. ◊

When regression residuals are autocorrelated, statistical inferences based upon their variance are degraded in the same way, and for the same reasons, that were discussed in Section 5.2.4 (Bloomfield and Nychka 1992; Matalas and Sankarasubramanian 2003;

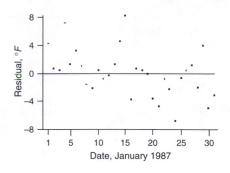

FIGURE 6.9   Residuals from the regression, Equation 6.20, plotted as a function of date. A strong serial correlation is not apparent, but the tendency for a negative slope suggests that the relationship between Ithaca and Canandaigua temperatures may be changing through the month.

Santer et al. 2000; Zheng et al. 1997). In particular, positive serial correlation of the residuals leads to inflation of the variance of the sampling distribution of their sum or average, because these quantities are less consistent from batch to batch of size n. When a first-order autoregression (Equation 8.16) is a reasonable representation for these correlations (characterized by r_1) it is appropriate to apply the same variance inflation factor, $(1 + r_1)/(1 - r_1)$ (bracketed quantity in Equation 5.13), to the variance s_e^2 in, for example, Equations 6.17b and 6.18b (Matalas and Sankarasubramanian 2003; Santer et al. 2000). The net effect is that the variance of the resulting sampling distribution is (appropriately) increased, relative to what would be calculated assuming independent regression residuals.

6.2.7 Prediction Intervals

Many times it is of interest to calculate confidence intervals around forecast values of the predictand (i.e., around the regression function). When it can be assumed that the residuals follow a Gaussian distribution, it is natural to approach this problem using the unbiasedness property of the residuals (Equation 6.8), together with their estimated variance $MSE = s_e^2$. Using Gaussian probabilities (see Table B.1), we expect a 95% confidence interval for a future residual, or specific future forecast, to be approximately bounded by $\hat{y} \pm 2s_e$.

The $\pm 2s_e$ rule of thumb is often a quite good approximation to the width of a true 95% confidence interval, especially when the sample size is large. However, because both the sample mean of the predictand and the slope of the regression are subject to sampling variations, the prediction variance for future data, not used in the fitting of the regression, is somewhat larger than the MSE. For a forecast of y using the predictor value x_0, this prediction variance is given by

$$s_{\hat{y}}^2 = s_e^2 \left[1 + \frac{1}{n} + \frac{(x_0 - \bar{x})^2}{\sum_{i=1}^{n}(x_i - \bar{x})^2} \right]. \tag{6.22}$$

That is, the prediction variance is proportional to the MSE, but is larger to the extent that the second and third terms inside the square brackets are appreciably larger than zero. The second term derives from the uncertainty in estimating the true mean of the predictand from a finite sample of size n (compare Equation 5.4), and becomes much smaller than one for large sample sizes. The third term derives from the uncertainty in estimation of the slope (it is similar in form to Equation 6.18b), and indicates that predictions far removed from the center of the data used to fit the regression will be more uncertain than predictions made near the sample mean. However, even if the numerator in this third term is fairly large, the term itself will tend to be much smaller than one if a large data sample was used to construct the regression equation, since there are n nonnegative terms in the denominator.

It is sometimes also of interest to compute confidence intervals for the regression function itself. These will be narrower than the confidence intervals for predictions, reflecting a smaller variance in the same way that the variance of a sample mean is smaller than the variance of the underlying data values. The variance for the sampling distribution of the regression function, or equivalently the variance of the conditional mean of the predictand given a particular predictor value x_0, is

$$s_{\bar{y}|x_0}^2 = s_e^2 \left[\frac{1}{n} + \frac{(x_0 - \bar{x})^2}{\sum_{i=1}^{n}(x_i - \bar{x})^2} \right]. \tag{6.23}$$

This expression is similar to Equation 6.22, but is smaller by the amount s_e^2. That is, there are contributions to this variance due to uncertainty in the mean of the predictand (or, equivalently the vertical position of the regression line, or the intercept), attributable to the first of the two terms in the square brackets; and to uncertainty in the slope, attributable to the second term. There is no contribution to Equation 6.23 reflecting scatter of data around the regression line, which is the difference between Equations 6.22 and 6.23. The extension of Equation 6.23 for multiple regression is given in Equation 9.41.

Figure 6.10 compares confidence intervals computed using Equations 6.22 and 6.23, in the context of the regression in Example 6.1. Here the regression (Equation 6.20) fit to the 31 data points (dots) is shown by the heavy solid line. The 95% prediction interval around the regression computed as $\pm 1.96 s_{\hat{y}}$, using the square root of Equation 6.22, is indicated by the pair of slightly curved solid black lines. As noted earlier, these bounds are only slightly wider than those given by the simpler approximation $\hat{y} \pm 1.96 s_e$ (dashed lines), because the second and third terms in the square brackets of Equation 6.22 are relatively small, even for moderate n. The pair of grey curved lines locate the 95% confidence interval for the conditional mean of the predictand. These are much narrower than the prediction interval because they account only for sampling variations in the regression parameters, without direct contributions from the prediction variance s_e^2.

Equations 6.17 through 6.19 define the parameters of a bivariate normal distribution for the two regression parameters. Imagine using the methods outlined in Section 4.7 to generate pairs of intercepts and slopes according to that distribution, and therefore to generate realizations of plausible regression lines. One interpretation of the gray curves in Figure 6.10 is that they would contain 95% of those regression lines (or, equivalently, 95% of the regression lines computed from different samples of data of this kind with size $n = 31$). The minimum separation between the gray curves (at the average Ithaca $T_{min} = 13°F$) reflects the uncertainty in the intercept. Their spreading at more extreme temperatures reflects the fact that uncertainty in the slope (i.e., uncertainty in the angle

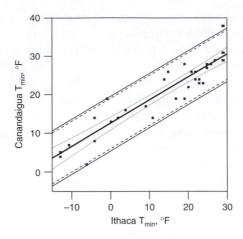

FIGURE 6.10   Confidence intervals around the regression derived in Example 6.1 (thick black line). Light solid lines indicate 95% confidence intervals for future predictions, computed using Equation 6.22, and the corresponding dashed lines simply locate the predictions $\pm 1.96 s_e$. Light gray lines locate 95% confidence intervals for the regression function (Equation 6.22). Data to which the regression was fit are also shown.

of the regression line) will produce more uncertainty in the conditional expected value of the predictand at the extremes than near the mean, because any regression line will pass through the point located by the two sample means.

The result of Example 6.2 is that the residuals for this regression can reasonably be regarded as independent. Also, some of the sample lag-1 autocorrelation of $r_1 = 0.191$ can be attributable to the time trend evident in Figure 6.9. However, if the residuals were significantly correlated, and that correlation was plausibly represented by a first-order autoregression (Equation 8.16), it would be appropriate to increase the residual variances s_e^2 in Equation 6.22 and 6.23 by multiplying them by the variance inflation factor $(1 + r_1)/(1 - r_1)$.

Special care is required when computing confidence intervals for regressions involving transformed predictands. For example, if the relationship shown in Figure 6.5a (involving a log-transformed predictand) were to be used in forecasting, dimensional values of the predictand would need to be recovered in order to make the forecasts interpretable. That is, the predictand $\ln(\hat{y})$ would need to be back-transformed, yielding the forecast $\hat{y} = \exp[\ln(\hat{y})] = \exp[a + bx]$. Similarly, the limits of the prediction intervals would also need to be back-transformed. For example the 95% prediction interval would be approximately $\ln(\hat{y}) \pm 1.96\, s_e$, because the regression residuals and their assumed Gaussian distribution pertain to the transformed predictand values. The lower and upper limits of this interval, when expressed on the original untransformed scale of the predictand, would be $\exp[a + bx - 1.96\, s_e]$ and $\exp[a + bx + 1.96\, s_e]$. These limits would not be symmetrical around $\hat{y}$, and would extend further for the larger values, consistent with the longer right tail of the predictand distribution.

Equations 6.22 and 6.23 are valid for simple linear regression. The corresponding equations for multiple regression are similar, but are more conveniently expressed in matrix algebra notation (e.g., Draper and Smith 1998; Neter et al. 1996). As is the case for simple linear regression, the prediction variance is quite close to the MSE for moderately large samples.

6.2.8 *Multiple Linear Regression*

Multiple linear regression is the more general (and more common) situation of linear regression. As in the case of simple linear regression, there is still a single predictand, y, but in distinction there is more than one predictor (x) variable. The preceding treatment of simple linear regression was relatively lengthy, in part because most of what was presented generalizes readily to the case of multiple linear regression.

Let K denote the number of predictor variables. Simple linear regression then reduces to the special case of $K = 1$. The prediction equation (corresponding to Equation 6.1) becomes

$$\hat{y} = b_0 + b_1 x_1 + b_2 x_2 + \cdots + b_K x_K. \tag{6.24}$$

Each of the K predictor variables has its own coefficient, analogous to the slope, b, in Equation 6.1. For notational convenience, the intercept (or regression constant) is denoted as b_0 rather than as a, as in Equation 6.1. These $K + 1$ regression coefficients often are called the regression parameters. In addition, the parameter b_0 is sometimes known as the regression constant.

Equation 6.2 for the residuals is still valid, if it is understood that the predicted value $\hat{y}$ is a function of a vector of predictors, $x_k, k = 1, \ldots, K$. If there are $K = 2$ predictor variables, the residual can still be visualized as a vertical distance. In this case, the regression function (Equation 6.24) is a surface rather than a line, and the residual corresponds geometrically to the distance above or below this surface along a line perpendicular to the (x_1, x_2) plane. The geometric situation is analogous for $K \geq 3$, but is not easily visualized. Also in common with simple linear regression, the average residual is guaranteed to be zero, so that the residual distributions are centered on the predicted values $\hat{y}_i$. Accordingly, these predicted values can be regarded as conditional means given particular values of a set of K predictors.

The $K + 1$ parameters in Equation 6.24 are found, as before, by minimizing the sum of squared residuals. This is achieved by simultaneously solving $K + 1$ equations analogous to Equation 6.5. This minimization is most conveniently done using matrix algebra, the details of which can be found in standard regression texts (e.g., Draper and Smith 1998; Neter *et al.* 1996). The basics of the process are outlined in Example 9.2. In practice, the calculations usually are done using statistical software. They are again summarized in an ANOVA table, of the form shown in Table 6.3. As before, SST is computed using Equation 6.12, SSR is computed using Equation 6.13a, and SSE is computed using the difference SST − SSR. The sample variance of the residuals is MSE = SSE/(n − K − 1). The coefficient of determination is computed according to Equation 6.16, although it is no longer the square of the Pearson correlation coefficient between the predictand and any of the predictor variables. The procedures presented previously for examination of residuals are applicable to multiple regression as well.

TABLE 6.3   Generic Analysis of Variance (ANOVA) table for multiple linear regression. Table 6.1 for simple linear regression can be viewed as a special case, with $K = 1$.

Source	df	SS	MS	
Total	$n - 1$	SST		
Regression	K	SSR	MSR = SSR/K	F = MSR/MSE
Residual	$n - K - 1$	SSE	MSE = SSE/(n − K − 1) = s_e^2	

6.2.9 *Derived Predictor Variables in Multiple Regression*

Multiple regression opens up the possibility of an essentially unlimited number of potential predictor variables. An initial list of potential predictor variables can be expanded manyfold by also considering mathematical transformations of these variables as potential predictors. Such derived predictors can be very useful in producing a good regression equation.

In some instances the forms of the predictor transformations may be suggested by the physics. In the absence of a strong physical rationale for particular variable transformations, the choice of a transformation or set of transformations may be made purely empirically, perhaps by subjectively evaluating the general shape of the point cloud in a scatterplot, or the nature of the deviation of a residual plot from its ideal form. For example, the curvature in the residual plot in Figure 6.6d suggests that addition of the derived predictor $x_2 = x_1^2$ might improve the regression relationship. It may happen that the empirical choice of a transformation of a predictor variable in regression leads to a greater physical understanding, which is a highly desirable outcome in a research setting. This outcome would be less important in a purely forecasting setting, where the emphasis is on producing good forecasts rather than knowing precisely why the forecasts are good.

Transformations such as $x_2 = x_1^2, x_2 = \sqrt{x_1}, x_2 = 1/x_1$, or any other power transformation of an available predictor can be regarded as another potential predictor. Similarly, trigonometric (sine, cosine, etc.), exponential or logarithmic functions, or combinations of these are useful in some situations. Another commonly used transformation is to a binary, or dummy variable. Binary variables take on one of two values (usually 0 and 1, although the particular choices do not affect the use of the equation), depending on whether the variable being transformed is above or below a threshold or cutoff, c. That is, a binary variable x_2 could be constructed from another predictor x_1 according to the transformation

$$ x_2 = \begin{cases} 1 & \text{if } x_1 > c \\ 0 & \text{if } x_1 \le c. \end{cases} \tag{6.25} $$

More than one binary predictor can be constructed from a single x_1 by choosing different values for the cutoff, c, for x_2, x_3, x_4, and so on.

Even though transformed variables may be nonlinear functions of other variables, the overall framework is still known as multiple linear regression. Once a derived variable has been defined it is just another variable, regardless of how the transformation was made. More formally, the linear in multiple linear regression refers to the regression equation being linear in the parameters, b_k.

EXAMPLE 6.3   A Multiple Regression with Derived Predictor Variables

Figure 6.11 is a scatterplot of a portion (1959–1988) of the famous Keeling monthly-averaged carbon dioxide (CO_2) concentration data from Mauna Loa in Hawaii. Representing the obvious time trend as a straight line yields the regression results shown in Table 6.4a. The regression line is also plotted (dashed) in Figure 6.11. The results indicate a strong time trend, with the calculated standard error for the slope being much smaller than the estimated slope. The intercept merely estimates the CO_2 concentration at $t = 0$, or December 1958, so the implied test for its difference from zero is of no interest. A literal interpretation of the MSE would suggest that a 95% prediction interval for measured CO_2 concentrations around the regression line would be about $\pm 2\sqrt{MSE} = 4.9$ ppm.

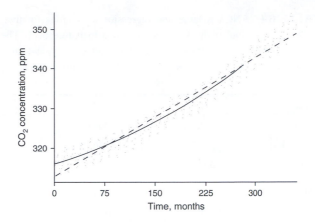

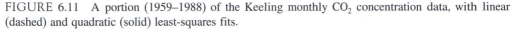

FIGURE 6.11 A portion (1959–1988) of the Keeling monthly CO_2 concentration data, with linear (dashed) and quadratic (solid) least-squares fits.

However, examination of a plot of the residuals versus time for this linear regression would reveal a bowing pattern similar to that in Figure 6.6d, with a tendency for positive residuals at the beginning and end of the record, and with negative residuals being more common in the central part of the record. This can be discerned from Figure 6.11 by noticing that most of the points fall above the dashed line at the beginning and end of the record, and fall below the line toward the middle. A plot of the residuals versus the predicted values would show this tendency for positive residuals at both high and low CO_2 concentrations, and negative residuals at intermediate concentrations.

This problem with the residuals can be alleviated (and the regression consequently improved) by fitting a quadratic curve to the time trend. To do this, a second predictor is added to the regression, and that predictor is simply the square of the time variable. That is, a multiple regression with $K = 2$ is fit using the predictors $x_1 = t$ and $x_2 = t^2$. Once defined, x_2 is just another predictor variable, taking on values between 1^2 and $360^2 = 129600$. The resulting least-squares quadratic curve is shown by the solid line in Figure 6.11, and the corresponding regression statistics are summarized in Table 6.4b.

Of course the SST in Tables 6.4a and 6.4b are the same since both pertain to the same predictand, the CO_2 concentrations. For the quadratic regression, both the coefficients $b_1 = 0.0501$ and $b_2 = 0.000136$ are substantially larger than their respective standard errors. The value of $b_0 = 312.9$ is again just the estimate of the CO_2 concentration at $t = 0$, and judging from the scatterplot this intercept is a better estimate of its true value than was obtained from the simple linear regression. The data points are fairly evenly scattered around the quadratic trend line throughout the time period, so residual plots would exhibit the desired horizontal banding. Consequently, an approximate 95% prediction interval of $\pm 2\sqrt{MSE} = 4.1$ ppm for CO_2 concentrations around the quadratic regression would be applied throughout the range of this data.

The quadratic function of time provides a reasonable approximation of the annual-average CO_2 concentration for the 30 years represented by the regression, although we can find periods of time where the point cloud wanders away from the curve. More importantly, however, a close inspection of the data points in Figure 6.11 reveals that they are not scattered randomly around the quadratic time trend. Rather, they execute a regular, nearly sinusoidal variation around the quadratic curve that is evidently an annual cycle. The resulting correlation in the residuals can easily be detected using the

TABLE 6.4   ANOVA table and regression summaries for three regressions fit to the 1959–1988 portion of the Keeling CO_2 data in Figure 6.11. The variable t (time) is a consecutive numbering of the months, with January 1959 = 1 and December 1988 = 360. There are $n = 357$ data points because February-April 1964 are missing.

(a) Linear Fit

Source	df	SS	MS	F
Total	356	39961.6		
Regression	1	37862.6	37862.6	6404
Residual	355	2099.0	5.913	

Variable	Coefficient	s.e.	t-ratio
Constant	312.9	0.2592	1207
t	0.0992	0.0012	90.0

(b) Quadratic Fit

Source	df	SS	MS	F
Total	356	39961.6		
Regression	2	38483.9	19242.0	4601
Residual	354	1477.7	4.174	

Variable	Coefficient	s.e.	t-ratio
Constant	315.9	0.3269	966
t	0.0501	0.0042	12.0
t^2	0.000136	0.0000	12.2

(c) Including quadratic trend, and harmonic terms to represent the annual cycle

Source	df	SS	MS	F
Total	356	39961.6		
Regression	4	39783.9	9946.0	19696
Residual	352	177.7	0.5050	

Variable	Coefficient	s.e.	t-ratio
Constant	315.9	0.1137	2778
t	0.0501	0.0014	34.6
t^2	0.000137	0.0000	35.2
$\cos(2\pi t/12)$	−1.711	0.0530	−32.3
$\sin(2\pi t/12)$	2.089	0.0533	39.2

Durbin-Watson statistic, $d = 0.334$ (compare Figure 6.8). The CO_2 concentrations are lower in late summer and higher in late winter as a consequence of the annual cycle of photosynthetic carbon uptake by northern hemisphere plants, and carbon release from the decomposing dead plant parts. As will be shown in Section 8.4.2, this regular 12-month variation can be represented by introducing two more derived predictor variables into the equation, $x_3 = \cos(2\pi t/12)$ and $x_4 = \sin(2\pi t/12)$. Notice that both of these derived variables are functions only of the time variable t.

Table 6.4c indicates that, together with the linear and quadratic predictors included previously, these two harmonic predictors produce a very close fit to the data. The resulting prediction equation is

$$[CO_2] = \underset{(0.1137)}{315.9} + \underset{(0.0014)}{0.0501t} + \underset{(0.0000)}{0.000137\,t^2} - \underset{(0.0530)}{1.711}\cos\left(\frac{2\pi t}{12}\right) + \underset{(0.0533)}{2.089}\sin\left(\frac{2\pi t}{12}\right), \quad (6.26)$$

with all regression coefficients being much larger than their respective standard errors. The near equality of SST and SSR indicate that the predicted values are nearly coincident with the observed CO_2 concentrations (compare Equations 6.12 and 6.13a). The resulting coefficient of determination is $R^2 = 39783.9/39961.6 = 99.56\%$, and the approximate 95% prediction interval implied by $\pm 2\sqrt{MSE}$ is only 1.4 ppm. A graph of Equation 6.26 would wiggle up and down around the solid curve in Figure 6.11, passing rather close to each of the data points. $\Diamond$

6.3 Nonlinear Regression

Although linear, least-squares regression accounts for the overwhelming majority of regression applications, it is also possible to fit regression functions that are nonlinear (in the regression parameters). Nonlinear regression can be appropriate when a nonlinear relationship is dictated by nature of the physical problem at hand, and/or the usual assumptions of Gaussian residuals with constant variance are untenable. In these cases the fitting procedure is usually iterative and based on maximum likelihood methods (see Section 4.6). This section introduces two such models.

6.3.1 Logistic Regression

One important advantage of statistical over (deterministic) dynamical forecasting methods is the capacity to produce probability forecasts. Inclusion of probability elements into the forecast format is advantageous because it provides an explicit expression of the inherent uncertainty or state of knowledge about the future weather, and because probabilistic forecasts allow users to extract more value from them when making decisions (e.g., Thompson 1962; Murphy 1977; Krzysztofowicz 1983; Katz and Murphy 1997). In a sense, ordinary linear regression produces probability information about a predictand, for example by constructing a 95% confidence interval around the regression function through application of the $\pm 2\sqrt{MSE}$ rule. More narrowly, however, probability forecasts are forecasts for which the predictand is a probability, rather than the value of a physical meteorological variable.

Most commonly, systems for producing probability forecasts are developed in a regression setting by first transforming the *predictand* to a binary (or dummy) variable, taking on the values zero and one. That is, regression procedures are implemented after applying Equation 6.25 to the predictand, *y*, rather than to a predictor. In a sense, zero and one can be viewed as probabilities of the dichotomous event not occurring or occurring, respectively, after it has been observed.

The simplest approach to regression when the predictand is binary is to use the machinery of ordinary multiple regression as described in the previous section. In the meteorological literature this is called Regression Estimation of Event Probabilities (REEP) (Glahn 1985). The main justification for the use of REEP is that it is no more computationally demanding than the fitting of any other linear regression, and so has been extensively used when computational resources have been limiting. The resulting predicted values are usually between zero and one, and it has been found through operational experience that these predicted values can usually be treated as specifications of probabilities for the event $\{Y = 1\}$. However, one obvious problem with REEP is that some of the resulting forecasts may not lie on the unit interval, particularly when the predictands are near the limits, or outside, of their ranges in the training data. This logical inconsistency usually causes little difficulty in an operational setting because multiple-regression forecast equations with many predictors rarely produce such nonsense probability estimates. When the problem does occur the forecast probability is usually near zero or one, and the operational forecast can be issued as such.

Two other difficulties associated with forcing a linear regression onto a problem with a binary predictand are that the residuals are clearly not Gaussian, and their variances are not constant. Because the predictand can take on only one of two values, a given regression residual can also take on only one of two values, and so the residual distributions are Bernoulli (i.e., binomial, Equation 4.1, with $N = 1$). Furthermore, the variance of the residuals is not constant, but depends on the ith predicted probability p_i according to $(p_i)(1 - p_i)$. It is possible to simultaneously bound the regression estimates for binary predictands on the interval $(0, 1)$, and to accommodate the Bernoulli distributions for the regression residuals, using a technique known as logistic regression. Some recent examples of logistic regression in the atmospheric science literature are Applequist *et al.* (2002), Buishand *et al.* (2004), Hilliker and Fritsch (1999), Lehmiller *et al.* (1997), Mazany *et al.* (2002), and Watson and Colucci (2002).

Logistic regressions are fit to binary predictands, according to the nonlinear equation

$$p_i = \frac{\exp(b_0 + b_1 x_1 + \cdots + b_K x_K)}{1 + \exp(b_0 + b_1 x_1 + \cdots + b_K x_K)} = \frac{1}{1 + \exp(-b_0 - b_1 x_1 - \cdots - b_K x_K)}, \quad (6.27a)$$

or

$$\ln\left(\frac{p_i}{1 - p_i}\right) = b_0 + b_1 x_1 + \cdots + b_K x_K. \quad (6.27b)$$

Here the predicted value p_i results from the ith set of predictors $(x_1, x_2, \ldots, x_K)$ of n such sets. Geometrically, logistic regression is most easily visualized for the single-predictor case $(K = 1)$, for which Equation 6.27a is an S-shaped curve that is a function of x_1. In the limits, $b_0 + b_1 x_1 \to +\infty$ results in the exponential function in the first equality of Equation 6.27a becoming arbitrarily large so that the predicted value p_i approaches one. As $b_0 + b_1 x_1 \to -\infty$, the exponential function approaches zero and thus so does the predicted value. Depending on the parameters b_0 and b_1, the function rises gradually or abruptly from zero to one (or falls, for $b_1 < 0$, from one to zero) at intermediate values of x_1. Thus it is guaranteed that logistic regression will produce properly bounded probability estimates. The logistic function is convenient mathematically, but it is not the only function that could be used in this context. Another alternative yielding a very similar shape involves using the Gaussian CDF for the form of the nonlinear regression; that is, $p_i = \Phi(b_0 + b_1 x_1 + \cdots)$, which is known as probit regression.

Equation 6.27b is a rearrangement of Equation 6.27a, and shows that logistic regression can be viewed as linear in terms of the logarithm of the odds ratio $p_i/(1 - p_i)$, also known as the logit transformation. Superficially it appears that Equation 6.27b could be fit using ordinary linear regression, except that the predictand is binary, so the left-hand side will be either $\ln(0)$ or $\ln(\infty)$. However, fitting the regression parameters can be accomplished using the method of maximum likelihood, recognizing that the residuals are Bernoulli variables. Assuming that Equation 6.27a is a reasonable model for the smooth changes in the probability of the binary outcome as a function of the predictors, the probability distribution function for the ith residual is Equation 4.1, with $N = 1$, and p_i as specified by Equation 6.27a. The corresponding likelihood is of the same functional form, except that the values of the predictand y and the predictors x are fixed, and the probability p_i is the variable. If the ith residual corresponds to a success (i.e., the event occurs, so $y_i = 1$), the likelihood is $\Lambda = p_i$ (as specified in Equation 6.27a), and otherwise $\Lambda = 1 - p_i = 1/(1 + \exp[b_0 + b_1 x_1 + \cdots])$. If the n sets of observations (predictand and predictor(s)) are independent, the joint likelihood for the $K + 1$ regression parameters is simply the product of the n individual likelihoods, or

$$\Lambda(\mathbf{b}) = \prod_{i=1}^{n} \frac{y_i \exp(b_0 + b_1 x_1 + \cdots + b_K x_K) + (1 - y_i)}{1 + \exp(b_0 + b_1 x_1 + \cdots + b_K x_K)}. \tag{6.28}$$

Since the y's are binary [0, 1] variables, each factor in Equation 6.28 for which $y_i = 1$ is equal to p_i (Equation 6.27a), and the factors for which $y_i = 0$ are equal to $1 - p_i$. As usual, it is more convenient to estimate the regression parameters by maximizing the log-likelihood

$$L(\mathbf{b}) = \ln[\Lambda(\mathbf{b})] = \sum_{i=1}^{n} \{y_i(b_0 + b_1 x_1 + \cdots + b_K x_K)$$
$$- \ln[1 + \exp(b_0 + b_1 x_1 + \cdots + b_K x_K)]\}. \tag{6.29}$$

Usually statistical software will be used to find the values of the b's maximizing this function, using the iterative methods such as those in Section 4.6.2 or 4.6.3.

Some software will display information relevant to the strength of the maximum likelihood fit using what is called the analysis of deviance table, which is analogous to the ANOVA table (see Table 6.3) for linear regression. More about analysis of deviance can be learned from sources such as Healy (1988) or McCullagh and Nelder (1989), although the idea underlying an analysis of deviance table is the likelihood ratio test (Equation 5.19). As more predictors and thus more regression parameters are added to Equation 6.27, the log-likelihood will progressively increase as more latitude is provided to accommodate the data. Whether that increase is sufficiently large to reject the null hypothesis that a particular, smaller, regression equation is adequate, is judged in terms of twice the difference of the log-likelihoods relative to the χ^2 distribution, with degrees-of-freedom ν equal to the difference in numbers of parameters between the null-hypothesis regression and the more elaborate regression being considered.

The likelihood ratio test is appropriate when a single candidate logistic regression is being compared to a null model. Often H_0 will specify that all the regression parameters except b_0 are zero, in which case the question being addressed is whether the predictors x being considered are justified in favor of the constant (no-predictor) model with $b_0 = \ln[\Sigma y_i / n / (1 - \Sigma y_i / n)]$. However, if multiple alternative logistic regressions are being entertained, computing the likelihood ratio test for each alternative raises the problem of test multiplicity (see Section 5.4.1). In such cases it is better to compute the Bayesian Information Criterion (BIC) statistic (Schwarz 1978)

$$\text{BIC} = -2L(\mathbf{b}) + (K + 1) \ln(n) \tag{6.30}$$

for each candidate model. The BIC statistic consists of twice the negative of the log-likelihood plus a penalty for the number of parameters fit, and the preferred regression will be the one with the smallest BIC.

EXAMPLE 6.4   Comparison of REEP and Logistic Regression

Figure 6.12 compares the results of REEP (dashed) and logistic regression (solid) for some of the January 1987 data from Table A.1. The predictand is daily Ithaca precipitation, transformed to a binary variable using Equation 6.25 with $c = 0$. That is, $y = 0$ if the precipitation is zero, and $y = 1$ otherwise. The predictor is the Ithaca minimum temperature for the same day. The REEP (linear regression) equation has been fit using ordinary least squares, yielding $b_0 = 0.208$ and $b_1 = 0.0212$. This equation specifies negative probability of precipitation if the temperature predictor is less than about $-9.8°F$, and specifies probability of precipitation greater than one if the minimum temperature is greater than about $37.4°F$. The parameters for the logistic regression, fit using maximum likelihood, are $b_0 = -1.76$ and $b_1 = 0.117$. The logistic regression curve produces probabilities that are similar to the REEP specifications through most of the temperature range, but are constrained by the functional form of Equation 6.27 to lie between zero and one, even for extreme values of the predictor.

Maximizing Equation 6.29 for logistic regression with a single ($K = 1$) is simple enough that the Newton-Raphson method (see Section 4.6.2) can be implemented easily and is reasonably robust to poor initial guesses for the parameters. The counterpart to Equation 4.73 for this problem is

$$
\begin{bmatrix} b_0^* \\ b_1^* \end{bmatrix} = \begin{bmatrix} b_0 \\ b_1 \end{bmatrix} - \begin{bmatrix} \sum_{i=1}^{n}(p_i^2 - p_i) & \sum_{i=1}^{n} x_i(p_i^2 - p_i) \\ \sum_{i=1}^{n} x_i(p_i^2 - p_i) & \sum_{i=1}^{n} x_i^2(p_i^2 - p_i) \end{bmatrix}^{-1} \begin{bmatrix} \sum_{i=1}^{n}(y_i - p_i) \\ \sum_{i=1}^{n} x_i(y_i - p_i) \end{bmatrix},
\tag{6.31}
$$

where p_i is a function of the regression parameters b_0 and b_1, and depends also on the predictor data x_i, as shown in Equation 6.27a. The first derivatives of the log-likelihood (Equation 6.29) with respect to b_0 and b_1 are in the vector enclosed by the rightmost square brackets, and the second derivatives are contained in the matrix to be inverted. Beginning with an initial guess for the parameters (b_0, b_1), updated parameters (b_0^*, b_1^*) are computed and then resubstituted into the right-hand side of Equation 6.31 for the next iteration. For example, assuming initially that the Ithaca minimum temperature is unrelated to the binary precipitation outcome, so $b_0 = -0.645$ (the log of the

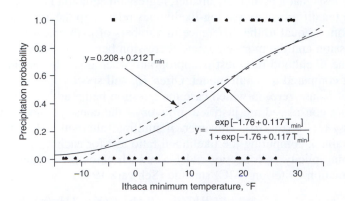

FIGURE 6.12   Comparison of regression probability forecasting using REEP (dashed) and logistic regression (solid) using the January 1987 data set in Table A.1. The linear function was fit using least squares, and the logistic curve was fit using maximum likelihood, to the data shown by the dots. The binary predictand $y = 1$ if Ithaca precipitation is greater than zero, and $y = 0$ otherwise.

observed odds ratio, for constant $p = 15/31$) and $b_1 = 0$; the updated parameters for the first iteration are $b_0^* = -0.645 - (-0.251)(-0.000297) - (0.00936)(118.0) = -1.17$, and $b_1^* = 0 - (0.00936)(-0.000297) - (-0.000720)(118.0) = 0.085$. These updated parameters increase the log-likelihood from -21.47 for the constant model (calculated using Equation 6.29, imposing $b_0 = -0.645$ and $b_1 = 0$), to -16.00. After four iterations the algorithm has converged, with a final (maximized) log-likelihood of -15.67.

Is the logistic relationship between Ithaca minimum temperature and the probability of precipitation statistically significant? This question can be addressed using the likelihood ratio test (Equation 5.19). The appropriate null hypothesis is that $b_1 = 0$, so $L(H_0) = -21.47$, and $L(H_A) = -15.67$ for the fitted regression. If H_0 is true then the observed test statistic $\Lambda^* = 2[L(H_A) - L(H_0)] = 11.6$ is a realization from the χ^2 distribution with $\nu = 1$ (the difference in the number of parameters between the two regressions), and the test is 1-tailed because small values of the test statistic are favorable to H_0. Referring to the first row of Table B.3, it is clear that the regression is significant at the 0.1% level. ◊

6.3.2 Poisson Regression

Another regression setting where the residual distribution may be poorly represented by the Gaussian is the case where the predictand consists of counts; that is, each of the y's is a nonnegative integer. Particularly if these counts tend to be small, the residual distribution is likely to be asymmetric, and we would like a regression predicting these data to be incapable of implying nonzero probability for negative counts.

A natural probability model for count data is the Poisson distribution (Equation 4.11). Recall that one interpretation of a regression function is as the conditional mean of the predictand, given specific value(s) of the predictor(s). If the outcomes to be predicted by a regression are Poisson-distributed counts, but the Poisson parameter μ may depend on one or more predictor variables, we can structure a regression to specify the Poisson mean as a nonlinear function of those predictors,

$$\mu_i = \exp[b_0 + b_1 x_1 + \cdots + b_K x_K], \qquad (6.32a)$$

or

$$\ln(\mu_i) = b_0 + b_1 x_1 + \cdots + b_K x_K. \qquad (6.32b)$$

Equation 6.32 is not the only function that could be used for this purpose, but framing the problem in this way makes the subsequent mathematics quite tractable, and the logarithm in Equation 6.32b ensures that the predicted Poisson mean is nonnegative. Some applications of Poisson regression are described in Elsner and Schmertmann (1993), McDonnell and Holbrook (2004), Paciorek *et al.* (2002), and Solow and Moore (2000).

Having framed the regression in terms of Poisson distributions for the y_i conditional on the corresponding set of predictor variables $x_i = \{x_1, x_2, \ldots, x_K\}$, the natural approach to parameter fitting is to maximize the Poisson log-likelihood, written in terms of the regression parameters. Again assuming independence, the log-likelihood is

$$L(\mathbf{b}) = \sum_{i=1}^{n} \{y_i(b_0 + b_1 x_1 + \cdots + b_K x_K) - \exp(b_0 + b_1 x_1 + \cdots + b_K x_K)\}, \qquad (6.33)$$

where the term involving $y!$ from the denominator of Equation 4.11 has been omitted because it does not involve the unknown regression parameters, and so will not influence the process of locating the maximum of the function. An analytic maximization of Equation 6.33 in general is not possible, so that statistical software will approximate the maximum iteratively, typically using one of the methods outlined in Sections 4.6.2 or 4.6.3. For example, if there is a single ($K = 1$) predictor, the Newton-Raphson method (see Section 4.6.2) iterates the solution according to

$$\begin{bmatrix} b_0^* \\ b_1^* \end{bmatrix} = \begin{bmatrix} b_0 \\ b_1 \end{bmatrix} - \begin{bmatrix} -\sum_{i=1}^n \mu_i & -\sum_{i=1}^n x_i\mu_i \\ -\sum_{i=1}^n x_i\mu_i & -\sum_{i=1}^n x_i^2\mu_i \end{bmatrix}^{-1} \begin{bmatrix} \sum_{i=1}^n (y_i - \mu_i) \\ \sum_{i=1}^n x_i(y_i - \mu_i) \end{bmatrix}, \qquad (6.34)$$

where μ_i is the conditional mean as a function of the regression parameters as defined in Equation 6.32a. Equation 6.34 is the counterpart of Equation 4.73 for fitting the gamma distribution, and Equation 6.31 for logistic regression.

EXAMPLE 6.5   A Poisson Regression

Consider again the annual counts of tornados reported in New York state for 1959–1988, in Table 4.3. Figure 6.13 shows a scatterplot of these as a function of average July temperatures at Ithaca in the corresponding years. The solid curve is a Poisson regression function, and the dashed line shows the ordinary least-squares linear fit. The nonlinearity of the Poisson regression is quite modest over the range of the training data, although the regression function would remain positive regardless of the magnitude of the predictor variable.

The relationship is weak, but slightly negative. The significance of the Poisson regression usually would be judged using the likelihood ratio test (Equation 5.19). The maximized log-likelihood (Equation 6.33) is 74.26 for $K = 1$, whereas the log-likelihood with only the intercept $b_0 = \ln(\Sigma y/n) = 1.526$ is 72.60. Comparing $\Lambda^* = 2(74.26 - 72.60) = 3.32$ to χ^2 distribution quantiles in Table B.3 with $\nu = 1$ (the difference in the number of fitted parameters) indicates that b_1 would be judged significantly different from zero at the 10% level, but not at the 5% level. For the linear regression, the t ratio for the slope parameter b_1 is -1.86, implying a two-tailed p value of 0.068, which is an essentially equivalent result.

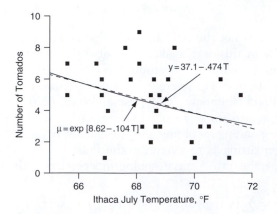

FIGURE 6.13   New York tornado counts, 1959–1988 (Table 4.3), as a function of average Ithaca July temperature in the same year. Solid curve shows the Poisson regression fit using maximum likelihood (Equation 6.34), and dashed line shows ordinary least-squares linear regression.

The primary difference between the Poisson and linear regressions in Figure 6.13 is in the residual distributions, and therefore in the probability statements about the specified predicted values. Consider, for example, the number of tornados specified when $T = 70°$ F. For the linear regression, $\hat{y} = 3.92$ tornados, with a Gaussian $\sigma_e = 2.1$. Rounding to the nearest integer (i.e., using a continuity correction), the linear regression assuming Gaussian residuals implies that the probability for a negative number of tornados is $\Phi[(-0.5 - 3.92)/2.1] = \Phi[-2.10] = 0.018$, rather than the true value of zero. On the other hand, conditional on a temperature of $70°$ F, the Poisson regression specifies that the number of tornados will be distributed as a Poisson variable with mean $\mu = 3.82$. Using to this mean, Equation 4.11 yields $\Pr\{Y < 0\} = 0$, $\Pr\{Y = 0\} = 0.022$, $\Pr\{Y = 1\} = 0.084$, $\Pr\{Y = 2\} = 0.160$, and so on. $\Diamond$

6.4 Predictor Selection

6.4.1 Why is Careful Predictor Selection Important?

There are almost always more potential predictors available than can be used in a statistical prediction procedure, and finding good subsets of these in particular cases is more difficult than we at first might imagine. The process is definitely not as simple as adding members of the list of potential predictors until an apparently good relationship is achieved. Perhaps surprisingly, there are dangers associated with including too many predictor variables in a forecast equation.

EXAMPLE 6.6   An Overfit Regression

To illustrate the dangers of too many predictors, Table 6.5 shows total winter snowfall at Ithaca (inches) for the seven winters beginning in 1980 through 1986 and four potential predictors arbitrarily taken from an almanac (Hoffman 1988): the U.S. federal deficit (in billions of dollars), the number of personnel in the U.S. Air Force, the sheep population of the U.S. (in thousands), and the average Scholastic Aptitude Test (SAT) scores of college-bound high-school students. Obviously these are nonsense predictors, which bear no real relationship to the amount of snowfall at Ithaca.

Regardless of their lack of relevance, we can blindly offer these predictors to a computer regression package, and it will produce a regression equation. For reasons that will be made clear shortly, assume that the regression will be fit using only the six winters

TABLE 6.5   A small data set to illustrate the dangers of overfitting. Nonclimatological data were taken from Hoffman (1988).

Winter Beginning	Ithaca Snowfall (in.)	U.S. Federal Deficit ($ $\times 10^9$)	U.S. Air Force Personnel	U.S. Sheep ($\times 10^3$)	Average SAT Scores
1980	52.3	59.6	557969	12699	992
1981	64.9	57.9	570302	12947	994
1982	50.2	110.6	582845	12997	989
1983	74.2	196.4	592044	12140	963
1984	49.5	175.3	597125	11487	965
1985	64.7	211.9	601515	10443	977
1986	65.6	220.7	606500	9932	1001

beginning in 1980 through 1985. That portion of available data used to produce the forecast equation is known as the developmental sample, dependent sample, or training sample. For the developmental sample of 1980–1985, the resulting equation is

$$\text{Snow} = 1161771 - 601.7(\text{yr}) - 1.733(\text{deficit}) + 0.0567(\text{AF pers.})$$
$$- 0.3799(\text{sheep}) + 2.882(\text{SAT}).$$

The ANOVA table accompanying this equation indicated MSE$= 0.0000$, $R^2 = 100.00\%$, and $F = \infty$; that is, a perfect fit!

Figure 6.14 shows a plot of the regression-specified snowfall totals (line segments) and the observed data (circles). For the developmental portion of the record, the regression does indeed represent the data exactly, as indicated by the ANOVA statistics, even though it is obvious from the nature of the predictor variables that the specified relationship is not meaningful. In fact, essentially any five predictors would have produced exactly the same perfect fit (although with different regression coefficients, b_k) to the six developmental data points. More generally, any $K = n - 1$ predictors will produce a perfect regression fit to any predictand for which there are n observations. This concept is easiest to see for the case of $n = 2$, where a straight line can be fit using any $K = 1$ predictor (simple linear regression), since a line can be found that will pass through any two points in the plane, and only an intercept and a slope are necessary to define a line. The problem, however, generalizes to any sample size.

This example illustrates an extreme case of overfitting the data. That is, so many predictors have been used that an excellent fit has been achieved on the dependent data, but the fitted relationship falls apart when used with independent, or verification data— data not used in the development of the equation. Here the data for 1986 has been reserved for a verification sample. Figure 6.14 indicates that the equation performs very poorly outside of the training sample, producing a meaningless forecast for negative snowfall

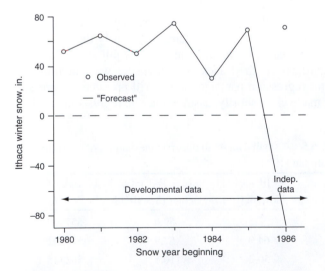

FIGURE 6.14   Forecasting Ithaca winter snowfall using the data in Table 6.5. The number of predictors is one fewer than the number of observations of the predictand in the developmental data, yielding perfect correspondence between the values specified by the regression and the data for this portion of the record. The relationship falls apart completely when used with the 1986 data, which was not used in equation development. The regression equation has been grossly overfit.

during 1986–1987. Clearly, issuing forecasts equal to the climatological average total snowfall, or the snowfall for the previous winter, would yield better results than the equation produced earlier. Note that the problem of overfitting is *not* limited to cases where nonsense predictors are used in a forecast equation, and will be a problem when too many meaningful predictors are included as well. ◊

As ridiculous as it may seem, several important lessons can be drawn from Example 6.6:

- Begin development of a regression equation by choosing only physically reasonable or meaningful potential predictors. If the predictand of interest is surface temperature, for example, then temperature-related predictors such as the 1000-700 mb thickness (reflecting the mean virtual temperature in the layer), the 700 mb relative humidity (perhaps as an index of clouds), or the climatological average temperature for the forecast date (as a representation of the annual cycle of temperature) could be sensible candidate predictors. Understanding that clouds will form only in saturated air, a binary variable based on the 700 mb relative humidity also might be expected to contribute meaningfully to the regression. One consequence of this lesson is that a statistically literate person with insight into the physical problem (domain expertise) may be more successful than a statistician at devising a forecast equation.
- A tentative regression equation needs to be tested on a sample of data not involved in its development. One way to approach this important step is simply to reserve a portion (perhaps a quarter, a third, or half) of the available data as the independent verification set, and fit the regression using the remainder as the training set. The performance of the resulting equation will nearly always be better for the dependent than the independent data, since (in the case of least-squares regression) the coefficients have been chosen specifically to minimize the squared residuals in the developmental sample. A very large difference in performance between the dependent and independent samples would lead to the suspicion that the equation had been overfit.
- We need a reasonably large developmental sample if the resulting equation is to be stable. Stability is usually understood to mean that the fitted coefficients are also applicable to independent (i.e., future) data, so that the coefficients would be substantially unchanged if based on a different sample of the same kind of data. The number of coefficients that can be estimated with reasonable accuracy increases as the sample size increases, although in forecasting practice it often is found that there is little to be gained from including more than about a dozen predictor variables in a final regression equation (Glahn 1985). In that kind of forecasting application there are typically thousands of observations of the predictand in the developmental sample. Unfortunately, there is not a firm rule specifying a minimum ratio of sample size (number of observations of the predictand) to the number of predictor variables in the final equation. Rather, testing on an independent data set is relied upon in practice to ensure stability of the regression.

6.4.2 Screening Predictors

Suppose the set of potential predictor variables for a particular problem could be assembled in a way that all physically relevant predictors were included, with exclusion of all irrelevant ones. This ideal can rarely, if ever, be achieved. Even if it could be, however, it generally would not be useful to include all the potential predictors in a final equation.

This is because the predictor variables are almost always mutually correlated, so that the full set of potential predictors contains redundant information. Table 3.5, for example, shows substantial correlations among the six variables in Table A.1. Inclusion of predictors with strong mutual correlation is worse than superfluous, because this condition leads to poor estimates (high-variance sampling distributions) for the estimated parameters. As a practical matter, then, we need a method to choose among potential predictors, and of deciding how many and which of them are sufficient to produce a good prediction equation.

In the jargon of statistical weather forecasting, the problem of selecting a good set of predictors from a pool of potential predictors is called screening regression, since the predictors must be subjected to some kind of screening, or filtering procedure. The most commonly used screening procedure is known as forward selection or stepwise regression in the broader statistical literature.

Suppose there are some number, M, of potential predictors. For linear regression we begin the process of forward selection with the uninformative prediction equation $\hat{y} = b_0$. That is, only the intercept term is in the equation, and this intercept is necessarily the sample mean of the predictand. On the first forward selection step, all M potential predictors are examined for the strength of their linear relationship to the predictand. In effect, all the possible M simple linear regressions between the available predictors and the predictand are computed, and that predictor whose linear regression is best among all candidate predictors is chosen as x_1. At this stage of the screening procedure, then, the prediction equation is $\hat{y} = b_0 + b_1 x_1$. Note that the intercept b_0, in general, no longer will be the average of the y values.

At the next stage of the forward selection, trial regressions are again constructed using all remaining $M - 1$ predictors. However, all these trial regressions also contain the variable selected on the previous step as x_1. That is, given the particular x_1 chosen on the previous step, that predictor variable yielding the best regression $\hat{y} = b_0 + b_1 x_1 + b_2 x_2$ is chosen as x_2. This new x_2 will be recognized as best because it produces that regression equation with $K = 2$ predictors that also includes the previously chosen x_1, having the highest R^2, the smallest MSE, and the largest F ratio.

Subsequent steps in the forward selection procedure follow this pattern exactly: at each step, that member of the potential predictor pool not yet in the regression is chosen that produces the best regression in conjunction with the $K - 1$ predictors chosen on previous steps. In general, when these regression equations are recomputed the regression coefficients for the intercept and for the previously chosen predictors will change. These changes will occur because the predictors usually are correlated to a greater or lesser degree, so that information about the predictand is spread around among the predictands differently as more predictors are added to the equation.

EXAMPLE 6.7 Equation Development Using Forward Selection

The concept of variable selection can be illustrated with the January 1987 temperature and precipitation data in Table A.1. As in Example 6.1 for simple linear regression, the predictand is Canandaigua minimum temperature. The potential predictor pool consists of maximum and minimum temperatures at Ithaca, maximum temperature at Canandaigua, the logarithms of the precipitation amounts plus 0.01 in. (in order for the logarithm to be defined for zero precipitation) for both locations, and the day of the month. The date predictor is included on the basis of the trend in the residuals apparent in Figure 6.9. Note that this example is somewhat artificial with respect to statistical weather forecasting, since the predictors (other than the date) will not be known in advance of the time that

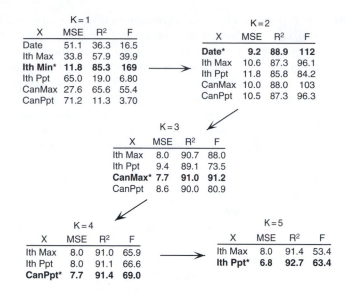

FIGURE 6.15 Diagram of the forward selection procedure for development of a regression equation for Canandaigua minimum temperature using as potential predictors the remaining variables in data set A1, plus the date. At each step the variable is chosen (bold, starred) whose addition would produce the largest decrease in MSE or, equivalently, the largest increase in R^2 or F. At the final ($K = 6$) stage, only IthMax remains to be chosen, and its inclusion would produce MSE $= 6.8$, $R^2 = 93.0\%$, and $F = 52.8$.

the predictand (minimum temperature at Canandaigua) will be observed. However, this small data set serves perfectly well to illustrate the principles.

Figure 6.15 diagrams the process of choosing predictors using forward selection. The numbers in each table summarize the comparisons being made at each step. For the first ($K = 1$) step, no predictors are yet in the equation, and all six potential predictors are under consideration. At this stage the predictor producing the best simple linear regression is chosen, as indicated by the smallest MSE, and the largest R^2 and F ratio among the six. This best predictor is the Ithaca minimum temperature, so the tentative regression equation is exactly Equation 6.20.

Having chosen the Ithaca minimum temperature in the first stage there are five potential predictors remaining, and these are listed in the $K = 2$ table. Of these five, the one producing the best predictions in an equation that also includes the Ithaca minimum temperature is chosen. Summary statistics for these five possible two-predictor regressions are also shown in the $K = 2$ table. Of these, the equation including Ithaca minimum temperature and the date as the two predictors is clearly best, producing MSE $= 9.2°F^2$ for the dependent data.

With these two predictors now in the equation, there are only four potential predictors left at the $K = 3$ stage. Of these, the Canandaigua maximum temperature produces the best predictions in conjunction with the two predictors already in the equation, yielding MSE $= 7.7°F^2$ on the dependent data. Similarly, the best predictor at the $K = 4$ stage is Canandaigua precipitation, and the better predictor at the $K = 5$ stage is Ithaca precipitation. For $K = 6$ (all predictors in the equation) the MSE for the dependent data is $6.8°F^2$, with $R^2 = 93.0\%$. ◊

An alternative approach to screening regression is called backward elimination. The process of backward elimination is analogous but opposite to that of forward selection.

Here the initial point is a regression containing all M potential predictors, $\hat{y} = b_0 + b_1 x_1 + b_2 x_2 + \cdots + b_M x_M$, so backward elimination will not be computationally feasible if $M \geq n$. Usually this initial equation will be grossly overfit, containing many redundant and some possibly useless predictors. At each step of the backward elimination procedure, the least important predictor variable is removed from the regression equation. That variable will be the one whose coefficient is smallest in absolute value, relative to its estimated standard error. In terms of the sample regression output tables presented earlier, the removed variable will exhibit the smallest (absolute) t ratio. As in forward selection, the regression coefficients for the remaining variables require recomputation if (as is usually the case) the predictors are mutually correlated.

There is no guarantee that forward selection and backward elimination will choose the same subset of the potential predictor pool for the final regression equation. Other variable selection procedures for multiple regression also exist, and these might select still different subsets. The possibility that a chosen variable selection procedure might not select the "right" set of predictor variables might be unsettling at first, but as a practical matter this is not usually an important problem in the context of producing an equation for use as a forecast tool. Correlations among the predictor variables result in the situation that essentially the same information about the predictand can be extracted from different subsets of the potential predictors. Therefore, if the aim of the regression analysis is only to produce reasonably accurate forecasts of the predictand, the black box approach of empirically choosing a workable set of predictors is quite adequate. However, we should not be so complacent in a research setting, where one aim of a regression analysis could be to find specific predictor variables most directly responsible for the physical phenomena associated with the predictand.

6.4.3 Stopping Rules

Both forward selection and backward elimination require a stopping criterion, or stopping rule. Without such a rule, forward selection would continue until all M candidate predictor variables were included in the regression equation, and backward elimination would continue until all predictors had been eliminated. It might seem that finding the stopping point would be a simple matter of evaluating the test statistics for the regression parameters and their nominal p values as supplied by the computer regression package. Unfortunately, because of the way the predictors are selected, these implied hypothesis tests are not quantitatively applicable. At each step (either in selection or elimination) predictor variables are not chosen randomly for entry or removal. Rather, the best or worst, respectively, among the available choices is selected. Although this may seem like a minor distinction, it can have very major consequences.

The problem is illustrated in Figure 6.16, taken from the study of Neumann *et al.* (1977). The specific problem represented in this figure is the selection of exactly $K = 12$ predictor variables from pools of potential predictors of varying sizes, M, when there are $n = 127$ observations of the predictand. Ignoring the problem of nonrandom predictor selection would lead us to declare as significant any regression for which the F ratio in the ANOVA table is larger than the nominal critical value of 2.35. Naïvely, this value would correspond to the minimum F ratio necessary to reject the null hypothesis of no real relationship between the predictand and the twelve predictors, at the 1% level. The curve labeled empirical F ratio was arrived at using a resampling test, in which the same meteorological predictor variables were used in a forward selection procedure to predict 100 artificial data sets of $n = 127$ independent Gaussian random numbers each.

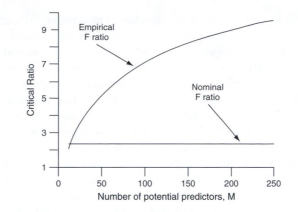

FIGURE 6.16   Comparison of the nominal and empirically (resampling) estimated critical $(p = 0.01)F$ ratios for overall significance of in a particular regression problem, as a function of the number of potential predictor variables, M. The sample size is $n = 127$, with $K = 12$ predictor variables to be included in each final regression equation. The nominal F ratio of 2.35 is applicable only for the case of $M = K$. When the forward selection procedure can choose from among more than K potential predictors the true critical F ratio is substantially higher. The difference between the nominal and actual values widens as M increases. From Neumann *et al.* (1977).

This procedure simulates the null hypothesis that the predictors bear no real relationship to the predictand, while automatically preserving the correlations among this particular set of predictors.

Figure 6.16 indicates that the nominal value gives the correct answer only in the case of $K = M$, for which there is no ambiguity in the predictor selection since all the $M = 12$ potential predictors must be used to construct the $K = 12$ predictor equation. When the forward selection procedure has available some larger number $M > K$ potential predictor variables to choose from, the true critical F ratio is higher, and sometimes by a substantial amount. Even though none of the potential predictors in the resampling procedure bears any real relationship to the artificial (random) predictand, the forward selection procedure chooses those predictors exhibiting the highest chance correlations with the predictand, and these relationships result in apparently high F ratio statistics. Put another way, the p value associated with the nominal critical $F = 2.35$ is larger (less significant) than the true p value, by an amount that increases as more potential predictors are offered to the forward selection procedure. To emphasize the seriousness of the problem, the nominal F ratio in the situation of Figure 6.16 for the very stringent 0.01% level test is only about 3.7. The practical result of relying literally on the nominal critical F ratio is to allow more predictors into the final equation than are meaningful, with the danger that the regression will be overfit.

Unfortunately, the results in Figure 6.16 apply only to the specific data set from which they were derived. In order to use this approach to estimate the true critical F-ratio using resampling methods it must be repeated for each regression to be fit, since the statistical relationships among the potential predictor variables will be different in different data sets. In practice, other less rigorous stopping criteria usually are employed. For example, we might stop adding predictors in a forward selection when none of the remaining predictors would reduce the R^2 by a specified amount, perhaps 0.05%.

The stopping criterion can also be based on the MSE. This choice is intuitively appealing because, as the standard deviation of the residuals around the regression function, $\sqrt{MSE}$ directly reflects the anticipated precision of a regression. For example, if a regression equation were being developed to forecast surface temperature, little would be gained by adding more predictors if the MSE were already $0.01°F^2$, since this would indicate a $\pm 2\sigma$ (i.e., approximately 95%) confidence interval around the forecast value of about $\pm 2\sqrt{0.01°F^2} = 0.2°F$. So long as the number of predictors K is substantially less than the sample size n, adding more predictor variables (even meaningless ones) will decrease the MSE for the developmental sample. This concept is illustrated schematically in Figure 6.17. Ideally the stopping criterion would be at the point where the MSE does not decline appreciably with the addition of more predictors, at perhaps $K = 12$ predictors in the hypothetical case shown in Figure 6.17.

Figure 6.17 indicates that the MSE for an independent data set will be larger than that achieved for the developmental data. This result should not be surprising, since the least-squares fitting procedure operates by selecting precisely those parameter values that minimize MSE for the developmental data. This underestimation of the operational MSE provided by a forecast equation on developmental data is an expression of what is sometimes called artificial skill (Davis 1976, Michaelson 1987). The precise magnitudes of the differences in MSE between developmental and independent data sets is not determinable solely from the regression output using the developmental data. That is, having seen only the regressions fit to the developmental data, we cannot know the value of the minimum MSE for independent data. Neither can we know if it will occur at a similar point (at around $K = 12$ in Figure 6.17), or whether the equation has been overfit and the minimum MSE for the independent data will be for a substantially smaller K. This situation is unfortunate, because the purpose of developing a forecast equation is to specify future, unknown values of the predictand using observations of the predictors that have yet to occur.

Figure 6.17 also indicates that, for forecasting purposes, the exact stopping point is not usually critical as long as it is approximately right. Again, this is because the MSE tends to change relatively little through a range of K near the optimum, and for purposes of forecasting it is the minimization of the MSE rather than the specific identities of the predictors that is important. That is, for forecasting it is acceptable to use the regression equation as a black box. By contrast, if the purpose of the regression analysis is scientific understanding, the specific identities of chosen predictor variables can be critically important, and the magnitudes of the resulting regression coefficients can lead to significant physical insight. In this case it is not reduction of prediction MSE, *per se*, that

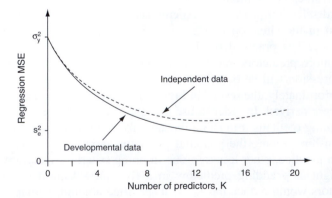

FIGURE 6.17   Schematic illustration of the regression MSE as a function of the number of predictor variables in the equation, K, for developmental data (solid), and for an independent verification set (dashed). After Glahn (1985).

is desired, but rather that causal relationships between particular variables be suggested by the analysis.

6.4.4 Cross Validation

Usually regression equations to be used for weather forecasting are tested on a sample of independent data that has been held back during the development of the forecast equation. In this way, once the number K and specific identities of the predictors have been fixed, an estimate of the distances between the solid and dashed MSE lines in Figure 6.17 can be estimated directly from the reserved data. If the deterioration in forecast precision (i.e., the unavoidable increase in MSE) is judged to be acceptable, the equation can be used operationally.

This procedure reserving an independent verification data set is actually a special case of a technique known as cross validation (Efron and Gong 1985; Efron and Tibshirani 1993; Elsner and Schmertmann 1994; Michaelson 1987). Cross validation simulates prediction for future, unknown data by repeating the fitting procedure on data subsets, and then examining the predictions on the data portions left out of each of these subsets. The most frequently used procedure is known as leave-one-out cross validation, in which the fitting procedure is repeated n times, each time with a sample of size $n-1$, because one of the predictand observations and its corresponding predictor set are left out. The result is n (often only slightly) different prediction equations.

The cross-validation estimate of the prediction MSE is computed by forecasting each omitted observation using the equation developed from the remaining $n-1$ data values, computing the squared difference between the prediction and predictand for each of these equations, and averaging the n squared differences. Thus, leave-one-out cross validation uses all n observations of the predictand to estimate the prediction MSE in a way that allows each observation to be treated, one at a time, as independent data. Cross validation can also be carried out for any number m of withheld data points, and developmental data sets of size $n-m$ (Zhang 1994). In this more general case, all $(n!)/[(m!)(n-m!)]$ possible partitions of the full data set could be employed. Particularly when the sample size n is small and the predictions will be evaluated using a correlation measure, leaving out $m > 1$ values at a time can be advantageous (Barnson and van den Dool 1993).

Cross validation requires some special care when the data are serially correlated. In particular, data records adjacent to or near the omitted observation(s) will tend to be more similar to them than randomly selected ones, so the omitted observation(s) will be more easily predicted than the uncorrelated future observations they are meant to simulate. A solution to this problem is to leave out blocks of an odd number of consecutive observations, L, so the fitting procedure is repeated $n-L+1$ times on samples of size $n-L$ (Burman et al. 1994; Elsner and Schmertmann 1994). The blocklength L is chosen to be large enough for the correlation between its middle value and the nearest data used in the cross-validation fitting to be small, and the cross-validation prediction is made only for that middle value. For $L = 1$ this moving-blocks cross validation reduces to leave-one-out cross validation.

It should be emphasized that each repetition of the cross-validation exercise is a repetition of the *fitting algorithm*, not of the specific statistical model derived from the full data set. In particular, different prediction variables must be allowed to enter for different data subsets. Any data transformations (e.g., standardizations with respect to climatological values) also need to be defined (and therefore possibly recomputed) without any reference to the withheld data in order for them to have no influence on the equation that will be used to predict them in the cross-validation exercise. However, the ultimate product equation, to be used for operational forecasts, would be fit using all the data after we are satisfied with the cross-validation results.

EXAMPLE 6.8   Protecting against Overfitting Using Cross Validation

Having used all the available developmental data to fit the regressions in Example 6.7, what can be done to ensure that these prediction equations have not been overfit? Fundamentally, what is desired is a measure of how the regressions will perform when used on data not involved in the fitting. Cross validation is an especially appropriate tool for this purpose in the present example, because the small ($n = 31$) sample would be inadequate if a substantial portion of it had to be reserved for a validation sample.

Figure 6.18 compares the regression MSE for the six regression equations obtained with the forward selection procedure outlined in Figure 6.15. This figure shows real results in the same form as the idealization of Figure 6.17. The solid line indicates the MSE achieved on the developmental sample, obtained by adding the predictors in the order shown in Figure 6.15. Because a regression chooses precisely those coefficients minimizing this quantity for the developmental data, the MSE is expected to be higher when the equations are applied to independent data. An estimate of how much higher is given by the average MSE from the cross-validation samples (dashed line). Because these data are autocorrelated, a simple leave-one-out cross validation is expected to underestimate the prediction MSE. Here the cross validation has been carried out omitting blocks of length $L = 7$ consecutive days. Since the lag-1 autocorrelation for the predictand is approximately $r_1 = 0.6$ and the autocorrelation function exhibits approximately exponential decay (similar to that in Figure 3.19), the correlation between the predictand in the centers of the seven-day moving blocks and the nearest data used for equation fitting is $0.6^4 = 0.13$, corresponding to $R^2 = 1.7\%$, indicating near-independence.

Each cross-validation point in Figure 6.18 represents the average of $25 (= 31 - 7 + 1)$ squared differences between an observed value of the predictand at the center of a block, and the forecast of that value produced by regression equations fit to all the data except those in that block. Predictors are added to each of these equations according to the usual

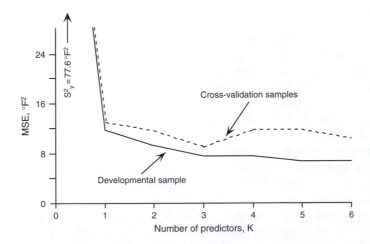

FIGURE 6.18   Plot of residual mean-squared error as a function of the number of regression predictors specifying Canandaigua minimum temperature, using the January 1987 data in Appendix A. Solid line shows MSE for developmental data (starred predictors in Figure 6.15). Dashed line shows MSE achievable on independent data, as estimated through cross-validation, leaving out blocks of seven consecutive days. This plot is a real-data example corresponding to the idealization in Figure 6.17.

forward selection algorithm. The order in which the predictors are added is often the same as that indicated in Figure 6.15 for the full data set, but this order is not forced onto the cross-validation samples, and indeed is different for some of the data partitions.

The differences between the dashed and solid lines in Figure 6.18 are indicative of the expected prediction errors for future independent data (dashed), and those that would be inferred from the MSE on the dependent data as provided by the ANOVA table (solid). The minimum cross-validation MSE at $K = 3$ suggests that the best regression for these data may be the one with three predictors, and that it should produce prediction MSE on independent data of around $91°F^2$, yielding $\pm2\sigma$ confidence limits of $\pm6.0°F$. ◊

Before leaving the topic of cross validation it is worthwhile to note that the procedure is sometimes mistakenly referred to as "jackknifing." The confusion is understandable because the jackknife is a statistical method that is computationally analogous to leave-one-out cross validation (e.g., Efron 1982; Efron and Tibshirani 1993). Its purpose, however, is to estimate the bias and/or standard deviation of a sampling distribution nonparametrically, and using only the data in a single sample. Accordingly it is similar in spirit to the bootstrap (see Section 5.3.4). Given a sample of n independent observations, the idea in jackknifing is to recompute a statistic of interest n times, omitting a different one of the data values each time. Attributes of the sampling distribution for the statistic can then be inferred from the resulting n-member jackknife distribution. The jackknife and leave-one-out cross validation share the mechanics of repeated recomputation on reduced samples of size $n - 1$, but cross validation seeks to infer future forecasting performance, whereas the jackknife seeks to nonparametrically characterize the sampling distribution of a sample statistic.

6.5 Objective Forecasts Using Traditional Statistical Methods

6.5.1 Classical Statistical Forecasting

Construction of weather forecasts through purely statistical means—that is, without the benefit of information from numerical (i.e., dynamical) weather prediction (NWP) models—has come to be known as classical statistical forecasting. This name reflects the long history of the use of purely statistical forecasting methods, dating from the time before the availability of NWP information. The accuracy of NWP forecasts has advanced sufficiently that pure statistical forecasting is used in practical settings only for very short lead times or for very long lead times.

Very often classical forecast products are based on multiple regression equations of the kinds described in Sections 6.2 and 6.3. These statistical forecasts are objective in the sense that a particular set of inputs or predictors will always produce the same forecast for the predictand, once the forecast equation has been developed. However, many subjective decisions necessarily go into the development of the forecast equations.

The construction of a classical statistical forecasting procedure follows from a straightforward implementation of the ideas presented in the previous sections of this chapter. Required developmental data consist of past values of the predictand to be forecast, and a matching collection of potential predictors whose values will be known prior to the forecast time. A forecasting procedure is developed using this set of historical data, which can then be used to forecast future values of the predictand on the basis of future observations of the predictor variables. It is a characteristic of classical statistical weather forecasting that the time lag is built directly into the forecast equation through the time-lagged relationships between the predictors and the predictand.

For lead times up to a few hours, purely statistical forecasts still find productive use. This short-lead forecasting niche is known as *nowcasting*. NWP-based forecasts are not practical for nowcasting because of the delays introduced by the processes of gathering weather observations, data assimilation (calculation of initial conditions for the NWP model), the actual running of the forecast model, and the post-processing and dissemination of the results. One very simple statistical approach that can produce competitive nowcasts is conditional climatology; that is, historical statistics subsequent to (conditional on) analogous weather situations in the past. The result could be a conditional frequency distribution for the predictand, or a single-valued forecast corresponding to the expected value (mean) of that conditional distribution. A more sophisticated approach is to construct a regression equation to forecast a few hours ahead. For example, Vislocky and Fritsch (1997) compare these two approaches for forecasting airport ceiling and visibility at lead times of one, three, and six hours.

At lead times beyond perhaps 10 days to two weeks, statistical forecasts are again competitive with dynamical NWP forecasts. At these long lead times the sensitivity of NWP models to the unavoidable small errors in their initial conditions, described in Section 6.6, renders explicit forecasting of specific weather events impossible. Although long-lead forecasts for seasonally averaged quantities currently are made using dynamical NWP models (e.g., Barnston *et al.* 2003), comparable or even better predictive accuracy at substantially lower cost is still obtained through statistical methods (Anderson *et al.* 1999; Barnston and Smith 1996; Barnston *et al.* 1999; Gershunov and Cayan 2003; Landsea and Knaff 2000; Moura and Hastenrath 2004). Often the predictands in these seasonal forecasts are spatial patterns, and so the forecasts involve multivariate statistical methods that are more elaborate than those described in Sections 6.2 and 6.3 (e.g., Barnston 1994; Mason and Mimmack 2002; Ward and Folland 1991; see Sections 12.2.3 and 13.4). However, regression methods are still appropriate and useful for single-valued predictands. For example, Knaff and Landsea (1997) used ordinary least-squares regression for seasonal forecasts of tropical sea-surface temperatures using observed sea-surface temperatures as predictors, and Elsner and Schmertmann (1993) used Poisson regression for seasonal prediction of hurricane numbers.

EXAMPLE 6.9   A Set of Classical Statistical Forecast Equations

The flavor of classical statistical forecast methods can be appreciated by looking at the NHC-67 procedure for forecasting hurricane movement (Miller *et al.* 1968). This relatively simple set of regression equations was used as part of the operational suite of forecast models at the U.S. National Hurricane Center until 1988 (Sheets 1990). Since hurricane movement is a vector quantity, each forecast consists of two equations: one for northward movement and one for westward movement. The two-dimensional forecast displacement is then computed as the vector sum of the northward and westward displacements.

The predictands were stratified according to two geographical regions: north and south of 27.5°N latitude. That is, separate forecast equations were developed to predict storms on either side of this latitude, on the basis of the subjective experience of the developers regarding the responses of hurricane movement to the larger-scale flow. Separate forecast equations were also developed for slow vs. fast storms. The choice of these two stratifications was also made subjectively, on the basis of the experience of the developers. Separate equations are also needed for each forecast projection (0–12 h, 12–24 h, 24–36 h, and 36–48 h, and 48–72 h), yielding a total of 2 (displacement directions) × 2 (regions) × 2 (speeds) × 5 (projections) = 40 separate regression equations in the NHC-67 package.

The available developmental data set consisted of 236 northern cases and 224 southern cases. Candidate predictor variables were derived primarily from 1000-, 700-, and 500-mb heights at each of 120 gridpoints in a $5° \times 5°$ coordinate system that follows the storm. Predictors derived from these $3 \times 120 = 360$ geopotential height predictors, including 24-h height changes at each level, geostrophic winds, thermal winds, and Laplacians of the heights, were also included as candidate predictors. Additionally, two persistence predictors, observed northward and westward storm displacements in the previous 12 hours were included.

With vastly more potential predictors than observations, some screening procedure is clearly required. Here forward selection was used, with the (subjectively determined) stopping rule that no more than 15 predictors would be in any equation, and new predictors would be included that increased the regression R^2 by at least 1%. This second criterion was apparently sometimes relaxed for regressions with few predictors.

Table 6.6 shows the results for the 0–12 h westward displacement of slow southern storms in NHC-67. The five predictors are shown in the order they were chosen by the forward selection procedure, together with the R^2 value achieved on the developmental data at each step. The coefficients are those for the final ($K = 5$) equation. The most important single predictor was the persistence variable (P_x), reflecting the tendency of hurricanes to change speed and direction fairly slowly. The 500 mb height at a point north and west of the storm (Z_{37}) corresponds physically to the steering effects of midtropospheric flow on hurricane movement. Its coefficient is positive, indicating a tendency for westward storm displacement given relatively high heights to the northwest, and slower or eastward (negative westward) displacement of storms located southwest of 500 mb troughs. The final two or three predictors appear to improve the regression only marginally—the predictor Z_3 increases the R^2 by less than 1%—and we would suspect that the $K = 2$ or $K = 3$ predictor models might have been chosen if cross validation had been computationally feasible for the developers, and might have been equally accurate for independent data. Remarks in Neumann *et al.* (1977) concerning the fitting of the similar NHC-72 regressions, in relation to Figure 6.16, are also consistent with the idea that the equation represented in Table 6.6 may have been overfit. ◊

TABLE 6.6   Regression results for the NHC-67 hurricane model for the 0–12 h westward displacement of slow southern zone storms, indicating the order in which the predictors were selected and the resulting R^2 at each step. The meanings of the symbols for the predictors are P_X = westward displacement in the previous 12 h, Z_{37} = 500 mb height at the point 10° north and 5° west of the storm, P_Y = northward displacement in the previous 12 h, Z_3 = 500 mb height at the point 20° north and 20° west of the storm, and P_{51} = 1000 mb height at the point 5° north and 5° west of the storm. Distances are in nautical miles, and heights are in meters. From Miller *et al.* (1968).

Predictor	Coefficient	Cumulative R^2
Intercept	−2909.5	—
P_X	0.8155	79.8%
Z_{37}	0.5766	83.7%
P_Y	−0.2439	84.8%
Z_3	−0.1082	85.6%
P_{51}	−0.3359	86.7%

6.5.2 *Perfect Prog and MOS*

Pure classical statistical weather forecasts for projections in the range of a few days are generally no longer employed, since current dynamical NWP models allow more accurate forecasts at this time scale. However, two types of statistical weather forecasting are in use that can improve on aspects of conventional NWP forecasts, essentially by post-processing the raw NWP output. Both of these methods use large multiple regression equations in a way that is analogous to the classical approach, so that many of the same technical considerations pertaining to equation fitting apply. The differences between these two approaches and classical statistical forecasting have to do with the range of available predictor variables. In addition to conventional predictors such as current meteorological observations, the date, or climatological values of a particular meteorological element, predictor variables taken from the output of the NWP models are also used.

There are three reasons why statistical reinterpretation of dynamical NWP output is useful for practical weather forecasting:

- There are important differences between the real world and its representation in NWP models. Figure 6.19 illustrates some of these. The NWP models necessarily simplify and homogenize surface conditions, by representing the world as an array of gridpoints to which the NWP output pertains. As implied by Figure 6.19, small-scale effects (e.g., of topography or small bodies of water) important to local weather may not be included in the NWP model. Also, locations and variables for which forecasts are needed may not be represented explicitly by the NWP model. However, statistical relationships can be developed between the information provided by the NWP models and desired forecast quantities to help alleviate these problems.
- The NWP models are not complete and true representations of the workings of the atmosphere, and their forecasts are subject to errors. To the extent that these errors are systematic, statistical forecasts based on the NWP information can compensate and correct forecast biases.
- The NWP models are deterministic. That is, even though the future state of the weather is inherently uncertain, a single NWP integration is capable of producing only a single forecast for any meteorological element, given a set of initial model conditions. Using NWP information in conjunction with statistical methods allows quantification and expression of the uncertainty associated with different forecast situations. In particular, it is possible to derive probability forecasts, using REEP or logistic regression, using predictors taken from a deterministic NWP integration.

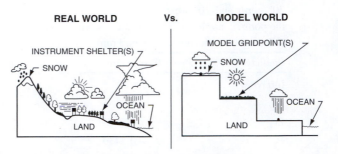

FIGURE 6.19   Cartoon illustration of differences between the real world, and the world as represented by numerical weather prediction models. From Karl *et al.* (1989).

The first statistical approach to be developed for taking advantage of the deterministic dynamical forecasts from NWP models (Klein *et al.* 1959) is called "perfect prog," which is short for perfect prognosis. As the name implies, the perfect-prog technique makes no attempt to correct for possible NWP model errors or biases, but takes their forecasts for future atmospheric variables at face value—assuming them to be perfect.

Development of perfect-prog regression equations is similar to the development of classical regression equations, in that observed predictors are used to specify observed predictands. That is, only historical climatological data is used in the development of a perfect-prog forecasting equation. The primary difference between development of classical and perfect-prog equations is in the time lag. Classical equations incorporate the forecast time lag by relating predictors available before the forecast must be issued (say, today) to values of the predictand to be observed at some later time (say, tomorrow). Perfect-prog equations do not incorporate any time lag. Rather, simultaneous values of predictors and predictands are used to fit the regression equations. That is, the equations specifying tomorrow's predictand are developed using tomorrow's predictor values.

At first, it might seem that this would not be a productive approach to forecasting. Tomorrow's 1000-850 mb thickness may be an excellent predictor for tomorrow's temperature, but tomorrow's thickness will not be known until tomorrow. However, in implementing the perfect-prog approach, it is the NWP forecasts of the predictors (e.g., the NWP forecast for tomorrow's thickness) that are substituted into the regression equation as predictor values. Therefore, the forecast time lag in the perfect-prog approach is contained entirely in the NWP model. Of course quantities not forecast by the NWP model cannot be included as potential predictors unless they will be known today. If the NWP forecasts for tomorrow's predictors really are perfect, the perfect-prog regression equations should provide very good forecasts.

The Model Output Statistics (MOS) approach (Glahn and Lowry 1972; Carter *et al.* 1989) is the second, and usually preferred, approach to incorporating NWP forecast information into traditional statistical weather forecasts. Preference for the MOS approach derives from its capacity to include directly in the regression equations the influences of specific characteristics of different NWP models at different projections into the future.

Although both the MOS and perfect-prog approaches use quantities from NWP output as predictor variables, the two approaches use the information differently. The perfect-prog approach uses the NWP forecast predictors only when making forecasts, but the MOS approach uses these predictors in both the development and implementation of the forecast equations. Think again in terms of today as the time at which the forecast must be made and tomorrow as the time to which the forecast pertains. MOS regression equations are developed for tomorrow's predictand using NWP forecasts for tomorrow's values of the predictors. The true values of tomorrow's predictors are unknown today, but the NWP forecasts for those quantities are known today. For example, in the MOS approach, one important predictor for tomorrow's temperature could be tomorrow's 1000-850 mb thickness *as forecast today* by a particular NWP model. Therefore, to develop MOS forecast equations it is necessary to have a developmental data set including historical records of the predictand, together with archived records of the forecasts produced by the NWP model for the same days on which the predictand was observed.

In common with the perfect-prog approach, the time lag in MOS forecasts is incorporated through the NWP forecast. Unlike perfect prog, the implementation of a MOS forecast equation is completely consistent with its development. That is, in both development and implementation, the MOS statistical forecast for tomorrow's predictand is made using the dynamical NWP forecast for tomorrow's predictors, which are available today. Also unlike the perfect-prog approach, separate MOS forecast equations must be developed for different forecast projections. This is because the error characteristics of the NWP forecasts are different at different projections, producing, for example, different statistical relationships between observed temperature and forecast thicknesses for 24 h versus 48 h in the future.

The classical, perfect-prog, and MOS approaches are all based on multiple regression, exploiting correlations between a predictand and available predictors. In the classical approach it is the correlations between today's values of the predictors and tomorrow's predictand that forms the basis of the forecast. For the perfect-prog approach it is the simultaneous correlations between today's values of both predictand and predictors that are the statistical basis of the prediction equations. In the case of MOS forecasts, the prediction equations are constructed on the basis of correlations between NWP forecasts as predictor variables, and the subsequently observed value of tomorrow's predictand.

These distinctions can be expressed mathematically, as follows. In the classical approach, the forecast predictand at some future time, t, is expressed in the regression function f_C using a vector of (i.e., multiple) predictor variables, $\mathbf{x}_0$ according to

$$\hat{y}_t = f_C(\mathbf{x}_0). \qquad (6.35)$$

The subscript 0 on the predictors indicates that they pertain to values observed at or before the time that the forecast must be formulated, which is earlier than the time t to which the forecast pertains. This equation emphasizes that the forecast time lag is built into the regression. It is applicable to both to the development and implementation of a Classical statistical forecast equation.

By contrast, the perfect-prog (PP) approach operates differently for development versus implementation of the forecast equation, and this distinction can be expressed as

$$\hat{y}_0 = f_{PP}(\mathbf{x}_0) \quad \text{in development,} \qquad (6.36a)$$

and

$$\hat{y}_t = f_{PP}(\mathbf{x}_t) \quad \text{in implementation.} \qquad (6.36b)$$

The perfect-prog regression function, f_{PP} is the same in both cases, but it is developed entirely with observed data having no time lag with respect to the predictand. In implementation it operates on forecast values of the predictors for the future time t, as obtained from the NWP model.

Finally, the MOS approach uses the same equation in development and implementation,

$$\hat{y}_t = f_{MOS}(\mathbf{x}_t). \qquad (6.37)$$

It is derived using the NWP forecast predictors $\mathbf{x}_t$, pertaining to the future time t (but known at time 0 when the forecast will be issued), and is implemented in the same way. In common with the perfect-prog approach, the time lag is carried by the NWP forecast, not the regression equation.

Given the comparatively high accuracy of dynamical NWP forecasts for weather forecasts, classical statistical forecasting is not competitive for forecasts in the range of a few hours to perhaps two weeks. Since the perfect-prog and MOS approaches both draw on NWP information, it is worthwhile to compare their advantages and disadvantages.

There is nearly always a large developmental sample for perfect-prog equations, since these are fit using only historical climatological data. This is an advantage over the MOS approach, since fitting MOS equations requires an archived record of forecasts from the same NWP model that will ultimately be used as input to the MOS equations. Typically, several years of archived NWP forecasts are required to develop a stable set of MOS forecast equations (Jacks *et al.* 1990). This requirement can be a substantial limitation, because the dynamical NWP models are not static. Rather, these models regularly undergo changes aimed at improving their performance. Minor changes in the NWP model leading to reductions in the magnitudes of random errors will not substantially degrade the performance of a set of MOS equations (e.g., Erickson *et al.* 1991). However, modifications to the NWP model that change—even substantially reduce—systematic errors in the NWP model will require redevelopment of accompanying MOS forecast equations. Since it is a change in the NWP model that will have necessitated the redevelopment of a set of MOS forecast equations, it is likely that a sufficiently long developmental sample of predictors from the improved NWP model will not be immediately available. By contrast, since the perfect-prog equations are developed without any NWP information, changes in the NWP models should not require changes in the perfect-prog regression equations. Furthermore, improving either the random or systematic error characteristics of the NWP model should improve the statistical forecasts produced by a perfect-prog equation.

Similarly, the same perfect-prog regression equations in principle can be used with any NWP model, or for any forecast projection provided by a single NWP model. Since the MOS equations are tuned to the particular error characteristics of the model for which they were developed, different MOS equations will, in general, be required for use with different dynamical NWP models. Analogously, since the error characteristics of the NWP model change with increasing projection, different MOS equations are required for forecasts of the same atmospheric variable for different projection times into the future. Note, however, that potential predictors for a perfect-prog equation must be variables that are well predicted by the NWP model with which they will be used. It may be possible to find an atmospheric predictor variable that relates closely to a predictand of interest, but which is badly forecast by a particular NWP model. Such a variable might well be selected for inclusion in a perfect-prog equation on the basis of the relationship of its observed values to the predictand, but would be ignored in the development of a MOS equation if the NWP forecast of that predictor bore little relationship to the predictand.

The MOS approach to statistical forecasting has two advantages over the perfect-prog approach that make MOS the method of choice when practical. The first of these is that model-calculated, but unobserved, quantities such as vertical velocity can be used as predictors. However, the dominating advantage of MOS over perfect prog is that systematic errors exhibited by the dynamical NWP model are accounted for in the process of developing the MOS equations. Since the perfect-prog equations are developed without reference to the characteristics of the NWP model, they cannot account for or correct any type of NWP forecast error. The MOS development procedure allows compensation for these systematic errors when forecasts are constructed. Systematic errors include such problems as progressive cooling or warming biases in the NWP model with increasing forecast projection, a tendency for modeled synoptic features to move too slowly or too quickly in the NWP model, and even the unavoidable decrease in forecast accuracy at increasing projections.

The compensation for systematic errors in a dynamical NWP model that is accomplished by MOS forecast equations is easiest to see in relation to a simple bias in an

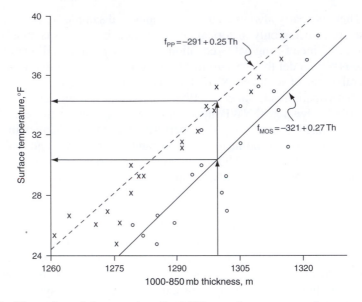

FIGURE 6.20   Illustration of the capacity of a MOS equation to correct for systematic bias in a hypothetical NWP model. The x's represent observed, and the circles represent NWP-forecast 1000-850 mb thicknesses, in relation to hypothetical surface temperatures. The bias in the NWP model is such that the forecast thicknesses are too large by about 15 m, on average. The MOS equation (solid line) is calibrated for this bias, and produces a reasonable temperature forecast (lower horizontal arrow) when the NWP forecast thickness is 1300 m. The perfect-prog equation (dashed line) incorporates no information regarding the attributes of the NWP model, and produces a surface temperature forecast (upper horizontal arrow) that is too warm as a consequence of the thickness bias.

important predictor. Figure 6.20 illustrates a hypothetical case, where surface temperature is to be forecast using the 1000-850 mb thickness. The x's in the figure represent the (unlagged, or simultaneous) relationship of a set of observed thicknesses with observed temperatures, and the circles represent the relationship between NWP-forecast thicknesses with the same temperature data. As drawn, the hypothetical NWP model tends to forecast thicknesses that are too large by about 15 m. The scatter around the perfect-prog regression line derives from the fact that there are influences on surface temperature other than those captured by the 1000-850 mb thickness. The scatter around the MOS regression line is greater, because in addition it reflects errors in the NWP model.

The observed thicknesses (x's) in Figure 6.20 appear to forecast the simultaneously observed surface temperatures reasonably well, yielding an apparently good perfect-prog regression equation (dashed line). The relationship between forecast thickness and observed temperature represented by the MOS equation (solid line) is substantially different, because it includes the tendency for this NWP model to systematically overforecast thickness. If the NWP model produces a thickness forecast of 1300 m (vertical arrows), the MOS equation corrects for the bias in the thickness forecast and produces a reasonable temperature forecast of about 30° F (lower horizontal arrow). Loosely speaking, the MOS knows that when this NWP model forecasts 1300 m, a more reasonable expectation for the true future thickness is closer to 1285 m, which in the climatological data (x's) corresponds to a temperature of about 30° F. The perfect-prog equation, on the other hand, operates under the assumption that the NWP model will forecast the future thickness perfectly. It therefore yields a temperature forecast that is too warm (upper horizontal arrow) when supplied with a thickness forecast that is too large.

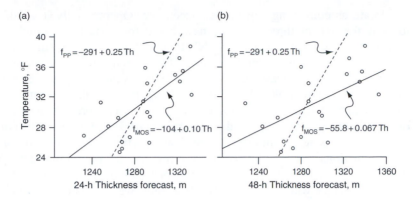

FIGURE 6.21   Illustration of the capacity of a MOS equation to account for the systematic tendency of NWP forecasts to become less accurate at longer projections. The points in these panels are simulated thickness forecasts, constructed from the x's in Figure 6.20 by adding random errors to the thickness values. As the forecast accuracy degrades at longer projections, the perfect-prog equation (dashed line, reproduced from Figure 6.20) is increasingly overconfident, and tends to forecast extreme temperatures too frequently. At longer projections (b) the MOS equations increasingly provide forecasts near the average temperature (30.8°F in this example).

A more subtle systematic error exhibited by all NWP models is the degradation of forecast accuracy at increasing projections. The MOS approach accounts for this type of systematic error as well. The situation is illustrated in Figure 6.21, which is based on the hypothetical observed data in Figure 6.20. The panels in Figure 6.21 simulate the relationships between forecast thicknesses from an unbiased NWP model at 24- and 48-h projections and the surface temperature, and have been constructed by adding random errors to the observed thickness values (x's) in Figure 6.20. These random errors exhibit $\sqrt{\text{MSE}} = 20\,\text{m}$ for the 24-h projection and $\sqrt{\text{MSE}} = 30\,\text{m}$ at the 48-h projection. The increased scatter of points for the simulated 48-h projection illustrates that the regression relationship is weaker when the NWP model is less accurate.

The MOS equations (solid lines) fit to the two sets of points in Figure 6.21 reflect the progressive loss of predictive accuracy in the NWP model at longer lead times. As the scatter of points increases at longer projections the slopes of the MOS forecast equations becomes more horizontal, leading to temperature forecasts that are more like the climatological temperature, on average. This characteristic is reasonable and desirable, since as the NWP model provides less information about the future state of the atmosphere, temperature forecasts differing substantially from the climatological average temperature are less justified. In the limit of an arbitrarily long forecast projection, an NWP model will really provide no more information than will the climatological value of the predictand, the slope of the corresponding MOS equation would be zero, and the appropriate temperature forecast consistent with this (lack of) information would simply be the climatological average temperature. Thus, it is sometimes said that MOS converges toward the climatology. By contrast, the perfect-prog equation (dashed lines, reproduced from Figure 6.20) take no account of the decreasing accuracy of the NWP model at longer projections, and continue to produce temperature forecasts as if the thickness forecasts were perfect. Figure 6.21 emphasizes that the result is overconfident temperature forecasts, with both very warm and very cold temperatures forecast much too frequently.

Although MOS postprocessing of NWP output is strongly preferred to perfect prog and to the raw NWP forecasts, the pace of changes made to NWP models continues to accelerate as computing capabilities accelerate. Operationally it would not be practical to wait for two or three years of new NWP forecasts to accumulate before deriving a new MOS system, even if the NWP model were to remain static for that period of time. One option for maintaining MOS systems in the face of this reality is to retrospectively re-forecast weather for previous years using the current updated NWP model (Jacks *et al.* 1990; Hamill *et al.* 2004). Because daily weather data are typically strongly autocorrelated, the reforecasting process is more efficient if several days are omitted between the reforecast days (Hamill *et al.* 2004). Even if the computing capacity to reforecast is not available, a significant portion of the benefit of fully calibrated MOS equations can be achieved using a few months of training data (Mao *et al.* 1999; Neilley *et al.* 2002). Alternative approaches include using longer developmental data records together with whichever NWP version was current at the time, and weighting the more recent forecasts more strongly. This can be done either by downweighting forecasts made with older NWP model versions (Wilson and Valée 2002, 2003), or by gradually downweighting older data, usually through an approach called the Kalman filter (Crochet 2004; Galanis and Anadranistakis 2002; Homleid 1995; Kalnay 2003; Mylne *et al.* 2002b; Valée *et al.* 1996), although other approaches are also possible (Yuval and Hsieh 2003).

6.5.3 *Operational MOS Forecasts*

Interpretation and extension of dynamical NWP forecasts using MOS systems has been implemented at a number of national meteorological centers, including those in the Netherlands (Lemcke and Kruizinga 1988), Britain (Francis *et al.* 1982), Italy (Conte *et al.* 1980), China (Lu 1991), Spain (Azcarraga and Ballester 1991), Canada (Brunet *et al.* 1988), and the U.S. (Carter *et al.* 1989), among others.

MOS forecast products can be quite extensive, as illustrated by Table 6.7, which shows the FOUS14 panel (Dallavalle *et al.* 1992) of 12 March 1993 for Albany, New York. This is one of hundreds of such panels for locations in the United States, for which these forecasts are issued twice daily and posted on the Web by the U.S. National Weather Service. Forecasts for a wide variety of weather elements are provided, at projections up to 60 h and at intervals as close as 3 h. After the first few lines indicating the dates and times (UTC) are forecasts for daily maximum and minimum temperatures; temperatures, dew point temperatures, cloud coverage, wind speed, and wind direction at 3 h intervals; probabilities of measurable precipitation at 6- and 12-h intervals; forecasts for precipitation amount; thunderstorm probabilities; precipitation type; freezing and frozen precipitation probabilities; snow amount; and ceiling, visibility, and obstructions to visibility. Similar panels, for several different NWP models, are also produced and posted.

The MOS equations underlying the FOUS14 forecasts are seasonally stratified, usually with separate forecast equations for the warm season (April through September) and cool season (October through March). This two-season stratification allows the MOS forecasts to incorporate different relationships between predictors and predictands at different times of the year. A finer stratification (three-month seasons, or separate month-by-month equations) would probably be preferable if sufficient developmental data were available.

The forecast equations for all elements except temperatures, dew points, and winds are regionalized. That is, developmental data from groups of nearby and climatically similar

TABLE 6.7 Example MOS forecasts produced by the U.S. National Meteorological Center for Albany, New York, shortly after 0000 UTC on 12 March 1993. A variety of weather elements are forecast, at projections up to 60h and at intervals as close as 3 h. Detailed explanations of all the forecast elements are given in Dallavalle *et al.* (1992).

FOUS14 KWBC 120000
ALB EC NGM MOS GUIDANCE 3/12/93 0000 UTC

DAY	/MAR 12						/MAR 13								/MAR 14				
HOUR	06	09	12	15	18	21	00	03	06	09	12	15	18	21	00	03	06	09	12
MX/MN							32				14				33				23
TEMP	20	16	14	22	30	31	23	18	17	17	19	28	31	32	30	26	24	25	25
DEWPT	10	7	4	3	3	4	7	7	9	10	12	18	21	23	26	26	24	25	25
CLDS	BK	SC	CL	CL	SC	SC	CL	SC	SC	OV	OV	OV	OV	OV	OV	OV	OV	OV	OV
WDIR	30	30	31	31	30	30	31	32	00	12	08	06	04	03	36	34	34	33	32
WSPD	08	08	07	09	09	07	03	01	00	03	03	05	10	18	13	23	32	29	25
POP06			0		0		0		7		42		72		98		100		92
POP12							0				37				100				100
QPF			0/		0/		0/0		0/		1/1		4/		5/6		2/		2/3
TSV06			0/3		0/3		0/3		1/4		0/6		4/0		14/0		21/1		23/0
TSV12					0/4				1/5				3/5				26/0		
PTYPE	S			S	S	S	S	S	S	S	S	S	S	S					
POZP	0	3	3	0	1	0	2	0	0	1	0	3	3	4					
POSN	94	100	97	99	100	99	100	98	100	99	98	96	96	98	90	86	95		
SNOW			0/		0/		0/0		0/		1/1		2/		4/6		2/		2/4
CIG	7	7	7	7	7	7	7	7	7	6	5	4	3						
VIS	5	5	5	5	5	5	5	5	4	4	2	1	1						
OBVIS	N	N	N	N	N	N	N	N	N	N	N	N	F	F					

stations were composited in order to increase the sample size when deriving the forecast equations. For each regional group, then, forecasts are made with the same equations and the same regression coefficients. This does not mean that the forecasts for all the stations in the group are the same, however, since interpolation of the NWP output to the different forecast locations yields different values for the predictors from the NWP model. Some of the MOS equations also contain predictors representing local climatological values, which introduces further differences in the forecasts for the nearby stations. Regionalization is especially valuable for producing good forecasts of rare events.

In order to enhance consistency among the forecasts, some of the MOS equations are developed simultaneously for several of the forecast elements. This means that the same predictor variables, although with different regression coefficients, are forced into prediction equations for related predictands in order to enhance the consistency of the forecasts. For example, it would be physically unreasonable and clearly undesirable for the forecast dew point to be higher than the forecast temperature. To help ensure that such inconsistencies appear in the forecasts as rarely as possible, the MOS equations for maximum temperature, minimum temperature, and the 3-h temperatures and dew points all contain the same predictor variables. Similarly, the four groups of forecast equations for wind speeds and directions, the 6- and 12-h precipitation probabilities, the 6- and 12-h thunderstorm probabilities, and the probabilities for precipitation types were also developed simultaneously to enhance their consistency.

Because MOS forecasts are made for a large number of locations, it is possible to view them as maps, which are also posted on the Web. Some of these maps display selected quantities from the MOS panels such as the one shown in Table 6.7. Figure 6.22 shows a forecast map for a predictand not currently included in the tabular forecast products: probabilities of at least 0.25 in. of (liquid-equivalent) precipitation, for the 24-h period ending 31 May 2004, at 72 h to 96 h lead time.

FIGURE 6.22 Example MOS forecasts in map form. The predictand is the probability of at least 0.25 in. of precipitation during the 24-h period between 72 and 96 h after forecast initialization. The contour interval is 0.10. From www.nws.noaa.gov/mdl.

6.6 Ensemble Forecasting

6.6.1 Probabilistic Field Forecasts

In Section 1.3 it was asserted that dynamical chaos ensures that the future behavior of the atmosphere cannot be known with certainty. Because the atmosphere can never be completely observed, either in terms of spatial coverage or accuracy of measurements, a fluid-dynamical numerical weather prediction (NWP) model of its behavior will always begin calculating forecasts from a state at least slightly different from that of the real atmosphere. These models (and other nonlinear dynamical systems, including the real atmosphere) exhibit the property that solutions (forecasts) started from only slightly different initial conditions will yield quite different results for projections sufficiently far into the future. For synoptic-scale predictions using NWP models, sufficiently far is a matter of days or (at most) weeks, and for mesoscale forecasts this window is even shorter, so that the problem of sensitivity to initial conditions is of practical importance.

NWP models are the mainstay of weather forecasting, and the inherent uncertainty of their results must be appreciated and quantified if their information is to be utilized most effectively. For example, a single deterministic forecast of the hemispheric 500 mb height field two days in the future is only one member of an essentially infinite collection of 500 mb height fields that could plausibly occur. Even if this deterministic forecast is the best possible single forecast that can be constructed, its usefulness and value will be enhanced if aspects of the probability distribution of which it is a member can be estimated and communicated. This is the problem of probabilistic field forecasting.

Probability forecasts for a scalar quantity, such as a maximum daily temperature at a single location, are relatively straightforward. Many aspects of producing such forecasts have been discussed in this chapter, and the uncertainty of such forecasts can be expressed using univariate probability distributions of the kind described in Chapter 4. However, producing a probability forecast for a field, such as the hemispheric 500 mb heights, is a much bigger and more difficult problem. A single atmospheric field might be represented by the values of thousands of 500 mb heights at regularly spaced locations, or gridpoints (e.g., Hoke *et al.* 1989). Construction of forecasts including probabilities for all these heights and their relationships (e.g., correlations) with heights at the other gridpoints is a very big task, and in practice only approximations to their complete probability description have been achieved. Expressing and communicating aspects of the large amounts of information in a probabilistic field forecast pose further difficulties.

6.6.2 Stochastic Dynamical Systems in Phase Space

Much of the conceptual basis for probabilistic field forecasting is drawn from Gleeson (1961, 1970), who noted analogies to quantum and statistical mechanics; and Epstein (1969c), who presented both theoretical and practical approaches to the problem of uncertainty in (simplified) NWP forecasts. In this approach, which Epstein called stochastic dynamic prediction, the physical laws governing the motions and evolution of the atmosphere are regarded as deterministic. However, in practical problems the equations that describe these laws must operate on initial values that are not known with certainty, and which can therefore be described by a joint probability distribution. Conventional deterministic forecasts use the governing equations to describe the future evolution of a single initial state that is regarded as the true initial state. The idea behind stochastic dynamic forecasts is to allow the deterministic governing equations to operate on the probability distribution describing the uncertainty about the initial state of the atmosphere. In principle this process yields, as forecasts, probability distributions describing uncertainty about the future state of the atmosphere. (But actually, since NWP models are not perfect representations of the real atmosphere, their imperfections further contribute to forecast uncertainty.)

Visualizing or even conceptualizing the initial and forecast probability distributions is difficult, especially when they involve joint probabilities pertaining to large numbers of variables. This visualization, or conceptualization, is most commonly and easily done using the concept of phase space. A phase space is a geometrical representation of the hypothetically possible states of a dynamical system, where each of the coordinate axes defining this geometry pertains to one of the forecast variables of the system.

For example, a simple dynamical system that is commonly encountered in textbooks on physics or differential equations is the swinging pendulum. The state of a pendulum can be completely described by two variables: its angular position and its velocity. At the extremes of the pendulum's arc, its angular position is maximum (positive or negative) and its velocity is zero. At the bottom of its arc the angular position of the swinging pendulum is zero and its speed is maximum. When the pendulum finally stops, both its angular position and velocity are zero. Because the motions of a pendulum can be described by two variables, its phase space is two-dimensional. That is, its phase space is a phase-plane. The changes through time of the state of the pendulum system can be described by a path, known as an orbit, or trajectory, on this phase-plane.

Figure 6.23 shows the trajectory of a hypothetical pendulum in its phase space. That is, this figure is a graph in phase space of the motions of a pendulum, and their changes through time. The trajectory begins at the single point corresponding to the initial state of the pendulum: it is dropped from the right with zero initial velocity. As it drops it accelerates and acquires leftward velocity, which increases until the pendulum passes through the vertical position. The pendulum then decelerates, slowing until it stops at its maximum left position. As the pendulum drops again it moves to the right, stopping

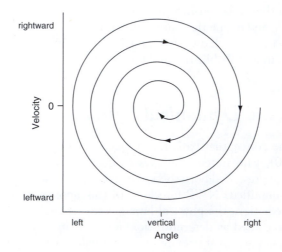

FIGURE 6.23   Trajectory of a swinging pendulum in its two-dimensional phase space, or phase-plane. The pendulum has been dropped from a position on the right, from which point it swings in arcs of decreasing angle. Finally, it slows to a stop, with zero velocity in the vertical position.

short of its initial position because of friction. The pendulum continues to swing back and forth until it finally comes to rest in the vertical position.

The phase space of an atmospheric model has many more dimensions than that of the pendulum system. Epstein (1969c) considered a highly simplified model of the atmosphere having only eight variables. Its phase space was therefore eight-dimensional, which is small but still much too big to imagine explicitly. The phase spaces of operational NWP models typically have millions of dimensions, each corresponding to one of the millions of variables ((horizontal gridpoints) × (vertical levels) × (prognostic variables)) represented. The trajectory of the atmosphere or a model of the atmosphere is also qualitatively more complicated than that of the pendulum because it is not attracted to a single point in the phase space, as is the trajectory in Figure 6.23. In addition, the pendulum dynamics do not exhibit the sensitivity to initial conditions that has come to be known as chaotic behavior. Releasing the pendulum slightly further to the right or left relative to Figure 6.23, or with a slight upward or downward push, would produce a very similar trajectory that would track the spiral in Figure 6.23 very closely, and arrive at the same place in the center of the diagram at nearly the same time. The corresponding behavior of the atmosphere, or of a realistic mathematical model of it, would be quite different. Nevertheless, the changes in the flow in a model atmosphere through time can still be imagined abstractly as a trajectory through its multidimensional phase space.

The uncertainty about the initial state of the atmosphere, from which an NWP model is initialized, can be conceived of as a probability distribution in its phase space. In a two-dimensional phase space like the one shown in Figure 6.23, we might imagine a bivariate normal distribution (Section 4.4.2), with ellipses of constant probability describing the spread of plausible initial states around the best guess, or mean value. Alternatively, we can imagine a cloud of points around the mean value, whose density decreases with distance from the mean. In a three-dimensional phase space the distribution might be imagined as a cigar- or egg-shaped cloud of points, again with density decreasing with distance from the mean value. Higher-dimensional spaces cannot be visualized explicitly, but probability distributions within them can be imagined by analogy.

In concept, a stochastic dynamic forecast moves the probability distribution of the initial state through the phase space as the forecast is advanced in time, according to the laws of fluid dynamics represented in the NWP model equations. However, trajectories in the phase space of a NWP model (or of the real atmosphere) are not nearly as smooth and regular as the pendulum trajectory shown in Figure 6.23. As a consequence, the shape of the initial distribution is stretched and distorted as the forecast is advanced. It will also become more dispersed at longer forecast projections, reflecting the increased uncertainty of forecasts further into the future. Furthermore, these trajectories are not attracted to a single point as are pendulum trajectories in the phase space of Figure 6.23. Rather, the attractor, or set of points in the phase space that can be visited after an initial transient period, is a rather complex geometrical object. A single point in phase space corresponds to a unique weather situation, and the collection of these possible points that constitutes the attractor can be interpreted as the climate of the NWP model. This set of allowable states occupies only a small fraction of the (hyper-) volume of the phase space, as many combinations of atmospheric variables will be physically impossible or dynamically inconsistent.

Equations describing the evolution of the initial-condition probability distribution can be derived through introduction of a continuity, or conservation equation for probability (Ehrendorfer 1994; Gleeson 1970). However, the dimensionality of phase spaces for problems of practical forecasting interest are too large to allow direct solution of these equations. Epstein (1969c) introduced a simplification that rests on a restrictive assumption about the shapes of probability distributions in phase space, which is expressed in terms of the moments of the distribution. However, even this approach is impractical for all but the simplest atmospheric models.

6.6.3 Ensemble Forecasts

The practical solution to the analytic intractability of sufficiently detailed stochastic dynamic equations is to approximate these equations using Monte-Carlo methods, as proposed by Leith (1974), and now called ensemble forecasting. These Monte-Carlo solutions bear the same relationship to stochastic dynamic forecast equations as the Monte-Carlo resampling tests described in Section 5.3 bear to the analytical tests they approximate. (Recall that resampling tests are appropriate and useful in situations where the underlying mathematics are difficult or impossible to evaluate analytically.) Reviews of current operational use of the ensemble forecasting approach can be found in Cheung (2001) and Kalnay (2003).

The ensemble forecast procedure begins in principle by drawing a finite sample from the probability distribution describing the uncertainty of the initial state of the atmosphere. Imagine that a few members of the point cloud surrounding the mean estimated atmospheric state in phase space are picked randomly. Collectively, these points are called the ensemble of initial conditions, and each represents a plausible initial state of the atmosphere consistent with the uncertainties in observation and analysis. Rather than explicitly predicting the movement of the entire initial-state probability distribution through phase space, that movement is approximated by the collective trajectories of the ensemble of sampled initial points. It is for this reason that the Monte-Carlo approximation to stochastic dynamic forecasting is known as ensemble forecasting. Each of the points in the initial ensemble provides the initial conditions for a separate run of the NWP model. At this initial time, all the ensemble members are very similar to each other. The distribution in phase space of this ensemble of points after the forecasts have been advanced to a future time then approximates how the full true initial probability distribution would have been transformed by the governing physical laws that are expressed in the NWP model.

Figure 6.24 illustrates the nature of ensemble forecasting in an idealized two-dimensional phase space. The circled X in the initial-time ellipse represents the single best initial value, from which a conventional deterministic NWP forecast would begin. Recall that, for a real model of the atmosphere, this initial point defines a full set of meteorological maps for all of the variables being forecast. The evolution of this single forecast in the phase space, through an intermediate forecast projection and to a final forecast projection, is represented by the heavy solid lines. However, the position of this point in phase space at the initial time represents only one of the many plausible initial states of the atmosphere consistent with errors in the analysis. Around it are other plausible states, which sample the probability distribution for states of the atmosphere at the initial time. This distribution is represented by the small ellipse. The open circles in this ellipse represent eight other members of this distribution. This ensemble of nine initial states approximates the variations represented by the full distribution from which they were drawn.

The Monte-Carlo approximation to a stochastic dynamic forecast is constructed by repeatedly running the NWP model, once for each of the members of the initial ensemble. The trajectories through the phase space of each of the ensemble members are only modestly different at first, indicating that all nine NWP integrations in Figure 6.24 are producing fairly similar forecasts at the intermediate projection. Accordingly, the

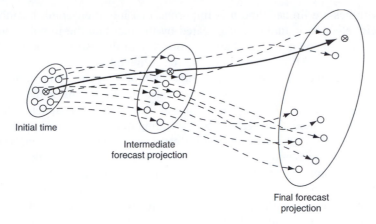

Initial time

Intermediate
forecast projection

Final forecast
projection

FIGURE 6.24   Schematic illustration of some concepts in ensemble forecasting, plotted in terms of an idealized two-dimensional phase space. The heavy line represents the evolution of the single best analysis of the initial state of the atmosphere, corresponding to the more traditional single deterministic forecast. The dashed lines represent the evolution of individual ensemble members. The ellipse in which they originate represents the probability distribution of initial atmospheric states, which are very close to each other. At the intermediate projection, all the ensemble members are still reasonably similar. By the time of the final projection some of the ensemble members have undergone a regime change, and represent qualitatively different flows. Any of the ensemble members, including the solid line, are plausible trajectories for the evolution of the real atmosphere, and there is no way of knowing in advance which will represent the real atmosphere most closely.

probability distribution describing uncertainty about the state of the atmosphere at the intermediate projection would not be a great deal larger than at the initial time. However, between the intermediate and final projections the trajectories diverge markedly, with three (including the one started from the mean value of the initial distribution) producing forecasts that are similar to each other, and the remaining six members of the ensemble predicting rather different atmospheric states at that time. The underlying distribution of uncertainty that was fairly small at the initial time has been stretched substantially, as represented by the large ellipse at the time of the final projection. The dispersion of the ensemble members at that time allows the nature of that distribution to be estimated, and is indicative of the uncertainty of the forecast, assuming that the NWP model includes only negligible errors in the representations of the governing physical processes. If only the single forecast started from the best initial condition had been made, this information would not be available.

6.6.4 Choosing Initial Ensemble Members

Ideally, we would like to produce ensemble forecasts based on a large number of possible initial atmospheric states drawn randomly from the PDF of initial-condition uncertainty in phase space. However, each member of an ensemble of forecasts is produced by a complete rerunning of the NWP model, each of which requires a substantial amount of computing. As a practical matter, computer time is a limiting factor at operational forecast centers, and each center must make a subjective judgment balancing the number of ensemble members to include in relation to the spatial resolution of the NWP model used to integrate them forward in time. Consequently, the sizes of operational forecast ensembles are limited, and it is important that initial ensemble members be chosen well. Their selection is further complicated by the fact that the initial-condition PDF in phase space is unknown, and it changes from day to day, so that the ideal of simple random samples from this distribution cannot be achieved in practice.

The simplest, and historically first, method of generating initial ensemble members is to begin with a best analysis, assumed to be the mean of the probability distribution representing the uncertainty of the initial state of the atmosphere. Variations around this mean state can be easily generated, by adding random numbers characteristic of the errors or uncertainty in the instrumental observations underlying the analysis (Leith 1974). For example, these random values might be Gaussian variates with zero mean, implying an unbiased combination of measurement and analysis errors. In practice, however, simply adding random numbers to a single initial field has been found to yield ensembles whose members are too similar to each other, probably because much of the variation introduced in this way is dynamically inconsistent, so that the corresponding energy is quickly dissipated in the model (Palmer *et al.* 1990). The consequence is that the variability of the resulting forecast ensemble underestimates the uncertainty in the forecast.

As of the time of this writing (2004), there are two dominant methods of choosing initial ensemble members in operational practice. In the United States, the National Centers for Environmental Prediction use the breeding method (Ehrendorfer 1997; Kalnay 2003; Toth and Kalnay 1993, 1997). In this approach, differences in the three-dimensional patterns of the predicted variables, between the ensemble members and the single "best" (control) analysis, are chosen to look like differences between recent forecast ensemble members and the forecast from the corresponding previous control analysis. The patterns are then scaled to have magnitudes appropriate to analysis uncertainties. These bred patterns are different from day to day, and emphasize features with respect to which the ensemble members are diverging most rapidly.

In contrast, the European Centre for Medium-Range Weather Forecasts generates initial ensemble members using singular vectors (Buizza 1997; Ehrendorfer 1997; Kalnay 2003; Molteni *et al.* 1996). Here the fastest growing characteristic patterns of differences from the control analysis in a linearized version of the full forecast model are calculated, again for the specific weather situation of a given day. Linear combinations of these patterns, with magnitudes reflecting an appropriate level of analysis uncertainty, are then added to the control analysis to define the ensemble members. Ehrendorfer and Tribbia (1997) present theoretical support for the use of singular vectors to choose initial ensemble members, although its use requires substantially more computation than does the breeding method.

In the absence of direct knowledge about the PDF of initial-condition uncertainty, how best to define initial ensemble members is the subject of some controversy (Errico and Langland 1999a, 1999b; Toth *et al.* 1999) and ongoing research. Interesting recent work includes computing ensembles of atmospheric analyses, with each analysis ensemble member leading to a forecast ensemble member. Some recent papers outlining current thinking on this subject are Anderson (2003), Hamill (2006), and Houtekamer *et al.* (1998).

6.6.5 *Ensemble Average and Ensemble Dispersion*

One simple application of ensemble forecasting is to average the members of the ensemble to obtain a single forecast. The motivation is to obtain a forecast that is more accurate than the single forecast initialized with the best estimate of the initial state of the atmosphere.

Epstein (1969a) pointed out that the time-dependent behavior of the ensemble mean is different from the solution of forecast equations using the initial mean value, and concluded that in general the best forecast is not the single forecast initialized with the best estimate of initial conditions. This conclusion should not be surprising, since a NWP model is in effect a highly nonlinear function that transforms a set of initial atmospheric conditions to a set of forecast atmospheric conditions.

In general, the average of a nonlinear function over some set of particular values of its argument is not the same as the function evaluated at the average of those values. That is, if the function $f(x)$ is nonlinear,

$$\frac{1}{n}\sum_{i=1}^{n} f(x_i) \neq f\left(\frac{1}{n}\sum_{i=1}^{n} x_i\right). \tag{6.38}$$

To illustrate simply, consider the three values $x_1 = 1$, $x_2 = 2$, and $x_3 = 3$. For the nonlinear function $f(x) = x^2 + 1$, the left side of Equation 6.38 is 5 2/3, and the right side of that equation is 5. We can easily verify that the inequality of Equation 6.38 holds for other nonlinear functions (e.g., $f(x) = \log(x)$ or $f(x) = 1/x$) as well. (By contrast, for the linear function $f(x) = 2x + 1$ the two sides of Equation 6.38 are both equal to 5.)

Extending this idea to ensemble forecasting, we might like to know the atmospheric state corresponding to the center of the ensemble in phase space for some time in the future. This central value of the ensemble will approximate the center of the stochastic dynamic probability distribution at that future time, after the initial distribution has been transformed by the forecast equations. The Monte Carlo approximation to this future value is the ensemble average forecast. The ensemble average forecast is obtained simply by averaging together the ensemble members for the lead time of interest, which corresponds to the left side of Equation 6.38. By contrast, the right side of Equation 6.38 represents the single forecast started from the average initial value of the ensemble members. Depending on the nature of the initial distribution and on the dynamics of the NWP model, this single forecast may or may not be close to the ensemble average forecast.

In the context of weather forecasts, the benefits of ensemble averaging appear to derive primarily from averaging out elements of disagreement among the ensemble members, while emphasizing features that generally are shared by the members of the forecast ensemble. Particularly for longer lead times, ensemble average maps tend to be smoother than instantaneous snapshots, and so may seem unmeteorological, or more similar to smooth climatic averages. Palmer (1993) suggests that ensemble averaging will improve the forecast only until a regime change, or a change in the long-wave pattern, and he illustrates this concept nicely using the simple Lorenz (1963) model. This problem also is illustrated in Figure 6.24, where a regime change is represented by the bifurcation of the trajectories of the ensemble members between the intermediate and final forecast projections. At the intermediate projection, before some of the ensemble members undergo this regime change, the center of the distribution of ensemble members is well represented by the ensemble average, which is a better central value than the single member of the ensemble started from the "best" initial condition. At the final forecast projection the distribution of states has been distorted into two distinct groups. Here the ensemble average will be located somewhere in the middle, but near none of the ensemble members.

A particularly important aspect of ensemble forecasting is its capacity to yield information about the magnitude and nature of the uncertainty in a forecast. In principle the forecast uncertainty is different on different forecast occasions, and this notion can be thought of as state-dependent predictability. The value to forecast users of communicating the different levels of forecast confidence that exist on different occasions was recognized early in the twentieth century (Cooke 1906b; Murphy 1998). Qualitatively, we have more confidence that the ensemble mean is close to the eventual state of the atmosphere if the dispersion of the ensemble is small. Conversely, if the ensemble members are all very different from each other the future state of the atmosphere is very uncertain. One approach to "forecasting forecast skill" (Ehrendorfer 1997; Kalnay and Dalcher 1987; Palmer and Tibaldi 1988) is to anticipate the accuracy of a forecast as being inversely related to the dispersion of ensemble members. Operationally, forecasters do this informally when comparing the results from different NWP models, or when comparing successive forecasts for a particular time in the future that were initialized on different days.

More formally, the spread-skill relationship for a collection of ensemble forecasts often is characterized by the correlation, over a collection of forecast occasions, between the variance or standard deviation of the ensemble members around their ensemble mean on each occasion, and the predictive accuracy of the ensemble mean on that occasion. The accuracy is often characterized using either the mean squared error (Equation 7.28) or its square root, although other measures have been used in some studies. These spread-skill correlations generally have been found to be fairly modest, and rarely exceed 0.5, which corresponds to accounting for 25% or less of the accuracy variations (e.g., Atger 1999; Barker 1991; Grimit and Mass 2002; Hamill *et al.* 2004; Molteni *et al.* 1996; Whittaker and Loughe 1998). Alternative approaches to the spread-skill problem continue to be investigated. Moore and Kleeman (1998) calculate probability distributions for forecast skill, conditional on ensemble spread. Toth *et al.* (2001) present an interesting alternative characterization of the ensemble dispersion, in terms of counts of ensemble forecasts between climatological deciles for the predictand. Some other promising alternative characterizations of the ensemble spread have been proposed by Ziehmann (2001).

6.6.6 *Graphical Display of Ensemble Forecast Information*

A prominent attribute of ensemble forecast systems is that they generate large amounts of multivariate information. As noted in Section 3.6, the difficulty of gaining even an initial understanding of a new multivariate data set can be reduced through the use of well-designed graphical displays. It was recognized early in the development of what is now ensemble forecasting that graphical display would be an important means of conveying the resulting complex information to forecasters (Epstein and Fleming 1971; Gleeson 1967), and operational experience is still accumulating regarding the most effective means of doing so. This section summarizes current practice according to three general types of graphics: displays of raw ensemble output or selected elements of the raw output, displays of statistics summarizing the ensemble distribution, and displays of ensemble relative frequencies for selected predictands. Displays based on more sophisticated statistical analysis of an ensemble are also possible (e.g., Stephenson and Doblas-Reyes 2000).

Perhaps the most direct way to visualize an ensemble of forecasts is to plot them simultaneously. Of course, for even modestly sized ensembles each element (corresponding to one ensemble member) of such a plot must be small in order for all the ensemble members to be viewed simultaneously. Such collections are called stamp maps, because each of its individual component maps is sized approximately like a postage stamp, allowing only the broadest features to be discerned. Legg *et al.* (2002) and Palmer (2002) show stamp maps of ensemble forecasts for sea-level pressure fields. Although fine details of the forecasts are difficult if not impossible to discern from the small images in a stamp map, a forecaster with experience in the interpretation of this kind of display can get an overall sense of the outcomes that are plausible, according to this sample of ensemble members.

A further step that sometimes is taken with a collection of stamp maps is to group them objectively into subsets of similar maps using a cluster analysis (see Section 14.2).

Part of the difficulty in interpreting a collection of stamp maps is that the many individual displays are difficult to comprehend simultaneously. Superposition of a set of stamp maps would alleviate this difficulty if not for the problem that the resulting plot would be too cluttered to be useful. However, seeing each contour of each map is not necessary to form a general impression of the flow, and indeed seeing only one or two well-chosen pressure or height contours is often sufficient to define the main features, since typically the contours roughly parallel each other. Superposition of one or two well-selected contours from each of the stamp maps does yield a sufficiently uncluttered composite to be interpretable, which is known as the "spaghetti plot." Figure 6.25 shows three spaghetti plots for the 5520-m contour of the 500 mb surface over North America, as forecast 12, 36, and 84 hours after the initial time of 0000 UTC, 14 March 1995. In Figure 6.25a the 17 ensemble members generally agree quite closely for the 12-hour forecast, and even with only the 5520-m contour shown the general nature of the flow is clear: the trough over the eastern Pacific and the cutoff low over the Atlantic are clearly indicated.

At the 36-hour projection (see Figure 6.25b) the ensemble members are still generally in close agreement about the forecast flow, except over central Canada, where some

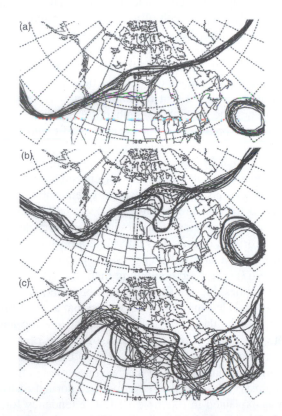

FIGURE 6.25   Spaghetti plots for the 5520-m contour of the 500 mb height field over North America forecast by the National Centers for Environmental Prediction, showing forecasts for (a) 12 h, (b) 36 h, and (c) 84 h after the initial time of 0000 UTC, 14 March 1995. Light lines show the contours produced by each of the 17 ensemble members, dotted line shows the control forecast, and the heavy lines in panels (b) and (c) indicate the verifying analyses. From Toth *et al.*, 1997.

ensemble members produce a short-wave trough. The 500 mb field over most of the domain would be regarded as fairly certain except in this area, where the interpretation would be a substantial but not dominant probability of a short-wave feature, which was missed by the single forecast from the control analysis (dotted). The heavy line in this panel indicates the subsequent analysis at the 36-hour projection. At the 84-hour projection (see Figure 6.25c) there is still substantial agreement about (and thus relatively high probability would be inferred for) the trough over the Eastern Pacific, but the forecasts for the continent and the Atlantic have begun to diverge quite strongly, suggesting the pasta dish for which this kind of plot is named. Spaghetti plots have proven to be quite useful in visualizing the evolution of the forecast flow, simultaneously with the dispersion of the ensemble. The effect is even more striking when a series of spaghetti plots is animated, which can be appreciated at some operational forecast center Web sites.

Stamp maps and spaghetti plots are the most frequently seen displays of raw ensemble output, but other useful and easily interpretable possibilities exist. For example, Legg *et al.* (2002) show ensembles of the tracks of low-pressure centers over a 48-hour forecast period, with the forecast cyclone intensities indicated by a color scale.

It can be informative to condense the large amount of information from an ensemble forecast into a small number of summary statistics, and to plot maps of these. By far the most common such plot, suggested initially by Epstein and Fleming (1971), is simultaneous display of the ensemble mean field and the field of the standard deviations of the ensemble. That is, at each of a number of gridpoints the average of the ensembles is calculated, as well as the standard deviation of the ensemble members around this average. Figure 6.26 is one such plot, for a 12-hour forecast of sea-level pressure (mb) valid at 0000 UTC, 29 January 1999. Here the solid contours represent the ensemble-mean field, and the shading indicates the field of ensemble standard deviation. These standard deviations indicate that the anticyclone over eastern Canada is predicted quite consistently among the ensemble members (ensemble standard deviations generally less than 1 mb), and the pressures in the eastern Pacific and east of Kamchatka, where large gradients are forecast, are somewhat less certain (ensemble standard deviations greater than 3 mb).

The third class of graphical display for ensemble forecasts portrays ensemble relative frequencies for selected predictands. Ideally, ensemble relative frequency would correspond closely to forecast probability; but because of nonideal initial sampling of initial ensemble members, together with inevitable deficiencies in the NWP models used to integrate them forward in time, this interpretation is not literally warranted (Hansen 2002; Krzysztofowicz 2001; Smith 2001).

Gleeson (1967) suggested combining maps of forecast u and v wind components with maps of probabilities that the forecasts will be within 10 knots of the eventual observed values. Epstein and Fleming (1971) suggested that a probabilistic depiction of a horizontal wind field could take the form of Figure 6.27. Here the lengths and orientations of the lines indicate the mean of the forecast distributions of wind vectors, blowing from the gridpoints the ellipses. The probability is 0.50 that the true wind vectors will terminate within the corresponding ellipse. It has been assumed in this figure that the uncertainty in the wind forecasts is described by the bivariate normal distribution, and the ellipses have been drawn as explained in Example 10.1. The tendency for the ellipses to be oriented in a north-south direction indicates that the uncertainties of the meridional winds are greater than those for the zonal winds, and the tendency for the larger velocities to be associated with larger ellipses indicates that these wind values are more uncertain.

Ensemble forecasts for surface weather elements at a single location can be concisely summarized by time series of boxplots for selected predictands, in a plot called a

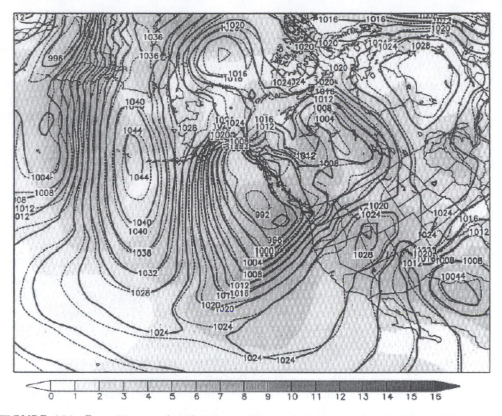

FIGURE 6.26   Ensemble mean (solid) and ensemble standard deviation (shading) for a 12-hour forecast of sea-level pressure, valid 0000 UTC, 29 January 1999. Dashed contours indicate the single control
forecast. From Toth *et al.* (2001).

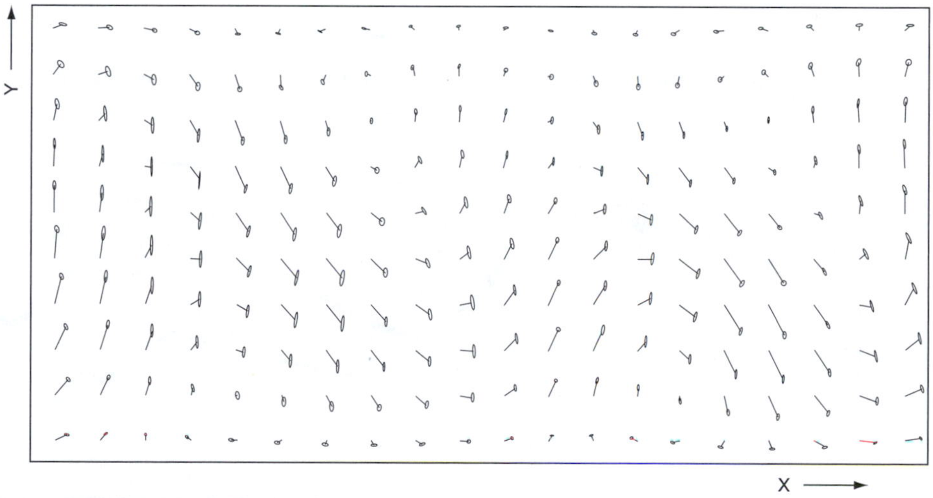

FIGURE 6.27   A forecast wind field from an idealized modeling experiment, expressed in probabilistic
form. Line lengths and orientations show forecast mean wind vectors directed from the gridpoint locations
to the ellipses. The ellipses indicate boundaries containing the observed wind with probability 0.50.
From Epstein and Fleming (1971).

meteogram (Palmer 2002). Each of these boxplots displays the dispersion of the ensemble for one predictand at a particular forecast projection, and jointly they show the time evolutions of the forecast central tendencies and uncertainties, through the forecast period. Palmer (2002) includes an example meteogram with frequency distributions over 50 ensemble members for fractional cloud cover, accumulated precipitation, 10 m wind speed, and 2 m temperature; all of which are described by boxplots at six-hour intervals.

Figure 6.28 shows an alternative to boxplots for portraying the time evolution of the ensemble distribution for a predictand. In this plume graph the contours indicate heights of the ensemble dispersion, expressed as a PDF, as a function of time for forecast 500 mb heights over southeast England. The ensemble is seen to be quite compact early in the forecast, and expresses a large degree of uncertainty by the end of the period.

Finally, information from ensemble forecasts is very commonly displayed as maps of ensemble relative frequencies for dichotomous events, which are often defined according to a threshold for a continuous variable. Figure 6.29 shows an example of a very common plot of this kind, for ensemble relative frequency of more than 2 mm of precipitation over 12 hours, at lead times of (a) 7 days, (b) 5 days, and (c) 3 days ahead of the observed event (d). As the lead time decreases, the areas with appreciable forecast probability become more compactly defined, and exhibit the generally larger relative frequencies indicative of greater confidence in the event outcome.

Other kinds of probabilistic field maps are also possible, many of which may be suggested by the needs of particular forecast applications. Figure 6.30, showing relative frequencies of forecast 1000-500 mb thickness over North America being less than 5400 m, is one such possibility. Forecasters often use this thickness value as the expected dividing line between rain and snow. At each grid point, the fraction of ensemble members predicting 5400 m thickness or less has been tabulated and plotted. Clearly, similar maps for other thickness values could be constructed as easily. Figure 6.30 indicates a relatively high confidence that the cold-air outbreak in the eastern United States will bring air sufficiently cold to produce snow as far south as the Gulf coast.

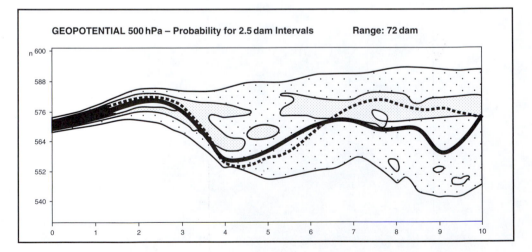

FIGURE 6.28   A plume graph, indicating probability density as a function of time, for a 10-day forecast of 500 mb height over southeast England, initiated 1200 UTC, 26 August 1999. The dashed line shows the high-resolution control forecast, and the solid line indicates the lower-resolution ensemble member begun from the same initial condition. From Young and Carroll (2002).

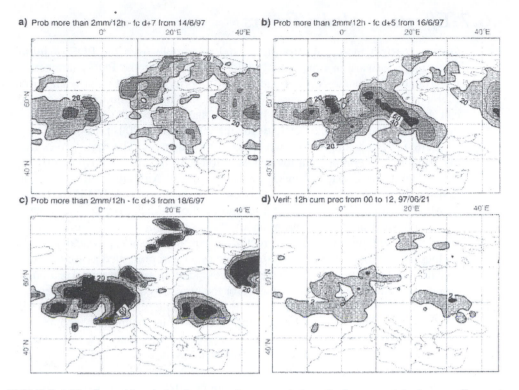

FIGURE 6.29   Ensemble relative frequency for accumulation of >2 mm precipitation over Europe in 12-hour periods (a) 7 days, (b) 5 days, and (c) 3 days ahead of (d) the observed events, on 21 June 1997. Contour interval in (a) – (c) is 0.2. From Buizza *et al.* (1999a).

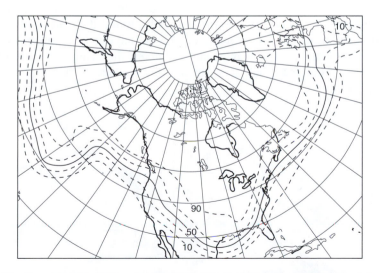

FIGURE 6.30   Forecast probabilities that the 1000-500 mb thicknesses will be less than 5400 m over North America, as estimated from the relative frequencies of thicknesses in 14 ensemble forecast members. From Tracton and Kalnay (1993).

6.6.7 *Effects of Model Errors*

Given a perfect model, integrating a random sample from the PDF of initial-condition uncertainty forward in time would yield a sample from the PDF characterizing forecast uncertainty. Of course NWP models are not perfect, so that even if an initial-condition PDF could be known and correctly sampled from, the distribution of a forecast ensemble can at best be only an approximation to a sample from the true PDF for the forecast uncertainty (Hansen 2002; Krzysztofowicz 2001; Smith 2001).

Leith (1974) distinguished two kinds of model errors. The first derives from the models inevitably operating at a lower resolution than the real atmosphere or, equivalently, occupying a phase space of much lower dimension. Although still significant, this problem has been addressed and ameliorated over the history of NWP through progressive increases in model resolution. The second kind of model error derives from the fact that certain physical processes—prominently those operating at scales smaller than the model resolution—are represented incorrectly. In particular, such physical processes (known colloquially in this context as physics) generally are represented as some relatively simple function of the explicitly resolved variables, known as a parameterization. Figure 6.31 shows a parameterization (solid curve) for the unresolved part of the tendency (dX/dt) of a resolved variable X, as a function of X itself, in the highly idealized Lorenz (1996) model (Wilks 2005). The individual points in Figure 6.31 are a sample of the actual unresolved tendencies, which are summarized by the regression function. In a real NWP model there are a number of such parameterizations for various unresolved physical processes, and the effects of these processes on the resolved variables are included in the model as functions of the resolved variables through these parameterizations. It is evident from Figure 6.31 that the parameterization (smooth curve) does not fully capture the range of behaviors for the parameterized process that are actually possible (scatter of points around the curve). One way of looking at this kind of model error is that the parameterized physics are not fully determined by the resolved variables. That is, they are uncertain.

One way of representing the errors, or uncertainties, in the model physics is to extend the idea of the ensemble to include simultaneously a collection of different initial conditions *and* multiple NWP models (each of which has a different collection of parameterizations). Harrison *et al.* (1999) found that forecasts using all four possible combinations of two sets of initial conditions and two NWP models differed significantly, with members of each of the four ensembles clustering relatively closely together, and distinctly from the other three, in the phase space. Other studies (e.g., Hansen 2002;

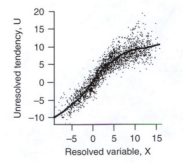

FIGURE 6.31   Scatterplot of the unresolved time tendency, U, of a resolved variable, X, as a function of the resolved variable; together with a regression function representing the average dependence of the tendency on the resolved variable. From Wilks (2005).

Houtekamer *et al.* 1996; Mylne *et al.* 2002a; Mullen *et al.* 1999; Stensrud *et al.* 2000) have found that using such multimodel ensembles improves the resulting ensemble forecasts. A substantial part of this improvement derives from the multimodel ensembles exhibiting larger ensemble dispersion, so that the ensemble members are less like each other than if a single NWP model is used for all forecast integrations. Typically the dispersion of forecast ensembles is too small (e.g., Buizza 1997; Stensrud *et al.* 1999; Toth and Kalnay 1997), and so express too little uncertainty about forecast outcomes (cf., Section 7.7).

Another approach to capturing uncertainties in NWP model physics is suggested by the scatter around the regression curve in Figure 6.31. From the perspective of Section 6.2, the regression residuals that are differences between the actual (points) and parameterized (regression curve) behavior of the modeled system are random variables. Accordingly, the effects of parameterized processes can be more fully represented in an NWP model if random numbers are added to the deterministic parameterization function, making the NWP model explicitly stochastic (Palmer 2001). This idea is not new, having been proposed in the 1970s for NWP (Lorenz 1975; Moritz and Sutera 1981; Pitcher 1977). However, the use of stochastic parameterizations in realistic atmospheric models is relatively recent (Buizza *et al.* 1999b; Garratt *et al.* 1990; Lin and Neelin 2000, 2002; Williams *et al.* 2003). Particularly noteworthy is the current operational use of a stochastic representation of the effects of unresolved processes in the forecast model at the European Centre for Medium-Range Forecasts, which they call "stochastic physics", and which results in improved forecasts relative to the conventional deterministic parameterizations (Buizza *et al.* 1999b; Mullen and Buizza 2001).

Stochastic parameterizations also have been used in simplified climate models, to represent atmospheric variations on the time scale of weather, since the 1970s (Hasselmann 1976). Some relatively recent papers applying this idea to prediction of the El Niño phenomenon are Penland and Sardeshmukh (1995), Saravanan and McWilliams (1998), and Thompson and Battisti (2001).

6.6.8 *Statistical Postprocessing: Ensemble MOS*

From the outset of ensemble forecasting (Leith 1974) it was anticipated that use of finite ensembles would yield errors in the forecast ensemble mean that could be statistically corrected using a database of previous errors—essentially a MOS postprocessing for the ensemble mean. In practice forecast errors benefiting from statistical postprocessing also derive from model deficiencies, as described in the previous section. However, though ensemble forecast methods continue to be investigated intensively, both in research and operational settings, work on their statistical postprocessing is still in its initial stages.

One class of ensemble MOS consists of interpretations or adjustments based only on the ensemble mean. Hamill and Colucci (1998) derived forecast distributions for precipitation amounts using gamma distributions with parameters specified according to the ensemble mean precipitation. Figure 6.32 shows the results, in terms of the logarithms of the two gamma distribution parameters as functions of the ensemble mean precipitation, which has been power-transformed similarly to Equation 3.18b using the exponent $\lambda = -0.3$ (and adding 0.01 in. to all ensemble means in order to avoid exponentiation of zero). Variations in the shape parameter α are relatively modest, whereas the scale parameter β increases very strongly with increasing ensemble mean.

Hamill *et al.* (2004) report very effective MOS postprocessing of the ensemble means for both surface temperature anomalies and accumulated precipitation, at 6–10 day and 8–14 day lead times. Consistent with current operational practice at the U.S. Climate

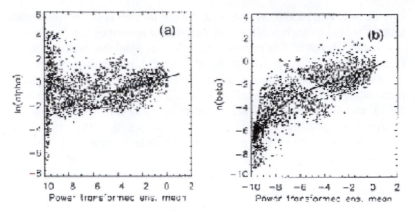

FIGURE 6.32   Relationships of the logarithms of gamma distribution parameters for forecast precipitation distributions, with the power-transformed ensemble mean. From Hamill and Colucci (1998).

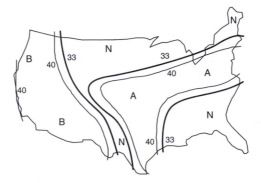

FIGURE 6.33   Example probabilistic forecast map for accumulated precipitation in the one-week period 23–29 June, 2004, at lead time 8–14 days, expressed as probability shifts among three equiprobable climatological classes. Adapted from http://www.cpc.ncep.noaa.gov/products/predictions/814day/.

Prediction Center (see Figure 6.33), the predictands are probabilities of these temperature or precipitation outcomes being either above the upper tercile, or below the lower tercile, of the respective climatological distributions. The approach taken was to produce these MOS probabilities through logistic regressions (see Section 6.3.1) that use as predictors only the respective ensemble means interpolated to each forecast location. These MOS forecasts, even for the 8–14 day lead time, were of higher quality that the 6–10 day operational forecasts. Adjusting the dispersion of the ensemble according to its historical error statistics can allow information on possible state- or flow-dependent predictability to be included also in an ensemble MOS procedure. Hamill and Colucci (1997, 1998) approached this problem through a forecast verification graph known as the verification rank histogram, described in Section 7.7.2. Their method is being used operationally at the U.K. Met Office (Mylne *et al.* 2002b).

Another approach to ensemble MOS involves defining a probability distribution around either the ensemble mean, or around each of the ensemble members. This process has been called ensemble dressing (Roulston and Smith 2003, Wang and Bishop 2005), because a forecast point has a probability distribution draped over it. Atger (1999) used Gaussian distributions around the ensemble mean for 500 mb height, with standard deviations proportional to the forecast ensemble standard deviation. Roulston and Smith

(2003) describe postprocessing a forecast ensemble by dressing each ensemble member with a probability distribution, in a manner similar to kernel density smoothing (see Section 3.3.6). This is an ensemble MOS procedure because the distributions that are superimposed are derived from historical error statistics of the ensemble prediction system being postprocessed. Because individual ensemble members rather than the ensemble mean are dressed, the procedure yields state-dependent uncertainty information even if the spread of the added error distributions is not conditional on the ensemble spread.

Bremnes (2004) forecasts probability distributions for precipitation using a two-stage ensemble MOS procedure that uses selected quantiles of the forecast ensemble precipitation distribution as predictors. First, the probability of nonzero precipitation is forecast using a probit regression, which is similar to logistic regression (Equation 6.27), but using the CDF of the standard Gaussian distribution to constrain the linear function of the predictors to the unit interval. That is $p_i = \Phi(b_0 + b_1 x_1 + b_2 x_2 + b_3 x_3)$, where the three predictors are the ensemble minimum, the ensemble median, and the ensemble maximum. Second, conditional on the occurrence of nonzero precipitation, the 5th, 25th, 50th, 75th, and 95th percentiles of the precipitation amount distributions are specified with separate regression equations, which each use the two ensemble quartiles as predictors. The final postprocessed precipitation probabilities then are obtained through the multiplicative law of probability (Equation 2.11), where E_1 is the event that nonzero precipitation occurs, and E_2 is a precipitation amount event defined by some combination of the forecast percentiles produced by the second regression step.

6.7 Subjective Probability Forecasts

6.7.1 The Nature of Subjective Forecasts

Most of this chapter has dealt with objective forecasts, or forecasts produced by means that are automatic. Objective forecasts are determined unambiguously by the nature of the forecasting procedure and the values of the variables that are used to drive it. However, objective forecasting procedures necessarily rest on a number of subjective judgments made during their development. Nevertheless, some people feel more secure with the results of objective forecasting procedures, seemingly taking comfort from their lack of contamination by the vagaries of human judgment. Apparently, such individuals feel that objective forecasts are in some way less uncertain than human-mediated forecasts.

One very important—and perhaps irreplaceable—role of human forecasters in the forecasting process is in the subjective integration and interpretation of objective forecast information. These objective forecast products often are called forecast guidance, and include deterministic forecast information from NWP integrations, and statistical guidance from MOS systems or other interpretive statistical products. Human forecasters also use, and incorporate into their judgments, available atmospheric observations (surface maps, radar images, etc.), and prior information ranging from persistence or simple climatological statistics, to their individual previous experiences with similar meteorological situations. The result is (or should be) a forecast reflecting, to the maximum practical extent, the forecaster's state of knowledge about the future evolution of the atmosphere.

Human forecasters can rarely, if ever, fully describe or quantify their personal forecasting processes. Thus, the distillation by a human forecaster of disparate and sometimes conflicting information is known as subjective forecasting. A subjective forecast is one formulated on the basis of the judgment of one or more individuals. Making a subjective weather forecast is a challenging process precisely because future states of the atmosphere are inherently uncertain. The uncertainty will be larger or smaller in different circumstances—some forecasting situations are more difficult than others—but it will never really be absent. Doswell (2004) provides some informed perspectives on the formation of subjective judgments in weather forecasting.

Since the future states of the atmosphere are inherently uncertain, a key element of a good and complete subjective weather forecast is the reporting of some measure of the forecaster's uncertainty. It is the forecaster who is most familiar with the atmospheric situation, and it is therefore the forecaster who is in the best position to evaluate the uncertainty associated with a given forecasting situation. Although it is common for nonprobabilistic forecasts (i.e., forecasts containing no expression of uncertainty) to be issued, such as "tomorrow's maximum temperature will be 27°F," an individual issuing this forecast would not seriously expect the temperature to be exactly 27°F. Given a forecast of 27°F, temperatures of 26 or 28°F would generally be regarded as nearly as likely, and in this situation the forecaster would usually not really be surprised to see tomorrow's maximum temperature anywhere between 25 and 30°F.

Although uncertainty about future weather can be reported verbally using phrases such as "chance" or "likely," such qualitative descriptions are open to different interpretations by different people (e.g., Murphy and Brown 1983). Even worse, however, is the fact that such qualitative descriptions do not precisely reflect the forecasters uncertainty about, or degree of belief in, the future weather. The forecaster's state of knowledge is most accurately reported, and the needs of the forecast user are best served, if the intrinsic uncertainty is quantified in probability terms. Thus, the Bayesian view of probability as the degree of belief of an individual holds a central place in subjective forecasting. Note that since different forecasters have somewhat different information on which to base their judgments (e.g., different sets of experiences with similar past forecasting situations), it is perfectly reasonable to expect that their probability judgments may differ somewhat as well.

6.7.2 The Subjective Distribution

Before a forecaster reports a subjective degree of uncertainty as part of a forecast, he or she needs to have an image of that uncertainty. The information about an individual's uncertainty can be thought of as residing in their subjective distribution for the event in question. The subjective distribution is a probability distribution in the same sense as the parametric distributions described in Chapter 4. Sometimes, in fact, one of the distributions specifically described in Chapter 4 may provide a very good approximation to our subjective distribution. Subjective distributions are interpreted from a Bayesian perspective as the quantification of an individual's degree of belief in each of the possible outcomes for the variable being forecast.

Each time a forecaster prepares to make a forecast, he or she internally develops a subjective distribution. The possible weather outcomes are subjectively weighed, and an internal judgment is formed as to their relative likelihoods. This process occurs whether or not the forecast is to be a probability forecast, or indeed whether or not the forecaster is even aware of the process. However, unless we believe that uncertainty can somehow be expunged from the process of weather forecasting, it should be clear that better forecasts will result when forecasters think explicitly about their subjective distributions and the uncertainty that those distributions describe.

It is easiest to approach the concept of subjective probabilities with a familiar but simple example. Subjective probability-of-precipitation (PoP) forecasts have been routinely issued in the United States. since 1965. These forecasts specify the probability that measurable precipitation (i.e., at least 0.01 in.) will occur at a particular location during a specified time period. The forecaster's subjective distribution for this event is so simple that we might not notice that it is a probability distribution. However, the events "precipitation" and "no precipitation" divide the sample space into two MECE events. The distribution of probabilities over these events is discrete, and consists of two elements: one probability for the event "precipitation" and another probability for the event "no precipitation." This distribution will be different for different forecasting situations, and perhaps for different forecasters assessing the same situation. However, the only thing about a forecaster's subjective distribution for the PoP that can change from one forecasting occasion to another is the probability, and this will be different to the extent that the forecaster's degree of belief regarding future precipitation occurrence is different. The PoP ultimately issued by the forecaster should be the forecaster's subjective probability for the event "precipitation," or perhaps a suitably rounded version of that probability. That is, it is the forecaster's job is to evaluate the uncertainty associated with the possibility of future precipitation occurrence, and to report that uncertainty to the users of the forecasts.

An operational subjective probability forecasting format that is only slightly more complicated than the format for PoP forecasts is that for the 6–10 and 8–14 day outlooks for temperature and precipitation issued by the Climate Prediction Center of the U.S. National Weather Service. Average temperature and total precipitation at a given location over an upcoming forecast period are continuous variables, and complete specifications of the forecaster's uncertainty regarding their values would require continuous subjective probability distribution functions. The format for these forecasts is simplified considerably by defining a three-event MECE partition of the sample spaces for the temperature and precipitation outcomes, transforming the problem into one of specifying a discrete probability distribution for a random variable with three possible outcomes.

The three events are defined with respect to the local climatological distributions for each of the periods to which these forecasts pertain, in a manner similar to that used in reporting recent climate anomalies (see Example 4.9). In the case of the temperature forecasts, a cold outcome is one that would fall in the lowest 1/3 of the climatological temperature distribution, a near-normal outcome is one that would fall in the middle 1/3 of that climatological distribution, and a warm outcome would fall in the upper 1/3 of the climatological distribution. Thus, the continuous temperature scale is divided into three discrete events according to the terciles of the local climatological temperature distribution for average temperature during the forecast period. Similarly, the precipitation outcomes are defined with respect to the same quantiles of the local precipitation distribution, so dry, near-normal, and wet precipitation outcomes correspond to the driest 1/3, middle 1/3, and wettest 1/3 of the climatological distribution, respectively. The same format also is used to express the uncertainty in seasonal forecasts (e.g., Barnston *et al.* 1999, Mason *et al.* 1999).

It is often not realized that forecasts presented in this format are probability forecasts. That is, the forecast quantities are probabilities for the events defined in the previous paragraph rather than temperatures or precipitation amounts per se. Through experience the forecasters have found that future periods with other than the climatological relative frequency of the near-normal events are difficult to discern. That is, regardless of the probability that forecasters have assigned to the near-normal event, the subsequent relative frequencies are nearly 1/3. Therefore, operationally the probability of 1/3 is generally assigned to the middle category. Since the probabilities for the three MECE events must all add to one, this restriction on the near-normal outcomes implies that the full forecast distribution can be specified by a single probability. In effect, the forecaster's subjective distribution for the upcoming month or season is collapsed to a single probability. Forecasting a 10% chance that average temperature will be cool implies a 57% chance that the outcome will be warm, since the format of the forecasts always carries the implication that the chance of near-normal is 1/3.

One benefit of this restriction is that the full probability forecast of the temperature or precipitation outcome for an entire region can be displayed on a single map. Figure 6.33 shows an example 8 to14 day precipitation forecast for the conterminous United States. The mapped quantities are probabilities in percentage terms for the more likely of the two extreme events (dry or wet). The heavy contours define the boundaries of areas (labeled N, for normal) where the climatological probabilities of 1/3–1/3–1/3 are forecast, indicating that this is the best information the forecaster has to offer for these locations in this particular case. Probability contours labeled 40 indicate 40% chance that the precipitation accumulation will be in either the wetter 1/3 (areas labeled A, for above) or drier 1/3 (areas labeled B, for below) of the local climatological distributions. The implied probability for the other extreme outcome is then 0.27.

6.7.3 *Central Credible Interval Forecasts*

It has been argued here that inclusion of some measure of the forecaster's uncertainty should be included in any weather forecast. Historically, resistance to this idea has been based in part on the practical consideration that the forecast format should be compact and easily understandable. In the case of PoP forecasts, the subjective distribution is sufficiently simple that it can be reported with a single number, and is no more cumbersome than issuing a nonprobabilistic forecast of "precipitation" or "no precipitation." When the subjective distribution is continuous, however, some approach to sketching its main features is a practical necessity if its probability information is to be conveyed succinctly in a publicly issued forecast. Discretizing the subjective distribution, as is done in Figure 6.33, is one approach to simplifying a continuous subjective probability distribution in terms of one or a few easily expressible quantities. Alternatively, if the forecaster's subjective distribution on a given occasion can be reasonably well approximated by one of the theoretical distributions described in Chapter 4, another approach to simplifying its communication could be to report the parameters of the approximating distribution. There is no guarantee, however, that subjective distributions will always (or even ever) correspond to a familiar theoretical form.

One very attractive and workable alternative for introducing probability information into forecasts for continuous meteorological variables is the use of credible interval forecasts. This forecast format has been used operationally in Sweden (Ivarsson *et al.* 1986), but to date has been used only experimentally in the United States (Murphy and Winkler 1974; Peterson *et al.* 1972; Winkler and Murphy 1979). In unrestricted form, a credible interval forecast requires specification of three quantities: two points defining an interval of the continuous forecast variable, and a probability (according to the forecaster's subjective distribution) that the forecast quantity will fall in the designated interval. Often the requirement is made that the credible interval be located in the middle of the subjective distribution. In this case the specified probability is distributed equally on either side of the subjective median, and the forecast is called a central credible interval forecast.

There are two special cases of the credible-interval forecast format, each requiring that only two quantities be forecast. The first is the fixed-width credible interval forecast. As the name implies, the width of the credible interval is the same for all forecasting situations and is specified in advance for each predictand. Thus the forecast includes a location for the interval, generally specified as its midpoint, and a probability that the outcome will occur in the forecast interval. For example, the Swedish credible interval forecasts for temperature are of the fixed-width type, with the interval size specified to be ±3°C around the midpoint temperature. These forecasts thus include a forecast temperature, together with a probability that the subsequently observed temperature will be within 3°C of the forecast temperature. The two forecasts 15°C, 90% and 15°C, 60% would both indicate that the forecaster expects the temperature to be about 15°C, but the inclusion of probabilities in the forecasts shows that much more confidence can be placed in the former as opposed to the latter of the two forecasts of 15°C. Because the forecast interval is central, these two forecasts would also imply 5%, and 20% chances, respectively, for the temperature to be colder than 12° and warmer than 18°.

Some forecast users would find the unfamiliar juxtaposition of a temperature and a probability in a fixed-width credible interval forecast to be somewhat jarring. An alternative forecast format that could be implemented more subtly is the fixed-probability credible interval forecast. In this format, it is the probability contained in the forecast interval, rather than the width of the interval, that is specified in advance and is constant from forecast to forecast. This format makes the probability part of the credible interval forecast implicit, so the forecast consists of two numbers having the same physical dimensions as the quantity being forecast.

Figure 6.34 illustrates the relationship of 75% central credible intervals for two subjective distributions having the same location. The shorter, broader distribution represents a relatively uncertain forecasting situation, where events fairly far away from the center of the distribution are regarded as having substantial probability. A relatively wide interval is therefore required to subsume 75% of this distribution's probability. On the other hand, the tall and narrow distribution describes considerably less uncertainty, and a much narrower forecast interval contains 75% of its density. If the variable being forecast is temperature, the 75% credible interval forecasts for these two cases might be 10° to 20°, and 14° to 16°, respectively.

A strong case can be made for operational credible-interval forecasts (Murphy and Winkler 1974, 1979). Since nonprobabilistic temperature forecasts are already often specified as ranges, fixed-probability credible interval forecasts could be introduced into forecasting operations quite unobtrusively. Forecast users not wishing to take advantage of the implicit probability information would notice little difference from the present

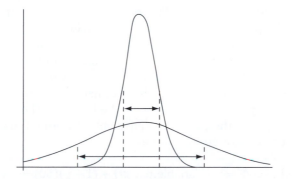

FIGURE 6.34   Two hypothetical subjective distributions shown as probability density functions. The two distributions have the same location, but reflect different degrees of uncertainty. The tall, narrow distribution represents an easier (less uncertain) forecasting situation, and the broader distribution represents a more difficult forecast problem. Arrows delineate 75% central credible intervals in each case.

forecast format, whereas those understanding the meaning of the forecast ranges would derive additional benefit. Even forecast users unaware that the forecast range is meant to define a particular interval of fixed probability might notice over time that the interval widths were related to the precision of the forecasts.

6.7.4 Assessing Discrete Probabilities

Experienced weather forecasters are able to formulate subjective probability forecasts that apparently quantify their uncertainty regarding future weather quite successfully. Examination of the error characteristics of such forecasts (see Chapter 7) reveals that they are largely free of the biases and inconsistencies sometimes exhibited in the subjective probability assessments made by less experienced individuals. Commonly, inexperienced forecasters produce probability forecasts exhibiting overconfidence (Murphy 1985), or biases due to such factors as excessive reliance on recently acquired information (Spetzler and Staël von Holstein 1975).

Individuals who are experienced at assessing their subjective probabilities can do so in a seemingly unconscious or automatic manner. People who are new to the practice often find it helpful to use physical or conceptual devices that allow comparison of the uncertainty to be assessed with an uncertain situation that is more concrete and familiar. For example, Spetzler and Staël von Holstein (1975) describe a physical device called a probability wheel, which consists of a spinner of the sort that might be found in a child's board game, on a background that has the form of a pie chart. This background has two colors, blue and orange, and the proportion of the background covered by each of the colors can be adjusted. The probability wheel is used to assess the probability of a dichotomous event (e.g., a PoP forecast) by adjusting the relative coverages of the two colors until the forecaster feels the probability of the event to be forecast is about equal to the probability of the spinner stopping in the orange sector. The subjective probability forecast is then read as the angle subtended by the orange section, divided by 360.

Conceptual devices can also be employed to assess subjective probabilities. For many people, comparison of the uncertainty surrounding the future weather is most easily assessed in the context of lottery games or betting games. Such conceptual devices translate the probability of an event to be forecast into more concrete terms by posing hypothetical questions such as "would you prefer to be given $2 if precipitation occurs tomorrow, or $1 for sure (regardless of whether or not precipitation occurs)?" Individuals preferring the sure $1 in this lottery situation evidently feel that the relevant PoP is less than 0.5, whereas individuals who feel the PoP is greater than 0.5 would prefer to receive $2 on the chance of precipitation. A forecaster can use this lottery device by adjusting the variable payoff relative to the certainty equivalent (the sum to be received for sure) until the point of indifference, where either choice would be equally attractive. That is, the variable payoff is adjusted until the expected (i.e., probability-weighted average) payment is equal to the certainty equivalent. Denoting the subjective probability as p, the procedure can be written formally as

$$\text{Expected payoff} = p\,(\text{Variable payoff}) + (1-p)(\$0) = \text{Certainty equivalent} \quad (6.39a)$$

which leads to

$$p = \frac{\text{Certainty equivalent}}{\text{Variable payoff}}. \quad (6.39b)$$

The same kind of logic can be applied in an imagined betting situation. Here the forecasters ask themselves whether receiving a specified payment should the weather event to be forecast occurs, or suffering some other monetary loss if the event does not occur, is preferable. In this case the subjective probability as assessed by finding monetary amounts for the payment and loss such that the bet is a fair one, implying that the forecaster would be equally happy to be on either side of it. Since the expected payoff from a fair bet is zero, the betting game situation can be represented as

$$\text{Expected payoff} = p(\$\,\text{payoff}) + (1-p)(-\$\,\text{loss}) = 0, \qquad (6.40\text{a})$$

leading to

$$p = \frac{\$\,\text{loss}}{\$\,\text{loss} + \$\,\text{payoff}}. \qquad (6.40\text{b})$$

Many betting people think in terms of odds in this context. Equation 6.40a can be expressed alternatively as

$$\text{odds ratio} = \frac{p}{1-p} = \frac{\$\,\text{loss}}{\$\,\text{payoff}}. \qquad (6.41)$$

Thus, a forecaster being indifferent to an even-money bet (1:1 odds) harbors an internal subjective probability of $p = 0.5$. Indifference to being on either side of a 2:1 bet implies a subjective probability of 2/3, and indifference at 1:2 odds is consistent with an internal probability of 1/3.

6.7.5 Assessing Continuous Distributions

The same kinds of lotteries or betting games just described can also be used to assess points on a subjective continuous probability distribution using the method of successive subdivision. Here the approach is to identify quantiles of the subjective distribution by comparing event probabilities that they imply with the reference probabilities derived from conceptual money games. Use of this method in an operational setting is described in Krzysztofowicz *et al.* (1993).

The easiest quantile to identify is the median. Suppose the distribution to be identified is for tomorrow's maximum temperature. Since the median divides the subjective distribution into two equally probable halves, its location can be assessed by evaluating a preference between, say, $1 for sure and $2 if tomorrow's maximum temperature is warmer than 14°C. The situation is the same as that described in Equation 6.39. Preferring the certainty of $1 implies a subjective probability for the event {maximum temperature warmer than 14°C} that is smaller than 0.5. A forecaster preferring the chance at $2 evidently feels that the probability for this event is larger than 0.5. Since the cumulative probability, p, for the median is fixed at 0.5, we can locate the threshold defining the event {outcome above median} by adjusting it to the point of indifference between the certainty equivalent and a variable payoff equal to twice the certainty equivalent.

The quartiles can be assessed in the same way, except that the ratios of certainty equivalent to variable payoff must correspond to the cumulative probabilities of the quartiles; that is, 1/4 or 3/4. At what temperature T_{lq} are we indifferent to the alternatives of receiving $1 for sure, or $4 if tomorrow's maximum temperature is below T_{lq}? The temperature T_{lq} then estimates the forecaster's subjective lower quartile. Similarly, the temperature T_{uq}, at which we are indifferent to the alternatives of $1 for sure or $4 if the temperature is above T_{uq}, estimates the upper quartile.

Especially when someone is inexperienced at probability assessments, it is a good idea to perform some consistency checks. In the method just described, the quartiles are assessed independently, but together define a range—the 50% central credible interval—in which half the probability should lie. Therefore a good check on their consistency would be to verify that we are indifferent to the choices between $1 for sure, and $2 if $T_{lq} \leq T \leq T_{uq}$. If we prefer the certainty equivalent in this comparison the quartile estimates T_{lq} and T_{uq} are apparently too close. If we prefer the chance at the $2 they apparently delineate too much probability. Similarly, we could verify indifference between the certainty equivalent, and four times the certainty equivalent if the temperature falls between the median and one of the quartiles. Any inconsistencies discovered in checks of this type indicate that some or all of the previously estimated quantiles need to be reassessed.

6.8 Exercises

6.1. a. Derive a simple linear regression equation using the data in Table A.3, relating June temperature (as the predictand) to June pressure (as the predictor).
   b. Explain the physical meanings of the two parameters.
   c. Formally test whether the fitted slope is significantly different from zero.
   d. Compute the R^2 statistic.
   e. Estimate the probability that a predicted value corresponding to $x_0 = 1013$ mb will be within 1°C of the regression line, using Equation 6.22.
   f. Repeat (e), assuming the prediction variance equals the MSE.

6.2. Consider the following ANOVA table, describing the results of a regression analysis:

Source	df	SS	MS	F
Total	23	2711.60		
Regression	3	2641.59	880.53	251.55
Residual	20	70.01	3.50	

   a. How many predictor variables are in the equation?
   b. What is the sample variance of the predictand?
   c. What is the R^2 value?
   d. Estimate the probability that a prediction made by this regression will be within ±2 units of the actual value.

6.3. Derive an expression for the maximum likelihood estimate of the intercept b_0 in logistic regression (Equation 6.27), for the constant model in which $b_1 = b_2 = \cdots = b_K = 0$.

6.4 The 19 nonmissing precipitation values in Table A.3 can be used to fit the regression equation:

$$\ln[(\text{Precipitation}) + 1\,\text{mm}] = 499.4 - 0.512(\text{Pressure}) + 0.796(\text{Temperature})$$

The MSE for this regression is 0.701. (The constant 1 mm has been added to ensure that the logarithm is defined for all data values.)
   a. Estimate the missing precipitation value for 1956 using this equation.
   b. Construct a 95% confidence interval for the estimated 1956 precipitation.

6.5. Explain how to use cross-validation to estimate the prediction mean squared error, and the sampling distribution of the regression slope, for the problem in Exercise 6.1. If the appropriate computing resources are available, implement your algorithm.

6.6. Hurricane Zeke is an extremely late storm in a very busy hurricane season. It has recently formed in the Caribbean, the 500 mb height at gridpoint 37 (relative to the storm) is 5400 m, the 500 mb height at gridpoint 3 is 5500 m, and the 1000 mb height at gridpoint 51 is −200 m (i.e., the surface pressure near the storm is well below 1000 mb).

a. Use the NHC 67 model (see Table 6.6) to forecast the westward component of its movement over the next 12 hours, if storm has moved 80 n.mi. due westward in the last 12 hours.

b. What would the NHC 67 forecast of the westward displacement be if, in the previous 12 hours, the storm had moved 80 n.mi. westward *and* 30 n.mi. northward (i.e., $P_y = 30$ n.mi.)?

6.7. The fall (September, October, November) LFM-MOS equation for predicting maximum temperature (in °F) at Binghamton, New York, at the 60-hour projection was

$$\text{MAX T} = -363.2 + 1.541(850 \text{ mb T}) - .1332(\text{SFC} - 490 \text{ mb RH}) - 10.3(\text{COS DOY})$$

where:

(850 mb T) is the 48-hour LFM forecast of temperature (°K) at 850 mb
(SFC-490 mb RH) is the 48-hour LFM-forecast lower tropospheric RH in %
(COS DOY) is the cosine of the Julian date transformed to radians or degrees; that is,
$\quad = \cos(2\pi t/365)$ or $= \cos(360°t/365)$
and t is the Julian date of the valid time (the Julian date for January 1 is 1, and for October 31 it is 304)

Calculate what the 60-hour MOS maximum temperature forecast would be for the following:

Valid time	48-hr 850 mb T fcst	48-hr mean RH fcst
a. September 4	278°K	30%
b. November 28	278°K	30%
c. November 28	258°K	30%
d. November 28	278°K	90%

6.8. A MOS equation for 12–24 hour PoP in the warm season might look something like:

$$\text{PoP} = 0.25 + .0063(\text{Mean RH}) - .163(0 - 12 \text{ ppt[bin@0.1 in.]})$$

$$-.165(\text{Mean RH[bin@70\%]})$$

where:

Mean RH is the same variable as in Exercise 6.7 (in %) for the appropriate projection
0–12 ppt is the model-forecast precipitation amount in the first 12 hours of the forecast
[bin @ xxx] indicates use as a binary variable:
$\quad = 1$ if the predictor is $\leq$ xxx
$\quad = 0$ otherwise

Evaluate the MOS PoP forecasts for the following conditions:

	12-hour mean RH	0–12 ppt
a.	90%	0.00 in.
b.	65%	0.15 in.
c.	75%	0.15 in.
d.	75%	0.09 in.

6.9. Explain why the slopes of the solid lines decrease, from Figure 6.20 to Figure 6.21a, to Figure 6.21b. What would the corresponding MOS equation be for an arbitrarily long projection into the future?

6.10. A forecaster is equally happy with the prospect of receiving $1 for sure, or $5 if freezing temperatures occur on the following night. What is the forecaster's subjective probability for frost?

6.11. A forecaster is indifferent between receiving $1 for sure and any of the following: $8 if tomorrow's rainfall is greater than 55 mm, $4 if tomorrow's rainfall is greater than 32 mm, $2 if tomorrow's rainfall is greater than 12 mm, $1.33 if tomorrow's rainfall is greater than 5 mm, and $1.14 if tomorrow's precipitation is greater than 1 mm.

a. What is the median of this subjective distribution?

b. What would be a consistent 50% central credible interval forecast? A 75% central credible interval forecast?

c. In this forecaster's view, what is the probability of receiving more than one but no more than 32 mm of precipitation?

CHAPTER ◆ 7

Forecast Verification

7.1 Background

7.1.1 *Purposes of Forecast Verification*

Forecast verification is the process of assessing the quality of forecasts. This process perhaps has been most fully developed in the atmospheric sciences, although parallel developments have taken place within other disciplines as well (e.g., Stephenson and Jolliffe, 2003), where the activity is sometimes called validation, or evaluation. Verification of weather forecasts has been undertaken since at least 1884 (Muller 1944; Murphy 1996). In addition to this chapter, other reviews of forecast verification can be found in Jolliffe and Stephenson (2003), Livezey (1995b), Murphy (1997), Murphy and Daan (1985), and Stanski *et al.* (1989).

Perhaps not surprisingly, there can be differing views of what constitutes a good forecast (Murphy 1993). A wide variety of forecast verification procedures exist, but all involve measures of the relationship between a forecast or set of forecasts, and the corresponding observation(s) of the predictand. Any forecast verification method thus necessarily involves comparisons between matched pairs of forecasts and the observations to which they pertain.

On a fundamental level, forecast verification involves investigation of the properties of the joint distribution of forecasts and observations (Murphy and Winkler 1987). That is, any given verification data set consists of a collection of forecast/observation pairs whose joint behavior can be characterized in terms of the relative frequencies of the possible combinations of forecast/observation outcomes. A parametric joint distribution such as the bivariate normal (see Section 4.4.2) can sometimes be useful in representing this joint distribution for a particular data set, but the empirical joint distribution of these quantities (more in the spirit of Chapter 3) more usually forms the basis of forecast verification measures. Ideally, the association between forecasts and the observations to which they pertain will be reasonably strong, and the nature and strength of this association will be reflected in their joint distribution.

Objective evaluations of forecast quality are undertaken for a variety of reasons. Brier and Allen (1951) categorized these as serving administrative, scientific, and economic purposes. In this view, administrative use of forecast verification pertains to ongoing monitoring of operational forecasts. For example, it is often of interest to examine trends of forecast performance through time. Rates of forecast improvement, if any, for different locations or lead times can be compared. Verification of forecasts from different sources for the same events can also be compared. Here forecast verification techniques allow comparison of the relative merits of competing forecasters or forecasting systems. This is the purpose to which forecast verification is often put in scoring student "forecast games" at colleges and universities.

Analysis of verification statistics and their components can also help in the assessment of specific strengths and weaknesses of forecasters or forecasting systems. Although classified by Brier and Allen as scientific, this application of forecast verification is perhaps better regarded as diagnostic verification (Murphy *et al.* 1989; Murphy and Winkler 1992). Here specific attributes of the relationship between forecasts and the subsequent events are investigated, which can highlight strengths and deficiencies in a set of forecasts. Human forecasters can be given feedback on the performance of their forecasts in different situations, which hopefully will lead to better forecasts in the future. Similarly, forecast verification measures can point to problems in forecasts produced by objective means, possibly leading to better forecasts through methodological improvements.

Ultimately, the justification for any forecasting enterprise is that it supports better decision making. The usefulness of forecasts to support decision making clearly depends on their error characteristics, which are elucidated through forecast verification methods. Thus the economic motivations for forecast verification are to provide the information necessary for users to derive full economic value from forecasts, and to enable estimation of that value. However, since the economic value of forecast information in different decision situations must be evaluated on a case-by-case basis (e.g., Katz and Murphy, 1997a), forecast value cannot be computed from the verification statistics alone. Similarly, although it is sometimes possible to guarantee the economic superiority of one forecast source over another for all forecast users on the basis of a detailed verification analysis, which is a condition called sufficiency (Ehrendorfer & Murphy 1988; Krzysztofowicz and Long 1990, 1991; Murphy 1997; Murphy and Ye 1990), superiority with respect to a single verification measure does not necessarily imply superior forecast value for all users.

7.1.2 *The Joint Distribution of Forecasts and Observations*

The joint distribution of the forecasts and observations is of fundamental interest with respect to the verification of forecasts. In practical settings, both the forecasts and observations are discrete variables. That is, even if the forecasts and observations are not already discrete quantities, they are rounded operationally to one of a finite set of values. Denote the forecast by y_i, which can take on any of the I values $y_1, y_2, \ldots, y_I$; and the corresponding observation as o_j, which can take on any of the J values $o_1, o_2, \ldots, o_J$. Then the joint distribution of the forecasts and observations is denoted

$$p(y_i, o_j) = \Pr\{y_i, o_j\} = \Pr\{y_i \cap o_j\}; \quad i = 1, \ldots, I; \ j = 1, \ldots, J. \tag{7.1}$$

This is a discrete bivariate probability distribution function, associating a probability with each of the $I \times J$ possible combinations of forecast and observation.

Even in the simplest cases, for which $I = J = 2$, this joint distribution can be difficult to use directly. From the definition of conditional probability (Equation 2.10) the joint distribution can be factored in two ways that are informative about different aspects of the verification problem. From a forecasting standpoint, the more familiar and intuitive of the two is

$$p(y_i, o_j) = p(o_j|y_i)\, p(y_i); \quad i = 1, \ldots, I;\ j = 1, \ldots, J; \tag{7.2}$$

which is called the calibration-refinement factorization (Murphy and Winkler 1987). One part of this factorization consists of a set of the I conditional distributions, $p(o_j|y_i)$, each of which consists of probabilities for all the J outcomes o_j, given one of the forecasts y_i. That is, each of these conditional distributions specifies how often each possible weather event occurred on those occasions when the single forecast y_i was issued, or how well each forecast y_i is calibrated. The other part of this factorization is the unconditional (marginal) distribution $p(y_i)$, which specifies the relative frequencies of use of each of the forecast values y_i, or how often each of the I possible forecast values were used. This marginal distribution is sometimes called the predictive distribution, or the refinement distribution of the forecasts. The refinement of a set of forecasts refers to the dispersion of the distribution $p(y_i)$. A refinement distribution with a large spread implies refined forecasts, in that different forecasts are issued relatively frequently, and so have the potential to discern a broad range of conditions. Conversely, if most of the forecasts f_i are the same or very similar, $p(f_i)$ is narrow, which indicates a lack of refinement. This attribute of forecast refinement often is referred to as sharpness in the sense that refined forecasts are called sharp.

The other factorization of the joint distribution of forecasts and observations is the likelihood-base rate factorization (Murphy and Winkler 1987),

$$p(y_i, o_j) = p(y_i|o_j)\, p(o_j); \quad i = 1, \ldots, I;\ j = 1, \ldots, J. \tag{7.3}$$

Here the conditional distributions $p(y_i|o_j)$ express the likelihoods that each of the allowable forecast values y_i would have been issued in advance of each of the observed weather events o_j. Although this concept may seem logically reversed, it can reveal useful information about the nature of forecast performance. In particular, these conditional distributions relate to how well a set of forecasts are able to discriminate among the events o_j, in the same sense of the word used in Chapter 13. The unconditional distribution $p(o_j)$ consists simply of the relative frequencies of the J weather events o_j in the verification data set, or the underlying rates of occurrence of each of the events o_j in the verification data sample. This distribution usually is called the sample climatological distribution, or simply the sample climatology.

Both the likelihood-base rate factorization (Equation 7.3) and the calibration-refinement factorization (Equation 7.2) can be calculated from the full joint distribution $p(y_i, o_j)$. Conversely, the full joint distribution can be reconstructed from either of the two factorizations. Accordingly, the full information content of the joint distribution $p(y_i, o_j)$ is included in either pair of distributions, Equation 7.2 or Equation 7.3. Forecast verification approaches based on these distributions are sometimes known as distributions-oriented (Murphy 1997) approaches, in distinction to potentially incomplete summaries based on one or a few scalar verification measures, known as measures-oriented approaches.

Although the two factorizations of the joint distribution of forecasts and observations can help organize the verification information conceptually, neither reduces the dimensionality (Murphy 1991), or degrees of freedom, of the verification problem. That is, since all the probabilities in the joint distribution (Equation 7.1) must add to 1, it is completely specified by any $(I \times J) - 1$ of these probabilities. The factorizations of Equations 7.2 and 7.3 reexpress this information differently and informatively, but $(I \times J) - 1$ distinct probabilities are still required to completely specify each factorization.

7.1.3 Scalar Attributes of Forecast Performance

Even in the simplest case of $I = J = 2$, complete specification of forecast performance requires a $(I \times J) - 1 = 3$- dimensional set of verification measures. This minimum level of dimensionality is already sufficient to make understanding and comparison of forecast evaluation statistics difficult. The difficulty is compounded in the many verification situations where $I > 2$ and/or $J > 2$, and such higher-dimensional verification situations may be further complicated if the sample size is not large enough to obtain good estimates for the required $(I \times J) - 1$ probabilities. As a consequence, it is traditional to summarize forecast performance using one or several scalar (i.e., one-dimensional) verification measures. Many of the scalar summary statistics have been found through analysis and experience to provide very useful information about forecast performance, but some of the information in the full joint distribution of forecasts and observations is inevitably discarded when the dimensionality of the verification problem is reduced.

The following is a partial list of scalar aspects, or attributes, of forecast quality. These attributes are not uniquely defined, so that each of these concepts may be expressible by more than one function of a verification data set.

1) *Accuracy* refers to the average correspondence between individual forecasts and the events they predict. Scalar measures of accuracy are meant to summarize, in a single number, the overall quality of a set of forecasts. Several of the more common measures of accuracy will be presented in subsequent sections. The remaining forecast attributes in this list can often be interpreted as components, or aspects, of accuracy.

2) *Bias*, or unconditional bias, or systematic bias, measures the correspondence between the average forecast and the average observed value of the predictand. This concept is different from accuracy, which measures the average correspondence between individual pairs of forecasts and observations. Temperature forecasts that are consistently too warm or precipitation forecasts that are consistently too wet both exhibit bias, whether or not the forecasts are otherwise reasonably accurate or quite inaccurate.

3) *Reliability* , or calibration, or conditional bias, pertains to the relationship of the forecast to the average observation, for specific values of (i.e., conditional on) the forecast. Reliability statistics sort the forecast/observation pairs into groups according to the value of the forecast variable, and characterize the conditional distributions of the observations given the forecasts. Thus, measures of reliability summarize the I conditional distributions $p(o_j|y_i)$ of the calibration-refinement factorization (Equation 7.2).

4) *Resolution* refers to the degree to which the forecasts sort the observed events into groups that are different from each other. It is related to reliability, in that both are concerned with the properties of the conditional distributions of the observations given the forecasts, $p(o_j|y_i)$. Therefore, resolution also relates to the calibration-refinement factorization of the joint distribution of forecasts and observations. However, resolution pertains to the differences between the conditional averages of the observations for different values of the forecast, whereas reliability compares the conditional averages of the observations with the forecast values themselves. If average temperature outcomes following forecasts of, say, 10°C and 20°C are very different, the forecasts can resolve these different temperature outcomes, and are said to exhibit resolution. If the temperature outcomes following forecasts of 10°C and 20°C are nearly the same on average, the forecasts exhibit almost no resolution.

5) *Discrimination* is the converse of resolution, in that it pertains to differences between the conditional averages of the forecasts for different values of the observation. Measures of discrimination summarize the conditional distributions of the forecasts

given the observations, $p(y_i|o_j)$, in the likelihood-base rate factorization (Equation 7.3). The discrimination attribute reflects the ability of the forecasting system to produce different forecasts for those occasions having different realized outcomes of the predictand. If a forecasting system forecasts $y =$ snow with equal frequency when $o =$ snow and $o =$ sleet, the two conditional probabilities of a forecast of snow are equal, and the forecasts are not able to discriminate between snow and sleet events.

6) *Sharpness*, or refinement, is an attribute of the forecasts alone, without regard to their corresponding observations. Measures of sharpness characterize the unconditional distribution (relative frequencies of use) of the forecasts, $p(y_i)$ in the calibration-refinement factorization (Equation 7.2). Forecasts that rarely deviate much from the climatological value of the predictand exhibit low sharpness. In the extreme, forecasts consisting only of the climatological value of the predictand exhibit no sharpness. By contrast, forecasts that are frequently much different from the climatological value of the predictand are sharp. Sharp forecasts exhibit the tendency to "stick their neck out." Sharp forecasts will be accurate only if they also exhibit good reliability, or calibration: anyone can produce sharp forecasts, but the difficult task is to ensure that these forecasts correspond well to the subsequent observations.

7.1.4 Forecast Skill

Forecast skill refers to the relative accuracy of a set of forecasts, with respect to some set of standard control, or reference, forecasts. Common choices for the reference forecasts are climatological average values of the predictand, persistence forecasts (values of the predictand in the previous time period), or random forecasts (with respect to the climatological relative frequencies of the forecast events o_j). Yet other choices for the reference forecasts can be more appropriate in some cases. For example, when evaluating the performance of a new forecasting system, it might be appropriate to compute skill relative to the forecasts that this new system might replace.

Forecast skill is usually presented as a skill score, which is interpreted as a percentage improvement over the reference forecasts. In generic form, the skill score for forecasts characterized by a particular measure of accuracy A, with respect to the accuracy A_{ref} of a set of reference forecasts, is given by

$$SS_{ref} = \frac{A - A_{ref}}{A_{perf} - A_{ref}} \times 100\%, \qquad (7.4)$$

where A_{perf} is the value of the accuracy measure that would be achieved by perfect forecasts. Note that this generic skill score formulation gives consistent results whether the accuracy measure has a positive (larger values of A are better) or negative (smaller values of A are better) orientation. If $A = A_{perf}$ the skill score attains its maximum value of 100%. If $A = A_{ref}$ then $SS_{ref} = 0\%$, indicating no improvement over the reference forecasts. If the forecasts being evaluated are inferior to the reference forecasts with respect to the accuracy measure A, $SS_{ref} < 0\%$.

The use of skill scores often is motivated by a desire to equalize effects of intrinsically more or less difficult forecasting situations, when comparing forecasters or forecast systems. For example, forecasting precipitation in a very dry climate is generally relatively easy, since forecasts of zero, or the climatological average (which will be very near zero), will exhibit good accuracy on most days. If the accuracy of the reference forecasts (A_{ref} in Equation 7.4) is relatively high, a higher accuracy A is required to achieve a given skill level than would be the case in a more difficult forecast situation, in which A_{ref} would be smaller. Some of the effects of the intrinsic ease or difficulty of different forecast situations can be equalized through use of skill scores such as Equation 7.4, but unfortunately skill scores have not been found to be fully effective for this purpose (Glahn and Jorgensen 1970; Winkler 1994, 1996).

7.2 Nonprobabilistic Forecasts of Discrete Predictands

Forecast verification is perhaps easiest to understand with reference to nonprobabilistic forecasts of discrete predictands. Nonprobabilistic indicates that the forecast consists of an unqualified statement that a single outcome will occur. Nonprobabilistic forecasts contain no expression of uncertainty, in distinction to probabilistic forecasts. A discrete predictand is an observable variable that takes on one and only one of a finite set of possible values. This is in distinction to a continuous predictand, which (at least conceptually) may take on any value on the relevant portion of the real line.

Verification for nonprobabilistic forecasts of discrete predictands has been undertaken since the nineteenth century (Murphy 1996), and during this considerable time a variety of sometimes conflicting terminology has been used. For example, nonprobabilistic forecasts have been called categorical, in the sense their being firm statements that do not admit the possibility of alternative outcomes. However, more recently the term categorical has come be understood as relating to a predictand belonging to one of a set of MECE categories; that is, a discrete variable. In an attempt to avoid confusion, the term categorical will be avoided here, in favor of the more explicit terms, nonprobabilistic and discrete. Other instances of the multifarious nature of forecast verification terminology will also be noted in this chapter.

7.2.1 The 2×2 Contingency Table

There is typically a one-to-one correspondence between allowable nonprobabilistic forecast values and the discrete observable predictand values to which they pertain. In terms of the joint distribution of forecasts and observations (Equation 7.1), $I = J$. The simplest possible situation is for the dichotomous $I = J = 2$ case, or verification of nonprobabilistic yes/no forecasts. Here there are $I = 2$ possible forecasts, either that the event will ($i = 1$, or y_1) or will not ($i = 2$, or y_2) occur. Similarly, there are $J = 2$ possible outcomes: either the event subsequently occurs (o_1) or it does not (o_2). Despite the simplicity of this verification setting, a surprisingly large body of work on the 2×2 verification problem has developed.

Conventionally, nonprobabilistic verification data is displayed in an $I \times J$ contingency table of absolute frequencies, or counts, of the $I \times J$ possible combinations of forecast and event pairs. If these counts are transformed to relative frequencies, by dividing each tabulated entry by the sample size (total number of forecast/event pairs), the (sample) joint distribution of forecasts and observations (Equation 7.1) is obtained. Figure 7.1 illustrates the essential equivalence of the contingency table and the joint distribution of forecasts and observations for the simple $I = J = 2$ case. The boldface portion in Figure 7.1a shows the arrangement of the four possible combinations of forecast/event pairs as a square contingency table, and the corresponding portion of Figure 7.1b shows these counts transformed to joint relative frequencies.

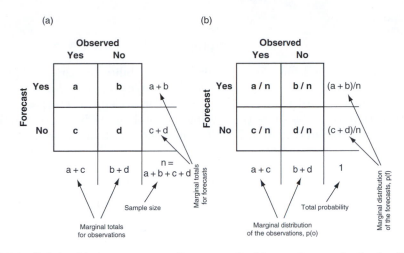

FIGURE 7.1   Relationship between counts (letters $a - d$) of forecast/event pairs for the dichotomous nonprobabilistic verification situation as displayed in a 2×2 contingency table (bold, panel a), and the corresponding joint distribution of forecasts and observations $[p(y, o)]$ (bold, panel b). Also shown are the marginal totals, indicating how often each of the two events were forecast and observed in absolute terms; and the marginal distributions of the observations $[p(o)]$ and forecasts $[p(y)]$, which indicates the same information in relative frequency terms.

In terms of Figure 7.1, the event in question was successfully forecast to occur a times out of n total forecasts. These a forecast-observation pairs usually are called *hits*, and their relative frequency, a/n, is the sample estimate of the corresponding joint probability $p(y_1, o_1)$ in Equation 7.1. Similarly, on b occasions, called *false alarms*, the event was forecast to occur but did not, and the relative frequency b/n estimates the joint probability $p(y_1, o_2)$. There are also c instances of the event of interest occurring despite not being forecast, called *misses*, the relative frequency of which estimates the joint probability $p(y_2, o_1)$; and d instances of the event not occurring after a forecast that it would not occur, sometimes called a *correct rejection* or *correct negative*, the relative frequency of which corresponds to the joint probability $p(y_2, o_2)$.

It is also common to include what are called *marginal totals* with a contingency table of counts. These are simply the row and column totals yielding, in this case, the numbers of times each yes or no forecast, or observation, respectively, occurred. These are shown in Figure 7.1a in normal typeface, as is the sample size, $n = a + b + c + d$. Expressing the marginal totals in relative frequency terms, again by dividing through by the sample size, yields the marginal distribution of the forecasts, $p(y_i)$, and the marginal distribution of the observations, $p(o_j)$. The marginal distribution $p(y_i)$ is the refinement distribution, of the calibration-refinement factorization (Equation 7.2) of the 2×2 joint distribution in Figure 7.1b. Since there are $I = 2$ possible forecasts, there are two calibration distributions $p(o_j|y_i)$, each of which consists of $J = 2$ probabilities. Therefore, in addition to the refinement distribution $p(y_1) = (a + b)/n$ and $p(y_2) = (c + d)/n$, the calibration-refinement factorization in the 2×2 verification setting consists of the conditional probabilities

$$p(o_1|y_1) = a/(a + b) \tag{7.5a}$$

$$p(o_2|y_1) = b/(a + b) \tag{7.5b}$$

$$p(o_1|y_2) = c/(c + d) \tag{7.5c}$$

and

$$p(o_2|y_2) = d/(c+d). \tag{7.5d}$$

In terms of the definition of conditional probability (Equation 2.10), Equation 7.5a (for example) would be obtained as $[a/n]/[(a+b)/n] = a/(a+b)$.

Similarly, the marginal distribution $p(o_j)$, with elements $p(o_1) = (a+c)/n$ and $p(o_2) = (b+d)/n$, is the base-rate (i.e., sample climatological) distribution in the likelihood-base rate factorization (Equation 7.3). The remainder of that factorization consists of the four conditional probabilities

$$p(y_1|o_1) = a/(a+c) \tag{7.6a}$$

$$p(y_2|o_1) = c/(a+c) \tag{7.6b}$$

$$p(y_1|o_2) = b/(b+d) \tag{7.6c}$$

and

$$p(y_2|o_2) = d/(b+d). \tag{7.6d}$$

7.2.2 Scalar Attributes Characterizing 2×2 Contingency Tables

Even though the 2×2 contingency table summarizes verification data for the simplest possible forecast setting, its dimensionality is 3. That is, the forecast performance information contained in the contingency table cannot fully be expressed with fewer than three parameters. It is perhaps not surprising that a wide variety of these scalar attributes have been devised and used to characterize forecast performance, over the long history of the verification of forecasts of this type. Unfortunately, a similarly wide variety of nomenclature also has appeared in relation to these attributes. This section lists scalar attributes of the 2×2 contingency table that have been most widely used, together with much of the synonymy associated with them. The organization follows the general classification of attributes in Section 7.1.3.

Accuracy

Accuracy measures reflect correspondence between pairs of forecasts and the events they are meant to predict. Perfectly accurate forecasts in the 2×2 nonprobabilistic forecasting situation will clearly exhibit $b = c = 0$, with all yes forecasts for the event followed by the event and all no forecasts for the event followed by nonoccurrence. For real, imperfect forecasts, accuracy measures characterize degrees of this correspondence. Several scalar accuracy measures are in common use, with each reflecting somewhat different aspects of the underlying joint distribution.

Perhaps the most direct and intuitive measure of the accuracy of nonprobabilistic forecasts for discrete events is the proportion correct proposed by Finley (1884). This is simply the fraction of the n forecast occasions for which the nonprobabilistic forecast correctly anticipated the subsequent event or non event. In terms of the counts Figure 7.1a, the proportion correct is given by

$$PC = \frac{a+d}{n}. \tag{7.7}$$

The proportion correct satisfies the principle of equivalence of events, since it credits correct yes and no forecasts equally. As Example 7.1 will show, however, this is not always a desirable attribute, particularly when the yes event is rare, so that correct no forecasts can be made fairly easily. The proportion correct also penalizes both kinds of errors (false alarms and misses) equally. The worst possible proportion correct is zero. The best possible proportion correct is one. Sometimes PC in Equation 7.7 is multiplied by 100%, and referred to as the percent correct, or percentage of forecasts correct. Because the proportion correct does not distinguish between correct forecasts of the event, a, and correct forecasts of the nonevent, d, this fraction of correct forecasts has also been called the hit rate. However, in current usage the term hit rate usually is reserved for the discrimination measure given in Equation 7.12.

An alternative to the proportion correct that is particularly useful when the event to be forecast (as the yes event) occurs substantially less frequently than the nonoccurrence (no), is the threat score (TS), or critical success index (CSI). In terms of Figure 7.1a, the threat score is computed as

$$TS = CSI = \frac{a}{a+b+c}. \tag{7.8}$$

The threat score is the number of correct yes forecasts divided by the total number of occasions on which that event was forecast and/or observed. It can be viewed as a proportion correct for the quantity being forecast, after removing correct no forecasts from consideration. The worst possible threat score is zero, and the best possible threat score is one. When originally proposed (Gilbert 1884) it was called the ratio of verification, and denoted as v, and so Equation 7.8 is sometimes called the Gilbert Score (as distinct from the Gilbert Skill Score, Equation 7.18). Very often, each of the counts in a 2×2 contingency table pertains to a different forecasting occasion (as illustrated in Example 7.1), but the threat score (and the skill score based on it, Equation 7.18) often is used to asses simultaneously issued spatial forecasts, for example severe weather warnings (e.g., Doswell *et al.* 1990; Ebert and McBride 2000; Schaefer 1990; Stensrud and Wandishin 2000). In this setting, a represents the intersection of the areas over which the event was forecast and subsequently occurred, b represents the area over which the event was forecast but failed to occur, and c is the area over which the event occurred but was not forecast to occur.

A third approach to characterizing forecast accuracy in the 2×2 situation is in terms of odds, or the ratio of a probability to its complementary probability, $p/(1-p)$. In the context of forecast verification the ratio of the conditional odds of a hit, given that the event occurs, to the conditional odds of a false alarm, given that the event does not occur, is called the odds ratio,

$$\theta = \frac{p(y_1|o_1)/[1-p(y_1|o_1)]}{p(y_1|o_2)/[1-p(y_1|o_2)]} = \frac{p(y_1|o_1)/p(y_2|o_1)}{p(y_1|o_2)/p(y_2|o_2)} = \frac{a\,d}{b\,c}. \tag{7.9}$$

The conditional distributions making up the odds ratio are all likelihoods from Equation 7.6. In terms of the 2×2 contingency table, the odds ratio is the product of the numbers of correct forecasts to the product of the numbers of incorrect forecasts. Clearly larger values of this ratio indicate more accurate forecasts. No-information forecasts, for which the forecasts and observations are statistically independent (i.e. $p(y_i, o_j) = p(y_i)p(o_j)$, cf. Equation 2.12), yield $\theta = 1$. The odds ratio was introduced into meteorological forecast verification by Stephenson (2000), although it has a longer history of use in medical statistics.

Bias

The bias, or comparison of the average forecast with the average observation, usually is represented as a ratio for verification of contingency tables. In terms of the 2×2 table in Figure 7.1a the bias ratio is

$$B = \frac{a+b}{a+c}. \tag{7.10}$$

The bias is simply the ratio of the number of yes forecasts to the number of yes observations. Unbiased forecasts exhibit $B = 1$, indicating that the event was forecast the same number of times that it was observed. Note that bias provides no information about the correspondence between the forecasts and observations of the event on particular occasions, so that the bias is not an accuracy measure. Bias greater than one indicates that the event was forecast more often than observed, which is called overforecasting. Conversely, bias less than one indicates that the event was forecast less often than observed, or was underforecast.

Reliability and Resolution

Equation 7.5 shows four reliability attributes for the 2×2 contingency table. That is, each quantity in Equation 7.5 is a conditional relative frequency for event occurrence or nonoccurrence, given either a yes or no forecast, in the sense of the calibration distributions $p(o_j|y_i)$ of Equation 7.2. Actually, Equation 7.5 indicates two calibration distributions, one conditional on the yes forecasts (Equations 7.5a and 7.5b), and the other conditional on the no forecasts (Equations 7.5c and 7.5d). Each of these four conditional probabilities is a scalar reliability statistic for the 2×2 contingency table, and all four have been given names (e.g., Doswell *et al.* 1990). By far the most commonly used of these conditional relative frequencies is Equation 7.5b, which is called the false alarm ratio (FAR). In terms of Figure 7.1a, the false alarm ratio is computed as

$$FAR = \frac{b}{a+b}. \tag{7.11}$$

That is, FAR is the fraction of yes forecasts that turn out to be wrong, or that proportion of the forecast events that fail to materialize. The FAR has a negative orientation, so that smaller values of FAR are to be preferred. The best possible FAR is zero, and the worst possible FAR is one. The FAR has also been called the false alarm *rate*, although this rather similar term is now generally reserved for the discrimination measure in Equation 7.13.

Discrimination

Two of the conditional probabilities in Equation 7.6 are used frequently to characterize 2×2 contingency tables, although all four of them have been named (e.g. Doswell *et al.* 1990). Equation 7.6a is commonly known as the hit rate,

$$H = \frac{a}{a+c}. \tag{7.12}$$

Regarding only the event o_1 as "the" event of interest, the hit rate is the ratio of correct forecasts to the number of times this event occurred. Equivalently this statistic can be regarded as the fraction of those occasions when the forecast event occurred on which it was also forecast, and so is also called the probability of detection (POD).

Equation 7.6c is called the false alarm rate,

$$F = \frac{b}{b+d},$$

(7.13)

which is the ratio of false alarms to the total number of nonoccurrences of the event o_1, or the conditional relative frequency of a wrong forecast given that the event does not occur. The false alarm rate is also known as the probability of false detection (POFD). Because the forecasts summarized in 2×2 tables are for dichotomous events, the hit rate and false alarm rate fully determine the four conditional probabilities in Equation 7.6. Jointly they provide both the conceptual and geometrical basis for the signal detection approach for verifying probabilistic forecasts (Section 7.4.6).

7.2.3 Skill Scores for 2×2 Contingency Tables

Commonly, forecast verification data in contingency tables are characterized using relative accuracy measures, or skill scores in the general form of Equation 7.4. A large number of such skill scores have been developed for the 2×2 verification situation, and many of these are presented by Muller (1944), Mason (2003), Murphy and Daan (1985), Stanski et al. (1989), and Woodcock (1976). A number of these skill measures date from the earliest literature on forecast verification (Murphy 1996), and have been rediscovered and (unfortunately) renamed on multiple occasions. In general the different skill scores perform differently, and sometimes inconsistently. This situation can be disconcerting if we hope to choose among alternative skill scores, but should not really be surprising given that all these skill scores are scalar measures of forecast performance in what is intrinsically a higher-dimensional setting. Scalar skill scores are used because they are conceptually convenient, but they are necessarily incomplete representations of forecast performance.

One of the most frequently used skill scores for summarizing square contingency tables was originally proposed by Doolittle (1888), but because it is nearly universally known as the Heidke Skill Score (Heidke 1926) this latter name will be used here. The Heidke Skill Score (HSS) is a skill score following the form of Equation 7.4, based on the proportion correct (Equation 7.7) as the basic accuracy measure. Thus, perfect forecasts receive $HSS = 1$, forecasts equivalent to the reference forecasts receive zero scores, and forecasts worse than the reference forecasts receive negative scores.

The reference accuracy measure in the Heidke score is the proportion correct that would be achieved by random forecasts that are statistically independent of the observations. In the 2×2 situation, the marginal probability of a yes forecast is $p(y_1) = (a+b)/n$, and the marginal probability of a yes observation is $p(o_1) = (a+c)/n$. Therefore, the probability of a correct yes forecast by chance is

$$p(y_1)p(o_1) = \frac{(a+b)}{n}\frac{(a+c)}{n} = \frac{(a+b)(a+c)}{n^2},$$

(7.14a)

and similarly the probability of a correct "no" forecast by chance is

$$p(y_2)p(o_2) = \frac{(b+d)}{n}\frac{(c+d)}{n} = \frac{(b+d)(c+d)}{n^2}.$$

(7.14b)

Thus, for the 2×2 verification setting, the Heidke Skill Score is

$$
\begin{aligned}
\text{HSS} &= \frac{(a+d)/n - [(a+b)(a+c)+(b+d)(c+d)]/n^2}{1 - [(a+b)(a+c)+(b+d)(c+d)]/n^2} \\
&= \frac{2(ad-bc)}{(a+c)(c+d)+(a+b)(b+d)},
\end{aligned}
\tag{7.15}
$$

where the second equality is easier to compute.

Another popular skill score for contingency-table forecast verification has been redis-covered many times since being first proposed by Peirce (1884). It is also commonly referred to as the Hanssen-Kuipers discriminant (Hanssen and Kuipers 1965) or Kuipers' performance index (Murphy and Daan 1985), and is sometimes also called the true skill statistic (TSS) (Flueck 1987). Gringorten's (1967) skill score contains equivalent information, as it is a linear transformation of the Peirce Skill Score.

The Peirce Skill Score is formulated similarly to the Heidke score, except that the reference hit rate in the denominator is that for random forecasts that are constrained to be unbiased. That is, the imagined random reference forecasts in the denominator have a marginal distribution that is equal to the (sample) climatology, so that $p(y_1) = p(o_1)$ and $p(y_2) = p(o_2)$. In the 2×2 situation of Figure 7.1, the Peirce Skill Score is computed as

$$
\begin{aligned}
\text{PSS} &= \frac{(a+d)/n - [(a+b)(a+c)+(b+d)(c+d)]/n^2}{1 - [(a+c)^2 + (b+d)^2]/n^2} \\
&= \frac{ad-bc}{(a+c)(b+d)},
\end{aligned}
\tag{7.16}
$$

where again the second equality is computationally more convenient. The PSS can also be understood as the difference between two conditional probabilities in the likelihood-base rate factorization of the joint distribution (Equation 7.6), namely the hit rate (Equation 7.12) and the false alarm rate (Equation 7.13); that is, $\text{PSS} = H - F$. Perfect forecasts receive a score of one (because $b = c = 0$; or in an alternative view, $H = 1$ and $F = 0$), random forecasts receive a score of zero (because $H = F$), and forecasts inferior to the random forecasts receive negative scores. Constant forecasts (i.e., always forecasting one or the other of y_1 or y_2) are also accorded zero skill. Furthermore, unlike the Heidke score, the contribution made to the Peirce Skill Score by a correct no or yes forecast increases as the event is more or less likely, respectively. Thus, forecasters are not discouraged from forecasting rare events on the basis of their low climatological probability alone.

The Clayton (1927, 1934) Skill Score can be formulated as the difference of the conditional probabilities in Equation 7.5a and 7.5c, relating to the calibration-refinement factorization of the joint distribution; that is,

$$
\text{CSS} = \frac{a}{a+b} - \frac{c}{c+d} = \frac{ad-bc}{(a+b)(c+d)}.
\tag{7.17}
$$

The CSS indicates positive skill to the extent that the event occurs more frequently when forecast than when not forecast, so that the conditional relative frequency of the yes outcome given yes forecasts is larger than the conditional relative frequency given no forecasts. Clayton (1927) originally called this difference of conditional relative frequencies (multiplied by 100%) the percentage of skill, where he understood skill in a modern sense of accuracy relative to climatological expectancy. Perfect forecasts exhibit $b = c = 0$, yielding $\text{CSS} = 1$. Random forecasts (Equation 7.14) yield $\text{CSS} = 0$.

A skill score in the form of Equation 7.4 can be constructed using the threat score (Equation 7.8) as the basic accuracy measure, using TS for random (Equation 7.14) forecasts as the reference. In particular, $TS_{ref} = a_{ref}/(a+b+c)$, where Equation 7.14a implies $a_{ref} = (a+b)(a+c)/n$. Since $TS_{perf} = 1$, the resulting skill score is

$$GSS = \frac{a/(a+b+c) - a_{ref}/(a+b+c)}{1 - a_{ref}/(a+b+c)} = \frac{a - a_{ref}}{a - a_{ref} + b + c}. \tag{7.18}$$

This skill score, called the Gilbert Skill Score (GSS) originated with Gilbert (1884), who referred to it as the ratio of success. It is also commonly called the Equitable Threat Score (ETS). Because the sample size n is required to compute a_{ref}, the GSS depends on the number of correct no forecasts, unlike the TS.

The odds ratio (Equation 7.9) can also be used as the basis of a skill score,

$$Q = \frac{\theta - 1}{\theta + 1} = \frac{ad/bc - 1}{ad/bc + 1} = \frac{ad - bc}{ad + bc}. \tag{7.19}$$

This skill score originated with Yule (1900), and is called Yule's Q (Woodcock 1976), or the Odds Ratio Skill Score (ORSS) (Stephenson 2000). Random (Equation 7.14) forecasts exhibit $\theta = 1$, yielding $Q = 0$; and perfect forecasts exhibit $b = c = 0$, producing $Q = 1$. However, an apparently perfect skill of $Q = 1$ is also obtained for imperfect forecasts, if either one or the other of b or c is zero.

All the skill scores listed in this section depend only on the four counts a, b, c, and d in Figure 7.1, and are therefore necessarily related. Notably, HSS, PSS, CSS, and Q are all proportional to the quantity $ad - bc$. Some specific mathematical relationships among the various skill scores are noted in Mason (2003), Murphy (1996), Stephenson (2000), and Wandishin and Brooks (2002).

EXAMPLE 7.1   The Finley Tornado Forecasts

A historical set of 2×2 forecast verification data set that often is used to illustrate evaluation of forecasts in this format is the collection of Finley's tornado forecasts (Finley 1884). John Finley was a sergeant in the U.S. Army who, using telegraphed synoptic information, formulated yes/no tornado forecasts for 18 regions of the United States east of the Rocky Mountains. The data set and its analysis were instrumental in stimulating much of the early work on forecast verification (Murphy 1996). The contingency table for Finley's $n = 2803$ forecasts is presented in Table 7.1a.

TABLE 7.1   Contingency tables for verification of the Finley tornado forecasts, from 1884. The forecast event is occurrence of a tornado, with separate forecasts for 18 regions of the United States east of the Rocky Mountains. (a) The table for the forecasts as originally issued; and (b) data that would have been obtained if no tornados had always been forecast.

(a)		Tornados Observed		(b)		Tornados Observed	
		Yes	No			Yes	No
Tornados	Yes	28	72	Tornados	Yes	0	0
Forecast	No	23	2680	Forecast	No	51	2752
		$n = 2803$				$n = 2803$	

Finley chose to evaluate his forecasts using the proportion correct (Equation 7.7), which for his data is PC = (28 + 2680)/2803 = 0.966. On the basis of this proportion correct, Finley claimed 96.6% accuracy. However, the proportion correct for this data set is dominated by the correct no forecasts, since tornados are relatively rare. Very shortly after Finley's paper appeared, Gilbert (1884) pointed out that always forecasting no would produce an even higher proportion correct. The contingency table that would be obtained if tornados had never been forecast is shown in Table 7.1b. These hypothetical forecasts yield a proportion correct of PC = (0 + 2752)/2803 = 0.982, which is indeed higher than the proportion correct for the actual forecasts.

Employing the threat score gives a more reasonable comparison, because the large number of easy, correct no forecasts are ignored. For Finley's original forecasts, the threat score is TS = 28/(28 + 72 + 23) = 0.228, whereas for the obviously useless no forecasts in Table 7.1b the threat score is TS = 0/(0 + 0 + 51) = 0. Clearly the threat score would be preferable to the proportion correct in this instance, but it is still not completely satisfactory. Equally useless would be a forecasting system that always forecast yes for tornados. For constant yes forecasts the threat score would be TS = 51/(51 + 2752 + 0) = 0.018, which is small, but not zero. The odds ratio for the Finley forecasts is $\theta = (28)(2680)/(72)(23) = 45.3 > 1$, suggesting better than random performance for the forecasts in Table 7.1a. The odds ratio is not computable for the forecasts in Table 7.1b.

The bias ratio for the Finely tornado forecasts is B = 1.96, indicating that approximately twice as many tornados were forecast as actually occurred. The false alarm ratio is FAR = 0.720, which expresses the fact that a fairly large fraction of the forecast tornados did not eventually occur. On the other hand, the hit rate is H = 0.549 and the false alarm rate is F = 0.0262; indicating that more than half of the actual tornados were forecast to occur, whereas a very small fraction of the nontornado cases falsely warned of a tornado.

The various skill scores yield a very wide range of results for the Finely tornado forecasts: HSS = 0.355, PSS = 0.523, CSS = 0.271, GSS = 0.216, and Q = 0.957. Zero skill is attributed to the constant no forecasts in Figure 7.1b by HSS, PSS and GSS, but CSS and Q cannot be computed for $a = b = 0$. ◊

7.2.4 Which Score?

The wide range of skills attributed to the Finley tornado forecasts in Example 7.1 may be somewhat disconcerting, but should not be surprising. The root of the problem is that, even in this simplest of all possible forecast verification settings, the dimensionality (Murphy 1991) of the problem is $I \times J - 1 = 3$, but the collapse of this three-dimensional information into a single number by any scalar verification measure necessarily involves a loss of information. Put another way, there are a variety of ways for forecasts to go right and for forecasts to go wrong, and different mixtures of these are combined by different scalar attributes and skill scores. There is no single answer to the question posed in the heading for this section.

Because the dimensionality of the 2×2 problem is 3, the full information in the 2×2 contingency table can be captured fully by three well-chosen scalar attributes. Using the likelihood-base rate factorization (Equation 7.6), the full joint distribution can be summarized by (and recovered from) the hit rate H (Equations 7.12 and 7.6a), the false alarm rate F (Equation 7.13 and 7.6c), and the base rate (or sample climatological relative frequency) $p(o_1) = (a + c)/n$. Similarly, using the calibration-refinement factorization (Equation 7.5), forecast performance depicted in a 2×2 contingency table can be fully captured using the false alarm ratio FAR (Equations 7.11 and 7.5b), its counterpart in

Equation 7.5d, and the probability $p(y_1) = (a+b)/n$ defining the calibration distribution. Other triplets of verification measures can also be used jointly to illuminate the data in a 2×2 contingency table (although not any three scalar statistics calculated from a 2×2 table will fully represent its information content). For example, Stephenson (2000) suggests use of H and F together with the bias ratio B, calling this the BHF representation. He also notes that, jointly, the likelihood ratio θ and Peirce Skill Score PSS represent the same information as H and F, so that these two statistics together with either $p(o_1)$ or B will also fully represent the 2×2 table.

It is sometimes necessary to choose a single scalar summary of forecast performance, accepting that the summary will necessarily be incomplete. For example, competing forecasters in a contest must be evaluated in a way that produces an unambiguous ranking of their performances. Choosing a single score for such a purpose involves investigating and comparing relevant properties of competing candidate verification statistics, a process that is called *metaverification* (Murphy 1996). Which property or properties might be most relevant may depend on the specific situation, but one reasonable choice is that a chosen verification statistic should be equitable (Gandin and Murphy 1992). An equitable skill score rates random forecasts, and all constant forecasts (such as no tornados in Example 7.1), equally. Usually this score is set to zero, and equitable scores are scaled such that perfect forecasts have unit skill. Equitability also implies that correct forecasts of less frequent events (such as tornados in Example 7.1) are weighted more strongly than correct forecasts of more common events, which discourages distortion of forecasts toward the more common event in order to artificially inflate the resulting score. In the 2×2 verification setting these considerations lead to the use of the Peirce skill score (Equation 7.16), if it can be assumed that false alarms $\{y_1 \cap o_2\}$ and misses $\{y_2 \cap o_1\}$ are equally undesirable. However, if these two kinds of errors are not equally severe, the notion of equitability does not fully inform the choice of a scoring statistic, even for 2×2 contingency tables (Marzban and Lakshmanan 1999). Derivation of equitable skill scores is described more fully in Section 7.2.6.

7.2.5 Conversion of Probabilistic to Nonprobabilistic Forecasts

The MOS system from which the nonprobabilistic precipitation amount forecasts in Table 6.7 were taken actually produces probability forecasts for discrete precipitation amount classes. The publicly issued precipitation amount forecasts were then derived by converting the underlying probabilities to the nonprobabilistic format by choosing one and only one of the possible categories. This unfortunate procedure is practiced with distressing frequency, and advocated under the rationale that nonprobabilistic forecasts are easier to understand. However, the conversion from probabilities inevitably results in a loss of information, to the detriment of the users of the forecasts.

For a dichotomous predictand, the conversion from a probabilistic to a nonprobabilistic format requires selection of a threshold probability, above which the forecast will be "yes," and below which the forecast will be "no." This procedure seems simple enough; however, the proper threshold to choose depends on the user of the forecast and the particular decision problem(s) to which that user will apply the forecast. Naturally, different decision problems will require different threshold probabilities, and this is the crux of the information-loss issue. In a real sense, the conversion from a probabilistic to a nonprobabilistic format amounts to the forecaster making decisions for the forecast

TABLE 7.2   Verification data for subjective 12-24h projection probability-of-precipitation forecasts for the United States during October 1980–March 1981, expressed in the form of the calibration-refinement factorization (Equation 7.2) of the joint distribution of these forecasts and observations. There are $I = 12$ allowable values for the forecast probabilities, y_i, and $J = 2$ events ($j = 1$ for precipitation and $j = 2$ for no precipitation). The sample climatological relative frequency is 0.162, and the sample size is $n = 12, 402$. From Murphy and Daan (1985).

y_i	0.00	0.05	0.10	0.20	0.30	0.40	0.50	0.60	0.70	0.80	0.90	1.00
$p(o_1|y_i)$	.006	.019	.059	.150	.277	.377	.511	.587	.723	.799	.934	.933
$p(y_i)$	.4112	.0671	.1833	.0986	.0616	.0366	.0303	.0275	.0245	.0220	.0170	.0203

users, but without knowing the particulars of their decision problems. Necessarily, then, the conversion from a probabilistic to a nonprobabilistic forecast is arbitrary.

EXAMPLE 7.2   Effects of Different Thresholds on Conversion to Nonprobabilistic Forecasts

It is instructive to examine the procedures used to convert probabilistic to nonprobabilistic forecasts. Table 7.2 contains a verification data set of probability of precipitation forecasts, issued for the United States during the period October 1980 through March 1981. Here the joint distribution of the $I = 12$ possible forecasts and the $J = 2$ possible observations is presented in the form of the calibration-refinement factorization (Equation 7.2). For each allowable forecast probability, y_i, the conditional probability $p(o_1|y_i)$ indicates the relative frequency of the event $j = 1$ (precipitation occurrence) for these $n = 12, 402$ forecasts. The marginal probabilities $p(y_i)$ indicate the relative frequencies with which each of the $I = 12$ possible forecast values were used.

These precipitation occurrence forecasts were issued as probabilities. If it had been intended to convert them first to a nonprobabilistic rain/no rain format, a threshold probability would have been chosen in advance. There are many possibilities for this choice, each of which give different results. The two simplest approaches are used rarely, if ever, in operational practice. The first procedure is to forecast the more likely event, which corresponds to selecting a threshold probability of 0.50. The other simple approach is to use the climatological relative frequency of the event being forecast as the threshold probability. For the data set in Table 7.2 this relative frequency is $\Sigma_i p(o_j|y_i)p(y_i) = 0.162$ (cf. 2.14), although in practice this probability threshold would need to have been estimated in advance using historical climatological data. Forecasting the more likely event turns out to maximize the expected values of both the proportion correct (Equation 7.7) and the Heidke Skill Score (Equation 7.15), and using the climatological relative frequency for the probability threshold maximizes the expected Peirce Skill Score (Equation 7.16) (Mason 1979).

The two methods for choosing the threshold probability that are most often used operationally are based on the threat score (Equation 7.8) and the bias ratio (Equation 7.10) for 2×2 contingency tables. For each possible choice of a threshold probability, a different 2×2 contingency table, in the form of Figure 7.1a, results, and therefore different values of TS and B are obtained. When using the threat score to choose the threshold, that threshold producing the maximum TS is selected. When using the bias ratio, choose that threshold producing, as nearly as possible, no bias ($B = 1$).

Figure 7.2 illustrates the dependence of the bias ratio and threat score on the threshold probability for the data given in Table 7.2. Also shown are the hit rates H and false alarm ratios FAR that would be obtained. The threshold probabilities that would be chosen according to the climatological relative frequency (Clim), the maximum threat

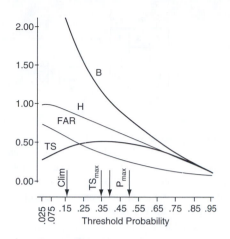

FIGURE 7.2   Derivation of candidate threshold probabilities for converting the probability-of-precipitation forecasts in Table 7.2 to nonprobabilistic rain/no rain forecasts. The Clim threshold indicates a forecast of rain if the probability is higher than the climatological probability of precipitation, TS_{max} is the threshold that would maximize the threat score of the resulting nonprobabilistic forecasts, the $B = 1$ threshold would produce unbiased forecasts, and the p_{max} threshold would produce nonprobabilistic forecasts of the more likely of the two events. Also shown (lighter lines) are the hit rates H and false alarm ratios FAR for the resulting 2×2 contingency tables.

score (TS_{max}), unbiased nonprobabilistic forecasts ($B = 1$), and maximum probability (p_{max}) are indicated by the arrows at the bottom of the figure. For example, choosing the relative frequency of precipitation occurrence, 0.162, as the threshold results in forecasts of PoP = 0.00, 0.05, and 0.10 being converted to no rain, and the other forecasts being converted to rain. This would have resulted in $n[p(y_1) + p(y_2) + p(y_3)] = 12,402[0.4112 + 0.0671 + 0.1833] = 8205$ no forecasts, and $12,402 - 8205 = 4197$ yes forecasts. Of the 8205 no forecasts, we can compute, using the multiplicative law of probability (Equation 2.11), that the proportion of occasions that no was forecast but precipitation occurred was $p(o_1|y_1)p(y_1) + p(o_1|y_2)p(y_2) + p(o_1|y_3)p(y_3) = (.006)(.4112) + (.019)(.0671) + (.059)(.1833) = 0.0146$. This relative frequency is c/n in Figure 7.1, so that $c = (0.0146)(12,402) = 181$, and $d = 8205 - 181 = 8024$. Similarly, we can compute that, for this cutoff, $a = 12,402[(0.150)(0.0986) + \cdots + (0.933)(0.203)] = 1828$ and $b = 2369$. The resulting 2×2 table yields $B = 2.09$, and $TS = 0.417$. By contrast, the threshold maximizing the threat score is near 0.35, which also would have resulted in overforecasting of precipitation occurrence. ◊

7.2.6 Extensions for Multicategory Discrete Predictands

Nonprobabilistic forecasts for discrete predictands are not limited to the 2×2 format, although that simple situation is the most commonly encountered and the easiest to understand. In some settings it is natural or desirable to consider and forecast more than two discrete MECE events. The left side of Figure 7.3, in boldface type, shows a generic contingency table for the case of $I = J = 3$ possible forecasts and events. Here the counts for each of the nine possible forecast/event pair outcomes are denoted by the letters r through z, yielding a total sample size $n = r + s + t + u + v + w + x + y + z$. As before,

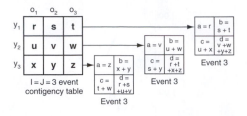

FIGURE 7.3   Contingency table for the $I = J = 3$ nonprobabilistic forecast verification situation (bold), and its reduction to three 2×2 contingency tables. Each 2×2 contingency table is constructed by regarding one of the three original events as the event being forecast, and the remaining two original events combined as complementary; i.e., not the forecast event. For example, the 2×2 table for Event 1 lumps Event 2 and Event 3 as the single event "not Event 1." The letters a, b, c, and d are used in the same sense as in Figure 7.1a. Performance measures specific to the 2×2 contingency tables can then be computed separately for each of the resulting tables. This procedure generalizes easily to square forecast verification contingency tables with arbitrarily many forecast and event categories.

dividing each of the nine counts in this 3×3 contingency table by the sample size yields a sample estimate of the joint distribution of forecasts and observations, $p(y_i, o_j)$.

Of the accuracy measures listed in Equations 7.7 through 7.9, only the proportion correct (Equation 7.7) generalizes directly to situations with more than two forecast and event categories. Regardless of the size of I and J, the proportion correct is still given by the number of correct forecasts divided by the total number of forecasts, n. This number of correct forecasts is obtained by adding the counts along the diagonal from the upper left to the lower right corners of the contingency table. In Figure 7.3, the numbers r, v, and z represent the numbers of occasions when the first, second, and third events were correctly forecast, respectively. Therefore in the 3×3 table represented in this figure, the proportion correct would be $\text{PC} = (r + v + z)/n$.

The other attributes listed in Section 7.7.2 pertain only to the dichotomous, yes/no forecast situation. In order to apply these to nonprobabilistic forecasts that are not dichotomous, it is necessary to collapse the $I = J > 2$ contingency table into a series of 2×2 contingency tables. Each of these 2×2 tables is constructed, as indicated in Figure 7.3, by considering the forecast event in distinction to the complementary, not the forecast event. This complementary event simply is constructed as the union of the $J - 1$ remaining events. In Figure 7.3, the 2×2 contingency table for Event 1 lumps Events 2 and 3 as "not Event 1." Thus, the number of times Event 1 is correctly forecast is still $a = r$, but the number of times it is incorrectly forecast is $b = s + t$. From the standpoint of this collapsed 2×2 contingency table, whether the incorrect forecast of Event 1 was followed by Event 2 or Event 3 is unimportant. Similarly, the number of times the event not Event 1 is correctly forecast is $d = v + w + y + z$, and includes cases where Event 2 was forecast but Event 3 occurred, and Event 3 was forecast but Event 2 occurred.

Attributes for 2×2 contingency tables can be computed for any or all of the 2×2 tables constructed in this way from larger square tables. For the 3×3 contingency table in Figure 7.3, the bias (Equation 7.10) for forecasts of Event 1 would be $B_1 = (r + s + t)/(r + u + x)$, the bias for forecasts of Event 2 would be $B_2 = (u + v + w)/(s + v + y)$, and the bias for forecasts of Event 3 would be $B_3 = (x + y + z)/(t + w + z)$.

EXAMPLE 7.3   A Set of Multicategory Forecasts

The left-hand side of Table 7.3 shows a 3×3 verification contingency table for forecasts of freezing rain (y_1), snow (y_2), and rain (y_3) from Goldsmith (1990). These are nonprobabilistic forecasts, conditional on the occurrence of some form of precipitation,

TABLE 7.3   Nonprobabilistic MOS forecasts for freezing rain (y_1), snow (y_2), and rain (y_3), conditional on occurrence of some form of precipitation, for the eastern region of the United States during cool seasons of 1983/1984 through 1988/1989. The verification data is presented as a 3×3 contingency table on the left, and then as three 2×2 contingency tables for each of the three precipitation types. Also shown are scalar attributes from Section 7.2.2 for each of the 2×2 tables. The sample size is $n = 6340$. From Goldsmith (1990).

Full 3×3 Contingency Table				Freezing Rain			Snow			Rain	
	o_1	o_2	o_3		o_1	not o_1		o_2	not o_2		o_3	not o_3
y_1	50	91	71	y_1	50	162	y_2	2364	217	y_3	3288	259
y_2	47	2364	170	not y_1	101	6027	not y_2	296	3463	not y_3	241	2552
y_3	54	205	3288									
				TS $= 0.160$			TS $= 0.822$			TS $= 0.868$		
				$\theta = 18.4$			$\theta = 127.5$			$\theta = 134.4$		
				B $= 1.40$			B $= 0.97$			B $= 1.01$		
				FAR $= 0.764$			FAR $= 0.084$			FAR $= 0.073$		
				H $= 0.331$			H $= 0.889$			H $= 0.932$		
				F $= 0.026$			F $= 0.059$			F $= 0.092$		

for the eastern region of the United States, for the months of October through March of 1983/1984 through 1988/1989. They are MOS forecasts of the form of the PTYPE forecasts in Table 6.7, but produced by an earlier MOS system. For each of the three precipitation types, a 2×2 contingency table can be constructed, following Figure 7.3, that summarizes the performance of forecasts of that precipitation type in distinction to the other two precipitation types together. Table 7.3 also includes forecast attributes from Section 7.2.2 for each 2×2 decomposition of the 3×3 contingency table. These are reasonably consistent with each other for a given 2×2 table, and indicate that the rain forecasts were slightly superior to the snow forecasts, but that the freezing rain forecasts were substantially less successful, with respect to most of these measures. ◊

The Heidke and Peirce Skill Scores can be extended easily to verification problems where there are more than $I = J = 2$ possible forecasts and events. The formulae for these scores in the more general case can be written most easily in terms of the joint distribution of forecasts and observations, $p(y_i, o_j)$, and the marginal distributions of the forecasts, $p(y_i)$ and of the observations, $p(o_j)$. For the Heidke Skill Score this more general form is

$$\text{HSS} = \frac{\sum_{i=1}^{I} p(y_i, o_i) - \sum_{i=1}^{I} p(y_i)p(o_i)}{1 - \sum_{i=1}^{I} p(y_i)p(o_i)}, \tag{7.20}$$

and the higher-dimensional generalization of the Peirce Skill Score is

$$\text{PSS} = \frac{\sum_{i=1}^{I} p(y_i, o_i) - \sum_{i=1}^{I} p(y_i)p(o_i)}{1 - \sum_{j=1}^{J} [p(o_j)]^2}. \tag{7.21}$$

Equation 7.20 reduces to Equation 7.15, and Equation 7.21 reduces to Equation 7.16, for $I = J = 2$.

Using Equation 7.20, the Heidke score for the 3×3 contingency table in Table 7.3 would be computed as follows. The proportion correct, $PC = \Sigma_i p(y_i, o_i) = (50/6340) + (2364/6340) + (3288/6340) = 0.8994$. The proportion correct for the random reference forecasts would be $\Sigma_i p(y_i)p(o_i) = (.0334)(.0238) + (.4071)(.4196) + (.5595)(.5566) = 0.4830$. Here, for example, the marginal probability $p(y_1) = (50 + 91 + 71)/6340 = 0.0344$. The proportion correct for perfect forecasts is of course one, yielding $HSS = (.8944 - .4830)/(1 - .4830) = 0.8054$. The computation for the Peirce Skill Score, Equation 7.21, is the same except that a different reference proportion correct is used in the denominator only. This is the unbiased random proportion $\Sigma_i [p(o_i)^2] = .0238^2 + .4196^2 + .5566^2 = 0.4864$. The Peirce Skill Score for this 3×3 contingency table is then $PSS = (.8944 - .4830)/(1 - .4864) = 0.8108$. The difference between the HSS and the PSS for these data is small, because the forecasts exhibit little bias.

There are many more degrees of freedom in the general $I \times J$ contingency table setting than in the simpler 2×2 problem. In particular $I \times J - 1$ elements are necessary to fully specify the contingency table, so that a scalar score must summarize much more even in the 3×3 setting as compared to the 2×2 problem. Accordingly, the number of possible scalar skill scores that are reasonable candidates increases rapidly with the size of the verification table. The notion of equitability for skill scores describing performance of nonprobabilistic forecasts of discrete predictands was proposed by Gandin and Murphy (1992) to define a restricted set of these yielding equal (zero) scores for random or constant forecasts.

When three or more events having a natural ordering are being forecast, it is usually required in addition that multiple-category forecast misses are scored as worse forecasts than single-category misses. Equations 7.20 and 7.21 both fail this requirement, as they depend only on the proportion correct. Gerrity (1992) has suggested a family of equitable (in the sense of Gandin and Murphy, 1992) skill scores that are also sensitive to distance in this sense and appear to provide generally reasonable results for rewarding correct forecasts and penalizing incorrect ones (Livezey 2003). The computation of Gandin-Murphy skill scores involves first defining a set of scoring weights $s_{i,j}, i = 1, \ldots, I, j = 1, \ldots, J$; each of which is applied to one of the joint probabilities $p(y_j, o_j)$, so that in general a Gandin-Murphy Skill Score is computed as

$$GMSS = \sum_{i=1}^{I} \sum_{j=1}^{J} p(y_i, o_j)s_{i,j}. \qquad (7.22)$$

As noted in Section 7.2.4 for the simple case of $I = J = 2$, when the scoring weights are derived according to the equitability criteria and equal penalties are assessed for the two types of errors (i.e., $s_{i,j} = s_{j,i}$), the result is the Peirce Skill Score (Equation 7.16). For larger verification problems, more constraints are required, and Gerrity (1992) suggested the following approach to defining the scoring weights based on the sample climatology $p(o_j)$. First, define the sequence of $J - 1$ odds ratios

$$D(j) = \frac{1 - \sum_{r=1}^{j} p(o_r)}{\sum_{r=1}^{j} p(o_r)}, \quad j = 1, \ldots, J - 1; \qquad (7.23)$$

where r is a dummy summation index. The scoring weights for correct forecasts are then

$$s_{j,j} = \frac{1}{J-1}\left[\sum_{r=1}^{j-1}\frac{1}{D(r)} + \sum_{r=j}^{J-1}D(r)\right], \quad j=1,\dots,J; \quad (7.24a)$$

and the weights for the incorrect forecasts are

$$s_{i,j} = \frac{1}{J-1}\left[\sum_{r=1}^{i-1}\frac{1}{D(r)} + \sum_{r=j}^{J-1}D(r) - (j-i)\right], \quad 1 \le i < j \le J. \quad (7.24b)$$

The summations in Equation 7.24 are taken to be equal to zero if the lower index is larger than the upper index. These two equations fully define the $I \times J$ scoring weights when symmetric errors are penalized equally; that is, when $s_{i,j} = s_{j,i}$. Equation 7.24a gives more credit for correct forecasts of rarer events and less credit for correct forecasts of common events. Equation 7.24b also accounts for the intrinsic rarity of the J events, and increasingly penalizes errors for greater differences between the forecast category i and the observed category j, through the penalty term $(j-i)$. Each scoring weight in Equation 7.24 is used together with the corresponding member of the joint distribution $p(y_j, o_j)$ in Equation 7.22 to compute the skill score. When the weights for the Gandin-Murphy Skill Score are computed according to Equations 7.23 and 7.24, the result is sometimes called the Gerrity skill score.

EXAMPLE 7.4 Gerrity Skill Score for a 3×3 Verification Table

Table 7.3 includes a 3×3 contingency table for nonprobabilistic forecasts of freezing rain, snow, and rain, conditional on the occurrence of precipitation of some kind. Figure 7.4a shows the corresponding joint probability distribution $p(y_i, o_j)$, calculated by dividing the counts in the contingency table by the sample size, $n = 6340$. Figure 7.4a also shows the sample climatological distribution $p(o_j)$, computed by summing the columns of the joint distribution.

The Gerrity (1992) scoring weights for the Gandin-Murphy Skill Score (Equation 7.22) are computed from these sample climatological relative frequencies using Equations 7.23 and 7.24. First, Equation 7.23 yields the $J - 1 = 2$ likelihood ratios $D(1) = (1 - .0238)/.0238 = 41.02$, and $D(2) = [1 - (.0238 + .4196)]/(.0238 + .4196) = 1.25$.

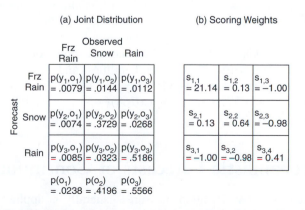

(a) Joint Distribution

	Observed		
Forecast	Frz Rain	Snow	Rain
Frz Rain	$p(y_1,o_1)$ = .0079	$p(y_1,o_2)$ = .0144	$p(y_1,o_3)$ = .0112
Snow	$p(y_2,o_1)$ = .0074	$p(y_2,o_2)$ = .3729	$p(y_2,o_3)$ = .0268
Rain	$p(y_3,o_1)$ = .0085	$p(y_3,o_2)$ = .0323	$p(y_3,o_3)$ = .5186

$p(o_1)$ = .0238 $p(o_2)$ = .4196 $p(o_3)$ = .5566

(b) Scoring Weights

$s_{1,1}$ = 21.14	$s_{1,2}$ = 0.13	$s_{1,3}$ = −1.00
$s_{2,1}$ = 0.13	$s_{2,2}$ = 0.64	$s_{2,3}$ = −0.98
$s_{3,1}$ = −1.00	$s_{3,2}$ = −0.98	$s_{3,4}$ = 0.41

FIGURE 7.4   (a) Joint distribution of forecasts and observations for the 3×3 contingency table in Table 7.3, with the marginal probabilities for the three observations (the sample climatological probabilities). (b) The Gerrity (1992) scoring weights computed from the sample climatological probabilities.

The rather large value for $D(1)$ reflects the fact that freezing rain was observed rarely, on only approximately 2% of the precipitation days during the period considered. The scoring weights for the three possible correct forecasts, computed using Equation 7.24a, are

$$s_{1,1} = \frac{1}{2}(41.02 + 1.25) = 21.14, \tag{7.25a}$$

$$s_{2,2} = \frac{1}{2}\left(\frac{1}{41.02} + 1.25\right) = 0.64, \tag{7.25b}$$

and

$$s_{3,3} = \frac{1}{2}\left(\frac{1}{41.02} + \frac{1}{1.25}\right) = 0.41; \tag{7.25c}$$

and the weights for the incorrect forecasts are

$$s_{1,2} = s_{2,1} = \frac{1}{2}(1.25 - 1) = 0.13, \tag{7.26a}$$

$$s_{2,3} = s_{3,2} = \frac{1}{2}\left(\frac{1}{41.02} - 1\right) = -0.98, \tag{7.26b}$$

and

$$s_{3,1} = s_{1,3} = \frac{1}{2}(-2) = -1.00. \tag{7.26c}$$

These scoring weights are arranged in Figure 7.4b in positions corresponding to the joint probabilities in Figure 7.4a to which they pertain.

The scoring weight $s_{1,1} = 21.14$ is much larger than the others in order to reward correct forecasts of the rare freezing rain events. Correct forecasts of snow and rain are credited with much smaller positive values, with $s_{3,3} = 0.41$ for rain being smallest because rain is the most common event. The scoring weight $s_{2,3} = -1.00$ is the minimum value according to the Gerrity algorithm, produced because the $(j - i) = 2$-category error (cf. Equation 7.24b) is the most severe possible when there is a natural ordering among the three outcomes. The penalty of an incorrect forecast of snow when rain occurs, or of rain when snow occurs (Equation 7.26b), is almost as large because these two events are relatively common. Mistakenly forecasting freezing rain when snow occurs, or vice versa, actually receives a small positive score because the frequency $p(o_1)$ is so small.

Finally, the Gandin-Murphy Skill Score in Equation 7.22 is computed by summing the products of pairs of joint probabilities and scoring weights in corresponding positions in Figure 7.4; that is, GMSS $= (.0079)(21.14) + (.0144)(.13) + (.0112)(-1) + (.0074)(.13) + (.3729)(.64) + (.0268)(-.98) + (.0085)(-1) + (.0323)(-.98) + (.5186)(.41) = 0.54.$ ◊

7.3 Nonprobabilistic Forecasts of Continuous Predictands

A different set of verification measures generally is applied to forecasts of continuous atmospheric variables. Continuous variables in principal can take on any value in a specified segment of the real line, rather than being limited to a finite number of discrete points. Temperature is an example of a continuous variable. In practice, however, forecasts and observations of continuous atmospheric variables are made using a finite number of discrete values. For example, temperature forecasts usually are rounded to integer degrees. It would be possible to deal with this kind of forecast verification data in discrete form, but there are usually so many allowable values of forecasts and observations that the resulting contingency tables would become unwieldy and possibly quite sparse. Just as discretely reported observations of continuous atmospheric variables were treated as continuous quantities in Chapter 4, it is convenient and useful to treat the verification of (operationally discrete) forecasts of continuous quantities in a continuous framework as well.

Conceptually, the joint distribution of forecasts and observations is again of fundamental interest. This distribution will be the continuous analog of the discrete joint distribution of Equation 7.1. Because of the finite nature of the verification data, however, explicitly using the concept of the joint distribution in a continuous setting generally requires that a parametric distribution such as the bivariate normal (Equation 4.33) be assumed and fit. Parametric distributions and other statistical models occasionally are assumed for the joint distribution of forecasts and observations or their factorizations (e.g., Bradley *et al.* 2003; Katz *et al.* 1982; Krzysztofowicz and Long 1991; Murphy and Wilks 1998), but it is far more common that scalar performance and skill measures, computed using individual forecast/observation pairs, are used in verification of continuous nonprobabilistic forecasts.

7.3.1 *Conditional Quantile Plots*

It is possible and quite informative to graphically represent certain aspects of the joint distribution of nonprobabilistic forecasts and observations for continuous variables. The joint distribution contains a large amount of information that is most easily absorbed from a well-designed graphical presentation. For example, Figure 7.5 shows conditional quantile plots for a sample of daily maximum temperature forecasts issued during the winters of 1980/1981 through 1985/1986 for Minneapolis, Minnesota. Panel (a) represents the performance of objective (MOS) forecasts, and panel (b) represents the performance of the corresponding subjective forecasts. These diagrams contain two parts, representing the two factors in the calibration-refinement factorization of the joint distribution of forecasts and observations (Equation 7.2). The conditional distributions of the observations given the forecasts are represented in terms of selected quantiles, in comparison to the 1:1 diagonal line representing perfect forecasts. Here it can be seen that the MOS forecasts (panel a) exhibit a small degree of overforecasting (the conditional medians of the observed temperatures are consistently colder than the forecasts), but that the subjective forecasts are essentially unbiased. The histograms in the lower parts of the panels represent the frequency of use of the forecasts, or $p(y_i)$. Here it can be seen that the subjective forecasts are somewhat sharper, or more refined, with more extreme temperatures being forecast more frequently, especially on the left tail.

Figure 7.5a shows the same data that is displayed in the glyph scatterplot in Figure 3.21, and the bivariate histogram in Figure 3.22. However, these latter two figures show the data in terms of their joint distribution, whereas the calibration-refinement factorization plotted in Figure 7.5a allows an easy visual separation between the frequencies of use of each of the possible forecasts, and the distributions of temperature outcomes conditional on each forecast. The conditional quantile plot is an example of a diagnostic verification technique, allowing diagnosis of particular strengths and weakness of a set of forecasts through exposition of the full joint distribution of the forecasts and observations.

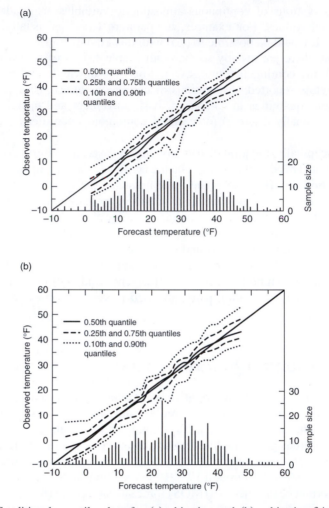

FIGURE 7.5   Conditional quantile plots for (a) objective and (b) subjective 24-h nonprobabilistic maximum temperature forecasts, for winter seasons of 1980 through 1986 at Minneapolis, Minnesota. Main body of the figures delineate smoothed quantiles from the conditional distributions $p(o_j|y_i)$ (i.e., the calibration distributions) in relation to the 1:1 line, and the lower parts of the figures show the unconditional distributions of the forecasts, $p(y_i)$ (the refinement distributions). From Murphy *et al.* (1989).

7.3.2 Scalar Accuracy Measures

Only two scalar measures of forecast accuracy for continuous predictands are in common use. The first is the Mean Absolute Error,

$$\text{MAE} = \frac{1}{n} \sum_{k=1}^{n} |y_k - o_k|. \tag{7.27}$$

Here (y_k, o_k) is the kth of n pairs of forecasts and observations. The MAE is the arithmetic average of the absolute values of the differences between the members of each pair. Clearly the MAE is zero if the forecasts are perfect (each $y_k = o_k$), and increases as

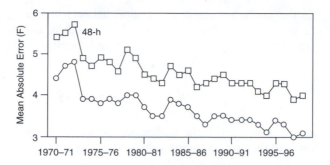

FIGURE 7.6   Year-by-year MAE for October-March objective maximum temperature forecasts at the 24- and 48-h projections, for approximately 95 locations in the United States. Forecasts for 1970–1971 through 1972–1973 were produced by perfect prog equations; those for 1973–1974 onward were produced by MOS equations. From www.nws.noaa.gov/tdl/synop.

discrepancies between the forecasts and observations become larger. We can interpret the MAE as a typical magnitude for the forecast error in a given verification data set.

The MAE often is used as a verification measure for temperature forecasts in the United States. Figure 7.6 shows MAE for objective maximum temperature forecasts at approximately 90 stations in the United States during the cool seasons (October-March) 1970/1971 through 1987/1988. Temperature forecasts with a 24-h lead time are more accurate than those for a 48-h lead time, exhibiting smaller average absolute errors. A clear trend of forecast improvement through time is also evident, as the MAE for the 48-h forecasts in the 1980s is comparable to the MAE for the 24-h forecasts in the early 1970s. The substantial reduction in error between 1972/1973 and 1973/1974 coincided with a change from perfect prog to MOS forecasts.

The other common accuracy measure for continuous nonprobabilistic forecasts is the Mean Squared Error,

$$\mathrm{MSE} = \frac{1}{n} \sum_{k=1}^{n} (y_k - o_k)^2. \tag{7.28}$$

The MSE is the average squared difference between the forecast and observation pairs. This measure is similar to the MAE except that the squaring function is used rather than the absolute value function. Since the MSE is computed by squaring forecast errors, it will be more sensitive to larger errors than will the MAE, and so will also be more sensitive to outliers. Squaring the errors necessarily produces positive terms in Equation 7.28, so the MSE increases from zero for perfect forecasts through larger positive values as the discrepancies between forecasts and observations become increasingly large. Sometimes the MSE is expressed as its square root, $\mathrm{RMSE} = \sqrt{\mathrm{MSE}}$, which has the same physical dimensions as the forecasts and observations, and can also be thought of as a typical magnitude for forecast errors.

Initially, we might think that the correlation coefficient (Equation 3.22) could be another useful accuracy measure for nonprobabilistic forecasts of continuous predictands. However, although the correlation does reflect linear association between two variables (in this case, forecasts and observations), it is sensitive to outliers, and is not sensitive to biases that may be present in the forecasts. This latter problem can be appreciated by considering an algebraic manipulation of the MSE (Murphy 1988):

$$\mathrm{MSE} = (\bar{y} - \bar{o})^2 + s_y^2 + s_o^2 - 2s_y s_o r_{yo}. \tag{7.29}$$

Here r_{yo} is the product-moment correlation between the forecasts and observations, s_y and s_o are the standard deviations of the marginal distributions of the forecasts and observations, respectively, and the first term in Equation 7.29 is the square of the Mean Error,

$$\text{ME} = \frac{1}{n}\sum_{k=1}^{n}(y_k - o_k) = \bar{y} - \bar{o}. \tag{7.30}$$

The Mean Error is simply the difference between the average forecast and average observation, and therefore expresses the bias of the forecasts. Equation 7.30 differs from Equation 7.28 in that the individual forecast errors are not squared before they are averaged. Forecasts that are, on average, too high will exhibit ME > 0 and forecasts that are, on average, too low will exhibit ME < 0. It is important to note that the bias gives no information about the typical magnitude of individual forecast errors, and is therefore not in itself an accuracy measure.

Returning to Equation 7.29, it can be seen that forecasts that are more highly correlated with the observations will exhibit lower MSE, other factors being equal. However, since the MSE can be written with the correlation r_{yo} and the bias (ME) in separate terms, we can imagine forecasts that may be highly correlated with the observations, but with sufficiently severe bias that they would be useless at face value. A set of temperature forecasts could exist, for example, that are exactly half of the subsequently observed temperature. For convenience, imagine that these temperatures are nonnegative. A scatterplot of the observed temperatures versus the corresponding forecasts would exhibit all points falling perfectly on a straight line ($r_{yo} = 1$), but the slope of that line would be 2. The bias, or mean error, would be $\text{ME} = n^{-1}\Sigma_k(f_k - o_k) = n^{-1}\Sigma_k(0.5o_k - o_k)$, or the negative of half of the average observation. This bias would be squared in Equation 7.29, leading to a very large MSE. A similar situation would result if all the forecasts were exactly 10° colder than the observed temperatures. The correlation r_{yo} would still be one, the points on the scatterplot would fall on a straight line (this time with unit slope), the Mean Error would be $-10°$, and the MSE would be inflated by $(10°)^2$. The definition of correlation (Equation 3.23) shows clearly why these problems would occur: the means of the two variables being correlated are separately subtracted, and any differences in scale are removed by separately dividing by the two standard deviations, before calculating the correlation. Therefore, any mismatches between either location or scale between the forecasts and observations are not reflected in the result.

7.3.3 Skill Scores

Skill scores, or relative accuracy measures, of the form of Equation 7.4, can easily be constructed using the MAE, MSE, or RMSE as the underlying accuracy statistics. Usually the reference, or control, forecasts are provided either by the climatological values of the predictand or by persistence (i.e., the previous value in a sequence of observations). For the MSE, the accuracies of these two references are, respectively,

$$\text{MSE}_{\text{Clim}} = \frac{1}{n}\sum_{k=1}^{n}(\bar{o} - o_k)^2 \tag{7.31a}$$

and

$$\text{MSE}_{\text{Pers}} = \frac{1}{n}\sum_{k=1}^{n}(o_{k-1} - o_k)^2. \tag{7.31b}$$

Completely analogous equations can be written for the MAE, in which the squaring function would be replaced by the absolute value function.

In Equation 7.31a, it is implied that the climatological average value does not change from forecast occasion to forecast occasion (i.e., as a function of the index, k). If this implication is true, then MSE_{Clim} in Equation 7.31a is an estimate of the sample variance of the predictand (compare Equation 3.6). In some applications the climatological value of the predictand will be different for different forecasts. For example, if daily temperature forecasts at a single location were being verified over the course of several months, the index k would represent time, and the climatological average temperature usually would change smoothly as a function of the date. In this case the quantity being summed in Equation 7.31a would be $(c_k - o_k)^2$, with c_k being the climatological value of the predictand on day k. Failing to account for a time-varying climatology would produce an unrealistically large MSE_{clim}, because the correct seasonality for the predictand would not be reflected. The MSE for persistence in 7.31b implies that the index k represents time, so that the reference forecast for the observation o_k at time k is just the observation of the predictand during the previous time period, o_{k-1}.

Either of the reference measures of accuracy in Equations 7.31a or 7.31b, or their MAE counterparts, can be used in Equation 7.4 to calculate skill. Murphy (1992) advocates use of the more accurate reference forecasts to standardize the skill. For skill scores based on MSE, Equation 7.31a is more accurate (i.e., is smaller) if the lag-1 autocorrelation (Equation 3.30) of the time series of observations is smaller than 0.5, and Equation 7.31b is more accurate when the autocorrelation of the observations is larger than 0.5. For the MSE using climatology as the control forecasts, the skill score (in proportion rather than percentage terms) becomes

$$ SS_{Clim} = \frac{MSE - MSE_{Clim}}{0 - MSE_{Clim}} = 1 - \frac{MSE}{MSE_{Clim}}. \tag{7.32} $$

Notice that perfect forecasts have MSE or MAE $= 0$, which allows the rearrangement of the skill score in Equation 7.32. By virtue of this second equality in Equation 7.32, SS_{clim} based on MSE is sometimes called the reduction of variance (RV), because the quotient being subtracted is the average squared error (or residual, in the nomenclature of regression) divided by the climatological variance (cf. Equation 6.16).

EXAMPLE 7.5   Skill of the Temperature Forecasts in Figure 7.6

The counterpart of Equation 7.32 for the MAE can be applied to the temperature forecast accuracy data in Figure 7.6. Assume the reference MAE is $MAE_{Clim} = 8.5°F$. This value will not depend on the forecast projection, and should be different for different years only to the extent that the average MAE values plotted in the figure are for slightly different collections of stations. However, in order for the resulting skill score not to be artificially inflated, the climatological values used to compute MAE_{clim} must be different for the different locations and different dates. Otherwise skill will be credited for correctly forecasting that January will be colder than October, or that high-latitude locations will be colder than low-latitude locations (Juras 2000).

For 1986/1987 the MAE for the 24-h projection is 3.5°F, yielding a skill score of $SS_{Clim} = 1 - (3.5°F)/(8.5°F) = 0.59$, or a 59% improvement over climatological forecasts. For the 48-h projection the MAE is 4.3°F, yielding $SS_{Clim} = 1 - (4.3°F)/(8.5°F) = 0.49$, or a 49% improvement over climatology. Not surprisingly, the forecasts for the 24-h projection are more skillful than those for the 48-h projection. ◊

The skill score for the MSE in Equation 7.32 can be manipulated algebraically in a way that yields some insight into the determinants of forecast skill as measured by the MSE, with respect to climatology as the reference (Equation 7.31a). Rearranging Equation 7.32, and substituting an expression for the Pearson product-moment correlation between the forecasts and observations, r_{yo}, yields (Murphy 1988)

$$SS_{Clim} = r_{yo}^2 - \left[r_{yo} - \frac{s_y}{s_o}\right]^2 - \left[\frac{\bar{y} - \bar{o}}{s_o}\right]^2. \tag{7.33}$$

Equation 7.33 indicates that the skill in terms of the MSE can be regarded as consisting of a contribution due to the correlation between the forecasts and observations, and penalties relating to the reliability and bias of the forecasts.

The first term in Equation 7.33 is the square of the product-moment correlation coefficient, and is a measure of the proportion of variability in the observations that is (linearly) accounted for by the forecasts. Here the squared correlation is similar to the R^2 in regression (Equation 6.16), although least-squares regressions are constrained to be unbiased by construction, whereas forecasts in general may not be.

The second term in Equation 7.33 is a measure of reliability, or conditional bias, of the forecasts. This is most easily appreciated by imagining a linear regression between the observations and the forecasts. The slope, b, of a linear regression equation can be expressed in terms of the correlation and the standard deviations of the predictor and predictand as $b = (s_o/s_y)r_{yo}$. This relationship can be verified by substituting Equations 3.6 and 3.23 into Equation 6.7a. If this slope is smaller than $b = 1$, then the predictions made with this regression are too large (positively biased) for smaller forecasts, and too small (negatively biased) for larger forecasts. However, if $b = 1$, there will be no conditional bias, and substituting $b = (s_o/s_y)r_{yo} = 1$ into the second term in Equation 7.33 yields a zero penalty for conditional bias.

The third term in Equation 7.33 is the square of the unconditional bias, as a fraction of the standard deviation of the observations, s_o. If the bias is small compared to the variability of the observations as measured by s_o the reduction in skill will be modest, whereas increasing bias of either sign progressively degrades the skill.

Thus, if the forecasts are completely reliable and unbiased, the second two terms in Equation 7.33 are both zero, and the skill score is exactly r_{yo}^2. To the extent that the forecasts are biased or not completely reliable (exhibiting conditional biases), then the square of the correlation coefficient will overestimate skill. Squared correlation is accordingly best regarded as measuring potential skill.

7.4 Probability Forecasts of Discrete Predictands

7.4.1 The Joint Distribution for Dichotomous Events

Formulation and verification of probability forecasts for weather events have a long history, dating at least to Cooke (1906a) (Murphy and Winkler 1984). Verification of probability forecasts is somewhat more subtle than verification of nonprobabilistic forecasts. Since nonprobabilistic forecasts contain no expression of uncertainty, it is clear whether an individual forecast is correct or not. However, unless a probability forecast is either 0.0 or 1.0, the situation is less clear-cut. For probability values between these two (certainty) extremes a single forecast is neither right nor wrong, so that meaningful assessments can only be made using collections of forecast and observation pairs.

Again, it is the joint distribution of forecasts and observations that contains the relevant information for forecast verification.

The simplest setting for probability forecasts is in relation to dichotomous predictands, which are limited to $J = 2$ possible outcomes. The most familiar example of probability forecasts for a dichotomous event is the probability of precipitation (PoP) forecast. Here the event is either the occurrence (o_1) or nonoccurrence (o_2) of measurable precipitation. The joint distribution of forecasts and observations is more complicated than for the case of nonprobabilistic forecasts of binary predictands, however, because more than $I = 2$ probability values can allowably be forecast. In theory any real number between zero and one is an allowable probability forecast, but in practice the forecasts usually are rounded to one of a reasonably small number of values.

Table 7.4a contains a hypothetical joint distribution for probability forecasts of a dichotomous predictand, where the $I = 11$ possible forecasts might have been obtained by rounding continuous probability assessments to the nearest tenth. Thus, this joint distribution of forecasts and observations contains $I \times J = 22$ individual probabilities. For example on 4.5% of the forecast occasions a zero forecast probability was nevertheless followed by occurrence of the event, and on 25.5% of the occasions zero probability forecasts were correct in that the event o_1 did not occur.

Table 7.4b shows the same joint distribution in terms of the Calibration-Refinement factorization (Equation 7.2). That is, for each possible forecast probability, y_i, Table 7.4b shows the relative frequency with which that forecast value was used, $p(y_i)$, and the conditional probability that the event o_1 occurred given the forecast y_i, $p(o_1|y_i)$, $i = 1, \ldots, I$. For example, $p(y_1) = p(y_1, o_1) + p(y_1, o_2) = .045 + .255 = .300$, and (using the definition of conditional probability, Equation 2.10) $p(o_1|y_1) = p(y_1, o_1)/p(y_1) = .045/.300 = .150$. Because the predictand is binary it is not necessary to specify the conditional probabilities for the complementary event, o_2, given each of the forecasts. That is, since the two predictand values represented by o_1 and o_2 constitute a MECE partition of the sample space, $p(o_2|y_i) = 1 - p(o_1|y_i)$. Not all the $J = 11$ probabilities in

TABLE 7.4   A hypothetical joint distribution of forecasts and observations (a) for probability forecasts (rounded to tenths) of a dichotomous event, with (b) its Calibration-Refinement factorization, and (c) its Likelihood-Base Rate factorization.

y_i	(a) Joint Distribution		(b) Calibration-Refinement		(c) Likelihood-Base Rate	
	$p(y_i, o_1)$	$p(y_i, o_2)$	$p(y_i)$	$p(o_1|y_i)$	$p(y_i|o_1)$	$p(y_i|o_2)$
0.0	.045	.255	.300	.150	.152	.363
0.1	.032	.128	.160	.200	.108	.182
0.2	.025	.075	.100	.250	.084	.107
0.3	.024	.056	.080	.300	.081	.080
0.4	.024	.046	.070	.350	.081	.065
0.5	.024	.036	.060	.400	.081	.051
0.6	.027	.033	.060	.450	.091	.047
0.7	.025	.025	.050	.500	.084	.036
0.8	.028	.022	.050	.550	.094	.031
0.9	.030	.020	.050	.600	.101	.028
1.0	.013	.007	.020	.650	.044	.010
					$p(o_1) = .297$	$p(o_2) = .703$

the refinement distribution $p(y_i)$ can be specified independently either, since $\Sigma_j p(y_j) = 1$. Thus the joint distribution can be completely specified with $I \times J - 1 = 21$ of the 22 probabilities given in either Table 7.4a or Table 7.4b, which is the dimensionality of this verification problem.

Similarly, Table 7.4c shows the Likelihood-Base Rate factorization (Equation 7.3) for the joint distribution in Table 7.4a. Since there are $J = 2$ MECE events, there are two conditional distributions $p(y_i|o_j)$, each of which include $I = 11$ probabilities. Since these 11 probabilities must sum to 1, each conditional distribution is fully specified by any 10 of them. The refinement (i.e., sample climatological) distribution consists of the two complementary probabilities $p(o_1)$ and $p(o_2)$, and so can be completely defined by either. Therefore the Likelihood-Base Rate factorization is also fully specified by $10 + 10 + 1 = 21$ probabilities. The information in any of the three portions of Table 7.4 can be recovered fully from either of the others. For example, $p(o_1) = \Sigma_i p(y_i, o_1) = .297$, and $p(y_1|o_1) = p(y_1, o_1)/p(o_1) = .045/.297 = .152$.

7.4.2 The Brier Score

Given the generally high dimensionality of verification problems involving probability forecasts even for dichotomous predictands (e.g., $I \times J - 1 = 21$ for Table 7.4), it is not surprising that forecast performance is often assessed with a scalar summary measure. Although attractive from a practical standpoint, such simplifications necessarily will give incomplete pictures of forecast performance. A number of scalar accuracy measures for verification of probabilistic forecasts of dichotomous events exist (Murphy and Daan 1985; Toth *et al.* 2003), but by far the most common is the Brier score (BS). The Brier score is essentially the mean squared error of the probability forecasts, considering that the observation is $o_1 = 1$ if the event occurs, and that the observation is $o_2 = 0$ if the event does not occur. The score averages the squared differences between pairs of forecast probabilities and the subsequent binary observations,

$$BS = \frac{1}{n} \sum_{k=1}^{n} (y_k - o_k)^2, \tag{7.34}$$

where the index k again denotes a numbering of the n forecast-event pairs. Comparing the Brier score with Equation 7.28 for the mean squared error, it can be seen that the two are completely analogous. As a mean-squared-error measure of accuracy, the Brier score is negatively oriented, with perfect forecasts exhibiting $BS = 0$. Less accurate forecasts receive higher Brier scores, but since individual forecasts and observations are both bounded by zero and one, the score can take on values only in the range $0 \leq BS \leq 1$.

The Brier score as expressed in Equation 7.34 is nearly universally used, but it differs from the score as originally introduced by Brier (1950) in that it averages only the squared differences pertaining to one of the two binary events. The original Brier score also included squared differences for the complementary (or non-) event in the average, with the result that Brier's original score is exactly twice that given by Equation 7.34. The confusion is unfortunate, but the usual present-day understanding of the meaning of Brier score is that given in Equation 7.34. In order to distinguish this from the original formulation, the Brier Score in Equation 7.34 sometimes is referred to as the half-Brier score.

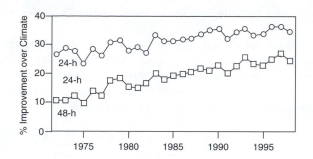

FIGURE 7.7 Trends in the skill of United States subjective PoP forecasts, measured in terms of the Brier score relative to climatological probabilities, April–September 1972–1998. From www.nws.noaa.gov/tdl/synop.

Skill scores of the form of Equation 7.4 often are computed for the Brier score, yielding the Brier Skill Score

$$BSS = \frac{BS - BS_{ref}}{0 - BS_{ref}} = 1 - \frac{BS}{BS_{ref}}, \tag{7.35}$$

since $BS_{perf} = 0$. The BSS is the conventional skill-score form using the Brier score as the underlying accuracy measure. Usually the reference forecasts are the relevant climatological relative frequencies that may vary with location and/or time of year (Juras 2000). Skill scores with respect to the climatological probabilities for subjective PoP forecasts during the warm seasons of 1972 through 1998 are shown in Figure 7.7. The labeling of the vertical axis as % improvement over climate indicates that it is the skill score in Equation 7.35, using climatological probabilities as the reference forecasts, that is plotted in the figure. According to this score, forecasts made at the 48-hour projection in the 1990s exhibited skill equivalent to 24-hour forecasts made in the 1970s.

7.4.3 Algebraic Decomposition of the Brier Score

An instructive algebraic decomposition of the Brier score (Equation 7.34) has been derived by Murphy (1973b). It relates to the calibration-refinement factorization of the joint distribution, Equation 7.2, in that it pertains to quantities that are conditional on particular values of the forecasts.

As before, consider that a verification data set contains forecasts taking on any of a discrete number, I, of forecast values y_i. For example, in the verification data set in Table 7.4, there are $I = 11$ allowable forecast values, ranging from $y_1 = 0.0$ to $y_{11} = 1.0$. Let N_i be the number of times each forecast y_i is used in the collection of forecasts being verified. The total number of forecast-event pairs is simply the sum of these subsample, or conditional sample, sizes,

$$n = \sum_{i=1}^{I} N_i. \tag{7.36}$$

The marginal distribution of the forecasts—the refinement—in the calibration-refinement factorization consists simply of the relative frequencies

$$p(y_i) = \frac{N_i}{n}. \tag{7.37}$$

The first column in Table 7.4b shows these relative frequencies for the data set represented there.

For each of the subsamples delineated by the I allowable forecast values there is a relative frequency of occurrence of the forecast event. Since the observed event is dichotomous, a single conditional relative frequency defines the conditional distribution of observations given each forecast y_i. It is convenient to think of this relative frequency as the subsample relative frequency, or conditional average observation,

$$\bar{o}_i = p(o_1|y_i) = \frac{1}{N_i} \sum_{k \in N_i} o_k, \tag{7.38}$$

where $o_k = 1$ if the event occurs for the kth forecast-event pair, $o_k = 0$ if it does not, and the summation is over only those values of k corresponding to occasions when the forecast y_i was issued. The second column in Table 7.4b shows these conditional relative frequencies. Similarly, the overall (unconditional) relative frequency, or sample climatology, of the observations is given by

$$\bar{o} = \frac{1}{n} \sum_{k=1}^{n} o_k = \frac{1}{n} \sum_{i=1}^{I} N_i \bar{o}_i. \tag{7.39}$$

After some algebra, the Brier score in Equation 7.34 can be expressed in terms of the quantities just defined as the sum of the three terms

$$BS = \frac{1}{n} \sum_{i=1}^{I} N_i (y_i - \bar{o}_i)^2 - \frac{1}{n} \sum_{i=1}^{I} N_i (\bar{o}_i - \bar{o})^2 + \bar{o}(1 - \bar{o}). \tag{7.40}$$

("Reliability")   ("Resolution")   ("Uncertainty")

As indicated in this equation, these three terms are known as reliability, resolution, and uncertainty. Since more accurate forecasts are characterized by smaller values of BS, a forecaster would like the reliability term to be as small as possible, and the resolution term to be as large (in absolute value) as possible. Equation 7.39 indicates that the uncertainty term depends only on the sample climatological relative frequency, and is unaffected by the forecasts.

The reliability and resolution terms in Equation 7.40 sometimes are used individually as scalar measures of these two aspects of forecast quality, and called REL and RES, respectively. Sometimes these two measures are normalized by dividing each by the uncertainty term (Kharin and Zwiers 2003a; Toth et al. 2003), so that their sum equals the Brier skill score BSS (cf. Equation 7.41).

The reliability term in Equation 7.40 summarizes the calibration, or conditional bias, of the forecasts. It consists of a weighted average of the squared differences between the forecast probabilities y_i and the relative frequencies of the forecast event in each subsample. For forecasts that are perfectly reliable, the subsample relative frequency is exactly equal to the forecast probability in each subsample. The relative frequency of the forecast event should be small on occasions when $y_1 = 0.0$ is forecast, and should be large when $y_I = 1.0$ is forecast. On those occasions when the forecast probability is 0.5, the relative frequency of the event should be near 1/2. For reliable, or well-calibrated forecasts, all the squared differences in the reliability term will be near zero, and their weighted average will be small.

The resolution term in Equation 7.40 summarizes the ability of the forecasts to discern subsample forecast periods with different relative frequencies of the event. The forecast probabilities y_i do not appear explicitly in this term, yet it still depends on the forecasts through the sorting of the events making up the subsample relative frequencies (Equation 7.38). Mathematically, the resolution term is a weighted average of the squared differences between these subsample relative frequencies, and the overall sample climatological relative frequency. Thus, if the forecasts sort the observations into subsamples having substantially different relative frequencies than the overall sample climatology, the resolution term will be large. This is a desirable situation, since the resolution term is subtracted in Equation 7.40. Conversely, if the forecasts sort the events into subsamples with very similar event relative frequencies, the squared differences in the summation of the resolution term will be small. In this case the forecasts resolve the event only weakly, and the resolution term will be small.

The uncertainty term in Equation 7.40 depends only on the variability of the observations, and cannot be influenced by anything the forecaster may do. This term is identical to the variance of the Bernoulli (binomial, with $N = 1$) distribution (see Table 4.4), exhibiting minima at zero when the climatological probability is either zero or one, and a maximum when the climatological probability is 0.5. When the event being forecast almost never happens, or almost always happens, the uncertainty in the forecasting situation is small. In these cases, always forecasting the climatological probability will give generally good results. When the climatological probability is close to 0.5 there is substantially more uncertainty inherent in the forecasting situation, and the third term in Equation 7.40 is commensurately larger.

The algebraic decomposition of the Brier score in Equation 7.40 is interpretable in terms of the calibration-refinement factorization of the joint distribution of forecasts and observations (Equation 7.2), as will become clear in Section 7.4.4. Murphy and Winkler (1987) also proposed a different three-term algebraic decomposition of the mean-squared error (of which the Brier score is a special case), based on the likelihood-base rate factorization (Equation 7.3), which has been applied to the Brier score for the data in Table 7.2 by Bradley *et al.* (2003).

7.4.4 *The Reliability Diagram*

Single-number summaries of forecast performance such as the Brier score can provide a convenient quick impression, but a comprehensive appreciation of forecast quality can be achieved only through the full joint distribution of forecasts and observations. Because of the typically large dimensionality ($= I \times J - 1$) of these distributions their information content can be difficult to absorb from numerical tabulations such as those in Tables 7.2 or 7.4, but becomes conceptually accessible when presented in a well-designed graphical format. The reliability diagram is a graphical device that shows the full joint distribution of forecasts and observations for probability forecasts of a binary predictand, in terms of its calibration-refinement factorization (Equation 7.2). Accordingly, it is the counterpart of the conditional quantile plot (see Section 7.3.1) for nonprobabilistic forecasts of continuous predictands. The fuller picture of forecast performance portrayed in the reliability diagram as compared to a scalar summary, such as BSS, allows diagnosis of particular strengths and weaknesses in a verification set.

The two elements of the calibration-refinement factorization are the calibration distributions, or conditional distributions of the observation given each of the I allowable values of the forecast, $p(o_j|y_i)$; and the refinement distribution $p(f_i)$, expressing the frequency of use of each of the possible forecasts. Each of the calibration distributions is a Bernoulli (binomial, with $N = 1$) distribution, because there is a single binary outcome O on each forecast occasion, and for each forecast y_i the probability of the outcome o_1 is the conditional probability $p(o_1|y_i)$. This probability fully defines the corresponding

Bernoulli distribution, because $p(o_2|y_i) = 1 - p(o_1|y_i)$. Taken together, these I calibration probabilities $p(o_1|y_i)$ define a calibration function, which expresses the conditional probability of the event o_1 as a function of the forecast y_i.

The first element of a reliability diagram is a plot of the calibration function, usually as I points connected by line segments for visual clarity. Figure 7.8a shows five characteristic forms for this portion of the reliability diagram, which allows immediate visual diagnosis of unconditional and conditional biases that may be exhibited by the forecasts in question. The center panel in Figure 7.8a shows the characteristic signature of

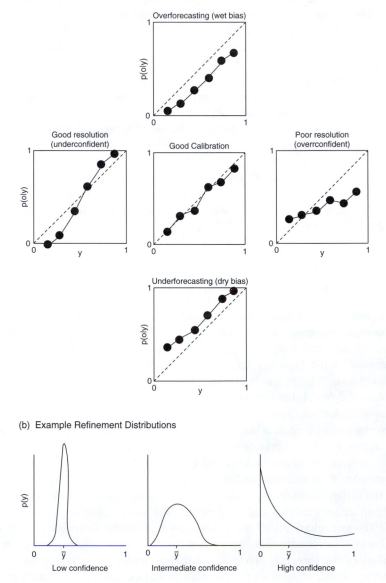

FIGURE 7.8   Example characteristic forms for the two elements of the reliability diagram. (a) Calibration functions, showing calibration distributions $p(o|y)$ (i.e., conditional Bernoulli probabilities), as functions of the forecast y. (b) Refinement distributions, $p(y)$, reflecting aggregate forecaster confidence.

well-calibrated forecasts, in which the conditional event relative frequency is essentially equal to the forecast probability; that is, $p(o_1|y_i) \approx y_i$, so that the I dots fall along the dashed 1:1 line except for deviations consistent with sampling variability. Well-calibrated probability forecasts mean what they say, in the sense that subsequent event relative frequencies are essentially equal to the forecast probabilities. In terms of the algebraic decomposition of the Brier score (Equation 7.40), such forecasts exhibit excellent reliability, because the squared differences in the reliability term correspond to squared vertical distances between the dots and the 1:1 line in the reliability diagram. These distances are all small for well-calibrated forecasts, yielding a small reliability term, which is a weighted average of these squared vertical distances.

The top and bottom panels in Figure 7.8a show characteristic forms of the calibration function for forecasts exhibiting unconditional biases. In the top panel, the calibration function is entirely to the right of the 1:1 line, indicating the forecasts are consistently too large relative to the conditional event relative frequencies, so that the average forecast is larger than the average observation (Equation 7.38). This pattern is the signature of overforecasting, or if the predictand is precipitation occurrence, a wet bias. Similarly, the bottom panel in Figure 7.8a shows the characteristic signature of underforecasting, or a dry bias, because the calibration function being entirely to the left of the 1:1 line indicates that the forecast probabilities are consistently too small relative to the corresponding conditional event relative frequencies given by $p(o_1|y_i)$, and so the average forecast is smaller than the average observation. Forecasts that are unconditionally biased in either of these two ways are miscalibrated, or not reliable, in the sense that the conditional event probabilities $p(o_1|y_i)$ do not correspond well to the stated probabilities y_i. The vertical distances between the points and the dashed 1:1 line are nonnegligible, leading to substantial squared differences in the first summation of Equation 7.40, and thus to a large reliability term in that equation.

The deficiencies in forecast performance indicated by the calibration functions in the left and right panels of Figure 7.8a are more subtle, and indicate conditional biases. That is, the sense and/or magnitudes of the biases exhibited by forecasts having these types of calibration functions depend on the forecasts themselves. In the left (good resolution) panel, there are overforecasting biases associated with smaller forecast probabilities and underforecasting biases associated with larger forecast probabilities, and the reverse is true of the calibration function in the right (poor resolution) panel.

The calibration function in the right panel of Figure 7.8a is characteristic of forecasts showing poor resolution in the sense that the conditional outcome relative frequencies $p(o_1|y_i)$ depend only weakly on the forecasts, and are all near the climatological probability. (That the climatological relative frequency is somewhere near the center of the vertical locations of the points in this panel can be appreciated from the law of total probability (Equation 2.14), which expresses the unconditional climatology as a weighted average of these conditional relative frequencies.) Because the differences in this panel between the calibration probabilities $p(o_1|y_i)$ (Equation 7.38) and the overall sample climatology are small, the resolution term in Equation 7.40 is small, reflecting the fact that these forecasts resolve the event o_1 poorly. Because the sign of this term in Equation 7.40 is negative, poor resolution leads to larger (worse) Brier scores.

Conversely, the calibration function in the left panel of Figure 7.8a indicates good resolution, in the sense that the weighted average of the squared vertical distances between the points and the sample climatology in the resolution term of Equation 7.40 is large. Here the forecasts are able to identify subsets of forecast occasions for which the outcomes are quite different from each other. For example, small but nonzero forecast probabilities have identified a subset of forecast occasions when the event o_1 did not occur at all.

However, the forecasts are conditionally biased, and so mislabeled, and therefore not well calibrated. Their Brier score would be penalized for this miscalibration through a substantial positive value for the reliability term in Equation 7.40.

The labels underconfident and overconfident in the left and right panels of Figure 7.8a can be understood in relation to the other element of the reliability diagram, namely the refinement distribution $p(y_i)$. The dispersion of the refinement distribution reflects the overall confidence of the forecaster, as indicated in Figure 7.8b. Forecasts that deviate rarely and quantitatively little from their average value (left panel) exhibit little confidence. Forecasts that are frequently extreme—that is, specifying probabilities close to the certainty values $y_1 = 0$ and $y_1 = 1$ (right panel)—exhibit high confidence. However, the degree to which a particular level of forecaster confidence may be justified will be evident only from inspection of the calibration function for the same forecasts. The forecast probabilities in the right-hand (overconfident) panel of Figure 7.8a are mislabeled in the sense that the extreme probabilities are too extreme. Outcome relative frequency following probability forecasts near 1 are substantially smaller than 1, and outcome relative frequencies following forecasts near 0 are substantially larger than 0. A calibration-function slope that is shallower than the 1:1 reference line is diagnostic of overconfident forecasts, because correcting the forecasts to bring the calibration function into the correct orientation would require adjusting extreme probabilities to be less extreme, thus shrinking the dispersion of the refinement distribution, which would connote less confidence. Conversely, the underconfident forecasts in the left panel of Figure 7.8a could achieve reliability (calibration function aligned with the 1:1 line) by adjusting the forecast probabilities to be more extreme, thus increasing the dispersion of the refinement distribution and connoting greater confidence.

A reliability diagram consists of plots of both the calibration function and the refinement distribution, and so is a full graphical representation of the joint distribution of the forecasts and observations, through its calibration-refinement factorization. Figure 7.9 shows two reliability diagrams, for seasonal (three-month) forecasts for (a) average temperatures and (b) total precipitation above the climatological terciles (outcomes in the warm and wet 1/3 of the respective local climatological distributions), for global land areas equatorward of 30° (Mason *et al.* 1999). The most prominent feature of Figure 7.9 is the substantial cold (underforecasting) bias evident for the temperature forecasts. The period 1997 through 2000 was evidently substantially warmer than the preceding several decades that defined the reference climate, so that the relative frequency of the observed warm outcome was about 0.7 (rather than the long-term climatological value of 1/3), but that warmth was not anticipated by these forecasts, in aggregate. There is also an indication of conditional bias in the temperature forecasts, with the overall calibration slope being slightly shallower than 45°, and so reflecting some forecast overconfidence. The precipitation forecasts (see Figure 7.9b) are better calibrated, showing only a slight overforecasting (wet) bias and a more nearly correct overall slope for the calibration function. The refinement distributions (insets, with logarithmic vertical scales) show much more confidence (more frequent use of more extreme probabilities) for the temperature forecasts.

The reliability diagrams in Figure 7.9 include some additional features that help interpret the results. The light lines through the calibration functions show weighted (to make points with larger subsample size N_i more influential) least-squares regressions (Murphy and Wilks 1998) that help guide the eye through the irregularities that are due at least in part to sampling variations. In order to emphasize the better-estimated portions of the calibration function, the line segments connecting points based on larger sample sizes have been drawn more heavily. Finally, the average forecasts are indicated by the triangles

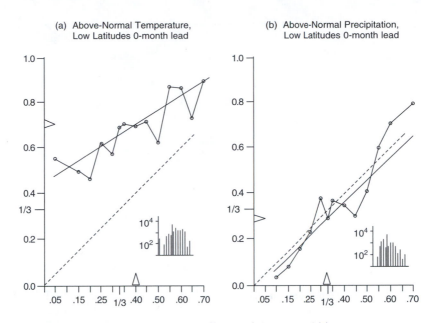

FIGURE 7.9   Reliability diagrams for seasonal (3-month) forecasts of (a) average temperature warmer than the climatological upper tercile, and (b) total precipitation wetter than the climatological upper tercile, for global land areas equatorward of 30°, during the period 1997–2000. From Wilks and Godfrey (2002).

on the horizontal axes, and the average observations are indicated by the triangles on the vertical axes, which emphasize the strong underforecasting of temperature in Figure 7.9a.

Another elaboration of the reliability diagram includes reference lines related to the algebraic decomposition of the Brier score (Equation 7.40) and the Brier skill score (Equation 7.35), in addition to plots of the calibration function and the refinement distribution. This version of the reliability diagram is called the attributes diagram (Hsu and Murphy 1986), an example of which (for the joint distribution in Table 7.2) is shown in Figure 7.10.

The horizontal no-resolution line in the attributes diagram relates to the resolution term in Equation 7.40. Geometrically, the ability of a set of forecasts to identify event subsets with different relative frequencies produces points in the attributes diagram that are well removed, vertically, from the level of the overall sample climatology, which is indicated by the no-resolution line. Points falling on the no-resolution line indicate forecasts y_i that are unable to resolve occasions where the event is more or less likely than the overall climatological probability. The weighted average making up the resolution term is of the squares of the vertical distances between the points (the subsample relative frequencies) and the no-resolution line. These distances will be large for forecasts exhibiting good resolution, in which case the resolution term will contribute to a small (i.e., good) Brier score. The forecasts summarized in Figure 7.10 exhibit a substantial degree of resolution, with forecasts that are most different from the sample climatological probability of 0.162 making the largest contributions to the resolution term.

Another interpretation of the uncertainty term in Equation 7.40 emerges from imagining the attributes diagram for climatological forecasts; that is, constant forecasts of the sample climatological relative frequency, Equation 7.39. Since only a single forecast value is ever used in this case, there is only $I = 1$ dot on the diagram. The horizontal position of this dot is at the constant forecast value, and the vertical position of the single

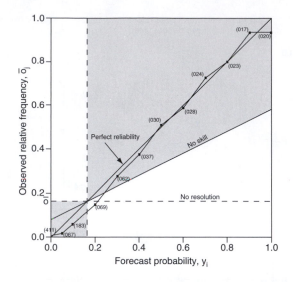

FIGURE 7.10   Attributes diagram for the PoP forecasts summarized in Table 7.2. Solid dots show observed relative frequency of precipitation occurrence, conditional on each of the $I = 12$ possible probability forecasts. Forecasts not defining event subsets with different relative frequencies of the forecast event would exhibit all points on the dashed no-resolution line, which is plotted at the level of the sample climatological probability. Points in the stippled region bounded by the dotted line labeled "no skill" contribute positively to forecast skill, according to Equation 7.35. Relative frequency of use of each of the forecast values, $p(y_i)$, are shown parenthetically, although they could also have been indicated graphically.

dot will be at the same sample climatological relative frequency. This single point will be located at the intersection of the 1:1 (perfect reliability), no-skill and no-resolution lines. Thus, climatological forecasts have perfect (zero, in Equation 7.40) reliability, since the forecast and the conditional relative frequency (Equation 7.38) are both equal to the climatological probability (Equation 7.39). Similarly, the climatological forecasts have zero resolution since the existence of only $I = 1$ forecast category precludes discerning different subsets of forecasting occasions with differing relative frequencies of the outcomes. Since the reliability and resolution terms in Equation 7.40 are both zero, it is clear that the Brier score for climatological forecasts is exactly the uncertainty term in Equation 7.40.

This observation of the equivalence of the uncertainty term and the BS for climatological forecasts has interesting consequences for the Brier skill score in Equation 7.35. Substituting Equation 7.40 for BS into Equation 7.35, and uncertainty for BS_{Ref} yields

$$BSS = \frac{\text{"Resolution"} - \text{"Reliability"}}{\text{"Uncertainty"}}. \tag{7.41}$$

Since the uncertainty term is always positive, the probability forecasts will exhibit positive skill in the sense of Equation 7.35 if the resolution term is larger in absolute value than the reliability term. This means that subsamples of the forecasts identified by the forecasts y_i will contribute positively to the overall skill when their resolution term is larger than their reliability term. Geometrically, this corresponds to points on the attributes diagram being closer to the 1:1 perfect-reliability line than to the horizontal no-resolution line. This condition defines the no-skill line, which is midway between the perfect-reliability and no-resolution lines, and delimits the stippled region, in which subsamples contribute positively to forecast skill. In Figure 7.10 only the subsample for $y_4 = 0.2$, which is nearly equal the climatological probability, fails to contribute positively to the overall forecast skill.

7.4.5 The Discrimination Diagram

The joint distribution of forecasts and observations can also be displayed graphically through the likelihood-base rate factorization (Equation 7.3). For probability forecasts of dichotomous ($J = 2$) predictands, this factorization consists of two conditional likelihood distributions $p(y_i|o_j)$, $j = 1, 2$; and a base rate (i.e., sample climatological) distribution $p(o_j)$ consisting of the relative frequencies for the two dichotomous events in the verification sample.

The discrimination diagram consists of superimposed plots of the two likelihood distributions, as functions of the forecast probability y_i, together with a specification of the sample climatological probabilities $p(o_1)$ and $p(o_2)$. Together, these quantities completely represent the information in the full joint distribution. Therefore, the discrimination diagram presents the same information as the reliability diagram, but in a different format.

Figure 7.11 shows an example discrimination diagram, for the probability-of-precipitation forecasts whose calibration-refinement factorization is displayed in Table 7.2 and whose attributes diagram is shown in Figure 7.10. The probabilities in the two likelihood distributions calculated from their joint distribution are shown in Table 13.2. Clearly the conditional probabilities given the no precipitation event o_2 are greater for the smaller forecast probabilities, and the conditional probabilities given the precipitation event o_1 are greater for the intermediate and larger probability forecasts. Forecasts that discriminated perfectly between the two events would exhibit no overlap in their likelihoods. The two likelihood distributions in Figure 7.11 overlap somewhat, but exhibit substantial separation, indicating substantial discrimination by the forecasts of dry and wet events.

The separation of the two likelihood distributions in a discrimination diagram can be summarized by the difference between their means, called the discrimination distance,

$$d = |\mu_{y|o_1} - \mu_{y|o_2}|. \tag{7.42}$$

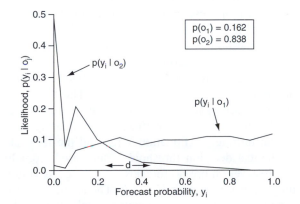

FIGURE 7.11 Discrimination diagram for the data in Table 7.2, which is shown in likelihood-base rate form in Table 13.2. The discrimination distance d (Equation 7.42) is also indicated.

For the two conditional distributions in Figure 7.11 this difference is $d = |0.448 - 0.229| = 0.219$, which is also plotted in the figure. This distance is zero if the two likelihood distributions are the same (i.e., if the forecasts cannot discriminate the event at all), and increases as the two likelihood distributions become more distinct. In the limit $d = 1$ for perfect forecasts, which have all probability concentrated at $p(1|o_1) = 1$ and $p(0|o_2) = 1$.

There is a connection between the likelihood distributions in the discrimination diagram, and statistical discrimination as discussed in Chapter 13. In particular, the two likelihood distributions in Figure 7.11 could be used together with the sample climatological probabilities, as in Section 13.3.3, to recalibrate these probability forecasts by calculating posterior probabilities for the two events given each of the possible forecast probabilities (cf. Exercise 13.3).

7.4.6 The ROC Diagram

The ROC (Relative Operating Characteristic, or Receiver Operating Characteristic) diagram is another discrimination-based graphical forecast verification display, although unlike the reliability diagram and discrimination diagram it does not include the full information contained in the joint distribution of forecasts and observations. The ROC diagram was first introduced into the meteorological literature by Mason (1982), although it has a longer history of use in such disciplines as psychology (Swets 1973) and medicine (Swets 1979), and arose from signal detection theory in electrical engineering.

One way to view the ROC diagram and the ideas behind it is in relation to the class of idealized decision problems outlined in Section 7.8. Here hypothetical decision makers must choose between two alternatives on the basis of a probability forecast for a dichotomous variable, with one of the decisions (say, action A) being preferred if the event o_1 does not occur, and the other (action B) being preferable if the event does occur. As explained in Section 7.8, the probability threshold determining which of the two decisions will be optimal depends on the decision problem, and in particular on the relative undesirability of having taken action A when the event occurs versus action B when the event does not occur. Therefore different probability thresholds for the choice between actions A and B will be appropriate for different decision problems.

If the forecast probabilities y_i have been rounded to I discrete values, there are $I - 1$ such thresholds, excluding the trivial cases of always taking action A or always taking action B. Operating on the joint distribution of forecasts and observations (e.g., Table 7.4a) consistent with each of these probability thresholds yields $I - 1$ 2×2 contingency tables of the kind treated in Section 7.2: a yes forecast is imputed if the probability y_i is above the threshold in question (sufficient probability to warrant a nonprobabilistic forecast of the event, for those decision problems appropriate to that probability threshold), and a no forecast is imputed if the forecast probability is below the threshold (insufficient probability for a nonprobabilistic forecast of the event). The mechanics of constructing these 2×2 contingency tables are exactly as illustrated in Example 7.2. As a discrimination-based technique, ROC diagrams are constructed by evaluating each of these $I - 1$ contingency tables using the hit rate H (Equation 7.12) and the false alarm rate F (Equation 7.13). As the hypothetical decision threshold is increased from lower to higher probabilities there are progressively more no forecasts and progressively fewer yes forecasts, yielding corresponding decreases in both H and F. The resulting $I - 1$ point pairs (F_i, H_i) are then plotted and connected with line segments to each other; and connected to the point $(0, 0)$ corresponding to never forecasting the event (i.e., always choosing action A), and to the point $(1,1)$ corresponding to always forecasting the event (always choosing action B).

The ability of a set of probability forecasts to discriminate a dichotomous event can be easily appreciated from its ROC diagram. Consider first the ROC diagram for perfect forecasts, which use only $I = 2$ probabilities, $y_1 = 0.00$ and $y_2 = 1.00$. For such forecasts there is only one probability threshold from which to calculate a 2×2 contingency table. That table for perfect forecasts exhibits $F = 0.0$ and $H = 1.0$, so its ROC curve consists of two line segments coincident with the left boundary and the upper boundary of the ROC diagram. At the other extreme of forecast performance, random forecasts consistent with the sample climatological probabilities $p(o_1)$ and $p(o_2)$ will exhibit $F_i = H_i$ regardless of how many or how few different probabilities y_i are used, and so their ROC curve will consist of the 45° diagonal connecting the points $(0, 0)$ and $(1,1)$. ROC curves for real forecasts generally fall between these two extremes, lying above and to the left of the 45° diagonal. Forecast with better discrimination exhibit ROC curves approaching the upper-left corner of the ROC diagram more closely, whereas forecasts with very little ability to discriminate the event o_1 exhibit ROC curves very close to the $H = F$ diagonal.

It can be convenient to summarize a ROC diagram using a single scalar value, and the usual choice for this purpose is the area under the ROC curve, A. Since ROC curves for perfect forecasts pass through the upper-left corner, the area under a perfect ROC curve includes the entire unit square, so $A_{\text{perf}} = 1$. Similarly ROC curves for random forecasts lie along the 45° diagonal of the unit square, yielding the area $A_{\text{rand}} = 0.5$. The area A under a ROC curve of interest can also be expressed in standard skill-score form (Equation 7.4), as

$$SS_{\text{ROC}} = \frac{A - A_{\text{rand}}}{A_{\text{perf}} - A_{\text{rand}}} = \frac{A - 1/2}{1 - 1/2} = 2\,A - 1. \tag{7.43}$$

Marzban (2004) describes some characteristics of forecasts that can be diagnosed from the shapes of their ROC curves, based on analysis of some simple idealized discrimination diagrams. Symmetrical ROC curves result when the two likelihood distributions $p(y_i|o_1)$ and $p(y_i|o_2)$ have similar dispersion, or widths, so the ranges of the forecasts y_i corresponding to each of the two outcomes are comparable. On the other hand asymmetrical ROC curves, which might intersect either the vertical or horizontal axis at either $H \approx 0.5$ or $F \approx 0.5$, respectively, are indicative of one or the other of the two likelihoods being substantially more concentrated than the other. Marzban (2004) also finds that A (or, equivalently, SS_{ROC}) is a reasonably good discriminator among relatively low-quality forecasts, but that relatively good forecasts tend to be characterized by quite similar (near-unit) areas under their ROC curves.

EXAMPLE 7.6   Two Example ROC Curves

Example 7.2 illustrated the conversion of the probabilistic forecasts summarized by the joint distribution in Table 7.2 to nonprobabilistic yes/no forecasts, using a probability threshold between $y_3 = 0.1$ and $y_4 = 0.2$. The resulting 2×2 contingency table consists of (cf. Figure 7.1a) $a = 1828$, $b = 2369$, $c = 181$, and $d = 8024$; yielding $F = 2369/(2369 + 8024) = 0.228$ and $H = 1828/(1828 + 181) = 0.910$. This point is indicated by the dot on the ROC curve for the Table 7.2 data in Figure 7.12. The entire ROC curve for the Table 7.2 data consists of this and all other partitions of these forecasts into yes/no forecasts using different probability thresholds. For example, the point just to the left of $(0.228, 0.910)$ on this ROC curve is obtained by moving the threshold between $y_4 = 0.2$ and $y_5 = 0.3$. This partition produces $a = 1644$, $b = 1330$, $c = 364$, and $d = 9064$, defining the point $(F, H) = (0.128, 0.819)$.

Summarizing ROC curves according to the areas underneath them requires summation of the areas under each of the I trapezoids defined by the point pairs

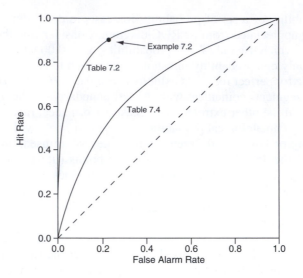

FIGURE 7.12   ROC diagrams for the PoP forecasts in Table 7.2 (upper solid curve), and the hypothetical forecasts in Table 7.4 (lower solid curve). Solid dot locates the (F, H) pair corresponding to the probability threshold in Example 7.2.

(F_i, H_i), $i = 1, \ldots, I-1$, together with the two endpoints $(0, 0)$ and $(1, 1)$. For example, the trapezoid defined by the dot in Figure 7.12 and the point just to its left has area $0.5(0.910 + 0.819)(0.228 - 0.128) = 0.08645$. This area, together with the areas of the other $I - 1 = 11$ trapezoids defined by the segments of the ROC curve for these data yield the area $A = 0.922$.

The ROC curve, and the area under it, can also be computed directly from the joint probabilities in $p(y_i, o_j)$; that is, without knowing the sample size n. Table 7.5 summarizes the conversion of the hypothetical joint distribution in Table 7.4a to the $I - 1 = 10$ sets of 2×2 tables, by operating directly on the joint probabilities. Note that these data have one fewer forecast value y_i than those in Table 7.2, because in Table 7.2 the forecast $y_2 = 0.05$ has been allowed. For example, for the first probability threshold in Table 7.5, 0.05, only

TABLE 7.5   The $I - 1 = 10$ 2×2 tables derived from successive partitions of the joint distribution in Table 7.4, and the corresponding values for H and F.

Threshold	a/n	b/n	c/n	d/n	H	F
0.05	.252	.448	.045	.255	.848	.637
0.15	.220	.320	.077	.383	.741	.455
0.25	.195	.245	.102	.458	.657	.348
0.35	.171	.189	.126	.514	.576	.269
0.45	.147	.143	.150	.560	.495	.203
0.55	.123	.107	.174	.596	.414	.152
0.65	.096	.074	.201	.629	.323	.105
0.75	.071	.049	.226	.654	.239	.070
0.85	.043	.027	.254	.676	.145	.038
0.95	.013	.007	.284	.696	.044	.010

the forecasts $y_1 = 0.0$ are converted to "no" forecasts, so the entries of the resulting 2×2 joint distribution (cf. Figure 7.1b) are $a/n = 0.032 + 0.025 + \cdots + 0.013 = 0.252$, $b/n = 0.128 + 0.075 + \cdots + 0.007 = 0.448$, $c/n = p(y_1, o_1) = 0.045$, and $d/n = p(y_1, o_1) = 0.255$. For the second probability threshold, 0.15, both the forecasts $y_1 = 0.0$ and $y_2 = 0.1$ are converted to "no" forecasts, so the resulting 2×2 joint distribution contains the four probabilities $a/n = 0.025 + 0.024 + \cdots + 0.013 = 0.220$, $b/n = 0.075 + 0.056 + \cdots + 0.007 = 0.320$, $c/n = 0.045 + 0.032 = 0.077$, and $d/n = 0.255 + 0.128 = 0.383$.

Table 7.5 also shows the hit rate H and false alarm rate F for each of the 10 partitions of the joint distribution in Table 7.4a. These pairs define the lower ROC curve in Figure 7.12, with the points corresponding to the smaller probability thresholds occurring in the upper right portion of the ROC diagram, and points corresponding to the larger probability thresholds occurring in the lower left portion. Proceeding from left to right, the areas under the $I = 11$ trapezoids defined by these points together with the points at the corners of the ROC diagram are $0.5(0.044 + 0.000)(0.010 - 0.000) = 0.00022$, $0.5(0.145 + 0.044)(0.038 - 0.010) = 0.00265$, $0.5(0.239 + 0.145)(0.070 - 0.038) = 0.00614, \ldots, 0.5(1.000 + 0.848)(1.000 - 0.637) = 0.33541$; yielding a total area of $A = 0.698$.

Figure 7.12 shows clearly that the forecasts in Table 7.2 exhibit greater event discrimination than those in Table 7.4, because the arc of the corresponding ROC curve for the former is everywhere above that for the latter, and approaches more closely the upper left-hand corner of the ROC diagram. This difference in discrimination is summarized by the differences in the areas under the two ROC curves; that is, $A = 0.922$ versus $A = 0.698$. ◊

ROC diagrams have been used increasingly in recent years to evaluate probability forecasts for binary predictands, so it is worthwhile to reiterate that (unlike the reliability diagram and the discrimination diagram) they do *not* provide a full depiction of the joint distribution of forecasts and observations. The primary deficiency of the ROC diagram can be appreciated by recalling the mechanics of its construction, as outlined in Example 7.6. In particular, the calculations behind the ROC diagrams are carried out without regard to the specific values for the probability labels, $p(y_i)$. That is, the actual forecast probabilities are used only to sort the elements of the joint distribution into a sequence of 2×2 tables, but otherwise their actual numerical values are immaterial. For example, Table 7.4b shows that the forecasts defining the lower ROC curve in Figure 7.12 are poorly calibrated, and in particular they exhibit strong conditional (overconfidence) bias. However this and other biases are not reflected in the ROC diagram, because the specific numerical values for the forecast probabilities $p(y_i)$ do not enter into the ROC calculations, and so ROC diagrams are insensitive to such conditional and unconditional biases (e.g., Kharin and Zwiers 2003b; Wilks 2001). In fact, if the forecast probabilities $p(y_i)$ had corresponded exactly to the corresponding conditional event probabilities $p(o_1|y_i)$, or even if the probability labels on the forecasts in Tables 7.2 or 7.4 had been assigned values that were allowed to range outside the [0, 1] interval (while maintaining the same ordering, and so the same groupings of event outcomes), the resulting ROC curves would be identical!

The insensitivity of ROC diagrams and ROC areas to both conditional and unconditional forecast biases—that they are independent of calibration—is sometimes cited as an advantage. This property is an advantage only in the sense that ROC diagrams reflect potential skill (which would be actually achieved only if the forecasts were correctly calibrated), in much the same way that the correlation coefficient reflects potential skill (cf. Equation 7.33). However, this property is not an advantage for forecast users who do not have access to the historical forecast data necessary to correct miscalibrations, and who therefore have no choice but to take forecast probabilities at face value. On the other hand, when forecasts underlying ROC diagrams are correctly calibrated, dominance of one ROC curve over another (i.e., one curve lying entirely above and to the left of another) implies statistical sufficiency for the dominating forecasts, so that these will be of greater economic value for all rational forecast users (Krzysztofowicz and Long 1990).

7.4.7 *Hedging, and Strictly Proper Scoring Rules*

When forecasts are evaluated quantitatively, it is natural for forecasters to want to achieve the best scores they can. Depending on the evaluation measure, it may be possible to improve scores by hedging, or gaming, which implies forecasting something other than our true beliefs about future weather events in order to achieve a better score. In the setting of a forecast contest in a college or university, if the evaluation of our performance can be improved by playing the score, then it is entirely rational to try to do so. Conversely, if we are responsible for assuring that forecasts are of the highest possible quality, evaluating those forecasts in a way that penalizes hedging is desirable.

A forecast evaluation procedure that awards a forecaster's best expected score only when his or her true beliefs are forecast is called strictly proper. That is, strictly proper scoring procedures cannot be hedged. One very appealing attribute of the Brier score is that it is strictly proper, and this is one strong motivation for using the Brier score to evaluate the accuracy of probability forecasts for dichotomous predictands. Of course it is not possible to know in advance what Brier score a given forecast will achieve, unless we can make perfect forecasts. However, it is possible on each forecasting occasion to calculate the expected, or probability-weighted, score using our subjective probability for the forecast event.

Suppose a forecaster's subjective probability for the event being forecast is y^*, and that the forecaster must publicly issue a forecast probability, y. The expected Brier score is simply

$$E[BS] = y^*(y-1)^2 + (1-y^*)(y-0)^2, \tag{7.44}$$

where the first term is the score received if the event occurs multiplied by the subjective probability that it will occur, and the second term is the score received if the event does not occur multiplied by the subjective probability that it will not occur. Consider that the forecaster has decided on a subjective probability y^*, and is weighing the problem of what forecast y to issue publicly. Regarding y^* as constant, it is easy to minimize the expected Brier score by differentiating Equation 7.44 by y, and setting the result equal to zero. Then,

$$\frac{\partial E[BS]}{\partial y} = 2y^*(y-1) + 2(1-y^*)y = 0, \tag{7.45}$$

yielding

$$2y\,y^* - 2y^* + 2y - 2y\,y^* = 0,$$
$$2y = 2y^*,$$

and

$$y = y^*.$$

That is, regardless of the forecaster's subjective probability, the minimum expected Brier score is achieved only when the publicly issued forecast corresponds exactly to the subjective probability. By contrast, the absolute error (linear) score, $LS = |y - o|$ is minimized by forecasting $y = 0$ when $y^* < 0.5$, and forecasting $y = 1$ when $y^* > 0.5$.

Equation 7.45 proves that the Brier score is strictly proper. Often Brier scores are expressed in the skill-score format of Equation 7.35. Unfortunately, even though the Brier score itself is strictly proper, this standard skill score based upon it is not. However, for moderately large sample sizes (perhaps $n > 100$) the BSS closely approximates a strictly proper scoring rule (Murphy 1973a).

7.4.8 *Probability Forecasts for Multiple-Category Events*

Probability forecasts may be formulated for discrete events having more than two (yes vs. no) possible outcomes. These events may be nominal, for which there is not a natural ordering; or ordinal, where it is clear which of the outcomes are larger or smaller than others. The approaches to verification of probability forecasts for nominal and ordinal predictands differ, because the magnitude of the forecast error is not a meaningful quantity in the case of nominal events, but is potentially quite important for ordinal events. The usual approach to verifying forecasts for nominal predictands is to collapse them to a sequence of binary predictands. Having done this, Brier scores, reliability diagrams, and so on, can be used to evaluate each of the derived binary forecasting situations.

Verification of probability forecasts for multicategory ordinal predictands presents a more difficult problem. First, the dimensionality of the verification problem increases exponentially with the number of outcomes over which the forecast probability is distributed. For example, consider a $J = 3$-event situation for which the forecast probabilities are constrained to be one of the 11 values $0.0, 0.1, 0.2, \ldots, 1.0$. The dimensionality of the problem is not simply $33 - 1 = 32$, as might be expected by extension of the dimensionality for the dichotomous forecast problem, because the forecasts are now vector quantities. For example, the forecast vector $(0.2, 0.3, 0.5)$ is a different and distinct forecast from the vector $(0.3, 0.2, 0.5)$. Since the three forecast probabilities must sum to 1.0, only two of them can vary freely. In this situation there are $I = 66$ possible three-dimensional forecast vectors, yielding a dimensionality for the forecast problem of $(66 \times 3) - 1 = 197$. Similarly, the dimensionality for the four-category ordinal verification situation with the same restriction on the forecast probabilities would be $(286 \times 4) - 1 = 1143$. As a practical matter, because of their high dimensionality, probability forecasts for ordinal predictands primarily have been evaluated using scalar performance measures, even though such approaches will necessarily be incomplete.

For ordinal predictands, collapsing the verification problem to a series of $I \times 2$ tables will result in the loss of potentially important information related to the ordering of the outcomes. For example, the probability forecasts for precipitation shown in Figure 6.33 distribute probability among three MECE outcomes: dry, near-normal, and wet. If we were to verify the dry events in distinction to not dry events composed of both the near-normal and wet categories, information pertaining to the magnitudes of the forecast errors would be thrown away. That is, the same error magnitude would be assigned to the difference between dry and wet as to the difference between dry and near-normal.

Verification that is sensitive to distance usually is preferred for probability forecasts of ordinal predictands. That is, the verification should be capable of penalizing forecasts increasingly as more probability is assigned to event categories further removed from the actual outcome. In addition, we would like the verification measure to be strictly proper (see Section 7.4.7), so that forecasters are encouraged to report their true beliefs. The most commonly used such measure is the ranked probability score (RPS) (Epstein 1969b; Murphy 1971). Many strictly proper scalar scores that are sensitive to distance exist (Murphy and Daan 1985; Staël von Holstein and Murphy 1978), but of these the ranked probability score usually is preferred (Daan 1985).

The ranked probability score is essentially an extension of the Brier score (Equation 7.34) to the many-event situation. That is, it is a squared-error score with respect to the observation 1 if the forecast event occurs, and 0 if the event does not occur. However, in order for the score to be sensitive to distance, the squared errors are computed with respect to the *cumulative* probabilities in the forecast and observation vectors. This characteristic introduces some notational complications.

As before, let J be the number of event categories, and therefore also the number of probabilities included in each forecast. For example, the precipitation forecasts in Figure 6.33 have $J = 3$ events over which to distribute probability. If the forecast is 20% chance of dry, 40% chance of near-normal, and 40% chance of wet; then $y_1 = 0.2$, $y_2 = 0.4$, and $y_3 = 0.4$. Each of these components y_j pertains to one of the J events being forecast. That is, y_1, y_2, and y_3, are the three components of a forecast vector y, and if all probabilities were to be rounded to tenths this forecast vector would be one of $I = 66$ possible forecasts y_i.

Similarly, the observation vector has three components. One of these components, corresponding to the event that occurs, will equal 1, and the other $J - 1$ components will equal zero. In the case of Figure 6.33, if the observed precipitation outcome is in the wet category, then $o_1 = 0$, $o_2 = 0$, and $o_3 = 1$.

The cumulative forecasts and observations, denoted Y_m and O_m, are defined as functions of the components of the forecast vector and observation vector, respectively, according to

$$Y_m = \sum_{j=1}^{m} y_j, \quad m = 1, \ldots, J;$$ (7.46a)

and

$$O_m = \sum_{j=1}^{m} o_j, \quad m = 1, \ldots, J.$$ (7.46b)

In terms of the foregoing hypothetical example, $Y_1 = y_1 = 0.2$, $Y_2 = y_1 + y_2 = 0.6$, and $Y_3 = y_1 + y_2 + y_3 = 1.0$; and $O_1 = o_1 = 0$, $O_2 = o_1 + o_2 = 0$, and $O_3 = o_1 + o_2 + o_3 = 1$. Notice that since Y_m and O_m are both cumulative functions of probability components that must add to one, the final sums Y_J and O_J are always both equal to one by definition.

The ranked probability score is the sum of squared differences between the components of the cumulative forecast and observation vectors in Equation 7.46a and 7.46b, given by

$$RPS = \sum_{m=1}^{J} (Y_m - O_m)^2,$$ (7.47a)

or, in terms of the forecast and observed vector components y_j and o_j,

$$RPS = \sum_{m=1}^{J} \left[\left(\sum_{j=1}^{m} y_j \right) - \left(\sum_{j=1}^{m} o_j \right) \right]^2.$$ (7.47b)

A perfect forecast would assign all the probability to the single y_j corresponding to the event that subsequently occurs, so that the forecast and observation vectors would be the same. In this case, RPS $= 0$. Forecasts that are less than perfect receive scores that are positive numbers, so the RPS has a negative orientation. Notice also that the final $(m = J)$ term in Equation 7.47 is always zero, because the accumulations in Equations 7.46 ensure that $Y_J = O_J = 1$. Therefore, the worst possible score is $J - 1$. For $J = 2$, the ranked probability score reduces to the Brier score, Equation 7.34. Note that since the last term, for $m = J$, is always zero, in practice it need not actually be computed.

EXAMPLE 7.7   Illustration of the Mechanics of the Ranked Probability Score

Table 7.6 demonstrates the mechanics of computing the RPS, and illustrates the property of sensitivity to distance, for two hypothetical probability forecasts for precipitation amounts. Here the continuum of precipitation has been divided into $J = 3$ categories, <0.01 in., $0.01 - 0.24$ in., and ≥ 0.25 in. Forecaster 1 has assigned the probabilities $(0.2, 0.5, 0.3)$ to the three events, and Forecaster 2 has assigned the probabilities $(0.2, 0.3, 0.5)$. The two forecasts are similar, except that Forecaster 2 has allocated more probability to the ≥ 0.25 in. category at the expense of the middle category. If no precipitation falls on this occasion the observation vector will be that indicated in the table. For most purposes, Forecaster 1 should receive a better score, because this forecaster has assigned more probability closer to the observed category than did Forecaster 2. The score for Forecaster 1 is RPS $= (0.2 - 1)^2 + (0.7 - 1)^2 = 0.73$, and for Forecaster 2 it is RPS $= (0.2 - 1)^2 + (0.5 - 1)^2 = 0.89$. The lower RPS for Forecaster 1 indicates a more accurate forecast.

If, on the other hand, some amount of precipitation larger than 0.25 in. had fallen, Forecaster 2's probabilities would have been closer, and would have received the better score. The score for Forecaster 1 would have been RPS $= (0.2 - 0)^2 + (0.7 - 0)^2 = 0.53$, and the score for Forecaster 2 would have been RPS $= (0.2 - 0)^2 + (0.5 - 0)^2 = 0.29$. Note that in both of these examples, only the first $J - 1 = 2$ terms in Equation 7.47 were needed to compute the RPS. $\lozenge$

Equation 7.47 yields the ranked probability score for a single forecast-event pair. Jointly evaluating a collection of n forecasts using the ranked probability score requires nothing more than averaging the RPS values for each forecast-event pair,

$$<RPS> = \frac{1}{n} \sum_{k=1}^{n} RPS_k. \tag{7.48}$$

TABLE 7.6   Comparison of two hypothetical probability forecasts for precipitation amount, divided into $J = 3$ categories. The three components of the observation vector indicate that the observed precipitation was in the smallest category.

Event	Forecaster 1		Forecaster 2		Observed	
	y_j	Y_m	y_j	Y_m	o_j	O_m
<0.01 in.	0.2	0.2	0.2	0.2	1	1
$0.01 - 0.24$ in.	0.5	0.7	0.3	0.5	0	1
≥ 0.25 in.	0.3	1.0	0.5	1.0	0	1

Similarly, the skill score for a collection of RPS values relative to the RPS computed from the climatological probabilities can be computed as

$$\text{SS}_{\text{RPS}} = \frac{<\text{RPS}> - <\text{RPS}_{\text{Clim}}>}{0 - <\text{RPS}_{\text{Clim}}>} = 1 - \frac{<\text{RPS}>}{<\text{RPS}_{\text{Clim}}>}. \tag{7.49}$$

7.5 Probability Forecasts for Continuous Predictands

7.5.1 Full Continuous Forecast Probability Distributions

It is usually logistically difficult to provide a full continuous PDF $f(y)$, or CDF $F(y)$, for a probability forecast for a continuous predictand y, unless a conventional parametric form (see Section 4.4) is assumed. In that case a particular forecast PDF or CDF can be summarized with a few specific values for the distribution parameters.

Regardless of how a forecast probability distribution is expressed, providing a full forecast probability distribution is both a conceptual and a mathematical extension of multicategory probability forecasting (see Section 7.4.8), to forecasts for an infinite number of predictand classes of infinitesimal width. A natural approach to evaluating this kind of forecast is to extend the ranked probability score to the continuous case, replacing the summations in Equation 7.47 with integrals. The result is the Continuous Ranked Probability Score (Hersbach 2000; Matheson and Winkler 1976; Unger 1985),

$$\text{CRPS} = \int_{-\infty}^{\infty} [F(y) - F_o(y)]^2 dy, \tag{7.50a}$$

where

$$F_o(y) = \begin{cases} 0, & y < \text{observed value} \\ 1, & y \geq \text{observed value} \end{cases} \tag{7.50b}$$

is a cumulative-probability step function that jumps from 0 to 1 at the point where the forecast variable y equals the observation. The squared difference between continuous CDFs in Equation 7.50a is analogous to the same operation applied to the cumulative discrete variables in Equation 7.47a. Like the discrete RPS, the CRPS is also strictly proper (Matheson and Winkler 1976).

The CRPS has a negative orientation (smaller values are better), and it rewards concentration of probability around the step function located at the observed value. Figure 7.13 illustrates the CRPS with a hypothetical example. Figure 7.13a shows three forecast PDFs $f(y)$ in relation to the single observed value of the continuous predictand y. Forecast Distribution 1 is centered on the eventual observation and strongly concentrates its probability around the observation. Distribution 2 is equally sharp (i.e., expresses the same degree of confidence in distributing probability), but is centered well away from the observation. Distribution 3 is centered on the observation but exhibits low confidence (distributes probability more diffusely than the other two forecast distributions). Figure 7.13b shows the same three forecast distributions expressed as CDFs, $F(y)$, together with the step-function CDF $F_0(y)$ (thick line) that jumps from 0 to 1 at the observed value (Equation 7.50b). Since the CRPS is the integrated squared difference between the CDF and the step function, CDFs that approximate the step function (Distribution 1) produce relatively small integrated squared differences, and so good scores. Distribution 2 is equally sharp, but

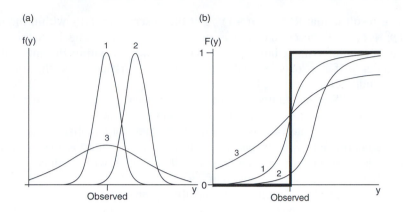

FIGURE 7.13   Schematic illustration of the continuous ranked probability score. Three forecast PDFs are shown in relation to the observed outcome in (a). The corresponding CDFs are shown in (b), together with the step-function CDF for the observation $F_0(y)$ (heavy line). Distribution 1 would produce a small (good) CRPS because its CDF is the closest approximation to the step function. Distribution 2 concentrates probability away from the observation, and Distribution 3 is penalized for lack of sharpness even though it is centered on the observation.

its displacement away from the observation produces large discrepancies with the step function, and therefore also large integrated squared differences. Distribution 3 is centered on the observation, but its diffuse assignment of forecast probability means that it is nevertheless a poor approximation to the step function, and so also yields large integrated squared differences.

Hersbach (2000) notes that the CRPS can also be computed as the Brier score for dichotomous events, integrated over all possible division points of the continuous variable y into the dichotomous variable above and below the division point. Accordingly the CRPS has an algebraic decomposition into reliability, resolution, and uncertainty components that is analogous to an integrated form of Equation 7.40.

7.5.2 Central Credible Interval Forecasts

The burden of communicating a full probability distribution is reduced considerably if the forecast distribution is merely sketched, using the central credible interval (CCI) format (see Section 6.7.3). In full form, a central credible interval forecast consists of a range of the predictand that is centered in a probability sense, together with the probability covered by the forecast range within the forecast distribution. Usually central credible interval forecasts are abbreviated in one of two ways: either the interval width is constant on every forecast occasion but the location of the interval and the probability it subtends are allowed to vary (fixed-width CCI forecasts), or the probability within the interval is constant on every forecast occasion but the interval location and width may both change (fixed-probability CCI forecasts).

The ranked probability score (Equation 7.47) is an appropriate scalar accuracy measure for fixed-width CCI forecasts (Baker 1981; Gordon 1982). In this case there are three categories (below, within, and above the forecast interval) among which the forecast probability is distributed. The probability p pertaining to the forecast interval is specified as part of the forecast, and because the forecast interval is located in the probability center of the distribution, probabilities for the two extreme categories are each $(1 - p)/2$.

The result is that RPS $= (p-1)^2/2$ if the observation falls within the interval, or RPS $= (p^2+1)/2$ if the observation is outside the interval. The RPS thus reflects a balance between preferring a large p if the observation is within the interval, but preferring a smaller p if it is outside, and that balance is satisfied when the forecaster reports their true judgment.

The RPS is not an appropriate accuracy measure for fixed-probability CCI forecasts. For this forecast format, small (i.e., better) RPS can be achieved by always forecasting extremely wide intervals, because the RPS does not penalize vague forecasts that include wide central intervals. In particular, forecasting an interval that is sufficiently wide that the observation is nearly certain to fall within it will produce a smaller RPS than a verification outside the interval if $(p-1)^2/2 < (p^2+1)/2$. A little algebra shows that this inequality is satisfied for any positive probability p.

Fixed-probability CCI forecasts are appropriately evaluated using Winkler's score (Winkler 1972a; Winkler and Murphy 1979),

$$W = \begin{cases} (b-a+1)+k(a-o), & o < a \\ (b-a+1), & a \le o \le b \\ (b-a+1)+k(o-b), & b < o \end{cases} \qquad (7.51)$$

Here the forecast interval ranges from a to b, inclusive, and the value of the observed variable is o. Regardless of the actual observation, a forecast is charged penalty points equal to the width of the forecast interval, which is $b-a+1$ to account for both endpoints when (as usual) the interval is specified in terms of integer units of the predictand. Additional penalty points are added if the observation falls outside the specified interval, and the magnitudes of these "miss" penalties are proportional to the distance from the interval. Winkler's score thus expresses a tradeoff between short intervals to reduce the fixed penalty (and thus encouraging sharp forecasts), versus sufficiently wide intervals to avoid incurring the additional penalties too frequently. This tradeoff is balanced by the constant k, which depends on the fixed probability to which the forecast CCI pertains, and increases as the implicit probability for the interval increases, because outcomes outside the interval should occur increasingly rarely for larger interval probabilities. In particular, $k=4$ for 50% CCI forecasts, and $k=8$ for 75% CCI forecasts. More generally, $k=1/F(a)$, where $F(a)=1-F(b)$ is the cumulative probability associated with the lower interval boundary according to the forecast CDF.

Winkler's score is equally applicable to fixed-width CCI forecasts, and to unabbreviated CCI forecasts for which the forecaster is free to choose both the interval width and the subtended probability. In these two cases, where the stated probability may change from forecast to forecast, the penalty function for observations falling outside the forecast interval will also change, according to $k=1/F(a)$.

7.6 Nonprobabilistic Forecasts of Fields

7.6.1 General Considerations for Field Forecasts

An important problem in forecast verification is characterization of the quality of forecasts of atmospheric fields; that is, spatial arrays of atmospheric variables. Forecasts for such fields as surface pressures, geopotential heights, temperatures, and so on, are produced routinely by weather forecasting centers worldwide. Often these forecasts are nonprobabilistic, without expressions of uncertainty as part of the forecast format. An example of

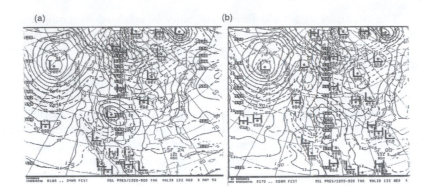

FIGURE 7.14 Forecast (a) and subsequently analyzed (b) sea-level pressures (solid) and 1000-500 mb thicknesses (dashed) over a portion of the northern hemisphere for 4 May 1993.

this kind of forecast is shown in Figure 7.14a, which displays 24-h forecasts of sea-level pressures and 1000-500 mb thicknesses over a portion of the northern hemisphere, made on the morning of 4 May 1993 by the U.S. National Meteorological Center. Figure 7.14b shows the same fields as analyzed 24 hours later. A subjective visual assessment of the two pairs of fields indicates that the main features correspond well, but that some discrepancies exist in their locations and magnitudes.

Objective, quantitative methods of verification for forecasts of atmospheric fields allow more rigorous assessments of forecast quality to be made. In practice, such methods operate on gridded fields, or collections of values of the field variable sampled at, or interpolated to, a grid in the spatial domain. Usually this geographical grid consists of regularly spaced points either in distance, or in latitude and longitude.

Figure 7.15 illustrates the gridding process for a hypothetical pair of forecast and observed fields in a small spatial domain. Each of the fields can be represented in map form as contours, or isolines, of the mapped quantity. The grid imposed on each map is a regular array of points at which the fields are sampled. Here the grid consists of four rows in the north-south direction, and five columns in the east-west direction. Thus the gridded forecast field consists of the $M = 20$ discrete values y_m, which sample the smoothly varying continuous forecast field. The gridded observed field consists of the $M = 20$ discrete values o_m, which represent the smoothly varying observed field at these same locations.

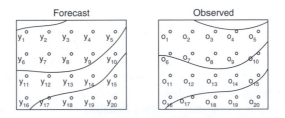

FIGURE 7.15 Hypothetical forecast (left) and observed (right) atmospheric fields represented as contour maps over a small rectangular domain. Objective assessment of the accuracy of the forecast begins with gridding both the forecast and observed fields; i.e., interpolating them to the same geographical grid (small circles). Here the grid has four rows in the north-south direction, and five columns in the east-west direction, so the forecast and observed fields are represented by the $M = 20$ discrete values y_m and o_m, respectively.

The accuracy of a field forecast usually is assessed by computing measures of the correspondence between the values y_m and o_m. If a forecast is perfect, then $y_m = o_m$ for each of the M gridpoints. Of course there are many ways that gridded forecast and observed fields can be different, even when there are only a small number of gridpoints. Put another way, the verification of field forecasts is a problem of very high dimensionality, even for small grids. Although examination of the joint distribution of forecasts and observation is in theory the preferred approach to verification of field forecasts, its large dimensionality suggests that this ideal may not be practically realizable. Rather, the correspondence between forecast and observed fields generally has been characterized using scalar summary measures. These scalar accuracy measures are necessarily incomplete, but are useful in practice.

7.6.2 The S1 Score

The S1 score is an accuracy measure that is primarily of historical interest. It was designed to reflect the accuracy of forecasts of gradients of pressure or geopotential height, in consideration of the relationship of these gradients to the wind field at the same level.

Rather than operating on individual gridded values, the S1 score operates on the differences between gridded values at adjacent gridpoints. Denote the differences between the gridded values at any particular pair adjacent gridpoints as Δy for points in the forecast field, and Δo for points in the observed field. In terms of Figure 7.15, for example, one possible value of Δy is $y_3 - y_2$, which would be compared to the corresponding gradient in the observed field, $\Delta o = o_3 - o_2$. Similarly, the difference $\Delta y = y_9 - y_4$, would be compared to the observed difference $\Delta o = o_9 - o_4$. If the forecast field reproduces the signs and magnitudes of the gradients in the observed field exactly, each Δy will equal its corresponding Δo.

The S1 score summarizes the differences between the $(\Delta y, \Delta o)$ pairs according to

$$
S1 = \frac{\sum\limits_{\text{adjacent pairs}} |\Delta y - \Delta o|}{\sum\limits_{\text{adjacent pairs}} \max\{|\Delta y|, |\Delta o|\}} \times 100. \tag{7.52}
$$

Here the numerator consists of the sum of the absolute errors in forecast gradient over all adjacent pairs of gridpoints. The denominator consists of the sum, over the same pairs of points, of the larger of the absolute value of the forecast gradient, $|\Delta y|$, or the absolute value of the observed gradient, $|\Delta o|$. The resulting ratio is multiplied by 100 as a convenience.

Clearly perfect forecasts will exhibit $S1 = 0$, with poorer gradient forecasts being characterized by increasingly larger scores. The S1 score exhibits some undesirable characteristics that have resulted in its going out of favor. The most obvious is that the actual magnitudes of the forecast pressures or heights are unimportant, since only pairwise gridpoint differences are scored. Thus the S1 score does not reflect bias. Summer scores tend to be larger (apparently worse) because of generally weaker gradients, producing a smaller denominator in Equation 7.52. Finally, the score depends on the size of the domain and the spacing of the grid, so that it is difficult to compare S1 scores not pertaining to the same domain and grid.

Equation 7.52 yields the S1 score for a single pair of forecast-observed fields. When the aggregate skill of a series of field forecasts is to be assessed, the S1 scores for each forecast occasion are simply averaged. This averaging smoothes sampling variations,

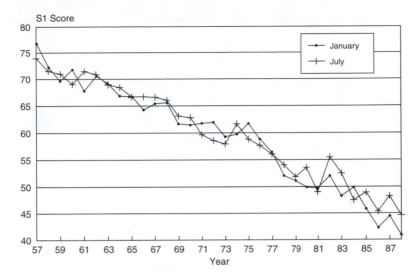

FIGURE 7.16   Average S1 scores for 36-h hemispheric forecasts of 500 mb heights for January and July, 1957–1988, produced by the Canadian Meteorological Centre. From Stanski *et al.* (1989).

and allows trends through time of forecast performance to be assessed more easily. For example, Figure 7.16 (Stanski *et al.* 1989) shows average S1 scores for 36-h hemispheric 500 mb height forecasts, for January and July 1957–1988. The steady decline through time indicates improved forecast performance.

The S1 score has limited operational usefulness for current forecasts, but its continued tabulation has allowed forecast centers to examine very long-term trends in their field-forecast accuracy. Decades-old forecast maps may not have survived, but summaries of their accuracy in terms of the S1 score have often been retained. Kalnay (2003) shows results comparable to those in Figure 7.16, for forecasts made for the United States from 1948 through 2001.

7.6.3 Mean Squared Error

The mean squared error, or MSE, is a much more common accuracy measure for field forecasts. The MSE operates on the gridded forecast and observed fields by spatially averaging the individual squared differences between the two at each of the M gridpoints. That is,

$$\mathrm{MSE} = \frac{1}{M} \sum_{m=1}^{M} (y_m - o_m)^2. \qquad (7.53)$$

This formulation is mathematically the same as Equation 7.28, with the mechanics of both equations centered on averaging squared errors. The difference in application between the two equations is that the MSE in Equation 7.53 is computed over the gridpoints of a *single* pair of forecast/observation fields—that is, to $n = 1$ pair of maps—whereas Equation 7.28 pertains to the average over n pairs of scalar forecasts and observations. Clearly the MSE for a perfectly forecast field is zero, with larger MSE indicating decreasing accuracy of the forecast.

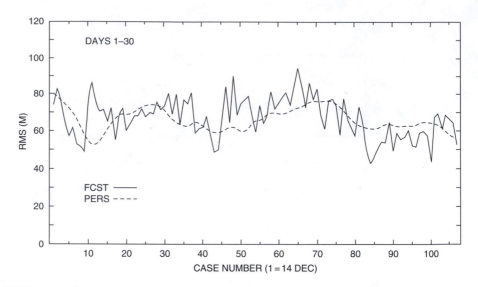

FIGURE 7.17   Root-mean squared error (RMSE) for dynamical 30-day forecasts of 500 mb heights for the northern hemisphere between 20° and 80° N (solid), and persistence of the previous 30-day average 500 mb field (dashed), for forecasts initialized 14 December 1986 through 31 March 1987. From Tracton *et al.* (1989).

Often the MSE is expressed as its square root, the root-mean squared error, RMSE = √MSE. This form of expression has the advantage that it retains the units of the forecast variable and is thus more easily interpretable as a typical error magnitude. To illustrate, the solid line in Figure 7.17 shows RMSE in meters for 30-day forecasts of 500 mb heights initialized on 108 consecutive days during 1986–1987 (Tracton *et al.* 1989). There is considerable variation in forecast accuracy from day to day, with the most accurate forecast fields exhibiting RMSE near 45 *m*, and the least accurate forecast fields exhibiting RMSE around 90 *m*. Also shown in Figure 7.17 are RMSE values of 30-day forecasts of persistence, obtained by averaging observed 500 mb heights for the most recent 30 days prior to the forecast. Usually the persistence forecast exhibits slightly higher RMSE than the 30-day dynamical forecasts, but it is apparent from the figure that there are many days when the reverse is true, and that at this extended range the accuracy of these persistence forecasts was competitive with that of the dynamical forecasts.

The plot in Figure 7.17 shows accuracy of individual field forecasts, but it is also possible to express the aggregate accuracy of a collection of field forecasts by averaging the MSEs for each of a collection of paired comparisons. This average of MSE values across many forecast maps can then be converted to an average RMSE as before, or expressed as a skill score in the same form as Equation 7.32. Since the MSE for perfect field forecasts is zero, the skill score following the form of Equation 7.4 is computed using

$$SS = \frac{\sum_{k=1}^{n} MSE(k) - \sum_{k=1}^{n} MSE_{ref}(k)}{0 - \sum_{k=1}^{n} MSE_{ref}(k)} = 1 - \frac{\sum_{k=1}^{n} MSE(k)}{\sum_{k=1}^{n} MSE_{ref}(k)}, \qquad (7.54)$$

where the aggregate skill of *n* individual field forecasts is being summarized. When this skill score is computed, the reference field forecast is usually either the climatological average field (in which case it may be called the reduction of variance, in common with Equation 7.32) or individual persistence forecasts as shown in Figure 7.17.

The MSE skill score in Equation 7.54, when calculated with respect to climatological forecasts as the reference, allows interesting interpretations for field forecasts when algebraically decomposed in the same way as in Equation 7.33. When applied to field forecasts, this decomposition is conventionally expressed in terms of the differences (anomalies) of forecasts and observations with the corresponding climatological values at each gridpoint (Murphy and Epstein 1989),

$$y'_m = y_m - c_m \tag{7.55a}$$

and

$$o'_m = o_m - c_m, \tag{7.55b}$$

where c_m is the climatological value at gridpoint m. The resulting MSE and skill scores are identical, because the climatological values c_m can be both added to and subtracted from the squared terms in Equation 7.53 without changing the result; that is,

$$\text{MSE} = \frac{1}{M} \sum_{m=1}^{M} (y_m - o_m)^2 = \frac{1}{M} \sum_{m=1}^{M} ([y_m - c_m] - [o_m - c_m])^2 = \frac{1}{M} \sum_{m=1}^{M} (y'_m - o'_m)^2. \tag{7.56}$$

When expressed in this way, the algebraic decomposition of MSE skill score in Equation 7.33 becomes

$$\text{SS}_{\text{Clim}} = \left\{ r_{y'o'}^2 - \left[r_{y'o'} - \frac{s_{y'}}{s_{o'}} \right]^2 - \left[\frac{\bar{y}' - \bar{o}'}{s_{o'}} \right]^2 + \left[\frac{\bar{o}'}{s_{o'}} \right]^2 \right\} \bigg/ \left(1 + \left[\frac{\bar{o}'}{s_{o'}} \right]^2 \right) \tag{7.57a}$$

$$\approx r_{y'o'}^2 - \left[r_{y'o'} - \frac{s_{y'}}{s_{o'}} \right]^2 - \left[\frac{\bar{y}' - \bar{o}'}{s_{o'}} \right]^2. \tag{7.57b}$$

The difference between this decomposition and Equation 7.33 is the normalization factor involving the average differences between the observed and climatological gridpoint values, in both the numerator and denominator of Equation 7.57a. This factor depends only on the observed field. Murphy and Epstein (1989) note that this normalization factor is likely to be small if the skill is being evaluated over a sufficiently large spatial domain, because positive and negative differences with the gridpoint climatological values will tend to balance. Neglecting this term leads to the approximate algebraic decomposition of the skill score in Equation 7.57b, which is identical to Equation 7.33 except that it involves the differences y' and o' with the gridpoint climatological values. It is worthwhile to work with these climatological anomalies when investigating skill of field forecasts in this way, in order to avoid ascribing spurious skill to forecasts for merely forecasting a correct climatology.

Livezey *et al.* (1995) have provided physical interpretations of the three terms in Equation 7.57b. They call the first term phase association, and refer to its complement $1 - r_{y'o'}^2$ as phase error. Of course $r_{y'o'}^2 = 1$ if the fields of forecast and observed anomalies are exactly equal, but because correlations are not sensitive to bias the phases association will also be 1 if the fields of forecast and observed anomalies are proportional according

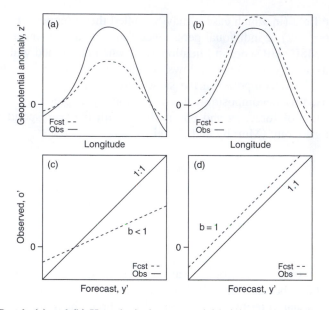

FIGURE 7.18   Panels (a) and (b) Hypothetical geopotential height anomaly forecasts (dashed) along a portion of a latitude circle, exhibiting excellent phase association with the corresponding observed feature (solid). Panels (c) and (d) show corresponding scatterplots of forecast and observed height anomalies.

to any positive constant—that is, if the locations, shapes, and *relative* magnitudes of the forecast features are correct. Figure 7.18a illustrates this concept for a hypothetical geopotential height forecast along a portion of a latitude circle: the forecast (dashed) height feature is located correctly with respect to the observations (solid), thus exhibiting good phase association, and therefore small phase error. Similarly, Figure 7.18b shows another hypothetical forecast with excellent phase association, but a different forecast bias. Offsetting either of these dashed forecast patterns to the left or right, putting them out of phase with the respective solid curve, would decrease the squared correlation and increase the phase error.

The second term in Equation 7.57b is a penalty for conditional bias, or deficiencies in reliability. In terms of errors in a forecast map, Livezey *et al.* (1995) refer to this term as amplitude error. A straightforward way to understand the structure of this term is in relation to a regression equation in which the predictor is y' and the predictand is o'. A little study of Equation 6.7a reveals that another way to express the regression slope is

$$b = \frac{\sum[(y'-\bar{y}')(o'-\bar{o}')]}{\sum(y'-\bar{y}')^2} = \frac{n\,\mathrm{cov}(y',o')}{n\,\mathrm{var}(y')} = \frac{n\,s_{y'}\,s_{o'}\,r_{y'o'}}{n\,s_{y'}^2} = \frac{s_{o'}}{s_{y'}}\,r_{y'o'}. \qquad (7.58)$$

If the forecasts are conditionally unbiased this regression slope will be 1, whereas forecasting features with excessive amplitudes will yield $b > 1$ and forecasting features with insufficient amplitude will yield $b < 1$. If $b = 1$ then $r_{y'o'}^2 = s_{y'}/s_{o'}$, yielding a zero amplitude error in the second term of Equation 7.57b. The dashed forecast in Figure 7.18a exhibits excellent phase association but insufficient amplitude, yielding $b < 1$ (see Figure 7.18c), and therefore a nonzero squared difference in the amplitude error term in Equation 7.57b. Because the amplitude error term is squared, penalties are subtracted for both insufficient and excessive forecast amplitudes.

Finally, the third term in Equation 7.57b is a penalty for unconditional bias, or map-mean error. It is the square of the difference between the overall map averages of the gridpoint forecasts and observations, scaled in units of the standard deviation of the observations. This third term will reduce the MSE skill score to the extent that the forecasts are consistently too high or too low, on average. Figure 7.18b shows a hypothetical forecast (dashed) exhibiting excellent phase association and the correct amplitude, but a consistent positive bias. Because the forecast amplitude is correct the corresponding regression slope (see Figure 7.18d) is $b = 1$, so there is no amplitude error penalty. However, the difference in overall mean between the forecast and observed field produces a map-mean error penalty in the third term of Equation 7.57b.

7.6.4 Anomaly Correlation

The anomaly correlation (AC) is another commonly used measure of association that operates on pairs of gridpoint values in the forecast and observed fields. To compute the anomaly correlation, the forecast and observed values are first converted to anomalies in the sense of Equation 7.55: the climatological average value of the observed field at each of M gridpoints is subtracted from both forecasts y_m and observations o_m.

There are actually two forms of anomaly correlation in use, and it is unfortunately not always clear which has been employed in a particular instance. The first form, called the centered anomaly correlation, was apparently first suggested by Glenn Brier in an unpublished 1942 U.S. Weather Bureau mimeo (Namias 1952). It is computed according to the usual Pearson correlation (Equation 3.22), operating on the M gridpoint pairs of forecasts and observations that have been referred to the climatological averages c_m at each gridpoint,

$$AC_C = \frac{\sum\limits_{m=1}^{M} (y'_m - \bar{y}')(o'_m - \bar{o}')}{\left[\sum\limits_{m=1}^{M} (y'_m - \bar{y}')^2 \sum\limits_{m=1}^{M} (o'_m - \bar{o}')^2\right]^{1/2}}. \tag{7.59}$$

Here the primed quantities are the anomalies relative to the climatological averages (Equation 7.55), and the overbars refer to these anomalies averaged over a given map of M gridpoints. The square of Equation 7.59 is thus exactly $r^2_{y'o'}$ in Equation 7.57.

The other form for the anomaly correlation differs from Equation 7.59 in that the map-mean anomalies are not subtracted, yielding the uncentered anomaly correlation

$$AC_U = \frac{\sum\limits_{m=1}^{M} [(y_m - c_m)(o_m - c_m)]}{\left[\sum\limits_{m=1}^{M} (y_m - c_m)^2 \sum\limits_{m=1}^{M} (o_m - c_m)^2\right]^{1/2}} = \frac{\sum\limits_{m=1}^{M} y'_m o'_m}{\left[\sum\limits_{m=1}^{M} (y'_m)^2 \sum\limits_{m=1}^{M} (o'_m)^2\right]^{1/2}}. \tag{7.60}$$

This form was apparently first suggested by Miyakoda *et al.* (1972). Superficially, the AC_U in Equation 7.60 resembles the Pearson product-moment correlation coefficient (Equations 3.22 and 7.59), in that that both are bounded by ± 1, and that neither are sensitive to biases in the forecasts. However, the centered and uncentered anomaly correlations are equivalent only if the averages over the M gridpoints of the two anomalies are zero; that is, only if $\sum_m (y_m - c_m) = 0$ and $\sum_m (o_m - c_m) = 0$. These conditions may be approximately true if the forecast and observed fields are being compared over a large (e.g., hemispheric) domain, but will almost certainly not hold if the fields are compared over a relatively small area.

The anomaly correlation is designed to detect similarities in the patterns of departures (i.e., anomalies) from the climatological mean field, and is sometimes referred to as a pattern correlation. This usage is consistent with the square of AC_C being interpreted as phase association in the algebraic decomposition of the MSE skill score in Equation 7.57. However, as Equation 7.57 makes clear, the anomaly correlation does not penalize either conditional or unconditional biases. Accordingly, it is reasonable to regard the anomaly correlation as reflecting potential skill (that might be achieved in the absence of conditional and unconditional biases), but it is incorrect to regard the anomaly correlation (or, indeed, any correlation) as measuring actual skill (e.g., Murphy 1995).

The anomaly correlation often is used to evaluate extended-range (beyond a few days) forecasts. The AC is designed to reward good forecasts of the pattern of the observed field, with less sensitivity to the correct magnitudes of the field variable. Figure 7.19 shows anomaly correlation values for the same 30-day dynamical and persistence forecasts of 500 mb height that are verified in terms of the RMSE in Figure 7.17. Since the anomaly correlation has a positive orientation (larger values indicate more accurate forecasts) and the RMSE has a negative orientation (smaller values indicate more accurate forecasts), we must mentally "flip" one of these two plots vertically in order to compare them. When this is done, it can be seen that the two measures usually rate a given forecast map similarly, although some differences are apparent. For example, in this data set the anomaly correlation values in Figure 7.19 show a more consistent separation between the performance of the dynamical and persistence forecasts than do the RMSE values in Figure 7.17.

As is also the case for the MSE, aggregate performance of a collection of field forecasts can be summarized by averaging anomaly correlations across many forecasts. However, skill scores of the form of Equation 7.4 usually are not calculated for the anomaly correlation. For the uncentered anomaly correlation, AC_U is undefined for climatological

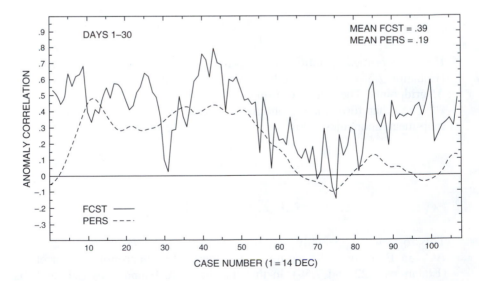

FIGURE 7.19   Anomaly correlations for dynamical 30-day forecasts of 500 mb heights for the Northern Hemisphere between 20° and 80° N (solid), and persistence of the previous 30-day average 500 mb field (dashed), for forecasts initialized 14 December 1986 through 31 March 1987. The same forecasts are evaluated in Figure 7.17 using the RMSE. From Tracton *et al.* (1989).

forecasts, because the denominator of Equation 7.60 is zero. Rather, skill with respect to the anomaly correlation generally is evaluated relative to the reference values AC = 0.6 or AC = 0.5. Individuals working operationally with the anomaly correlation have found, subjectively, that AC = 0.6 seems to represent a reasonable lower limit for delimiting field forecasts that are synoptically useful (Hollingsworth *et al.* 1980). Murphy and Epstein (1989) have shown that if the average forecast and observed anomalies are zero, and if the forecast field exhibits a realistic level of variability (i.e., the two summations in the denominator of Equation 7.59 are of comparable magnitude), then $AC_C = 0.5$ corresponds to the skill score for the MSE in Equation 7.54 being zero. Under these same restrictions, $AC_C = 0.6$ corresponds to the MSE skill score being 0.20.

Figure 7.20 illustrates the use of the subjective AC = 0.6 reference level. Panel (a) shows average AC values for 500 mb height forecasts made during the winters (December-February) of 1981/1982 through 1989/1990. For projection zero days into the future (i.e., initial time), AC = 1 since $y_m = o_m$ at all grid points. The average AC declines progressively for longer forecast lead times, falling below AC = 0.6 for projections between five and seven days. The curves for the later years tend to lie above the curves for the earlier years, reflecting, at least in part, improvements made to the forecast model during this decade. One measure of this overall improvement is the increase in the average lead time at which the AC curve crosses the 0.6 line. These times are plotted in Figure 7.20b, and range from five days in the early and mid-1980s, to seven days in the late 1980s. Also plotted in this panel are the average projections at which anomaly correlations for persistence forecasts fall below 0.4 and 0.6. The increase for the AC = 0.4 threshold in later winters indicates that some of the apparent improvement for the dynamical forecasts may be attributable to more persistent synoptic patterns. The crossover time at the AC = 0.6 threshold for persistence forecasts is consistently about two days. Thus, imagining the average correspondence between observed 500 mb maps separated by 48 hour intervals allows a qualitative appreciation of the level of forecast performance represented by the AC = 0.6 threshold.

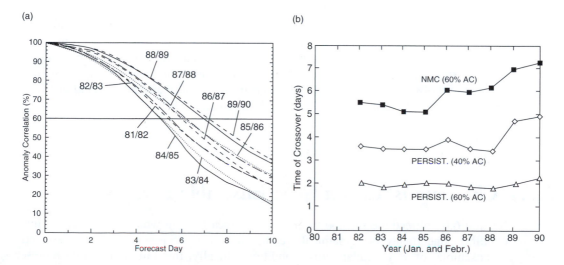

FIGURE 7.20   (a) Average anomaly correlations as a function of forecast projection for 1981/1982 through 1989/1990 winter 500 mb heights between 20° N and 80° N. Accuracy decreases as forecast projection increases, but there are substantial differences between winters. (b) Average projection at which forecast anomaly correlations cross the 0.6 level, and persistence forecasts cross the 0.4 and 0.6 levels, for Januarys and Februarys of these nine winters. From Kalnay *et al.* (1990).

7.6.5 Recent Ideas in Nonprobabilistic Field Verification

Because the numbers of gridpoints M typically used to represent meteorological fields is relatively large, and the numbers of allowable values for forecasts and observations of continuous predictands defined on these grids is also large, the dimensionality of verification problems for field forecasts is typically huge. Using scalar scores such as MSE or AC to summarize forecast performance in these settings may at times be a welcome relief from the inherent complexity of the verification problem, but necessarily masks very much relevant detail. Forecasters and forecast evaluators often are dissatisfied with the correspondence between single-number performance summaries and their subjective perceptions about the goodness of a forecast. For example, a modest error in the advection of a relatively small-scale meteorological feature may produce a large phase error in Equation 7.57, and thus result in a poor MSE skill score, even though the feature itself may otherwise have been well forecast.

Recent and still primarily experimental work has been undertaken to address such concerns, by attempting to design verification methods for fields that may be able to quantify aspects of forecast performance that reflect human visual reactions to map features more closely. One such new approach involves scale decompositions of the forecast and observed fields, allowing a separation of the verification for features of different sizes. Briggs and Levine (1997) proposed this general approach using *wavelets*, which are a particular kind of mathematical basis function. Casati *et al.* (2004) have extended the wavelet approach to both position and intensity of rainfall features, by considering a series of binary predictands defined according to a sequence of precipitation-amount thresholds. Denis *et al.* (2002) and de Elia *et al.* (2002) consider a similar approach based on more conventional spectral basis functions. Zepeda-Arce *et al.* (2000) address scale dependence in conventional verification measures such as the threat score through different degrees of spatial aggregation.

An approach to field verification based specifically on forecast features, or objects, was proposed by Hoffman *et al.* (1995). Here a feature is a forecast or observation defined by a closed contour for the predictand in the spatial domain. Errors in forecasting features may be expressed as a decomposition into displacement, amplitude, and residual components. Location error is determined by horizontal translation of the forecast field until the best match with the observed feature is obtained, where best may be interpreted through such criteria as minimum MSE, maximal area overlap, or alignment of the forecast and observed centroids. Applications of this basic idea can be found in Du *et al.* (2000), Ebert and McBride (2000), and Nehrkorn *et al.* (2003).

7.7 Verification of Ensemble Forecasts

7.7.1 Characteristics of a Good Ensemble Forecast

Section 6.6 outlined the method of ensemble forecasting, in which the effects of initial-condition uncertainty on dynamical weather forecasts are represented by a collection, or ensemble, of very similar initial conditions. Ideally, this initial ensemble represents a random sample from the PDF quantifying initial-condition uncertainty, defined on the phase space of the dynamical model. Integrating the forecast model forward in time from each of these initial conditions individually thus becomes a Monte-Carlo approach to estimating the effects of the initial-condition uncertainty on uncertainty for the quantities being predicted. That is, if the initial ensemble members have been chosen as a random sample from the initial-condition uncertainty PDF, and if the forecast model contains an accurate representation of the physical dynamics, the dispersion of the ensemble after being integrated forward in time represents a random sample from the PDF of forecast uncertainty. If this ideal situation could be obtained, the true state of the atmosphere would be just one more member of the ensemble, at the initial time and throughout the integration period, and should be statistically indistinguishable from the forecast ensemble. This condition, that the actual future atmospheric state behaves like a random draw from the same distribution that produced the ensemble, is called consistency of the ensemble (Anderson 1997).

In light of this background, it should be clear that ensemble forecasts are probability forecasts that are expressed as a discrete approximation to a full forecast PDF. According to this approximation, ensemble relative frequency should estimate actual probability. Depending on what the predictand(s) of interest may be, the formats for these probability forecasts can vary widely. Probability forecasts can be obtained for simple predictands, such as continuous scalars (e.g., temperature or precipitation at a single location), or discrete scalars (possibly constructed by thresholding a continuous variable, e.g., zero or trace precipitation vs. nonzero precipitation, at a given location); or quite complicated multivariate predictands such as entire fields (e.g., the joint distribution of 500 mb heights at the global set of horizontal gridpoints).

In any of these cases, the probability forecasts from an ensemble will be good (i.e., appropriately express the forecast uncertainty) to the extent that the consistency condition has been met, so that the observation being predicted looks statistically like just another member of the forecast ensemble. A necessary condition for ensemble consistency is an appropriate degree of ensemble dispersion. If the ensemble dispersion is consistently too small, then the observation will often be an outlier in the distribution of ensemble members, implying that ensemble relative frequency will be a poor approximation to probability. This condition of ensemble underdispersion, in which the ensemble members look too much like each other and not enough like the observation, is illustrated hypothetically in Figure 7.21a. If the ensemble dispersion is consistently too large, as in Figure 7.21c, then the observation may too often be in the middle of the ensemble distribution. The result will again be that ensemble relative frequency will be a poor approximation to probability. If the ensemble distribution is appropriate, as illustrated by the hypothetical example in Figure 7.21b, then the observation may have an equal chance of occurring at any quantile of the distribution that is estimated by the ensemble.

The empirical frequency distribution of a forecast ensemble, as expressed for example using histograms as in Figure 7.21, provides a direct estimate of the forecast PDF for a scalar continuous predictand. These raw ensemble distributions could also be smoothed

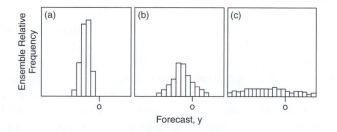

FIGURE 7.21   Histograms of hypothetical ensembles predicting a continuous scalar, y, exhibiting relatively (a) too little dispersion, (b) an appropriate degree of dispersion, and (c) excessive dispersion, in comparison to a typical observation, o.

using kernel density estimates, as in Section 3.3.6 (Roulston and Smith 2003), or by fitting parametric probability distributions (Hannachi and O'Neill 2001; Stephenson and Doblas-Reyes 2000; Wilks 2002b). Probability forecasts for discrete predictands are constructed from these ensemble distributions through the corresponding empirical cumulative frequency distribution (see Section 3.3.7), which will approximate $\Pr\{Y \leq y\}$ on the basis of the ranks $y_{(i)}$ of the ensemble members within the ensemble distribution. A probability forecast for the occurrence of the predictand at or below some threshold y can then be obtained directly from this function. Using the simple plotting position $p(y) = i/n_{\text{ens}}$, where i is the rank of the order statistic $y_{(i)}$ within the ensemble distribution of size n_{ens}, probability would be estimated directly as ensemble relative frequency. That is, $\Pr\{Y \leq y\}$ would be estimated by the relative frequency of ensemble members below the level y, and this is the basis upon which forecast probability is often equated to ensemble relative frequency. In practice it may be better to use one of the more sophisticated plotting positions in Table 3.2 to estimate the cumulative probabilities.

Regardless of how probability forecasts are estimated from a forecast ensemble, the appropriateness of these probability assignments can be investigated through techniques of forecast verification for probabilistic forecasts (see Sections 7.4 and 7.5). Often ensembles are used to produce probability forecasts for dichotomous predictands, for example using ensemble relative frequency (i.e., $\Pr\{Y \leq y_{(i)}\} \approx i/n_{\text{ens}}$), and in these cases standard verification tools such as the Brier score, the reliability diagram, and the ROC diagram are routinely used (e.g., Atger 1999; Legg *et al.* 2002). However, additional verification tools have been developed specifically for ensemble forecasts, many of which are oriented toward investigating the plausibility of the consistency condition that provides the underpinning for ensemble-based probability forecasting, namely that the ensemble members and the corresponding observation are samples from the same underlying population.

7.7.2 The Verification Rank Histogram

Construction of a verification rank histogram, sometimes called simply the rank histogram, is the most common approach to evaluating whether a collection of ensemble forecasts for a scalar predictand satisfy the consistency condition. That is, the rank histogram is used to evaluate whether the ensembles apparently include the observations being predicted as equiprobable members. The rank histogram is a graphical approach that was devised independently by Anderson (1996), Hamill and Colucci (1997), and Talagrand *et al.* (1997), and is sometimes also called the Talagrand Diagram.

Consider the evaluation of n ensemble forecasts, each of which consists of n_{ens} ensemble members, in relation to the n corresponding observed values for the predictand. For each of these n sets, if the n_{ens} members and the single observation all have been drawn from the same distribution, then the rank of the observation within these $n_{\text{ens}} + 1$ values is equally likely to take on any of the values $i = 1, 2, 3, \ldots, n_{\text{ens}} + 1$. For example, if the observation is smaller than all n_{ens} ensemble members, then its rank is $i = 1$. If it is larger than all the ensemble members (as in Figure 7.21a), then its rank is $i = n_{\text{ens}} + 1$. For each of the n forecasting occasions, the rank of the verification (i.e., the observation) within this $n_{\text{ens}} + 1$ -member distribution is tabulated. Collectively these n verification ranks are plotted in the form of a histogram to produce the verification rank histogram. (Equality of the observation with one or more of the ensemble members requires a slightly more elaborate procedure; see Hamill and Colucci 1997, 1998.) If the consistency condition has been met this histogram of verification ranks will be uniform, reflecting equiprobability

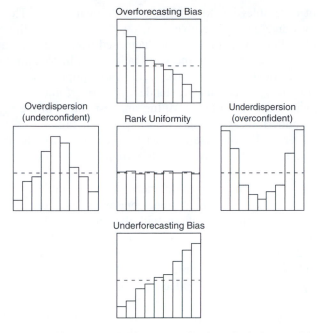

FIGURE 7.22 Example verification rank histograms for hypothetical ensembles of size $n_{ens} = 8$, illustrating characteristic ensemble dispersion and bias errors. Perfect rank uniformity is indicated by the horizontal dashed lines. The arrangement of the panels corresponds to the calibration portions of the reliability diagrams in Figure 7.8a.

of the observations within their ensemble distributions, except for departures that are small enough to be attributable to sampling variations.

Departures from the ideal of rank uniformity can be used to diagnose aggregate deficiencies of the ensembles (Hamill 2001). Figure 7.22 shows four problems that can be discerned from the rank histogram, together with a rank histogram (center panel) that shows only small sampling departures from a uniform distribution of ranks, or rank uniformity. The horizontal dashed lines in Figure 7.22 indicate the relative frequency $[= (n_{ens} + 1)^{-1}]$ attained by a uniform distribution for the ranks, which is often plotted as a reference as part of the rank histogram. The hypothetical rank histograms in Figure 7.22 each have $n_{ens} + 1 = 9$ bars, and so would pertain to ensembles of size $n_{ens} = 8$.

Overdispersed ensembles produce rank histograms with relative frequencies concentrated in the middle ranks (left-hand panel in Figure 7.22). In this situation, corresponding to Figure 7.21c, excessive dispersion produces ensembles that range beyond the verification more frequently than would occur by chance if the ensembles exhibited consistency. The verification is accordingly an extreme member (of the $n_{ens} + 1$ -member ensemble + verification collection) too infrequently, so that the extreme ranks are underpopulated; and is near the center of the ensemble too frequently, producing overpopulation of the middle ranks. Conversely, a set of n underdispersed ensembles produce a U-shaped rank histogram (right-hand panel in Figure 7.22) because the ensemble members tend to be too much like each other, and different from the verification, as in Figure 7.21a. The result is that the verification is too frequently an outlier among the collection of $n_{ens} + 1$ values, so the extreme ranks are overpopulated; and occurs too rarely as a middle value, so the central ranks are underpopulated.

An appropriate degree of ensemble dispersion is a necessary condition for a set of ensemble forecasts to exhibit consistency, but it is not sufficient. It is also necessary for consistent ensembles not to exhibit unconditional biases. That is, consistent ensembles will not be centered either above or below their corresponding verifications, on average. Unconditional ensemble bias can be diagnosed from overpopulation of either the smallest ranks, or the largest ranks, in the verification rank histogram. Forecasts that are centered above the verification, on average, exhibit overpopulation of the smallest ranks (upper panel in Figure 7.22) because the tendency for overforecasting leaves the verification too frequently as the smallest or one of the smallest values of the $n_{ens} + 1$ -member collection. Similarly, underforecasting bias (lower panel in Figure 7.22) produces overpopulation of the higher ranks, because a consistent tendency for the ensemble to be below the verification leaves the verification too frequently as the largest or one of the largest members.

The rank histogram reveals deficiencies in ensemble calibration, or reliability. That is, either conditional or unconditional biases produce deviations from rank uniformity. Accordingly, there are connections with the calibration function $p(o_j|y_i)$ that is plotted as part of the reliability diagram (see Section 7.4.4), which can be appreciated by comparing Figures 7.22 and 7.8a. The five pairs of panels in these two figures bear a one-to-one correspondence for forecast ensembles yielding probabilities for a dichotomous variable defined by a fixed threshold applied to a continuous predictand. That is, the yes component of a dichotomous outcome occurs if the value of the continuous predictand y is at or above a threshold. For example, the event "precipitation occurs" corresponds to the value of a continuous precipitation variable being at or above a detection limit, such as 0.01 in. In this setting, forecast ensembles that would produce each of the five reliability diagrams in Figure 7.8a would exhibit rank histograms having the forms in the corresponding positions in Figure 7.22.

Correspondences between the unconditional bias signatures in these two figures are easiest to understand. Ensemble overforecasting (upper panels) yields average probabilities that are larger than average outcome relative frequencies in Figure 7.8a, because ensembles that are too frequently centered above the verification will exhibit a majority of members above a given threshold more frequently that the verification is above that threshold (or, equivalently, more frequently than the corresponding probability of being above the threshold, according to the climatological distribution). Conversely, underforecasting (lower panels) simultaneously yields average probabilities for dichotomous events that are smaller than the corresponding average outcome relative frequencies in Figure 7.8a, and overpopulation of the higher ranks in Figure 7.22.

In underdispersed ensembles, most or all ensemble members will fall too frequently on one side or the other of the threshold defining a dichotomous event. The result is that probability forecasts from underdispersed ensembles will be excessively sharp, and will use extreme probabilities more frequently than justified by the ability of the ensemble to resolve the event being forecast. The probability forecasts will be overconfident; that is, too little uncertainty is communicated, so that the conditional event relative frequencies are less extreme than the forecast probabilities. Reliability diagrams exhibiting conditional biases, in the form of the right-hand panel of Figure 7.8a, are the result. On the other hand, overdispersed ensembles will rarely have most members on one side or the other of the event threshold, so the probability forecasts derived from them will rarely be extreme. These probability forecasts will be underconfident, and produce conditional biases of the kind illustrated in the left-hand panel of Figure 7.8a, namely that the conditional event relative frequencies tend to be more extreme than the forecast probabilities.

Lack of uniformity in a rank histogram quickly reveals the presence of conditional and/or unconditional biases in a collection of ensemble forecasts, but unlike the reliability diagram it does not provide a complete picture of forecast performance in the sense of fully expressing the joint distribution of forecasts and observations. In particular, the rank histogram does not include an absolute representation of the refinement, or sharpness, of the ensemble forecasts. Rather, it indicates only if the forecast refinement is appropriate relative to the degree to which the ensemble can resolve the predictand. The nature of this incompleteness can be appreciated by imagining the rank histogram for ensemble forecasts constructed as random samples of size n_{ens} from the historical climatological distribution of the predictand. Such ensembles would be consistent, by definition, because the value of the predictand on any future occasion will have been drawn from the same distribution that generated the finite sample in each ensemble. The resulting rank histogram would be accordingly flat, but would not reveal that these forecasts exhibited so little refinement as to be useless. If these climatological ensembles were to be converted to probability forecasts for a discrete event according to a fixed threshold of the predictand, their reliability diagram would consist of a single point, located on the 1:1 diagonal, at the magnitude of the climatological relative frequency. This abbreviated reliability diagram immediately would communicate the fact that the forecasts underlying it exhibited no sharpness, because the same event probability would have been forecast on each of the n occasions.

Distinguishing between true deviations from uniformity and mere sampling variations usually is approached through the chi-square goodness-of-fit test (see Section 5.2.5). Here the null hypothesis is a uniform rank histogram, so the expected number of counts in each bin is $n/(n_{ens}+1)$, and the test is evaluated using the chi-square distribution with $\nu = n_{ens}$ degrees of freedom (because there are $n_{ens}+1$ bins). This approach assumes independence of the n ensembles being evaluated, and so is not appropriate in unmodified form, for example, to ensemble forecasts on consecutive days, or at nearby gridpoints. To the extent that a rank histogram may be nonuniform, reflecting conditional and/or unconditional biases, the forecast probabilities can be recalibrated on the basis of the rank histogram, as described by Hamill and Colucci (1997, 1998).

7.7.3 Recent Ideas in Verification of Ensemble Forecasts

Ensemble forecasting was proposed by Leith (1974), but became computationally practical only much more recently. Both the practice of ensemble forecasting, and the verification of ensemble forecasts, are still evolving methodologically. This section contains brief descriptions of some methods that have been proposed, but not yet widely used, for ensemble verification.

The verification rank histogram (Section 7.7.2) is used to investigate ensemble consistency for a single scalar predictand. The concept behind the rank histogram can be extended to simultaneous forecast for multiple predictands, using the minimum spanning tree (MST) histogram. This idea was proposed by Smith (2001), and explored more fully by Smith and Hansen (2004) and Wilks (2004). The MST histogram is constructed from an ensemble of K-dimensional vector forecasts $\boldsymbol{y}_i, i, = 1, \ldots, n_{ens}$, and the corresponding vector observation $\boldsymbol{o}$. Each of these vectors defines a point in a K-dimensional space, the coordinate axes of which corresponds to the K variables in the vectors $\boldsymbol{y}$ and $\boldsymbol{o}$. In general these vectors will not have a natural ordering in the same way that a set of $n_{ens}+1$ scalars would, so the conventional verification rank histogram is not applicable. The minimum spanning tree for n_{ens} members $\boldsymbol{y}_i$ of a particular ensemble is the set of line segments (in the K-dimensional space of these vectors) that connect all the points $\boldsymbol{y}_i$ in an arrangement having no closed loops, and for which the sum of the lengths of these line segments is minimized. The solid lines in Figure 7.23 show a minimum spanning tree for a hypothetical $n_{ens} = 10$-member forecast ensemble, labeled $A - J$.

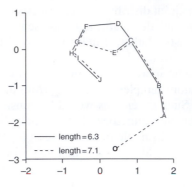

FIGURE 7.23   Hypothetical example minimum spanning trees in $K = 2$ dimensions. The $n_{ens} = 10$ ensemble members are labeled $A - J$, and the corresponding observation is O. Solid lines indicate MST for the ensemble as forecast, and dashed lines indicate the MST that results from the observation being substituted for ensemble member D. From Wilks (2004).

If each ensemble member is replaced in turn with the observation vector o, the lengths of the minimum spanning trees for each of these substitutions make up a set of n_{ens} reference MST lengths. The dashed lines in Figure 7.23 show the MST obtained when ensemble member D is replaced by the observation, O. To the extent that the ensemble consistency condition has been satisfied, the observation vector is statistically indistinguishable from any of the forecast vectors y_i, implying that the length of the MST connecting only the n_{ens} vectors y_i has been drawn from the same distribution of MST lengths as those obtained by substituting the observation for each of the ensemble members in turn. The MST histogram investigates the plausibility of this proposition, and thus the plausibility of ensemble consistency for the n K-dimensional ensemble forecasts, by tabulating the ranks of the MST lengths for the ensemble as forecast within each group of $n_{ens} + 1$ MST lengths. This concept is similar to that underlying the rank histogram for scalar ensemble forecasts, but it is not a multidimensional generalization of the rank histogram, and the interpretations of the MST histograms are different. In raw form, it is unable to distinguish between ensemble underdispersion and bias (the outlier observation in Figure 7.23 could be the result of either of these problems), and deemphasizes variables in the forecast and observation vectors with small variance. However, useful diagnostics can be obtained from MST histograms of debiased and rescaled forecast and observation vectors, and if the n ensembles are independent the chi-square test is again appropriate to evaluate rank uniformity for the MST lengths (Wilks 2004).

Ensemble consistency through time—that is, as the projection of a forecast into the future increases—can also be investigated. If an initial ensemble has been well chosen, its members are consistent with the initial observation, or analysis. The question of how far into the future there are time trajectories within the forecast ensemble that are statistically indistinguishable from the true state being predicted is the question of ensemble shadowing; that is, how long the ensemble shadows the truth (Smith 2001). Smith (2001) suggests using the geometrical device of the bounding box to approximate ensemble shadowing. A vector observation o is contained by the bounding box defined by an ensemble $y_i, i, = 1, \ldots, n_{ens}$; if for each of the K dimensions of these vectors the element o_k of the observation vector is no larger than at least one of its counterparts in the ensemble, and no smaller than at least one of its other counterparts in the ensemble. The observation in Figure 7.23 is not within the $K = 2$-dimensional bounding box defined by the ensemble: even though its value in the horizontal dimension is not extreme, it is smaller than all of the ensemble members with respect to the vertical dimension. The shadowing properties of a set of n ensemble forecasts could be evaluated by tabulating the relative frequencies of lead times at which the bounding box from the ensemble first fails to contain the corresponding observation. The multidimensional scaling plots in Stephenson and Doblas-Reyes (2000) offer a way to visualize approximate shadowing in two dimensions, regardless of the dimensionality K of the forecast vectors.

Wilson *et al.* (1999) have proposed a score based on the likelihood of the verifying observation in the context of the ensemble distribution, in relation to its likelihood relative to the climatological (or other reference) distribution. Their idea is to compare the conditional likelihoods $f\{$observation|ensemble$\}$ and $f\{$observation|climatology$\}$. The former should be larger, to the extent that the ensemble distribution is sharper (lower-variance) and/or centered closer to the observation than the climatological distribution. Wilson *et al.* (1999) also suggest expressing this difference in the form of a skill score (Equation 7.4), in which $f\{$observation|perfect$\} = 1$, since a perfect forecast distribution would concentrate all its probability exactly at the (discrete) observation. They also suggest evaluating the probabilities using parametric distributions fitted to the ensemble and climatological distributions, in order to reduce the effects of sampling errors on the calculations. Wilson *et al.* (1999) propose the method for evaluating ensemble forecasts of scalar predictands, but the method apparently could be applied to higher-dimensional ensemble forecasts as well.

7.8 Verification Based on Economic Value

7.8.1 *Optimal Decision Making and the Cost/Loss Ratio Problem*

The practical justification for effort expended in developing forecast systems and making forecasts is that these forecasts should result in better decision making in the face of weather uncertainty. Often such decisions have direct economic consequences, or their consequences can be mapped onto an economic (i.e., monetary) scale. There is substantial literature in the fields of economics and statistics on the use and value of information for decision making under uncertainty (e.g., Clemen 1996; Johnson and Holt 1997), and the concepts and methods in this body of knowledge have been extended to the context of optimal use and economic value of weather forecasts (e.g., Katz and Murphy 1997a; Winkler and Murphy 1985). Forecast verification is an essential component of this extension, because it is the joint distribution of forecasts and observations (Equation 7.1) that will determine the economic value of forecasts (on average) for a particular decision problem. It is therefore natural to consider characterizing forecast goodness (i.e., computing forecast verification measures) in terms of the mathematical transformations of the joint distribution that define forecast value for particular decision problems.

The reason that economic value of weather forecasts must be calculated for particular decision problems—that is, on a case-by-case basis—is that the value of a particular set of forecasts will be different for different decision problems (e.g., Roebber and Bosart 1996, Wilks 1997a). However, a useful and convenient prototype, or "toy," decision model is available, called the cost/loss ratio problem (Katz and Murphy 1997b; Murphy 1977). This simplified decision model apparently originated with Anders Angstrom, in a 1922 paper (Liljas and Murphy 1994), and has been frequently used since that time. Despite its simplicity, the cost/loss problem nevertheless can reasonably approximate some simple real-world decision problems (Roebber and Bosart 1996).

The cost/loss decision problem relates to a hypothetical decision maker for whom some kind of adverse weather may or may not occur, and who has the option of either protecting or not protecting against the possibility of the adverse weather. That is, this decision maker must choose one of two alternatives in the face of an uncertain dichotomous weather outcome. Because there are only two possible actions and two possible outcomes, this is the simplest possible decision problem: no decision would be needed if there was only one course of action, and no uncertainty would be involved if only one weather outcome was possible. The protective action available to the decision maker is assumed to be completely effective, but requires payment of a cost C, regardless of whether or not the adverse weather subsequently occurs. If the adverse weather occurs in the absence of the protective action being taken, the decision maker suffers a loss L. The economic effect is zero loss if protection is not taken and the event does not occur. Figure 7.24a shows the loss function for the four possible combinations of decisions and outcomes in this problem.

Probability forecasts for the dichotomous weather event are assumed to be available and, depending on their quality, better decisions (in the sense of improved economic outcomes, on average) may be possible. Taking these forecasts at face value (i.e., assuming that they are calibrated, so $p(o_1|y_i) = y_i$ for all forecasts y_i), the optimal decision on any particular occasion will be the one yielding the smallest expected (i.e., probability-weighted average) expense. If the decision is made to protect, the expense will be C with probability 1, and if no protective action is taken the expected loss will be $y_i L$ (because no loss is incurred, with probability $1 - y_i$). Therefore, the smaller expected expense will be associated with the protection action whenever

$$C < y_i L, \tag{7.61a}$$

or

$$\frac{C}{L} < y_i. \tag{7.61b}$$

Protection is the optimal action when the probability of the adverse event is larger than the ratio of the cost C to the loss L, which is the origin of the name cost/loss ratio. Different decision makers face problems involving different costs and losses, and so their optimal thresholds for action will be different. Clearly this situation is meaningful only if $C < L$, because otherwise the protective action offers no potential gains, so that meaningful cost/loss ratios are confined to the unit interval, $0 < C/L < 1$.

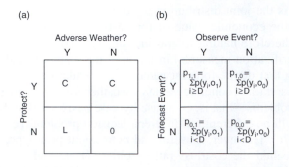

FIGURE 7.24   (a) Loss function for the 2×2 cost/loss ratio situation. (b) Corresponding 2×2 verification table resulting from probability forecasts characterized by the joint distribution $p(y_i, o_j)$ being transformed to nonprobabilistic forecasts according to a particular decision maker's cost/loss ratio. Adapted from Wilks (2001).

Mathematically explicit decision problems of this kind not only prescribe optimal actions, but also provide a way to calculate expected economic outcomes associated with forecasts having particular characteristics. For the simple cost/loss ratio problem these expected economic expenses are the probability-weighted average costs and losses, according the probabilities in the joint distribution of the forecasts and observations, $p(y_i, o_j)$. If only climatological forecasts are available (i.e., if the climatological relative frequency is forecast on each occasion), the optimal action will be to protect if this climatological probability is larger than C/L, and not to protect otherwise. Accordingly, the expected expense associated with the climatological forecast depends on its magnitude relative to the cost/loss ratio:

$$EE_{dim} = \begin{cases} C, & \text{if } C/L < \bar{o} \\ \bar{o}L, & \text{otherwise.} \end{cases} \tag{7.62}$$

Similarly, if perfect forecasts were available the hypothetical decision maker would incur the protection cost only on the occasions when the adverse weather was about to occur, so the corresponding expected expense would be

$$EE_{perf} = \bar{o}C. \tag{7.63}$$

The expressions for expected expenses in Equation 7.62 and 7.63 are simple because the joint distributions of forecasts and observations for climatological and perfect forecasts are also very simple. More generally, a set of probability forecasts for a dichotomous event would be characterized by a joint distribution of the kind shown in Table 7.4a. A cost/loss decision maker with access to probability forecasts that may range throughout the unit interval has an optimal decision threshold, D, corresponding to the cost/loss ratio, C/L. That is, the decision threshold D is that value of the index i corresponding to the smallest probability y_i that is larger than C/L. In effect, the hypothetical cost/loss decision maker transforms probability forecasts summarized by a joint distribution $p(y_i, o_j)$ into nonprobabilistic forecasts for the dichotomous event adverse weather, in the same way that was described in Sections 7.2.5 and 7.4.6: probabilities y_i for which $i \geq D$ are transformed to yes forecasts and forecasts for which $i < D$ are transformed to no forecasts. Figure 7.24b illustrates the 2×2 joint distribution (corresponding to Figure 7.1b) for the resulting nonprobabilistic forecasts of the binary event, in terms of the joint distribution of forecasts and observations for the probability forecasts. Here $p_{1,1}$ is the joint frequency that the probability forecast y_i is above the decision threshold D and the event subsequently occurs, $p_{1,0}$ is the joint frequency that the forecast is above the probability threshold but the event does not occur, $p_{0,1}$ is the joint frequency of forecasts below the threshold and the event occurring, and $p_{0,0}$ is the joint frequency of the probability forecasts being below threshold and the event not occurring.

Because the hypothetical decision maker has constructed yes/no forecasts using the decision threshold D that is customized to a particular cost/loss ratio of interest, there is a one-to-one correspondence between the joint probabilities in Figures 7.24b and the loss function in Figure 7.24a. Combining these leads to the expected expense associated with the forecasts characterized by the joint distribution $p(y_i, o_j)$,

$$EE_f = (p_{1,1} + p_{1,0})C + p_{0,1}L \tag{7.64a}$$

$$= C \sum_{j=0}^{1} \sum_{i \geq D} p(y_i, o_j) + L \sum_{i < D} p(y_i, o_1). \tag{7.64b}$$

This expected expense depends both on the particular nature of the decision maker's circumstances, through the cost/loss ratio that defines the decision threshold D; and on the quality of the probability forecasts available to the decision maker, as summarized in the joint distribution of forecasts and observations $p(y_i|o_j)$.

7.8.2 The Value Score

Economic value as calculated in the simple cost/loss ratio decision problem is, for a given cost/loss ratio, a rational and meaningful single-number summary of the quality of probabilistic forecasts for a dichotomous event summarized by the joint distribution $p(y_i, o_j)$. However, this measure of forecast quality is different for different decision makers (i.e., different values of C/L). Richardson (2000) proposed using economic value, plotted as a function of the cost/loss ratio, as a graphical verification device for probabilistic forecasts for dichotomous events, after a transformation that ensures calibration of the forecasts (i.e., $y_i \equiv p(o_1|y_i)$). The ideas are similar to those behind the ROC diagram (see Section 7.4.6), in that forecasts are evaluated through a function that is based on reducing probability forecasts to yes/no forecasts at all possible probability thresholds y_D, and also because conditional and unconditional biases are not penalized. The result is a strictly nonnegative measure of potential (not necessarily actual) economic value in the simplified decision problem, as a function of C/L, for $0 < C/L < 1$.

This basic procedure can be extended to reflect potentially important forecast deficiencies by computing the economic expenses using the original, uncalibrated forecasts (Wilks 2001). A forecast user without the information necessary to recalibrate the forecasts would need to take them at face value and, to the extent that they might be miscalibrated (i.e., that the probability labels y_i might be inaccurate), make suboptimal decisions. Whether or not the forecasts are preprocessed to remove biases, the calculated expected expense (Equation 7.64) can be expressed in the form of a standard skill score (Equation 7.4), relative to the expected expenses associates with climatological (Equation 7.62) and perfect (Equation 7.63) forecasts, called the value score:

$$VS = \frac{EE_f - EE_{clim}}{EE_{perf} - EE_{clim}} \tag{7.65a}$$

$$= \begin{cases} \dfrac{(C/L)(p_{1,1}+p_{1,0}-1)+p_{0,1}}{(C/L)(\bar{o}-1)}, & \text{if } C/L < \bar{o} \\[2ex] \dfrac{(C/L)(p_{1,1}+p_{1,0})+p_{0,1}-\bar{o}}{\bar{o}[(C/L)-1]}, & \text{if } C/L > \bar{o}. \end{cases} \tag{7.65b}$$

The advantage of this rescaling of EE_f is that sensitivity to particular values of C and L are removed, so that (unlike Equations 7.62–7.64) Equation 7.65 depends only on their ratio, C/L. Perfect forecasts exhibit $VS = 1$, and climatological forecasts exhibit $VS = 0$, for all cost/loss ratios. If the forecasts are recalibrated before calculation of the value score, it will be nonnegative for all cost/loss ratios. Richardson (2001) called this score, for recalibrated forecasts, the potential value, V. However, in the more realistic case that the forecasts are scored at face value, $VS < 0$ is possible if some or all of the hypothetical decision makers would be better served on average by adopting the climatological decision rule, leading to EE_{clim} in Equation 7.62. Mylne (2002) has extended this framework for 2×2 decision problems in which protection against the adverse event is only partially effective.

Figure 7.25 shows VS curves for MOS (dashed) and subjective (solid) probability-of-precipitation forecasts for (a) October–March, and (b) April–September, 1980–1987,

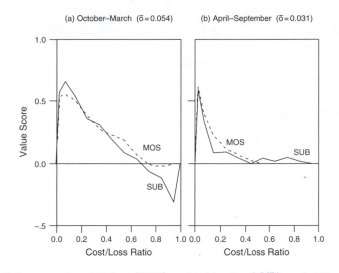

FIGURE 7.25   VS curves for objective (MOS) and subjective (SUB) probability of precipitation forecasts at Las Vegas, Nevada, for the period April 1980–March 1987. From Wilks (2001).

at a desert location. The larger values for smaller cost/loss ratios indicate that all these forecasts would be of greater utility in decisions for which the potential losses are large relative to the costs of protecting against them. Put another way, these figures indicate that the forecasts would be of greatest economic value for problems having relatively small probability thresholds, y_D. Figure 7.25a shows that decision makers whose problems during the cool-season months involve large relative protection costs would have made better decisions on average based only on the climatological probability of 0.054 (i.e., by never protecting), especially if only the subjective forecasts had been available. These negative values derive from miscalibration of these forecasts, in particular that the event relative frequencies conditional on the higher forecast probabilities were substantially smaller than the forecasts (i.e., the forecasts exhibited substantial overconfidence for the high probabilities). Recalibrating the forecasts before computing VS would remove the scoring penalties for this overconfidence.

Brier skill scores (Equation 7.35) are higher for the MOS as compared to the subjective forecasts in both panels of Figure 7.25, but the figure reveals that there are potential forecast users for whom one or the other of the forecasts would have been more valuable. The VS curve thus provides a more complete perspective on forecast quality than is possible with the scalar Brier skill score, or indeed than would be possible with any single-number measure. The warm-season forecasts (see Figure 7.25b) are particularly interesting, because the MOS system never forecast probabilities greater than 0.5. The human forecasters were able to successfully forecast some larger probabilities, but this was apparently done at the expense of forecast quality for the smaller probabilities.

7.8.3 Connections with Other Verification Approaches

Just as ROC curves are sometimes characterized in terms of the area beneath them, value score curves also can be collapsed to scalar summary statistics. The simple unweighted integral of VS over the full unit interval of C/L is one such summary. This simple function of VS turns out to be equivalent to evaluation of the full set of forecasts using the Brier score, because the expected expense in the cost/lost ratio situation (Equation 7.64) is a linear function of BS (Equation 7.34) (Murphy 1966). That is, ranking competing forecasts according to their Brier scores, or Brier skill scores, yields the same result as a ranking based on the unweighted integrals of their VS curves. To the extent that the expected forecast user community might have a nonuniform distribution of cost/loss ratios (for example, a preponderance of forecast users for whom the protection option is relatively inexpensive), single-number weighted-averages of VS also can be computed as statistical expectations of VS with respect to the probability density function for C/L among users of interest (Richardson 2001; Wilks 2001).

The VS curve is constructed through a series of 2×2 verification tables, and there are accordingly connections both with scores used to evaluate nonprobabilistic forecasts of binary predictands, and with the ROC curve. For correctly calibrated forecasts, maximum economic value in the cost/loss decision problem is achieved for decision makers for whom C/L is equal to the climatological event relative frequency, because for these individuals the optimal action is least clear from the climatological information alone. Lev Gandin called these ideal users, recognizing that such individuals will benefit most from forecasts. Interestingly, this maximum (potential, because calibrated forecasts are assumed) economic value is given by the Peirce skill score (Equation 7.16), evaluated for the 2×2 table appropriate to this "ideal" cost/loss ratio (Richardson 2000; Wandishin and Brooks 2002). Furthermore, the odds ratio (Equation 7.9) for this same 2×2 table is equal to the width of the range of cost/loss ratios over which decision makers can potentially realize economic value from the forecasts, so that $\theta > 1$ for this table is a necessary condition for economic value to be imparted for at least one cost/loss ratio decision problem (Richardson 2003; Wandishin and Brooks 2002). The range of cost/loss ratios for which positive potential economic value can be realized for a given 2×2 verification table is given by its Clayton skill score (Equation 7.17) (Wandishin and Brooks 2002). Additional connections between VS and attributes of the ROC diagram are provided in Mylne (2002) and Richardson (2003).

7.9 Sampling and Inference for Verification Statistics

Practical forecast verification is necessarily concerned with finite samples of forecast-observation pairs. The various verification statistics that can be computed from a particular verification data set are no less subject to sampling variability than are any other sort of statistics. If a different sample of the same kind of forecasts and observations were hypothetically to become available, the value of verification statistic(s) computed from it likely would be at least somewhat different. To the extent that the sampling distribution for a verification statistic is known or can be estimated, confidence intervals around it can be obtained, and formal tests (for example, against a null hypothesis of zero skill) can be constructed. Relatively little work on the sampling characteristics of forecast verification statistics has appeared to date. With a few exceptions, the best or only means of characterizing the sampling properties of a verification statistic may be through a resampling approach (see Section 7.9.4).

7.9.1 Sampling Characteristics of Contingency Table Statistics

In principle, the sampling characteristics of many 2×2 contingency table statistics follow from a fairly straightforward application of binomial sampling (Agresti 1996).

For example, such measures as the false alarm ratio (Equation 7.11), the hit rate (Equation 7.12), and the false alarm rate (Equation 7.13) are all proportions that estimate (conditional) probabilities. If the contingency table counts (see Figure 7.1a) have been produced independently from stationary (i.e., constant-p) forecast and observation systems, those counts are (conditional) binomial variables, and the corresponding proportions (such as FAR, H and F) are sample estimates of the corresponding binomial probabilities.

A direct approach to finding confidence intervals for sample proportions x/N that estimate the binomial parameter p is to use the binomial probability distribution function (Equation 4.1). A $1 - \alpha$ confidence interval for the underlying probability that is consistent with the observed proportion x/N can be defined by the extreme values of x on each tail that include probabilities of at least $1 - \alpha$ between them, inclusive. Unfortunately the result, called the Clopper-Pearson exact interval, generally will be inaccurate to a degree (and, specifically, too wide) because of the discreteness of the binomial distribution (Agresti and Coull 1998). Another simple approach to calculation of confidence intervals for sample proportions is to invert the Gaussian approximation to the binomial distribution (Equation 5.2). Since equation 5.2b is the standard deviation σ_x for the number of binomial successes X, the corresponding variance for the estimated proportion $\hat{p} = x/N$ is $\sigma_p^2 = \sigma_x^2/N^2 = \hat{p}(1 - \hat{p})/N$ (using Equation 4.16). The resulting $1 - \alpha$ confidence interval is then

$$p = \hat{p} \pm z_{(1-\alpha/2)}[\hat{p}(1 - \hat{p})/N]^{1/2}, \qquad (7.66)$$

where $z_{(1-\alpha/2)}$ is the $(1 - \alpha/2)$ quantile of the standard Gaussian distribution (e.g., $z_{(1-\alpha/2)} = 1.96$ for $\alpha = 0.05$).

Equation 7.66 can be quite inaccurate, in the sense that the actual probability of including the true p is substantially smaller than $1 - \alpha$, unless N is very large. However, this bias can be corrected using the modification (Agresti and Coull 1998) to Equation 7.66,

$$p = \frac{\hat{p} + \dfrac{z_{(1-\alpha/2)}^2}{2N} \pm z_{(1-\alpha/2)}\sqrt{\dfrac{\hat{p}(1 - \hat{p})}{N} + \dfrac{z_{(1-\alpha/2)}^2}{4N^2}}}{1 + \dfrac{z_{(1-\alpha/2)}^2}{N}}. \qquad (7.67)$$

The differences between Equations 7.67 and 7.66 are in the three terms involving $z_{(1-\alpha/2)}^2/N$, which approach zero for large N. Standard errors according to Equation 7.67 are tabulated for ranges of $\hat{p}$ and N in Thornes and Stephenson (2001).

Another relevant result from the statistics of contingency tables (Agresti 1996), is that the sampling distribution of the logarithm of the odds ratio (Equation 7.9) is approximately Gaussian-distributed for sufficiently large $n = a + b + c + d$, with estimated standard deviation

$$\hat{s}_{\ln(\theta)} = \left[\frac{1}{a} + \frac{1}{b} + \frac{1}{c} + \frac{1}{d}\right]^{1/2}. \qquad (7.68)$$

Thus, a floor on the magnitude of the sampling uncertainty for the odds ratio is imposed by the smallest of the four counts in Table 7.1a. When the null hypothesis of independence between forecasts and observations (i.e., $\theta = 1$) is of interest, it could be rejected if the observed $\ln(\theta)$ is sufficiently far from $\ln(1) = 0$, with respect to Equation 7.68.

EXAMPLE 7.8   Inferences for Selected Contingency Table Verification Measures

The hit and false alarm rates for the Finley tornado forecasts in Table 7.1a are $H = 28/51 = 0.549$ and $F = 72/2752 = 0.026$, respectively. These proportions are sample estimates of the conditional probabilities of tornados having been forecast, given either that tornados were or were not subsequently reported. Using Equation 7.67, $1 - \alpha = 95\%$ confidence intervals for the true underlying conditional probabilities can be estimated as

$$H = \frac{0.549 + \dfrac{1.96^2}{(2)(51)} \pm 1.96 \sqrt{\dfrac{0.549(1 - 0.549)}{51} + \dfrac{1.96^2}{(4)(51)^2}}}{1 + \dfrac{1.96^2}{51}}$$

$$= 0.546 \pm 0.132 = \{0.414, 0.678\}, \tag{7.69a}$$

and

$$F = \frac{0.026 + \dfrac{1.96^2}{(2)(2752)} \pm 1.96 \sqrt{\dfrac{0.026(1 - 0.026)}{2752} + \dfrac{1.96^2}{(4)(2752)^2}}}{1 + \dfrac{1.96^2}{2752}}$$

$$= 0.0267 \pm 0.00598 = \{0.0207, 0.0326\}. \tag{7.69b}$$

The precision of the estimated false alarm rate is much better (its standard error is much smaller) in part because the overwhelming majority of observations $(b + d)$ were no tornado; but also in part because $p(1 - p)$ is small for extreme values, and larger for intermediate values of p. Assuming independence of the forecasts and observations (in the sense illustrated in Equation 7.14), plausible useless-forecast benchmarks for the hit and false alarm rates might be $H_0 = F_0 = (a + b)/n = 100/2803 = 0.0357$. Neither of the 95% confidence intervals in Equation 7.69 include this value, leading to the inference that H and F for the Finley forecasts are better than would have been achieved by chance.

Stephenson (2000) notes that, because the Peirce skill score (Equation 7.16) can be calculated as the difference between H and F, confidence intervals for it can be calculated using simple binomial sampling considerations if it can be assumed that H and F are mutually independent. In particular, since the sampling distributions of both H and F are Gaussian for sufficiently large sample sizes, under these conditions the sampling distribution of the PSS will be Gaussian, with standard deviation

$$\hat{s}_{PSS} = \sqrt{\hat{s}_H^2 + \hat{s}_F^2}. \tag{7.70}$$

For the Finley tornado forecasts, $PSS = 0.523$, so that a 95% confidence interval around this value could be constructed as $0.523 \pm 1.96 \hat{s}_{PSS}$. Taking numerical values from Equation 7.69, or interpolating from the table in Thornes and Stephenson (2001), this interval would be $0.523 \pm (0.132^2 + 0.00598^2)^{1/2} = 0.523 \pm 0.132 = \{0.391, 0.655\}$. Since this interval does not include zero, a reasonable inference would be that these forecasts exhibited significant skill.

Finally the odds ratio for the Finley forecasts is $\theta = (28)(2680)/(23)(72) = 45.31$, and the standard deviation of the (approximately Gaussian) sampling distribution of its logarithm (Equation 7.68) is $(1/28 + 1/72 + 1/23 + 1/2680)^{1/2} = 0.306$. The null hypothesis that the forecasts and observations are independent (i.e., $\theta_0 = 1$) produces the t-statistic $[\ln(45.31) - \ln(1)]/0.306 = 12.5$, which would lead to emphatic rejection of that null hypothesis. ◊

The calculations in this section rely on the assumptions that the verification data are independent and, for the sampling distribution of proportions, that the underlying probability p is stationary (i.e., constant). The independence assumption might be violated, for example, if the data set consists of a sequence of daily forecast-observation pairs. The stationarity assumption might be violated if the data set includes a range of locations with different climatologies for the forecast variable. In cases where either of these assumptions might be violated, inferences for contingency-table verification measures still can be made, by estimating their sampling distributions using resampling approaches (see Section 7.9.4).

7.9.2 ROC Diagram Sampling Characteristics

Because confidence intervals around sample estimates for the hit rate H and the false alarm rate F can be calculated using Equation 7.67, confidence regions around individual (F, H) points in a ROC diagram can also be calculated and plotted. A complication is that, in order to define a joint, *simultaneous* $1 - \alpha$ confidence region around a sample (F, H) point, each of the two individual confidence intervals must cover its corresponding true value with a probability that is somewhat larger than $1 - \alpha$. Essentially, this adjustment is necessary in order to make valid simultaneous inference in a multiple testing situation (cf. Section 5.4.1). If H and F are at least approximately independent, a reasonable approach to deciding the appropriate sizes of the two confidence intervals is to use the Bonferroni inequality (Equation 10.53). In the present case of the ROC diagram, where there are $K = 2$ dimensions to the joint confidence region, Equation 10.53 says that the rectangular region defined by two $1 - \alpha/2$ confidence intervals for F and H will jointly enclose the true (F, H) pair with probability at least as large as $1 - \alpha$. For example, a joint 95% rectangular confidence region will be defined by two 97.5% confidence intervals, calculated using $z_{1-\alpha/4} = z_{.9875} = 2.24$, in Equation 7.67.

Mason and Graham (2002) have pointed out that a test for the statistical significance of the area A under the ROC curve, against the null hypothesis that the forecasts and observations are independent (i.e., that $A_0 = 1/2$), is available. In particular, the sampling distribution of the ROC area, given the null hypothesis of no relationship between forecasts and observations is proportional to the distribution of the Mann-Whitney U (Equations 5.22 and 5.23), and this test for the ROC area is equivalent to the Wilcoxon-Mann-Whitney test applied to the two likelihood distributions $p(y_i|o_1)$ and $p(y_i|o_2)$ (cf. Figure 7.11). In order to calculate this test, the ROC area A is transformed to a Mann-Whitney U variable according to

$$U = n_1 n_2 (1 - A). \tag{7.71}$$

Here $n_1 = a + c$ is the number of yes observations, and $n_2 = b + d$ is the number of no observations. Notice that, under the null hypothesis $A_0 = 1/2$, Equation 7.71 is exactly the mean of the Gaussian approximation to the sampling distribution of U in Equation 5.23a. This null hypothesis is rejected for sufficiently small U, or equivalently for sufficiently large ROC area A.

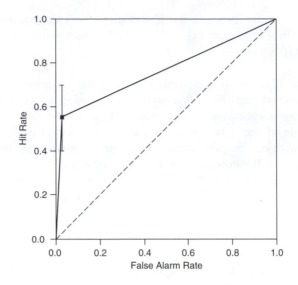

FIGURE 7.26   ROC diagram for the Finley tornado forecasts (Table 7.1a), with the 95% simulta-neous Bonferroni (Equation 10.53) confidence intervals for the single (F, H) point, calculated using Equation 7.67.

EXAMPLE 7.9   Confidence and Significance Statements about a ROC Diagram

Figure 7.26 shows the ROC diagram for the Finley tornado forecasts (see Table 7.1a), together with the 97.5% confidence intervals for F and H. These are $0.020 \leq F \leq 0.034$ and $0.396 \leq H \leq 0.649$, and were calculated from Equation 7.67 using $z_{1-\alpha/4} = z_{.9875} = 2.24$. The confidence interval for F is only about as wide as the dot locating the sample (F, H) pair, both because the number of no tornado observations is large, and because the proportion of false alarms is quite small. These two 97.5% confidence intervals define a rectangular region that contains the true (F, H) pair with at least 95% probability, according to the Bonferroni inequality (Equation 10.53). This region does not include the dashed 1:1 line, indicating that it is improbable for these forecasts to have been generated by a process that was independent of the observations.

The area under the ROC curve in Figure 7.26 is 0.761. If the true ROC curve for the process from which these forecast-observation pairs were sampled is the dashed 1:1 diag-onal line, what is the probability that a ROC area A this large or larger could have been achieved by chance, given $n_1 = 51$ yes observations and $n_2 = 2752$ no observations? Equa-tion 7.71 yields $U = (51)(2752)(0.761) = 33544$, the unusualness of which can be eval-uated in the context of the (null) Gaussian distribution with mean $\mu_U = (51)(2752)/2 = 70176$ (Equation 5.23a) and standard deviation $\sigma_U = [(51)(2752)(51 + 2752 + 1)/12]^{1/2} = 5727$ (Equation 5.23b). The resulting test statistic is $z = (33544 - 70176)/5727 = -6.4$, so that the null hypothesis of no association between the forecasts and observations would be strongly rejected. ◊

7.9.3 *Reliability Diagram Sampling Characteristics*

The calibration-function portion of the reliability diagram consists of I conditional out-come relative frequencies that estimate the conditional probabilities $p(o_1|y_i)$, $i = 1, \ldots, I$.

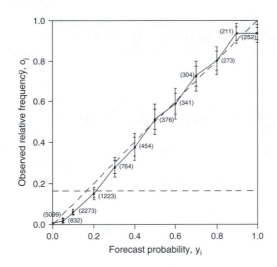

FIGURE 7.27   Reliability diagram for the probability-of-precipitation data in Table 7.2, with 95% confidence intervals on each conditional probability estimate, calculated using Equation 7.66. Inner confidence limits pertain to single points, and outer bounds are joint Bonferroni (Equation 10.53) confidence limits. Raw subsample sizes N_i are shown parenthetically. The 1:1 perfect reliability and horizontal no resolution lines are dashed.

If independence and stationarity are reasonably approximated, then confidence intervals around these points can be computed using either Equation 7.66 or Equation 7.67. To the extent that these intervals include the 1:1 perfect reliability diagonal, a null hypothesis that the forecaster(s) or forecast system produce calibrated forecasts would not be rejected. To the extent that these intervals do not include the horizontal no resolution line, a null hypothesis that the forecasts are no better than climatological guessing would be rejected.

Figure 7.27 shows the reliability diagram for the forecasts summarized in Table 7.2, with 95% confidence intervals drawn around each of the $I = 12$ conditional relative frequencies. The stationarity assumption for these estimated probabilities is reasonable, because the forecasters have sorted the forecast-observation pairs according to their judgments about those probabilities. The independence assumption is less well justified, because these data are simultaneous forecast-observation pairs for about one hundred locations in the United States, so that positive spatial correlations among both the forecasts and observations would be expected. Accordingly the confidence intervals drawn in Figure 7.27 are possibly too narrow.

Because the sample sizes (shown parenthetically in Figure 7.27) are large, Equation 7.66 was used to compute the confidence intervals. For each point, two confidence intervals are shown. The inner, narrower intervals are ordinary individual confidence intervals, computed using $z_{1-\alpha/2} = 1.96$, for $\alpha = 0.05$ in Equation 7.66. An interval of this kind would be appropriate if confidence statements about a single one of these points is of interest. The outer, wider confidence intervals are joint $1 - \alpha = 95\%$ Bonferroni (Equation 10.53) intervals, computed using $z_{1-[\alpha/12]/2} = 2.87$, again for $\alpha = 0.05$. The meaning of these outer, Bonferroni, intervals is that the probability is at least 0.95 that all $I = 12$ of the conditional probabilities being estimated are simultaneously within their respective individual confidence intervals. Thus, a (joint) null hypothesis that all of the forecast probabilities are calibrated would be rejected if any one of them fails to include the diagonal 1:1 line (dashed), which in fact does occur for $y_1 = 0.0$, $y_2 = 0.05$, $y_3 = 0.1$, $y_4 = 0.2$, and $y_{12} = 1.0$. On the other hand it is clear that these forecasts are overall much better than random climatological guessing, since the Bonferroni confidence intervals overlap the dashed horizontal no resolution line only for $y_4 = 0.2$, and are in general quite far from it.

7.9.4 *Resampling Verification Statistics*

Often the sampling characteristics for verification statistics with unknown sampling distributions are of interest. Or, sampling characteristics of verification statistics discussed previously in this section are of interest, but the assumption of independent sampling cannot be supported. In either case, statistical inference for forecast verification statistics can be addressed through resampling tests, as described in Sections 5.3.2 through 5.3.4. These procedures are very flexible, and the resampling algorithm used in any particular case will depend on the specific setting.

For problems where the sampling distribution of the verification statistic is unknown, but independence can reasonably be assumed, implementation of conventional permutation (see Section 5.3.3) or bootstrap (see Section 5.3.4) tests are straightforward. Illustrative examples of the bootstrap in forecast verification can be found in Roulston and Smith (2003) and Wilmott *et al.* (1985). Bradley *et al.* (2003) use the bootstrap to evaluate the sampling distributions of the reliability and resolution terms in Equation 7.4, using the probability-of-precipitation data in Table 7.2. Déqué (2003) illustrates permutation tests for a variety of verification statistics.

Special problems occur when the data to be resampled exhibit spatial and/or temporal correlation. A typical cause of spatial correlation is the occurrence of simultaneous data at multiple locations; that is, maps of forecasts and observations. Hamill (1999) describes a permutation test for a paired comparison of two forecasting systems, in which problems of nonindependence of forecast errors have been obviated by spatial pooling. Livezey (2003) notes that the effects of spatial correlation on resampled verification statistics can be accounted for automatically if the resampled objects are entire maps, rather than individual locations resampled independently of each other. Similarly, the effects of time correlation in the forecast verification statistics can be accounted for using the moving-blocks bootstrap (see Section 5.3.4). The moving-blocks bootstrap is equally applicable to scalar data (e.g., individual forecast-observation pairs at single locations, which are autocorrelated), or to entire autocorrelated maps of forecasts and observations (Wilks 1997b).

7.10 Exercises

7.1. For the forecast verification data in Table 7.2,

a. Reconstruct the joint distribution, $p(y_i, o_j), i = 1, \ldots, 12, j = 1, 2$.
b. Compute the unconditional (sample climatological) probability $p(o_1)$.

7.2. Construct the 2×2 contingency table that would result if the probability forecasts in Table 7.2 had been converted to nonprobabilistic rain/no rain forecasts, with a threshold probability of 0.25.

7.3. Using the 2×2 contingency table from Exercise 7.2, compute

a. The proportion correct.
b. The threat score.

TABLE 7.7 A 4×4 contingency table for snow amount forecasts in the eastern region of the United States during the winters 1983/1984 through 1988/1989. The event o_1 is $0 - 1$ in., o_2 is $2 - 3$ in., o_3 is $3 - 4$ in., and o_4 is ≥ 6 in. From Goldsmith (1990).

	o_1	o_2	o_3	o_4
y_1	35, 915	477	80	28
y_2	280	162	51	17
y_3	50	48	34	10
y_4	28	23	185	34

   c. The Heidke skill score.
   d. The Peirce skill score.
   e. The Gilbert skill score.

7.4. For the event o_3 (3 to 4 in. of snow) in Table 7.7 find

   a. The threat score.
   b. The hit rate.
   c. The false alarm ratio.
   d. The bias ratio.

7.5. Using the 4×4 contingency table in Table 7.7, compute

   a. The joint distribution of the forecasts and the observations.
   b. The proportion correct.
   c. The Heidke skill score.
   d. The Peirce skill score.

7.6. For the persistence forecasts for the January 1987 Ithaca maximum temperatures in Table A.1 (i.e., the forecast for 2 January is the observed temperature on 1 January, etc.), compute

   a. The MAE
   b. The RMSE
   c. The ME (bias)
   d. The skill, in terms of RMSE, with respect to the sample climatology.

7.7. Using the collection of hypothetical PoP forecasts summarized in Table 7.8,

   a. Calculate the Brier Score.
   b. Calculate the Brier Score for (the sample) climatological forecast.
   c. Calculate the skill of the forecasts with respect to the sample climatology.
   d. Draw the reliability diagram.

7.8. For the hypothetical forecast data in Table 7.8,

   a. Compute the likelihood-base rate factorization of the joint distribution $p(y_i, o_j)$.
   b. Draw the discrimination diagram.

TABLE 7.8 Hypothetical verification data for 1000 probability-of-precipitation forecasts.

forecast probability, y_i	0.00	0.10	0.20	0.30	0.40	0.50	0.60	0.70	0.80	0.90	1.00
number of times forecast	293	237	162	98	64	36	39	26	21	14	10
number of precipitation occurrences	9	21	34	31	25	18	23	18	17	12	9

TABLE 7.9   Hypothetical verification for 500 probability forecasts of precipitation amounts.

Forecast Probabilities for			Number of Forecast Periods Verifying as		
< 0.01 in.	0.01 in. − 0.24 in.	≥ 0.25 in.	< 0.01 in.	0.01 in. − 0.24 in.	≥ 0.25 in.
.8	.1	.1	263	24	37
.5	.4	.1	42	37	12
.4	.4	.2	14	16	10
.2	.6	.2	4	13	6
.2	.3	.5	4	6	12

  c. Draw the ROC curve.
  d. Test whether the area under the ROC curve is significantly greater than 1/2.

7.9. Using the hypothetical probabilistic three-category precipitation amount forecasts in Table 7.9,

  a. Calculate the average RPS.
  b. Calculate the skill of the forecasts with respect to the sample climatology.

7.10. For the hypothetical forecast and observed 500 mb fields in Figure 7.28,

  a. Calculate the S1 score, comparing the 24 pairs of gradients in the north-south and east-west directions.
  b. Calculate the MSE.
  c. Calculate the skill score for the MSE with respect to the climatological field.
  d. Calculate the centered AC.
  e. Calculate the uncentered AC.

7.11. Table 7.10 shows a set of 20 hypothetical ensemble forecasts, each with five members, and corresponding observations.

  a. Plot the verification rank histogram.
  b. Qualitatively diagnose the performance of this sample of forecast ensembles.

7.12. Using the results from Exercise 7.1, construct the VS curve for the verification data in Table 7.2.

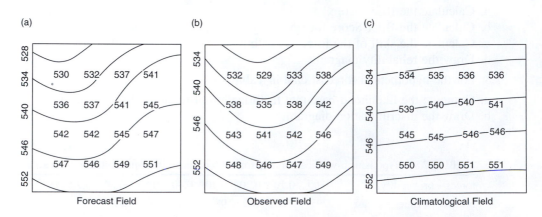

FIGURE 7.28   Hypothetical forecast (a), observed (b), and climatological average (c) fields of 500 mb heights (dam) over a small domain, and interpolations onto 16-point grids.

TABLE 7.10 A set of 20 hypothetical ensemble forecasts, of ensemble size 5, and corresponding observations.

Case	Member 1	Member 2	Member 3	Member 4	Member 5	Observation
1	7.9	7.3	5.5	6.9	8.3	7.7
2	7.4	5.6	8.2	5.8	6.1	9.4
3	9.5	8.3	10.5	8.9	6.1	8.7
4	6.1	7.8	5.1	10.4	4.9	3.4
5	6.3	5.8	5.1	6.0	4.1	7.3
6	8.1	6.8	1.8	6.7	10.5	8.2
7	4.4	5.6	7.7	6.0	7.0	4.3
8	5.9	3.0	4.4	7.2	9.1	7.0
9	5.2	5.7	5.3	6.0	7.5	4.1
10	2.7	6.6	5.8	7.5	5.1	8.3
11	6.6	5.2	5.3	5.5	3.2	4.7
12	6.7	6.0	8.6	7.7	4.8	8.7
13	8.9	1.3	5.9	7.3	6.3	8.5
14	8.5	5.0	4.6	7.6	1.4	4.8
15	9.2	4.4	8.9	5.3	6.5	9.5
16	2.7	8.7	3.4	7.6	5.1	4.3
17	4.1	7.0	7.5	7.2	7.0	5.4
18	7.7	4.7	5.7	5.7	6.8	2.1
19	6.7	7.4	6.2	5.3	5.8	3.3
20	4.4	3.3	1.9	5.4	6.6	7.4

CHAPTER ◆ 8

Time Series

8.1 Background

This chapter presents methods for characterizing and analyzing the time variations of data series. Often we encounter data sets consisting of consecutive measurements of atmospheric variables. When the ordering of the data in time is important to their information content, summarization and analysis using time series methods are appropriate.

As has been illustrated earlier, atmospheric observations separated by relatively short times tend to be similar, or correlated. Analyzing and characterizing the nature of these temporal correlations, or relationships through time, can be useful both for understanding atmospheric processes and for forecasting future atmospheric events. Accounting for these correlations is also necessary if valid statistical inferences about time-series data are to be made (see Chapter 5).

8.1.1 Stationarity

Of course, we do not expect the future values of a data series to be identical to some past series of existing observations. However, in many instances it may be very reasonable to assume that their statistical properties will be the same. The idea that past and future values of a time series will be similar statistically is an informal expression of what is called *stationarity*. Usually, the term stationarity is understood to mean weak stationarity, or covariance stationarity. In this sense, stationarity implies that the mean and autocovariance function (Equation 3.33) of the data series do not change through time. Different time slices of a stationary data series (for example, the data observed to date and the data to be observed in the future) can be regarded as having the same underlying mean, variance, and covariances. Furthermore, the correlations between variables in a stationary series are determined only by their separation in time (i.e., their lag, k, in Equation 3.31), and not their absolute positions in time. Qualitatively, different portions of a stationary time series look alike statistically, even though the individual data values may be very different. Covariance stationarity is a less restrictive assumption than strict stationarity, which implies that the full joint distribution of the variables in the series does not change through time. More technical expositions of the concept of stationarity can be found in, for example, Fuller (1996) or Kendall and Ord (1990).

Most methods for analyzing time series assume stationarity of the data. However, many atmospheric processes are distinctly not stationary. Obvious examples of nonstationary atmospheric data series are those exhibiting annual or diurnal cycles. For example, temperatures typically exhibit very strong annual cycles in mid- and high-latitude climates, and we expect the average of the distribution of January temperature to be very different from that for July temperature. Similarly, time series of wind speeds often exhibit a diurnal cycle, which derives physically from the tendency for diurnal changes in static stability, imposing a diurnal cycle on downward momentum transport.

There are two approaches to dealing with nonstationary series. Both aim to process the data in a way that will subsequently allow stationarity to be reasonably assumed. The first approach is to mathematically transform the nonstationary data to approximate stationarity. For example, subtracting a periodic mean function from data subject to an annual cycle would produce a transformed data series with constant (zero) mean. In order to produce a series with both constant mean and variance, it might be necessary to further transform these anomalies to standardized anomalies (Equation 3.21)—that is, to divide the values in the anomaly series by standard deviations that also vary through an annual cycle. Not only do temperatures tend to be colder in winter, but the variability of temperature tends to be higher. Data that become stationary after such annual cycles have been removed are said to exhibit cyclostationarity. A possible approach to transforming a monthly cyclostationary temperature series to (at least approximate) stationarity could be to compute the 12 monthly mean values and 12 monthly standard deviation values, and then to apply Equation 3.21 using the different means and standard deviations for the appropriate calendar month. This was the first step used to construct the time series of SOI values in Figure 3.14.

The alternative to data transformation is to stratify the data. That is, we can conduct separate analyses of subsets of the data record that are short enough to be regarded as nearly stationary. We might analyze daily observations for all available January records at a given location, assuming that each 31-day data record is a sample from the same physical process, but not necessarily assuming that process to be the same as for July, or even for February, data.

8.1.2 Time-Series Models

Characterization of the properties of a time series often is achieved by invoking mathematical models for the observed data variations. Having obtained a time-series model for an observed data set, that model might then be viewed as a generating process, or algorithm, that could have produced the data. A mathematical model for the time variations of a data set can allow compact representation of the characteristics of that data in terms of a few parameters. This approach is entirely analogous to the fitting of parametric probability distributions, which constitute another kind of probability model, in Chapter 4. The distinction is that the distributions in Chapter 4 are used without regard to the ordering of the data, whereas the motivation for using time-series models is specifically to characterize the nature of the ordering. Time-series methods are thus appropriate when the ordering of the data values in time is important to a given application.

Regarding an observed time series as having been generated by a theoretical (model) process is convenient because it allows characteristics of future, yet unobserved, values of a time series to be inferred from the inevitably limited data in hand. That is, characteristics of an observed time series are summarized by the parameters of a time-series model. Invoking the assumption of stationarity, future values of the time series should then also exhibit the statistical properties implied by the model, so that the properties of the model generating process can be used to infer characteristics of yet unobserved values of the series.

8.1.3 Time-Domain vs. Frequency-Domain Approaches

There are two fundamental approaches to time series analysis: time domain analysis and frequency domain analysis. Although these two approaches proceed very differently and may seem quite distinct, they are not independent. Rather, they are complementary methods that are linked mathematically.

Time-domain methods seek to characterize data series in the same terms in which they are observed and reported. A primary tool for characterization of relationships between data values in the time-domain approach is the autocorrelation function. Mathematically, time-domain analyses operate in the same space as the data values. Separate sections in this chapter describes different time-domain methods for use with discrete and continuous data. Here discrete and continuous are used in the same sense as in Chapter 4: discrete random variables are allowed to take on only a finite (or possibly countably infinite) number of values, and continuous random variables may take on any of the infinitely many real values within their range.

Frequency-domain analysis represents data series in terms of contributions occurring at different time scales, or characteristic frequencies. Each time scale is represented by a pair of sine and cosine functions. The overall time series is regarded as having arisen from the combined effects of a collection of sine and cosine waves oscillating at different rates. The sum of these waves reproduces the original data, but it is often the relative strengths of the individual component waves that are of primary interest. Frequency-domain analyses take place in the mathematical space defined by this collection of sine and cosine waves. That is, frequency-domain analysis involves transformation of the n original data values into coefficients that multiply an equal number of periodic (the sine and cosine) functions. At first exposure this process can seem very strange, and is sometimes difficult to grasp. However, frequency-domain methods very commonly are applied to atmospheric time series, and important insights can be gained from frequency-domain analyses.

8.2 Time Domain—I. Discrete Data

8.2.1 Markov Chains

Recall that a discrete random variable is one that can take on only values from among a defined, finite or countably infinite set. The most common class of model, or stochastic process, used to represent time series of discrete variables is known as the Markov chain. A Markov chain can be imagined as being based on collection of states of a model system. Each state corresponds to one of the elements of the MECE partition of the sample space describing the random variable in question.

For each time period, the length of which is equal to the time separation between observations in the time series, the Markov chain can either remain in the same state or change to one of the other states. Remaining in the same state corresponds to two successive observations of the same value of the discrete random variable in the time series, and a change of state implies two successive values of the time series that are different.

The behavior of a Markov chain is governed by a set of probabilities for these transitions, called the transition probabilities. The transition probabilities specify probabilities for the system being in each of its possible states during the next time period. The most common form is called a first-order Markov chain, for which the transition probabilities controlling the next state of the system depend only on the current state of the system. That is, knowing the current state of the system and the full sequence of states leading up to the current state, provides no more information about the probability distribution for the states at the next observation time than does knowledge of the current state alone. This characteristic of first-order Markov chains is known as the Markovian property, which can be expressed more formally as

$$\Pr\{X_{t+1}|X_t, X_{t-1}, X_{t-2}, \cdots, X_1\} = \Pr\{X_{t+1}|X_t\}. \tag{8.1}$$

The probabilities of future states depend on the present state, but they do not depend on the particular way that the model system arrived at the present state. In terms of a time series of observed data the Markovian property means, for example, that forecasts of tomorrow's data value can be made on the basis of today's observation, but also knowing yesterday's data value provides no additional information.

The transition probabilities of a Markov chain are conditional probabilities. That is, there is a conditional probability distribution pertaining to each possible current state, and each of these distributions specifies probabilities for the states of the system in the next time period. To say that these probability distributions are conditional allows for the possibility that the transition probabilities can be different, depending on the current state. The fact that these distributions can be different is the essence of the capacity of a Markov chain to represent the serial correlation, or persistence, often exhibited by atmospheric variables. If probabilities for future states are the same, regardless of the current state, then the time series consists of independent values. In that case the probability of occurrence of any given state in the upcoming time period is not affected by the occurrence or nonoccurrence of a particular state in the current time period. If the time series being modeled exhibits persistence, the probability of the system staying in a given state will be higher than the probabilities of arriving at that state from other states, and higher than the corresponding unconditional probability.

If the transition probabilities of a Markov chain do not change through time and none of them are zero, then the resulting time series will be stationary. Modeling nonstationary data series exhibiting, for example, an annual cycle can require allowing the transition probabilities to vary through an annual cycle as well. One way to capture this kind of nonstationarity is to specify that the probabilities vary according to some smooth periodic curve, such as a cosine function. Alternatively, separate transition probabilities can be used for nearly stationary portions of the cycle, for example four three-month seasons or 12 calendar months.

Certain classes of Markov chains are described more concretely, but relatively informally, in the following sections. More formal and comprehensive treatments can be found in, for example, Feller (1970), Karlin and Taylor (1975), or Katz (1985).

8.2.2 Two-State, First-Order Markov Chains

The simplest kind of discrete random variable pertains to dichotomous (yes/no) events. The behavior of a stationary sequence of independent (exhibiting no serial correlation) values of a dichotomous discrete random variable is described by the binomial distribution

TABLE 8.1 Time series of a dichotomous random variable derived from the January 1987 Ithaca precipitation data in Table A.1. Days on which nonzero precipitation was reported yield $x_t = 1$, and days with zero precipitation yield $x_t = 0$.

Date, t	1	2	3	4	5	6	7	8	9	10	11	12	13	14	15	16	17	18	19	20	21	22	23	24	25	26	27	28	29	30	31
x_t	0	1	1	0	0	0	0	1	1	1	1	1	1	1	1	1	0	0	0	0	1	0	0	1	0	0	0	0	1	1	1

(Equation 4.1). That is, for serially independent events, the ordering in time is of no importance from the perspective of specifying probabilities for future events, so that a time-series model for their behavior does not provide more information than does the simple binomial distribution.

A two-state Markov chain is a statistical model for the persistence of binary events. The occurrence or nonoccurrence of rain on a given day is a simple meteorological example of a binary random event, and a sequence of daily observations of "rain" and "no rain" for a particular location would constitute a time series of that variable. Consider a series where the random variable takes on the values $x_t = 1$ if precipitation occurs on day t and $x_t = 0$ if it does not. For the January 1987 Ithaca precipitation data in Table A.1, this time series would consist of the values shown in Table 8.1. That is, $x_1 = 0, x_2 = 1, x_3 = 1, x_4 = 0, \ldots,$ and $x_{31} = 1$. It is evident from looking at this series of numbers that the 1's and 0's tend to cluster in time. As was illustrated in Example 2.2, this clustering is an expression of the serial correlation present in the time series. That is, the probability of a 1 following a 1 is apparently higher than the probability of a 1 following a 0, and the probability of a 0 following a 0 is apparently higher than the probability of a 0 following a 1.

A common and often quite good stochastic model for data of this kind is a first-order, two-state Markov chain. A two-state Markov chain is natural for dichotomous data since each of the two states will pertain to one of the two possible data values. A first-order Markov chain has the property that the transition probabilities governing each observation in the time series depend only on the value of the previous member of the time series.

Figure 8.1 illustrates schematically the nature of a two-state first-order Markov chain. In order to help fix ideas, the two states are labeled in a manner consistent with the data in Table 8.1. For each value of the time series, the stochastic process is either in state 0 (no precipitation occurs and $x_t = 0$), or in state 1 (precipitation occurs and $x_t = 1$). At each time step the process can either stay in the same state or switch to the other state. Therefore four distinct transitions are possible, corresponding to a dry day following a dry day (p_{00}), a wet day following a dry day (p_{01}), a dry day following a wet day (p_{10}),

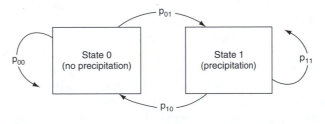

FIGURE 8.1   Schematic representation of a two-state, first-order Markov chain, illustrated in terms of daily precipitation occurrence or nonoccurrence. The two states are labeled 0 for no precipitation, and 1 for precipitation occurrence. For a first-order Markov chain, there are four transition probabilities controlling the state of the system in the next time period. Since these four probabilities are pairs of conditional probabilities, $p_{00} + p_{01} = 1$ and $p_{10} + p_{11} = 1$. For quantities like day-to-day precipitation occurrence that exhibit positive serial correlation, $p_{01} < p_{00}$, and $p_{01} < p_{11}$.

and a wet day following a wet day (p_{11}). Each of these four transitions is represented in Figure 8.1 by arrows, labeled with the appropriate transition probabilities. Here the notation is such that the first subscript on the probability is the state at time t, and the second subscript is the state at time $t+1$.

The transition probabilities are conditional probabilities for the state at time $t+1$ (e.g., whether precipitation will occur tomorrow) given the state at time t (e.g., whether or not precipitation occurred today). That is,

$$p_{00} = \Pr\{X_{t+1} = 0 | X_t = 0\} \tag{8.2a}$$

$$p_{01} = \Pr\{X_{t+1} = 1 | X_t = 0\} \tag{8.2b}$$

$$p_{10} = \Pr\{X_{t+1} = 0 | X_t = 1\} \tag{8.2c}$$

$$p_{11} = \Pr\{X_{t+1} = 1 | X_t = 1\}. \tag{8.2d}$$

Together, Equations 8.2a and 8.2b constitute the conditional probability distribution for the value of the time series at time $t+1$, given that $X_t = 0$ at time t. Similarly, Equations 8.2c and 8.2d express the conditional probability distribution for the next value of the time series given that the current value is $X_t = 1$.

Notice that the four probabilities in Equation 8.2 provide some redundant information. Given that the Markov chain is in one state or the other at time t, the sample space for X_{t+1} consists of only two MECE events. Therefore, $p_{00} + p_{01} = 1$ and $p_{10} + p_{11} = 1$, so that it is really only necessary to focus on one of each of the pairs of transition probabilities, say p_{01} and p_{11}. In particular, it is sufficient to estimate only two parameters for a two-state first-order Markov chain, since the two pairs of conditional probabilities must sum to 1. The parameter estimation procedure consists simply of computing the conditional relative frequencies, which yield the maximum likelihood estimators (MLEs)

$$\hat{p}_{01} = \frac{\text{\# of 1's following 0's}}{\text{Total \# of 0's}} = \frac{n_{01}}{n_{0\bullet}} \tag{8.3a}$$

and

$$\hat{p}_{11} = \frac{\text{\# of 1's following 1's}}{\text{Total \# of 1's}} = \frac{n_{11}}{n_{1\bullet}}. \tag{8.3b}$$

Here n_{01} is the number of transitions from State 0 to State 1, n_{11} is the number of pairs of time steps in which there are two consecutive 1's in the series, $n_{0\bullet}$ is the number of 0's in the series followed by another data point, and $n_{1\bullet}$ is the number of 1's in the series followed by another data point. That is, the subscript $\bullet$ indicates the total over all values of the index replaced by this symbol, so that $n_{1\bullet} = n_{10} + n_{11}$ and $n_{0\bullet} = n_{00} + n_{01}$. Equations 8.3 state that the parameter p_{01} is estimated by looking at the conditional relative frequency of the event $X_{t+1} = 1$ considering only those points in the time series following data values for which $X_t = 0$. Similarly, p_{11} is estimated as the fraction of points for which $X_t = 1$ that are followed by points with $X_{t+1} = 1$. These somewhat labored definitions of $n_{0\bullet}$ and $n_{1\bullet}$ are necessary to account for the edge effect in a finite sample. The final point in the time series is not counted in the denominator of Equation 8.3a or 8.3b, whichever is appropriate, because there is no available data value following it to be incorporated into the counts in one of the numerators. These definitions also cover cases of missing values, and stratified samples such as 30 years of January data, for example.

Equation 8.3 suggests that parameter estimation for a two-state first-order Markov chain is equivalent to fitting two Bernoulli distributions (i.e., binomial distributions with

$N = 1$). One of these binomial distributions pertains to points in the time series preceded by 0's, and the other describes the behavior of points in the time series preceded by 1's. Knowing that the process is currently in state 0 (e.g., no precipitation today), the probability distribution for the event $X_{t+1} = 1$ (precipitation tomorrow) is simply binomial (Equation 4.1) with $p = p_{01}$. The second binomial parameter is $N = 1$, because there is only one data point in the series for each time step. Similarly, if $X_t = 1$, then the distribution for the event $X_{t+1} = 1$ is binomial with $N = 1$ and $p = p_{11}$. The conditional dichotomous events of a stationary Markov chain satisfy the requirements listed in Chapter 4 for the binomial distribution. For a stationary process the probabilities do not change through time, and conditioning on the current value of the time series satisfies the independence assumption for the binomial distribution because of the Markovian property. It is the fitting of two Bernoulli distributions that allows the time dependence in the data series to be represented.

Certain properties are implied for a time series described by a Markov chain. These properties are controlled by the values of the transition probabilities, and can be computed from them. First, the long-run relative frequencies of the events corresponding to the two states of the Markov chain are called the stationary probabilities. For a Markov chain describing the daily occurrence or nonoccurrence of precipitation, the stationary probability for precipitation, π_1, corresponds to the (unconditional) climatological probability of precipitation. In terms of the transition probabilities p_{01} and p_{11},

$$\pi_1 = \frac{p_{01}}{1 + p_{01} - p_{11}}, \tag{8.4}$$

with the stationary probability for state 0 being simply $\pi_0 = 1 - \pi_1$. In the usual situation of positive serial correlation or persistence, we find $p_{01} < \pi_1 < p_{11}$. Applied to daily precipitation occurrence, this relationship means that the conditional probability of a wet day following a dry day is less than the overall climatological relative frequency, which in turn is less than the conditional probability of a wet day following a wet day.

The transition probabilities also imply a specific degree of serial correlation, or persistence, for the binary time series. In terms of the transition probabilities, the lag-1 autocorrelation (Equation 3.30) of the binary time series is simply

$$r_1 = p_{11} - p_{01}. \tag{8.5}$$

In the context of Markov chains, r_1 is sometimes known as the persistence parameter. As the correlation r_1 increases, the difference between p_{11} and p_{01} widens, so that state 1 is more and more likely to follow state 1, and less and less likely to follow state 0. That is, there is an increasing tendency for 0's and 1's to cluster in time, or occur in runs. A time series exhibiting no autocorrelation would be characterized by $r_1 = p_{11} - p_{01} = 0$, or $p_{11} = p_{01} = \pi_1$. In this case the two conditional probability distributions specified by Equation 8.2 are the same, and the time series is simply a string of independent Bernoulli realizations. The Bernoulli distribution can be viewed as defining a two-state, *zero*-order Markov chain.

Once the state of a Markov chain has changed, the number of time periods it will remain in the new state is a random variable, with a probability distribution function. Because the conditional independence implies conditional Bernoulli distributions, this probability distribution function for numbers of consecutive time periods in the same state, or spell lengths, will be the geometric distribution (Equation 4.5), with $p = p_{01}$ for sequences of 0's (dry spells), and $p = p_{10} = 1 - p_{11}$ for sequences of 1's (wet spells).

The full autocorrelation function, Equation 3.31, for the first-order Markov chain follows easily from the lag-1 autocorrelation r_1. Because of the Markovian property, the autocorrelation between members of the time series separated by k time steps is simply the lag-1 autocorrelation multiplied by itself k times,

$$r_k = (r_1)^k. \tag{8.6}$$

A common misconception is that the Markovian property implies independence of values in a first-order Markov chain that are separated by more than one time period. Equation 8.6 shows that the correlation, and hence the statistical dependence, among elements of the time series tails off at increasing lags, but it is never exactly zero unless $r_1 = 0$. Rather, the Markovian property implies *conditional* independence of data values separated by more than one time period, as expressed by Equation 8.1. Given a particular value for x_t, the different possible values for $x_{t-1}, x_{t-2}, x_{t-3}$, and so on, do not affect the probabilities for x_{t+1}. However, for example, $\Pr\{x_{t+1} = 1 | x_{t-1} = 1\} \neq \Pr\{x_{t+1} = 1 | x_{t-1} = 0\}$, indicating statistical dependence among members of a Markov chain separated by more than one time period. Put another way, it is not that the Markov chain has no memory of the past, but rather that it is only the recent past that matters.

8.2.3 Test for Independence vs. First-Order Serial Dependence

Even if a series of binary data is generated by a mechanism producing serially independent values, the sample lag-one autocorrelation (Equation 8.5) computed from a finite sample is unlikely to be exactly zero. A formal test, similar to the χ^2 goodness-of-fit test (Equation 5.14), can be conducted to investigate the statistical significance of the sample autocorrelation for a binary data series. The null hypothesis for this test is that the data series is serially independent (i.e., the data are independent Bernoulli variables), with the alternative being that the series was generated by a first-order Markov chain.

The test is based on a contingency table of the observed transition counts n_{00}, n_{01}, n_{10}, and n_{11}, in relation to the numbers of transitions expected under the null hypothesis. The corresponding expected counts, e_{00}, e_{01}, e_{10}, and e_{11}, are computed from the observed transition counts under the constraint that the marginal totals of the expected counts are the same as for the observed transitions. The comparison is illustrated in Figure 8.2, which shows generic contingency tables for the observed transition counts (a) and those

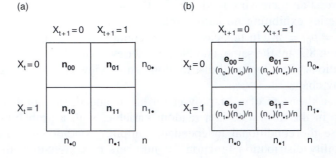

FIGURE 8.2   Contingency tables of observed transition counts n_{ij} (a) for a binary time series, and (b) transition counts e_{ij} expected if the time series actually consists of serially independent values with the same marginal totals. The transition counts are shown in boldface, and the marginal totals are in plain type.

expected under the null hypothesis of independence (b). For example, the transition count n_{00} specifies the number of consecutive pairs of 0's in the time series. This is related to the joint probability $\Pr\{X_t = 0 \cap X_{t+1} = 0\}$. Under the null hypothesis of independence this joint probability is simply the product of the two event probabilities, or in relative frequency terms, $\Pr\{X_t = 0\}\Pr\{X_{t+1} = 0\} = (n_{0\bullet}/n)(n_{\bullet 0}/n)$. Thus, the corresponding number of expected transition counts is simply this product multiplied by the sample size, or $e_{00} = (n_{0\bullet})(n_{\bullet 0})/n$. More generally,

$$e_{ij} = \frac{n_{i\bullet}n_{\bullet j}}{n}. \tag{8.7}$$

The test statistic is computed from the observed and expected transition counts using

$$\chi^2 = \sum_i \sum_j \frac{(n_{ij} - e_{ij})^2}{e_{ij}}, \tag{8.8}$$

where, for the 2×2 contingency table appropriate for dichotomous data, the summations are for $i = 0$ to 1 and $j = 0$ to 1. That is, there is a separate term in Equation 8.8 for each of the four pairs of contingency table cells in Figure 8.2. Note that Equation 8.8 is analogous to Equation 5.14, with the n_{ij} being the observed counts, and the e_{ij} being the expected counts. Under the null hypothesis, the test statistic follows the χ^2 distribution with $\nu = 1$ degree of freedom. This value of the degrees-of-freedom parameter is appropriate because, given that the marginal totals are fixed, arbitrarily specifying one of the transition counts completely determines the other three.

The fact that the numerator in Equation 8.8 is squared implies that values of the test statistic on the left tail of the null distribution are favorable to H_0, because small values of the test statistic are produced by pairs of observed and expected transition counts of similar magnitudes. Therefore, the test is one-tailed. The p value associated with a particular test can be assessed using the χ^2 quantiles in Table B.3.

EXAMPLE 8.1   Fitting a Two-State, First Order Markov Chain

Consider summarizing the time series in Table 8.1, derived from the January 1987 Ithaca precipitation series in Table A.1, using a first-order Markov chain. The parameter estimates in Equation 8.3 are obtained easily from the transition counts. For example, the number of 1's following 0's in the time series of Table 8.1 is $n_{01} = 5$. Similarly, $n_{00} = 11$, $n_{10} = 4$, and $n_{11} = 10$. The resulting sample estimates for the transition probabilities (Equations 8.3) are $p_{01} = 5/16 = 0.312$, and $p_{11} = 10/14 = 0.714$. Note that these are identical to the conditional probabilities computed in Example 2.2.

Whether the extra effort of fitting the first-order Markov chain to the data in Table 8.1 is justified can be investigated using the χ^2 test in Equation 8.8. Here the null hypothesis is that these data resulted from an independent (i.e., Bernoulli) process, and the expected transition counts e_{ij} that must be computed are those consistent with this null hypothesis. These are obtained from the marginal totals $n_{0\bullet} = 11 + 5 = 16$, $n_{1\bullet} = 4 + 10 = 14$, $n_{\bullet 0} = 11 + 4 = 15$, and $n_{\bullet 1} = 5 + 10 = 15$. The expected transition counts follow easily as $e_{00} = (16)(15)/30 = 8$, $e_{01} = (16)(15)/30 = 8$, $e_{10} = (14)(15)/30 = 7$, and $e_{11} = (14)(15)/30 = 7$. Note that usually the expected transition counts will be different from each other, and need not be integer values.

Computing the test statistic in Equation 8.8, we find $\chi^2 = (11 - 8)^2/8 + (5 - 8)^2/8 + (4 - 7)^2/7 + (10 - 7)^2/7 = 4.82$. The degree of unusualness of this result with reference to the null hypothesis can be assessed with the aid of Table B.3. Looking on the $\nu = 1$

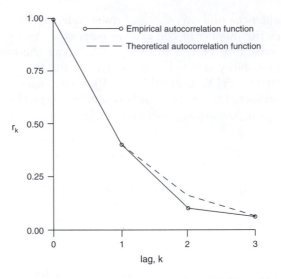

FIGURE 8.3   Sample autocorrelation function for the January 1987 Ithaca binary precipitation occurrence series, Table 8.1 (solid, with circles), and theoretical autocorrelation function (dashed) specified by the fitted first-order Markov chain model (Equation 8.6). The correlations are 1.00 for $k = 0$, since the unlagged data are perfectly correlated with themselves.

row, we find that the result lies between the 95th and 99th percentiles of the appropriate χ^2 distribution. Thus, even for this rather small sample size, the null hypothesis of serial independence would be rejected at the 5% level, although not at the 1% level.

The degree of persistence exhibited by this data sample can be summarized using the persistence parameter, which is also the lag-one autocorrelation, $r_1 = p_{11} - p_{01} = 0.714 - 0.312 = 0.402$. This value is obtained by operating on the series of 0's and 1's in Table 8.1, using Equation 3.30. This lag-1 autocorrelation is fairly large, indicating substantial serial correlation in the time series. It also implies the full autocorrelation function, through Equation 8.6. Figure 8.3 shows that the implied theoretical correlation function for this Markov process, shown as the dashed line, agrees very closely with the sample autocorrelation function shown by the solid line, for the first few lags. This agreement provides qualitative support for the first-order Markov chain as an appropriate model for the data series.

Finally, the stationary (i.e., climatological) probability for precipitation implied for this data by the Markov chain model is, using Equation 8.4, $\pi_1 = 0.312/(1+0.312-0.714) = 0.522$. This value agrees closely with the relative frequency $16/30 = 0.533$, obtained by counting the number of 1's in the last 30 values of the series in Table 8.1. ◊

8.2.4 Some Applications of Two-State Markov Chains

One interesting application of the Markov chain model is in the computer generation of synthetic rainfall series. Time series of random binary numbers, statistically resembling real rainfall occurrence data, can be generated using the Markov chain as an algorithm. This procedure is an extension of the ideas presented in Section 4.7, to time-series data. To generate sequences of numbers statistically resembling those in Table 8.1, for example, the parameters $p_{01} = 0.312$ and $p_{11} = 0.714$, estimated in Example 8.1, would be used together with a uniform [0, 1] random number generator (see Section 4.7.1). The synthetic time series would begin using the stationary probability $\pi_1 = 0.522$. If the first uniform number generated were less than π_1, then $x_1 = 1$, meaning that the first simulated day would be wet. For subsequent values in the series, each new uniform random number would be compared to the appropriate transition probability, depending on whether the most recently generated number, corresponding to day t, was wet or dry. That is, the transition probability p_{01} would be used to generate x_{t+1} if $x_t = 0$, and p_{11} would be used if $x_t = 1$. A wet day ($x_{t+1} = 1$) is simulated if the next uniform random number is less than the transition probability, and a dry day ($x_{t+1} = 0$) is generated if it is not. Since typically $p_{11} > p_{01}$ for daily precipitation occurrence data, simulated wet days are more likely to follow wet days than dry days, as is the case in the real data series.

The Markov chain approach for simulating precipitation occurrence can be extended to include simulation of daily precipitation amounts. This is accomplished by adopting a statistical model for the nonzero rainfall amounts, yielding a sequence of random variables defined on the Markov chain, called a chain-dependent process (Katz 1977; Todorovic and Woolhiser 1975). Commonly a gamma distribution (see Chapter 4) is fit to the precipitation amounts on wet days in the data record (e.g., Katz 1977; Richardson 1981; Stern and Coe 1984), although the mixed exponential distribution (Equation 4.66) often provides a better fit to nonzero daily precipitation data (e.g., Foufoula-Georgiou and Lettenmaier 1987; Wilks 1999a; Woolhiser and Roldan 1982). Computer algorithms are available to generate random variables drawn from gamma distributions (e.g., Bratley et al. 1987; Johnson 1987), or together Example 4.14 and Section 4.7.5 can be used to simulate from the mixed exponential distribution, to produce synthetic precipitation amounts on days when the Markov chain calls for a wet day. The tacit assumption that precipitation amounts on consecutive wet days are independent has turned out to be a reasonable approximation in most instances where it has been investigated (e.g., Katz 1977; Stern and Coe 1984), but may not adequately simulate extreme multiday precipitation events that could arise, for example, from a slow-moving landfalling hurricane (Wilks 2002a). Generally both the Markov chain transition probabilities and the parameters of the distributions describing precipitation amounts change through the year. These seasonal cycles can be handled by fitting separate sets of parameters for each of the 12 calendar months (e.g., Wilks 1989), or by representing them using smoothly varying sine and cosine functions (Stern and Coe 1984).

Properties of longer-term precipitation quantities resulting from simulated daily series (e.g., the monthly frequency distributions of numbers of wet days in a month, or of total monthly precipitation) can be calculated from the parameters of the chain-dependent process that governs the daily precipitation series. Since observed monthly precipitation statistics are computed from individual daily values, it should not be surprising that the statistical characteristics of monthly precipitation quantities will depend directly on the statistical characteristics of daily precipitation occurrences and amounts. Katz (1977, 1985) gives equations specifying some of these relationships, which can be used in a variety of ways (e.g., Katz and Parlange 1993; Wilks 1992, 1999b; Wilks and Wilby 1999).

Finally, another interesting perspective on the Markov chain model for daily precipitation occurrence is in relation to forecasting precipitation probabilities. Recall that forecast skill is assessed relative to a set of benchmark, or reference forecasts (Equation 7.4). Usually one of two reference forecasts are used: either the climatological probability of the forecast event, in this case π_1; or persistence forecasts specifying unit probability if precipitation occurred in the previous period, or zero probability if the event did not occur. Neither of these reference forecasting systems is particularly sophisticated, and both are relatively easy to improve upon, at least for short-range forecasts. A more challenging, yet still fairly simple alternative is to use the transition probabilities of a two-state Markov chain as the reference forecasts. If precipitation did not occur in the preceding period, the reference forecast would be p_{01}, and the conditional forecast probability for precipitation following a day with precipitation would be p_{11}. Note that for meteorological quantities exhibiting persistence, $0 < p_{01} < \pi_1 < p_{11} < 1$, so that reference forecasts consisting of Markov chain transition probabilities constitute a compromise between the persistence (either 0 or 1) and climatological (π_1) probabilities. Furthermore, the balance of this compromise depends on the strength of the persistence exhibited by the climatological data on which the estimated transition probabilities are based. A weakly persistent quantity would be characterized by transition probabilities differing little from π_1, whereas strong serial correlation will produce transition probabilities much closer to 0 and 1.

8.2.5 *Multiple-State Markov Chains*

Markov chains are also useful for representing the time correlation of discrete variables that can take on more than two values. For example, a three-state, first-order Markov chain is illustrated schematically in Figure 8.4. Here the three states are arbitrarily labeled 1, 2, and 3. At each time t, the random variable in the series can take on one of the three values $x_t = 1$, $x_t = 2$, or $x_t = 3$, and each of these values corresponds to a state. First-order time dependence implies that the transition probabilities for x_{t+1} depend only on the state x_t, so that there are $3^2 =$ nine transition probabilities, p_{ij}. In general, for a first-order, s-state Markov chain, there are s^2 transition probabilities.

As is the case for the two-state Markov chain, the transition probabilities for multiple-state Markov chains are conditional probabilities. For example, the transition probability p_{12} in Figure 8.4 is the conditional probability that state 2 will occur at time $t + 1$, given that state 1 occurred at time t. Therefore, in a s-state Markov chain it must be the case that the probabilities for the s transitions emanating from each state must sum to one, or $\Sigma_j p_{ij} = 1$ for each value of i.

Estimation of the transition probabilities for multiple-state Markov chains is a straightforward generalization of the formulas in Equations 8.3 for two-state chains. Each of these

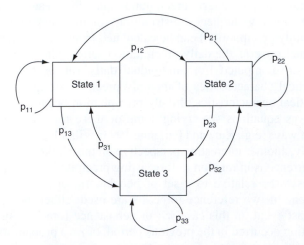

FIGURE 8.4   Schematic illustration of a three-state, first-order Markov chain. There are nine possible transitions among the three states, including the possibility that two consecutive points in the time series will be in the same state. First-order time dependence implies that the transition probabilities depend only on the current state of the system, or present value of the time series.

estimates are simply obtained from the conditional relative frequencies of the transition counts,

$$\hat{p}_{ij} = \frac{n_{ij}}{n_{i\bullet}}; i, j = 1, \ldots, s. \qquad (8.9)$$

As before, the dot indicates summation over all values of the replaced subscript so that, for example, $n_{1\bullet} = \Sigma_j n_{1j}$. For the $s = 3$-state Markov chain represented in Figure 8.4, for example, $\hat{p}_{12} = n_{12}/(n_{11} + n_{12} + n_{13})$. In general, a contingency table of transition counts, corresponding to Figure 8.2a for the $s = 2$-state case, will contain s^2 entries.

Testing whether the observed degree of serial correlation is significantly different from zero in a multiple-state situation can be done using the χ^2 test in Equation 8.8. Here the summations are over all s possible states, and will include s^2 terms. As before, the expected numbers of transition counts e_{ij} are computed using Equation 8.7. Under the null hypothesis of no serial correlation, the distribution of the test statistic in Equation 8.8 is χ^2 with $\nu = (s - 1)^2$ degrees of freedom.

Three-state Markov chains have been used to characterize transitions between below-normal, near-normal, and above-normal months, as defined by the U.S. Climate Prediction Center (see Example 4.9), by Preisendorfer and Mobley (1984) and Wilks (1989). Mo and Ghil (1987) used a five-state Markov chain to characterize transitions between persistent hemispheric 500-mb flow types.

8.2.6 Higher-Order Markov Chains

First-order Markov chains often provide good representations of daily precipitation occurrence, but it is not obvious just from inspection of the series in Table 8.1, for example, that this simple model will be adequate to capture the observed correlation structure. More generally, an mth order Markov chain is one where the transition probabilities depend on the states in the previous m time periods. Formally, the extension of the Markovian property expressed in Equation 8.1 to the mth order Markov chain is

$$\Pr\{X_{t+1}|X_t, X_{t-1}, X_{t-2}, \cdots, X_1\} = \Pr\{X_{t+1}|X_t, X_{t-1}, \cdots, X_{t-m}\}. \qquad (8.10)$$

Consider, for example a second-order Markov chain. Second-order time dependence means that the transition probabilities depend on the states (values of the time series) at lags of both one and two time periods. Notationally, then, the transition probabilities for a second-order Markov chain require three subscripts: the first denotes the state at time $t - 1$, the second denotes the state in time t, and the third specifies the state at (the future) time $t + 1$. The notation for the transition probabilities of a second-order Markov chain can be defined as

$$p_{hij} = \{X_{t+1} = j|X_t = i, X_{t-1} = h\}. \qquad (8.11)$$

In general, the notation for a mth order Markov chain requires $m + 1$ subscripts on the transition counts and transition probabilities. If Equation 8.11 is being applied to a binary time series such as that in Table 8.1, the model would be a two-state, second-order Markov chain, and the indices h, i, and j could take on either of the $s = 2$ values of the time series, say 0 and 1. However, Equation 8.11 is equally applicable to discrete time series with larger numbers ($s > 2$) of states.

TABLE 8.2   Arrangement of the $2^{2+1} = 8$ transition counts for a two-state, second-order Markov chain in a table of the form of Figure 8.2a. Determining these counts from an observed time series requires examination of successive triplets of data values.

X_{t-1}	X_t	$X_{t+1} = 0$	$X_{t+1} = 1$	Marginal Totals
0	0	n_{000}	n_{001}	$n_{00\bullet} = n_{000} + n_{001}$
0	1	n_{010}	n_{011}	$n_{01\bullet} = n_{010} + n_{011}$
1	0	n_{100}	n_{101}	$n_{10\bullet} = n_{100} + n_{101}$
1	1	n_{110}	n_{111}	$n_{11\bullet} = n_{110} + n_{111}$

As is the case for first-order Markov chains, transition probability estimates are obtained from relative frequencies of observed transition counts. However, since data values further back in time now need to be considered, the number of possible transitions increases exponentially with the order, m, of the Markov chain. In particular, for an s-state, m^{th} order Markov chain, there are $s^{(m+1)}$ distinct transition counts and transition probabilities. The arrangement of the resulting transition counts, in the form of Figure 8.2a, is shown in Table 8.2 for a $s = 2$ state, $m =$ second-order Markov chain. The transition counts are determined from the observed data series by examining consecutive groups of $m + 1$ data points. For example, the first three data points in Table 8.1 are $x_{t-1} = 0$, $x_t = 1$, $x_{t+1} = 1$, and this triplet would contribute one to the transition count n_{011}. Overall the data series in Table 8.1 exhibits three transitions of this kind, so the final transition count $n_{011} = 3$ for this data set. The second triplet in the data set in Table 8.1 would contribute one count to n_{110}. There is only one other triplet in this data for which $x_{t-1} = 1$, $x_t = 1$, $x_{t+1} = 0$, so the final count for $n_{110} = 2$.

The transition probabilities for a second-order Markov chain are obtained from the conditional relative frequencies of the transition counts

$$\hat{p}_{hij} = \frac{n_{hij}}{n_{hi\bullet}}. \tag{8.12}$$

That is, given that the value of the time series at time $t - 1$ was $x_{t-1} = h$ and the value of the time series at time t was $x_t = i$, the probability that the future value of the time series $x_{t+1} = j$ is p_{hij}, and the sample estimate of this probability is given in Equation 8.12. Just as the two-state first-order Markov chain consists essentially of two conditional Bernoulli distributions, a two-state second-order Markov chain amounts to four conditional Bernoulli distributions, with parameters $p = p_{hi1}$, for each of the four distinct combinations of the indices h and i.

Note that the small data set in Table 8.1 is really too short to fit a second-order Markov chain. Since there are no triplets in this series for which $x_{t-1} = 1$, $x_t = 0$, $x_{t+1} = 1$ (i.e., a single dry day following and followed by a wet day) the transition count $n_{101} = 0$. This zero transition count would lead to the sample estimate for the transition probability $\hat{p}_{101} = 0$, even though there is no physical reason why that particular sequence of wet and dry days could not or should not occur.

8.2.7 Deciding among Alternative Orders of Markov Chains

How are we to know what order m is appropriate for a Markov chain to represent a particular data series? One approach is to use a hypothesis test. For example, the χ^2

test in Equation 8.8 can be used to assess the significance of a first-order Markov chain model versus a zero-order, or binomial model. The mathematical structure of this test can be modified to investigate the suitability of, say, a first-order versus a second-order, or a second-order versus a third-order Markov chain, but the overall significance of a collection of such tests would be difficult to evaluate. This difficulty arises in part because of the issue of test multiplicity. As discussed in Section 5.5, the overall significance level of a collection of simultaneous, correlated tests is difficult if not impossible to evaluate.

Two criteria are in common use for choosing among alternative orders of Markov chain models. These are the Akaike Information Criterion (AIC) (Akaike 1974; Tong 1975) and the Bayesian Information Criterion (BIC) (Schwarz 1978; Katz 1981). Both are based on the log-likelihood functions for the transition probabilities of the fitted Markov chains. These log-likelihoods depend on the transition counts and the estimated transition probabilities. The log-likelihoods for s-state Markov chains of order 0, 1, 2, and 3 are

$$L_0 = \sum_{j=0}^{s-1} n_j \ln(\hat{p}_j) \tag{8.13a}$$

$$L_1 = \sum_{i=0}^{s-1}\sum_{j=0}^{s-1} n_{ij} \ln(\hat{p}_{ij}) \tag{8.13b}$$

$$L_2 = \sum_{h=0}^{s-1}\sum_{i=0}^{s-1}\sum_{j=0}^{s-1} n_{hij} \ln(\hat{p}_{hij}) \tag{8.13c}$$

and

$$L_3 = \sum_{g=0}^{s-1}\sum_{h=0}^{s-1}\sum_{i=0}^{s-1}\sum_{j=0}^{s-1} n_{ghij} \ln(\hat{p}_{ghij}), \tag{8.13d}$$

with obvious extension for fourth-order and higher Markov chains. Here the summations are over all s states of the Markov chain, and so will include only two terms each for two-state (binary) time series. Equation 8.13a is simply the log-likelihood for the independent binomial model.

EXAMPLE 8.2   Likelihood Ratio Test for the Order of a Markov Chain

To illustrate the application of Equations 8.13, consider a likelihood ratio test of first-order dependence of the binary time series in Table 8.1, versus the null hypothesis of zero serial correlation. The test involves computation of the log-likelihoods in Equations 8.13a and 8.13b. The resulting two log-likelihoods are compared using the test statistic given by Equation 5.19.

In the last 30 data points in Table 8.1, there are $n_0 = 14$ 0's and $n_1 = 16$ 1's, yielding the unconditional relative frequencies of rain and no rain $\hat{p}_0 = 14/30 = 0.467$ and $\hat{p}_1 = 16/30 = 0.533$, respectively. The last 30 points are used because the first-order Markov chain amounts to two conditional Bernoulli distributions, given the previous day's value, and the value for 31 December 1986 is not available in Table A.1. The log-likelihood in Equation 8.13a for these data is $L_0 = 14\ln(0.467) + 16\ln(0.533) = -20.73$. Values of n_{ij} and $\hat{p}_{ij}$ were computed previously, and can be substituted into Equation 8.13b to yield $L_1 = 11\ln(0.688) + 5\ln(0.312) + 4\ln(0.286) + 10\ln(0.714) = -18.31$. Necessarily, $L_1 \geq L_0$ because the greater number of parameters in the more elaborate first-order

Markov model provide more flexibility for a closer fit to the data at hand. The statistical significance of the difference in log-likelihoods can be assessed knowing that the null distribution of $\Lambda = 2(L_1 - L_0) = 4.83$ is χ^2, with $\nu = (s^{m(H_A)} - s^{m(H_0)})(s-1)$ degrees of freedom. Since the time series being tested in binary, $s = 2$. The null hypothesis is that the time dependence is zero-order, so $m(H_0) = 0$, and the alternative hypothesis is first-order serial dependence, or $m(H_A) = 1$. Thus, $\nu = (2^1 - 2^0)(2-1) = 1$ degree of freedom. In general the appropriate degrees-of-freedom will be the difference in dimensionality between the competing models. This likelihood test result is consistent with the χ^2 goodness-of-fit test conducted in Example 8.1, which is not surprising because the χ^2 test conducted there is an approximation to the likelihood ratio test◊

Both the AIC and BIC criteria attempt to find the most appropriate model order by striking a balance between goodness of fit, as reflected in log-likelihoods, and a penalty that increases with the number of fitted parameters. The two approaches differ only in the form of the penalty function. The AIC and BIC statistics are computed for each trial order m, using

$$\text{AIC}(m) = -2 \, L_m + 2 \, s^m(s-1), \tag{8.14}$$

or

$$\text{BIC}(m) = -2 \, L_m + s^m(\ln n), \tag{8.15}$$

respectively. The order m is chosen as appropriate that minimizes either Equation 8.14 or 8.15. The BIC criterion tends to be more conservative, generally picking lower orders than the AIC criterion when results of the two approaches differ. Use of the BIC statistic may be preferable for sufficiently long time series, although "sufficiently long" may range from around $n = 100$ to over $n = 1000$, depending on the nature of the serial correlation (Katz 1981).

8.3 Time Domain—II. Continuous Data

8.3.1 *First-Order Autoregression*

The Markov chain models described in the previous section are not suitable for describing time series of data that are continuous, in the sense of the data being able to take on infinitely many values on all or part of the real line. As discussed in Chapter 4, atmospheric variables such as temperature, wind speed, geopotential height, and so on, are continuous variables in this sense. The correlation structure of such time series often can be represented successfully using a class of time series models known as Box-Jenkins models, after the classic text by Box and Jenkins (1994).

The simplest Box-Jenkins model is the first-order autoregression, or AR(1) model. It is the continuous analog of the first-order Markov chain. As the name suggests, one way of viewing the AR(1) model is as a simple linear regression (see Section 6.2.1), where the predictand is the value of the time series at time $t+1$, x_{t+1}, and the predictor is the current value of the time series, x_t. The AR(1) model can be written as

$$x_{t+1} - \mu = \phi(x_t - \mu) + \varepsilon_{t+1}, \tag{8.16}$$

where μ is the mean of the time series, ϕ is the autoregressive parameter, and ε_{t+1} is a random quantity corresponding to the residual in ordinary regression. The right-hand side of Equation 8.16 consists of a deterministic part in the first term, and a random part in the second term. That is, the next value of the time series x_{t+1} is given by the function of x_t in the first term, plus the random shock or innovation ε_{t+1}.

The time series of x is assumed to be stationary, so that its mean μ is the same for each interval of time. The data series also exhibits a variance, σ_x^2, the sample counterpart of which is just the ordinary sample variance computed from the values of the time series by squaring Equation 3.6. The ε's are mutually independent random quantities having mean $\mu_\varepsilon = 0$ and variance σ_ε^2. Very often it is further assumed that the ε's follow a Gaussian distribution.

As illustrated in Figure 8.5, the autoregressive model in Equation 8.16 can represent the serial correlation of a time series. This is a scatterplot of minimum temperatures at Canandaigua, New York, during January 1987, from Table A.1. Plotted on the horizontal axis are the first 30 data values, for 1–30 January. The corresponding temperatures for the following days, 2–31 January, are plotted on the vertical axis. The serial correlation, or persistence, is evident from the appearance of the point cloud, and from the positive slope of the regression line. Equation 8.16 can be viewed as a prediction equation for x_{t+1} using x_t as the predictor. Rearranging Equation 8.16 to more closely resemble the simple linear regression Equation 6.3 yields the intercept $a = \mu(1 - \phi)$, and slope $b = \phi$.

Another way to look at Equation 8.16 is as an algorithm for generating synthetic time series of values of x, in the same sense as Section 4.7. Beginning with an initial value, x_0, we would subtract the mean value (i.e., construct the corresponding anomaly), multiply by the autoregressive parameter ϕ, and then add a randomly generated variable ε_1 drawn from a Gaussian distribution (see Section 4.7.4) with mean zero and variance σ_ε^2. The first value of the time series, x_1, would then be produced by adding back the mean μ. The next time series value x_2, would then be produced in a similar way, by operating on

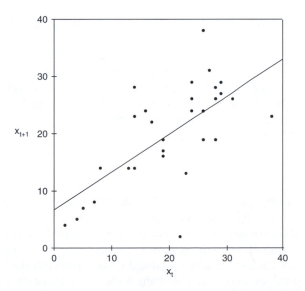

FIGURE 8.5 Scatterplot of January 1–30, 1987 minimum temperatures (°F) at Canandaigua, New York (X_t, horizontal) paired with minimum temperatures for the following days, January 2–31 (x_{t+1}, vertical). The data are from Table A.1. The regression line corresponding to the first term of the AR(1) time series model (Equation 8.16) is also shown.

x_1 and adding a new random Gaussian quantity ε_2. For positive values of the parameter ϕ, synthetic time series constructed in this way will exhibit positive serial correlation because each newly generated data value x_{t+1} includes some information carried forward from the preceding value x_t. Since x_t was in turn generated in part from x_{t-1}, and so on, members of the time series separated by more than one time unit will be correlated, although this correlation becomes progressively weaker as the time separation increases.

The first-order autoregression is sometimes called the Markov process, or Markov scheme. It shares with the first-order Markov chain the property that the full history of the time series prior to x_t provides no additional information regarding x_{t+1}, once x_t is known. This property can be expressed formally as

$$\Pr\{X_{t+1} \leq x_{t+1} | X_t \leq x_t, X_{t-1} \leq x_{t-1}, \ldots, X_1 \leq x_1\} = \Pr\{X_{t+1} \leq x_{t+1} | X_t \leq x_t\}. \quad (8.17)$$

Here the notation for continuous random variables has been used to express essentially the same idea as in Equation 8.1 for a series of discrete events. Again, Equation 8.17 does not imply that values of the time series separated by more than one time step are independent, but only that the influence of the prior history of the time series on its future values is fully contained in the current value x_t, regardless of the particular path by which the time series arrived at x_t.

Equation 8.16 is also sometimes known as a red noise process, because a positive value of the parameter ϕ averages or smoothes out short-term fluctuations in the serially independent series of innovations, ε, while affecting the slower random variations much less strongly. The resulting time series is called red noise by analogy to visible light depleted in the shorter wavelengths, which appears reddish. This topic will be discussed further in Section 8.5, but the effect can be appreciated by looking at Figure 5.4. This figure compares a series of uncorrelated Gaussian values, ε_t (panel a), with an autocorrelated series generated from them using Equation 8.16 and the value $\phi = 0.6$ (panel b). It is evident that the most erratic point-to-point variations in the uncorrelated series have been smoothed out, but the slower random variations are essentially preserved. In the time domain this smoothing is expressed as serial correlation. From a frequency perspective, the resulting series is "reddened."

Parameter estimation for the first-order autoregressive model is straightforward. The estimated mean of the time series, μ, is simply the usual sample average (Equation 3.2) of the data set, provided that the series can be considered to be stationary. Nonstationary series must first be dealt with in one of the ways sketched in Section 8.1.1.

The estimated autoregressive parameter is simply equal to the sample lag-1 autocorrelation coefficient, Equation 3.30:

$$\hat{\phi} = r_1. \quad (8.18)$$

For the resulting probability model to be stationary, it is required that $-1 < \phi < 1$. As a practical matter this presents no problem for the first-order autoregression, because the correlation coefficient also is bounded by the same limits. For most atmospheric time series the parameter ϕ will be positive, reflecting persistence. Negative values of ϕ are possible, but correspond to very jagged (anti-correlated) time series with a tendency for alternating values above and below the mean. Because of the Markovian property, the full (theoretical, or population) autocorrelation function for a time series governed by a first-order autoregressive process can be written in terms of the autoregressive parameter as

$$\rho_k = \phi^k. \quad (8.19)$$

Thus, the autocorrelation function for an AR(1) process decays exponentially from $\rho_0 = 1$, approaching zero as $k \to \infty$.

A series of truly independent data would have $\phi = 0$. However, a finite sample of independent data generally will exhibit a nonzero sample estimate of the autoregressive parameter. For a sufficiently long data series the sampling distribution of $\hat{\phi}$ is approximately Gaussian, with $\mu_{\hat{\phi}} = \hat{\phi}$ and variance $\sigma_{\hat{\phi}}^2 = (1 - \phi^2)/n$. Therefore, a test for the sample estimate of the autoregressive parameter, corresponding to Equation 5.3 with the null hypothesis that $\phi = 0$, can be carried out using the test statistic

$$z = \frac{\hat{\phi} - 0}{[\text{Var}(\hat{\phi})]^{1/2}} = \frac{\hat{\phi}}{[1/n]^{1/2}}, \tag{8.20}$$

because $\phi = 0$ under the null hypothesis. Statistical significance would be assessed approximately using standard Gaussian probabilities. This test is virtually identical to the t test for the slope of a regression line.

The final parameter of the statistical model in Equation 8.16 is the residual variance, or innovation variance, σ_ε^2. This quantity is sometimes also known as the white-noise variance, for reasons that are explained in Section 8.5. This parameter expresses the variability or uncertainty in the time series not accounted for by the serial correlation or, put another way, the uncertainty in x_{t+1} given that x_t is known. The brute-force approach to estimating σ_ε^2 is to estimate ϕ using Equation 8.18, compute the time series ε_{t+1} from the data using a rearrangement of Equation 8.16, and then to compute the ordinary sample variance of these ε values. Since the variance of the data is often computed as a matter of course, another way to estimate the white-noise variance is to use the relationship between the variances of the data series and the innovation series in the AR(1) model,

$$\sigma_\varepsilon^2 = (1 - \phi^2)\sigma_x^2. \tag{8.21}$$

Equation 8.21 implies $\sigma_\varepsilon^2 \leq \sigma_x^2$, with equality only for independent data, for which $\phi = 0$. Equation 8.21 implies that knowing the current value of an autocorrelated time series decreases uncertainty about the next value of the time series. In practical settings we work with sample estimates of the autoregressive parameter and of the variance of the data series, so that the corresponding sample estimate of the white-noise variance is

$$s_\varepsilon^2 = \frac{1 - \hat{\phi}^2}{n - 2} \sum_{t=1}^{n} (x_t - \bar{x})^2 = \frac{n - 1}{n - 2} (1 - \hat{\phi}^2) \, s_x^2. \tag{8.22}$$

The difference between Equations 8.22 and 8.21 is appreciable only if the data series is relatively short.

EXAMPLE 8.3   A First-Order Autoregression

Consider fitting an AR(1) process to the series of January 1987 minimum temperatures from Canandaigua, New York, in Table A.1. As indicated in the table, the average of these 31 values is $20.23°F$, and this would be adopted as the estimated mean of the time series, assuming stationarity. The sample lag-1 autocorrelation coefficient, from Equation 3.23, is $r_1 = 0.67$, and this value would be adopted as the estimated autoregressive parameter according to Equation 8.18.

The scatterplot of this data against itself lagged by one time unit in Figure 8.5 suggests the positive serial correlation typical of daily temperature data. A formal test of the estimated autoregressive parameter versus the null hypothesis that it is really zero would use the test statistic in Equation 8.20, $z = 0.67/[1/31]^{1/2} = 3.73$. This test provides extremely strong evidence that the observed nonzero sample autocorrelation did not arise by chance from a sequence of 31 independent values.

The sample standard deviation of the 31 Canandaigua minimum temperatures in Table A.1 is 8.81°F. Using Equation 8.22, the estimated white-noise variance for the fitted autoregression would be $s_\varepsilon^2 = (30/29)(1 - 0.67^2)(8.81^2) = 44.24$°F^2, corresponding to a standard deviation of 6.65°F. By comparison, the brute-force sample standard deviation of the series of sample residuals, each computed from the rearrangement of Equation 8.16 as $e_{t+1} = (x_{t+1} - \mu) - \phi(x_t - \mu)$ is 6.55°F.

The computations in this example have been conducted under the assumption that the time series being analyzed is stationary, which implies that the mean value does not change through time. This assumption is not exactly satisfied by this data, as illustrated in Figure 8.6. Here the time series of the January 1987 Canandaigua minimum temperature data is shown together with the climatological average temperatures for the period 1961–1990 (dashed line), and the linear trend fit to the 31 data points for 1987 (solid line).

Of course, the dashed line in Figure 8.6 is a better representation of the long-term (population) mean minimum temperatures at this location, and it indicates that early January is slightly warmer than late January on average. Strictly speaking, the data series is not stationary, since the underlying mean value for the time series is not constant through time. However, the change through the month represented by the dashed line is sufficiently minor (in comparison to the variability around this mean function) that generally we would be comfortable in pooling data from a collection of Januaries and assuming stationarity. In fact, the preceding results for the 1987 data sample are not very much different if the January 1987 mean minimum temperature of 20.23°F, or the long-term climatological temperatures represented by the dashed line, are assumed. In the latter case, we find $\phi = 0.64$, and $s_\varepsilon^2 = 6.49$°F.

Because the long-term climatological minimum temperature declines so slowly, it is clear that the rather steep negative slope of the solid line in Figure 8.6 results mainly from sampling variations in this short example data record. Normally an analysis of this kind would be carried out using a much longer time series. However, if no other information about the January minimum temperature climate of this location were available, it would be sensible to produce a stationary series before proceeding further, by subtracting the mean values represented by the solid line from the data points, provided the estimated

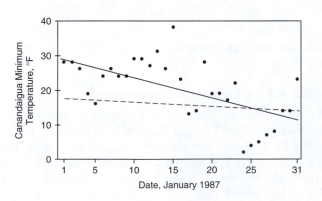

FIGURE 8.6   Time series of the January 1987 Canandaigua minimum temperature data. Solid line is the least-squares linear trend in the data, and the dashed line represents the climatological average minimum temperatures for the period 1961–1990.

slope is significantly different from zero (and accounting for the serial correlation in the data). The regression equation for this line is $\mu(t) = 29.6 - 0.584t$, where t is the date, and the slope is indeed significant. Hypothetically, the autoregressive process in Equation 8.16 would then be fit using the time series of the anomalies $x'_t = x_t - \mu(t)$. For example, $x'_1 = 28°F - (29.6 - 0.584) = -1.02°F$. Since the average residual from a least-squares regression line is zero (see Section 6.2.2), the mean of this series of anomalies x'_t will be zero. Fitting Equation 8.16 to this anomaly series yields $\hat{\phi} = 0.47$, and $s_\varepsilon^2 = 39.95°F^2$. ◊

8.3.2 Higher-Order Autoregressions

The first-order autoregression in Equation 8.16 generalizes readily to higher orders. That is, the regression equation predicting x_{t+1} can be expanded to include data values progressively further back in time as predictors. The general autoregressive model of order K, or AR(K) model is

$$x_{t+1} - \mu = \sum_{k=1}^{K} \phi_k (x_{t-k+1} - \mu) + \varepsilon_{t+1}. \tag{8.23}$$

Here the anomaly for the next time point, $x_{t+1} - \mu$, is a weighted sum of the previous K anomalies plus the random component ε_{t+1}, where the weights are the autoregressive coefficients ϕ_k. As before, the εs are mutually independent, with zero mean and variance σ_ε^2. Stationarity of the process implies that μ and σ_ε^2 do not change through time. For $K = 1$, Equation 8.23 is identical to Equation 8.16.

Estimation of the K autoregressive parameters ϕ_k is most easily done using the set of equations relating them to the autocorrelation function, which are known as the Yule-Walker equations. These are

$$\left.\begin{array}{rcl}
r_1 &=& \hat{\phi}_1 + \hat{\phi}_2 r_1 + \hat{\phi}_3 r_2 + \cdots + \hat{\phi}_K r_{K-1} \\
r_2 &=& \hat{\phi}_1 r_1 + \hat{\phi}_2 + \hat{\phi}_3 r_1 + \cdots + \hat{\phi}_K r_{K-2} \\
r_3 &=& \hat{\phi}_1 r_2 + \hat{\phi}_2 r_1 + \hat{\phi}_3 + \cdots + \hat{\phi}_K r_{K-3} \\
\vdots && \vdots \quad \vdots \quad \vdots \quad \cdots \quad \vdots \\
r_K &=& \hat{\phi}_1 r_{K-1} + \hat{\phi}_2 r_{K-2} + \hat{\phi}_3 r_{K-3} + \cdots + \hat{\phi}_K
\end{array}\right\}. \tag{8.24}$$

Here $\phi_k = 0$ for $k > K$. The Yule-Walker equations arise from Equation 8.23, by multiplying by x_{t-k}, applying the expected value operator, and evaluating the result for different values of k (e.g., Box and Jenkins 1994). These equations can be solved simultaneously for the ϕ_k. Alternatively, a method to use these equations recursively for parameter estimation—that is, to compute ϕ_1 and ϕ_2 to fit the AR(2) model knowing ϕ for the AR(1) model, and then to compute ϕ_1, ϕ_2, and ϕ_3 for the AR(3) model knowing ϕ_1 and ϕ_2 for the AR(2) model, and so on—is given in Box and Jenkins (1994) and Katz (1982). Constraints on the autoregressive parameters necessary for Equation 8.23 to be stationary are given in Box and Jenkins (1994).

The theoretical autocorrelation function corresponding to a particular set of the ϕ_ks can be determined by solving Equation 8.24 for the first K autocorrelations, and then applying

$$\rho_m = \sum_{k=1}^{K} \phi_k \rho_{m-k}. \tag{8.25}$$

Equation 8.25 holds for lags $m \geq k$, with the understanding that $\rho_0 \equiv 1$. Finally, the generalization of Equation 8.21 for the relationship between the white-noise variance and the variance of the data values themselves is

$$\sigma_\varepsilon^2 = \left(1 - \sum_{k=1}^{K} \phi_k \rho_k\right) \sigma_x^2. \tag{8.26}$$

8.3.3 The AR(2) Model

A common and important higher-order autoregressive model is the AR(2) process. It is reasonably simple, requiring the fitting of only two parameters in addition to the sample mean and variance of the series, yet it can describe a variety of qualitatively quite different behaviors of time series. The defining equation for AR(2) processes is

$$x_{t+1} - \mu = \phi_1(x_t - \mu) + \phi_2(x_{t-1} - \mu) + \varepsilon_{t+1}, \tag{8.27}$$

which is easily seen to be a special case of Equation 8.23. Using the first $K = 2$ of the Yule-Walker Equations (8.24),

$$r_1 = \hat{\phi}_1 + \hat{\phi}_2 r_1 \tag{8.28a}$$

$$r_2 = \hat{\phi}_1 r_1 + \hat{\phi}_2, \tag{8.28b}$$

the two autoregressive parameters can be estimated as

$$\hat{\phi}_1 = \frac{r_1(1 - r_2)}{1 - r_1^2} \tag{8.29a}$$

and

$$\hat{\phi}_2 = \frac{r_2 - r_1^2}{1 - r_1^2}. \tag{8.29b}$$

Here the estimation equations 8.29 have been obtained simply by solving Equations 8.28 for $\hat{\phi}_1$ and $\hat{\phi}_2$.

The white-noise variance for a fitted AR(2) model can be estimated in several ways. For very large samples, Equation 8.26 with $K = 2$ can be used with the sample variance of the time series, s_x^2. Alternatively, once the autoregressive parameters have been fit using Equations 8.29 or some other means, the corresponding estimated time series of the random innovations ε can be computed from a rearrangement of Equation 8.27 and their sample variance computed, as was done in Example 8.3 for the fitted AR(1) process. Another possibility is to use the recursive equation given by Katz (1982),

$$s_\varepsilon^2(m) = [1 - \hat{\phi}_m^2(m)]s_\varepsilon^2(m - 1). \tag{8.30}$$

Here the autoregressive models AR(1), AR(2), ... are fitted successively, $s_\varepsilon^2(m)$ is the estimated white-noise variance of the mth (i.e., current) autoregression, $s_\varepsilon^2(m - 1)$ is the estimated white-noise variance for the previously fitted (one order smaller) model, and $\hat{\phi}_m^2(m)$ is the estimated autoregressive parameter for the highest lag in the current model. For the AR(2) model, Equation 8.30 can be used with the expression for $s_\varepsilon^2(1)$ in Equation 8.22 to yield

$$s_\varepsilon^2(2) = \left(1 - \hat{\phi}_2^2\right) \frac{n-1}{n-2} \left(1 - r_1^2\right) s_x^2, \tag{8.31}$$

since $\hat{\phi} = r_1$ for the AR(1) model.

For an AR(2) process to be stationary, its two parameters must satisfy the constraints

$$\left. \begin{aligned} \phi_1 + \phi_2 &< 1 \\ \phi_2 - \phi_1 &< 1 \\ -1 < \phi_2 &< 1 \end{aligned} \right\}, \tag{8.32}$$

which define the triangular region in the (ϕ_1, ϕ_2) plane shown in Figure 8.7. Note that substituting $\phi_2 = 0$ into Equation 8.32 yields the stationarity condition $-1 < \phi_1 < 1$ applicable to the AR(1) model. Figure 8.7 includes AR(1) models as special cases on the horizontal $\phi_2 = 0$ line, for which that stationarity condition applies.

The first two values of the theoretical autocorrelation function for a particular AR(2) process can be obtained by solving Equations 8.28 as

$$\rho_1 = \frac{\phi_1}{1 - \phi_2} \tag{8.33a}$$

and

$$\rho_2 = \phi_2 + \frac{\phi_1^2}{1 - \phi_2}, \tag{8.33b}$$

and subsequent values of the autocorrelation function can be calculated using Equation 8.25. Figure 8.7 indicates that a wide range of types of autocorrelation functions, and thus a wide range of time correlation behaviors, can be represented by AR(2) processes.

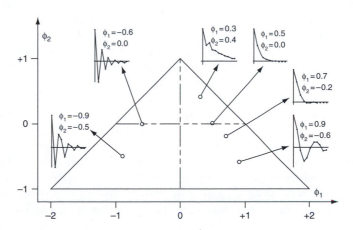

FIGURE 8.7 The allowable parameter space for stationary AR(2) processes, with insets showing autocorrelation functions for selected AR(2) models. The horizontal $\phi_2 = 0$ line locates the AR(1) models as special cases, and autocorrelation functions for two of these are shown. AR(2) models appropriate to atmospheric time series usually exhibit $\phi_1 > 0$.

First, AR(2) models include the simpler AR(1) models as special cases. Two AR(1) auto-correlation functions are shown in Figure 8.7. The autocorrelation function for the model with $\phi_1 = 0.5$ and $\phi_2 = 0.0$ decays exponentially toward zero, following Equation 8.19. Autocorrelation functions for many atmospheric time series exhibit this kind of behavior, at least approximately. The other AR(1) model for which an autocorrelation function is shown is for $\phi_1 = -0.6$ and $\phi_2 = 0.0$. Because of the negative lag-one autocorrelation, the autocorrelation function exhibits oscillations around zero that are progressively damped at longer lags (again, compare Equation 8.19). That is, there is a tendency for the anomalies of consecutive data values to have opposite signs, so that data separated by even numbers of lags are positively correlated. This kind of behavior rarely is seen in atmospheric data series, and most AR(2) models for atmospheric data have $\phi_1 > 0$.

The second autoregressive parameter allows many other kinds of behaviors in the autocorrelation function. For example, the autocorrelation function for the AR(2) model with $\phi_1 = 0.3$ and $\phi_2 = 0.4$ exhibits a larger correlation at two lags than at one lag. For $\phi_1 = 0.7$ and $\phi_2 = -0.2$ the autocorrelation function declines very quickly, and is almost zero for lags $k \geq 4$. The autocorrelation function for the AR(2) model with $\phi_1 = 0.9$ and $\phi_2 = -0.6$ is very interesting in that it exhibits a slow damped oscillation around zero. This characteristic reflects what are called *pseudoperiodicities* in the corresponding time series. That is, time series values separated by very few lags exhibit fairly strong positive correlation, those separated by a few more lags exhibit negative correlation, and values separated by a few more lags yet exhibit positive correlation again. The qualitative effect is for time series to exhibit oscillations around the mean resembling an irregular cosine curve with an average period that is approximately equal to the number of lags at the first positive hump in the autocorrelation function. Thus, AR(2) models can represent data that are approximately but not strictly periodic, such as barometric pressure variations resulting from the movement of midlatitude synoptic systems.

Some properties of autoregressive models are illustrated by the four example synthetic time series in Figure 8.8. Series (a) is simply a sequence of 50 independent Gaussian variates with $\mu = 0$. Series (b) is a realization of the AR(1) process generated using Equation 8.16 or, equivalently, Equation 8.27 with $\mu = 0$, $\phi_1 = 0.5$ and $\phi_2 = 0.0$. The apparent similarity between series (a) and (b) arises because series (a) has been used as the ε_{t+1} series forcing the autoregressive process in Equation 8.27. The effect of the parameter $\phi = \phi_1 > 0$ is to smooth out step-to-step variations in the white-noise series (a), and to give the resulting time series a bit of memory. The relationship of the series in these two panels is analogous to that in Figure 5.4, in which $\phi = 0.6$.

Series (c) in Figure 8.8 is a realization of the AR(2) process with $\mu = 0$, $\phi_1 = 0.9$ and $\phi_2 = -0.6$. It resembles qualitatively some atmospheric series (e.g., midlatitude sea-level pressures), but has been generated using Equation 8.27 with series (a) as the forcing white noise. This series exhibits pseudoperiodicities. That is, peaks and troughs in this time series tend to recur with a period near six or seven time intervals, but these are not so regular that a cosine function or the sum of a few cosine functions would represent them very well. This feature is the expression in the data series of the positive hump in the autocorrelation function for this autoregressive model shown in the inset of Figure 8.7, which occurs at a lag interval of six or seven time periods. Similarly, the peak-trough pairs tend to be separated by perhaps three or four time intervals, corresponding to the minimum in the autocorrelation function at these lags shown in the inset in Figure 8.7.

The autoregressive parameters $\phi_1 = 0.9$ and $\phi_2 = 0.11$ for series (d) in Figure 8.8 fall outside the triangular region in Figure 8.7 that define the limits of stationary AR(2) processes. The series is therefore not stationary, and the nonstationarity can be seen as a drifting of the mean value in the realization of this process shown in Figure 8.8.

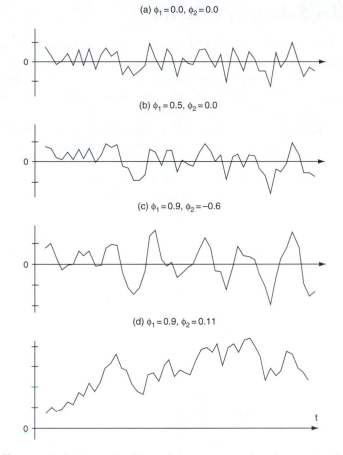

FIGURE 8.8   Four synthetic time series illustrating some properties of autoregressive models. Series (a) consists of independent Gaussian variables (white noise). Series (b) is a realization of the AR(1) process with $\phi_1 = 0.5$, and series (c) is a realization of the AR(2) process with $\phi_1 = 0.9$ and $\phi_2 = -0.6$, both of whose autocorrelation functions are shown in Figure 8.7. Series (d) is nonstationary because its parameters lie outside the triangle in Figure 8.7, and this nonstationarity can be seen as a drifting in the mean value. The series (b)–(d) were constructed using Equation 8.27 with $\mu = 0$ and the εs from series (a).

Finally, series (a) through (c) in Figure 8.8 illustrate the nature of the relationship between the variance of the time series, σ_x^2, and the white-noise variance, σ_ε^2, of an autoregressive process.

Series (a) consists simply of independent Gaussian variates, or white noise. Formally, it can be viewed as a special case of an autoregressive process, with all the ϕ_k's $= 0$. Using Equation 8.26 it is clear that $\sigma_x^2 = \sigma_\varepsilon^2$ for this series. Since series (b) and (c) were generated using series (a) as the white-noise forcing ε_{t+1}, σ_ε^2 for all three of these series are equal. Time series (c) gives the visual impression of being more variable than series (b), which in turn appears to exhibit more variability than series (a). Using Equations 8.33 with Equation 8.26 it is easy to compute that σ_x^2 for series (b) is 1.33 times larger than the common σ_ε^2, and for series (c) it is 2.29 times larger. The equations on which these computations are based pertain only to stationary autoregressive series, and so cannot be applied meaningfully to the nonstationary series (d).

8.3.4 Order Selection Criteria

The Yule-Walker equations (8.24) can be used to fit autoregressive models to essentially arbitrarily high order. At some point, however, expanding the complexity of the model will not appreciably improve its representation of the data. Arbitrarily adding more terms in Equation 8.23 will eventually result in the model being overfit, or excessively tuned to the data used for parameter estimation.

The BIC (Schwarz 1978) and AIC (Akaike 1974) statistics, applied to Markov chains in Section 8.2, are also often used to decide among potential orders of autoregressive models. Both statistics involve the log-likelihood plus a penalty for the number of parameters, with the two criteria differing only in the form of the penalty function. Here the likelihood function involves the estimated white-noise variance.

For each candidate order m, the order selection statistics

$$BIC(m) = n \ln\left[\frac{n}{n-m-1}s_\varepsilon^2(m)\right] + (m+1)\ln n, \qquad (8.34)$$

or

$$AIC(m) = n \ln\left[\frac{n}{n-m-1}s_\varepsilon^2(m)\right] + 2(m+1), \qquad (8.35)$$

are computed, using $s_\varepsilon^2(m)$ from Equation 8.30. Better fitting models will exhibit smaller white-noise variance, implying less residual uncertainty. Arbitrarily adding more parameters (fitting higher- and higher-order autoregressive models) will not increase the white-noise variance estimated from the data sample, but neither will the estimated white-noise variance decrease much if the extra parameters are not effective in describing the behavior of the data. Thus, the penalty functions serve to guard against overfitting. That order m is chosen as appropriate, which minimizes either Equation 8.34 or 8.35.

EXAMPLE 8.4   Order Selection among Autoregressive Models

Table 8.3 summarizes the results of fitting successively higher autoregressive models to the January 1987 Canandaigua minimum temperature data, assuming that they are stationary without removal of a trend. The second column shows the sample autocorrelation function up to seven lags. The estimated white-noise variance for autoregressions of orders one through seven, computed using the Yule-Walker equations and Equation 8.30, are shown in the third column. Notice that $s_\varepsilon^2(0)$ is simply the sample variance of the time series itself, or s_x^2. The estimated white-noise variances decrease progressively as more terms are added to Equation 8.23, but toward the bottom of the table adding yet more terms has little further effect.

The BIC and AIC statistics for each candidate autoregression are shown in the last two columns. Both indicate that the AR(1) model is most appropriate for this data, as this produces the minimum in both order selection statistics. Similar results also are obtained for the other three temperature series in Table A.1. Note, however, that with a larger sample size, higher-order autoregressions could be chosen by both criteria. For the estimated residual variances shown in Table 8.3, using the AIC statistic would lead to the choice of the AR(2) model for n greater than about 290, and the AR(2) model would minimize the BIC statistic for n larger than about 430. ◊

TABLE 8.3 Illustration of order selection for autoregressive models to represent the January 1987 Canandaigua minimum temperature series, assuming stationarity. Presented are the autocorrelation function for the first seven lags m, the estimated white-noise variance for each AR(m) model, and the BIC and AIC statistics for each trial order. For $m = 0$ the autocorrelation function is 1.00, and the white-noise variance is equal to the sample variance of the series. The AR(1) model is selected by both the BIC and AIC criteria.

Lag, m	r_m	$s_\varepsilon^2(m)$	BIC(m)	AIC(m)
0	1.000	77.58	138.32	136.89
1	0.672	42.55	125.20	122.34
2	0.507	42.11	129.41	125.11
3	0.397	42.04	133.91	128.18
4	0.432	39.72	136.76	129.59
5	0.198	34.39	136.94	128.34
6	0.183	33.03	140.39	130.35
7	0.161	33.02	145.14	133.66

8.3.5 *The Variance of a Time Average*

An important application of time series models in atmospheric data analysis is estimation of the sampling distribution of the average of a correlated time series. Recall that a sampling distribution characterizes the batch-to-batch variability of a statistic computed from a finite data sample. If the data values making up a sample average are independent, the variance of the sampling distribution of that average is given by the variance of the data, s_x^2, divided by the sample size (Equation 5.4).

Since atmospheric data are often positively correlated, using Equation 5.4 to calculate the variance of (the sampling distribution of) a time average leads to an underestimate. This discrepancy is a consequence of the tendency for nearby values of correlated time series to be similar, leading to less batch-to-batch consistency of the sample average. The phenomenon is illustrated in Figure 5.4. As discussed in Chapter 5, underestimating the variance of the sampling distribution of the mean can lead to serious problems for hypothesis tests, leading in general to unwarranted rejection of null hypotheses.

The effect of serial correlation on the variance of a time average over a sufficiently large sample can be accounted for through a variance inflation factor, V, modifying Equation 5.4:

$$\mathrm{Var}[\bar{x}] = \frac{V \sigma_x^2}{n}. \tag{8.36}$$

If the data series is uncorrelated, $V = 1$ and Equation 8.36 corresponds to Equation 5.4. If the data exhibit positive serial correlation, $V > 1$ and the variance of the time average is inflated above what would result from independent data. Note, however, that even if the underlying data are correlated, the mean of the sampling distribution of the time average is the same as the underlying mean of the data being averaged,

$$E[\bar{x}] = \mu_{\bar{x}} = E[x_t] = \mu_x. \tag{8.37}$$

For large sample size, the variance inflation factor depends on the autocorrelation function according to

$$V = 1 + 2 \sum_{k=1}^{\infty} \rho_k. \tag{8.38}$$

However, the variance inflation factor can be estimated with greater ease and precision if a data series is well represented by an autoregressive model. In terms of the parameters of an $AR(K)$ model, the large-sample variance inflation factor in Equation 8.38 is

$$V = \frac{1 - \sum_{k=1}^{K} \phi_k \rho_k}{\left[1 - \sum_{k=1}^{K} \phi_k \right]^2}. \tag{8.39}$$

Note that the theoretical autocorrelations ρ_k in Equation 8.39 can be expressed in terms of the autoregressive parameters by solving the Yule-Walker Equations (8.24) for the correlations. In the special case of an $AR(1)$ model being appropriate for a time series of interest, Equation 8.39 reduces to

$$V = \frac{1 + \phi_1}{1 - \phi_1}, \tag{8.40}$$

which was used to estimate the effective sample size in Equation 5.12, and the variance of the sampling distribution of a sample mean in Equation 5.13. Equations 8.39 and 8.40 are convenient large-sample approximations to the formula for the variance inflation factor based on sample autocorrelation estimates

$$V = 1 + 2 \sum_{k=1}^{n} \left(1 - \frac{k}{n} \right) r_k. \tag{8.41}$$

Equation 8.41 approaches Equations 8.39 and 8.40 for large sample size n, when the autocorrelations r_k are expressed in terms of the autoregressive parameters (Equation 8.24), but yields $V = 1$ for $n = 1$. Usually either Equation 8.39 or 8.40, as appropriate, would be used to compute the variance inflation factor.

EXAMPLE 8.5 Variances of Time Averages of Different Lengths

The relationship between the variance of a time average and the variance of the individual elements of a time series in Equation 8.36 can be useful in surprising ways. Consider, for example, the average winter (December–February) geopotential heights, and the standard deviations of those averages, for the northern hemisphere shown in Figures 8.9a and b, respectively. Figure 8.9a shows the average field (Equation 8.37), and Figure 8.9b shows the standard deviation of 90-day averages of winter 500 mb heights, representing the interannual variability. That is, Figure 8.9b shows the square root of Equation 8.36, with s_x^2 being the variance of the daily 500 mb height measurements and $n = 90$.

Suppose, however, that the sampling distribution of 500 mb heights averaged over a different length of time were needed. We might be interested in the variance of 10-day averages of 500 mb heights at selected locations, for use as a climatological

(a)

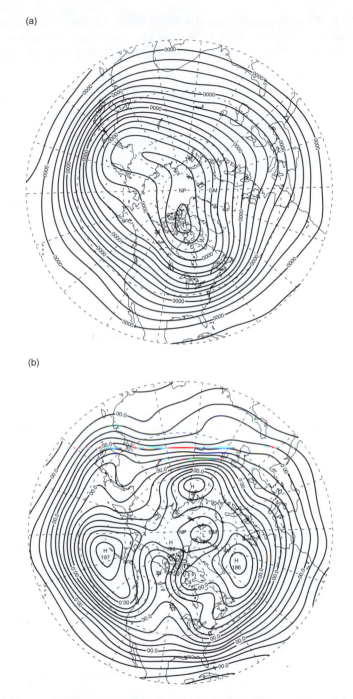

(b)

FIGURE 8.9   Average 500 mb height field for the northern hemisphere winter (a), and the field of standard deviations of that average, reflecting winter-to-winter variations (b). From Blackmon (1976).

reference for calculating the skill of forecasts of 10-day averages of this quantity, using Equation 7.32. (Note that the variance of the climatological distribution is exactly the mean-squared error of the climatological reference forecast.) Assuming that time series of winter 500 mb heights are stationary, the variance of an average over some different time period can be approximated without explicitly knowing the variance inflation factor in either Equations 8.38 or 8.39, and therefore without necessarily having the daily data. The ratio of the variances of 10-day and 90-day averages can be constructed from Equation 8.36,

$$\frac{\mathrm{Var}[\bar{x}_{10}]}{\mathrm{Var}[\bar{x}_{90}]} = \frac{Vs_x^2/10}{Vs_x^2/90}, \tag{8.42a}$$

leading to

$$\mathrm{Var}[\bar{x}_{10}] = \frac{90}{10}\mathrm{Var}[\bar{x}_{90}]. \tag{8.42b}$$

Regardless of the averaging period, the variance inflation factor V and the variance of the daily observations s_x^2 are the same because they are characteristics of the underlying daily time series. Thus, the variance of a 10-day average is approximately nine times larger than the variance of a 90-day average, and a map of hemispheric 10-day standard deviations of winter 500 mb heights would be qualitatively very similar to Figure 8.9b, but exhibiting magnitudes about $\sqrt{9} = 3$ times larger. ◊

8.3.6 Autoregressive-Moving Average Models

Autoregressive models actually constitute a subset of a broader class of time-domain models, known as autoregressive-moving average, or ARMA, models. The general ARMA(K, M) model has K autoregressive terms, as in the AR(K) process in Equation 8.23, and in addition contains M moving average terms that comprise a weighted average of the M previous values of the εs. The ARMA(K, M) model thus contains K autoregressive parameters ϕ_k and M moving average parameters θ_m that affect the time series according to

$$x_{t+1} - \mu = \sum_{k=1}^{K} \phi_k (x_{t-k+1} - \mu) + \varepsilon_{t+1} - \sum_{m=1}^{M} \theta_m \varepsilon_{t-m+1}. \tag{8.43}$$

The AR(K) process in Equation 8.23 is a special case of the ARMA(K, M) model in Equation 8.43, with all the $\theta_m = 0$. Similarly, a pure moving average process of order M, or MA(M) process, would be a special case of Equation 8.43, with all the $\phi_k = 0$.

Parameter estimation and derivation of the autocorrelation function for the general ARMA(K, M) process is more difficult than for the simpler AR(K) models. Parameter estimation methods are given in Box and Jenkins (1994), and many time-series computer packages will fit ARMA models. An important and common ARMA model is the ARMA(1,1) process,

$$x_{t+1} - \mu = \phi_1 (x_t - \mu) + \varepsilon_{t+1} - \theta_1 \varepsilon_t. \tag{8.44}$$

Computing parameter estimates even for this simple ARMA model is somewhat complicated, although Box and Jenkins (1994) present an easy graphical technique that allows estimation of ϕ_1 and θ_1 using the first two sample lag correlations r_1 and r_2. The autocorrelation function for the ARMA (1,1) process can be calculated from the parameters using

$$\rho_1 = \frac{(1 - \phi_1 \theta_1)(\phi_1 - \theta_1)}{1 + \theta_1^2 - 2\phi_1 \theta_1} \tag{8.45a}$$

and

$$\rho_k = \phi_1 \rho_{k-1}, \quad k > 1. \tag{8.45b}$$

The autocorrelation function of an ARMA(1, 1) process decays exponentially from its value at ρ_1, which depends on both ϕ_1 and θ_1. This differs from the autocorrelation function for an AR(1) process, which decays exponentially from $\rho_0 = 1$. The relationship between the time-series variance and the white-noise variance of an ARMA(1,1) process is

$$\sigma_\varepsilon^2 = \frac{1 - \phi_1^2}{1 + \theta_1^2 + 2\phi_1\theta_1} \sigma_x^2. \tag{8.46}$$

Equations 8.45 and 8.46 also apply to the simpler AR(1) and MA(1) processes, for which $\theta_1 = 0$ or $\phi_1 = 0$, respectively.

8.3.7 Simulation and Forecasting with Continuous Time-Domain Models

An important application of time-domain models is in the simulation of synthetic (i.e., random-number, as in Section 4.7) series having statistical characteristics that are similar to observed atmospheric data. These Monte-Carlo simulations are useful for investigating impacts of atmospheric variability in situations where the record length of the observed data is known or suspected to be insufficient to include representative sequences of the relevant variable(s). Here it is necessary to choose the type and order of time-series model carefully, so that the simulated time series will represent the variability of the real atmosphere well.

Once an appropriate time series model has been identified and its parameters estimated, its defining equation can be used as an algorithm to generate synthetic time series. For example, if an AR(2) model is representative of the data, Equation 8.27 would be used, whereas Equation 8.44 would be used as the generation algorithm for ARMA(1,1) models. The simulation method is similar to that described earlier for sequences of binary variables generated using the Markov chain model. Here, however, the noise or innovation series, ε_{t+1}, usually is assumed to consist of independent Gaussian variates with $\mu_\varepsilon = 0$ and variance σ_ε^2, which is estimated from the data as described earlier.

At each time step, a new Gaussian ε_{t+1} is chosen (see Section 4.7.4) and substituted into the defining equation. The next value of the synthetic time series x_{t+1} is then computed using the previous K values of x (for AR models), the previous M values of ε (for MA models), or both (for ARMA models). The only real difficulty in implementing the process is at the beginning of each synthetic series, where there are no prior values of x and/or ε that can be used. A simple solution to this problem is to substitute the corresponding averages (expected values) for the unknown previous values. That is, we assume $(x_t - \mu) = 0$ and $\varepsilon_t = 0$ for $t \leq 0$.

A better procedure is to generate the first values in a way that is consistent with the structure of the time-series model. For example, with an AR(1) model we could choose x_0 from a Gaussian distribution with variance $\sigma_x^2 = \sigma_\varepsilon^2/(1 - \phi^2)$ (cf. Equation 8.21). Another very workable solution is to begin with $(x_t - \mu) = 0$ and $\varepsilon_t = 0$, but generate a longer time series than needed. The first few members of the resulting time series, which are most influenced by the initial values, are then discarded.

EXAMPLE 8.6   Statistical Simulation with an Autoregressive Model

The time series in Figures 8.8b–d were produced according to the procedure just described, using the independent Gaussian series in Figure 8.8a as the series of εs. The first and last few values of this independent series, and of the two series plotted in Figures 8.8b and c, are given in Table 8.4. For all three series, $\mu = 0$ and $\sigma_\varepsilon^2 = 1$. Equation 8.16 has been used to generate the values of the AR(1) series, with $\phi_1 = 0.5$, and Equation 8.27 was used to generate the AR(2) series, with $\phi_1 = 0.9$ and $\phi_2 = -0.6$.

Consider the more difficult case of generating the AR(2) series. Calculating x_1 and x_2 in order to begin the series presents an immediate problem, because x_0 and x_{-1} do not exist. This simulation was initialized by assuming the expected values $E[x_0] = E[x_{-1}] = \mu = 0$. Thus, since $\mu = 0$, $x_1 = \phi_1 x_0 + \phi_2 x_{-1} + \varepsilon_1 = (0.9)(0) - (0.6)(0) + 1.526 = 1.562$. Having generated x_1 in this way, it is then used to obtain $x_2 = \phi_1 x_1 + \phi_2 x_0 + \varepsilon_2 = (0.9)(1.526) - (0.6)(0) + 0.623 = 1.996$. For values of the AR(2) series at times $t = 3$ and larger, the computation is a straightforward application of Equation 8.27. For example, $x_3 = \phi_1 x_2 + \phi_2 x_1 + \varepsilon_3 = (0.9)(1.996) - (0.6)(1.526) - 0.272 = 0.609$. Similarly, $x_4 = \phi_1 x_3 + \phi_2 x_2 + \varepsilon_4 = (0.9)(0.609) - (0.6)(1.996) + 0.092 = -0.558$. If this synthetic series to be used as part of a larger simulation, the first portion would generally be discarded, so that the retained values would have negligible memory of the initial condition $x_{-1} = x_0 = 0$. ◊

Purely statistical forecasts of the future evolution of time series can be produced using time-domain models. These are accomplished by simply extrapolating the most recently observed value(s) into the future using the defining equation for the appropriate model, on the basis of parameter estimates fitted from the previous history of the series. Since the future values of the εs cannot be known, the extrapolations are made using their expected values; that is, $E[\varepsilon] = 0$. Probability bounds on these extrapolations can be calculated as well.

The nature of this kind of forecast is most easily appreciated for the AR(1) model, the defining equation for which is Equation 8.16. Assume that the mean μ and the autoregressive parameter ϕ have been estimated from a time series of observations, the most recent of which is x_t. A nonprobabilistic forecast for x_{t+1} could be made by setting the unknown future ε_{t+1} to zero, and rearranging Equation 8.16 to yield $x_{t+1} = \mu + \phi(x_t - \mu)$. Note that, in common with the forecasting of a binary time series using a Markov chain model, this forecast is a compromise between persistence ($x_{t+1} = x_t$, which would result

TABLE 8.4   Values of the time series plotted in Figure 8.8a–c. The AR(1) and AR(2) series have been generated from the independent Gaussian series using Equations 8.16 and 8.27, respectively, as the algorithm.

t	Independent Gaussian Series, ε_t, (Figure 8.8a)	AR(1) Series, X_t, (Figure 8.8b)	AR(2) Series, X_t, (Figure 8.8c)
1	1.526	1.526	1.526
2	0.623	1.387	1.996
3	−0.272	0.421	0.609
4	0.092	0.302	−0.558
5	0.823	0.974	−0.045
⋮	⋮	⋮	⋮
49	−0.505	−1.073	−3.172
50	−0.927	−1.463	−2.648

if $\phi = 1$) and climatology ($x_{t+1} = \mu$, which would result if $\phi = 0$). Further projections into the future would be obtained by extrapolating the previously forecast values, e.g., $x_{t+2} = \mu + \phi(x_{t+1} - \mu)$, and $x_{t+3} = \mu + \phi(x_{t+2} - \mu)$. For the AR(1) model and $\phi > 0$, this series of forecasts would exponentially approach $x_\infty = \mu$.

The same procedure is used for higher order autoregressions, except that the most recent K values of the time series are needed to extrapolate an AR(K) process (Equation 8.23). Forecasts derived from an AR(2) model, for example, would be made using the previous two observations of the time series, or $x_{t+1} = \mu + \phi_1(x_t - \mu) + \phi_2(x_{t-1} - \mu)$. Forecasts using ARMA models are only slightly more difficult, requiring that the last M values of the ε series be back-calculated before the projections begin.

Forecasts made using time-series models are of course uncertain, and the forecast uncertainty increases for longer projections into the future. This uncertainty also depends on the nature of the appropriate time-series model (e.g., the order of an autoregression and its parameter values) and on the intrinsic uncertainty in the random noise series that is quantified by the white-noise variance σ_ε^2. The variance of a forecast made only one time step into the future is simply equal to the white-noise variance. Assuming the εs follow a Gaussian distribution, a 95% confidence interval on a forecast of x_{t+1} is then approximately $x_{t+1} \pm 2\sigma_\varepsilon$. For very long extrapolations, the variance of the forecasts approaches the variance of the time series itself, σ_x^2, which for AR models can be computed from the white-noise variance using Equation 8.26.

For intermediate time projections, calculation of forecast uncertainty is more complicated. For a forecast j time units into the future, the variance of the forecast is given by

$$\sigma^2(x_{t+j}) = \sigma_\varepsilon^2 \left[1 + \sum_{i=1}^{j-1} \psi_i^2 \right]. \tag{8.47}$$

Here the weights ψ_i depend on the parameters of the time series model, so that Equation 8.47 indicates the variance of the forecast increases with both the white-noise variance and the projection, and that the increase in uncertainty at increasing lead time depends on the specific nature of the time-series model. For the $j = 1$ time step forecast, there are no terms in the summation in Equation 8.47, and the forecast variance is equal to the white-noise variance, as noted earlier.

For AR(1) models, the ψ weights are simply

$$\psi_i = \phi^i, \quad i > 0, \tag{8.48}$$

so that, for example, $\psi_1 = \phi$, $\psi_2 = \phi^2$, and so on. More generally, for AR(K) models, the ψ weights are computed recursively, using

$$\psi_i = \sum_{k=1}^{K} \phi_k \psi_{i-k}, \tag{8.49}$$

where it is understood that $\psi_0 = 1$ and $\psi_i = 0$ for $i < 0$. For AR(2) models, for example, $\psi_1 = \phi_1$, $\psi_2 = \phi_1^2 + \phi_2$, $\psi_3 = \phi_1(\phi_1^2 + \phi_2) + \phi_2\phi_1$, and so on. Equations that can be used to compute the ψ weights for MA and ARMA models are given in Box and Jenkins (1994).

EXAMPLE 8.7 Forecasting with an Autoregressive Model

Figure 8.10 illustrates forecasts using the AR(2) model with $\phi_1 = 0.9$ and $\phi_2 = -0.6$. The first six points in the time series, shown by the circles connected by heavy lines, are

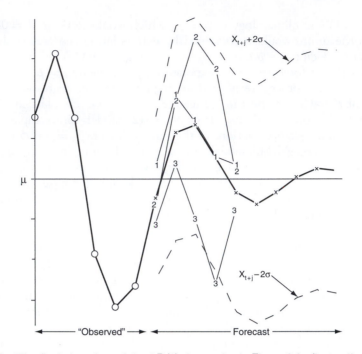

FIGURE 8.10   The final six points of the AR(2) time series in Figure 8.8c (heavy line, with circles), and its forecast evolution (heavy line, with xs) extrapolated using Equation 8.27 with all the $\varepsilon = 0$. The $\pm 2\sigma$ limits describing the uncertainty of the forecast values are shown with dashed lines. These standard deviations depend on the forecast lead time. For the 1-step ahead forecast, the width of the confidence interval is a function simply of the white-noise variance, $\pm 2\sigma_\varepsilon$. For very long lead times, the forecasts converge to the mean, μ, of the process, and the width of the confidence interval increases to $\pm 2\sigma_x$. Three example realizations of the first five points of the future evolution of the time series, computed using Equation 8.27 and particular random ε values, are also shown (thin lines connecting numbered points).

the same as the final six points in the time series shown in Figure 8.8c. The extrapolation of this series into the future, using Equation 8.27 with all $\varepsilon_{t+1} = 0$, is shown by the continuation of the heavy line connecting the xs. Note that only the final two observed values are used to extrapolate the series. The forecast series continues to show the pseudoperiodic behavior characteristic of this AR(2) model, but its oscillations damp out at longer projections as the forecast series approaches the mean μ.

The approximate 95% confidence intervals for the forecast time series values, given by $\pm 2\sigma(x_{t+j})$ as computed from Equation 8.47 are shown by the dashed lines. For the particular values of the autoregressive parameters $\phi_1 = 0.9$ and $\phi_2 = -0.6$, Equation 8.49 yields $\psi_1 = 0.90$, $\psi_2 = 0.21$, $\psi_3 = -0.35$, $\psi_4 = -0.44$, and so on. Note that the confidence band follows the oscillations of the forecast series, and broadens from $\pm 2\sigma_\varepsilon$ at a projection of one time unit to nearly $\pm 2\sigma_x$ at the longer projections.

Finally, Figure 8.10 shows the relationship between the forecast time series, and the first five points of three realizations of this AR(2) process, shown by the thin lines connecting points labeled 1, 2, and 3. Each of these three series was computed using Equation 8.27, starting from $x_t = -2.648$ and $x_{t-1} = -3.172$, but using different sequences of independent Gaussian εs. For the first two or three projections these remain reasonably close to the forecasts. Subsequently the three series begin to diverge as the influence of the final two points from Figure 8.8c diminishes and the accumulated influence of the new (and different) random εs increases. For clarity these series have not been plotted more than five time units into the future, although doing so would have shown each to oscillate irregularly, with progressively less relationship to the forecast series. $\Diamond$

8.4 Frequency Domain—I. Harmonic Analysis

Analysis in the frequency domain involves representing data series in terms of contributions made at different time scales. For example, a time series of hourly temperature data from a midlatitude location usually will exhibit strong variations both at the daily time scale (corresponding to the diurnal cycle of solar heating) and at the annual time scale (reflecting the march of the seasons). In the time domain, these cycles would appear as large positive values in the autocorrelation function for lags at and near 24 hours for the diurnal cycle, and $24 \times 365 = 8760$ hours for the annual cycle. Thinking about the same time series in the frequency domain, we speak of large contributions to the total variability of the time series at periods of 24 and 8760 hours, or at frequencies of $1/24 = 0.0417\,\text{h}^{-1}$ and $1/8760 = 0.000114\,\text{h}^{-1}$.

Harmonic analysis consists of representing the fluctuations or variations in a time series as having arisen from the adding together of a series of sine and cosine functions. These trigonometric functions are harmonic in the sense that they are chosen to have frequencies exhibiting integer multiples of the fundamental frequency determined by the sample size (i.e., length) of the data series. A common physical analogy is the musical sound produced by a vibrating string, where the pitch is determined by the fundamental frequency, but the aesthetic quality of the sound depends also on the relative contributions of the higher harmonics.

8.4.1 Cosine and Sine Functions

It is worthwhile to review briefly the nature of the cosine function $\cos(\alpha)$, and the sine function $\sin(\alpha)$. The argument in both is a quantity α, measured in angular units, which can be either degrees or radians. Figure 8.11 shows portions of the cosine (solid) and sine (dashed) functions, on the angular interval 0 to $5\pi/2$ radians (0° to 450°).

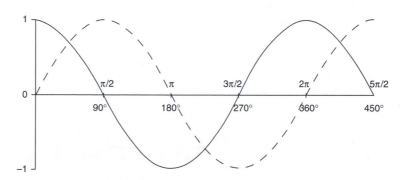

FIGURE 8.11   Portions of the cosine (solid) and sine (dashed) functions on the interval 0° to 450° or, equivalently, 0 to $5\pi/2$ radians. Each executes a full cycle every 360°, or 2π radians, and extends left to $-\infty$ and right to $+\infty$.

The cosine and sine functions extend through indefinitely large negative and positive angles. The same wave pattern repeats every 2π radians or $360°$, so that

$$\cos(2\pi k + \alpha) = \cos(\alpha), \tag{8.50}$$

where k is any integer. An analogous equation holds for the sine function. That is, both cosine and sine functions are periodic. Both functions oscillate around their average value of zero, and attain maximum values of $+1$ and minimum values of -1. The cosine function is maximized at $0°, 360°$, and so on, and the sine function is maximized at $90°, 450°$, and so on.

These two functions have exactly the same shape but are offset from each other by $90°$. Sliding the cosine function to the right by $90°$ produces the sine function, and sliding the sine function to the left by $90°$ produces the cosine function. That is,

$$\cos\left(\alpha - \frac{\pi}{2}\right) = \sin(\alpha) \tag{8.51a}$$

and

$$\sin\left(\alpha + \frac{\pi}{2}\right) = \cos(\alpha). \tag{8.51b}$$

8.4.2 *Representing a Simple Time Series with a Harmonic Function*

Even in the simple situation of time series having a sinusoidal character and executing a single cycle over the course of n observations, three small difficulties must be overcome in order to use a sine or cosine function to represent it. These are:

1) The argument that a trigonometric function is an angle, whereas the data series is a function of time.
2) Cosine and sine functions fluctuate between $+1$ and -1, but the data will generally fluctuate between different limits.
3) The cosine function is at its maximum value for $\alpha = 0$ and $\alpha = 2\pi$, and the sine function is at its mean value for $\alpha = 0$ and $\alpha = 2\pi$. Both the sine and cosine may thus be positioned arbitrarily in the horizontal with respect to the data.

The solution to the first problem comes through regarding the length of the data record, n, as constituting a full cycle, or the fundamental period. Since the full cycle corresponds to $360°$ or 2π radians in angular measure, it is easy to proportionally rescale time to angular measure, using

$$\alpha = \left(\frac{360°}{\text{cycle}}\right)\left(\frac{t \text{ time units}}{n \text{ time units/cycle}}\right) = \frac{t}{n}360° \tag{8.52a}$$

or

$$\alpha = \left(\frac{2\pi}{\text{cycle}}\right)\left(\frac{t \text{ time units}}{n \text{ time units/cycle}}\right) = 2\pi\frac{t}{n}. \tag{8.52b}$$

These equations can be viewed as specifying the angle that subtends proportionally the same part of the distance between 0 and 2π, as the point t is located in time between 0 and n. The quantity

$$\omega_1 = \frac{2\pi}{n} \tag{8.53}$$

is called the fundamental frequency. This quantity is an angular frequency, having physical dimensions of radians per unit time. The fundamental frequency specifies the fraction of the full cycle, spanning n time units, that is executed during a single time unit. The subscript "1" indicates that ω_1 pertains to the wave that executes one full cycle over the whole data series.

The second problem is overcome by shifting a cosine or sine function up or down to the general level of the data, and then stretching or compressing it vertically until its range corresponds to that of the data. Since the mean of a pure cosine or sine wave is zero, simply adding the mean value of the data series to the cosine function assures that it will fluctuate around that mean value. The stretching or shrinking is accomplished by multiplying the cosine function by a constant, C_1, known as the amplitude. Again, the subscript 1 indicates that this is the amplitude of the fundamental harmonic. Since the maximum and minimum values of a cosine function are ± 1, the maximum and minimum values of the function $C_1 \cos(\alpha)$ will be $\pm C_1$. Combining the solutions to these first two problems for a data series (call it y) yields

$$y_t = \bar{y} + C_1 \cos\left(\frac{2\pi t}{n}\right). \tag{8.54}$$

This function is plotted as the lighter curve in Figure 8.12. In this figure the horizontal axis indicates the equivalence of angular and time measure, through Equation 8.52, and the vertical shifting and stretching has produced a function fluctuating around the mean, with a range of $\pm C_1$.

Finally, it is usually necessary to shift a harmonic function laterally in order to have it match the peaks and troughs of a data series. This time-shifting is most conveniently

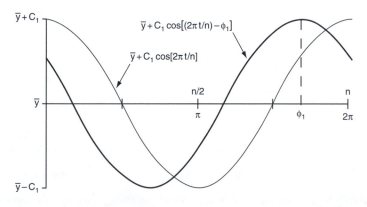

FIGURE 8.12   Transformation of a simple cosine function defined on 0 to 2π radians to a function representing a data series on the interval 0 to n time units. After changing from time to angular units, multiplying the cosine function by the amplitude C_1 stretches it so that it fluctuates through a range of $2C_1$. Adding the mean of the time series then shifts it to the proper vertical level, producing the lighter curve. The function can then be shifted laterally by subtracting the phase angle ϕ_1 that corresponds to the time of the maximum in the data series (heavier curve).

accomplished when the cosine function is used, because its maximum value is achieved when the angle on which it operates is zero. Shifting the cosine function to the right by the angle ϕ_1 results in a new function that is maximized at $\omega_1 t = 2\pi t/n = \phi_1$,

$$y_t = \bar{y} + C_1 \cos\left(\frac{2\pi t}{n} - \phi_1\right). \tag{8.55}$$

The angle ϕ_1 is called the phase angle, or phase shift. Shifting the cosine function to the right by this amount requires *subtracting* ϕ_1, so that the argument of the cosine function is zero when $(2\pi t/n) = \phi_1$. Notice that by using Equation 8.51 it would be possible to rewrite Equation 8.55 using the sine function. However, the cosine usually is used as in Equation 8.55, because the phase angle can then be easily interpreted as corresponding to the time of the maximum of the harmonic function. That is, the function in Equation 8.55 is maximized at time $t = \phi_1 n/2\pi$.

EXAMPLE 8.8   Transforming a Cosine Wave to Represent an Annual Cycle

Figure 8.13 illustrates the foregoing procedure using the 12 mean monthly temperatures (°F) for 1943–1989 at Ithaca, New York. Figure 8.13a is simply a plot of the 12 data points, with $t = 1$ indicating January, $t = 2$ indicating February, and so on. The overall annual average temperature of 46.1°F is indicated by the dashed horizontal line. These data appear to be at least approximately sinusoidal, executing a single full cycle over the course of the 12 months. The warmest mean temperature is 68.8°F in July and the coldest is 22.2°F in January.

The light curve at the bottom of Figure 8.13b is simply a cosine function with the argument transformed so that it executes one full cycle in 12 months. It is obviously a poor representation of the data. The dashed curve in Figure 8.13b shows this function lifted to the level of the average annual temperature, and stretched so that its range is

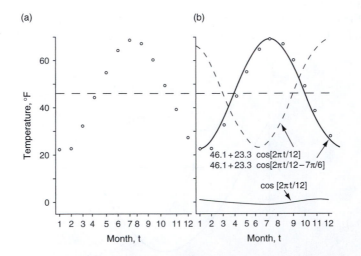

FIGURE 8.13   Illustration of the approximate matching of a cosine function to a data series. (a) Average monthly temperatures (°F) for Ithaca, New York for the years 1943–1989 (the data values are given in Table 8.5). The annual cycle of average temperature is evidently approximately sinusoidal. (b) Three cosine functions illustrating transformation from time to angular measure (light line at bottom), vertical positioning and stretching (dashed line), and lateral shifting (heavy line) yielding finally the function matching the data approximately. The horizontal dashed lines indicate the average of the 12 data points, 4.61°F.

similar to that of the data series (Equation 8.54). The stretching has been done only approximately, by choosing the amplitude C_1 to be half the difference between the July and January temperatures.

Finally, the cosine curve needs to be shifted to the right to line up well with the data. The maximum in the curve can be made to occur at $t = 7$ months (July) by introducing the phase shift, using Equation 8.52, of $\phi_1 = (7)(2\pi)/12 = 7\pi/6$. The result is the heavy curve in Figure 8.13, which is of the form of Equation 8.55. This function lines up with the data points, albeit somewhat roughly. The correspondence between the curve and the data can be improved by using better estimators for the amplitude and phase of the cosine wave. ◊

8.4.3 Estimation of the Amplitude and Phase of a Single Harmonic

The heavy curve in Figure 8.13 represents the associated temperature data reasonably well, but the correspondence will be improved if better choices for C_1 and ϕ_1 can be found. The easiest way to do this is to use the trigonometric identity

$$\cos(\alpha - \phi_1) = \cos(\phi_1)\cos(\alpha) + \sin(\phi_1)\sin(\alpha). \tag{8.56}$$

Substituting $\alpha_1 = 2\pi t/n$ from Equation 8.52 and multiplying both sides by the amplitude C_1 yields

$$C_1 \cos\left(\frac{2\pi t}{n} - \phi_1\right) = C_1 \cos(\phi_1)\cos\left(\frac{2\pi t}{n}\right) + C_1 \sin(\phi_1)\sin\left(\frac{2\pi t}{n}\right)$$

$$= A_1 \cos\left(\frac{2\pi t}{n}\right) + B_1 \sin\left(\frac{2\pi t}{n}\right), \tag{8.57}$$

where

$$A_1 = C_1 \cos(\phi_1) \tag{8.58a}$$

and

$$B_1 = C_1 \sin(\phi_1). \tag{8.58b}$$

Equation 8.57 says that it is mathematically equivalent to represent a harmonic wave either as a cosine function with amplitude C_1 and phase ϕ_1, or as the sum of an unshifted cosine and unshifted sine wave with amplitudes A_1 and B_1.

For the purpose of estimating one or the other of these pairs of parameters from a set of data, the advantage of representing the wave using the second line of Equation 8.57 rather than Equation 8.55 derives from the fact that the former is a linear function of the parameters. Notice that making the variable transformations $x_1 = \cos(2\pi t/n)$ and $x_2 = \sin(2\pi t/n)$, and substituting these into the second line of Equation 8.57, produces what looks like a two-predictor regression equation with $A_1 = b_1$ and $B_1 = b_2$. In fact, given a data series y_t we can use this transformation together with ordinary regression software to find least-squares estimates of the parameters A_1 and B_1, with y_t as the predictand. Furthermore, the regression package will also produce the average of the predictand values as the intercept, b_0. Subsequently, the operationally more convenient form of Equation 8.55 can be recovered by inverting Equations 8.58 to yield

$$C_1 = [A_1^2 + B_1^2]^{1/2} \tag{8.59a}$$

and

$$\phi_1 = \begin{cases} \tan^{-1}(B_1/A_1), & A_1 > 0 \\ \tan^{-1}(B_1/A_1) \pm \pi, & \text{or } \pm 180°, \quad A_1 < 0 \\ \pi/2, & \text{or } 90°, \quad A_1 = 0. \end{cases} \tag{8.59b}$$

Notice that since the trigonometric functions are periodic, effectively the same phase angle is produced by adding or subtracting a half-circle of angular measure if $A_1 < 0$. The alternative that produces $0 < \phi_1 < 2\pi$ is usually selected.

Finding the parameters A_1 and B_1 in Equation 8.57 using least-squares regression will work in the general case. For the special (although not too unusual) situation where the data values are equally spaced in time with no missing values, the properties of the sine and cosine functions allow the same least-squares parameter values to be obtained more easily and efficiently using

$$A_1 = \frac{2}{n} \sum_{t=1}^{n} y_t \cos\left(\frac{2\pi t}{n}\right) \tag{8.60a}$$

and

$$B_1 = \frac{2}{n} \sum_{t=1}^{n} y_t \sin\left(\frac{2\pi t}{n}\right). \tag{8.60b}$$

EXAMPLE 8.9   Harmonic Analysis of Average Monthly Temperatures

Table 8.5 shows the calculations necessary to obtain least-squares estimates for the parameters of the annual harmonic representing the Ithaca mean monthly temperatures plotted in Figure 8.13a, using Equations 8.60. The temperature data are shown in the column labeled y_t, and their average is easily computed as $552.9/12 = 46.1°F$. The $n = 12$ terms of the sums in Equations 8.60a and b are shown in the last two columns. Applying Equations 8.60 to these yields $A_1 = (2/12)(-110.329) = -18.39$, and $B_1 = (2/12)(-86.417) = -14.40$.

Equation 8.59 transforms these two amplitudes to the parameters of the amplitude-phase form of Equation 8.55. This transformation allows easier comparison to the heavy curve plotted in Figure 8.13b. The amplitude is $C_1 = [-18.39^2 - 14.40^2]^{1/2} = 23.36°F$, and the phase angle is $\phi_1 = \tan^{-1}(-14.40/-18.39) + 180° = 218°$. Here 180° has been added rather than subtracted, so that $0° < \phi_1 < 360°$. The least-squares amplitude of $C_1 = 23.36°F$ is quite close to the one used to draw Figure 8.13b, and the phase angle is 8° greater than the $(7)(360°)/12 = 210°$ angle that was eyeballed on the basis of the July mean being the warmest of the 12 months. The value of $\phi_1 = 218°$ is a better estimate, and implies a somewhat later (than mid-July) date for the time of the climatologically warmest temperature at this location. In fact, since there are very nearly as many degrees in a full cycle as there are days in one year, the results from Table 8.5 indicate that the heavy curve in Figure 8.13b should be shifted to the right by about one week. It is apparent that the result would be an improved correspondence with the data points. ◊

TABLE 8.5 Illustration of the mechanics of using Equations 8.60 to estimate the parameters of a fundamental harmonic. The data series y_t are the mean monthly temperatures at Ithaca for month t plotted in Figure 8.13a. Each of the 12 terms in Equations 8.60a and b, respectively, are shown in the last two columns.

t	y_t	$\cos(2\pi t/12)$	$\sin(2\pi t/12)$	$y_t \cos(2\pi t/12)$	$yt \sin(2\pi t/12)$
1	22.2	0.866	0.500	19.225	11.100
2	22.7	0.500	0.866	11.350	19.658
3	32.2	0.000	1.000	0.000	32.200
4	44.4	−0.500	0.866	−22.200	38.450
5	54.8	−0.866	0.500	−47.457	27.400
6	64.3	−1.000	0.000	−64.300	0.000
7	68.8	−0.866	−0.500	−59.581	−34.400
8	67.1	−0.500	−0.866	−33.550	−58.109
9	60.2	0.000	−1.000	0.000	−60.200
10	49.5	0.500	−0.866	24.750	−42.867
11	39.3	0.866	−0.500	34.034	−19.650
12	27.4	1.000	0.000	27.400	0.000
Sums:	552.9	0.000	0.000	−110.329	−86.417

EXAMPLE 8.10   Interpolation of the Annual Cycle to Average Daily Values

The calculations in Example 8.9 result in a smoothly varying representation of the annual cycle of mean temperature at Ithaca, based on the monthly values. Particularly if this were a location for which daily data were not available, it might be valuable to be able to use a function like this to represent the climatological average temperatures on a day-by-day basis. In order to employ the cosine curve in Equation 8.55 with time t in days, it would be necessary to use $n = 365$ days rather than $n = 12$ months. The amplitude can be left unchanged, although Epstein (1991) suggests a method to adjust this parameter that will produce a somewhat better representation of the annual cycle of daily values. In any case, however, it is necessary to make an adjustment to the phase angle.

Consider that the time $t = 1$ month represents all of January, and thus might be reasonably assigned to the middle of the month, perhaps the 15th. Thus, the $t = 0$ months point of this function corresponds to the middle of December. Therefore, simply substituting the Julian date (1 January = 1, 2 January = 2, ..., 1 February = 32, etc.) for the time variable will result in a curve that is shifted too far left by about two weeks. What is required is a new phase angle, say ϕ_1', consistent with a time variable t' in days, that will position the cosine function correctly.

On 15 December, the two time variables are $t = 0$ months, and $t' = -15$ days. On 31 December, they are $t = 0.5$ month $= 15$ days, and $t' = 0$ days. Thus, in consistent units, $t' = t - 15$ days, or $t = t' + 15$ days. Substituting $n = 365$ days and $t = t' + 15$ into Equation 8.55 yields

$$y_t = \bar{y} + C_1 \cos\left[\frac{2\pi t}{n} - \phi_1\right] = \bar{y} + C_1 \cos\left[\frac{2\pi(t'+15)}{365} - \phi_1\right]$$

$$= \bar{y} + C_1 \cos\left[\frac{2\pi t'}{365} + 2\pi\frac{15}{365} - \phi_1\right]$$

$$= \bar{y} + C_1 \cos \left[\frac{2\pi t'}{365} - \left(\phi_1 - 2\pi \frac{15}{365} \right) \right]$$

$$= \bar{y} + C_1 \cos \left[\frac{2\pi t'}{365} - \phi_1' \right]. \tag{8.61}$$

That is, the required new phase angle is $\phi_1' = \phi_1 - (2\pi)(15)/365.$ ◊

8.4.4 Higher Harmonics

The computations in Example 8.9 produced a single cosine function passing quite close to the 12 monthly mean temperature values. This very good fit results because the shape of the annual cycle of temperature at this location is approximately sinusoidal, with a single full cycle being executed over the $n = 12$ points of the time series. We do not expect that a single harmonic wave will represent every time series this well. However, just as adding more predictors to a multiple regression will improve the fit to a set of dependent data, adding more cosine waves to a harmonic analysis will improve the fit to any time series.

Any data series consisting of n points can be represented exactly, meaning that a harmonic function can be found that passes through each of the points, by adding together a series of $n/2$ harmonic functions,

$$y_t = \bar{y} + \sum_{k=1}^{n/2} \left\{ C_k \cos \left[\frac{2\pi kt}{n} - \phi_k \right] \right\} \tag{8.62a}$$

$$= \bar{y} + \sum_{k=1}^{n/2} \left\{ A_k \cos \left[\frac{2\pi kt}{n} \right] + B_k \sin \left[\frac{2\pi kt}{n} \right] \right\}. \tag{8.62b}$$

Notice that Equation 8.62b emphasizes that Equation 8.57 holds for any cosine wave, regardless of its frequency. The cosine wave comprising the $k = 1$ term of Equation 8.62a is simply the fundamental, or first harmonic, that was the subject of the previous section. The other $n/2 - 1$ terms in the summation of Equation 8.62 are higher harmonics, or cosine waves with frequencies

$$\omega_k = \frac{2\pi k}{n} \tag{8.63}$$

that are integer multiples of the fundamental frequency ω_1.

For example, the second harmonic is that cosine function that completes exactly two full cycles over the n points of the data series. It has its own amplitude C_2 and phase angle ϕ_2. Notice that the factor k inside the cosine and sine functions in Equation 8.62a is of critical importance. When $k = 1$, the angle $\alpha = 2\pi kt/n$ varies through a single full cycle of 0 to 2π radians as the time index increased from $t = 0$ to $t = n$, as described earlier. In the case of the second harmonic where $k = 2$, $\alpha = 2\pi kt/n$ executes one full cycle as t increases from 0, to $n/2$, and then executes a second full cycle between $t = n/2$ and $t = n$. Similarly, the third harmonic is defined by the amplitude C_3 and the phase angle ϕ_3, and varies through three cycles as t increases from 0 to n.

Equation 8.62b suggests that the coefficients A_k and B_k corresponding to particular data series y_t can be found using multiple regression methods, after the data transformations $x_1 = \cos(2\pi t/n)$, $x_2 = \sin(2\pi t/n)$, $x_3 = \cos(2\pi 2t/n)$, $x_4 = \sin(2\pi 2t/n)$,

$x_5 = \cos(2\pi 3t/n)$, and so on. This is, in fact, the case in general, but if the data series is equally spaced in time and contains no missing values, Equation 8.60 generalizes to

$$A_k = \frac{2}{n}\sum_{t=1}^{n} y_t \cos\left(\frac{2\pi kt}{n}\right) \tag{8.64a}$$

and

$$B_k = \frac{2}{n}\sum_{t=1}^{n} y_t \sin\left(\frac{2\pi kt}{n}\right). \tag{8.64b}$$

To compute a particular A_k, for example, these equations indicate than an n-term sum is formed, consisting of the products of the data series y_t with values of a cosine function executing k full cycles during the n time units. For relatively short data series these equations can be easily programmed and evaluated using a hand calculator or spreadsheet software. For larger data series the A_k and B_k coefficients usually are computed using a more efficient method that will be mentioned in Section 8.5.3. Having computed these coefficients, the amplitude-phase form of the first line of Equation 8.62 can be arrived at by computing, separately for each harmonic,

$$C_k = [A_k^2 + B_k^2]^{1/2} \tag{8.65a}$$

and

$$\phi_k = \begin{cases} \tan^{-1}(B_k/A_k), & A_k > 0 \\ \tan^{-1}(B_k/A_k) \pm \pi, \text{ or } \pm 180°, & A_k < 0 \\ \pi/2, \text{ or } 90°, & A_k = 0. \end{cases} \tag{8.65b}$$

Recall that a multiple regression function will pass through all the developmental data points, and exhibit $R^2 = 100\%$, if there are as many predictor values as data points. The series of cosine terms in Equation 8.62 is an instance of this overfitting principle, because there are two parameters (the amplitude and phase) for each harmonic term. Thus the $n/2$ harmonics in Equation 8.62 consist of n predictor variables, and any set of data, regardless of how untrigonometric it may look, can be represented exactly using Equation 8.62.

Since the sample mean in Equation 8.62 is effectively one of the estimated parameters, corresponding to the regression intercept b_0, an adjustment to Equation 8.62 is required if n is odd. In this case a summation over only $(n-1)/2$ harmonics is required to completely represent the function. That is, $(n-1)/2$ amplitudes plus $(n-1)/2$ phase angles plus the sample average of the data equals n. If n is even, there are $n/2$ terms in the summation, but the phase angle for the final and highest harmonic, $\phi_{n/2}$, is zero.

We may or may not want to use all $n/2$ harmonics indicated in Equation 8.62, depending on the context. Often for defining, say, an annual cycle of a climatological quantity, the first few harmonics may give a quite adequate representation from a practical standpoint. If the goal is to find a function passing exactly through each of the data points, then all $n/2$ harmonics would be used. Recall that Section 6.4 warned against overfitting in the context of developing forecast equations, because the artificial skill exhibited on the developmental data does not carry forward when the equation is used to forecast future independent data. In this latter case the goal would not be to forecast but rather to represent the data, so that the overfitting ensures that Equation 8.62 reproduces a particular data series exactly.

EXAMPLE 8.11   A More Complicated Annual Cycle

Figure 8.14 illustrates the use of a small number of harmonics to smoothly represent the annual cycle of a climatological quantity. Here the quantity is the probability (expressed as a percentage) of five consecutive days without measurable precipitation, for El Paso, Texas. The irregular curve is a plot of the individual daily relative frequencies computed using data for the years 1948–1983. These execute a regular but asymmetric annual cycle, with the wettest time of year being summer, and with dry springs and falls separated by a somewhat less dry winter. The figure also shows irregular, short-term fluctuations that have probably arisen mainly from sampling variations particular to the specific years analyzed. If a different sample of El Paso precipitation data had been used to compute the relative frequencies (say, 1900–1935), the same broad pattern would be evident, but the details of the individual "wiggles" would be different.

The annual cycle in Figure 8.14 is quite evident, yet it does not resemble a simple cosine wave. However, this cycle is reasonably well represented by the smooth curve, which is a sum of the first three harmonics. That is, the smooth curve is a plot of Equation 8.62 with three, rather than $n/2$, terms in the summation. The mean value for this data is 61.4%, and the parameters for the first two of these harmonics are $C_1 = 13.6\%$, $\phi_1 = 72° = 0.4\pi$, $C_2 = 13.8\%$, and $\phi_2 = 272° = 1.51\pi$. These values can be computed from the underlying data using Equations 8.64 and 8.65. Computing and plotting the sum of all possible $(365 - 1)/2 = 182$ harmonics would result in a function identical to the irregular curve in Figure 8.14.

Figure 8.15 illustrates the construction of the smooth curve representing the annual cycle in Figure 8.14. Panel (a) shows the first (dashed) and second (solid) harmonics plotted separately, both as a function of time (t) in days and as a function of the corresponding angular measure in radians. Also indicated are the magnitudes of the amplitudes C_k in the vertical, and the correspondence of the phase angles ϕ_k to the maxima of the two functions. Note that since the second harmonic executes two cycles during

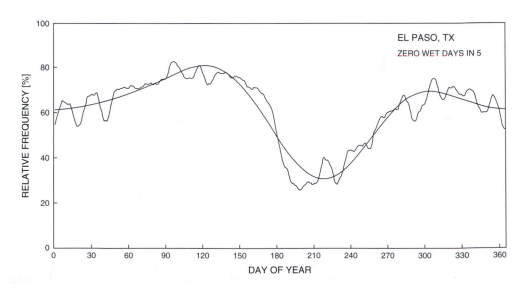

FIGURE 8.14   The annual cycle of the climatological probability that no measurable precipitation will fall during the five-day period centered on the date on the horizontal axis, for El Paso, Texas. Irregular line is the plot of the daily relative frequencies, and the smooth curve is a three-harmonic fit to the data. From Epstein and Barnston, 1988.

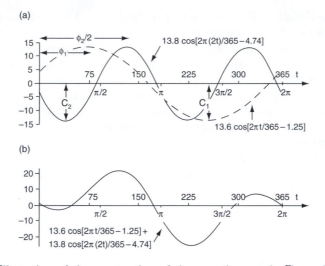

FIGURE 8.15   Illustration of the construction of the smooth curve in Figure 8.14. (a) The first (dashed) and second (solid) harmonics of the annual cycle plotted separately. These are defined by $C_1 = 13.6\%$, $\phi_1 = 72° = 0.4\pi$, $C_2 = 13.8\%$, and $\phi_2 = 272° = 1.51\pi$. The horizontal axis is labeled both in days and radians. (b) The smoothed representation of the annual cycle is produced by adding the values of the two functions in panel (a) for each time point. Subsequently adding the annual mean value of 61.4% produces a curve very similar to that in Figure 8.14. The small differences are accounted for by the third harmonic. Note that the two panels have different vertical scales.

the full 365 days of the year, there are two times of maximum, located at $\phi_2/2$ and $\pi + \phi_2/2$. (The maxima for the third harmonic would occur at $\phi_3/3$, $2\pi/3 + \phi_3/3$, and $4\pi/3 + \phi_3/3$, with a similar pattern holding for the higher harmonics.)

The curve in Figure 8.15b has been constructed by simply adding the values for the two functions in Figure 8.15a at each time point. Note that the two panels in Figure 8.15 have been plotted using different vertical scales. During times of the year where the two harmonics are of opposite sign but comparable magnitude, their sum is near zero. The maximum and minimum of the function in Figure 8.15b are achieved when its two components have relatively large magnitudes of the same sign. Adding the annual mean value of 61.4% to the lower curve results in a close approximation to the smooth curve in Figure 8.14, with the small differences between the two attributable to the third harmonic.◊

8.5 Frequency Domain—II. Spectral Analysis

8.5.1 The Harmonic Functions as Uncorrelated Regression Predictors

The second line of Equation 8.62 suggests the use of multiple regression to find best-fitting harmonics for a given data series y_t. But for equally spaced data with no missing values Equations 8.64 will produce the same least-squares estimates for the coefficients A_k and B_k as will multiple regression software. Notice, however, that Equations 8.64 do not depend on any harmonic other than the one whose coefficients are being computed. That is, these equations depend on the current value of k, but not $k - 1$, or $k - 2$, or any other harmonic index. This fact implies that the coefficients A_k and B_k for any particular harmonic can be computed independently of those for any other harmonic.

Recall that usually regression parameters need to be recomputed each time a new predictor variable is entered into a multiple regression equation, and each time a predictor variable is removed from a regression equation. As noted in Chapter 6, this recomputation is necessary in the general case of sets of predictor variables that are mutually correlated, because correlated predictors carry redundant information to a greater or lesser extent. It is a remarkable property of the harmonic functions that (for equally spaced and complete data) they are uncorrelated so, for example, the parameters (amplitude and phase) for the first or second harmonic are the same whether or not they will be used in an equation with the third, fourth, or any other harmonics.

This remarkable attribute of the harmonic functions is a consequence of what is called the orthogonality property of the sine and cosine functions. That is, for integer harmonic indices k and j,

$$\sum_{t=1}^{n} \cos\left(\frac{2\pi kt}{n}\right) \sin\left(\frac{2\pi jt}{n}\right) = 0, \quad \text{for any integer values of } k \text{ and } j; \quad (8.66a)$$

and

$$\sum_{t=1}^{n} \cos\left(\frac{2\pi kt}{n}\right) \cos\left(\frac{2\pi jt}{n}\right) = \sum_{t=1}^{n} \sin\left(\frac{2\pi kt}{n}\right) \sin\left(\frac{2\pi jt}{n}\right) = 0, \quad \text{for } k \neq j. \quad (8.66b)$$

Consider, for example, the two transformed predictor variables $x_1 = \cos[2\pi t/n]$ and $x_3 = \cos[2\pi(2t)/n]$. The Pearson correlation between these derived variables is given by

$$r_{x_1 x_3} = \frac{\sum_{t=1}^{n}(x_1 - \bar{x}_1)(x_3 - \bar{x}_3)}{\left[\sum_{t=1}^{n}(x_1 - \bar{x}_1)^2 \sum_{t=1}^{n}(x_3 - \bar{x}_3)^2\right]^{1/2}}, \quad (8.67a)$$

and since the averages $\bar{X}_1$ and $\bar{X}_3$ of cosine functions over integer numbers of cycles are zero,

$$r_{x_1 x_3} = \frac{\sum_{t=1}^{n} \cos\left(\frac{2\pi t}{n}\right) \cos\left(\frac{2\pi 2t}{n}\right)}{\left[\sum_{t=1}^{n} \cos^2\left(\frac{2\pi t}{n}\right) \sum_{t=1}^{n} \cos^2\left(\frac{2\pi 2t}{n}\right)\right]^{1/2}} = 0, \quad (8.67b)$$

because the numerator is zero by Equation 8.66b.

Since the relationships between harmonic predictor variables and the data series y_t do not depend on what other harmonic functions are also being used to represent the series, the proportion of the variance of y_t accounted for by each harmonic is also fixed. Expressing this proportion as the R^2 statistic commonly computed in regression, the R^2 for the kth harmonic is simply

$$R_k^2 = \frac{\frac{n}{2}C_k^2}{(n-1)s_y^2}. \quad (8.68)$$

In terms of the regression ANOVA table, the numerator of Equation 8.68 is the regression sum of squares for the kth harmonic. The factor s_y^2 is simply the sample variance of the data series, so the denominator of Equation 8.68 is the total sum of squares, SST. Notice that the strength of the relationship between the kth harmonic and the data series can be expressed entirely in terms of the amplitude C_k. The phase angle ϕ_k is necessary only to determine the positioning of the cosine curve in time. Furthermore, since each harmonic provides independent information about the data series, the joint R^2 exhibited by a regression equation with only harmonic predictors is simply the sum of the R_k^2 values for each of the harmonics,

$$R^2 = \sum_{\text{k in the equation}} R_k^2. \qquad (8.69)$$

If all the $n/2$ possible harmonics are used as predictors (Equation 8.62), then the total R^2 in Equation 8.69 will be exactly 1. Another perspective on Equations 8.68 and 8.69 is that the variance of the time-series variable y_t can be apportioned among the $n/2$ harmonic functions, each of which represents a different time scale of variation.

Equation 8.62 says that a data series y_t of length n can be specified completely in terms of the n parameters of $n/2$ harmonic functions. Equivalently, we can take the view that the data y_t are transformed into new set of quantities A_k and B_k according to Equations 8.64. For this reason, Equations 8.64 are called the discrete Fourier transform. Equivalently the data series can be represented as the n quantities C_k and ϕ_k, obtained from the A_ks and B_ks using the transformations in Equations 8.65. According to Equations 8.68 and 8.69, this data transformation accounts for all of the variation in the series y_t.

8.5.2 The Periodogram, or Fourier Line Spectrum

The foregoing suggests that a different way to look at a time series is as a collection of Fourier coefficients A_k and B_k that are a function of frequency ω_k (Equation 8.63), rather than as a collection of data points y_t measured as a function of time. The advantage of this new perspective is that it allows us to see separately the contributions to a time series that are made by processes varying at different speeds; that is, by processes operating at a spectrum of different frequencies. Panofsky and Brier (1958) illustrate this distinction with a nice analogy: "An optical spectrum shows the contributions of different wave lengths or frequencies to the energy of a given light source. The spectrum of a time series shows the contributions of oscillations with various frequencies to the variance of a time series." Even if the underlying physical basis for a data series y_t is not really well represented by a series of cosine waves, often much can still be learned about the data by viewing it from this perspective.

The characteristics of a time series that has been Fourier-transformed into the frequency domain are most often examined graphically, using a plot known as the periodogram, or Fourier line spectrum. This plot sometimes is called the power spectrum, or simply the spectrum, of the data series. In simplest form, this plot of a spectrum consists of the squared amplitudes C_k^2 as a function of the frequencies ω_k. The vertical axis is sometimes numerically rescaled, in which case the plotted points are proportional to the squared amplitudes. One choice for this proportional rescaling is that in Equation 8.68. Note that information contained in the phase angles ϕ_k is not portrayed in the spectrum. Therefore, the spectrum conveys the proportion of variation in the original data series accounted for by oscillations at the harmonic frequencies, but does not supply information about when in time these oscillations are expressed. A spectrum thus does not provide a full picture of the behavior of the time series from which it has been calculated, and is not sufficient to reconstruct the time series.

It is common for the vertical axis of a spectrum to be plotted on a logarithmic scale. Plotting the vertical axis logarithmically is particularly useful if the variations in the time series are dominated by harmonics of only a few frequencies. In this case a linear plot would result in the remaining spectral components being invisibly small. A logarithmic vertical axis also regularizes the representation of confidence limits for the spectral estimates.

The horizontal axis of the line spectrum consists of $n/2$ frequencies ω_k if n is even, and $(n-1)/2$ frequencies if n is odd. The smallest of these will be the lowest frequency $\omega_1 = 2\pi/n$ (the fundamental frequency), and this corresponds to the cosine wave that executes a single cycle over the n time points. The highest frequency, $\omega_{n/2} = \pi$, is called the Nyquist frequency. It is the frequency of the cosine wave that executes a full cycle over only two time intervals, and which executes $n/2$ cycles over the full data record. The Nyquist frequency depends on the time resolution of the original data series y_t, and imposes an important limitation on the information available from a spectral analysis.

The horizontal axis is often simply the angular frequency, ω, with units of radians/time. A common alternative is to use the frequencies

$$f_k = \frac{k}{n} = \frac{\omega_k}{2\pi}, \qquad (8.70)$$

which have dimensions of time^{-1}. Under this alternative convention, the allowable frequencies range from $f_1 = 1/n$ for the fundamental to $f_{n/2} = 1/2$ for the Nyquist frequency. The horizontal axis of a spectrum can also be scaled according to the reciprocal of the frequency, or the period of the kth harmonic

$$\tau_k = \frac{n}{k} = \frac{2\pi}{\omega_k} = \frac{1}{f_k}. \qquad (8.71)$$

The period τ_k specifies the length of time required for a cycle of frequency ω_k to be completed. Associating periods with the periodogram estimates can help visualize the time scales on which the important variations in the data are occurring.

EXAMPLE 8.12   Discrete Fourier Transform of a Small Data Set

Table 8.6 shows a simple data set and its discrete Fourier transform. The leftmost columns contain the observed average monthly temperatures at Ithaca, New York, for the two

TABLE 8.6   Average monthly temperatures, °F, at Ithaca, New York, for 1987–1988, and their discrete Fourier transform.

Month	1987	1988	k	τ_k, months	A_k	B_k	C_k
1	21.4	20.6	1	24.00	−0.14	0.44	0.46
2	17.9	22.5	2	12.00	−23.76	−2.20	23.86
3	35.9	32.9	3	8.00	−0.99	0.39	1.06
4	47.7	43.6	4	6.00	−0.46	−1.25	1.33
5	56.4	56.5	5	4.80	−0.02	−0.43	0.43
6	66.3	61.9	6	4.00	−1.49	−2.15	2.62
7	70.9	71.6	7	3.43	−0.53	−0.07	0.53
8	65.8	69.9	8	3.00	−0.34	−0.21	0.40
9	60.1	57.9	9	2.67	1.56	0.07	1.56
10	45.4	45.2	10	2.40	0.13	0.22	0.26
11	39.5	40.5	11	2.18	0.52	0.11	0.53
12	31.3	26.7	12	2.00	0.79	—	0.79

years 1987 and 1988. This is such a familiar type of data that, even without doing a spectral analysis, we know in advance that the primary feature will be the annual cycle of cold winters and warm summers. This expectation is validated by the plot of the data in Figure 8.16a, which shows these temperatures as a function of time. The overall impression is of a data series that is approximately sinusoidal with a period of 12 months, but that a single cosine wave with this period would not pass exactly through all the points.

Columns 4 to 8 of Table 8.6 shows the same data after being subjected to the discrete Fourier transform. Since $n = 24$ is an even number, the data are completely represented

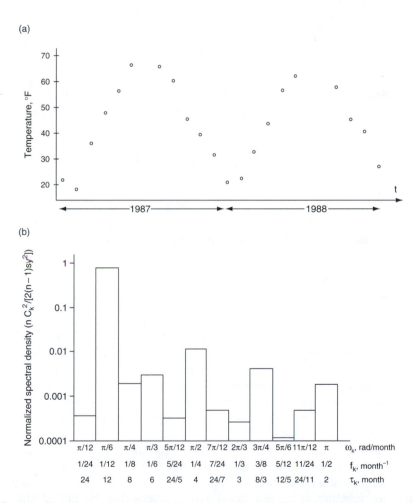

FIGURE 8.16   Illustration of the relationship between a simple time series and its spectrum. (a) Average monthly temperatures at Ithaca, New York for 1987–1988, from Table 8.6. The data are approximately sinusoidal, with a period of 12 months. (b) The spectrum of the data series in panel (a), plotted in the form of a histogram, and expressed in the normalized form of Equation 8.68. Clearly the most important variations in the data series are represented by the second harmonic, with period $\tau_2 = 12$ months, which is the annual cycle. Note that the vertical scale is logarithmic, so that the next most important harmonic accounts for barely more than 1% of the total variation. The horizontal scale is linear in frequency.

by $n/2 = 12$ harmonics. These are indicated by the rows labeled by the harmonic index, k. Column 5 of Table 8.6 indicates the period (Equation 8.71) of each of the 12 harmonics used to represent the data. The period of the fundamental frequency, $\tau_1 = 24$ months, is equal to the length of the data record. Since there are two annual cycles in the $n = 24$ month record, it is the $k = 2^{\text{nd}}$ harmonic with period $\tau_2 = 24/2 = 12$ months that is expected to be most important. The Nyquist frequency is $\omega_{12} = \pi$ radians/month, or $f_{12} = 0.5$ month^{-1}, corresponding to the period $\tau_{12} = 2$ months.

The coefficients A_k and B_k that could be used in Equation 8.62 to reconstruct the original data are shown in the next columns of the table. These constitute the discrete Fourier transform of the data series of temperatures. Notice that there are only 23 Fourier coefficients, because 24 independent pieces of information are necessary to fully represent the $n = 24$ data points, including the sample mean of 46.1°F. To use Equation 8.62 to reconstitute the data, we would substitute $B_{12} = 0$.

Column 8 in Table 8.6 shows the amplitudes C_k, computed according to Equation 8.65a. The phase angles could also be computed, using Equation 8.65b, but these are not needed to plot the spectrum. Figure 8.16b shows the spectrum for this temperature data, plotted in the form of a histogram. The vertical axis consists of the squared amplitudes C_k^2, normalized according to Equation 8.68 to show the R^2 attributable to each harmonic. The horizontal axis is linear in frequency, but the corresponding periods are also shown, to aid the interpretation. Clearly most of the variation in the data is described by the second harmonic, the R^2 for which is 97.5%. As expected, the variations of the annual cycle dominate this data, but the fact that the amplitudes of the other harmonics are not zero indicates that the data do not consist of a pure cosine wave with a frequency of $f_2 = 1$ year^{-1}. Notice, however, that the logarithmic vertical axis tends to deemphasize the smallness of these other harmonics. If the vertical axis were scaled linearly, the plot would consist of a spike at $k = 2$ and a small bump at $k = 6$, with the rest of the points being essentially indistinguishable from the horizontal axis. ◊

EXAMPLE 8.13   Another Sample Spectrum

A less trivial example spectrum is shown in Figure 8.17. This is a portion of the spectrum of the monthly Tahiti minus Darwin sea-level pressure time series for 1951–1979. That time series resembles the (normalized) SOI index shown in Figure 3.14, including the tendency for a quasiperiodic behavior. That the variations in the time series are not strictly periodic is evident from the irregular variations in Figure 3.14, and from the broad (i.e., spread over many frequencies) maximum in the spectrum. Figure 8.17 indicates that the typical length of one of these pseudocycles (corresponding to typical times between El Niño events) is something between $\tau = [(1/36)\text{mo}^{-1}]^{-1} = 3$ years and $\tau = [(1/84)\text{mo}^{-1}]^{-1} = 7$ years.

The vertical axis in Figure 8.17 has been plotted on a linear scale, but units have been omitted because they do not contribute to a qualitative interpretation of the plot. The horizontal axis is linear in frequency, and therefore nonlinear in period. Notice also that the labeling of the horizontal axis indicates that the full spectrum of the underlying data series is not presented in the figure. Since the data series consists of monthly values, the Nyquist frequency must be 0.5 month^{-1}, corresponding to a period of two months. Only the left-most one-eighth of the spectrum has been shown because it is these lower frequencies that reflect the physical phenomenon of interest, namely the El Niño-Southern Oscillation (ENSO) cycle. The estimated spectral density function for the omitted higher frequencies would comprise only a long, irregular and generally uninformative right tail. ◊

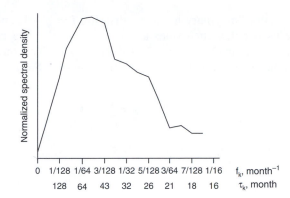

FIGURE 8.17   The low-frequency portion of the smoothed spectrum for the monthly time series of Tahiti minus Darwin sea-level pressures, 1951–1979. This underlying time series resembles that in Figure 3.14, and the tendency for oscillations to occur in roughly three- to seven-year cycles is reflected in the broad maximum of the spectrum in this range. After Chen (1982a).

8.5.3 Computing Spectra

One way to compute the spectrum of a data series is simply to apply Equations 8.64, and then to find the amplitudes using Equation 8.65b. This is a reasonable approach for relatively short data series, and can be programmed easily using, for example, spreadsheet software. These equations would be implemented only for $k = 1, 2, \ldots, (n/2 - 1)$. Because we want exactly n Fourier coefficients (A_k and B_k) to represent the n points in the data series, the computation for the highest harmonic, $k = n/2$, is done using

$$A_{n/2} = \begin{cases} \left(\frac{1}{2}\right)\left(\frac{2}{n}\right) \sum_{t=1}^{n} y_t \cos\left[\frac{2\pi(n/2)t}{n}\right] = \frac{1}{n} \sum_{t=1}^{n} y_t \cos[\pi t], & n \text{ even} \\ 0, & n \text{ odd} \end{cases} \qquad (8.72a)$$

and

$$B_{n/2} = 0, \qquad n \text{ even or odd.} \qquad (8.72b)$$

Although straightforward notationally, this method of computing the discrete Fourier transform is quite inefficient computationally. In particular, many of the calculations called for by Equation 8.64 are redundant. Consider, for example, the data for April 1987 in Table 8.6. The term for $t = 4$ in the summation in Equation 8.64b is $(47.7°F)\sin[(2\pi)(1)(4)/24] = (47.7°F)(0.866) = 41.31°F$. However, the term involving this same data point for $k = 2$ is exactly the same: $(47.7°F)\sin[(2\pi)(2)(4)/24] = (47.7°F)(0.866) = 41.31°F$. There are many other such redundancies in the computation of discrete Fourier transforms using Equations 8.64. These can be avoided through the use of clever algorithms known as Fast Fourier Transforms (FFTs). Most scientific software packages include one or more FFT routines, which give very substantial speed improvements, especially as the length of the data series increases. In comparison to computation of the Fourier coefficients using a regression approach, an FFT is approximately $n/\log_2(n)$ times faster; or about 15 times faster for $n = 100$, and about 750 times faster for $n = 10000$.

It is worth noting that FFTs usually are documented and implemented in terms of the Euler complex exponential notation,

$$e^{i\omega t} = \cos(\omega t) + i\sin(\omega t), \qquad (8.73)$$

where i is the unit imaginary number satisfying $i = \sqrt{-1}$, and $i^2 = -1$. Complex exponentials are used rather than sines and cosines purely as a notational convenience that makes some of the manipulations less cumbersome. The mathematics are still entirely the same. In terms of complex exponentials, Equation 8.62 becomes

$$y_t = \bar{y} + \sum_{k=1}^{n/2} H_k e^{i[2\pi k/n]t}, \tag{8.74}$$

where H_k is the complex Fourier coefficient

$$H_k = A_k + iB_k. \tag{8.75}$$

That is, the real part of H_k is the coefficient A_k, and the imaginary part of H_k is the coefficient B_k.

8.5.4 Aliasing

Aliasing is a hazard inherent in spectral analysis of discrete data. It arises because of the limits imposed by the sampling interval, or the time between consecutive pairs of data points. Since a minimum of two points are required to even sketch a cosine wave—one point for the peak and one point for the trough—the highest representable frequency is the Nyquist frequency, with $\omega_{n/2} = \pi$, or $f_{n/2} = 0.5$. A wave of this frequency executes one cycle every two data points, and thus a discrete data set can represent explicitly variations that occur no faster than this speed.

It is worth wondering what happens to the spectrum of a data series if it includes important physical process that vary faster than the Nyquist frequency. If so, the data series is said to be undersampled, which means that the points in the time series are spaced too far apart to properly represent these fast variations. However, variations that occur at frequencies higher than the Nyquist frequency do not disappear. Rather, their contributions are spuriously attributed to some lower but representable frequency, between ω_1 and $\omega_{n/2}$. These high-frequency variations are said to be aliased.

Figure 8.18 illustrates the meaning of aliasing. Imagine that the physical data-generating process is represented by the dashed cosine curve. The data series y_t is produced

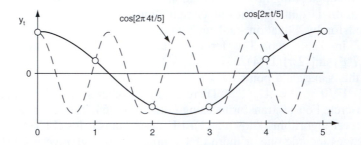

FIGURE 8.18 Illustration of the basis of aliasing. Heavy circles represent data points in a data series y_t. Fitting a harmonic function to them results in the heavy curve. However, if the data series actually had been produced by the process indicated by the light dashed curve, the fitted heavy curve would present the misleading impression that the source of the data was actually fluctuating at the lower frequency. The lighter curve has not been sampled densely enough because its frequency, $\omega = 8\pi/5$ (or $f = 4/5$), is higher than the Nyquist frequency of $\omega = \pi$ (or $f = 1/2$). Variations at the frequency of the dashed curve are said to be aliased into the frequency of the heavier curve.

by sampling this process at integer values of the time index t, resulting in the points indicated by the heavy circles. However, the frequency of the dashed curve ($\omega = 8\pi/5$, or $f = 4/5$) is higher than the Nyquist frequency ($\omega = \pi$, or $f = 1/2$), meaning that it oscillates too quickly to be adequately sampled at this time resolution. Rather, if only the information in the discrete time points is available, this data looks like the heavy cosine function, the frequency of which ($\omega = 2\pi/5$, or $f = 1/5$) is lower than the Nyquist frequency, and is therefore representable. Note that because the cosine functions are orthogonal, this same effect will occur whether or not variations of different frequencies are also present in the data.

The effect of aliasing on spectral analysis is that any energy (squared amplitudes) attributable to processes varying at frequencies higher than the Nyquist frequency will be erroneously added to that of one of the $n/2$ frequencies that are represented by the discrete Fourier spectrum. A frequency $f_A > 1/2$ will be aliased into one of the representable frequencies f (with $0 < f \leq 1/2$) if it differs by an integer multiple of 1 time^{-1}, that is, if

$$f_A = j \pm f, \quad j \text{ any positive integer.} \tag{8.76a}$$

In terms of angular frequency, variations at the aliased frequency ω_A appear to occur at the representable frequency ω if

$$\omega_A = 2\pi j \pm \omega, \quad j \text{ any positive integer.} \tag{8.76b}$$

These equations imply that the squared amplitudes for frequencies higher than the Nyquist frequency will be added to the representable frequencies in an accordion-like pattern, with each "fold" of the accordion occurring at an integer multiple of the Nyquist frequency. For this reason the Nyquist frequency is sometimes called the "folding" frequency. An aliased frequency f_A that is just slightly higher than the Nyquist frequency of $f_{n/2} = 1/2$ is aliased to a frequency slightly lower than 1/2. Frequencies only slightly lower than twice the Nyquist frequency are aliased to frequencies only slightly higher than zero. The pattern then reverses for $2f_{n/2} < f_A < 3f_{n/2}$. That is, frequencies just higher than $2f_{n/2}$ are aliased to very low frequencies, and frequencies almost as high as $3f_{n/2}$ are aliased to frequencies near $f_{n/2}$.

Figure 8.19 illustrates the effects of aliasing on a hypothetical spectrum. The lighter line represents the true spectrum, which exhibits a concentration of density at low frequencies, but also has a sharp peak at $f = 5/8$ and a broader peak at $f = 19/16$. These second two peaks occur at frequencies higher than the Nyquist frequency of $f = 1/2$, which means that the physical process that generated the data was not sampled often enough to resolve them explicitly. The variations actually occurring at the frequency $f_A = 5/8$ are aliased to (i.e., appear to occur at) the frequency $f = 3/8$. That is, according to Equation 8.76a, $f_A = 1 - f = 1 - 3/8 = 5/8$. In the spectrum, the squared amplitude for $f_A = 5/8$ is added to the (genuine) squared amplitude at $f = 3/8$ in the true spectrum. Similarly, the variations represented by the broader hump centered at $f_A = 19/16$ in the true spectrum are aliased to frequencies at and around $f = 3/16$ ($f_A = 1 + f = 1 + 3/16 = 19/16$). The dashed line in Figure 8.19 indicates the portions of the aliased spectral energy (the total area between the light and dark lines) contributed by frequencies between $f = 1/2$ and $f = 1$ (the area below the dashed line), and by frequencies between $f = 1$ and $f = 3/2$ (the area above the dashed line); emphasizing the fan-folded nature of the aliased spectral density.

Aliasing can be particularly severe when isolated segments of a time series are averaged and then analyzed, for example a time series of average January values in each of n years. This problem has been studied by Madden and Jones (2001), who conclude that badly aliased spectra are expected to result unless the averaging time is at least as

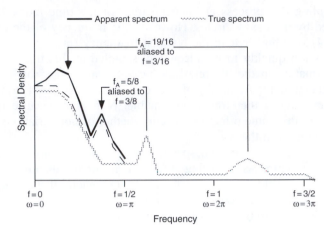

FIGURE 8.19   Illustration of aliasing in a hypothetical spectrum. The true spectrum (lighter line) exhibits a sharp peak at $f = 5/8$, and a broader peak at $f = 19/16$. Since both of these frequencies are higher than the Nyquist frequency $f = 1/2$, they are aliased in the spectrum (erroneously attributed) to the frequencies indicated. The aliasing follows an accordion-like pattern, with the area between the light line and the dashed line contributed by frequencies from $f = 1$ to $f = 1/2$, and the area between the dashed line and the heavy line contributed by frequencies between $f = 1$ and $f = 3/2$. The resulting apparent spectrum (heavy line) includes both the true spectral density values for frequencies between zero and 1/2, as well as the contributions from the aliased frequencies.

large as the sampling interval. For example, a spectrum for January averages is expected to be heavily aliased since the one-month averaging period is much shorter than the annual sampling interval.

Unfortunately, once a data series has been collected, there is no way to "de-alias" its spectrum. That is, it is not possible to tell from the data values alone whether appreciable contributions to the spectrum have been made by frequencies higher than $f_{n/2}$, or how large these contributions might be. In practice, it is desirable to have an understanding of the physical basis of the processes generating the data series, so that we can see in advance that the sampling rate is adequate. Of course in an exploratory setting this advice is of no help, since the point of an exploratory analysis is exactly to gain a better understanding of an unknown or a poorly known generating process. In this latter situation, we would like to see the spectrum approach zero for frequencies near $f_{n/2}$, which would give some hope that the contributions from higher frequencies are minimal. A spectrum such as the heavy line in Figure 8.19 would lead us to expect that aliasing might be a problem, since its not being essentially zero at the Nyquist frequency could well mean that the true spectrum is nonzero at higher frequencies as well.

8.5.5 Theoretical Spectra of Autoregressive Models

Another perspective on the time-domain autoregressive models described in Section 8.3 is provided by their spectra. The types of time dependence produced by different autoregressive models produce characteristic spectral signatures that can be related to the autoregressive parameters.

The simplest case is the AR(1) process, Equation 8.16. Here positive values of the single autoregressive parameter ϕ induce a memory into the time series that tends to smooth over short-term (high-frequency) variations in the ε series, and emphasize the slower (low-frequency) variations. In terms of the spectrum, these effects lead to more density at lower frequencies, and less density at higher frequencies. Furthermore, these effects are progressively stronger for ϕ closer to 1.

These ideas are quantified by the theoretical spectral density function for AR(1) processes,

$$S(f) = \frac{4\sigma_\varepsilon^2/n}{1+\phi^2-2\phi\cos(2\pi f)}, \quad 0 \le f \le 1/2. \tag{8.77}$$

This is a function that associates a spectral density with all frequencies in the range $0 \le f \le 1/2$. The shape of the function is determined by the denominator, and the numerator contains scaling constants that give the function numerical values that are comparable to the empirical spectrum of squared amplitudes, C_k^2. This equation also applies to negative values of the autoregressive parameter, which produce time series tending to oscillate quickly around the mean, and for which the spectral density is greatest at the high frequencies.

Note that, for zero frequency, Equation 8.77 is proportional to the variance of a time average. This can be appreciated by substituting $f = 0$, and Equations 8.21 and 8.39 into Equation 8.77, and comparing to Equation 8.36. Thus, the extrapolation of the spectrum to zero frequency has been used to estimate variances of time averages (e.g., Madden and Shea 1978).

Figure 8.20 shows theoretical spectra for the AR(1) processes having $\phi = 0.5, 0.3, 0.0,$ and -0.6. The autocorrelation functions for the first and last of these are shown as

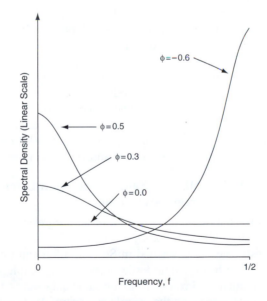

FIGURE 8.20   Theoretical spectral density functions for four AR(1) processes, computed using Equation 8.77. Autoregressions with $\phi > 0$ are red-noise processes, since their spectra are enriched at the lower frequencies and depleted at the higher frequencies. The spectrum for the $\phi = 0$ process (i.e., serially independent data) is flat, exhibiting no tendency to emphasize either high- or low-frequency variations. This is a white-noise process. The autoregression with $\phi = -0.6$ tends to exhibit rapid variations around its mean, which results in a spectrum enriched in the high frequencies, or a blue-noise process. Autocorrelation functions for the $\phi = 0.5$ and $\phi = -0.6$ processes are shown as insets in Figure 8.7.

insets in Figure 8.7. As might have been anticipated, the two processes with $\phi > 0$ show enrichment of the spectral densities in the lower frequencies and depletion in the higher frequencies, and these characteristics are stronger for the process with the larger autoregressive parameter. By analogy to the properties of visible light, AR(1) processes with $\phi > 0$ are sometimes referred to as red-noise processes.

The AR(1) process with $\phi = 0$ actually consists of the series of temporally uncorrelated data values $x_{t+1} = \mu + \varepsilon_{t+1}$ (compare Equation 8.16). These exhibit no tendency to emphasize either low-frequency or high-frequency variations, so their spectrum is constant, or flat. Again by analogy to visible light, this is called white noise because of the equal mixture of all frequencies. This analogy is the basis of the independent series of εs being called the white-noise forcing, and of the parameter σ_ε^2 being known as the white-noise variance.

Finally, the AR(1) process with $\phi = -0.6$ tends to produce erratic short-term variations in the time series, resulting in negative correlations at odd lags and positive correlations at even lags. (This kind of correlation structure is rare in atmospheric time series.) The spectrum for this process is thus enriched at the high frequencies and depleted at the low frequencies, as indicated in Figure 8.20. Such series are accordingly known as blue-noise processes.

Expressions for the spectra of other autoregressive processes, and for ARMA processes as well, are given in Box and Jenkins (1994). Of particular importance is the spectrum for the AR(2) process,

$$S(f) = \frac{4\sigma_\varepsilon^2/n}{1 + \phi_1^2 + \phi_2^2 - 2\phi_1(1 - \phi_2)\cos(2\pi f) - 2\phi_2\cos(4\pi f)}, \quad 0 \le f \le 1/2. \quad (8.78)$$

This equation reduces to Equation 8.77 for $\phi_2 = 0$, since an AR(2) process (Equation 8.27) with $\phi_2 = 0$ is simply an AR(1) process.

The AR(2) processes are particularly interesting because of their capacity to exhibit a wide variety of behaviors, including pseudoperiodicities. This diversity is reflected in the various forms of the spectra that are included in Equation 8.78. Figure 8.21 illustrates a few of these, corresponding to the AR(2) processes whose autocorrelation functions are shown as insets in Figure 8.7. The processes with $\phi_1 = 0.9$, $\phi_2 = -0.6$, and $\phi_1 = -0.9$, $\phi_2 = -0.5$, exhibit pseudoperiodicities, as indicated by the broad humps in their spectra at intermediate frequencies. The process with $\phi_1 = 0.3$, $\phi_2 = 0.4$ exhibits most of its variation at low frequencies, but also shows a smaller maximum at high frequencies. The spectrum for the process with $\phi_1 = 0.7$, $\phi_2 = -0.2$ resembles the red-noise spectra in Figure 8.20, although with a broader and flatter low-frequency maximum.

EXAMPLE 8.14   Smoothing a Sample Spectrum Using an Autoregressive Model

The equations for the theoretical spectra of autoregressive models can be useful in interpreting sample spectra from data series. The erratic sampling properties of the individual periodogram estimates as described in Section 8.5.6 can make it difficult to discern features of the true spectrum that underlies a particular sample spectrum. However, if a well-fitting time-domain model can be fit to the same data series, its theoretical spectrum can be used to guide the eye through the sample spectrum. Autoregressive models are sometimes fitted to time series for the sole purpose of obtaining smooth spectra. Chu and Katz (1989) show that the spectrum corresponding to a time-domain model fit using the SOI time series (see Figure 8.17) corresponds well to the spectrum computed directly from the data.

Consider the data series in Figure 8.8c, which was generated according to the AR(2) process with $\phi_1 = 0.9$ and $\phi_2 = -0.6$. The sample spectrum for this particular batch of 50 points is shown as the solid curve in Figure 8.22. Apparently the series exhibits

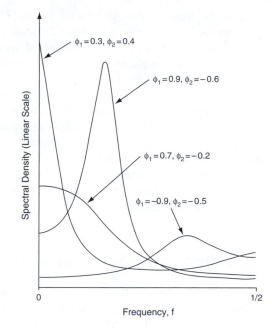

FIGURE 8.21   Theoretical spectral density functions for four AR(2) processes, computed using Equation 8.78. The diversity of the forms of the of spectra in this figure illustrates the flexibility of the AR(2) model. The autocorrelation functions for these autoregressions are shown as insets in Figure 8.7.

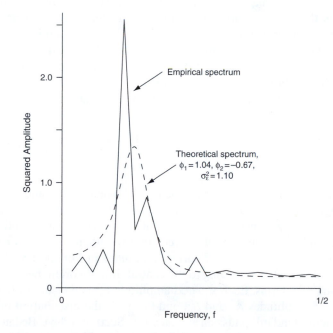

FIGURE 8.22   Illustration of the use of theoretical spectra for autoregressive models to guide the eye in interpreting sample spectra. The solid curve is the sample spectrum for the $n = 50$ data points shown in Figure 8.8c, generated by the AR(2) process with $\phi_1 = 0.9$, $\phi_2 = -0.6$, and $\sigma_\varepsilon^2 = 1.0$. A fuller perspective on this spectrum is provided by the dashed line, which is the theoretical spectrum of the AR(2) process fitted to this same series of 50 data points.

pseudoperiodicities in the frequency range around $f = 0.12$ through $f = 0.16$, but sampling variability makes the interpretation somewhat difficult. Although the empirical spectrum in Figure 8.22 somewhat resembles the theoretical spectrum for this AR(2) model shown in Figure 8.21, its nature might not be obvious from the empirical spectrum alone.

A fuller perspective on the spectrum in Figure 8.22 is gained when the dashed curve is provided to guide the eye. This is the theoretical spectrum for an AR(2) model fitted to the same data points from which the empirical spectrum was computed. The first two sample autocorrelations for these data are $r_1 = 0.624$ and $r_2 = -0.019$, which are near the theoretical values that would be obtained using Equation 8.33. Using Equation 8.29, the corresponding estimated autoregressive parameters are $\phi_1 = 1.04$ and $\phi_2 = -0.67$. The sample variance of the $n = 50$ data points is 1.69, which leads through Equation 8.30 to the estimated white-noise variance $\sigma_\varepsilon^2 = 1.10$. The resulting spectrum, according to Equation 8.78, is plotted as the dashed curve. ◊

8.5.6 *Sampling Properties of Spectral Estimates*

Since the data from which atmospheric spectra are computed are subject to sampling fluctuations, Fourier coefficients computed from these data will exhibit random batch-to-batch variations as well. That is, different data batches of size n from the same source will transform to somewhat different C_k^2 values, resulting in somewhat different sample spectra.

Each squared amplitude is an unbiased estimator of the true spectral density, which means that averaged over many batches the mean of the many C_k^2 values would closely approximate their true population counterpart. Another favorable property of raw sample spectra is that the periodogram estimates at different frequencies are uncorrelated with each other. Unfortunately, the sampling distribution for an individual C_k^2 is rather broad. In particular, the sampling distribution of suitably scaled squared amplitudes is the χ^2 distribution with $\nu = 2$ degrees of freedom, which is an exponential distribution, or a gamma distribution having $\alpha = 1$ (compare Figure 4.7).

The particular scaling of the raw spectral estimates that has this χ^2 sampling distribution is

$$\frac{\nu C_k^2}{S(f_k)} \sim \chi_\nu^2, \tag{8.79}$$

where $S(f_k)$ is the spectral density being estimated by C_k^2, and $\nu = 2$ degrees of freedom for a single spectral estimate C_k^2. Note that the various choices that can be made for multiplicative scaling of periodogram estimates will cancel in the ratio on the left-hand side of Equation 8.79. One way of appreciating the appropriateness of the χ^2 sampling distribution is to realize that the Fourier amplitudes in Equation 8.64 will be approximately Gaussian-distributed according to the central limit theorem, because they are each derived from sums of n terms. Each squared amplitude C_k^2 is the sum of the squares of its respective pair of amplitudes A_k^2 and B_k^2, and the χ^2 is the distribution of the sum of ν squared independent standard Gaussian variates (cf. Section 4.4.3). Because the sampling distributions of the squared Fourier amplitudes in Equation 8.64a are not *standard* Gaussian, the scaling constants in Equation 8.79 are necessary to produce a χ^2 distribution.

Because the sampling distribution of the periodogram estimates is exponential, these estimates are strongly positively skewed, and their standard errors (standard deviation of the sampling distribution) are equal to their means. An unhappy consequence of these properties is that the individual C_k^2 estimates represent the true spectrum rather poorly. The very erratic nature of raw spectral estimates is illustrated by the two sample spectra shown in Figure 8.23. The heavy and light lines are two sample spectra computed from different batches of $n = 30$ independent Gaussian random variables. Each of the two sample spectra vary rather wildly around the true spectrum, which is shown by the dashed horizontal line. In a real application, the true spectrum is, of course, not known in advance, and Figure 8.23 shows that the poor sampling properties of the individual spectral estimates can make it very difficult to discern much about the true spectrum if only a single sample spectrum is available.

Confidence limits for the underlying population quantities corresponding to raw spectral estimates are rather broad. Equation 8.79 implies that confidence interval widths are proportional to the raw periodogram estimates themselves, so that

$$\Pr\left[\frac{\nu C_k^2}{\chi_\nu^2 \left(1 - \frac{\alpha}{2}\right)} < S(f_k) \leq \frac{\nu C_k^2}{\chi_\nu^2 \left(\frac{\alpha}{2}\right)} \right] = 1 - \alpha, \tag{8.80}$$

where again $\nu = 2$ for a single raw periodogram estimate, and $\chi_k^2(\alpha)$ is the α quantile of the appropriate χ^2 distribution. For example, $\alpha = 0.05$ for a 95% confidence interval. The form of Equation 8.80 suggests one reason that it can be convenient to plot spectra on a logarithmic scale, since in that case the widths of the $(1 - \alpha) \times 100\%$ confidence intervals are constant across frequencies, regardless of the magnitudes of the estimated C_k^2.

The usual remedy in statistics for an unacceptably broad sampling distribution is to increase the sample size. For spectra, however, simply increasing the sample size does not give more precise information about any of the individual frequencies, but rather results in equally imprecise information about more frequencies. For example, the spectra in Figure 8.23 were computed from $n = 30$ data points, and thus consist of $n/2 = 15$ squared amplitudes. Doubling the sample size to $n = 60$ data values would result in a spectrum at $n/2 = 30$ frequencies, each point of which would exhibit the same large sampling variations as the individual C_k^2 values in Figure 8.23.

It is possible, however, to use larger data samples to obtain sample spectra that are more representative of the underlying population spectra. One approach is to compute

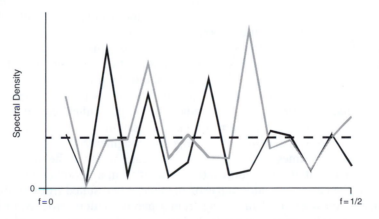

FIGURE 8.23   Illustration of the erratic sampling characteristics of estimated spectra. The solid and grey curves are two sample spectra, each computed using different batches of $n = 30$ independent Gaussian random variables. Both are quite erratic, with points of relative agreement being more fortuitous than meaningful. The true spectrum for the underlying serially independent data is shown by the horizontal dashed line. The vertical axis is linear.

replicate spectra from separate sections of the time series, and then to average the resulting squared amplitudes. In the context of Figure 8.23, for example, a time series of $n = 60$ could be split into two series of length $n = 30$. The two spectra in Figure 8.23 might be viewed as having resulted from such a process. Here averaging each of the $n/2 = 15$ pairs of C_k^2 values would result in a less erratic spectrum, that somewhat more faithfully represents the true spectrum. In fact the sampling distributions of each of these $n/2$ average spectral values would be proportional (Equation 8.79) to the χ^2 distribution with $\nu = 4$, or a gamma distribution with $\alpha = 2$, as each would be proportional to the sum of four squared Fourier amplitudes. This distribution is substantially less variable and less strongly skewed than the exponential distribution, having standard deviation of $1/\sqrt{2}$ of the averaged estimates, or about 70% of the previous individual ($\nu = 2$) estimates. If we had a data series with $n = 300$ points, 10 sample spectra could be computed whose average would smooth out a large fraction of the sampling variability evident in Figure 8.23. The sampling distribution for the averaged squared amplitudes in this case would have $\nu = 20$. The standard deviations for these averages would be smaller by the factor $1/\sqrt{10}$, or about one-third of the magnitudes of those for single squared amplitudes. Since the confidence interval widths are still proportional to the estimated squared amplitudes, a logarithmic vertical scale again results in plotted confidence interval widths not depending on frequency.

An essentially equivalent approach to obtaining a smoother and more representative spectrum using more data begins with computation of the discrete Fourier transform for the longer data series. Although this results at first in more spectral estimates that are equally variable, their sampling variability can be smoothed out by adding (not averaging) the squared amplitudes for groups of adjacent frequencies. The spectrum shown in Figure 8.17 has been smoothed in this way. For example, if we wanted to estimate the spectrum at the 15 frequencies plotted in Figure 8.23, these could be obtained by summing consecutive pairs of the 30 squared amplitudes obtained from the spectrum of a data record that was $n = 60$ observations long. If $n = 300$ observations were available, the spectrum at these same 15 frequencies could be estimated by adding the squared amplitudes for groups of 10 of the $n/2 = 150$ original frequencies. Here again the sampling distribution is χ^2 with ν equal to twice the number of pooled frequencies, or gamma with α equal to the number of pooled frequencies.

A variety of more sophisticated smoothing functions are commonly applied to sample spectra (e.g., Jenkins and Watts 1968; von Storch and Zwiers 1999). Note that, regardless of the specific form of the smoothing procedure, the increased smoothness and representativeness of the resulting spectra come at the expense of decreased frequency resolution and introduction of bias. Essentially, stability of the sampling distributions of the spectral estimates is obtained by smearing spectral information from a range of frequencies across a frequency band. Smoothing across broader bands produces less erratic estimates, but hides sharp contributions that may be made at particular frequencies. In practice, there is always a compromise to be made between sampling stability and frequency resolution, which is resolved as a matter of subjective judgment.

It is sometimes of interest to investigate whether the largest C_k^2 among K such squared amplitudes is significantly different from a hypothesized population value. That is, has the largest periodogram estimate plausibly resulted from sampling variations in the Fourier transform of data arising from a purely random process, or does it reflect a real periodicity that may be partially hidden by random noise in the time series? Addressing this question is complicated by two issues: choosing a null spectrum that is appropriate to the data series, and accounting for test multiplicity if the frequency f_k corresponding to the largest C_k^2 is chosen according to the test data rather than on the basis of external, prior information.

Initially we might adopt the white-noise spectrum (Equation 8.77, with $\phi = 0$) to define the null hypothesis. This could be an appropriate choice if there is little or no prior information about the nature of the data series, or if we expect in advance that the possible periodic signal is embedded in uncorrelated noise. However, most atmospheric time series are positively autocorrelated, and usually a null spectrum reflecting this tendency is a preferable null reference function (Gilman *et al.* 1963). Commonly it is the AR(1) spectrum (Equation 8.77) that is chosen for the purpose, with ϕ and σ_ε^2 fit to the data whose spectrum is being investigated. Using Equation 8.79, the null hypothesis that the squared amplitude C_k^2 at frequency f_k is significantly larger than the null (possibly red-noise) spectrum at that frequency, $S_0(f_k)$, would be rejected at the α level if

$$C_k^2 \geq \frac{S_0(f_k)}{\nu} \chi_\nu^2 (1 - \alpha), \tag{8.81}$$

where $\chi_\nu^2(1 - \alpha)$ denotes right-tail quantiles of the appropriate Chi-square distribution, given in Table B.3. The parameter ν may be greater than 2 if spectral smoothing has been employed.

The rejection rule given in Equation 8.81 is appropriate if the frequency f_k being tested has been defined out of sample; that is, by prior information, and is in no way dependent on the data used to calculate the C_k^2. When such prior information is lacking, testing the statistical significance of the largest squared amplitude is complicated by the problem of test multiplicity. Because, in effect, K independent hypothesis tests are conducted in the search for the most significant squared amplitude, direct application of Equation 8.81 results in a test that is substantially less stringent than the nominal level, α. Because the K spectral estimates being tested are uncorrelated, dealing with this multiplicity problem is reasonably straightforward, and involves choosing a nominal test level that is small enough that Equation 8.81 specifies the correct rejection rule when applied to the *largest* of the K squared amplitudes. Fisher (1929) provides the equation to compute the exact values,

$$\alpha^* = 1 - (1 - \alpha)^K, \tag{8.82}$$

attributing it without a literature citation to Gilbert Walker. The resulting nominal test levels, α, to be used in Equation 8.81 to yield a true probability α^* of falsely rejecting the null hypothesis that the *largest* of K periodogram estimates is significantly larger than the null spectral density at that frequency, are closely approximated by those calculated by the Bonferroni method (Section 10.5.3),

$$\alpha = \alpha^*/K. \tag{8.83}$$

In order to account for the test multiplicity it is necessary to choose a nominal test level α that is smaller that the actual test level α^*, and that reduction is proportional to the number of frequencies (i.e., independent tests) considered. The result is that a relatively large C_k^2 is required in order to reject the null hypothesis in the properly reformulated test.

EXAMPLE 8.15   Statistical Significance of the Largest Spectral Peak Relative to a Red-Noise H_0

Imagine a hypothetical time series of length $n = 200$ for which the sample estimates of the lag-1 autocorrelation and white-noise variance are $r_1 = 0.6$ and s_e^2, respectively. A reasonable candidate to describe the behavior of the data as a purely random series could be the AR(1) process with these two parameters. Substituting these values into Equation 8.77

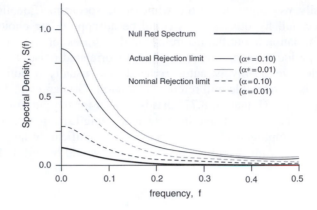

FIGURE 8.24   Red spectrum for $\phi_1 = 0.6$, $\sigma_\varepsilon^2 = 1.0$, and $n = 200$ (heavy curve) with minimum values necessary to conclude that the largest of $K = 100$ periodogram estimates is significantly larger (lighter solid curves) at the 0.10 (black) and 0.01 (grey) levels. Dashed curves show erroneous minimum values resulting when test multiplicity is not accounted for.

yields the spectrum for this process, shown as the heavy curve in Figure 8.24. A sample spectrum, C_k^2, $k = 1, \ldots 100$, can also be computed from this series. This spectrum will include squared amplitudes at $K = 100$ frequencies because $n = 200$ data points have been Fourier transformed. Whether or not the series also contains one or more periodic components the sample spectrum will be rather erratic, and it may be of interest to calculate how large the largest C_k^2 must be in order to infer that it is significantly different from the null red spectrum at that frequency. Equation 8.81 provides the decision criterion.

Because $K = 100$ frequencies are being searched for the largest squared amplitude, the standard of proof must be much more stringent than if a particular single frequency had been chosen for testing in advance of seeing the data. In particular, Equation 8.82 and Equation 8.83 both show that a test at the $\alpha^* = 0.10$ level requires that the largest of the 100 squared amplitudes trigger a test rejection at the nominal $\alpha = 0.10/100 = 0.001$ level, and a test at the $\alpha^* = 0.01$ level requires the nominal test level $\alpha = 0.01/100 = 0.0001$. Each squared amplitude in the unsmoothed sample spectrum follows a χ^2 distribution with $\nu = 2$ degrees of freedom, so the relevant right-tail quantiles $\chi_2^2(1 - \alpha)$ from the second line of Table B.3 are $\chi_2^2(0.999) = 13.816$ and $\chi_2^2(0.9999) = 18.421$, respectively. (Because $\nu = 2$ these probabilities can also be calculated using the quantile function for the exponential distribution, Equation 4.80, with $\beta = 2$.) Substituting these values into Equation 8.81, and using Equation 8.77 with $\phi_1 = 0.6$ and σ_ε^2 to define $S_0(f_k)$, yields the two light solid lines in Figure 8.24. If the largest of the $K = 100 C_k^2$ values does not rise above these curves, the null hypothesis that the series arose from a purely random AR(1) process cannot be rejected at the specified α^* levels.

The dashed curves in Figure 8.24 are the rejection limits computed in the same was as the solid curves, except that the nominal test levels α have been taken to be equal to the overall test levels α^*, so that $\chi_2^2(0.90) = 4.605$ and $\chi_2^2(0.99) = 9.210$ have been used in Equation 8.81. These dashed curves would be appropriate thresholds for rejecting the null hypothesis that the estimated spectrum *at a single frequency that had been chosen in advance without reference to the data being tested*, had resulted from sampling variations in the null red-noise process. If these thresholds were to be used to evaluate the largest among $K = 100$ squared amplitudes, the probabilities according to Equation 8.82 of falsely rejecting the null hypothesis if it were true would be $\alpha^* = 0.634$ and $\alpha^* = 0.99997$ (i.e., virtual certainty), at the nominal $\alpha = 0.01$ and $\alpha = 0.10$ levels, respectively.

Choice of the null spectrum can also have a large effect on the test results. If instead a white spectrum—Equation 8.77, with $\phi = 0$, implying $\sigma_x^2 = 1.5625$ (cf. Equation 8.21)— had been chosen as the baseline against which to judge potentially significant squared amplitudes, the null spectrum in Equation 8.81 would have been $S_0(f_k) = 0.031$ for all frequencies. In that case, the rejection limits would be parallel horizontal lines with magnitudes comparable to those at $f = 0.15$ in Figure 8.24. ◊

8.6 Exercises

8.1. Using the January 1987 precipitation data for Canandaigua in Table A.1,

   a. Fit a two-state, first-order Markov chain to represent daily precipitation occurrence.
   b. Test whether this Markov model provides a significantly better representation of the data than does the assumption of independence.
   c. Compare the theoretical stationary probability, π_1 with the empirical relative frequency.
   d. Graph the theoretical autocorrelation function, for the first three lags.
   e. Compute the probability according to the Markov model that a sequence of consecutive wet days will last at least three days.

8.2. Graph the autocorrelation functions up to five lags for

   a. The AR(1) process with $\phi = 0.4$.
   b. The AR(2) process with $\phi_1 = 0.7$ and $\phi_2 = -0.7$.

8.3. Compute sample lag correlations for a time series with $n = 100$ values, whose variance is 100, yields $r_1 = 0.80$, $r_2 = 0.60$, and $r_3 = 0.50$.

   a. Use the Yule-Walker equations to fit AR(1), AR(2), and AR(3) models to the data. Assume the sample size is large enough that Equation 8.26 provides a good estimate for the white-noise variance.
   b. Select the best autoregressive model for the series according to the BIC statistic.
   c. Select the best autoregressive model for the series according to the AIC statistic.

8.4. Given that the mean of the time series in Exercise 8.3 is 50, use the fitted AR(2) model to forecast the future values of the time series x_1, x_2, and x_3; assuming the current value is $x_0 = 76$ and the previous value is $x_{-1} = 65$.

8.5. The variance of a time series governed by the AR(1) model with $\phi = 0.8$, is 25. Compute the variances of the sampling distributions of averages of consecutive values of this time series, with lengths

   a. $n = 5$,
   b. $n = 10$,
   c. $n = 50$.

8.6. For the temperature data in Table 8.7,

   a. Calculate the first two harmonics.
   b. Plot each of the two harmonics separately.
   c. Plot the function representing the annual cycle defined by the first two harmonics. Also include the original data points in this plot, and visually compare the goodness of fit.

TABLE 8.7   Average monthly temperature data for New Delhi, India.

Month	J	F	M	A	M	J	J	A	S	O	N	D
Average Temperature, °F	57	62	73	82	92	94	88	86	84	79	68	59

8.7. Use the two-harmonic equation for the annual cycle from Exercise 8.6 to estimate the mean daily temperatures for

a. April 10.
b. October 27.

8.8. The amplitudes of the third, fourth, fifth, and sixth harmonics, respectively, of the data in Table 8.7 are 1.4907, 0.5773, 0.6311, and 0.0001°F.

a. Plot a periodogram for this data. Explain what it shows.
b. What proportion of the variation in the monthly average temperature data is described by the first two harmonics?

8.9. How many tic-marks for frequency are missing from the horizontal axis of Figure 8.17?

8.10. Suppose the minor peak in Figure 8.17 at $f = 13/256 = 0.0508 \, \text{mo}^{-1}$ resulted in part from aliasing.

a. Compute a frequency that could have produced this spurious signal in the spectrum.
b. How often would the underlying sea-level pressure data need to be recorded and processed in order to resolve this frequency explicitly?

8.11. Derive and plot the theoretical spectra for the two autoregressive processes in Exercise 8.2, assuming unit white-noise variance, and $n = 100$.

8.12. The largest squared amplitude in Figure 8.23 is $C_{11}^2 = 0.413$ (in the grey spectrum).

a. Compute a 95% confidence interval for the value of the underlying spectral density at this frequency.
b. Test whether this largest value is significantly different from the null white-noise spectral density at this frequency, assuming that the variance of the underlying data is 1, using the $\alpha^* = 0.015$ level.

PART · III

Multivariate Statistics

CHAPTER • 9

Matrix Algebra and Random Matrices

9.1 Background to Multivariate Statistics

9.1.1 Contrasts between Multivariate and Univariate Statistics

Much of the material in the first eight chapters of this book has pertained to analysis of univariate, or one-dimensional data. That is, the analysis methods presented were oriented primarily toward scalar data values and their distributions. However, we find in many practical situations that data sets are composed of vector observations. In this situation each data record consists of simultaneous observations of multiple quantities. Such data sets are known as *multivariate*. Examples of multivariate atmospheric data include simultaneous observations of multiple variables at one location, or an atmospheric field as represented by a set of gridpoint values at a particular time.

Univariate methods can be, and are, applied to individual scalar elements of multivariate data observations. The distinguishing attribute of multivariate methods is that both the joint behavior of the multiple simultaneous observations, as well as the variations of the individual data elements, are considered. The remaining chapters of this book present introductions to some of the multivariate methods that are used most commonly with atmospheric data. These include approaches to data reduction and structural simplification, characterization and summarization of multiple dependencies, prediction of one or more of the variables from the remaining ones, and grouping and classification of the multivariate observations.

Multivariate methods are more difficult to understand and implement than univariate methods. Notationally, they require use of matrix algebra to make the presentation tractable, and the elements of matrix algebra that are necessary to understand the subsequent material are presented briefly in Section 9.3. The complexities of multivariate data and the methods that have been devised to deal with them dictate that all but the very simplest multivariate analyses will be implemented using a computer. Enough detail is included here for readers comfortable with numerical methods to be able to implement the analyses themselves. However, many readers will use statistical software for this purpose, and the material in these final chapters should help them to understand what these computer programs are doing, and why.

9.1.2 *Organization of Data and Basic Notation*

In conventional univariate statistics, each datum or observation is a single number, or scalar. In multivariate statistics each datum is a collection of simultaneous observations of $K \geq 2$ scalar values. For both notational and computational convenience, these multivariate observations are arranged in an ordered list known as a vector, with a boldface single symbol being used to represent the entire collection, for example,

$$\mathbf{x}^T = [x_1, x_2, x_3, \ldots, x_K].$$ (9.1)

The superscript T on the left-hand side has a specific meaning that will be explained in Section 9.3, but for now we can safely ignore it. Because the K individual values are arranged horizontally, Equation 9.1 is called a row vector, and each of the positions within it corresponds to one of the K scalars whose simultaneous relationships will be considered. It can be convenient to visualize or (for higher dimensions, K) imagine a data vector geometrically, as a point in a K-dimensional space, or as an arrow whose tip position is defined by the listed scalars, and whose base is at the origin. Depending on the nature of the data, this abstract geometric space may correspond to a phase- or state-space (see Section 6.6.2), or some subset of the dimensions (a subspace) of such a space.

A univariate data set consists of a collection of n scalar observations $x_i, i = 1, \ldots, n$. Similarly, a multivariate data set consists of a collection of n data vectors $\mathbf{x}_i, i = 1, \ldots, n$. Again for both notational and computational convenience this collection of data vectors can be arranged into a rectangular array of numbers having n rows, each corresponding to one multivariate observation, and with each of the K columns containing all n observations of one of the variables. This arrangement of the $n \times K$ numbers in the multivariate data set is called a data matrix,

$$[X] = \begin{bmatrix} \mathbf{x}_1^T \\ \mathbf{x}_2^T \\ \mathbf{x}_3^T \\ \vdots \\ \mathbf{x}_n^T \end{bmatrix} = \begin{bmatrix} x_{1,1} & x_{1,2} & \cdots & x_{1,K} \\ x_{2,1} & x_{2,2} & \cdots & x_{2,K} \\ x_{3,1} & x_{3,2} & \cdots & x_{3,K} \\ \vdots & \vdots & & \vdots \\ x_{n,1} & x_{n,2} & \cdots & x_{n,K} \end{bmatrix}.$$ (9.2)

Here n row-vector observations of the form shown in Equation 9.1 have been stacked vertically to yield a rectangular array, called a matrix, with n rows and K columns. Conventionally, the first of the two subscripts of the scalar elements of a matrix denotes the row number, and the second indicates the column number so, for example, $x_{3,2}$ is the third of the n observations of the second of the K variables. In this book matrices will be denoted using square brackets, as a pictorial reminder that the symbol within represents a rectangular array.

The data matrix $[X]$ in Equation 9.2 corresponds exactly to a conventional data table or spreadsheet display, in which each column pertains to one of the variables considered, and each row represents one of the n observations. Its contents can also be visualized or imagined geometrically within an abstract K-dimensional space, with each of the n rows defining a single point. The simplest example is a data matrix for bivariate data, which has n rows and $K = 2$ columns. The pair of numbers in each of the rows locates a point on the Cartesian plane. The collection of these n points on the plane defines a scatterplot of the bivariate data.

9.1.3 *Multivariate Extensions of Common Univariate Statistics*

Just as the data vector in Equation 9.1 is the multivariate extension of a scalar datum, multivariate sample statistics can be expressed using the notation of vectors and matrices. The most common of these is the multivariate sample mean, which is just a vector of the K individual scalar sample means (Equation 3.2), arranged in the same order as the elements of the underlying data vectors,

$$\bar{\mathbf{x}}^T = \left[\frac{1}{n}\sum_{i=1}^{n} x_{i,1}, \frac{1}{n}\sum_{i=1}^{n} x_{i,2}, \dots, \frac{1}{n}\sum_{i=1}^{n} x_{i,K} \right] = [\bar{x}_1, \bar{x}_2, \dots, \bar{x}_K]. \qquad (9.3)$$

As before, the boldface symbol on the left-hand side of Equation 9.3 indicates a vector quantity, and the double-subscripted variables in the first equality are indexed according to the same convention as in Equation 9.2.

The multivariate extension of the sample standard deviation (Equation 3.6), or (much more commonly, its square) the sample variance, is a little more complicated because all pairwise relationships among the K variables need to be considered. In particular, the multivariate extension of the sample variance is the collection of covariances between all possible pairs of the K variables,

$$\text{Cov}(x_k, x_\ell) = s_{k,\ell} = \frac{1}{n-1}\sum_{i=1}^{n} (x_{i,k} - \bar{x}_k)(x_{i,\ell} - \bar{x}_\ell). \qquad (9.4)$$

which is equivalent to the numerator of Equation 3.22. If the two variables are the same, that is, if $k = \ell$, then Equation 9.4 defines the sample variance, $s_k^2 = s_{k,k}$, or the square of Equation 3.6. Although the notation $s_{k,k}$ for the sample variance of the kth variable may seem a little strange at first, it is conventional in multivariate statistics, and is also convenient from the standpoint of arranging the covariances calculated according to Equation 9.4 into a square array called the sample covariance matrix,

$$[S] = \begin{bmatrix} s_{1,1} & s_{1,2} & s_{1,3} & \cdots & s_{1,\ell} \\ s_{2,1} & s_{2,2} & s_{2,3} & \cdots & s_{2,\ell} \\ s_{3,1} & s_{3,2} & s_{3,3} & \cdots & s_{3,\ell} \\ \vdots & \vdots & \vdots & & \vdots \\ s_{k,1} & s_{k,2} & s_{k,3} & \cdots & s_{k,\ell} \end{bmatrix}. \qquad (9.5)$$

That is, the covariance $s_{k,\ell}$ is displayed in the kth row and ℓ^{th} column of the covariance matrix. The sample covariance matrix, or variance-covariance matrix, is directly analogous to the sample (Pearson) correlation matrix (see Figure 3.25), with the relationship between corresponding elements of the two matrices being given by Equation 3.22; that is, $r_{k,\ell} = s_{k,\ell}/[(s_{k,k})(s_{\ell,\ell})]^{1/2}$. The K covariances $s_{k,k}$ in the diagonal positions between the upper-left and lower-right corners of the sample covariance matrix are simply the K sample variances. The remaining, off-diagonal, elements are covariances among unlike variables, and the values below and to the left of the diagonal positions duplicate the values above and to the right.

The variance-covariance matrix is also known as the dispersion matrix, because it describes how the observations are dispersed around their (vector) mean in the K-dimensional space defined by the K variables. The diagonal elements are the individual variances, which index the degree to which the data are spread out in directions parallel to the K coordinate axes for this space, and the covariances in the off-diagonal positions describe the extent to which the cloud of data points is oriented at angles to these axes. The matrix [S] is the sample estimate of the population dispersion matrix [Σ], which appears in the probability density function for the multivariate normal distribution (Equation 10.1).

9.2 Multivariate Distance

It was pointed out in the previous section that a data vector can be regarded as a point in the K-dimensional geometric space whose coordinate axes correspond to the K variables being simultaneously represented. Many multivariate statistical approaches are based on, and/or can be interpreted in terms of, distances within this K-dimensional space. Any number of distance measures can be defined (see Section 14.1.2), but two of these are of particular importance.

9.2.1 Euclidean Distance

Perhaps the easiest and most intuitive distance measure is conventional Euclidean distance, because it corresponds to our ordinary experience in the three-dimensional world. Euclidean distance is easiest to visualize in two dimensions, where it can easily be seen as a consequence of the Pythagorean theorem, as illustrated in Figure 9.1. Here two points, $\mathbf{x}$ and $\mathbf{y}$, located by the dots, define the hypotenuse of a right triangle whose other two legs are parallel to the two data axes. The Euclidean distance $\|\mathbf{y}-\mathbf{x}\| = \|\mathbf{x}-\mathbf{y}\|$ is obtained by taking the square root of the sum of the squared lengths of the other two sides.

Euclidean distance generalizes directly to $K \geq 3$ dimensions even though the corresponding geometric space may be difficult or impossible to imagine. In particular,

$$\|\mathbf{x}-\mathbf{y}\| = \sqrt{\sum_{k=1}^{K}(x_k - y_k)^2}. \tag{9.6}$$

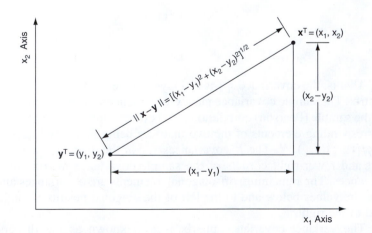

FIGURE 9.1   Illustration of the Euclidean distance between points $\mathbf{x}$ and $\mathbf{y}$ in $K = 2$ dimensions using the Pythagorean theorem.

Distance between a point x and the origin can also be calculated using Equation 9.6 by substituting a vector of K zeros (which locates the origin in the corresponding K-dimensional space) for the vector y.

It can be mathematically convenient to work in terms of squared distances. No information is lost in so doing, because distance ordinarily is regarded as necessarily nonnegative, so that squared distance is a monotonic and invertable transformation of ordinary dimensional distance (e.g., Equation 9.6). In addition, the square-root operation is avoided. Points at a constant squared distance $C^2 = \|x - y\|^2$ define a circle on the plane with radius C for $K = 2$ dimensions, a sphere in a volume with radius C for $K = 3$ dimensions, and a hypersphere with radius C within a K-dimensional hypervolume for $K > 3$ dimensions.

9.2.2 Mahalanobis (Statistical) Distance

Euclidean distance treats the separation of pairs of points in a K-dimensional space equally, regardless of their relative orientation. But it will be very useful to interpret distances between points in terms of statistical dissimilarity or unusualness, and in this sense point separations in some directions are more unusual than others. This context for unusualness is established by a (K-dimensional, joint) probability distribution for the data points, which may be characterized using the scatter of a finite sample, or using a parametric probability density.

Figure 9.2 illustrates the issues in $K = 2$ dimensions. Figure 9.2a shows a statistical context established by the scatter of points $x^T = [x_1, x_2]$. The distribution is centered on the origin; and the standard deviation of x_1 is approximately three times that of x_2; that is, $s_1 \approx 3 s_2$. The orientation of the point cloud along one of the axes reflects the fact that the two variables x_1 and x_2 are essentially uncorrelated (the points in fact have been drawn from a bivariate Gaussian distribution; see Section 4.4.2). Because of this difference in dispersion, horizontal distances are less unusual than vertical ones relative to this data scatter. Although point A is closer to the center of the distribution according to Euclidean distance, it is more unusual than point B in the context established by the point cloud, and so is statistically further from the origin.

Because the points in Figure 9.2a are uncorrelated, a distance measure that reflects unusualness in the context of the data scatter can be defined simply as

$$D^2 = \frac{(x_1 - \bar{x}_1)^2}{s_{1,1}} + \frac{(x_2 - \bar{x}_2)^2}{s_{2,2}}, \tag{9.7}$$

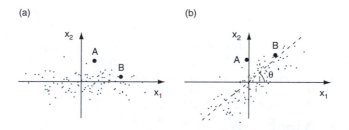

(a)   (b)

FIGURE 9.2   Distance in the context of data scatters centered at the origin. (a) The standard deviation of x_1 is approximately three times larger than the standard deviation of x_2. Point A is closer to the origin in terms of Euclidean distance, but point B is less unusual relative to the data scatter, and so is closer in statistical distance. (b) The same points rotated through an angle $\theta = 40°$.

which is a special case of the Mahalanobis distance between the point $x^T = [x_1, x_2]$ and the origin (because the two sample means are zero) when variations in the $K = 2$ dimensions are uncorrelated. For convenience Equation 9.7 is expressed as a squared distance, and it is equivalent to the ordinary squared Euclidean distance after the transformation that divides each element of the data vector by its respective standard deviation (recall that, for example, $s_{1,1}$ is the sample variance of x_1). Another interpretation of Equation 9.7 is as the sum of the two squared standardized anomalies, or z-scores (see Section 3.4.2). In either case, the importance ascribed to a distance along one of the axes is inversely proportional to the data scatter, or uncertainty, in that direction. Consequently point A is further from the origin than point B in Figure 9.2 when measured according to the Mahalanobis distance.

For a fixed Mahalanobis distance D^2, Equation 9.7 defines an ellipse of constant statistical distance on the plane, and that ellipse is also a circle if $s_{1,1} = s_{2,2}$. Generalizing Equation 9.7 to three dimensions by adding a third term for x_3, the set of points at a fixed distance D^2 constitute an ellipsoid that will be spherical if all three variances are equal, blimp-like if two variances are nearly equal but smaller than the third, and disk-like if two variances are nearly equal and larger than the third.

In general the variables within a multivariate data vector x will not be uncorrelated, and these correlations must also be accounted for when defining distances in terms of data scatter or probability density. Figure 9.2b illustrates the situation in two dimensions, in which the points from Figure 9.2a have been rotated around the origin through an angle θ, which results in the two variables being relatively strongly positively correlated. Again point B is closer to the origin in a statistical sense, although in order to calculate the actual Mahalanobis distances in terms of the variables x_1 and x_2 it would be necessary to use an equation of the form

$$D^2 = a_{1,1}(x_1 - \bar{x}_1)^2 + 2a_{1,2}(x_1 - \bar{x}_1)(x_2 - \bar{x}_2) + a_{2,2}(x_2 - \bar{x}_2)^2. \tag{9.8}$$

Analogous expressions of this kind for the Mahalanobis distance in K dimensions would involve $K(K+1)/2$ terms. Even in only two dimensions the coefficients $a_{1,1}$, $a_{1,2}$, and $a_{2,2}$, are fairly complicated functions of the rotation angle θ and the three covariances $s_{1,1}$, $s_{1,2}$, and $s_{2,2}$. For example,

$$a_{1,1} = \frac{\cos^2(\theta)}{\cos^2(\theta)s_{1,1} - 2\sin(\theta)\cos(\theta)s_{1,2} + \sin^2(\theta)s_{2,2}}$$
$$+ \frac{\sin^2(\theta)}{\cos^2(\theta)s_{2,2} + 2\sin(\theta)\cos(\theta)s_{1,2} + \sin^2(\theta)s_{1,1}}. \tag{9.9}$$

Do not study this equation at all closely. It is here to help convince you, if that is even required, that conventional scalar notation is hopelessly impractical for expressing the mathematical ideas necessary to multivariate statistics. Matrix notation and matrix algebra, which will be reviewed in the next section, are practical necessities for taking the development further. Section 9.4 will resume the statistical development using this notation, including a revisiting of the Mahalanobis distance in Section 9.4.4.

9.3 Matrix Algebra Review

The mathematics of dealing simultaneously with multiple variables and their mutual correlations is greatly simplified by use of matrix notation, and a set of computational rules called matrix algebra, or linear algebra. The notation for vectors and matrices was briefly introduced in Section 9.1.2. Matrix algebra is the toolkit used to mathematically manipulate these notational objects. A brief review of this subject, sufficient for the multivariate techniques described in the following chapters, is presented in this section. More complete introductions are readily available elsewhere (e.g., Golub and van Loan 1996; Lipschutz 1968; Strang 1988).

9.3.1 Vectors

The vector is a fundamental component of the notation of matrix algebra. It is essentially nothing more than an ordered list of scalar variables, or ordinary numbers, that are called the elements of the vector. The number of elements, also called the vector's dimension, will depend on the situation at hand. A familiar meteorological example is the two-dimensional horizontal wind vector, whose two elements are the eastward wind speed u, and the northward wind speed v.

Vectors already have been introduced in Equation 9.1, and as previously noted will be indicated using boldface type. A vector with only $K = 1$ element is just an ordinary number, or scalar. Unless otherwise indicated, vectors will be regarded as column vectors, which means that their elements are arranged vertically. For example, the column vector x would consist of the elements $x_1, x_2, x_3, \ldots, x_K$; arranged as

$$\mathbf{x} = \begin{bmatrix} x_1 \\ x_2 \\ x_3 \\ \vdots \\ x_K \end{bmatrix}. \tag{9.10}$$

These same elements can be arranged horizontally, as in Equation 9.1, which is a row vector. Column vectors are transformed to row vectors, and vice versa, through an operation called transposing the vector. The transpose operation is denoted by the superscript T, so that we can write the vector x in Equation 9.10 as the row vector x^T in Equation 9.1, which is pronounced x-transpose. The transpose of a column vector is useful for notational consistency with certain matrix operations. It is also useful for typographical purposes, as it allows a vector to be written on a horizontal line of text.

Addition of two or more vectors with the same dimension is straightforward. Vector addition is accomplished by adding the corresponding elements of the two vectors, for example

$$\mathbf{x} + \mathbf{y} = \begin{bmatrix} x_1 \\ x_2 \\ x_3 \\ \vdots \\ x_K \end{bmatrix} + \begin{bmatrix} y_1 \\ y_2 \\ y_3 \\ \vdots \\ y_K \end{bmatrix} = \begin{bmatrix} x_1 + y_1 \\ x_2 + y_2 \\ x_3 + y_3 \\ \vdots \\ x_K + y_K \end{bmatrix}. \tag{9.11}$$

Subtraction is accomplished analogously. This operation, and other operations with vectors, reduces to ordinary scalar addition or subtraction when the two vectors have dimension $K = 1$. Addition and subtraction of vectors with different dimensions is not defined.

Multiplying a vector by a scalar results in a new vector whose elements are simply the corresponding elements of the original vector multiplied by that scalar. For example, multiplying the vector x in Equation 9.10 by a constant c yields

$$c\mathbf{x} = \begin{bmatrix} cx_1 \\ cx_2 \\ cx_3 \\ \vdots \\ cx_K \end{bmatrix}. \tag{9.12}$$

Two vectors of the same dimension can be multiplied using an operation called the dot product, or inner product. This operation consists of multiplying together each of the K like pairs of vector elements, and then summing these K products. That is,

$$\mathbf{x}^T\mathbf{y} = [x_1, x_2, x_3, \ldots, x_K] \begin{bmatrix} y_1 \\ y_2 \\ y_3 \\ \vdots \\ y_K \end{bmatrix} = x_1y_1 + x_2y_2 + x_3y_3 + \cdots + x_Ky_K$$

$$= \sum_{k=1}^{K} x_k y_k. \tag{9.13}$$

This vector multiplication has been written as the product of a row vector on the left and a column vector on the right in order to be consistent with the operation of matrix multiplication, which will be presented in Section 9.3.2. As will be seen, the dot product is in fact a special case of matrix multiplication, and (unless $K = 1$) the order of vector and matrix multiplication is important: the multiplications $\mathbf{x}^T\mathbf{y}$ and $\mathbf{y}\mathbf{x}^T$ yield entirely different results. Equation 9.13 also shows that vector multiplication can be expressed in component form using summation notation. Expanding vector and matrix operations in component form can be useful if the calculation is to be programmed for a computer, depending on the programming language.

As noted previously, a vector can be visualized as a point in K-dimensional space. The Euclidean length of a vector in that space is the ordinary distance between the point and the origin. Length is a scalar quantity that can be computed using the dot product, as

$$\|\mathbf{x}\| = \sqrt{\mathbf{x}^T\mathbf{x}} = \left[\sum_{k=1}^{K} x_k^2 \right]^{1/2}. \tag{9.14}$$

Equation 9.14 is sometimes known as the Euclidean norm of the vector x. Figure 9.1, with $y = 0$ as the origin, illustrates that this length is simply an application of the Pythagorean theorem. A common application of Euclidean length is in the computation of the total horizontal wind speed from the horizontal velocity vector $v^T = [u, v]$, according to $v_H = (u^2 + v^2)^{1/2}$. However Equation 9.14 generalizes to arbitrarily high K as well.

The angle θ between two vectors is also computed using the dot product, using

$$\theta = \cos^{-1}\left[\frac{\mathbf{x}^T\mathbf{y}}{\|\mathbf{x}\|\|\mathbf{y}\|} \right]. \tag{9.15}$$

This relationship implies that two vectors are perpendicular if their dot product is zero, since $\cos[0] = 90°$. Mutually perpendicular vectors are also called orthogonal.

The magnitude of the projection (or "length of the shadow") of a vector x onto a vector y is also a function of the dot product, given by

$$L_{x,y} = \frac{\mathbf{x}^T\mathbf{y}}{\|\mathbf{y}\|}. \tag{9.16}$$

The geometric interpretations of these three computations of length, angle, and projection are illustrated in Figure 9.3, for the vectors $\mathbf{x}^T = [1, 1]$ and $\mathbf{y}^T = [2, 0.8]$. The length of $\mathbf{x}$ is simply $\|\mathbf{x}\| = (1^2 + 1^2)^{1/2} = \sqrt{2}$, and the length of $\mathbf{y}$ is $\|\mathbf{y}\| = (2^2 + 0.8^2)^{1/2} = 2.154$. Since the dot product of the two vectors is $\mathbf{x}^T\mathbf{y} = 1 \cdot 2 + 1 \cdot 0.8 = 2.8$, the angle between them is $\theta = \cos^{-1}[2.8/(\sqrt{2} \cdot 2.154)] = 23°$, and the length of the projection of $\mathbf{x}$ onto $\mathbf{y}$ is $2.8/2.154 = 1.302$.

9.3.2 Matrices

A matrix is a two-dimensional rectangular array of numbers having I rows and J columns. The dimension of a matrix is specified by these numbers of rows and columns. A matrix dimension is written $(I \times J)$, and pronounced I by J. Matrices are denoted here by uppercase letters surrounded by square brackets. Sometimes, for notational clarity and convenience, the parenthetical expression for the dimension of a matrix will be written directly below it. The elements of a matrix are the individual variables or numerical values occupying the rows and columns. The matrix elements are identified notationally by two subscripts; the first of these identifies the row number and the second identifies the column number. Equation 9.2 shows a $(n \times K)$ data matrix, and Equation 9.5 shows a $(K \times K)$ covariance matrix, with the subscripting convention illustrated.

A vector is a special case of a matrix, and matrix operations are applicable also to vectors. A K-dimensional row vector is a $(1 \times K)$ matrix, and a column vector is a $(K \times 1)$ matrix. Just as a $K = 1$-dimensional vector is also a scalar, so too is a (1×1) matrix.

A matrix with the same number of rows and columns, such as [S] in Equation 9.5, is called a square matrix. The elements of a square matrix for which $i = j$ are arranged on the diagonal between the upper left to the lower right corners, and are called diagonal elements. Correlation matrices [R] (see Figure 3.25) are square matrices having all 1s on the diagonal. A matrix for which $a_{i,j} = a_{j,i}$ for all values of i and j is called symmetric. Correlation and covariance matrices are symmetric because the correlation between

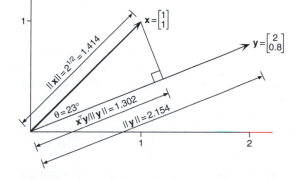

FIGURE 9.3   Illustration of the concepts of vector length (Equation 9.14), the angle between two vectors (Equation 9.15), and the projection of one vector onto another (Equation 9.16); for the two vectors $\mathbf{x}^T = [1, 1]$ and $\mathbf{y}^T = [2, 0.8]$.

variable i and variable j is identical to the correlation between variable j and variable i. Another important square, symmetric matrix is the identity matrix [I], consisting of 1s on the diagonal and zeros everywhere else,

$$[\mathbf{I}] = \begin{bmatrix} 1 & 0 & 0 & \cdots & 0 \\ 0 & 1 & 0 & \cdots & 0 \\ 0 & 0 & 1 & \cdots & 0 \\ \vdots & \vdots & \vdots & & \vdots \\ 0 & 0 & 0 & \cdots & 1 \end{bmatrix}. \tag{9.17}$$

An identity matrix can be constructed for any (square) dimension. When the identity matrix appears in an equation it can be assumed to be of appropriate dimension for the relevant matrix operations to be defined.

The transpose operation is defined for any matrix, including the special case of vectors. The transpose of a matrix is obtained in general by exchanging row and column indices, not by a 90° rotation as might have been anticipated from a comparison of Equations 9.1 and 9.10. Geometrically, the transpose operation is like a reflection across the matrix diagonal, that extends downward and to the right from the upper, left-hand element. For example, the relationship between the (3×4) matrix [B] and its transpose, the (4×3) matrix $[\mathbf{B}]^{\mathrm{T}}$, is illustrated by comparing

$$[\mathbf{B}] = \begin{bmatrix} b_{1,1} & b_{1,2} & b_{1,3} & b_{1,4} \\ b_{2,1} & b_{2,2} & b_{2,3} & b_{2,4} \\ b_{3,1} & b_{3,2} & b_{3,3} & b_{3,4} \end{bmatrix} \tag{9.18a}$$

and

$$[\mathbf{B}]^{\mathrm{T}} = \begin{bmatrix} b_{1,1} & b_{2,1} & b_{3,1} \\ b_{1,2} & b_{2,2} & b_{3,2} \\ b_{1,3} & b_{2,3} & b_{3,3} \\ b_{1,4} & b_{2,4} & b_{3,4} \end{bmatrix}. \tag{9.18b}$$

If a matrix [A] is symmetric, then $[\mathbf{A}]^{\mathrm{T}} = [\mathbf{A}]$.

Multiplication of a matrix by a scalar is the same as for vectors, and is accomplished by multiplying each element of the matrix by the scalar,

$$c[\mathbf{D}] = c \begin{bmatrix} d_{1,1} & d_{1,2} \\ d_{2,1} & d_{2,2} \end{bmatrix} = \begin{bmatrix} c\,d_{1,1} & c\,d_{1,2} \\ c\,d_{2,1} & c\,d_{2,2} \end{bmatrix}. \tag{9.19}$$

Similarly, matrix addition and subtraction are defined only for matrices of identical dimension, and are accomplished by performing these operations on the elements in corresponding row and column positions. For example, the sum of two (2×2) matrices would be computed as

$$[\mathbf{D}] + [\mathbf{E}] = \begin{bmatrix} d_{1,1} & d_{1,2} \\ d_{2,1} & d_{2,2} \end{bmatrix} + \begin{bmatrix} e_{1,1} & e_{1,2} \\ e_{2,1} & e_{2,2} \end{bmatrix} = \begin{bmatrix} d_{1,1}+e_{1,1} & d_{1,2}+e_{1,2} \\ d_{2,1}+e_{2,1} & d_{2,2}+e_{2,2} \end{bmatrix}. \tag{9.20}$$

Matrix multiplication is defined between two matrices if the number of columns in the left matrix is equal to the number of rows in the right matrix. Thus, not only is matrix multiplication not commutative (i.e., $[\mathbf{A}][\mathbf{B}] \neq [\mathbf{B}][\mathbf{A}]$), but multiplication of two matrices in reverse order is not even defined unless the two have complementary row and column dimensions. The product of a matrix multiplication is another matrix, the row dimension of which is the same as the row dimension of the left matrix, and the column dimension of which is the same as the column dimension of the right matrix. That is, multiplying a $(I \times J)$ matrix [A] and a $(J \times K)$ matrix [B] yields a $(I \times K)$ matrix [C]. In effect, the middle dimension J is "multiplied out."

Consider the case where $I = 2$, $J = 3$, and $K = 2$. In terms of the individual matrix elements, the matrix multiplication [A][B] = [C] expands to

$$
\begin{bmatrix} a_{1,1} & a_{1,2} & a_{1,3} \\ a_{2,1} & a_{2,2} & a_{2,3} \end{bmatrix}
\begin{bmatrix} b_{1,1} & b_{1,2} \\ b_{2,1} & b_{2,2} \\ b_{3,1} & b_{3,2} \end{bmatrix}
= \begin{bmatrix} c_{1,1} & c_{1,2} \\ c_{2,1} & c_{2,2} \end{bmatrix}, \tag{9.21a}
$$

$$\quad (2 \times 3) \qquad\qquad (3 \times 2) \qquad\qquad (2 \times 2)$$

where

$$
[C] = \begin{bmatrix} c_{1,1} & c_{1,2} \\ c_{2,1} & c_{2,2} \end{bmatrix} = \begin{bmatrix} a_{1,1}b_{1,1} + a_{1,2}b_{2,1} + a_{1,3}b_{3,1} & a_{1,1}b_{1,2} + a_{1,2}b_{2,2} + a_{1,3}b_{1,3} \\ a_{2,1}b_{1,1} + a_{2,2}b_{2,1} + a_{2,3}b_{3,1} & a_{2,1}b_{1,2} + a_{2,2}b_{2,2} + a_{2,3}b_{3,2} \end{bmatrix}. \tag{9.21b}
$$

The individual components of [C] as written out in Equation 9.21b may look confusing at first exposure. In understanding matrix multiplication, it is helpful to realize that each element of the product matrix [C] is simply the dot product, as defined in Equation 9.13, of one of the rows in the left matrix [A] and one of the columns in the right matrix [B]. In particular, the number occupying the ith row and k^{th} column of the matrix [C] is exactly the dot product between the row *vector* comprising the ith row of [A] and the column *vector* comprising the kth column of [B]. Equivalently, matrix multiplication can be written in terms of the individual matrix elements using summation notation,

$$
c_{i,k} = \sum_{j=1}^{J} a_{i,j} b_{j,k}; \qquad i = 1, \ldots, I; \quad k = 1, \ldots, K. \tag{9.22}
$$

Figure 9.4 illustrates the procedure graphically, for one element of the matrix [C] resulting from the multiplication [A][B] = [C].

The identity matrix (Equation 9.17) is so named because it functions as the multiplicative identity—that is, [A][I] = [A], and [I][A] = [A] regardless of the dimension of [A]—although in the former case [I] is a square matrix with the same number of columns as [A], and in the latter its dimension is the same as the number of rows in [A].

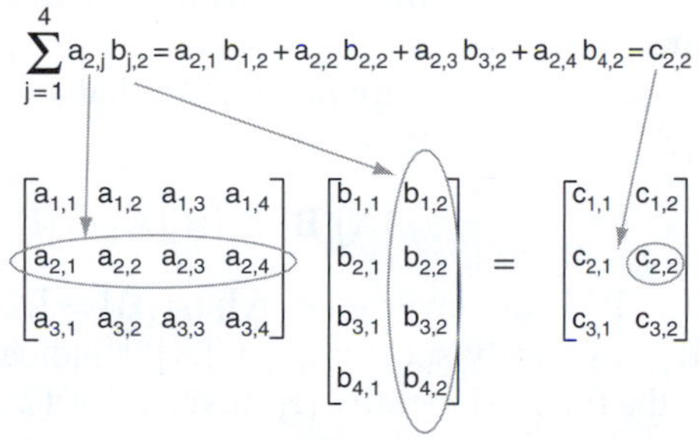

FIGURE 9.4   Graphical illustration of matrix multiplication as the dot product of the ith row of the left-hand matrix with the jth column of the right-hand matrix, yielding the element in the ith row and j^{th} column of the matrix product.

The dot product, or inner product (Equation 9.13) is one application of matrix multiplication to vectors. But the rules of matrix multiplication also allow multiplication of two vectors of the same dimension in the opposite order, which is called the outer product. In contrast to the inner product, which is a $(1 \times K) \times (K \times 1)$ matrix multiplication yielding a (1×1) scalar; the outer product of two vectors is a $(K \times 1) \times (1 \times K)$ matrix multiplication, yielding a $(K \times K)$ square matrix. For example, for $K = 3$,

$$\mathbf{x} \, \mathbf{y}^T = \begin{bmatrix} x_1 \\ x_2 \\ x_3 \end{bmatrix} [y_1 \; y_2 \; y_3] = \begin{bmatrix} x_1 y_1 & x_1 y_2 & x_1 y_3 \\ x_2 y_1 & x_2 y_2 & x_2 y_3 \\ x_3 y_1 & x_3 y_2 & x_3 y_3 \end{bmatrix}. \tag{9.23}$$

The trace of a square matrix is simply the sum of its diagonal elements; that is,

$$\text{tr}[A] = \sum_{k=1}^{K} a_{k,k}, \tag{9.24}$$

for the $(K \times K)$ matrix $[A]$. For the $(K \times K)$ identity matrix, $\text{tr}[I] = K$.

The determinant of a square matrix is a scalar quantity defined as

$$\det[A] = |A| = \sum_{k=1}^{K} a_{1,k} |A_{1,k}| (-1)^{1+k}, \tag{9.25}$$

where $[A_{1,k}]$ is the $(K-1 \times K-1)$ matrix formed by deleting the first row and k^{th} column of $[A]$. The absolute value notation for the matrix determinant suggests that this operation produces a scalar that is in some sense a measure of the magnitude of the matrix. The definition in Equation 9.25 is recursive, so for example computing the determinant of a $(K \times K)$ matrix requires that K determinants of reduced $(K-1 \times K-1)$ matrices be calculated first, and so on until reaching and $|A| = a_{1,1}$ for $K = 1$. Accordingly the process is quite tedious and usually best left to a computer. However, in the (2×2) case,

$$\det[A] = \det \begin{bmatrix} a_{1,1} & a_{1,2} \\ a_{2,1} & a_{2,2} \end{bmatrix} = a_{1,1} \, a_{2,2} - a_{1,2} \, a_{2,1}. \tag{9.26}$$

The analog of arithmetic division exists for square matrices that have a property known as full rank, or nonsingularity. This condition can be interpreted to mean that the matrix does not contain redundant information, in the sense that none of the rows can be constructed from linear combinations of the other rows. Considering each row of a nonsingular matrix as a vector, it is impossible to construct vector sums of rows multiplied by scalar constants, that equal any one of the other rows. These same conditions applied to the columns also imply that the matrix is nonsingular. Nonsingular matrices also have nonzero determinant.

Nonsingular square matrices are *invertible*; that a matrix $[A]$ is invertible means that another matrix $[B]$ exists such that

$$[A][B] = [B][A] = [I]. \tag{9.27}$$

It is then said that $[B]$ is the inverse of $[A]$, or $[B] = [A]^{-1}$; and that $[A]$ is the inverse of $[B]$, or $[A] = [B]^{-1}$. Loosely speaking, $[A][A]^{-1}$ indicates division of the matrix $[A]$ by itself, and yields the (matrix) identity $[I]$. Inverses of (2×2) matrices are easy to compute by hand, using

$$[A]^{-1} = \frac{1}{\det[A]} \begin{bmatrix} a_{2,2} & -a_{1,2} \\ -a_{2,1} & a_{1,1} \end{bmatrix} = \frac{1}{(a_{1,1} \, a_{2,2}) - (a_{2,1} \, a_{1,2})} \begin{bmatrix} a_{2,2} & -a_{1,2} \\ -a_{2,1} & a_{1,1} \end{bmatrix}. \tag{9.28}$$

TABLE 9.1   Some elementary properties of arithmetic operations with matrices.

Distributive multiplication by a scalar	$c([A][B]) = (c[A])[B]$
Distributive matrix multiplication	$[A]([B]+[C]) = [A][B]+[A][C]$
	$([A]+[B])[C] = [A][C]+[B][C]$
Associative matrix multiplication	$[A]([B][C]) = ([A][B])[C]$
Inverse of a matrix product	$([A][B])^{-1} = [B]^{-1}[A]^{-1}$
Transpose of a matrix product	$([A][B])^{T} = [B]^{T}[A]^{T}$
Combining matrix transpose and inverse	$([A]^{-1})^{T} = ([A]^{T})^{-1}$

This matrix is pronounced A inverse. Explicit formulas for inverting matrices of higher dimension also exist, but quickly become very cumbersome as the dimensions get larger. Computer algorithms for inverting matrices are widely available, and as a consequence matrices with dimension higher than two or three are rarely inverted by hand. An important exception is the inverse of a diagonal matrix, which is simply another diagonal matrix whose nonzero elements are the reciprocals of the diagonal elements of the original matrix. If [A] is symmetric (frequently in statistics, symmetric matrices are inverted), then $[A]^{-1}$ is also symmetric.

Table 9.1 lists some additional properties of arithmetic operations with matrices that have not been specifically mentioned in the foregoing.

EXAMPLE 9.1   Computation of the Covariance and Correlation Matrices

The covariance matrix [S] was introduced in Equation 9.5, and the correlation matrix [R] was introduced in Figure 3.25 as a device for compactly representing the mutual correlations among K variables. The correlation matrix for the January 1987 data in Table A.1 (with the unit diagonal elements and the symmetry implicit) is shown in Table 3.5. The computation of the covariances in Equation 9.4 and of the correlations in Equation 3.23 can also be expressed in notation of matrix algebra.

One way to begin the computation is with the $(n \times K)$ data matrix [X] (Equation 9.2). Each row of this matrix is a vector, consisting of one observation for each of K variables. The number of these rows is the same as the sample size, n, so [X] is just an ordinary data table such as Table A.1. In Table A.1 there are $K = 6$ variables (excluding the column containing the dates), each simultaneously observed on $n = 31$ occasions. An individual data element $x_{i,k}$ is the ith observation of the kth variable. For example, in Table A.1, $x_{4,6}$ would be the Canandaigua minimum temperature (19°F) observed on 4 January.

Define the $(n \times n)$ matrix [1], whose elements are all equal to 1. The $(n \times K)$ matrix of anomalies (in the meteorological sense of variables with their means subtracted), or centered data [X'] is then

$$[X'] = [X] - \frac{1}{n}[1][X]. \tag{9.29}$$

(Note that some authors use the prime notation in this context to indicate matrix transpose, but the superscript T has been used to indicate transpose throughout this book, to avoid confusion.) The second term in Equation 9.29 is a $(n \times K)$ matrix containing the sample means. Each of its n rows is the same, and consists of the K sample means in the same order as the corresponding variables appear in each row of [X].

Multiplying $[X']$ by the transpose of itself, and dividing by $n-1$, yields the covariance matrix,

$$[S] = \frac{1}{n-1}[X']^T[X'].\qquad(9.30)$$

This is the same symmetric $(K \times K)$ matrix as in Equation 9.5, whose diagonal elements are the sample variances of the K variables, and whose other elements are the covariances among all possible pairs of the K variables. The operation in Equation 9.30 corresponds to the summation in the numerator of Equation 3.22.

Now define the $(K \times K)$ diagonal matrix $[D]$, whose diagonal elements are the sample standard deviations of the K variables. That is, $[D]$ consists of all zeros except for the diagonal elements, whose values are the square roots of the corresponding elements of $[S]$: $d_{k,k} = \sqrt{s_{k,k}}$, $k = 1, \ldots, K$. The correlation matrix can then be computed from the covariance matrix using

$$[R] = [D]^{-1}[S][D]^{-1}.\qquad(9.31)$$

Since $[D]$ is diagonal, its inverse is the diagonal matrix whose elements are the reciprocals of the sample standard deviations on the diagonal of $[D]$. The matrix multiplication in Equation 9.31 corresponds to division by the standard deviations in Equation 3.23.

Note that the correlation matrix $[R]$ is equivalently the covariance matrix of the standardized variables (or standardized anomalies) z_k (Equation 3.21). That is, dividing the anomalies x'_k by their standard deviations $\sqrt{s_{k,k}}$ nondimensionalizes the variables, and results in their having unit variance (1's on the diagonal of $[R]$) and covariances equal to their correlations. In matrix notation this can be seen by substituting Equation 9.30 into Equation 9.31 to yield

$$[R] = \frac{1}{n-1}[D]^{-1}[X']^T[X'][D]^{-1}$$

$$= \frac{1}{n-1}[Z]^T[Z],\qquad(9.32)$$

where $[Z]$ is the $(n \times K)$ matrix whose rows are the vectors of standardized variables z, analogously to the matrix $[X']$ of the anomalies. The first line converts the matrix $[X']$ to the matrix $[Z]$ by dividing each element by its standard deviation, $d_{k,k}$. Comparing Equation 9.32 and 9.30 shows that $[R]$ is indeed the covariance matrix for the standardized variables z.

It is also possible to formulate the computation of the covariance and correlation matrices in terms of outer products of vectors. Define the ith of n (column) vectors of anomalies

$$\mathbf{x}'_i = \mathbf{x}_i - \bar{\mathbf{x}}_i,\qquad(9.33)$$

where the vector (sample) mean is the transpose of any of the rows of the matrix that is subtracted on the right-hand side of Equation 9.29. Also let the corresponding standardized anomalies (the vector counterpart of Equation 3.21) be

$$\mathbf{z}_i = [D]^{-1}\mathbf{x}'_i,\qquad(9.34)$$

where $[D]$ is again the diagonal matrix of standard deviations. Equation 9.34 is called the scaling transformation, and simply indicates division of all the values in a data vector by their respective standard deviations. The covariance matrix can then be computed in a way that is notationally analogous to the usual computation of the scalar variance (Equation 3.6, squared),

$$[S] = \frac{1}{n-1} \sum_{i=1}^{n} \mathbf{x}_i' \mathbf{x}_i'^{\,T}; \tag{9.35}$$

and, similarly, the correlation matrix is

$$[R] = \frac{1}{n-1} \sum_{i=1}^{n} \mathbf{z}_i' \mathbf{z}_i'^{\,T}. \tag{9.36}$$

◊

EXAMPLE 9.2 Multiple Linear Regression Expressed in Matrix Notation

The discussion of multiple linear regression in Section 6.2.8 indicated that the relevant mathematics are most easily expressed and solved using matrix algebra. In this notation, the expression for the predictand y as a function of the predictor variables x_i (Equation 6.24) becomes

$$\mathbf{y} = [X]\mathbf{b}, \tag{9.37a}$$

or

$$\begin{bmatrix} y_1 \\ y_2 \\ y_3 \\ \vdots \\ y_n \end{bmatrix} = \begin{bmatrix} 1 & x_{1,1} & x_{1,2} & \cdots & x_{1,K} \\ 1 & x_{2,1} & x_{2,2} & \cdots & x_{2,K} \\ 1 & x_{3,1} & x_{3,2} & \cdots & x_{3,K} \\ \vdots & \vdots & \vdots & & \vdots \\ 1 & x_{n,1} & x_{n,2} & \cdots & x_{n,K} \end{bmatrix} \begin{bmatrix} b_0 \\ b_1 \\ b_2 \\ \vdots \\ b_K \end{bmatrix}. \tag{9.37b}$$

Here $\mathbf{y}$ is a $(n \times 1)$ matrix (i.e., a vector) of the n observations of the predictand, $[X]$ is a $(n \times K+1)$ data matrix containing the values of the predictor variables, and $\mathbf{b}^T = [b_0, b_1, b_2, \ldots, b_K]$ is a $(K+1 \times 1)$ vector of the regression parameters. The data matrix in the regression context is similar to that in Equation 9.2, except that it has $K+1$ rather than K columns. This extra column is the leftmost column of $[X]$ in Equation 9.37, and consists entirely of 1's. Thus, Equation 9.37 is a vector equation, with dimension $(n \times 1)$ on each side. It is actually n repetitions of Equation 6.24, once each for the n data records.

The normal equations (presented in Equation 6.6 for the simple case of $K = 1$) are obtained by left-multiplying each side of Equation 9.37 by $[X]^T$,

$$[X]^T \mathbf{y} = [X]^T [X] \mathbf{b}, \tag{9.38a}$$

or

$$\begin{bmatrix} \Sigma y \\ \Sigma x_1 y \\ \Sigma x_2 y \\ \vdots \\ \Sigma x_K y \end{bmatrix} = \begin{bmatrix} n & \Sigma x_1 & \Sigma x_2 & \cdots & \Sigma x_K \\ \Sigma x_1 & \Sigma x_1^2 & \Sigma x_1 x_2 & \cdots & \Sigma x_1 x_K \\ \Sigma x_2 & \Sigma x_2 x_1 & \Sigma x_2^2 & \cdots & \Sigma x_2 x_K \\ \vdots & \vdots & \vdots & & \vdots \\ \Sigma x_K & \Sigma x_K x_1 & \Sigma x_K x_2 & \cdots & \Sigma x_K^2 \end{bmatrix} \begin{bmatrix} b_0 \\ b_1 \\ b_2 \\ \vdots \\ b_K \end{bmatrix}, \tag{9.38b}$$

where all the summations are over the n data points. The $[X]^T[X]$ matrix has dimension $(K+1 \times K+1)$. Each side of Equation 9.38 has dimension $(K+1 \times 1)$, and this equation actually represents $K+1$ simultaneous equations involving the $K+1$ unknown regression coefficients. Matrix algebra very commonly is used to solve sets of simultaneous linear equations such as these. One way to obtain the solution is to left-multiply both sides of Equation 9.38 by the inverse of the $[X]^T[X]$ matrix. This operation is analogous to dividing both sides by this quantity, and yields

$$([X]^T[X])^{-1}[X]^T\mathbf{y} = ([X]^T[X])^{-1}[X]^T[X]\mathbf{b}$$
$$= [I]\mathbf{b}$$
$$= \mathbf{b}, \tag{9.39}$$

which is the solution for the vector of regression parameters. If there are no linear dependencies among the predictor variables, then the matrix $[X]^T[X]$ is nonsingular, and its inverse will exist. Otherwise, regression software will be unable to compute Equation 9.39, and a suitable error message should be reported.

Variances for the joint sampling distribution of the $K+1$ regression parameters $\boldsymbol{b}^T$, corresponding to Equations 6.17b and 6.18b, can also be calculated using matrix algebra. The $(K+1 \times K+1)$ covariance matrix, jointly for the intercept and the K regression coefficients, is

$$[S_{\mathbf{b}}] = \begin{bmatrix} s_{b_0}^2 & s_{b_0,b_1} & \cdots & s_{b_0,b_K} \\ s_{b_1,b_0} & s_{b_1}^2 & \cdots & s_{b_1,b_K} \\ s_{b_2,b_0} & s_{b_2,b_2} & \cdots & s_{b_2,b_K} \\ \vdots & \vdots & & \vdots \\ s_{b_K,b_0} & s_{b_K,b_1} & \cdots & s_{b_K}^2 \end{bmatrix} = s_e^2([X]^T[X])^{-1}. \tag{9.40}$$

As before, s_e^2 is the estimated residual variance, or MSE (see Table 6.3). The diagonal elements of Equation 9.40 are the estimated variances of the sampling distributions of each of the elements of the parameter vector $\boldsymbol{b}$; and the off-diagonal elements are the covariances among them, corresponding to (for covariances involving the intercept, b_0), Equation 6.19. For sufficiently large sample sizes, the joint sampling distribution is multivariate normal (see Chapter 10) so Equation 9.40 fully defines its dispersion.

Similarly, the conditional variance of the sampling distribution of the multiple linear regression function, which is the multivariate extension of Equation 6.23, can be expressed in matrix form as

$$s_{\hat{y}|\mathbf{x}_0}^2 = s_e^2 \mathbf{x}_0^T ([X]^T[X])^{-1}\mathbf{x}_0. \tag{9.41}$$

As before, this quantity depends on the values of the predictor(s) for which the regression function is evaluated, $\mathbf{x}_0^T = (1, x_1, x_2, \ldots, x_K)$. $\Diamond$

A square matrix is called orthogonal if the vectors defined by its columns have unit lengths, and are mutually perpendicular (i.e., $\theta = 90°$ according to Equation 9.15), and the same conditions hold for the vectors defined by its columns. In that case,

$$[A]^T = [A]^{-1}, \tag{9.42a}$$

which implies

$$[A][A]^T = [A]^T[A] = [I]. \tag{9.42b}$$

Orthogonal matrices are also called unitary, with this latter term encompassing also matrices that may have complex elements.

An orthogonal transformation is achieved by multiplying a vector by an orthogonal matrix. Considering a vector to define a point in K-dimensional space, an orthogonal transformation corresponds to a rigid rotation of the coordinate axes (and also a reflection, if the determinant is negative), resulting in a new basis for the space. For example, consider $K = 2$ dimensions, and the orthogonal matrix

$$[T] = \begin{bmatrix} \cos(\theta) & \sin(\theta) \\ -\sin(\theta) & \cos(\theta) \end{bmatrix}. \tag{9.43}$$

The lengths of both rows and both columns of this matrix are $\sin^2(\theta) + \cos^2(\theta) = 1$ (Equation 9.14), and the angles between the two pairs of vectors are both 90° (Equation 9.15), so $[T]$ is an orthogonal matrix.

Multiplication of a vector x by this matrix corresponds to a rigid clockwise rotation of the coordinate axes through an angle θ. Consider the point $x^T = (1, 1)$ in Figure 9.5. Multiplying it by the matrix $[T]$, with $\theta = 72°$, yields the point in a new (dashed) coordinate system

$$\tilde{x} = \begin{bmatrix} \cos(72°) & \sin(72°) \\ -\sin(72°) & \cos(72°) \end{bmatrix} x$$

$$= \begin{bmatrix} 0.309 & 0.951 \\ -0.951 & 0.309 \end{bmatrix} \begin{bmatrix} 1 \\ 1 \end{bmatrix} = \begin{bmatrix} .309 + .951 \\ -.951 + .309 \end{bmatrix} = \begin{bmatrix} 1.26 \\ -0.64 \end{bmatrix}. \tag{9.44}$$

Because the rows and columns of an orthogonal matrix all have unit length, orthogonal transformations preserve length. That is, they do not compress or expand the (rotated) coordinate axes. In terms of (squared) Euclidean length (Equation 9.14),

$$\tilde{x}^T \tilde{x} = ([T]x)^T ([T]x)$$

$$= x^T [T]^T [T] x$$

$$= x^T [I] x \tag{9.45}$$

$$= x^T x.$$

The result for the transpose of a matrix product from Table 9.1 has been used in the second line, and Equation 9.42 has been used in the third.

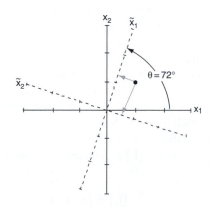

FIGURE 9.5 The point $x^T = (1, 1)$, when subjected to an orthogonal rotation of the coordinate axes through an angle of $\theta = 72°$, is transformed to the point $\tilde{x}^T = (1.26, -0.64)$ in the new basis (dashed coordinate axes).

9.3.3 *Eigenvalues and Eigenvectors of a Square Matrix*

An eigenvalue λ, and an eigenvector, e of a square matrix $[A]$ are a scalar and nonzero vector, respectively, satisfying the equation

$$[A]\mathbf{e} = \lambda\mathbf{e}, \tag{9.46a}$$

or equivalently

$$([A] - \lambda[I])\mathbf{e} = \mathbf{0}, \tag{9.46b}$$

where $\mathbf{0}$ is a vector consisting entirely of zeros. For every eigenvalue and eigenvector pair that can be found to satisfy Equation 9.46, any scalar multiple of the eigenvector, $c\mathbf{e}$, will also satisfy the equation together with that eigenvalue. Consequently, for definiteness it is usual to require that the eigenvectors have unit length,

$$\|\mathbf{e}\| = 1. \tag{9.47}$$

This restriction removes the ambiguity only up to a change in sign, since if a vector e satisfies Equation 9.46 then its negative, $-e$ will also.

If $[A]$ is nonsingular there will be K eigenvalue-eigenvector pairs λ_k and e_k with nonzero eigenvalues, where K is the number of rows and columns in $[A]$. Each eigenvector will be dimensioned $(K \times 1)$. If $[A]$ is singular at least one of its eigenvalues will be zero, with the corresponding eigenvectors being arbitrary. Synonymous terminology that is sometimes also used for eigenvalues and eigenvectors includes characteristic values and characteristic vectors, latent values and latent vectors, and proper values and proper vectors.

Because each eigenvector is defined to have unit length, the dot product of any eigenvector with itself is one. If, in addition, the matrix $[A]$ is symmetric, then its eigenvectors are mutually orthogonal, so that

$$\mathbf{e}_i^T\mathbf{e}_j = \begin{cases} 1, & i = j \\ 0, & i \neq j. \end{cases} \tag{9.48}$$

Orthogonal vectors of unit length are said to be orthonormal. (This terminology has nothing to do with the Gaussian, or "normal" distribution.) The orthonormality property is analogous to Equation 8.66, expressing the orthogonality of the sine and cosine functions.

For many statistical applications, eigenvalues and eigenvectors are calculated for real (not containing complex or imaginary numbers) symmetric matrices, such as covariance or correlation matrices. Eigenvalues and eigenvectors of such matrices have a number of important and remarkable properties. The first of these properties is that their eigenvalues and eigenvectors are real-valued. Also, as just noted, the eigenvectors of symmetric matrices are orthogonal. That is, their dot products with each other are zero, so that they are mutually perpendicular in K-dimensional space.

Often the $(K \times K)$ matrix $[E]$ is formed, the K columns of which are the eigenvectors e_k. That is,

$$[E] = \begin{bmatrix} \mathbf{e}_1 & \mathbf{e}_2 & \mathbf{e}_3 & \cdots & \mathbf{e}_K \end{bmatrix}. \tag{9.49}$$

Because of the orthogonality and unit length of the eigenvectors of symmetric matrices, the matrix $[E]$ is orthogonal, having the properties expressed in Equation 9.42. The orthogonal transformation $[E]^T x$ defines a rigid rotation of the K-dimensional coordinate axes of x, called an eigenspace. This space covers the same "territory" as the original coordinates, but using the different set of axes defined by the solutions to Equation 9.46.

The K eigenvalue-eigenvector pairs contain the same information as the matrix $[A]$ from which they were computed, and so can be regarded as a transformation of $[A]$. This equivalence can be expressed, again for $[A]$ symmetric, as the spectral decomposition, or Jordan decomposition,

$$[A] = [E][\Lambda][E]^T \tag{9.50a}$$

$$= [E] \begin{bmatrix} \lambda_1 & 0 & 0 & \cdots & 0 \\ 0 & \lambda_2 & 0 & \cdots & 0 \\ 0 & 0 & \lambda_3 & \cdots & 0 \\ \vdots & \vdots & \vdots & & \vdots \\ 0 & 0 & 0 & \cdots & \lambda_K \end{bmatrix} [E]^T, \tag{9.50b}$$

so that $[\Lambda]$ denotes a diagonal matrix whose nonzero elements are the K eigenvalues of $[A]$. It is illuminating to consider also the equivalent of Equation 9.50 in summation notation,

$$[A] = \sum_{k=1}^{K} \lambda_k e_k e_k^T \tag{9.51a}$$

$$= \sum_{k=1}^{K} \lambda_k [E_k]. \tag{9.51b}$$

The outer product of each eigenvector with itself in Equation 9.51a defines a matrix $[E_k]$. Equation 9.51b shows that the original matrix $[A]$ can be recovered as a weighted sum of these $[E_k]$ matrices, where the weights are the corresponding eigenvalues. Hence the spectral decomposition of a matrix is analogous to the Fourier decomposition of a function or data series (Equation 8.62a), with the eigenvalues playing the role of the Fourier amplitudes and the $[E_k]$ matrices corresponding to the cosine functions.

Other consequences of the equivalence of the information on the two sides of Equation 9.50 pertain to the eigenvalues. The first of these is

$$\text{tr}[A] = \sum_{k=1}^{K} a_{k,k} = \sum_{k=1}^{K} \lambda_k = \text{tr}[\Lambda]. \tag{9.52}$$

This relationship is particularly important when $[A]$ is a covariance matrix, in which case its diagonal elements $a_{k,k}$ are the K variances. Equation 9.52 says the sum of these variances is given by the sum of the eigenvalues of the covariance matrix.

The second consequence of Equation 9.50 for the eigenvalues is

$$\det[A] = \prod_{k=1}^{K} \lambda_k = \det[\Lambda], \tag{9.53}$$

which is consistent with the property that at least one of the eigenvalues of a singular matrix (that has zero determinant) will be zero. A real symmetric matrix with all eigenvalues nonnegative is called positive definite.

The matrix of eigenvectors [E] has the property that it diagonalizes the original symmetric matrix [A] from which the eigenvectors and eigenvalues were calculated. Left-multiplying Equation 9.50a by $[E]^{-1}$, right-multiplying by [E], and using the orthogonality of [E] yields

$$[E]^{-1}[A][E] = [\Lambda]. \tag{9.54}$$

Multiplication of [A] on the left by $[E]^{-1}$ and on the right by [E] produces the diagonal matrix of eigenvalues $[\Lambda]$.

There is also a strong connection between the eigenvalues λ_k and eigenvectors e_k of a nonsingular symmetric matrix, and the corresponding quantities λ_k^* and e_k^* of its inverse. The eigenvectors of matrix-inverse pairs are the same—that is, $e_k^* = e_k$—and the corresponding eigenvalues are reciprocals, $\lambda_k^* = \lambda_k^{-1}$. Therefore, the eigenvector of [A] associated with its largest eigenvalue is the same as the eigenvector of $[A]^{-1}$ associated with its smallest eigenvalue, and vice versa.

The extraction of eigenvalue-eigenvector pairs from matrices is a computationally demanding task, particularly as the dimensionality of the problem increases. It is possible but very tedious to do the computations by hand if $K = 2$, 3, or 4, using the equation

$$\det([A] - \lambda[I]) = 0. \tag{9.55}$$

This calculation requires first solving a Kth-order polynomial for the K eigenvalues, and then solving K sets of K simultaneous equations to obtain the eigenvectors. In general, however, widely available computer algorithms for calculating numerical approximations to eigenvalues and eigenvectors are used. These computations can also be done within the framework of the singular value decomposition (see Section 9.3.5).

EXAMPLE 9.3   Eigenvalues and Eigenvectors of a (2 × 2) Symmetric Matrix

The symmetric matrix

$$[A] = \begin{bmatrix} 185.47 & 110.84 \\ 110.84 & 77.58 \end{bmatrix} \tag{9.56}$$

has as its eigenvalues $\lambda_1 = 254.76$ and $\lambda_2 = 8.29$, with corresponding eigenvectors $e_1^T = [0.848, 0.530]$ and $e_2^T = [-0.530, 0.848]$. It is easily verified that both eigenvectors are of unit length. Their dot product is zero, which indicates that the two vectors are perpendicular, or orthogonal.

The matrix of eigenvectors is therefore

$$[E] = \begin{bmatrix} 0.848 & -0.530 \\ 0.530 & 0.848 \end{bmatrix}, \tag{9.57}$$

and the original matrix can be recovered using the eigenvalues and eigenvectors (Equations 9.50 and 9.51) as

$$[A] = \begin{bmatrix} 185.47 & 110.84 \\ 110.84 & 77.58 \end{bmatrix} = \begin{bmatrix} .848 & -.530 \\ .530 & .848 \end{bmatrix} \begin{bmatrix} 254.76 & 0 \\ 0 & 8.29 \end{bmatrix} \begin{bmatrix} .848 & .530 \\ -.530 & .848 \end{bmatrix} \tag{9.58a}$$

$$= 254.76 \begin{bmatrix} .848 \\ .530 \end{bmatrix} [.848 \ .530] + 8.29 \begin{bmatrix} -.530 \\ .848 \end{bmatrix} [-.530 .848] \tag{9.58b}$$

$$= 254.76 \begin{bmatrix} .719 & .449 \\ .449 & .281 \end{bmatrix} + 8.29 \begin{bmatrix} .281 & -.449 \\ -.449 & .719 \end{bmatrix}. \tag{9.58c}$$

Equation 9.58a expresses the spectral decomposition of [A] in the form of Equation 9.50, and Equations 9.58b and 9.58c show the same decomposition in the form of Equation 9.51. The matrix of eigenvectors diagonalizes the original matrix [A] according to

$$[E]^{-1}[A][E] = \begin{bmatrix} 0.848 & 0.530 \\ -0.530 & 0.848 \end{bmatrix} \begin{bmatrix} 185.47 & 110.84 \\ 110.84 & 77.58 \end{bmatrix} \begin{bmatrix} 0.848 & -0.530 \\ 0.530 & 0.848 \end{bmatrix}$$

$$= \begin{bmatrix} 254.76 & 0 \\ 0 & 8.29 \end{bmatrix} = [\Lambda]. \tag{9.59}$$

Because of the orthonormality of the eigenvectors, the inverse of [E] can be and has been replaced by its transpose in Equation 9.59. Finally, the sum of the eigenvalues, $254.76 + 8.29 = 263.05$, equals the sum of the diagonal elements of the original [A] matrix, $185.47 + 77.58 = 263.05$. ◊

9.3.4 Square Roots of a Symmetric Matrix

Consider two square matrices of the same order, [A] and [B]. If the condition

$$[A] = [B][B]^T \tag{9.60}$$

holds, then [B] multiplied by itself yields [A], so [B] is said to be a "square root" of [A], or $[B] = [A]^{1/2}$. Unlike the square roots of scalars, the square root of a symmetric matrix is not uniquely defined. That is, there are any number of matrices [B] that can satisfy Equation 9.60, although two algorithms are used most frequently to find solutions for it.

If [A] is of full rank, a lower-triangular matrix [B] satisfying Equation 9.60 can be found using the Cholesky decomposition of [A]. (A lower-triangular matrix has zeros above and to the right of the main diagonal; i.e., $b_{i,j} = 0$ for $i < j$.) Beginning with

$$b_{1,1} = \sqrt{a_{1,1}} \tag{9.61}$$

as the only nonzero element in the first row of [B], the Cholesky decomposition proceeds iteratively, by calculating the nonzero elements of each of the subsequent rows, i, of [B] in turn according to

$$b_{i,j} = \frac{a_{i,j} - \sum_{k=1}^{j-1} b_{i,k} b_{j,k}}{b_{j,j}}, \quad j = 1, \dots, i-1; \tag{9.62a}$$

and

$$b_{i,i} = \left[a_{i,i} - \sum_{k=1}^{i-1} b_{i,k}^2 \right]^{1/2}. \tag{9.62b}$$

It is a good idea to do these calculations in double precision in order to minimize the accumulation roundoff errors that can lead to a division by zero in Equation 9.62a for large matrix dimension K, even if [A] is of full rank.

The second commonly used method to find a square root of [A] uses its eigenvalues and eigenvectors, and is computable even if the symmetric matrix [A] is not of full rank. Using the spectral decomposition (Equation 9.50) for [B],

$$[B] = [A]^{1/2} = [E][\Lambda]^{1/2}[E]^T, \tag{9.63}$$

where [E] is the matrix of eigenvectors for *both* [A] and [B] (i.e., they are the same vectors). The matrix $[\Lambda]$ contains the eigenvalues of [A], which are the squares of the eigenvalues of [B] on the diagonal of $[\Lambda]^{1/2}$. That is, $[\Lambda]^{1/2}$ is a diagonal matrix with elements $\lambda_k^{1/2}$, where the λ_k are the eigenvalues of [A]. Equation 9.63 is still defined even if some of these eigenvalues are zero, so this method can be used to find a square-root for a matrix that is not of full rank. Note that $[\Lambda]^{1/2}$ also conforms to the definition of a square-root matrix, since $[\Lambda]^{1/2}([\Lambda]^{1/2})^{\mathrm{T}} = [\Lambda]^{1/2}[\Lambda]^{1/2} = [\Lambda]$. The square-root decomposition in Equation 9.63 produces a symmetric square-root matrix. It is more tolerant than the Cholesky decomposition to roundoff error when the matrix dimension is large, because (computationally, as well as truly) zero eigenvalues do not produce undefined arithmetic operations.

Equation 9.63 can be extended to find the square root of a matrix inverse, $[A]^{-1/2}$, if [A] is symmetric and of full rank. Because a matrix has the same eigenvectors as its inverse, so also will it have the same eigenvectors as the square root of its inverse. Accordingly,

$$[A]^{-1/2} = [E][\Lambda]^{-1/2}[E]^{\mathrm{T}}, \tag{9.64}$$

where $[\Lambda]^{-1/2}$ is a diagonal matrix with elements $\lambda_k^{-1/2}$, the reciprocals of the square roots of the eigenvalues of [A]. The implications of Equation 9.64 are those that would be expected; that is, $[A]^{-1/2}([A]^{-1/2})^{\mathrm{T}} = [A]^{-1}$, and $[A]^{-1/2}([A]^{1/2})^{\mathrm{T}} = [I]$.

EXAMPLE 9.4   Square Roots of a Matrix and its Inverse

The symmetric matrix [A] in Equation 9.56 is of full rank, since both of its eigenvalues are positive. Therefore a lower-triangular square-root matrix $[B] = [A]^{1/2}$ can be computed using the Cholesky decomposition. Equation 9.61 yields $b_{1,1} = (a_{1,1})^{1/2} = 185.47^{1/2} = 13.619$ as the only nonzero element of the first row $(i = 1)$ of [B]. Because [B] has only one additional row, Equation 9.62 needs to be applied only once each. Equation 9.62a yields $b_{2,1} = (a_{1,1} - 0)/b_{1,1} = 110.84/13.619 = 8.139$. Zero is subtracted in the numerator of Equation 9.62a for $b_{2,1}$ because there are no terms in the summation. (If [A] had been a (3×3) matrix, Equation 9.62a would be applied twice for the third $(i = 3)$ row: the first of these applications, for $b_{3,1}$, would again have no terms in the summation; but when calculating $b_{3,2}$ there would be one term corresponding to $k = 1$.) Finally, the calculation indicated by Equation 9.62b is $b_{2,2} = (a_{2,2} - b_{2,1}^2)^{1/2} = (77.58 - 8.139^2)^{1/2} = 3.367$. The Cholesky lower-triangular square-root matrix for [A] is thus

$$[B] = [A]^{1/2} = \begin{bmatrix} 13.619 & 0 \\ 8.139 & 3.367 \end{bmatrix}, \tag{9.65}$$

which can be verified as a valid square root of [A] through the matrix multiplication $[B][B]^{\mathrm{T}}$.

A symmetric square-root matrix for [A] can be computed using its eigenvalues and eigenvectors from Example 9.3, and Equation 9.63:

$$\begin{aligned}
[B] = [A]^{1/2} &= [E][\Lambda]^{1/2}[E]^{\mathrm{T}} \\
&= \begin{bmatrix} .848 & -.530 \\ .530 & .848 \end{bmatrix} \begin{bmatrix} \sqrt{254.76} & 0 \\ 0 & \sqrt{8.29} \end{bmatrix} \begin{bmatrix} .848 & .530 \\ -.530 & .848 \end{bmatrix} \\
&= \begin{bmatrix} 12.286 & 5.879 \\ 5.879 & 6.554 \end{bmatrix}.
\end{aligned} \tag{9.66}$$

This matrix also can be verified as a valid square root of [A] by calculating $[B][B]^{\mathrm{T}}$.

Equation 9.64 allows calculation of a square-root matrix for the inverse of [A],

$$
\begin{aligned}
[A]^{-1/2} &= [E][\Lambda]^{-1/2}[E]^T \\
&= \begin{bmatrix} .848 & -.530 \\ .530 & .848 \end{bmatrix} \begin{bmatrix} 1/\sqrt{254.76} & 0 \\ 0 & 1/\sqrt{8.29} \end{bmatrix} \begin{bmatrix} .848 & .530 \\ -.530 & .848 \end{bmatrix} \\
&= \begin{bmatrix} .1426 & -.1279 \\ -.1279 & .2674 \end{bmatrix}.
\end{aligned}
\tag{9.67}
$$

This is also a symmetric matrix. The matrix product $[A]^{-1/2}([A]^{-1/2})^T = [A]^{-1/2}[A]^{-1/2} = [A]^{-1}$. The validity of Equation 9.67 can be checked by comparing the product $[A]^{-1/2}[A]^{-1/2}$ with $[A]^{-1}$ as calculated using Equation 9.28, or by verifying $[A][A]^{-1/2}[A]^{-1/2} = [A][A]^{-1} = [I]$. ◊

9.3.5 Singular-Value Decomposition (SVD)

Equation 9.50 expresses the spectral decomposition of a symmetric square matrix. This decomposition can be extended to any $(n \times m)$ rectangular matrix [A] with at least as many rows as columns $(n \geq m)$ using the singular-value decomposition (SVD),

$$
\underset{(n \times m)}{[A]} = \underset{(n \times m)}{[L]} \; \underset{(m \times m)}{[\Omega]} \; \underset{(m \times m)}{[R]^T}, \quad n \geq m.
\tag{9.68}
$$

The m columns of [L] are called the left singular vectors, and the m columns of [R] are called the right singular vectors. (Note that, in the context of SVD, [R] does not denote a correlation matrix.) Both sets of vectors are mutually orthonormal, so $[L]^T[L] = [R]^T[R] = [R][R]^T = [I]$, with dimension $(m \times m)$. The matrix $[\Omega]$ is diagonal, and its nonnegative elements are called the singular values of [A]. Equation 9.68 is sometimes called the thin SVD, in contrast to an equivalent expression in which the dimension of [L] is $(n \times n)$, and the dimension of $[\Omega](n \times m)$, but with the last $n - m$ rows containing all zeros so that the last $n - m$ columns of [L] are arbitrary.

If [A] is square and symmetric, then Equation 9.68 reduces to Equation 9.50, with $[L] = [R] = [E]$, and $[\Omega] = [\Lambda]$. It is therefore possible to compute eigenvalues and eigenvectors for symmetric matrices using an SVD algorithm from a package of computer routines, which are widely available (e.g., Press *et al.* 1986). Analogously to Equation 9.51 for the spectral decomposition of a symmetric square matrix, Equation 9.68 can be expressed as a summation of weighted outer products of the left and right singular vectors,

$$
[A] = \sum_{i=1}^{m} \omega_i \ell_i \mathbf{r}_i^T.
\tag{9.69}
$$

Even if [A] is not symmetric, there is a connection between the SVD and the eigenvalues and eigenvectors of both $[A]^T[A]$ and $[A][A]^T$, both of which matrix products are square (with dimensions $(m \times m)$ and $(n \times n)$, respectively) and symmetric. Specifically, the columns of [R] are the $(m \times 1)$ eigenvectors of $[A]^T[A]$, the columns of [L] are the $(n \times 1)$ eigenvectors of $[A][A]^T$. The respective singular values are the square roots of the corresponding eigenvalues, i.e., $\omega_i^2 = \lambda_i$.

EXAMPLE 9.5   Eigenvalues and Eigenvectors of a Covariance Matrix Using SVD

Consider the (31×2) matrix $(30)^{-1/2}[X']$, where $[X']$ is the matrix of anomalies (Equation 9.29) for the minimum temperature data in Table A.1. The SVD of this matrix, in the form of Equation 9.68, is

$$\frac{1}{\sqrt{30}}[X'] = \begin{bmatrix} 1.09 & 1.42 \\ 2.19 & 1.42 \\ 1.64 & 1.05 \\ \vdots & \vdots \\ 1.83 & 0.51 \end{bmatrix} = \begin{bmatrix} .105 & .216 \\ .164 & .014 \\ .122 & .008 \\ \vdots & \vdots \\ .114 & -.187 \end{bmatrix} \begin{bmatrix} 15.961 & 0 \\ 0 & 2.879 \end{bmatrix} \begin{bmatrix} .848 & .530 \\ -.530 & .848 \end{bmatrix}. \quad (9.70)$$

$$(31 \times 2) \qquad\qquad (31 \times 2) \qquad\qquad (2 \times 2) \qquad\qquad (2 \times 2)$$

The reason for multiplying the anomaly matrix $[X']$ by $30^{-1/2}$ should be evident from Equation 9.30: the product $(30^{-1/2}[X']^T)(30^{-1/2}[X']) = (n-1)^{-1}[X']^T [X]$ yields the covariance matrix $[S]$ for these data, which is the same as the matrix $[A]$ in Equation 9.56. Because the matrix of right singular vectors $[R]$ contains the eigenvectors for the product of the matrix on the left-hand side of Equation 9.70, left-multiplied by its transpose, the matrix $[R]^T$ on the far right of Equation 9.70 is the same as the (transpose of) the matrix $[E]$ in Equation 9.57. Similarly the squares of the singular values in the diagonal matrix $[\Omega]$ in Equation 9.70 are the corresponding eigenvalues; for example, $\omega_1^2 = 15.961^2 = \lambda_1 = 254.7$.

The right-singular vectors of $(n-1)^{-1/2}[X] = [S]^{1/2}$ are the eigenvectors of the (2×2) covariance matrix $[S] = (n-1)^{-1}[X]^T[X]$. The left singular vectors in the matrix $[L]$ are eigenvectors of the (31×31) matrix $(n-1)^{-1}[X][X]^T$. This matrix actually has 31 eigenvectors, but only two of them (the two shown in Equation 9.70) are associated with nonzero eigenvalues. It is in this sense, of truncating the zero eigenvalues and their associated irrelevant eigenvectors, that Equation 9.70 is an example of a thin SVD. ◊

9.4 Random Vectors and Matrices

9.4.1 Expectations and Other Extensions of Univariate Concepts

Just as ordinary random variables are scalar quantities, a random vector (or random matrix) is a vector (or matrix) whose entries are random variables. The purpose of this section is to extend the rudiments of matrix algebra presented in Section 9.3 to include statistical ideas.

A vector x whose K elements are the random variables x_k is a random vector. The expected value of this random vector is also a vector, called the vector mean, whose K elements are the individual expected values (i.e., probability-weighed averages) of the corresponding random variables. If all the x_k are continuous variables,

$$\boldsymbol{\mu} = \begin{bmatrix} \int_{-\infty}^{\infty} x_1 f_1(x_1) dx_1 \\ \int_{-\infty}^{\infty} x_2 f_2(x_2) dx_2 \\ \vdots \\ \int_{-\infty}^{\infty} x_K f_K(x_K) dx_K \end{bmatrix}. \quad (9.71)$$

If some or all of the K variables in $\mathbf{x}$ are discrete, the corresponding elements of $\boldsymbol{\mu}$ will be sums in the form of Equation 4.12.

The properties of expectations listed in Equation 4.14 extend also to vectors and matrices in ways that are consistent with the rules of matrix algebra. If c is a scalar constant, [X] and [Y] are random matrices with the same dimensions (and which may be random vectors if one of their dimensions is 1), and [A] and [B] are constant (nonrandom) matrices,

$$E(c[X]) = c \, E([X]), \tag{9.72a}$$

$$E([X]+[Y]) = E([X]) + E([Y]), \tag{9.72b}$$

$$E([A][X][B]) = [A]E([X])[B], \tag{9.72c}$$

$$E([A][X]+[B]) = [A]E([X]) + [B]. \tag{9.72d}$$

The (population) covariance matrix, corresponding to the sample estimate [S] in Equation 9.5, is the matrix expected value

$$\underset{(K \times K)}{[\Sigma]} = E(\underset{(K \times 1)}{[\mathbf{x}-\boldsymbol{\mu}]} \underset{(1 \times K)}{[\mathbf{x}-\boldsymbol{\mu}]^T}) \tag{9.73a}$$

$$= E\left(\begin{bmatrix} (x_1-\mu_1)^2 & (x_1-\mu_1)(x_2-\mu_2) & \cdots & (x_1-\mu_1)(x_K-\mu_K) \\ (x_2-\mu_2)(x_1-\mu_1) & (x_2-\mu_2)^2 & \cdots & (x_2-\mu_2)(x_K-\mu_K) \\ \vdots & \vdots & & \vdots \\ (x_K-\mu_K)(x_1-\mu_1) & (x_K-\mu_K)(x_2-\mu_2) & \cdots & (x_K-\mu_K)^2 \end{bmatrix}\right) \tag{9.73b}$$

$$= \begin{bmatrix} \sigma_{1,1} & \sigma_{1,2} & \cdots & \sigma_{1,K} \\ \sigma_{2,1} & \sigma_{2,2} & \cdots & \sigma_{2,K} \\ \vdots & \vdots & & \vdots \\ \sigma_{K,1} & \sigma_{K,2} & \cdots & \sigma_{K,K} \end{bmatrix}. \tag{9.73c}$$

The diagonal elements of Equation 9.73 are the scalar (population) variances, which would be computed (for continuous variables) using Equation 4.20 with $g(x_k) = (x_k - \mu_k)^2$ or, equivalently, Equation 4.21. The off-diagonal elements are the covariances, which would be computed using the double integrals

$$\sigma_{k,\ell} = \int_{-\infty}^{\infty} \int_{-\infty}^{\infty} (x_k - \mu_k)(x_\ell - \mu_\ell) f_k(x_k) f_\ell(x_\ell) dx_\ell dx_k, \tag{9.74}$$

each of which is analogous to the summation in Equation 9.4 for the sample covariances. Analogously to Equation 4.21b for the scalar variance, an equivalent expression for the (population) covariance matrix is

$$[\Sigma] = E(\mathbf{x} \, \mathbf{x}^T) - \boldsymbol{\mu}\boldsymbol{\mu}^T. \tag{9.75}$$

9.4.2 Partitioning Vectors and Matrices

In some settings it is natural to define collections of variables that segregate into two or more groups. Simple examples are one set of L predictands together with a different set of $K - L$ predictors, or two or more sets of variables, each observed simultaneously at some large number of locations or gridpoints. In such cases is it often convenient and useful to maintain these distinctions notationally, by partitioning the corresponding vectors and matrices.

Partitions are indicated by dashed lines in the expanded representation of vectors and matrices. These indicators of partitions are imaginary lines, in the sense that they have no effect whatsoever on the matrix algebra as applied to the larger vectors or matrices. For example, consider a $(K \times 1)$ random vector $\mathbf{x}$ that consists of one group of L variables and another group of $K - L$ variables,

$$\mathbf{x}^T = [x_1 x_2 \cdots x_L | x_{L+1} x_{L+2} \cdots x_K], \qquad (9.76a)$$

which would have expectation

$$E[\mathbf{x}^T] = \boldsymbol{\mu}^T = [\mu_1 \mu_2 \cdots \mu_L | \mu_{L+1} \mu_{L+2} \cdots \mu_K], \qquad (9.76b)$$

exactly as Equation 9.71, except that both $\mathbf{x}$ and $\boldsymbol{\mu}$ are partitioned (i.e., composed of a concatenation of) a $(L \times 1)$ vector and a $(K - L \times 1)$ vector.

The covariance matrix of $\mathbf{x}$ in Equation 9.76 would be computed in exactly the same way as indicated in Equation 9.73, with the partitions being carried forward:

$$[\Sigma] = E([\mathbf{x} - \boldsymbol{\mu}][\mathbf{x} - \boldsymbol{\mu}]^T) \qquad (9.77a)$$

$$= \left[\begin{array}{cccc|cccc}
\sigma_{1,1} & \sigma_{2,1} & \cdots & \sigma_{1,L} & \sigma_{1,L+1} & \sigma_{1,L+2} & \cdots & \sigma_{1,K} \\
\sigma_{2,1} & \sigma_{2,2} & \cdots & \sigma_{2,L} & \sigma_{2,L+1} & \sigma_{2,L+2} & \cdots & \sigma_{2,K} \\
\vdots & \vdots & & \vdots & \vdots & \vdots & & \vdots \\
\sigma_{L,1} & \sigma_{L,2} & \cdots & \sigma_{L,L} & \sigma_{L,L+1} & \sigma_{L,L+2} & \cdots & \sigma_{L,K} \\
\hline
\sigma_{L+1,1} & \sigma_{L+1,2} & \cdots & \sigma_{L+1,L} & \sigma_{L+1,L+1} & \sigma_{L+1,L+2} & \cdots & \sigma_{L+1,K} \\
\sigma_{L+2,1} & \sigma_{L+2,2} & \cdots & \sigma_{L+2,L} & \sigma_{L+2,L+1} & \sigma_{L+2,L+2} & \cdots & \sigma_{L+2,K} \\
\vdots & \vdots & & \vdots & \vdots & \vdots & & \vdots \\
\sigma_{K,1} & \sigma_{K,2} & \cdots & \sigma_{K,L} & \sigma_{K,L+1} & \sigma_{K,L+2} & \cdots & \sigma_{K,K}
\end{array}\right] \qquad (9.77b)$$

$$= \left[\begin{array}{c|c}
[\Sigma_{1,1}] & [\Sigma_{2,1}] \\
\hline
[\Sigma_{2,1}] & [\Sigma_{2,2}]
\end{array}\right]. \qquad (9.77c)$$

The covariance matrix $[\Sigma]$ for a data vector $\mathbf{x}$ partitioned into two segments as in Equation 9.76, is itself partitioned into four submatrices. The $(L \times L)$ matrix $[\Sigma_{1,1}]$ is the covariance matrix for the first L variables, $[x_1 x_2 \ldots x_L]^T$, and the $(K - L \times K - L)$ matrix $[\Sigma_{2,2}]$ is the covariance matrix for the last $K - L$ variables, $[x_{L+1} x_{L+2} \ldots x_K]^T$. Both of these matrices have variances on the main diagonal, and covariances among the variables in its respective group in the other positions.

The $(K - L \times L)$ matrix $[\Sigma_{2,1}]$ contains the covariances among all possible pairs of variables with one member in the second group and the other member in the first group. Because it is not a full covariance matrix it does not contain variances along the main diagonal even if it is square, and in general it is not symmetric. The $(L \times K - L)$ matrix $[\Sigma_{1,2}]$ contains the same covariances among all possible pairs of variables with one member in the first group and the other member in the second group. Because the full covariance matrix $[\Sigma]$ is symmetric, $[\Sigma_{1,2}]^T = [\Sigma_{2,1}]$.

9.4.3 *Linear Combinations*

A linear combination is essentially a weighted sum of two or more variables $x_1, x_2, \ldots, x_K$. For example, the multiple linear regression in Equation 6.24 is a linear combination of the K regression predictors that yields a new variable, which in this case is the regression prediction. For simplicity, consider that the parameter $b_0 = 0$ in Equation 6.24. Then Equation 6.24 can be expressed in matrix notation as

$$y = \mathbf{b}^{\mathrm{T}}\mathbf{x}, \tag{9.78}$$

where $\mathbf{b}^{\mathrm{T}} = [b_1, b_2, \ldots, b_K]$ is the vector of parameters that are the weights in the weighted sum.

Usually in regression the predictors $\mathbf{x}$ are considered to be fixed constants rather than random variables. But consider now the case where $\mathbf{x}$ is a random vector with mean $\boldsymbol{\mu}_x$ and covariance $[\Sigma_x]$. The linear combination in Equation 9.78 will also be a random variable. Extending Equation 4.14c for vector $\mathbf{x}$, with $g_j(x) = b_j x_j$, the mean of y will be

$$\mu_y = \sum_{k=1}^{K} b_k \mu_k, \tag{9.79}$$

where $\mu_k = E(x_k)$. The variance of the linear combination is more complicated, both notationally and computationally, and involves the covariances among all pairs of the x's. For simplicity, suppose $K = 2$. Then,

$$
\begin{aligned}
\sigma_y^2 &= \mathrm{Var}(b_1 x_1 + b_2 x_2) = E\{[(b_1 x_1 + b_2 x_2) - (b_1 \mu_1 + b_2 \mu_2)]^2\} \\
&= E\{[(b_1(x_1 - \mu_1) + b_2(x_2 - \mu_2)]^2 \\
&= E\{b_1^2(x_1 - \mu_1)^2 + b_2^2(x_2 - \mu_2)^2 + 2b_1 b_2(x_1 - \mu_1)(x_2 - \mu_2)\} \\
&= b_1^2 E\{(x_1 - \mu_1)^2\} + b_2^2 E\{(x_2 - \mu_2)^2\} + 2b_1 b_2 E\{(x_1 - \mu_1)(x_2 - \mu_2)\} \\
&= b_1^2 \sigma_{1,1} + b_2^2 \sigma_{2,2} + 2b_1 b_2 \sigma_{1,2}.
\end{aligned} \tag{9.80}
$$

This scalar result is fairly cumbersome, even though the linear combination is of only two random variables, and the general extension to linear combinations of K random variables involves $K(K+1)/2$ terms. More generally, and much more compactly, in matrix notation Equations 9.79 and 9.80 become

$$\mu_y = \mathbf{b}^{\mathrm{T}}\boldsymbol{\mu} \tag{9.81a}$$

and

$$\sigma_y^2 = \mathbf{b}^{\mathrm{T}}[\Sigma_x]\mathbf{b}. \tag{9.81b}$$

The quantities on the left-hand side of Equation 9.81 are scalars, because the result of the single linear combination in Equation 9.78 is scalar. But consider simultaneously forming L linear combinations of the K random variables $\mathbf{x}$,

$$
\begin{aligned}
y_1 &= b_{1,1}x_1 + b_{1,2}x_2 + \cdots + b_{1,K}x_K \\
y_2 &= b_{2,1}x_1 + b_{2,2}x_2 + \cdots + b_{2,K}x_K \\
&\;\;\vdots \qquad \vdots \qquad \vdots \qquad\qquad \vdots \\
y_L &= b_{L,1}x_1 + b_{L,2}x_2 + \cdots + b_{L,K}x_K,
\end{aligned} \tag{9.82a}
$$

or

$$\underset{(L\times1)}{\mathbf{y}} = \underset{(L\times K)}{[B]}\ \underset{(K\times1)}{\mathbf{x}}\ . \tag{9.82b}$$

Here each row of [B] defines a single linear combination as in Equation 9.78, and collectively these L linear combinations define the random vector $\mathbf{y}$. Extending Equations 9.81 to the mean vector and covariance matrix of this collection of L linear combinations of $\mathbf{x}$,

$$\underset{(L\times1)}{\boldsymbol{\mu}_y} = \underset{(L\times K)}{[B]}\ \underset{(K\times1)}{\boldsymbol{\mu}_x} \tag{9.83a}$$

and

$$\underset{(L\times L)}{[\Sigma_y]} = \underset{(L\times K)}{[B]}\ \underset{(K\times K)}{[\Sigma_x]}\ \underset{(K\times L)}{[B]^{\mathrm{T}}}. \tag{9.83b}$$

Note that it is not actually necessary to compute the transformed variables in Equation 9.82 in order to find their mean and covariance, if the mean vector and covariance matrix of the x's are known.

EXAMPLE 9.6   Mean Vector and Covariance Matrix for a Pair of Linear Combinations

Example 9.5 showed that the matrix in Equation 9.56 is the covariance matrix for the Ithaca and Canandaigua minimum temperature data in Table A.1. The mean vector for these data is $\boldsymbol{\mu}^{\mathrm{T}} = [\mu_{\mathrm{Ith}}\ \mu_{\mathrm{Can}}] = [13.0\ 20.2]$. Consider now two linear combinations of these minimum temperature data in the form of Equation 9.43, with $\theta = 32°$. That is, each of the two rows of [T] defines a linear combination (Equation 9.78), which can be expressed jointly as in Equation 9.82. Together, these two linear combinations are equivalent to a transformation that corresponds to a clockwise rotation of the coordinate axes through the angle θ. That is, the vectors $\mathbf{y} = [T]\mathbf{x}$ would locate the same points, but in the framework of the rotated coordinate system.

One way to find the mean and covariance for the transformed points, $\boldsymbol{\mu}_y$ and $[\Sigma_y]$, would be to carry out the transformation for all $n = 31$ point pairs, and then to compute the mean vector and covariance matrix for the transformed data set. However, knowing the mean and covariance of the underlying x's it is straightforward and much easier to use Equation 9.83 to obtain

$$\boldsymbol{\mu}_y = \begin{bmatrix} \cos 32° & \sin 32° \\ -\sin 32° & \cos 32° \end{bmatrix} \boldsymbol{\mu}_x = \begin{bmatrix} .848 & .530 \\ -.530 & .848 \end{bmatrix} \begin{bmatrix} 13.0 \\ 20.2 \end{bmatrix} = \begin{bmatrix} 21.7 \\ 10.2 \end{bmatrix} \tag{9.84a}$$

and

$$[\Sigma_y] = [T][\Sigma_x][T]^{\mathrm{T}} = \begin{bmatrix} .848 & .530 \\ -.530 & .848 \end{bmatrix} \begin{bmatrix} 185.47 & 110.84 \\ 110.84 & 77.58 \end{bmatrix} \begin{bmatrix} .848 & -.530 \\ .530 & .848 \end{bmatrix}$$

$$= \begin{bmatrix} 254.76 & 0 \\ 0 & 8.29 \end{bmatrix}. \tag{9.84b}$$

The rotation angle $\theta = 32°$ is evidently a special one for these data, as it produces a pair of transformed variables $\mathbf{y}$ that are uncorrelated. In fact this transformation is exactly the same as in Equation 9.59, which was expressed in terms of the eigenvectors of $[\Sigma_x]$. ◊

9.4.4 *Mahalanobis Distance, Revisited*

Section 9.2.2 introduced the Mahalanobis, or statistical, distance as a way to gauge differences or unusualness within the context established by an empirical data scatter or an underlying multivariate probability density. If the K variables in the data vector x are mutually uncorrelated, the (squared) Mahalanobis distance takes the simple form of the sum of the squared standardized anomalies z_k, as indicated in Equation 9.7 for $K = 2$ variables. When some or all of the variables are correlated the Mahalanobis distance accounts for the correlations as well, although as noted in Section 9.2.2 the notation is prohibitively complicated in scalar form. In matrix notation, the Mahalanobis distance between points x and y in their K-dimensional space is

$$D^2 = [\mathbf{x} - \mathbf{y}]^T [\mathbf{S}]^{-1} [\mathbf{x} - \mathbf{y}], \tag{9.85}$$

where [S] is the covariance matrix in the context of which the distance is being calculated.

If the dispersion defined by [S] involves zero correlation among the K variables, it is not difficult to see that Equation 9.85 reduces to Equation 9.7 (in two dimensions, with obvious extension to higher dimensions). In that case, [S] is diagonal, its inverse is also diagonal with elements $(s_{k,k})^{-1}$, so Equation 9.85 would reduce to $D^2 = \Sigma (x_k - y_k)^2 / s_{k,k}$. This observation underscores one important property of the Mahalanobis distance, namely that different intrinsic scales of variability for the K variables in the data vector do not confound D^2, because each is divided by its standard deviation before squaring. If [S] is diagonal, the Mahalanobis distance is the same as the Euclidean distance after dividing each variable by its standard deviation.

The second salient property of the Mahalanobis distance is that it accounts for the redundancy in information content among correlated variables, in the calculation of statistical distances. Again, this concept is easiest to see in two dimensions. Two strongly correlated variables provide very nearly the same information, and ignoring strong correlations when calculating statistical distance (i.e., using Equation 9.7 when the correlation is not zero), effectively double-counts the contribution of the (nearly) redundant second variable. The situation is illustrated in Figure 9.6, which shows the standardized point $z^T = (1, 1)$ in the contexts of three very different point clouds. In Figure 9.6a the correlation in the point cloud is zero, so it is appropriate to use Equation 9.7 to calculate the Mahalanobis distance to the origin (which is also the vector mean of the point cloud), after having accounted for possibly different scales of variation for the two variables by dividing by the respective standard deviations. That distance is $D^2 = 2$ (corresponding to an ordinary Euclidean distance of $\sqrt{2} = 1.414$). The correlation between the two

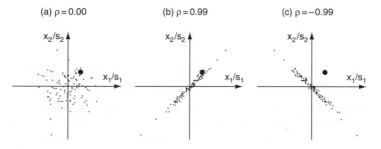

(a) $\rho = 0.00$ (b) $\rho = 0.99$ (c) $\rho = -0.99$

FIGURE 9.6 The point $z^T = (1, 1)$ (large dot) in the contexts of data scatters with (a) zero correlation, (b) correlation 0.99, and (c) correlation -0.99. Mahalanobis distances to the origin are drastically different in these three cases.

variables in Figure 9.6b is 0.99, so that one or the other of the two variables provides nearly the same information as both together: z_1 and z_2 are nearly the same variable. Using Equation 9.85 the Mahalanobis distance to the origin is $D^2 = 1.005$, which is only slightly more than if only one of the two nearly redundant variables had been considered alone, and substantially smaller than the distance appropriate to the context of the scatter in Figure 9.6a.

Finally, Figure 9.6c shows a very different situation, in which the correlation is -0.99. Here the point $(1, 1)$ is extremely unusual in the context of the data scatter, and using Equation 9.85 we find $D^2 = 200$. That is, it is extremely far from the origin relative to the dispersion of the point cloud, and this unusualness is reflected by the Mahalanobis distance. The point $(1, 1)$ in Figure 9.6c is a multivariate outlier. Visually it is well removed from the point scatter in two dimensions. But relative to either of the two univariate distributions it is a quite ordinary point that is relatively close to (one standard deviation from) the origin, so that it would not stand out as unusual according to standard EDA methods applied to the two variables individually. It is an outlier in the sense that it does not behave like the scatter of negatively correlated point cloud, in which large values of x_1/s_1 are associated with small values of x_2/s_2, and vice versa. The large Mahalanobis distance to the center of the point cloud identifies it as a multivariate outlier.

Equation 9.85 is an example of what is called a quadratic form. It is quadratic in the vector $x - y$, in the sense that this vector is multiplied by itself, together with scaling constants in the symmetric matrix $[S]^{-1}$. In $K = 2$ dimensions a quadratic form written in scalar notation is of the form of Equation 9.7 if the symmetric matrix of scaling constants is diagonal, and in the form of Equation 9.80 if it is not. Equation 9.85 emphasizes that quadratic forms can be interpreted as (squared) distances, and as such it is generally desirable for them to be nonnegative, and furthermore strictly positive if the vector being squared is not zero. This condition is met if all the eigenvalues of the symmetric matrix of scaling constants are positive, in which case that matrix is called positive definite.

Finally, it was noted in Section 9.2.2 that Equation 9.7 describes ellipses of constant distance D^2. These ellipses described by Equation 9.7, corresponding to zero correlations in the matrix $[S]$ in Equation 9.85, have their axes aligned with the coordinate axes. Equation 9.85 also describes ellipses of constant Mahalanobis distance D^2, whose axes are rotated away from the coordinate axes to the extent that some or all of the correlations in $[S]$ are nonzero. In these cases the axes of the ellipses of constant D^2 are aligned in the directions of the eigenvectors of $[S]$, as will be seen in Section 10.1.

9.5 Exercises

9.1. Calculate the matrix product $[A][E]$, using the values in Equations 9.56 and 9.57.

9.2. Derive the regression equation produced in Example 6.1, using matrix notation.

9.3. Calculate the angle between the two eigenvectors of the matrix $[A]$ in Equation 9.56.

9.4. Verify through matrix multiplication that $[T]$ in Equation 9.43 is an orthogonal matrix.

9.5. Show that Equation 9.63 produces a valid square root.

9.6. The eigenvalues and eigenvectors of the covariance matrix for the Ithaca and Canandaigua maximum temperatures in Table A.1 are $\lambda_1 = 118.8$ and $\lambda_2 = 2.60$, and $\mathbf{e}_1^T = [.700, .714]$ and $\mathbf{e}_2^T = [-.714, .700]$, where the first element of each vector corresponds to the Ithaca temperature.

   a. Find the covariance matrix $[S]$, using its spectral decomposition.

   b. Find $[S]^{-1}$ using its eigenvalues and eigenvectors.

   c. Find $[S]^{-1}$ using the result of part (a), and Equation 9.28.

   d. Find the $[S]^{1/2}$ that is symmetric.

   e. Find the Mahalanobis distance between the observations for 1 January and 2 January.

9.7. a. Use the Pearson correlations in Table 3.5 and the standard deviations from Table A.1 to compute the covariance matrix $[S]$ for the four temperature variables.

   b. Consider the average daily temperatures defined by the two linear combinations:

$$y_1 = 0.5 \text{ (Ithaca Max)} + 0.5 \text{ (Ithaca Min)}$$
$$y_2 = 0.5 \text{ (Canandaigua Max)} + 0.5 \text{ (Canandaigua Min)}$$

Find $\boldsymbol{\mu}_y$ and $[S_y]$ without actually computing the individual y values.

CHAPTER • 10

The Multivariate Normal (MVN) Distribution

10.1 Definition of the MVN

The Multivariate Normal (MVN) distribution is the natural generalization of the Gaussian, or normal distribution (Section 4.4.2) to multivariate, or vector data. The MVN is by no means the only known continuous parametric multivariate distribution (e.g., Johnson and Kotz 1972; Johnson 1987), but overwhelmingly it is the most commonly used. Some of the popularity of the MVN follows from its relationship to the multivariate central limit theorem, although it is also used in other settings without strong theoretical justification because of a number of convenient properties that will be outlined in this section. This convenience is often sufficiently compelling to undertake transformation of non-Gaussian multivariate data to approximate multinormality before working with them, and this has been a strong motivation for development of the methods described in Section 3.4.1.

The univariate Gaussian PDF (Equation 4.23) describes the individual, or marginal, distribution of probability density for a scalar Gaussian variable. The MVN describes the joint distribution of probability density collectively for the K variables in a vector x. The univariate Gaussian PDF is visualized as the bell curve defined on the real line (i.e., in a one-dimensional space). The MVN PDF is defined on the K-dimensional space whose coordinate axes correspond to the elements of x, in which multivariate distances were calculated in Sections 9.2 and 9.4.4.

The probability density function for the MVN is

$$f(\mathbf{x}) = \frac{1}{(2\pi)^{K/2}\sqrt{\det[\Sigma]}} \exp\left[-\frac{1}{2}(\mathbf{x}-\boldsymbol{\mu})^{\mathrm{T}}[\Sigma]^{-1}(\mathbf{x}-\boldsymbol{\mu})\right], \qquad (10.1)$$

where $\boldsymbol{\mu}$ is the K-dimensional mean vector, and $[\Sigma]$ is the $(K \times K)$ covariance matrix for the K variables in the vector x. In $K = 1$ dimension, Equation 10.1 reduces to Equation 4.23, and for $K = 2$ it reduces to the PDF for the bivariate normal distribution (Equation 4.33). The key part of the MVN PDF is the argument of the exponential function, and regardless of the dimension of x this argument is a squared, standardized distance (i.e., the difference between x and its mean, standardized by the (co-)variance). In the general multivariate form of Equation 10.1 this distance is the Mahalanobis distance, which is a positive-definite quadratic form when $[\Sigma]$ is of full rank, and not defined

435

otherwise because in that case $[\Sigma]^{-1}$ does not exist. The constants outside of the exponential in Equation 10.1 serve only to ensure that the integral over the entire K-dimensional space is 1,

$$\int_{-\infty}^{\infty}\int_{-\infty}^{\infty}\cdots\int_{-\infty}^{\infty} f(\mathbf{x})dx_1 dx_2 \cdots dx_K = 1, \tag{10.2}$$

which is the multivariate extension of Equation 4.17.

If each of the K variables in $\mathbf{x}$ are separately standardized according to 4.25, the result is the standardized MVN density,

$$\phi(\mathbf{z}) = \frac{1}{(2\pi)^{K/2}\sqrt{\det[R]}} \exp\left[\frac{-\mathbf{z}^{\mathrm{T}}[R]^{-1}\mathbf{z}}{2}\right] \tag{10.3}$$

where $[R]$ is the (Pearson) correlation matrix (e.g., Figure 3.25) for the K variables. Equation 10.3 is the multivariate generalization of Equation 4.24. The nearly universal notation for indicating that a random vector follows a K-dimensional MVN is

$$\mathbf{x} \sim N_K(\boldsymbol{\mu}, [\Sigma]) \tag{10.4a}$$

or, for standardized variables,

$$\mathbf{z} \sim N_K(\mathbf{0}, [R]), \tag{10.4b}$$

where $\mathbf{0}$ is the K-dimensional mean vector whose elements are all zero.

Because the only dependence of Equation 10.1 on the random vector $\mathbf{x}$ is through the Mahalanobis distance inside the exponential, contours of equal probability density are ellipsoids of constant D^2 from $\boldsymbol{\mu}$. These ellipsoidal contours centered on the mean enclose the smallest regions in the K-dimensional space containing a given portion of the probability mass, and the link between the size of these ellipsoids and the enclosed probability is the χ^2 distribution:

$$\Pr\{D^2 = (\mathbf{x} - \boldsymbol{\mu})^{\mathrm{T}}[\Sigma]^{-1}(\mathbf{x} - \boldsymbol{\mu}) \le \chi_K^2(\alpha)\} = \alpha. \tag{10.5}$$

Here $\chi_K^2(\alpha)$ denotes the α quantile of the χ^2 distribution with K degrees of freedom, associated with cumulative probability α (Table B.3). That is, the probability of an $\mathbf{x}$ being within a given Mahalanobis distance D^2 of the mean is the area to the left of D^2 under the χ^2 distribution with degrees of freedom $\nu = K$. As noted at the end of Section 9.4.4 the orientations of these ellipsoids are given by the eigenvectors of $[\Sigma]$, which are also the eigenvectors of $[\Sigma]^{-1}$. Furthermore, the elongation of the ellipsoids in the directions of each of these eigenvectors is given by the square root of the product of the respective eigenvalue of $[\Sigma]$ multiplied by the relevant χ^2 quantile. For a given D^2 the (hyper-) volume enclosed by one of these ellipsoids is proportional to the square root of the determinant of $[\Sigma]$,

$$V = \frac{2(\pi D^2)^{K/2}}{K\, \Gamma(K/2)}\sqrt{\det[\Sigma]}, \tag{10.6}$$

where $\Gamma(\cdot)$ denotes the gamma function (Equation 4.7). Here the determinant of $[\Sigma]$ functions as a scalar measure of the magnitude of the matrix, in terms of the volume occupied by the probability dispersion it describes. Accordingly, $\det[\Sigma]$ is sometimes called the generalized variance. The determinant, and thus also the volumes enclosed by constant-D^2 ellipsoids, increases as the K variances $\sigma_{k,k}$ increase; but also these volumes decrease as the correlations among the K variables increase, because larger correlations result in the ellipsoids being less spherical and more elongated.

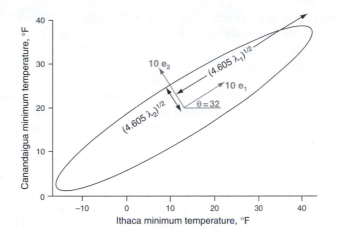

FIGURE 10.1 The 90% probability ellipse for the bivariate normal distribution representing the minimum temperature data in Table A.1, centered at the vector sample mean. Its major and minor axes are oriented in the directions of the eigenvectors (gray) of the covariance matrix in Equation 9.56, and stretched in these directions in proportion to the square roots of the respective eigenvalues. The constant of proportionality is the square root of the appropriate χ_2^2 quantile. The eigenvectors are drawn 10x larger than unit length for clarity.

EXAMPLE 10.1 Probability Ellipses for the Bivariate Normal Distribution

It is easiest to visualize multivariate ideas in two dimensions. Consider the MVN distribution fit to the Ithaca and Canandaigua minimum temperature data in Table A.1. Here $K = 2$, so this is a bivariate normal distribution with sample mean vector $[13.0, 20.2]^T$ and (2×2) covariance matrix as shown in Equation 9.56. Example 9.3 shows that this covariance matrix has eigenvalues $\lambda_1 = 254.76$ and $\lambda_2 = 8.29$, with corresponding eigenvectors $e_1^T = [0.848, 0.530]$ and $e_2^T = [-0.530, 0.848]$.

Figure 10.1 shows the 90% probability ellipse for this distribution. All the probability ellipses for this distribution are oriented 32° from the data axes, as shown in Example 9.6. (This angle between e_1 and the horizontal unit vector $[1, 0]^T$ can also be calculated using Equation 9.15.) The extent of this 90% probability ellipse in the directions of its two axes is determined by the 90% quantile of the χ^2 distribution with $\nu = K = 2$ degrees of freedom, which is $\chi_2^2(0.90) = 4.605$ from Table B.3. Therefore the ellipse extends to $(\chi_2^2(0.90)\lambda_k)^{1/2}$ in the directions of each of the two eigenvectors e_k; or the distances $(4.605 \cdot 254.67)^{1/2} = 34.2$ in the e_1 direction, and $(4.605 \cdot 8.29)^{1/2} = 6.2$ in the e_2 direction.

The volume enclosed by this ellipse is actually an area in two dimensions. From Equation 10.6 this area is $V = 2(\pi 4.605)^1 \sqrt{2103.26}/(2 \cdot 1) = 663.5$, since $\det[S] = 2103.26$. ◊

10.2 Four Handy Properties of the MVN

1) *All subsets of variables from a MVN distribution are themselves distributed MVN.* Consider the partition of a $(K \times 1)$ MVN random vector x into the vectors $x_1 = (x_1, x_2, \ldots, x_L)$, and $x_2 = (x_{L+1}, x_{L+2}, \ldots, x_K)$, as in Equation 9.76a. Then each of these two subvectors themselves follow MVN distributions, with $x_1 \sim N_L(\mu_1, [\Sigma_{1,1}])$ and $x_2 \sim N_{K-L}(\mu_2, [\Sigma_{2,2}])$. Here the two mean vectors comprise the corresponding

partition of the original mean vector as in Equation 9.76b, and the covariance matrices are the indicated submatrices in Equation 9.77b and 9.77c. Note that the original ordering of the elements of x is immaterial, and that a MVN partition can be constructed from any subset. If a subset of the MVN x contains only one element (e.g., the scalar x_1) its distribution is univariate Gaussian: $x_1 \sim N_1(\mu_1, \sigma_{1,1})$. That is, this first handy property implies that all the marginal distributions for the K elements of a MVN x are univariate Gaussian. The converse may not be true: it is not necessarily the case that the joint distribution of an arbitrarily selected set of K Gaussian variables will follow a MVN.

2) *Linear combinations of a MVN* **x** *are Gaussian.* If x is a MVN random vector, then a single linear combination in the form of Equation 9.78 will be univariate Gaussian with mean and variance given by Equations 9.81a and 9.81b, respectively. This fact is a consequence of the property that sums of Gaussian variables are themselves Gaussian, as noted in connection with the sketch of the Central Limit Theorem in Section 4.4.2. Similarly the result of L simultaneous linear transformations, as in Equation 9.82, will have an L-dimensional MVN distribution, with mean vector and covariance matrix given by Equations 9.83a and 9.83b, respectively, provided the covariance matrix $[\Sigma_y]$ is invertable. This condition will hold if $L \leq K$, and if none of the transformed variables y_ℓ can be expressed as an exact linear combination of the others. In addition, the mean of a MVN distribution can be shifted without changing the covariance matrix. If c is a $(K \times 1)$ vector of constants and $x \sim N_K(\mu, [\Sigma])$, then

$$\mathbf{x} + \mathbf{c} \sim N_K(\mu_X + \mathbf{c}, [\Sigma_X]). \tag{10.7}$$

3) *Independence implies zero correlation, and vice versa, for Gaussian distributions.* Again consider the partition of a MVN x as in Equation 9.76a. If x_1 and x_2 are independent then the off-diagonal matrices of cross-covariances in Equation 9.77 contain only zeros: $[\Sigma_{1,2}] = [\Sigma_{2,1}]^T = [0]$. Conversely, if $[\Sigma_{1,2}] = [\Sigma_{2,1}]^T = [0]$ then the MVN PDF can be factored as $f(x) = f(x_1)f(x_2)$, implying independence (cf. Equation 2.12), because the argument inside the exponential in Equation 10.1 then breaks cleanly into two factors.

4) *Conditional distributions of subsets of a MVN* **x**, *given fixed values for other subsets, are also MVN.* This is the multivariate generalization of Equations 4.37, which are illustrated in Example 4.7, expressing this idea for the bivariate normal distribution. Consider again the partition $x = [x_1, x_2]$ as defined in Equation 9.76b and used to illustrate properties (1) and (3). The conditional mean of one subset of the variables x_1 given particular values for the remaining variables $X_2 = x_2$ is

$$\mu_1 | \mathbf{x}_2 = \mu_1 + [\Sigma_{12}][\Sigma_{22}]^{-1}(\mathbf{x}_2 - \mu_2), \tag{10.8a}$$

and the conditional covariance matrix is

$$[\Sigma_1 | \mathbf{x}_2] = [\Sigma_{11}] - [\Sigma_{12}][\Sigma_{22}]^{-1}[\Sigma_{21}], \tag{10.8b}$$

where the submatrices of $[\Sigma]$ are again as defined in Equation 9.77. As was the case for the bivariate normal distribution, the conditional mean shift in Equation 10.8a depends on the particular value of the conditioning variable x_2, whereas the conditional covariance matrix in Equation 10.8b does not. If x_1 and x_2 are independent, then knowledge of one provides no additional information about the other. Mathematically, if $[\Sigma_{1,2}] = [\Sigma_{2,1}]^T = [0]$ then Equation 10.8a reduces to $\mu_1 | \mathbf{x}_2 = \mu_1$, and Equation 10.8b reduces to $[\Sigma_1 | \mathbf{x}_2] = [\Sigma_1]$.

EXAMPLE 10.2   Three-Dimensional MVN Distributions as Cucumbers

Imagine a three-dimensional MVN PDF as a cucumber, which is a solid, three-dimensional ovoid. Since the cucumber has a distinct edge, it would be more correct to imagine that it represents that part of a MVN PDF enclosed within a fixed-D^2 ellipsoidal surface. The cucumber would be an even better metaphor if its density increased toward the core and decreased toward the skin.

Figure 10.2a illustrates property (1), which is that all subsets of a MVN distribution are themselves MVN. Here are three hypothetical cucumbers floating above a kitchen cutting board in different orientations, and are illuminated from above. Their shadows represent the joint distribution of the two variables whose axes are aligned with the edges of the board. Regardless of the orientation of the cucumber relative to the board (i.e., regardless of the covariance structure of the three-dimensional distribution) each of these two-dimensional joint shadow distributions for x_1 and x_2 is bivariate normal, with probability contours within fixed Mahalanobis distances of the mean that are ovals in the plane of the board.

Figure 10.2b illustrates property (4), that conditional distributions of subsets given particular values for the remaining variables in a MVN, are themselves MVN. Here portions of two cucumbers are lying on the cutting board, with the long axis of the left cucumber (indicated by the direction of the arrow, or the corresponding eigenvector) oriented parallel to the x_1 axis of the board, and the long axis of the right cucumber has been placed diagonal to the edges of the board. The three variables represented by the left cucumber are thus mutually independent, whereas the two horizontal (x_1 and x_2) variables for the right cucumber are positively correlated. Each cucumber has been sliced perpendicularly to the x_1 axis of the cutting board, and the exposed faces represent the joint conditional distributions of the remaining two (x_2 and x_3) variables. Both faces are ovals, illustrating that both of the resulting conditional distributions are bivariate normal. Because the cucumber on the left is oriented parallel to the cutting board edges (coordinate axes) it represents independent variables and the exposed oval is a circle.

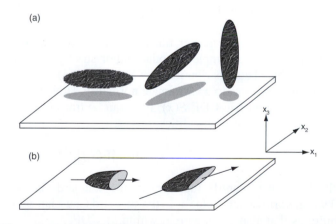

FIGURE 10.2   Three-dimensional MVN distributions as cucumbers on a kitchen cutting board. (a) Three cucumbers floating slightly above the cutting board and illuminated from above, illustrating that their shadows (the bivariate normal distributions representing the two-dimensional subsets of the original three variables in the plane of the cutting board) are ovals, regardless of the orientation (covariance structure) of the cucumber. (b) Two cucumbers resting on the cutting board, with faces exposed by cuts made perpendicularly to the x_1 coordinate axis; illustrating bivariate normality in the other two (x_2, x_3) dimensions, given the left-right location of the cut. Arrows indicate directions of the cucumber long-axis eigenvectors.

If parallel cuts had been made elsewhere on these cucumbers, the shapes of the exposed faces would have been the same, illustrating (as in Equation 10.8b) that the conditional covariance (shape of the exposed cucumber face) does not depend on the value of the conditioning variable (location left or right along the x_1 axis at which the cut is made). On the other hand, the conditional means (the centers of the exposed faces projected onto the $x_2 - x_3$ plane, Equation 10.8a) depend on the value of the conditioning variable (x_1), but only if the variables are correlated as in the right-hand cucumber: Making the cut further to the right shifts the location of the center of the exposed face toward the back of the board (the x_2 component of the conditional bivariate vector mean is greater). On the other hand, because the axes of the left cucumber ellipsoid are aligned with the coordinate axes, the location of the center of the exposed face in the $x_2 - x_3$ plane is the same regardless of where on the x_1 axis the cut has been made. ◊

10.3 Assessing Multinormality

It was noted in Section 3.4.1 that one strong motivation for transforming data to approximate normality is the ability to use the MVN to describe the joint variations of a multivariate data set. Usually either the Box-Cox power transformations (Equation 3.18), or the Yeo and Johnson (2000) generalization to possibly nonpositive data, are used. The Hinkley statistic (Equation 3.19), which reflects the degree of symmetry in a transformed univariate distribution, is the simplest way to decide among power transformations. However, when the goal is specifically to approximate a Gaussian distribution, as is the case when we hope that each of the transformed distributions will form one of the marginal distributions of a MVN, it is probably better to choose transformation exponents that maximize the Gaussian likelihood function (Equation 3.20). It is also possible to choose transformation exponents simultaneously for multiple elements of x, by choosing the corresponding vector of exponents λ that maximize the MVN likelihood function (Andrews *et al.* 1971), although this approach requires substantially more computation than fitting the individual exponents independently, and in most cases is probably not worth the additional effort.

Choices other than the power transforms are also possible, and may sometimes be more appropriate. For example bimodal and/or strictly bounded data, such as might be well described by a beta distribution (see Section 4.4.4) with both parameters less than 1, will not power-transform to approximate normality. However, if such data are adequately described by a parametric CDF $F(x)$, they can be transformed to approximate normality by matching cumulative probabilities; that is,

$$z_i = \Phi^{-1}[F(x_i)]. \tag{10.9}$$

Here $\Phi^{-1}[\bullet]$ is the quantile function for the standard Gaussian distribution, so Equation 10.9 transforms a data value x_i to the standard Gaussian z_i having the same cumulative probability as that associated with x_i within its CDF.

Methods for evaluating normality are necessary, both to assess the need for transformations, and to evaluate the effectiveness of candidate transformations. There is no single best approach to the problem for evaluating multinormality, and in practice we usually look at multiple indicators, which may include both quantitative formal tests and qualitative graphical tools.

Because all marginal distributions of a MVN are univariate Gaussian, goodness-of-fit tests are often calculated for the univariate distributions corresponding to each of the elements of the x whose multinormality is being assessed. A good choice for the specific purpose of testing Gaussian distribution is the Filliben test for the Gaussian Q-Q plot correlation (see Table 5.3). Gaussian marginal distributions are a necessary consequence of joint multinormality, but are not sufficient to guarantee it. In particular, looking only at marginal distributions will not identify the presence of multivariate outliers (e.g., Figure 9.6c), which are points that are not extreme with respect to any of the individual variables, but are unusual in the context of the overall covariance structure.

Two tests for multinormality (i.e., jointly for all K dimensions of x) with respect to multivariate skewness and kurtosis are available (Mardia 1970; Mardia *et al.* 1979). Both rely on the function of the point pair x_i and x_j given by

$$g_{i,j} = (x_i - \bar{x})[S]^{-1}(x_j - \bar{x}), \tag{10.10}$$

where $[S]$ is the sample covariance matrix. This function is used to calculate the multivariate skewness measure

$$b_{1,K} = \frac{1}{n^2} \sum_{i=1}^{n} \sum_{j=1}^{n} g_{i,j}^3, \tag{10.11}$$

which reflects high-dimensional symmetry, and will be near zero for MVN data. This test statistic can be evaluated using

$$\frac{n\, b_{1,K}}{6} \sim \chi_{\nu}^2, \tag{10.12a}$$

where the degrees-of-freedom parameter is

$$\nu = \frac{K(K+1)(K+2)}{6}, \tag{10.12b}$$

and the null hypothesis of multinormality, with respect to its symmetry, is rejected for sufficiently large values of $b_{1,K}$.

Multivariate kurtosis (appropriately heavy tails for the MVN relative to the center of the distribution) can be tested using the statistic

$$b_{2,K} = \frac{1}{n} \sum_{i=1}^{n} g_{i,i}^2, \tag{10.13}$$

which is equivalent to the average of $(D^2)^2$ because for this statistic $i = j$ in Equation 10.10. Under the null hypothesis of multinormality,

$$\left[\frac{b_{2,K} - K(K+2)}{8\,K(K+2)/n} \right]^{1/2} \sim N[0, 1]. \tag{10.14}$$

Scatterplots of variable pairs are valuable qualitative indicators of multinormality, since all subsets of variables from a MVN distribution are jointly normal also, and two-dimensional graphs are easy to plot and grasp. Thus looking at a scatterplot matrix (see Section 3.6.5) is typically a valuable tool in assessing multinormality. Point clouds that are elliptical or circular are indicative of multinormality. Outliers away from the main scatter in one or more of the plots may be multivariate outliers, as in Figure 9.6c. Similarly, it can be valuable to look at rotating scatterplots of various three-dimensional subsets of x.

Absence of evidence for multivariate outliers in all possible pairwise scatterplots does not guarantee that none exist in higher-dimensional combinations. An approach to exposing the possible existence of high-dimensional multivariate outliers, as well as to detect other possible problems, is to use Equation 10.5. This equation implies that if the data x are MVN, the (univariate) distribution for $D_i^2, i = 1, \ldots, n$, is χ_K^2. That is, the Mahalanobis distance D_i^2 from the sample mean for each x_i can be calculated, and the closeness of this distribution of D_i^2 values to the χ^2 distribution with K degrees of freedom can be evaluated. The easiest and most usual evaluation method is to visually inspect the Q-Q plot. It would also be possible to derive critical values to test the null hypothesis of multinormality according to the correlation coefficient for this kind of plot, using the method sketched in Section 5.2.5.

Because any linear combination of variables that are jointly multinormal will be univariate Gaussian, it can also be informative to look at and formally test linear combinations for Gaussian distribution. Often it is useful to look specifically at the linear combinations given by the eigenvectors of [S],

$$y_i = \mathbf{e}_k^T \mathbf{x}_i. \tag{10.15}$$

It turns out that the linear combinations defined by the elements of the eigenvectors associated with the smallest eigenvalues can be particularly useful in identifying multivariate outliers, either by inspection of the Q-Q plots, or by formally testing the Q-Q correlations. (The reason behind linear combinations associated with the smallest eigenvalues being especially powerful in exposing outliers relates to principal component analysis, as explained in Section 11.1.5.) Inspection of pairwise scatterplots of linear combinations defined by the eigenvectors of [S] can also be revealing.

EXAMPLE 10.3   Assessing Bivariate Normality for the Canandaigua Temperature Data

Are the January 1987 Canandaigua maximum and minimum temperature data in Table A.1 consistent with the proposition that they were drawn from a bivariate normal distribution? Figure 10.3 presents four plots indicating that this assumption is not unreasonable, considering the rather small sample size.

Figures 10.3a and 10.3b are Gaussian Q-Q plots for the maximum and minimum temperatures, respectively. The temperatures are plotted as functions of the standard Gaussian variables with the same cumulative probability, which has been estimated using a median plotting position (see Table 3.2). Both plots are close to linear, supporting the notion that each of the two data batches were drawn from univariate Gaussian distributions. Somewhat more quantitatively, the correlations of the points in these two panels are 0.984 for the maximum temperatures and 0.978 for the minimum temperatures. If these data were independent, we could refer to Table 5.3 and find that both are larger than 0.970, which is the 10% critical value for $n = 30$. Since these data are serially correlated, the Q-Q correlations provide even weaker evidence against the null hypotheses that these two marginal distributions are Gaussian.

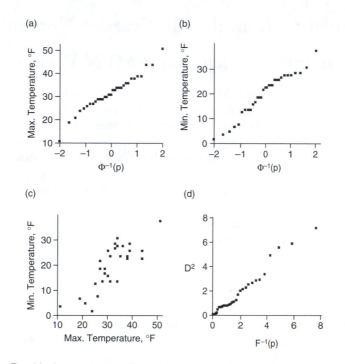

FIGURE 10.3   Graphical assessments of bivariate normality for the Canandaigua maximum and minimum temperature data. (a) Gaussian Q-Q plot for the maximum temperatures, (b) Gaussian Q-Q plot for the minimum temperatures, (c) scatterplot for the bivariate temperature data, and (d) Q-Q plot for Mahalanobis distances relative to the χ^2 distribution.

Figure 10.3c shows the scatterplot for the two variables jointly. The distribution of points appears to be reasonably elliptical, with greater density near the sample mean, $[31.77, 20.23]^T$, and less density at the extremes. This assessment is supported by Figure 10.3d, which is the Q-Q plot for the Mahalanobis distances of each of the points from the sample mean. If the data are bivariate normal, the distribution of these D_i^2 values will be χ^2, with two degrees of freedom, which is an exponential distribution (Equations 4.45 and 4.46), with $\beta = 2$. Values of its quantile function on the horizontal axis of Figure 10.3d have been calculated using Equation 4.80. The points in this Q-Q plot are also reasonably straight, with the largest bivariate outlier ($D^2 = 7.23$) obtained for 25 January. This is the leftmost point in Figure 10.3c, corresponding to the coldest maximum temperature. The second-largest D^2 of 6.00 results from the data for 15 January, which is the warmest day in both the maximum and minimum temperature data.

The correlation of the points in Figure 10.3d is 0.989, but it would be inappropriate to use Table 5.3 to judge its unusualness relative to a null hypothesis that the data were drawn from a bivariate normal distribution, for two reasons. First, Table 5.3 was derived for Gaussian Q-Q plot correlations, and the null distribution (under the hypothesis of MVN data) for the Mahalanobis distance is χ^2. In addition, these data are not independent. However, it would be possible to derive critical values analogous to those in Table 5.3, by synthetically generating a large number of samples from a bivariate normal distribution with (bivariate) time correlations that simulate those in the Canandaigua temperatures, calculating the D^2 Q-Q plot for each of these samples, and tabulating the distribution of the resulting correlations. Methods appropriate to constructing such simulations are described in the next section. ◊

10.4 Simulation from the Multivariate Normal Distribution

10.4.1 Simulating Independent MVN Variates

Statistical simulation of MVN variates is accomplished through an extension of the univariate ideas presented in Section 4.7. Generation of synthetic MVN values takes advantage of property (2) in Section 10.2, that linear combinations of MVN values are themselves MVN. In particular, realizations of K-dimensional MVN vectors $x \sim N_K(\mu, [\Sigma])$ are generated as linear combinations of K-dimensional standard MVN vectors $z \sim N_K(0, [I])$. These standard MVN realizations are in turn generated on the basis of uniform variates (see Section 4.7.1) transformed according to an algorithm such as that described in Section 4.7.4.

Specifically, the linear combinations used to generate MVN variates with a given mean vector and covariance matrix are given by the rows of a square-root matrix (see Section 9.3.4) for $[\Sigma]$, with the appropriate element of the mean vector added:

$$\mathbf{x}_i = [\Sigma]^{1/2}\mathbf{z}_i + \mu. \tag{10.16}$$

As a linear combination of the K standard Gaussian values in the vector z, the generated vectors x will have a MVN distribution. It is straightforward to see that they will also have the correct mean vector and covariance matrix:

$$\mathrm{E}(\mathbf{x}) = \mathrm{E}([\Sigma]^{1/2}\mathbf{z} + \mu) = [\Sigma]^{1/2}\mathrm{E}(\mathbf{z}) + \mu = \mu \tag{10.17a}$$

because $\mathrm{E}(\mathbf{z}) = 0$, and

$$[\Sigma_x] = [\Sigma]^{1/2}[\Sigma_z]([\Sigma]^{1/2})^{\mathrm{T}} = [\Sigma]^{1/2}[I]([\Sigma]^{1/2})^{\mathrm{T}}$$
$$= [\Sigma]^{1/2}([\Sigma]^{1/2})^{\mathrm{T}} = [\Sigma]. \tag{10.17b}$$

Different choices for the nonunique matrix $[\Sigma]^{1/2}$ will yield different simulated x vectors for a given input z, but Equation 10.17 shows that collectively, the resulting $x \sim N_K(\mu, [\Sigma])$ so long as $[\Sigma]^{1/2}([\Sigma]^{1/2})^{\mathrm{T}} = [\Sigma]$.

It is interesting to note that the transformation in Equation 10.16 can be inverted to produce standard MVN vectors $z \sim N_K(0, [I])$ corresponding to MVN vectors x of known distributions. Usually this manipulation is done to transform a sample of vectors x to the standard MVN according to their estimated mean and covariance of x, analogously to the standardized anomaly (Equation 3.21),

$$\mathbf{z}_i = [S]^{-1/2}(\mathbf{x}_i - \bar{\mathbf{x}}) = [S]^{-1/2}(\mathbf{x}_i'). \tag{10.18}$$

This relationship is called the Mahalanobis transformation. It is distinct from the scaling transformation (Equation 9.34), which produces a vector of standard Gaussian variates having unchanged covariance structure. It is straightforward to show that Equation 10.18 produces uncorrelated z_k values, each with unit variance:

$$[S_z] = [S_x]^{-1/2}[S_x]([S_x]^{-1/2})^{\mathrm{T}}$$
$$= [S_x]^{-1/2}[S_x]^{1/2}([S_x]^{1/2})^{\mathrm{T}}([S_x]^{-1/2})^{\mathrm{T}} = [I][I] = [I]. \tag{10.19}$$

10.4.2 Simulating Multivariate Time Series

The autoregressive processes for scalar time series described in Sections 8.3.1 and 8.3.2 can be generalized to stationary multivariate, or vector, time series. In this case the variable x is a vector quantity observed at discrete and regularly spaced time intervals. The multivariate generalization of the AR(p) process in Equation 8.23 is

$$\mathbf{x}_{t+1} - \boldsymbol{\mu} = \sum_{i=1}^{p} [\Phi_i](\mathbf{x}_{t-i+1} - \boldsymbol{\mu}) + [\mathrm{B}]\boldsymbol{\varepsilon}_{t+1}. \tag{10.20}$$

Here the elements of the vector x consist of a set of K correlated time series, $\boldsymbol{\mu}$ contains the corresponding mean vector, and the elements of the vector $\boldsymbol{\varepsilon}$ are mutually independent (and usually Gaussian) random variables with unit variance. The matrices of autoregressive parameters $[\Phi_i]$ correspond to the scalar autoregressive parameters ϕ_k in Equation 8.23. The matrix $[\mathrm{B}]$, operating on the vector $\boldsymbol{\varepsilon}_{t+1}$, allows the random components in Equation 10.20 to have different variances, and to be mutually correlated at each time step (although they are uncorrelated in time). Note that the order, p, of the autoregression was denoted as K in Chapter 8, and does not indicate the dimension of a vector there. Multivariate autoregressive-moving average models, extending the scalar models in Section 8.3.6 to vector data, can also be defined.

The most common special case of Equation 10.20 is the multivariate AR(1) process,

$$\mathbf{x}_{t+1} - \boldsymbol{\mu} = [\Phi](\mathbf{x}_t - \boldsymbol{\mu}) + [\mathrm{B}]\boldsymbol{\varepsilon}_{t+1}, \tag{10.21}$$

which is obtained from Equation 10.20 for the autoregressive order $p = 1$. It is the multivariate generalization of Equation 8.16, and will describe a stationary process if all the eigenvalues of $[\Phi]$ are between -1 and 1. Matalas (1967) and Bras and Rodríguez-Iturbe (1985) describe use of Equation 10.21 in hydrology, where the elements of x are simultaneously measured streamflows at different locations. This equation is also often used as part of a common synthetic weather generator formulation (Richardson 1981). In this second application x usually has three elements, corresponding to daily maximum temperature, minimum temperature, and solar radiation at a given location.

The two parameter matrices in Equation 10.21 are most easily estimated using the simultaneous and lagged covariances among the elements of x. The simultaneous covariances are contained in the usual covariance matrix $[\mathrm{S}]$, and the lagged covariances are contained in the matrix

$$[\mathrm{S}_1] = \frac{1}{n-1} \sum_{t=1}^{n-1} \mathbf{x}'_{t+1} \mathbf{x}'^{T}_{t} \tag{10.22a}$$

$$= \begin{bmatrix} s_1(1 \to 1) & s_1(2 \to 1) & \cdots & s_1(K \to 1) \\ s_1(1 \to 2) & s_1(2 \to 2) & \cdots & s_1(K \to 2) \\ \vdots & \vdots & & \vdots \\ s_1(1 \to K) & s_1(2 \to K) & \cdots & s_1(K \to K) \end{bmatrix}. \tag{10.22b}$$

This equation is similar to Equation 9.35 for $[\mathrm{S}]$, except that the pairs of vectors whose outer products are summed are data (anomalies) at pairs of successive time points. The diagonal elements of $[\mathrm{S}_1]$ are the lag-1 autocovariances (the autocorrelations in Equation 3.30 multiplied by the respective variances) for each of the K elements of x. The off-diagonal elements of $[\mathrm{S}_1]$ are the lagged covariances among unlike elements of x.

The arrow notation in this equation indicates the time sequence of the lagging of the variables. For example, $s_1(1 \rightarrow 2)$ denotes the correlation between x_1 at time t, and x_2 at time $t+1$, and $s_1(2 \rightarrow 1)$ denotes the correlation between x_2 at time t, and x_1 at time $t+1$. Notice that the matrix $[S]$ is symmetric, but that in general $[S_1]$ is not.

The matrix of autoregressive parameters $[\Phi]$ in Equation 10.21 is obtained from the lagged and unlagged covariance matrices using

$$[\Phi] = [S_1][S]^{-1}. \tag{10.23}$$

Obtaining the matrix $[B]$ requires finding a matrix square root (see Section 9.3.4) of

$$[B][B]^{T} = [S] - [\Phi][S_1]^{T}. \tag{10.24}$$

Having defined a multivariate autoregressive model, it is straightforward to simulate from it using the defining equation (e.g., Equation 10.21) together with an appropriate random number generator to provide time series of realizations for the random-forcing vector $\boldsymbol{\varepsilon}$. Usually these are taken to be standard Gaussian, in which case they can be generated using the algorithm described in Section 4.7.4. In any case the K elements of $\boldsymbol{\varepsilon}$ will have zero mean and unit variance, will be uncorrelated with each other at any one time t, and will be uncorrelated with other forcing vectors at different times $t+i$:

$$E[\boldsymbol{\varepsilon}_t] = \mathbf{0} \tag{10.25a}$$

$$E[\boldsymbol{\varepsilon}_t \boldsymbol{\varepsilon}_t^{T}] = [I] \tag{10.25b}$$

$$E[\boldsymbol{\varepsilon}_t \boldsymbol{\varepsilon}_{t+i}^{T}] = [0], \quad i \neq 0. \tag{10.25c}$$

If the $\boldsymbol{\varepsilon}$ vectors contain realizations of independent Gaussian variates, then the resulting $\boldsymbol{x}$ vectors will have a MVN distribution, because they are linear combinations of (standard) MVN vectors $\boldsymbol{\varepsilon}$. If the original data are clearly non-Gaussian they may be transformed before fitting the time series model.

EXAMPLE 10.4   Fitting and Simulating from a Bivariate Autoregression

Example 10.3 examined the Canandaigua maximum and minimum temperature data in Table A.1, and concluded that the MVN is a reasonable model for their joint variations. The first-order autoregression (Equation 10.21) is a reasonable model for the time dependence of these data, and fitting the parameter matrices $[\Phi]$ and $[B]$ will allow statistical simulation of synthetic bivariate series that statistically resemble these data. This process can be regarded as an extension of Example 8.3, which illustrated the univariate AR(1) model for the time series of Canandaigua minimum temperatures alone.

The sample statistics necessary to fit Equation 10.21 are easily computed from the Canandaigua temperature data in Table A.1 as

$$\bar{\mathbf{x}} = [31.77, \quad 20.23]^{T} \tag{10.26a}$$

$$[S] = \begin{bmatrix} 61.85 & 56.12 \\ 56.12 & 77.58 \end{bmatrix} \tag{10.26b}$$

and

$$[S_1] = \begin{bmatrix} s_{max \rightarrow max} & s_{min \rightarrow max} \\ s_{max \rightarrow min} & s_{min \rightarrow min} \end{bmatrix} = \begin{bmatrix} 37.32 & 44.51 \\ 42.11 & 51.33 \end{bmatrix}. \tag{10.26c}$$

The matrix of simultaneous covariances is the ordinary covariance matrix [S], and is of course symmetric. The matrix of lagged covariances (Equation 10.26c) is not symmetric. Using Equation 10.23, the estimated matrix of autoregressive parameters is

$$[\Phi] = [S_1][S]^{-1} = \begin{bmatrix} 37.32 & 44.51 \\ 42.11 & 51.33 \end{bmatrix} \begin{bmatrix} .04705 & -.03404 \\ -.03404 & .03751 \end{bmatrix}$$

$$= \begin{bmatrix} 0.241 & 0.399 \\ 0.234 & 0.492 \end{bmatrix}. \tag{10.27}$$

The matrix [B] can be anything satisfying (c.f. Equation 10.24)

$$[B][B]^T = \begin{bmatrix} 61.85 & 56.12 \\ 56.12 & 77.58 \end{bmatrix} - \begin{bmatrix} 0.241 & 0.399 \\ 0.234 & 0.492 \end{bmatrix} \begin{bmatrix} 37.32 & 42.11 \\ 44.51 & 51.33 \end{bmatrix}$$

$$= \begin{bmatrix} 35.10 & 25.49 \\ 25.49 & 42.47 \end{bmatrix}, \tag{10.28}$$

with one solution given by the Cholesky factorization (Equations 9.61 and 9.62),

$$[B] = \begin{bmatrix} 5.92 & 0 \\ 4.31 & 4.89 \end{bmatrix}. \tag{10.29}$$

Using the estimated values in Equations 10.27 and 10.29, and substituting the sample mean from Equation 10.26a for the mean vector, Equation 10.21 becomes an algorithm for simulating bivariate x_t series with the same (sample) first- and second-moment statistics as the Canandaigua temperatures in Table A.1. The Box-Muller algorithm (see Section 4.7.4) is especially convenient for generating the vectors ε_t in this case because it produces them in pairs. Figure 10.4a shows a 100-point realization of a bivariate time series generated in this way. Here the vertical lines connect the simulated maximum and minimum temperatures for a given day, and the light horizontal lines locate the two mean values (Equation 10.26a). These two time series statistically resemble the January 1987 Canandaigua temperature data to the extent that Equation 10.21 is capable of doing so. They are unrealistic in the sense that the underlying population statistics do not change through the 100 simulated days, since the underlying generating model is covariance stationary. That is, the means, variances, and covariances are constant throughout the 100 time points, whereas in nature these statistics would change over the course of a winter. Also, the time series is unrealistic in the sense that, for example, some of the minimum temperatures are warmer than the maxima for the preceding day, which could not occur if the values had been abstracted from a single continuous temperature record. Recalculating the simulation, but starting from a different random number seed, would yield a different series, but with the same statistical characteristics.

Figure 10.4b shows a scatterplot for the 100 point pairs, corresponding to the scatterplot of the actual data in the lower-right panel of Figure 3.26. Since the points were generated by forcing Equation 10.21 with synthetic Gaussian variates for the elements of ε, the resulting distribution for x is bivariate normal by construction. However, the points are not independent, and exhibit time correlation mimicking that found in the original data series. The result is that successive points do not appear at random within the scatterplot, but rather tend to cluster. The light grey line illustrates this time dependence by tracing a path from the first point (circled) to the tenth point (indicated by the arrow tip). ◊

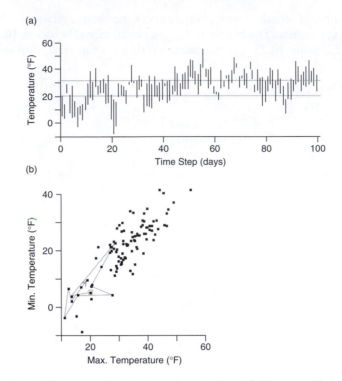

FIGURE 10.4   (a) A 100-point realization from the bivariate AR(1) process fit to the January 1987 Canandaigua daily maximum and minimum temperatures. Vertical lines connect the simulated maximum and minimum for each day, and light horizontal lines locate the two means. (b) Scatterplot of the 100 bivariate points. Light gray line segments connect the first 10 pairs of values.

Since the statistics underlying Figure 10.4a remained constant throughout the simulation, it is a realization of a stationary time series—in this case a perpetual January. Simulations of this kind can be made to be more realistic by allowing the parameters, based on the statistics in Equations 10.26, to vary periodically through an annual cycle. The result would be a cyclostationarity autoregression whose statistics are different for different dates, but the same on the same date in different years. Cyclostationary autoregressions are described in Richardson (1981), von Storch and Zwiers (1999), and Wilks and Wilby (1999), among others.

10.5 Inferences about a Multinormal Mean Vector

This section describes parametric multivariate hypothesis tests concerning mean vectors, based on the MVN. There are many instances where multivariate nonparametric approaches are more appropriate. Some of these multivariate nonparametric tests have been described, as extensions to their univariate counterparts, in Sections 5.3 and 5.4. The parametric tests described in this section require the invertability of the sample covariance matrix of x, $[S_x]$, and so will be infeasible if $n \leq K$. In that case nonparametric tests would be indicated. Even if $[S_x]$ is invertable, the resulting parametric test may have disappointing power unless $n >> K$, and this limitation can be another reason to choose a nonparametric alternative.

10.5.1 Multivariate Central Limit Theorem

The central limit theorem for univariate data was described briefly in Section 4.4.2, and again more quantitatively in Section 5.2.1. It states that the sampling distribution of the average of a sufficiently large number of random variables will be Gaussian, and that if the variables being averaged are mutually independent the variance of that sampling distribution will be smaller than the variance of the original variables by the factor $1/n$. The multivariate generalization of the central limit theorem states that the sampling distribution of the mean of n independent random $(K \times 1)$ vectors x with mean $\boldsymbol{\mu}_x$ and covariance matrix $[\Sigma_x]$ will be MVN with the same covariance matrix, again scaled by the factor $1/n$. That is,

$$\bar{x} \sim N_K \left(\boldsymbol{\mu}_x, \frac{1}{n}[\Sigma_x] \right) \tag{10.30a}$$

or, equivalently,

$$\sqrt{n}(\bar{x} - \boldsymbol{\mu}_x) \sim N_K(\mathbf{0}, [\Sigma_x]). \tag{10.30b}$$

If the random vectors x being averaged are themselves MVN, then the distributions indicated in Equations 10.30 are exact, because then the mean vector is a linear combination of the MVN vectors x. Otherwise, the multinormality for the sample mean is approximate, and that approximation improves as the sample size n increases.

Multinormality for the sampling distribution of the sample mean vector implies that the sampling distribution for the Mahalanobis distance between the sample and population means will be χ^2. That is, assuming that $[\Sigma_x]$ is known, Equation 10.5 implies that

$$(\bar{x} - \boldsymbol{\mu})^T \left(\frac{1}{n}[\Sigma_x] \right)^{-1} (\bar{x} - \boldsymbol{\mu}) \sim \chi_K^2, \tag{10.31a}$$

or

$$n(\bar{x} - \boldsymbol{\mu})^T [\Sigma_x]^{-1} (\bar{x} - \boldsymbol{\mu}) \sim \chi_K^2. \tag{10.31b}$$

10.5.2 Hotelling's T^2

Usually inferences about means must be made without knowing the population variance, and this is true in both univariate and multivariate settings. Substituting the estimated covariance matrix into Equation 10.31 yields the one-sample Hotelling T^2 statistic,

$$T^2 = (\bar{x} - \boldsymbol{\mu}_0)^T \left(\frac{1}{n}[S_x] \right)^{-1} (\bar{x} - \boldsymbol{\mu}_0) = n(\bar{x} - \boldsymbol{\mu}_0)^T [S_x]^{-1} (\bar{x} - \boldsymbol{\mu}_0). \tag{10.32}$$

Here $\boldsymbol{\mu}_0$ indicates the unknown population mean about which inferences will be made. Equation 10.32 is the multivariate generalization of (the square of) the univariate one-sample t statistic that is obtained by combining Equations 5.3 and 5.4. The univariate t is obtained from the square root of Equation 10.32 for scalar (i.e., $K = 1$) data. Both t and T^2 express differences between the sample mean being tested and its hypothesized true value under H_0, "divided by" an appropriate characterization of the dispersion of the null distribution. T^2 is a quadratic (and thus nonnegative) quantity, because the unambiguous ordering of univariate magnitudes on the real line that is expressed by the univariate t statistic does not generalize to higher dimensions. That is, the ordering of scalar magnitude is unambiguous (e.g., it is clear that $5 > 3$), whereas the ordering of vectors is not (e.g., is $[3, 5]^T$ larger or smaller than $[-5, 3]^T$?).

The one-sample T^2 is simply the Mahalanobis distance between the vectors $\boldsymbol{x}$ and $\boldsymbol{\mu}_0$, within the context established by the estimated covariance matrix for the sampling distribution of the mean vector, $(1/n)[S_x]$. Since $\boldsymbol{\mu}_0$ is unknown, a continuum of T^2 values are possible, and the probabilities for these outcomes are described by a PDF—the null distribution for T^2. Under the null hypothesis H_0: $E(\boldsymbol{x}) = \boldsymbol{\mu}_0$, an appropriately scaled version of T^2 follows what is known as the F distribution,

$$\frac{(n-K)}{(n-1)K}T^2 \sim F_{K,n-K}. \tag{10.33}$$

The F distribution is a two-parameter distribution whose quantiles are tabulated in most beginning statistics textbooks. Both parameters are referred to as degrees-of-freedom parameters, and in the context of Equation 10.33 they are $\nu_1 = K$ and $\nu_2 = n - K$, as indicated by the subscripts in Equation 10.33. Accordingly, a null hypothesis that $E(\boldsymbol{x}) = \boldsymbol{\mu}_0$ would be rejected at the α level if

$$T^2 > \frac{(n-1)K}{(n-K)}F_{K,n-K}(1-\alpha), \tag{10.34}$$

where $F_{K,n-K}(1 - \alpha)$ is the $1 - \alpha$ quantile of the F distribution with K and $n - K$ degrees of freedom.

One way of looking at the F distribution is as the multivariate generalization of the t distribution, which is the null distribution for the t statistic in Equation 5.3. The sampling distribution of Equation 5.3 is t rather than standard univariate Gaussian, and the distribution of T^2 is F rather than χ^2 (as might have been expected from Equation 10.31) because the corresponding dispersion measures (s^2 and $[S]$, respectively) are sample estimates rather than known population values. Just as the univariate t distribution converges to the univariate standard Gaussian as its degrees-of-freedom parameter increases (and the variance s^2 is estimated increasingly more precisely), the F distribution approaches proportionality to the χ^2 with $\nu_1 = K$ degrees of freedom as the sample size (and thus also ν_2) becomes large, because [S] is estimated more precisely:

$$\chi^2_K(1-\alpha) = K\, F_{K,\infty}(1-\alpha). \tag{10.35}$$

That is, the $(1 - \alpha)$ quantile of the χ^2 distribution with K degrees of freedom is exactly a factor of K larger than the $(1 - \alpha)$ quantile of the F distribution with $\nu_1 = K$ and $\nu_2 = \infty$ degrees of freedom. Since $(n - 1) \approx (n - K)$ for sufficiently large n, the large-sample counterparts of Equations 10.33 and 10.34 are

$$T^2 \sim \chi^2_K \tag{10.36a}$$

if the null hypothesis is true, leading to rejection at the α level if

$$T^2 > \chi^2_K(1-\alpha). \tag{10.36b}$$

Differences between χ^2 and F quantiles are about 5% for $n - K = 100$, so that this is a reasonable rule of thumb for appropriateness of Equations 10.36 as large-sample approximations to Equations 10.33 and 10.34.

The two-sample t-test statistic (Equation 5.5) is also extended in a straightforward way to inferences regarding the differences of two independent sample mean vectors:

$$T^2 = [(\bar{\mathbf{x}}_1 - \bar{\mathbf{x}}_2) - \boldsymbol{\delta}_0]^T [S_{\Delta\bar{\mathbf{x}}}]^{-1} [(\bar{\mathbf{x}}_1 - \bar{\mathbf{x}}_2) - \boldsymbol{\delta}_0], \tag{10.37}$$

where

$$\boldsymbol{\delta}_0 = E[\bar{\mathbf{x}}_1 - \bar{\mathbf{x}}_2] \tag{10.38}$$

is the difference between the two population mean vectors under H_0, corresponding to the second term in the numerator of Equation 5.5. If, is as often the case, the null hypothesis is that the two underlying means are equal, then $\boldsymbol{\delta}_0 = \mathbf{0}$ (corresponding to Equation 5.6). The two-sample Hotelling T^2 in Equation 10.37 is a Mahalanobis distance between the difference of the two sample mean vectors being tested, and the corresponding difference of their expected values under the null hypothesis. If the null hypothesis is $\boldsymbol{\delta}_0 = \mathbf{0}$, Equation 10.37 is reduced to a Mahalanobis distance between the two sample mean vectors.

The covariance matrix for the (MVN) sampling distribution of the difference of the two mean vectors is estimated differently, depending on whether the covariance matrices for the two samples, $[\Sigma_1]$ and $[\Sigma_2]$, can plausibly be assumed equal. If so, this matrix is estimated using a pooled estimate of that common covariance,

$$[S_{\Delta\bar{\mathbf{x}}}] = \left(\frac{1}{n_1} + \frac{1}{n_2} \right) [S_{\text{pool}}], \tag{10.39a}$$

where

$$[S_{\text{pool}}] = \frac{n_1 - 1}{n_1 + n_2 - 2} [S_1] + \frac{n_2 - 1}{n_1 + n_2 - 2} [S_2] \tag{10.39b}$$

is a weighted average of the two sample covariance matrices for the underlying data. If these two matrices cannot plausibly be assumed equal, and if in addition the sample sizes are relatively large, then the dispersion matrix for the sampling distribution of the difference of the sample mean vectors may be estimated as

$$[S_{\Delta\bar{\mathbf{x}}}] = \frac{1}{n_1} [S_1] + \frac{1}{n_2} [S_2], \tag{10.40}$$

which is numerically equal to Equation 10.39 for $n_1 = n_2$.

If the sample sizes are not large, the two-sample null hypothesis is rejected at the α level if

$$T^2 > \frac{(n_1 + n_2 - 2)K}{(n_1 + n_2 - K - 1)} F_{K, n_1 + n_2 - K - 1}(1 - \alpha). \tag{10.41}$$

That is, critical values are proportional to quantiles of the F distribution with $\nu_1 = K$ and $\nu_2 = n_1 + n_2 - K - 1$ degrees of freedom. For ν_2 sufficiently large (> 100, perhaps), Equation 10.36b can be used, as before.

Finally, if $n_1 = n_2$ and corresponding observations of x_1 and x_2 are linked physically—and correlated as a consequence—it is appropriate to account for the correlations between the pairs of observations by computing a one-sample test on their differences. Defining Δ_i as the difference between the ith observations of the vectors x_1 and x_2, analogously to

Equation 5.10, the one-sample Hotelling T^2 test statistic, corresponding to Equation 5.11, and of exactly the same form as Equation 10.32, is

$$T^2 = (\bar{\Delta} - \boldsymbol{\mu}_\Delta)^T \left(\frac{1}{n}[S_\Delta] \right)^{-1} (\bar{\Delta} - \boldsymbol{\mu}_\Delta) = n(\bar{\Delta} - \boldsymbol{\mu}_\Delta)^T [S_\Delta]^{-1} (\bar{\Delta} - \boldsymbol{\mu}_\Delta). \quad (10.42)$$

Here $n = n_1 = n_2$ is the common sample size, and $[S_\Delta]$ is the sample covariance matrix for the n vectors of differences $\boldsymbol{\Delta}_i$. The unusualness of Equation 10.42 in the context of the null hypothesis that the true difference of means is $\boldsymbol{\mu}_\Delta$ is evaluated using the F distribution (Equation 10.34) for relatively small samples, and the χ^2 distribution (Equation 10.36b) for large samples.

EXAMPLE 10.5 Two-Sample, and One-Sample Paired T^2 Tests

Table 10.1 shows January averages of daily maximum and minimum temperatures at New York City and Boston, for the 30 years 1971 through 2000. Because these are annual values, their serial correlations are quite small. As averages of 31 daily values each, the univariate distributions of these monthly values are expected to closely approximate the Gaussian. Figure 10.5 shows scatterplots for the values at each location. The ellipsoidal dispersions of the two point clouds suggests bivariate normality for both pairs of maximum and minimum temperatures. The two scatterplots overlap somewhat, but the visual separation is sufficiently distinct to suspect strongly that their distributions are different.

The two vector means, and their difference vector, are

$$\bar{x}_N = \begin{bmatrix} 38.68 \\ 26.15 \end{bmatrix}, \quad (10.43a)$$

$$\bar{x}_B = \begin{bmatrix} 36.50 \\ 22.13 \end{bmatrix}, \quad (10.43b)$$

and

$$\bar{\Delta} = \bar{x}_N - \bar{x}_B = \begin{bmatrix} 2.18 \\ 4.02 \end{bmatrix}. \quad (10.43c)$$

As might have been expected from its lower latitude, the average temperatures at New York are warmer. The sample covariance matrix for all four variables jointly is

$$[S] = \begin{bmatrix} [S_N] & | & [S_{N-B}] \\ - - - & & - - - \\ [S_{B-N}] & | & [S_B] \end{bmatrix} = \begin{bmatrix} 21.485 & 21.072 & | & 17.150 & 17.866 \\ 21.072 & 22.090 & & 16.652 & 18.854 \\ - - - - & - - - - & | & - - - - & - - - - \\ 17.150 & 16.652 & & 15.948 & 16.070 \\ 17.866 & 18.854 & | & 16.070 & 18.386 \end{bmatrix}. \quad (10.44)$$

Because the two locations are relatively close to each other and the data were taken in the same years, it is appropriate to treat them as paired values. This assertion is supported by the large cross-covariances in the submatrices $[S_{B-N}] = [S_{N-B}]^T$, corresponding to correlations ranging from 0.89 to 0.94: the data at the two locations are clearly not independent of each other. Nevertheless, it is instructive to first carry through T^2 calculations for differences of mean vectors as a two-sample test, ignoring these large cross-covariances for the moment.

TABLE 10.1 Average January maximum and minimum temperatures for New York City and Boston, 1971–2000, and the corresponding year-by year differences.

Year	New York		Boston		Differences	
	T_{max}	T_{min}	T_{max}	T_{min}	Δ_{max}	Δ_{min}
1971	33.1	20.8	30.9	16.6	2.2	4.2
1972	42.1	28.0	40.9	25.0	1.2	3.0
1973	42.1	28.8	39.1	23.7	3.0	5.1
1974	41.4	29.1	38.8	24.6	2.6	4.5
1975	43.3	31.3	41.4	28.4	1.9	2.9
1976	34.2	20.5	34.1	18.1	0.1	2.4
1977	27.7	16.4	29.8	16.7	−2.1	−0.3
1978	33.9	22.0	35.6	21.3	−1.7	0.7
1979	40.2	26.9	39.1	25.8	1.1	1.1
1980	39.4	28.0	35.6	23.2	3.8	4.8
1981	32.3	20.2	28.5	14.3	3.8	5.9
1982	32.5	19.6	30.5	15.2	2.0	4.4
1983	39.6	29.4	37.6	24.8	2.0	4.6
1984	35.1	24.6	32.4	20.9	2.7	3.7
1985	34.6	23.0	31.2	17.5	3.4	5.5
1986	40.8	27.4	39.6	23.1	1.2	4.3
1987	37.5	27.1	35.6	22.2	1.9	4.9
1988	35.8	23.2	35.1	20.5	0.7	2.7
1989	44.0	30.7	42.6	26.4	1.4	4.3
1990	47.5	35.2	43.3	29.5	4.2	5.7
1991	41.2	28.5	36.6	22.2	4.6	6.3
1992	42.5	28.9	38.2	23.8	4.3	5.1
1993	42.5	30.1	39.4	25.4	3.1	4.7
1994	33.2	17.9	31.0	13.4	2.2	4.5
1995	43.1	31.9	41.0	28.1	2.1	3.8
1996	37.0	24.0	37.5	22.7	−0.5	1.3
1997	39.2	25.1	36.7	21.7	2.5	3.4
1998	45.8	34.2	39.7	28.1	6.1	6.1
1999	40.8	27.0	37.5	21.5	3.3	5.5
2000	37.9	24.7	35.7	19.3	2.2	5.4

Regarding the Boston and New York temperatures as mutually independent, the appropriate test statistic would be Equation 10.37. If the null hypothesis is that the underlying vector means of the two distributions from which these data were drawn are equal, $\delta_0 = 0$. Both the visual impressions of the two data scatters in Figure 10.5, and the similarity of the covariance matrices $[S_N]$ and $[S_B]$ in Equation 10.44, suggest that assuming equality of covariance matrices would be reasonable. The appropriate covariance for the sampling distribution of the mean difference would then be calculated

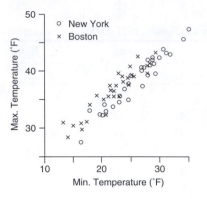

FIGURE 10.5   January average maximum and minimum temperatures, 1971–2000, for New York City (circles) and Boston (Xs).

using Equation 10.39, although because the sample sizes are equal the same numerical result is obtained with Equation 10.40:

$$[S_{\Delta\bar{x}}] = \left(\frac{1}{30} + \frac{1}{30}\right)\left(\frac{29}{58}[S_N] + \frac{29}{58}[S_B]\right) = \frac{1}{30}[S_N] + \frac{1}{30}[S_B]$$

$$= \begin{bmatrix} 1.248 & 1.238 \\ 1.238 & 1.349 \end{bmatrix}. \tag{10.45}$$

The test statistic (Equation 10.37) can now be calculated as

$$T^2 = [2.18, 4.02]\begin{bmatrix} 1.248 & 1.238 \\ 1.238 & 1.349 \end{bmatrix}^{-1}\begin{bmatrix} 2.18 \\ 4.02 \end{bmatrix} = 32.34. \tag{10.46}$$

The $1 - \alpha = .9999$ quantile of the F distribution with $\nu_1 = 2$ and $\nu_2 = 57$ degrees of freedom is 10.9 so the null hypothesis is rejected at the $\alpha = .0001$ level because $[(30 + 30 - 2)(2)/(30 + 30 - 2 - 1)]10.9 = 22.2 << T^2 = 32.34$ (cf. Equation 10.41). The actual p-value is smaller than 0.0001, but more extreme F-distribution quantiles are not commonly tabulated. Using the χ^2 distribution will provide only a moderately close approximation (Equation 10.35) because $\nu_2 = 57$, but the cumulative probability corresponding to $\chi_2^2 = 32.34$ can be calculated using Equation 4.46 (because χ_2^2 is the exponential distribution with $\beta = 2$) to be 0.99999991, corresponding to $\alpha = 0.00000001$ (Equation 10.36b).

Even though the two-sample T^2 test provides a definitive rejection of the null hypothesis, it underestimates the statistical significance, because it does not account for the positive covariances between the New York and Boston temperatures that are evident in the submatrices $[S_{N-B}]$ and $[S_{B-N}]$ in Equation 10.44. One way to account for these correlations is to compute the differences between the maximum temperatures as the linear combination $\mathbf{b}_1^T = [1, 0, -1, 0]$; compute the differences between the minimum temperatures as the linear combination $\mathbf{b}_2^T = [0, 1, 0, -1]$; and then use these two vectors as the rows of the transformation matrix $[B]$ in Equation 9.83b to compute the covariance $[S_\Delta]$ of the $n = 30$ vector differences, from the full covariance matrix $[S]$ in Equation 10.44.

Equivalently, we could compute this covariance matrix from the 30 data pairs in the last two columns of Table 10.1. In either case the result is

$$[S_\Delta] = \begin{bmatrix} 3.133 & 2.623 \\ 2.623 & 2.768 \end{bmatrix}. \tag{10.47}$$

The null hypothesis of equal mean vectors for New York and Boston implies $\mu_\Delta = 0$ in Equation 10.42, yielding the test statistic

$$T^2 = 30[2.18, 4.02]\begin{bmatrix} 3.133 & 2.623 \\ 2.623 & 2.768 \end{bmatrix}^{-1}\begin{bmatrix} 2.18 \\ 4.02 \end{bmatrix} = 298. \tag{10.48}$$

Because these temperature data are spatially correlated, much of the variability that was ascribed to sampling uncertainty for the mean vectors separately in the two-sample test is actually shared, and does not contribute to sampling uncertainty about the temperature *differences*. The numerical consequence is that the variances in the matrix $(1/30)\,[S_\Delta]$ are much smaller than their counterparts in Equation 10.45 for the two-sample test. Accordingly, T^2 for the paired test in Equation 10.48 is much larger than for the two-sample test in Equation 10.46. In fact it is huge, leading to the rough (because the sample sizes are only moderate) estimate, through Equation 4.46, of $\alpha \approx 2 \times 10^{-65}$.

Both the (incorrect) two-sample test, and the (appropriate) paired test yield strong rejections of the null hypothesis that the New York and Boston mean vectors are equal. But what can be concluded about the way(s) in which they are different? This question will be taken up in Example 10.7. ◊

The T^2 tests described so far are based on the assumption that the data vectors are mutually uncorrelated. That is, although the K elements of x may have nonzero correlations, each of the observations x_i, $i = 1, \ldots, n$, have been assumed to be mutually independent. As noted in Section 5.2.4, ignoring serial correlation leads to large errors in statistical inference, typically because the sampling distributions of the test statistics have greater dispersion (the test statistics are more variable from batch to batch of data) than would be the case if the underlying data were independent.

A simple adjustment (Equation 5.13) is available for scalar t tests if the serial correlation in the data is consistent with a first-order autoregression (Equation 8.16). The situation is more complicated for the multivariate T^2 test because, even if the time dependence for each of K elements of x is reasonably represented by an AR(1) process, their autoregressive parameters ϕ may not be the same, and the lagged correlations among the elements of x must also be accounted for. However, if the multivariate AR(1) process (Equation 10.21) can be assumed as reasonably representing the serial dependence of the data, and if the sample size is large enough to produce multinormality as a consequence of the Central Limit Theorem, the sampling distribution of the sample mean vector is

$$\bar{x} \sim N_K\left(\mu_x, \frac{1}{n}[\Sigma_\Phi]\right), \tag{10.49a}$$

where

$$[\Sigma_\Phi] = ([I] - [\Phi])^{-1}[\Sigma_x] + [\Sigma_x]([I] - [\Phi]^T)^{-1} - [\Sigma_x]. \tag{10.49b}$$

Equation 10.49 corresponds to Equation 10.30a for independent data, and $[\Sigma_\Phi]$ reduces to $[\Sigma_x]$ if $[\Phi] = [0]$ (i.e., if the x's are serially independent). For large n, sample counterparts of the quantities in Equation 10.49 can be substituted, and the matrix $[S_\Phi]$ used in place of $[S_x]$ in the computation of T^2 test statistics.

10.5.3 Simultaneous Confidence Statements

As noted in Section 5.1.7, a confidence interval is a region around a sample statistic, containing values that would not be rejected by a test whose null hypothesis is that the observed sample value is the true value. In effect, confidence intervals are constructed by working hypothesis tests in reverse. The difference in multivariate settings is that a confidence interval defines a region in the K-dimensional space of the data vector $\mathbf{x}$ rather than an interval on the one-dimensional space (the real line) of the scalar x. That is, multivariate confidence intervals are K-dimensional hypervolumes, rather than one-dimensional line segments.

Consider the one-sample T^2 test, Equation 10.32. Once the data $\mathbf{x}_i, i = 1, \ldots, n$, have been observed and their sample covariance matrix $[S_x]$ has been computed, a $(1 - \alpha) \times 100\%$ confidence region for the true vector mean consists of the set of points satisfying

$$n(\mathbf{x} - \bar{\mathbf{x}})^\mathrm{T}[S_x]^{-1}(\mathbf{x} - \bar{\mathbf{x}}) \le \frac{K(n-1)}{(n-K)}F_{K,n-K}(1-\alpha), \tag{10.50}$$

because these are the $\mathbf{x}$'s that would not trigger a rejection of the null hypothesis that the true mean is the observed sample mean. For sufficiently large $n - K$, the right-hand side of Equation 10.50 would be well approximated by $\chi_K^2(1 - \alpha)$. Similarly, for the two-sample T^2 test (Equation 10.37) a $(1 - \alpha) \times 100\%$ confidence region for the difference of the two means consists of the points δ satisfying

$$[\delta - (\bar{\mathbf{x}}_1 - \bar{\mathbf{x}}_2)]^\mathrm{T}[S_{\Delta\bar{\mathbf{x}}}]^{-1}[\delta - (\bar{\mathbf{x}}_1 - \bar{\mathbf{x}}_2)] \le \frac{K(n_1 + n_2 - 2)}{(n_1 + n_2 - K - 1)}F_{K,n_1+n_2-K-1}(1-\alpha), \tag{10.51}$$

where again the right-hand side is approximately $\chi_K^2(1 - \alpha)$ for large samples.

The points $\mathbf{x}$ satisfying Equation 10.50 are those whose Mahalanobis distance from $\bar{\mathbf{x}}$ is no larger than the scaled $(1 - \alpha)$ quantile of the F (or χ^2, as appropriate) distribution on the right-hand side, and similarly for the points δ satisfying Equation 10.51. Therefore the confidence regions defined by these equations are bounded by (hyper-) ellipsoids whose characteristics are defined by the covariance matrix for the sampling distribution of the respective test statistic; for example, $(1/n)[S_x]$ for Equation 10.50. Because the sampling distribution of $\bar{\mathbf{x}}$ approximates the MVN on the strength of the central limit theorem, the confidence regions defined by Equation 10.50 are confidence ellipsoids for the MVN with mean $\bar{\mathbf{x}}$ and covariance $(1/n)[S_x]$ (cf. Equation 10.5). Similarly, the confidence regions defined by Equation 10.51 are hyperellipsoids centered on the vector mean difference between the two sample means.

As illustrated in Example 10.1, the properties of these confidence ellipses, other than their center, are defined by the eigenvalues and eigenvectors of the covariance matrix for the sampling distribution in question. In particular, each axis of one of these ellipses will be aligned in the direction of one of the eigenvectors, and will be elongated in proportion to the square roots of the corresponding eigenvalues. In the case of the one-sample confidence region, for example, the limits of $\mathbf{x}$ satisfying Equation 10.50 in the directions of each of the axes of the ellipse are

$$\mathbf{x} = \bar{\mathbf{x}} \pm \mathbf{e}_k \left[\lambda_k \frac{K(n-1)}{(n-K)}F_{K,n-K}(1-\alpha) \right]^{1/2}, \quad k = 1, \ldots, K, \tag{10.52}$$

where λ_k and e_k are the kth eigenvalue-eigenvector pair of the matrix $(1/n)[S_x]$. Again, for sufficiently large n, the quantity in the square brackets would be well approximated by $[\lambda_k \chi^2_K (1 - \alpha)]$. Equation 10.52 indicates that the confidence ellipses are centered at the observed sample mean $\bar{x}$, and extend further in the directions associated with the largest eigenvalues. They also extend further for smaller α because these produce larger cumulative probabilities for the distribution quantiles $F(1 - \alpha)$ and $\chi^2_K (1 - \alpha)$.

It would be possible, and computationally simpler, to conduct K univariate t tests separately for the means of each of the elements of x rather than the T^2 test examining the vector mean $\bar{x}$. What is the relationship between an ellipsoidal multivariate confidence region of the kind just described, and a collection of K univariate confidence intervals? Jointly, these univariate confidence intervals would define a hyperrectangular region in the K-dimensional space of x; but the probability (or confidence) associated with outcomes enclosed by it will be substantially less than $1 - \alpha$, if the lengths of each of its K sides are the corresponding $(1 - \alpha) \times 100\%$ scalar confidence intervals. The problem is one of test multiplicity: if the K tests on which the confidence intervals are based are independent, the joint probability of all the elements of the vector x being simultaneously within their scalar confidence bounds will be $(1 - \alpha)^K$. To the extent that the scalar confidence interval calculations are not independent, the joint probability will be different, but difficult to calculate.

An expedient workaround for this multiplicity problem is to calculate the K one-dimensional Bonferroni confidence intervals, and use these as the basis of a joint confidence statement

$$\Pr \left\{ \bigcap_{k=1}^{K} \left[\bar{x}_k + z \left(\frac{\alpha/K}{2} \right) \sqrt{\frac{s_{k,k}}{n}} \leq \mu_k \leq \bar{x}_k + z \left(1 - \frac{\alpha/K}{2} \right) \sqrt{\frac{s_{k,k}}{n}} \right] \right\} \geq 1 - \alpha. \quad (10.53)$$

The expression inside the square bracket defines a univariate, $(1 - \alpha/K) \times 100\%$ confidence interval for the kth variable in x. Each of these confidence intervals is expanded relative to the nominal $(1 - \alpha) \times 100\%$ confidence interval, to compensate for the multiplicity in K dimensions simultaneously. For convenience, it has been assumed in Equation 10.53 that the sample size is adequate for standard Gaussian quantiles to be appropriate, although quantiles of the t distribution with $n - 1$ degrees of freedom usually would be used for n smaller than about 30.

There are two problems with using Bonferroni confidence regions in this context. First, Equation 10.53 is an inequality rather than an exact specification. That is, the probability that all the K elements of the hypothetical true mean vector μ are contained simultaneously in the respective one-dimensional confidence intervals is *at least* $1 - \alpha$, not exactly $1 - \alpha$. That is, in general the K-dimensional Bonferroni confidence region is too large, but exactly how much more probability than $1 - \alpha$ may be enclosed by it is not known.

The second problem is more serious. As a collection of univariate confidence intervals, the resulting K-dimensional (hyper-) rectangular confidence region ignores the covariance structure of the data. Bonferroni confidence statements can be reasonable if the correlation structure is weak, for example in the setting described in Section 8.5.6. But Bonferroni confidence regions are inefficient when the correlations among elements of x are strong, in the sense that they will include large regions of very low plausibility. As a consequence they are too large in a multivariate sense, and can lead to silly inferences.

EXAMPLE 10.6   Comparison of Unadjusted Univariate, Bonferroni, and MVN Confidence Regions

Assume that the covariance matrix in Equation 9.56, for the Ithaca and Canandaigua minimum temperatures, had been calculated from $n = 100$ independent temperature pairs. This many observations would justify large-sample approximations for the sampling distributions (standard Gaussian Z and χ^2, rather than t and F quantiles), and assuming independence obviates the need for the nonindependence adjustments in Equation 10.49.

What is the best two-dimensional confidence region for the true climatological mean vector, given the sample mean $[13.00, 20.23]^T$, and assuming the sample covariance matrix for the data in Equation 9.56? Relying on the multivariate normality for the sampling distribution of the sample mean implied by the Central Limit Theorem, Equation 10.50 defines an elliptical 95% confidence region when the right-hand side is the χ^2 quantile $\chi_2^2(0.95) = 5.991$. The result is the elliptical region shown in Figure 10.6, centered on the sample mean (+). Compare this ellipse to Figure 10.1, which is centered on the same mean and based on the same covariance matrix (although drawn to enclose slightly less probability). Figure 10.6 has exactly the same shape and orientation, but it is much more compact, even though it encloses somewhat more probability. Both ellipses have the same eigenvectors, $e_1^T = [0.848, 0.530]$ and $e_2^T = [-0.530, 0.848]$, but the eigenvalues for Figure 10.6 are 100-fold smaller; that is, $\lambda_1 = 2.5476$ and $\lambda_2 = 0.0829$. The difference is that Figure 10.1 represents one contour of the MVN distribution for the data, with covariance $[S_x]$ given by Equation 9.56, but Figure 10.6 shows one contour of the MVN with covariance $(1/n)[S_x]$, appropriate to Equation 10.50. This ellipse is the smallest region enclosing 95% of the probability of this distribution. Its elongation reflects the strong correlation between the minimum temperatures at the two locations, so that differences between the sample and true means due to sampling variations are much

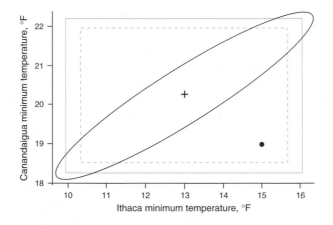

FIGURE 10.6   Hypothetical 95% joint confidence regions for the mean Ithaca and Canandaigua minimum temperatures, assuming that $n = 100$ independent bivariate observations had been used to calculate the covariance matrix in Equation 9.56. Ellipse encloses points within a Mahalanobis distance of $\chi^2 = 5.991$ of the sample mean (indicated by +) $[13.00, 20.23]^T$. Horizontal and vertical limits of the dashed rectangle are defined by two independent confidence intervals for the two variables, with $\pm z(0.025) = \pm 1.96$. Solid rectangle indicates corresponding Bonferroni confidence region, calculated with $\pm z(0.0125) = \pm 2.24$. The point $[15, 19]^T$ (large dot) is comfortably within both rectangular confidence regions, but is at Mahalanobis distance $\chi^2 = 1006$ from the mean relative to the joint covariance structure of the two variables, and is thus highly implausible.

more likely to involve differences of the same sign for both the Ithaca and Canandaigua means.

The solid rectangle outlines the 95% Bonferroni confidence region. It has been calculated using $\alpha = 0.05$ in Equation 10.53, and so is based on the 0.0125 and 0.9875 quantiles of the standard Gaussian distribution, or $z = \pm 2.24$. The resulting rectangular region encloses at least $(1 - \alpha) \times 100\% = 95\%$ of the probability of the joint sampling distribution. It occupies much more area in the plane than does the confidence ellipse, because the rectangle includes large regions in the upper left and lower right that contain very little probability. However, from the standpoint of univariate inference—that is, confidence intervals for one location without regard to the other—the Bonferroni limits are narrower.

The dashed rectangular region results jointly from the two standard 95% confidence intervals. The length of each side has been computed using the 0.025 and 0.975 quantiles of the standard Gaussian distribution, which are $z = \pm 1.96$. They are, of course, narrower than the corresponding Bonferroni intervals, and according to Equation 10.53 the resulting rectangle includes at least 90% of the probability of this sampling distribution. Like the Bonferroni confidence region, it claims large areas with very low probabilities as plausible.

The main difficulty with Bonferroni confidence regions is illustrated by the point $[15, 19]^{T}$, located by the large dot in Figure 10.6. It is comfortably within the solid rectangle delineating the Bonferroni confidence region, which carries the implication that this is a plausible value for the true mean vector. However, a Bonferroni confidence region is defined without regard to the multivariate covariance structure of the distribution that it purports to represent. In the case of Figure 10.6 the Bonferroni confidence region ignores the fact that sampling variations for these two positively correlated variables are much more likely to yield differences between the two sample and true means that are of the same sign. The Mahalanobis distance between the points $[15, 19]^{T}$ and $[13.00, 20.23]^{T}$, according to the covariance matrix $(1/n)[S_x]$, is 1006, implying an astronomically small probability for the separation of these two vectors (cf. Equation 10.31a). The vector $[15, 19]^{T}$ is an extremely implausible candidate for the true mean $\boldsymbol{\mu}_x$. $\Diamond$

10.5.4 Interpretation of Multivariate Statistical Significance

What can be said about multivariate mean differences if the null hypothesis for a T^2 test is rejected; that is, if Equation 10.34 or 10.41 (or their large-sample counterpart, Equation 10.36b) are satisfied? This question is complicated by the fact that there are many ways for multivariate means to differ from one another, including but not limited to one or more pairwise differences between the elements that would be detected by the corresponding univariate tests.

If a T^2 tests results in the rejection of its multivariate null hypothesis, the implication is that at least one scalar test for a linear combination $\boldsymbol{a}^T\boldsymbol{x}$ or $\boldsymbol{a}^T(\boldsymbol{x}_1 - \boldsymbol{x}_2)$, for one- and two-sample tests, respectively, will be statistically significant. In any case, the scalar linear combination providing the most convincing evidence against the null hypothesis (regardless of whether or not it is sufficiently convincing to reject at a given test level) will satisfy

$$\mathbf{a} \propto [\mathrm{S}]^{-1}(\bar{\mathbf{x}} - \boldsymbol{\mu}_0) \qquad (10.54a)$$

for one–sample tests, or

$$\mathbf{a} \propto [\mathbf{S}]^{-1}[(\bar{\mathbf{x}}_1 - \bar{\mathbf{x}}_2) - \boldsymbol{\delta}_0] \tag{10.54b}$$

for two-sample tests. At minimum, then, if a multivariate T^2 calculation results in a null hypothesis rejection, then linear combinations corresponding to the K-dimensional direction defined by the vector $\mathbf{a}$ in Equation 10.54 will lead to significant results also. It can be very worthwhile to interpret the meaning, in the context of the data, of the direction $\mathbf{a}$ defined by Equation 10.54. Of course, depending on the strength of the overall multivariate result, other linear combinations may also lead to scalar test rejections, and it is possible that all linear combinations will be significant. The direction $\mathbf{a}$ also indicates the direction that best discriminates between the populations from which $\mathbf{x}_1$ and $\mathbf{x}_2$ were drawn (see Section 13.2.2).

The reason that any linear combination $\mathbf{a}$ satisfying Equation 10.54 yields the same test result can be seen most easily in terms of the corresponding confidence interval. Consider for simplicity the confidence interval for a one-sample T^2 test, Equation 10.50. Using the results in Equation 9.81, the resulting scalar confidence interval is defined by

$$\mathbf{a}^{\mathrm{T}}\bar{\mathbf{x}} - c\sqrt{\frac{\mathbf{a}^{\mathrm{T}}[\mathbf{S}_x]\mathbf{a}}{n}} \leq \mathbf{a}^{\mathrm{T}}\boldsymbol{\mu} \leq \mathbf{a}^{\mathrm{T}}\bar{\mathbf{x}} + c\sqrt{\frac{\mathbf{a}^{\mathrm{T}}[\mathbf{S}_x]\mathbf{a}}{n}}, \tag{10.55}$$

where c^2 equals $[K(n-1)/(n-K)]F_{K,n-K}(1-\alpha)$, or χ_K^2, as appropriate. Even though the length of the vector $\mathbf{a}$ is arbitrary, so that the magnitude of the linear combination $\mathbf{a}^{\mathrm{T}}\mathbf{x}$ is also arbitrary, the quantity $\mathbf{a}^{\mathrm{T}}\boldsymbol{\mu}$ is scaled identically.

Another remarkable property of the T^2 test is that valid inferences about *any and all* linear combinations can be made, even though they may not have been specified *a priori*. The price that is paid for this flexibility is that inferences made using conventional scalar tests for linear combinations that *are* specified in advance, will be more precise. This point can be appreciated in the context of the confidence regions shown in Figure 10.6. If a test regarding the Ithaca minimum temperature only had been of interest, corresponding to the linear combination $\mathbf{a} = [1, 0]^{\mathrm{T}}$, the appropriate confidence interval would be defined by the horizontal extent of the dashed rectangle. The corresponding interval for this linear combination from the full T^2 test is substantially wider, being defined by the projection, or shadow, of the ellipse onto the horizontal axis. But what is gained from the multivariate test is the ability to make valid simultaneous probability statements regarding as many linear combinations as may be of interest.

EXAMPLE 10.7   Interpreting the New York and Boston Mean January Temperature Differences

Return now to the comparisons made in Example 10.5, between the vectors of average January maximum and minimum temperatures for New York City and Boston. The difference between the sample means was $[2.18, 4.02]^{\mathrm{T}}$, and the null hypothesis was that the true means were equal, so the corresponding difference $\boldsymbol{\delta}_0 = \mathbf{0}$. Even assuming, erroneously, that there is no spatial correlation between the two locations (or, equivalently for the purpose of the test, that the data for the two locations were taken in different years), T^2 in Equation 10.46 indicates that the null hypothesis should be strongly rejected.

Both means are warmer at New York, but Equation 10.46 does not necessarily imply significant differences between the average maxima or the average minima. Figure 10.5 shows substantial overlap between the data scatters for both maximum and minimum temperatures, with each scalar mean near the center of the corresponding data distribution for the other city. Computing the separate univariate tests (Equation 5.8) yields $z = 2.18/\sqrt{1.248} = 1.95$ for the maxima and $z = 4.02/\sqrt{1.349} = 3.46$ for the minima. Even leaving aside the problem that two simultaneous comparisons are being made, the result for the difference of the average maximum temperatures is not quite significant at the 5% level, although the difference for the minima is stronger.

The significant result in Equation 10.46 ensures that there is at least one linear combination $\boldsymbol{a}^T(\boldsymbol{x}_1 - \boldsymbol{x}_2)$ (and possibly others, although not necessarily the linear combinations resulting from $\boldsymbol{a}^T = [1, 0]$ or $[0, 1]$) for which there is a significant difference. According to Equation 10.54b, the vectors producing the most significant linear combinations are proportional to

$$\mathbf{a} \propto [S_{\Delta\bar{\mathbf{x}}}]^{-1}\bar{\Delta} = \begin{bmatrix} 1.248 & 1.238 \\ 1.238 & 1.349 \end{bmatrix}^{-1} \begin{bmatrix} 2.18 \\ 4.02 \end{bmatrix} = \begin{bmatrix} -13.5 \\ 15.4 \end{bmatrix}. \tag{10.56}$$

This linear combination of the mean differences, and the estimated variance of its sampling distribution, are

$$\mathbf{a}^T\bar{\Delta} = [-13.5, 15.4] \begin{bmatrix} 2.18 \\ 4.02 \end{bmatrix} = 32.5, \tag{10.57a}$$

and

$$\mathbf{a}^T[S_{\Delta\bar{\mathbf{x}}}]\mathbf{a} = [-13.5, 15.4] \begin{bmatrix} 1.248 & 1.238 \\ 1.238 & 1.349 \end{bmatrix} \begin{bmatrix} -13.5 \\ 15.4 \end{bmatrix} = 32.6, \tag{10.57b}$$

yielding the univariate test statistic for this linear combination of the differences $z = 32.5/\sqrt{32.6} = 5.69$. This is, not coincidentally, the square root of Equation 10.46. The appropriate benchmark against which to compare the unusualness of this result in the context of the null hypothesis is not the standard Gaussian or t distributions (because this linear combination was derived from the test data, not a priori), but rather the square roots of either χ_2^2 quantiles or of appropriately scaled $F_{2,30}$ quantiles. The result is still very highly significant, with $p \approx 10^{-7}$.

Equation 10.56 indicates that the most significant aspect of the difference between the New York and Boston mean vectors is not the warmer temperatures at New York relative to Boston (which would correspond to $\boldsymbol{a} \propto [1, 1]^T$). Rather, the elements of $\boldsymbol{a}$ are of opposite sign and of nearly equal magnitude, and so describe a contrast. Since $-\boldsymbol{a} \propto \boldsymbol{a}$, one way of interpreting this contrast is as the difference between the average maxima and minima; that is, choosing $\boldsymbol{a} \approx [1, -1]^T$. That is, the most significant aspect of the difference between the two mean vectors is closely approximated by the difference in the average diurnal range, with the range for Boston being larger. The null hypothesis that that the two diurnal ranges are equal can be tested specifically, using the contrast vector $\boldsymbol{a} = [1, -1]^T$ in Equation 10.57, rather than the linear combination defined by Equation 10.56. The result is $z = -1.84/\sqrt{0.121} = -5.29$. This test statistic is negative because the diurnal range at New York is smaller than the diurnal range at Boston. It is slightly smaller than the result obtained when using $\boldsymbol{a} = [-13.5, 15.4]$, because that is the most significant linear combination, although the result is almost the same because the two vectors are aligned in nearly the same direction. Comparing the result to the χ_2^2 distribution yields the very highly significant result $p \approx 10^{-6}$. Visually, the separation between the two point clouds in Figure 10.5 is consistent with this difference in diurnal range: The points for Boston tend to be closer to the upper left, and those for New York are closer to the lower right. On the other hand, the relative orientation of the two means is almost exactly opposite, with the New York mean closer to the upper right corner, and the Boston mean closer to the lower left. ◊

10.6 Exercises

10.1. Assume that the Ithaca and Canandaigua maximum temperatures in Table A.1 constitute a sample from a MVN distribution, and that their covariance matrix [S] has eigenvalues and eigenvectors as given in Exercise 9.6. Sketch the 50% and 95% probability ellipses of this distribution.

10.2. Assume that the four temperature variables in Table A.1 are MVN-distributed, with the ordering of the variables in x being $[\text{Max}_{\text{Ith}}, \text{Min}_{\text{Ith}}, \text{Max}_{\text{Can}}, \text{Min}_{\text{Can}}]^T$. The respective means are also given in Table A.1, and the covariance matrix [S] is given in the answer to Exercise 9.7a. Assuming the true mean and covariance are the same as the sample values,

   a. Specify the conditional distribution of $[\text{Max}_{\text{Ith}}, \text{Min}_{\text{Ith}}]^T$, given that $[\text{Max}_{\text{Can}}, \text{Min}_{\text{Can}}]^T = [31.77, 20.23]^T$ (i.e., the average values for Canandaigua).

   b. Consider the linear combinations $b_1 = [1, 0, -1, 0]$, expressing the difference between the maximum temperatures, and $b_2 = [1, -1 - 1, 1]$, expressing the difference between the diurnal ranges, as rows of a transformation matrix [B]. Specify the distribution of the transformed variables [B] x.

10.3. The eigenvector associated with the smallest eigenvalue of the covariance matrix [S] for the January 1987 temperature data referred to in Exercise 10.2 is $e_4^T = [-.665, .014, .738, -.115]$. Assess the normality of the linear combination $e_4^T x$,

   a. Graphically, with a Q-Q plot. For computational convenience, evaluate $\Phi(z)$ using Equation 4.29.

   b. Formally, with the Filliben test (see Table 5.3), assuming no autocorrelation.

10.4. a. Compute the 1-sample T^2 testing the linear combinations [B] $\bar{x}$ with respect to $H_0 : \mu_0 = 0$, where x and [B] are defined as in Exercise 10.2. Ignoring the serial correlation, evaluate the plausibility of H_0, assuming that the χ^2 distribution is an adequate approximation to the sampling distribution of the test statistic.

   b. Compute the most significant linear combination for this test.

10.5. Repeat Exercise 10.4, assuming spatial independence (i.e., setting all cross-covariances between Ithaca and Canandaigua variables to zero).

CHAPTER • 11

Principal Component (EOF) Analysis

11.1 Basics of Principal Component Analysis

Possibly the most widely used multivariate statistical technique in the atmospheric sciences is principal component analysis, often denoted as PCA. The technique became popular for analysis of atmospheric data following the paper by Lorenz (1956), who called the technique *empirical orthogonal function* (EOF) analysis. Both names are commonly used, and refer to the same set of procedures. Sometimes the method is incorrectly referred to as factor analysis, which is a related but distinct multivariate statistical method. This chapter is intended to provide a basic introduction to what has become a very large subject. Book-length treatments of PCA are given in Preisendorfer (1988), which is oriented specifically toward geophysical data; and in Jolliffe (2002), which describes PCA more generally. In addition, most textbooks on multivariate statistical analysis contain chapters on PCA.

11.1.1 Definition of PCA

PCA reduces a data set containing a large number of variables to a data set containing fewer (hopefully many fewer) new variables. These new variables are linear combinations of the original ones, and these linear combinations are chosen to represent the maximum possible fraction of the variability contained in the original data. That is, given multiple observations of a $(K \times 1)$ data vector x, PCA finds $(M \times 1)$ vectors u whose elements are linear combinations of the elements of the xs, which contain most of the information in the original collection of xs. PCA is most effective when this data compression can be achieved with $M << K$. This situation occurs when there are substantial correlations among the variables within x, in which case x contains redundant information. The elements of these new vectors u are called the principal components (PCs).

Data for atmospheric and other geophysical fields generally exhibit many large correlations among the variables x_k, and a PCA results in a much more compact representation of their variations. Beyond mere data compression, however, a PCA can be a very useful tool for exploring large multivariate data sets, including those consisting of geophysical fields. Here PCA has the potential for yielding substantial insights into both the spatial and temporal variations exhibited by the field or fields being analyzed, and new interpretations of the original data x can be suggested by the nature of the linear combinations that are most effective in compressing the data.

Usually it is convenient to calculate the PCs as linear combinations of the anomalies $x' = x - \bar{x}$. The first PC, u_1, is that linear combination of x' having the largest variance. The subsequent principal components u_m, $m = 2, 3, \ldots$, are the linear combinations having the largest possible variances, subject to the condition that they are uncorrelated with the principal components having lower indices. The result is that all the PCs are mutually uncorrelated.

The new variables or PCs—that is, the elements u_m of u that will account successively for the maximum amount of the joint variability of x' (and therefore also of x)—are uniquely defined by the eigenvectors of the covariance matrix of x, [S]. In particular, the mth principal component, u_m is obtained as the projection of the data vector x' onto the mth eigenvector, e_m,

$$u_m = e_m^T x' = \sum_{k=1}^{K} e_{km} x_k', \quad m = 1, \ldots, M. \tag{11.1}$$

Notice that each of the M eigenvectors contains one element pertaining to each of the K variables, x_k. Similarly, each realization of the mth principal component in Equation 11.1 is computed from a particular set of observations of the K variables x_k. That is, each of the M principal components is a sort of weighted average of the x_k values. Although the weights (the e_{km}s) do not sum to 1, their squares do because of the scaling convention $\|e_m\| = 1$. (Note that a fixed scaling convention for the weights e_m of the linear combinations in Equation 11.1 allows the maximum variance constraint defining the PCs to be meaningful.) If the data sample consists of n observations (and therefore of n data vectors x, or n rows in the data matrix [X]), there will be n values for each of the principal components, or new variables, u_m. Each of these constitutes a single-number index of the resemblance between the eigenvector e_m and the corresponding individual data vector x.

Geometrically, the first eigenvector, e_1, points in the direction (in the K-dimensional space of x') in which the data vectors jointly exhibit the most variability. This first eigenvector is the one associated with the largest eigenvalue, λ_1. The second eigenvector e_2, associated with the second-largest eigenvalue λ_2, is constrained to be perpendicular to e_1 (Equation 9.48), but subject to this constraint it will align in the direction in which the x' vectors exhibit their next strongest variations. Subsequent eigenvectors e_m, $m = 3, 4, \ldots, M$, are similarly numbered according to decreasing magnitudes of their associated eigenvalues, and in turn will be perpendicular to all the previous eigenvectors. Subject to this orthogonality constraint these eigenvectors will continue to locate directions in which the original data jointly exhibit maximum variability.

Put another way, the eigenvectors define a new coordinate system in which to view the data. In particular, the orthogonal matrix [E] whose columns are the eigenvectors (Equation 9.49) defines the rigid rotation

$$\mathbf{u} = [E]^T \mathbf{x}', \tag{11.2}$$

which is the simultaneous matrix-notation representation of $M = K$ linear combinations of the form of Equation 11.1 (i.e., here the matrix [E] is square, with K eigenvector columns). This new coordinate system is oriented such that each consecutively numbered axis is aligned along the direction of the maximum joint variability of the data, consistent with that axis being orthogonal to the preceding ones. These axes will turn out to be different for different data sets, because they are extracted from the sample covariance matrix $[S_x]$ particular to a given data set. That is, they are orthogonal functions, but are defined empirically according to the particular data set at hand. This observation is the basis for the eigenvectors being known in this context as empirical orthogonal functions (EOFs). The implied distinction is with theoretical orthogonal functions, such as Fourier harmonics or Tschebyschev polynomials, which also can be used to define alternative coordinate systems in which to view a data set.

It is a remarkable property of the principal components that they are uncorrelated. That is, the correlation matrix for the new variables u_m is simply $[I]$. This property implies that the covariances between pairs of the u_m's are all zero, so that the corresponding covariance matrix is diagonal. In fact, the covariance matrix for the principal components is obtained by the diagonalization of $[S_x]$ (Equation 9.54), and is thus simply the diagonal matrix $[\Lambda]$ of the eigenvalues of $[S]$:

$$[S_u] = \text{Var}([E]^T\mathbf{x}) = [E]^T[S_x][E] = [E]^{-1}[S_x][E] = [\Lambda]. \tag{11.3}$$

That is, the variance of the mth principal component u_m is the mth eigenvalue λ_m. Equation 9.52 then implies that each PC represents a share of the total variation in $\mathbf{x}$ that is proportional to its eigenvalue,

$$R_m^2 = \frac{\lambda_m}{\sum\limits_{k=1}^{K} \lambda_k} \times 100\% = \frac{\lambda_m}{\sum\limits_{k=1}^{K} s_{k,k}} \times 100\%. \tag{11.4}$$

Here R^2 is used in the same sense that is familiar from linear regression (see Section 6.2). The total variation exhibited by the original data is completely represented in (or accounted for by) the full set of K u_m's, in the sense that the sum of the variances of the centered data $\mathbf{x}'$ (and therefore also of the uncentered variables $\mathbf{x}$), $\Sigma_k s_{k,k}$, is equal to the sum of the variances $\Sigma_m \lambda_m$. of the principal component variables $\mathbf{u}$.

Equation 11.2 expresses the transformation of a $(K \times 1)$ data vector $\mathbf{x}'$ to a vector $\mathbf{u}$ of PCs. If $[E]$ contains all K eigenvectors of $[S_x]$ (assuming it is nonsingular) as its columns, the resulting vector $\mathbf{u}$ will also have dimension $(K \times 1)$. Equation 11.2 sometimes is called the analysis formula for $\mathbf{x}'$, expressing that the data can be analyzed, or summarized in terms of the principal components. Reversing the transformation in Equation 11.2, the data $\mathbf{x}'$ can be reconstructed from the principal components according to

$$\underset{(K \times 1)}{\mathbf{x}'} = \underset{(K \times K)(K \times 1)}{[E] \, \mathbf{u}}, \tag{11.5}$$

which is obtained from Equation 11.2 by multiplying on the left by $[E]$ and using the orthogonality property of this matrix (Equation 9.42). The reconstruction of $\mathbf{x}'$ expressed by Equation 11.5 is sometimes called the synthesis formula. If the full set of $M = K$ PCs is used in the synthesis, the reconstruction is complete and exact, since $\Sigma_m R_m^2 = 1$ (cf. Equation 11.4). If $M < K$ PCs (usually corresponding to the M largest eigenvalues) are used, the reconstruction is approximate,

$$\underset{(K \times 1)}{\mathbf{x}'} \approx \underset{(K \times M)(M \times 1)}{[E] \, \mathbf{u}}, \tag{11.6a}$$

or

$$x'_k \approx \sum_{m=1}^{M} e_{km} u_m, \quad k = 1, \ldots, K, \tag{11.6b}$$

but improves as the number M of PCs used (or, more accurately, as the sum of the corresponding eigenvalues, because of Equation 11.4) increases. Because [E] has only M columns, and operates on a truncated PC vector u of dimension $(M \times 1)$, Equation 11.6 is called the truncated synthesis formula. The original (in the case of Equation 11.5) or approximated (for Equation 11.6) uncentered data x can easily be obtained by adding back the vector of sample means; that is, by reversing Equation 9.33.

Because each principal component u_m is a linear combination of the original variables x_k (Equation 11.1), and vice versa (Equation 11.5), pairs of principal components and original variables will be correlated unless the eigenvector element $e_{k,m}$ relating them is zero. It can sometimes be informative to calculate these correlations, which are given by

$$r_{u,x} = \text{corr}(u_m, x_k) = e_{k,m} \sqrt{\frac{\lambda_m}{s_{k,k}}}. \tag{11.7}$$

EXAMPLE 11.1   PCA in Two Dimensions

The basics of PCA are most easily appreciated in a simple example where the geometry can be visualized. If $K = 2$ the space of the data is two-dimensional, and can be graphed on a page. Figure 11.1 shows a scatterplot of centered (at zero) January 1987

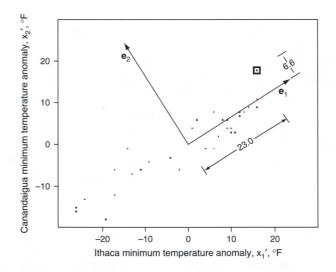

FIGURE 11.1   Scatterplot of January 1987 Ithaca and Canandaigua minimum temperatures (converted to anomalies, or centered), illustrating the geometry of PCA in two dimensions. The eigenvectors e_1 and e_2 of the covariance matrix [S] for these two variables, as computed in Example 9.3, have been plotted with exaggerated lengths for clarity. The data stretch out in the direction of e_1 to the extent that 96.8% of the joint variance of these two variables occurs along this axis. The coordinates u_1 and u_2, corresponding to the data point x'^T [16.0, 17.8], recorded on January 15 and indicated by the large square symbol, are shown by lengths in the directions of the new coordinate system defined by the eigenvectors. That is, the vector $u^T = [23.0, 6.6]$ locates the same point as $x'^T = [16.0, 17.8]$.

Ithaca minimum temperatures (x_1') and Canandaigua minimum temperatures (x_2') from Table A.1. This is the same scatterplot that appears in the middle of the bottom row of Figure 3.26. It is apparent that the Ithaca temperatures are more variable than the Canandaigua temperatures, with the two standard deviations being $\sqrt{s_{1,1}} = 13.62°F$ and $\sqrt{s_{2,2}} = 8.81°F$, respectively. The two variables are clearly strongly correlated, and have a Pearson correlation of $+0.924$ (see Table 3.5). The covariance matrix [S] for these two variables is given as [A] in Equation 9.56. The two eigenvectors of this matrix are $e_1^T = [0.848, 0.530]$ and $e_2^T = [-0.530, 0.848]$, so that the eigenvector matrix [E] is that shown in Equation 9.57. The corresponding eigenvalues are $\lambda_1 = 254.76$ and $\lambda_2 = 8.29$. These are the same data used to fit the bivariate normal probability ellipses shown in Figures 10.1 and 10.6.

The orientations of the two eigenvectors are shown in Figure 11.1, although their lengths have been exaggerated for clarity. It is evident that the first eigenvector is aligned in the direction that the data jointly exhibit maximum variation. That is, the point cloud is inclined at the same angle as is e_1, which is 32° from the horizontal (i.e., from the vector [1, 0]), according to Equation 9.15. Since the data in this simple example exist in only $K = 2$ dimensions, the constraint that the second eigenvector must be perpendicular to the first determines its direction up to sign (i.e., it could as easily be $-e_2^T = [0.530, -0.848]$). This last eigenvector locates the direction in which data jointly exhibit their smallest variations.

The two eigenvectors determine an alternative coordinate system in which to view the data. This fact may become more clear if you rotate this book 32° clockwise. Within this rotated coordinate system, each point is defined by a principal component vector $u^T = [u_1, u_2]$ of new transformed variables, whose elements consist of the projections of the original data onto the eigenvectors, according to the dot product in Equation 11.1. Figure 11.1 illustrates this projection for the 15 January data point $x'^T = [16.0, 17.8]$, which is indicated by the large square symbol. For this datum, $u_1 = (0.848)(16.0) + (0.530)(17.8) = 23.0$, and $u_2 = (-0.530)(16.0) + (0.848)(17.8) = 6.6$.

The sample variance of the new variable u_1 is an expression of the degree to which it spreads out along its axis (i.e., along the direction of e_1). This dispersion is evidently greater than the dispersion of the data along either of the original axes, and indeed it is larger than the dispersion of the data along any other axis in this plane. This maximum sample variance of u_1 is equal to the eigenvalue $\lambda_1 = 254.76°F^2$. The points in the data set tend to exhibit quite different values of u_1, whereas they have more similar values for u_2. That is, they are much less variable in the e_2 direction, and the sample variance of u_2 is only $\lambda_2 = 8.29°F^2$.

Since $\lambda_1 + \lambda_2 = s_{1,1} + s_{2,2} = 263.05°F^2$, the new variables retain all the variation exhibited by the original variables. However, the fact that the point cloud seems to exhibit no slope in the new coordinate frame defined by the eigenvectors indicates that u_1 and u_2 are uncorrelated. Their lack of correlation can be verified by transforming the 31 pairs of minimum temperatures in Table A.1 to principal components and computing the Pearson correlation, which is zero. The variance-covariance matrix for the principal components is therefore [Λ], shown in Equation 9.59.

The two original temperature variables are so strongly correlated that a very large fraction of their joint variance, $\lambda_1/(\lambda_1 + \lambda_2) = 0.968$, is represented by the first principal component. It would be said that the first principal component describes 96.8% of the total variance. The first principal component might be interpreted as reflecting the regional minimum temperature for the area including these two locations (they are about 50 miles apart), with the second principal component describing random variations departing from the overall regional value.

Since so much of the joint variance of the two temperature series is captured by the first principal component, resynthesizing the series using only the first principal component will yield a good approximation to the original data. Using the synthesis Equation 11.6 with only the first $(M = 1)$ principal component yields

$$\mathbf{x}'(t) = \begin{bmatrix} x_1'(t) \\ x_2'(t) \end{bmatrix} \approx \mathbf{e}_1 u_1(t) = \begin{bmatrix} .848 \\ .530 \end{bmatrix} u_1(t). \tag{11.8}$$

The temperature data $\mathbf{x}$ are time series, and therefore so are the principal components $\mathbf{u}$. The time dependence for both has been indicated explicitly in Equation 11.8. On the other hand, the eigenvectors are fixed by the covariance structure of the entire series, and do not change through time. Figure 11.2 compares the original series (black) and the reconstructions using the first principal component $u_1(t)$ only (gray) for the (a) Ithaca and (b) Canandaigua anomalies. The discrepancies are small because $R_1^2 = 96.8\%$. The residual differences would be captured by u_2. The two gray series are exactly proportional to each other, since each is a scalar multiple of the same first principal component time series. Since $\text{Var}(u_1) = \lambda_1 = 254.76$, the variances of the reconstructed series are $(0.848)^2 254.76 = 183.2$ and $(0.530)^2 254.76 = 71.6°F^2$, respectively, which are close to but smaller than the corresponding diagonal elements of the original covariance matrix (Equation 9.56). The larger variance for the Ithaca temperatures is also visually evident in Figure 11.2. Using Equation 11.7, the correlations between the first principal component series $u_1(t)$ and the original temperature variables are $0.848(254.76/185.47)^{1/2} = 0.994$ for Ithaca, and $0.530(254.76/77.58)^{1/2} = 0.960$ for Canandaigua. ◊

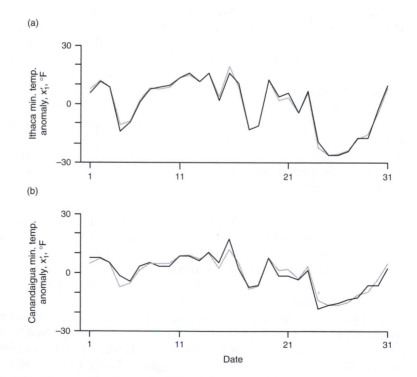

FIGURE 11.2   Time series of January 1987 (a) Ithaca and (b) Canandaigua minimum temperature anomalies (black), and their reconstruction using the first principal component only (grey), through the synthesis Equation 11.8.

11.1.2 PCA Based on the Covariance Matrix vs. the Correlation Matrix

PCA can be conducted as easily on the correlation matrix [R] as it can on the covariance matrix [S]. The correlation matrix is the variance-covariance matrix of the vector of standardized variables z (Equation 9.32). The vector of standardized variables z is related to the vectors of original variables x and their centered counterparts x' according to the scaling transformation (Equation 9.34). Therefore, PCA on the correlation matrix amounts to analysis of the joint variance structure of the standardized variables z_k, as computed using either Equation 9.34 or (in scalar form) Equation 3.21.

The difference between a PCA performed using the variance-covariance and correlation matrices will be one of emphasis. Since PCA seeks to find variables successively maximizing the proportion of the total variance ($\Sigma_k s_{k,k}$) represented, analyzing the covariance matrix [S] results in principal components that emphasize the x_ks having the largest variances. Other things equal, the tendency will be for the first few eigenvectors to align near the directions of the variables having the biggest variances. In Example 11.1, the first eigenvector points more toward the Ithaca minimum temperature axis because the variance of the Ithaca minimum temperatures is larger than the variance of the Canandaigua minimum temperatures. Conversely, PCA applied to the correlation matrix [R] weights all the standardized variables z_k equally, since all have equal (unit) variance.

If the PCA is conducted using the correlation matrix, the analysis formula, Equations 11.1 and 11.2, will pertain to the standardized variables, z_k and z, respectively. Similarly the synthesis formulae, Equations 11.5 and 11.6 will pertain to z and z_k rather than to x' and x'_k. In this case the original data x can be recovered from the result of the synthesis formula by reversing the standardization given by Equations 9.33 and 9.34; that is,

$$\mathbf{x} = [\mathbf{D}]\mathbf{z} + \bar{\mathbf{x}}. \tag{11.9}$$

Although z and x' can easily be obtained from each other using Equation 9.34, the eigenvalue–eigenvector pairs of [R] and [S] do not bear simple relationships to one another. In general, it is not possible to compute the principal components of one knowing only the principal components of the other. This fact implies that these two alternatives for PCA do not yield equivalent information, and that it is important to make an intelligent choice of one over the other for a given application. If an important goal of the analysis is to identify or isolate the strongest variations in a data set, the better alternative usually will be PCA using the covariance matrix, although the choice will depend on the judgment of the analyst and the purpose of the study. For example, in analyzing gridded numbers of extratropical cyclones, Overland and Preisendorfer (1982) found that PCA on their covariance matrix better identified regions having the highest variability in cyclone numbers, and that correlation-based PCA was more effective at locating the primary storm tracks.

However, if the analysis is of unlike variables—variables not measured in the same units—it will almost always be preferable to compute the PCA using the correlation matrix. Measurement in unlike physical units yields arbitrary relative scalings of the variables, which results in arbitrary relative magnitudes of the variances of these variables. To take a simple example, the variance of a set of temperatures measured in °F will be $(1.8)^2 = 3.24$ times as large as the variance of the same temperatures expressed in °C. If the PCA has been done using the correlation matrix, the analysis formula, Equation 11.2, pertains to the vector z rather than x'; and the synthesis in Equation 11.5 will yield the standardized variables z_k (or approximations to them if Equation 11.6 is used for the reconstruction). The summations in the denominators of Equation 11.4 will equal the number of standardized variables, since each has unit variance.

EXAMPLE 11.2   Correlation-versus Covariance-Based PCA for Arbitrarily Scaled Variables

The importance of basing a PCA on the correlation matrix when the variables being analyzed are not measured on comparable scales is illustrated in Table 11.1. This table summarizes PCAs of the January 1987 data in Table A.1 in (a) unstandardized (covariance matrix) and (b) standardized (correlation matrix) forms. Sample variances of the variables are shown, as are the six eigenvectors, the six eigenvalues, and the cumulative percentages of variance accounted for by the principal components. The (6×6) arrays in the upper-right portions of parts (a) and (b) of this table constitute the matrices [E] whose columns are the eigenvectors.

TABLE 11.1   Comparison of PCA computed using (a) the covariance matrix, and (b) the correlation matrix, of the data in Table A.1. The sample variances of each of the variables are shown, as are the six eigenvectors e_m arranged in decreasing order of their eigenvalues λ_m. The cumulative percentage of variance represented is calculated according to Equation 11.4. The much smaller variances of the precipitation variables in (a) is an artifact of the measurement units, but results in precipitation being unimportant in the first four principal components computed from the covariance matrix, which collectively account for 99.9% of the total variance of the data set. Computing the principal components from the correlation matrix ensures that variations of the temperature and precipitation variables are weighted equally.

(a) Covariance results:

Variable	Sample Variance	e_1	e_2	e_3	e_4	e_5	e_6
Ithaca ppt.	$0.059\,inch^2$	.003	.017	.002	−.028	.818	−.575
Ithaca T_{max}	$892.2°F^2$	.359	−.628	.182	−.665	−.014	−.003
Ithaca T_{min}	$185.5°F^2$	.717	.527	.456	.015	−.014	.000
Canandaigua ppt.	$0.028\,inch^2$	.002	.010	.005	−.023	.574	.818
Canandaigua T_{max}	$61.8°F^2$	.381	−.557	.020	.737	.037	.000
Canandaigua T_{min}	$77.6°F^2$	.459	.131	−.871	−.115	−.004	.003
Eigenvalues, λ_k		337.7	36.9	7.49	2.38	0.065	0.001
Cumulative % variance		87.8	97.4	99.3	99.9	100.0	100.0

(b) Correlation results:

Variable	Sample Variance	e_1	e_2	e_3	e_4	e_5	e_6
Ithaca ppt.	1.000	.142	.677	.063	−.149	−.219	.668
Ithaca T_{max}	1.000	.475	−.203	.557	.093	.587	.265
Ithaca T_{min}	1.000	.495	.041	−.526	.688	−.020	.050
Canandaigua ppt.	1.000	.144	.670	.245	.096	.164	−.658
Canandaigua T_{max}	1.000	.486	−.220	.374	−.060	−.737	−.171
Canandaigua T_{min}	1.000	.502	−.021	−.458	−.695	−.192	−.135
Eigenvalues, λ_k		3.532	1.985	0.344	0.074	0.038	0.027
Cumulative % variance		58.9	92.0	97.7	98.9	99.5	100.0

Because of the different magnitudes of the variations of the data in relation to their measurement units, the variances of the unstandardized precipitation data are tiny in comparison to the variances of the temperature variables. This is purely an artifact of the measurement unit for precipitation (inches) being relatively large in comparison to the range of variation of the data (about 1 in.), and the measurement unit for temperature (°F) being relatively small in comparison to the range of variation of the data (about 40°F). If the measurement units had been millimeters and °C, respectively, the differences in variances would have been much smaller. If the precipitation had been measured in micrometers, the variances of the precipitation variables would dominate the variances of the temperature variables.

Because the variances of the temperature variables are so much larger than the variances of the precipitation variables, the PCA calculated from the covariance matrix is dominated by the temperatures. The eigenvector elements corresponding to the two precipitation variables are negligibly small in the first four eigenvectors, so these variables make negligible contributions to the first four principal components. However, these first four principal components collectively describe 99.9% of the joint variance. An application of the truncated synthesis formula (Equation 11.6) with the leading $M = 4$ eigenvector therefore would result in reconstructed precipitation values very near their average values. That is, essentially none of the variation in precipitation would be represented.

Since the correlation matrix is the covariance matrix for comparably scaled variables z_k, each has equal variance. Unlike the analysis on the covariance matrix, this PCA does not ignore the precipitation variables when the correlation matrix is analyzed. Here the first (and most important) principal component represents primarily the closely intercorrelated temperature variables, as can be seen from the relatively larger elements of e_1 for the four temperature variables. However, the second principal component, which accounts for 33.1% of the total variance in the scaled data set, represents primarily the precipitation variations. The precipitation variations would not be lost in the truncated data representation including at least the first $M = 2$ eigenvectors, but rather would be very nearly completely reconstructed. ◊

11.1.3 *The Varied Terminology of PCA*

The subject of PCA is sometimes regarded as a difficult and confusing one, but much of this confusion derives from a proliferation of the associated terminology, especially in writings by analysts of atmospheric data. Table 11.2 organizes the more common of these in a way that may be helpful in deciphering the PCA literature.

Lorenz (1956) introduced the term empirical orthogonal function (EOF) into the literature as another name for the eigenvectors of a PCA. The terms modes of variation and pattern vectors also are used primarily by analysts of geophysical data, especially in relation to analysis of fields, to be described in Section 11.2. The remaining terms for the eigenvectors derive from the geometric interpretation of the eigenvectors as basis vectors, or axes, in the K-dimensional space of the data. These terms are used in the literature of a broader range of disciplines.

The most common name for individual elements of the eigenvectors in the statistical literature is loading, connoting the weight of the kth variable x_k that is borne by the mth eigenvector e_m through the individual element $e_{k,m}$. The term coefficient is also a usual one in the statistical literature. The term pattern coefficient is used mainly in relation to PCA of field data, where the spatial patterns exhibited by the eigenvector elements can be illuminating. Empirical orthogonal weights is a term that is sometimes used to be consistent with the naming of the eigenvectors as EOFs.

TABLE 11.2   A partial guide to synonymous terminology associated with PCA.

Eigenvectors, $\mathbf{e}_m$	Eigenvector Elements, $\mathbf{e}_{k,m}$	Principal Components, $\mathbf{u}_m$	Principal Component Elements, $u_{i,m}$
EOFs	Loadings	Empirical Orthogonal Variables	Scores
Modes of Variation	Coefficients		Amplitudes
Pattern Vectors	Pattern Coefficients		Expansion Coefficients
Principal Axes	Empirical Orthogonal Weights		Coefficients
Principal Vectors			
Proper Functions			
Principal Directions			

The new variables u_m defined with respect to the eigenvectors are almost universally called principal components. However, they are sometimes known as empirical orthogonal variables when the eigenvectors are called EOFs. There is more variation in the terminology for the individual values of the principal components $u_{i,m}$ corresponding to particular data vectors $\mathbf{x}_i'$. In the statistical literature these are most commonly called scores, which has a historical basis in the early and widespread use of PCA in psychometrics. In atmospheric applications, the principal component elements are often called amplitudes by analogy to the amplitudes of a Fourier series, that multiply the (theoretical orthogonal) sine and cosine functions. Similarly, the term expansion coefficient is also used for this meaning. Sometimes expansion coefficient is shortened simply to coefficient, although this can be the source of some confusion since it is more standard for the term coefficient to denote an eigenvector element.

11.1.4 Scaling Conventions in PCA

Another contribution to confusion in the literature of PCA is the existence of alternative scaling conventions for the eigenvectors. The presentation in this chapter has assumed that the eigenvectors are scaled to unit length; that is, $||\mathbf{e}_m|| \equiv 1$. Recall that vectors of any length will satisfy Equation 9.46 if they point in the appropriate direction, and as a consequence it is common for the output of eigenvector computations to be expressed with this scaling.

However, it is sometimes useful to express and manipulate PCA results using alternative scalings of the eigenvectors. When this is done, each element of an eigenvector is multiplied by the same constant, so their relative magnitudes and relationships remain unchanged. Therefore, the qualitative results of an exploratory analysis based on PCA do not depend on the scaling selected, but if different, related analyses are to be compared it is important to be aware of the scaling convention used in each.

Rescaling the lengths of the eigenvectors changes the magnitudes of the principal components by the same factor. That is, multiplying the eigenvector $\mathbf{e}_m$ by a constant requires that the principal component scores u_m be multiplied by the same constant in order for the analysis formulas that define the principal components (Equations 11.1 and 11.2) to remain valid. The expected values of the principal component scores for centered data $\mathbf{x}'$ are zero, and multiplying the principal components by a constant will produce rescaled principal components whose means are also zero. However, their variances will change by a factor of the square of the scaling constant.

TABLE 11.3 Three common eigenvector scalings used in PCA, and their consequences for the properties of the principal components, u_m; and their relationship to the original variables, x_k, and the standardized original variables, z_k.

Eigenvector Scaling	$E[u_m]$	$Var[u_m]$	$Corr[u_m, x_k]$	$Corr[u_m, z_k]$
$\|\|e_m\|\| = 1$	0	λ_m	$e_{k,m}(\lambda_m)^{1/2}/s_k$	$e_{k,m}(\lambda_m)^{1/2}$
$\|\|e_m\|\| = (\lambda_m)^{1/2}$	0	λ_m^2	$e_{k,m}/s_k$	$e_{k,m}$
$\|\|e_m\|\| = (\lambda_m)^{-1/2}$	0	1	$e_{k,m}\lambda_m/s_k$	$e_{k,m}\lambda_m$

Table 11.3 summarizes the effects of three common scalings of the eigenvectors on the properties of the principal components. The first row indicates their properties under the scaling convention $\|\|e_m\|\| \equiv 1$ adopted in this presentation. Under this scaling, the expected value (mean) of each of the principal components is zero, and the variance of each is equal to the respective eigenvalue, λ_m. This result is simply an expression of the diagonalization of the variance-covariance matrix (Equation 9.54) produced by adopting the geometric coordinate system defined by the eigenvectors. When scaled in this way, the correlation between a principal component u_m and a variable x_k is given by Equation 11.7. The correlation between u_m and the standardized variable z_k is given by the product of the eigenvector element and the square root of the eigenvalue, since the standard deviation of a standardized variable is one.

The eigenvectors sometimes are rescaled by multiplying each element by the square root of the corresponding eigenvalue. This rescaling produces vectors of differing lengths, $\|\|e_m\|\| \equiv (\lambda_m)^{1/2}$, but which point in exactly the same directions as the original eigenvectors with unit lengths. Consistency in the analysis formula implies that the principal components are also changed by the factor $(\lambda_m)^{1/2}$, with the result that the variance of each u_m increases to λ_m^2. A major advantage of this rescaling, however, is that the eigenvector elements are more directly interpretable in terms of the relationship between the principal components and the original data. Under this rescaling, each eigenvector element $e_{k,m}$ is numerically equal to the correlation $r_{u,z}$ between the mth principal component u_m and the kth standardized variable z_k.

The last scaling shown in Table 11.3, resulting in $\|\|e_m\|\| \equiv (\lambda_m)^{-1/2}$, is less commonly used. This scaling is achieved by dividing each element of the original unit-length eigenvectors by the square root of the corresponding eigenvalue. The resulting expression for the correlations between the principal components and the original data is more awkward, but this scaling has the advantage that all the principal components have equal, unit variance. This property can be useful in the detection of outliers.

11.1.5 Connections to the Multivariate Normal Distribution

The distribution of the data x whose sample covariance matrix $[S]$ is used to calculate a PCA need not be multivariate normal in order for the PCA to be valid. Regardless of the joint distribution of x, the resulting principal components u_m will uniquely be those uncorrelated linear combinations that successively maximize the represented fractions of the variances on the diagonal of $[S]$. However, if in addition $x \sim N_K(\mu_x, [\Sigma_x])$, then as linear combinations of the multinormal x, the joint distribution of the principal components will also have the multivariate normal distribution,

$$\mathbf{u} \sim N_M([E]^T\mu_x, [\Lambda]). \qquad (11.10)$$

Equation 11.10 is valid both when the matrix [E] contains the full number $M = K$ of eigenvectors as its columns, or some fewer number $1 \leq M < K$. If the principal components are calculated from the centered data x', then $\boldsymbol{\mu}_u = \boldsymbol{\mu}_{x'} = \mathbf{0}$.

If the joint distribution of x is multivariate normal, then the transformation of Equation 11.2 is a rigid rotation to the principal axes of the probability ellipses of the distribution of x, yielding the uncorrelated u_m. With this background it is not difficult to understand Equations 10.5 and 10.31, which say that the distribution of Mahalanobis distances to the mean of a multivariate normal distribution follow the χ^2_K distribution. One way to view the χ^2_K is as the distribution of K squared independent standard Gaussian variables z^2_k (see Section 4.4.3). Calculation of the Mahalanobis distance (or, equivalently, the Mahalanobis transformation, Equation 10.18) produces uncorrelated values with zero mean and unit variance, and a (squared) distance involving them is then simply the sum of the squared values.

It was noted in Section 10.3 that an effective way to search for multivariate outliers when assessing multivariate normality is to examine the distribution of linear combinations formed using eigenvectors associated with the smallest eigenvalues of [S] (Equation 10.15). These linear combinations are, of course, the last principal components. Figure 11.3 illustrates why this idea works, in the easily visualized $K = 2$ situation. The point scatter shows a strongly correlated pair of Gaussian variables, with one multivariate outlier. The outlier is not especially unusual within either of the two univariate distributions, but it stands out in two dimensions because it is inconsistent with the strong positive correlation of the remaining points. The distribution of the first principal component u_1, obtained geometrically by projecting the points onto the first eigenvector e_1, is Gaussian, and the projection of the outlier is a very ordinary member of this distribution. On the other hand, the Gaussian distribution of the second principal component u_2, obtained by projecting the points onto the second eigenvector e_2, is concentrated near the origin except for the single large outlier. This approach is effective in identifying the multivariate outlier because its existence has distorted the PCA only slightly, so that the leading eigenvector continues to be oriented in the direction of the main data scatter. Because a small number of outliers contribute only slightly to the full variability, it is the last (low-variance) principal components that represent them.

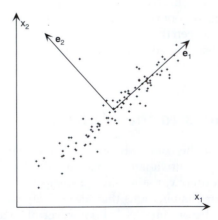

FIGURE 11.3   Identification of a multivariate outlier by examining the distribution of the last principal component. The projection of the single outlier onto the first eigenvector yields a quite ordinary value for its first principal component u_1, but its projection onto the second eigenvector yields a prominent outlier in the distribution of the u_2 values.

11.2 Application of PCA to Geophysical Fields

11.2.1 PCA for a Single Field

The overwhelming majority of applications of PCA to atmospheric data have involved analysis of fields (i.e., spatial arrays of variables) such as geopotential heights, temperatures, precipitation, and so on. In these cases the full data set consists of multiple observations of a field or set of fields. Frequently these multiple observations take the form of time series, for example a sequence of daily hemispheric 500 mb heights. Another way to look at this kind of data is as a collection of K mutually correlated time series that have been sampled at each of K gridpoints or station locations. The goal of PCA as applied to this type of data is usually to explore, or to express succinctly, the joint space/time variations of the many variables in the data set.

Even though the locations at which the field is sampled are spread over a two-dimensional (or possibly three-dimensional) space, the data from these locations at a given observation time are arranged in the one-dimensional vector x. That is, regardless of their geographical arrangement, each location is assigned a number (as in Figure 7.15) from 1 to K, which refers to the appropriate element in the data vector $x = [x_1, x_2, x_3, \cdots, x_K]^T$. In this most common application of PCA to fields, the data matrices [X] and [X'] are thus dimensioned $(n \times K)$, or (time × space), since data at K locations in space have been sampled at n different times.

To emphasize that the original data consists of K time series, the analysis equation (11.1 or 11.2) is sometimes written with an explicit time index:

$$\mathbf{u}(t) = [E]^T \mathbf{x}'(t), \tag{11.11a}$$

or, in scalar form,

$$u_m(t) = \sum_{k=1}^{K} e_{km} x'_k(t), \quad m = 1, \ldots, M. \tag{11.11b}$$

Here the time index t runs from 1 to n. The synthesis equations (11.5 or 11.6) can be written using the same notation, as was done in Equation 11.8. Equation 11.11 emphasizes that, if the data x consist of a set of time series, then the principal components u are also time series. The time series of one of the principal components, $u_m(t)$, may very well exhibit serial correlation (correlation with itself through time), and the principal component time series are sometimes analyzed using the tools presented in Chapter 8. However, each of the time series of principal components will be uncorrelated with the time series of all the other principal components.

When the K elements of x are measurements at different locations in space, the eigenvectors can be displayed graphically in a quite informative way. Notice that each eigenvector contains exactly K elements, and that these elements have a one-to-one correspondence with each of the K locations in the dot product from which the corresponding principal component is calculated (Equation 11.11b). Each eigenvector element e_{km} can be plotted on a map at the same location as its corresponding data value x'_k, and this field of eigenvector elements can itself be displayed with smooth contours in the same way as ordinary meteorological fields. Such maps depict clearly which locations are contributing most strongly to the respective principal components. Looked at another way, such maps indicate the geographic distribution of simultaneous data anomalies represented by the

corresponding principal component. These geographic displays of eigenvectors sometimes also are interpreted as representing uncorrelated modes of variability of the fields
from which the PCA was extracted. There are cases where this kind of interpretation can
be reasonable (but see Section 11.2.4 for a cautionary counterexample), particularly for
the leading eigenvector. However, because of the mutual orthogonality constraints on the
eigenvectors, strong interpretations of this sort are often not justified for the subsequent
EOFs (North, 1984).

Figure 11.4, from Wallace and Gutzler (1981), shows the first four eigenvectors of
a PCA of the correlation matrix for winter monthly-mean 500 mb heights at gridpoints
in the northern hemisphere. The numbers below and to the right of the panels show
the percentage of the total hemispheric variance (Equation 11.4) represented by each of
the corresponding principal components. Together, the first four principal components

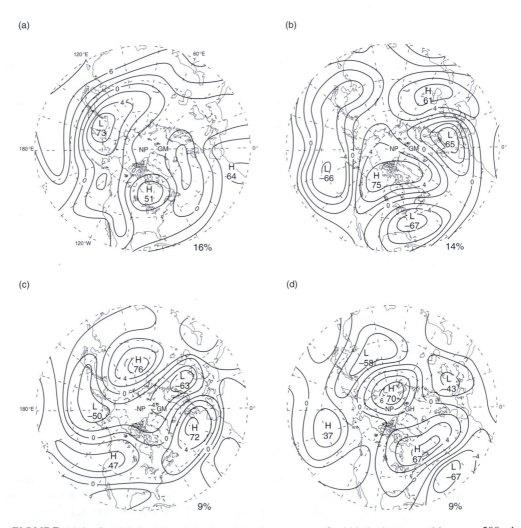

FIGURE 11.4   Spatial displays of the first four eigenvectors of gridded winter monthly-mean 500 mb
heights for the northern hemisphere, 1962–1977. This PCA was computed using the correlation matrix
of the height data, and scaled so that $||e_m|| = \lambda_m^{1/2}$. Percentage values below and to the right of each map
are proportion of total variance × 100% (Equation 11.4). The patterns resemble the teleconnectivity
patterns for the same data (Figure 3.28). From Wallace and Gutzler (1981).

account for nearly half of the (normalized) hemispheric winter height variance. These patterns resemble the teleconnectivity patterns for the same data shown in Figure 3.28, and apparently reflect the same underlying physical processes in the atmosphere. For example, Figure 11.4b evidently reflects the PNA pattern of alternating height anomalies stretching from the Pacific Ocean through northwestern North America to southeastern North America. A positive value of the second principal component of this data set corresponds to negative 500 mb height anomalies (troughs) in the northeastern Pacific and in the southeastern United States, and to positive height anomalies (ridges) in the western part of the continent, and over the central tropical Pacific. A negative value of the second principal component yields the reverse pattern of anomalies, and a more zonal 500 mb flow over North America.

11.2.2 Simultaneous PCA for Multiple Fields

It is also possible to apply PCA to vector-valued fields, which are fields with observations of more than one variable at each location or gridpoint. This kind of analysis is equivalent to simultaneous PCA of two or more fields. If there are L such variables at each of the K gridpoints, then the dimensionality of the data vector x is given by the product KL. The first K elements of x are observations of the first variable, the second K elements are observations of the second variable, and the last K elements of x will be observations of the Lth variable. Since the L different variables generally will be measured in unlike units, it will almost always be appropriate to base the PCA of such data on the correlation matrix. The dimension of [R], and of the matrix of eigenvectors [E], will then be $(KL \times KL)$. Application of PCA to this kind of correlation matrix will produce principal components successively maximizing the joint variance of the L variables in a way that considers the correlations both between and among these variables at the K locations. This joint PCA procedure is sometimes called combined PCA, or CPCA.

Figure 11.5 illustrates the structure of the correlation matrix (left) and the matrix of eigenvectors (right) for PCA of vector field data. The first K rows of [R] contain the correlations between the first of the L variables at these locations and all of the

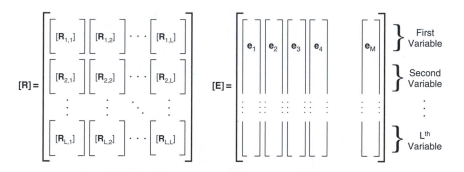

FIGURE 11.5   Illustration of the structures of the correlation matrix and of the matrix of eigenvectors for PCA of vector field data. The basic data consist of multiple observations of L variables at each of K locations, so the dimensions of both [R] and [E] are $(KL \times KL)$. The correlation matrix consists of $(K \times K)$ submatrices containing the correlations between sets of the L variables jointly at the K locations. The submatrices located on the diagonal of [R] are the ordinary correlation matrices for each of the L variables. The off-diagonal submatrices contain correlation coefficients, but are not symmetrical and will not contain 1s on the diagonals. Each eigenvector column of [E] similarly consists of L segments, each of which contains K elements pertaining to individual locations.

KL variables. Rows $K+1$ to $2K$ similarly contain the correlations between the second of the L variables and all the KL variables, and so on. Another way to look at the correlation matrix is as a collection of L^2 submatrices, each dimensioned $(K \times K)$, which contain the correlations between sets of the L variables jointly at the K locations. The submatrices located on the diagonal of [R] thus contain ordinary correlation matrices for each of the L variables. The off-diagonal submatrices contain correlation coefficients, but are not symmetric and will not contain 1s on their diagonals. However, the overall symmetry of [R] implies that $[R_{i,j}] = [R_{j,i}]^T$. Similarly, each column of [E] consists of L segments, and each of these segments contains the K elements pertaining to each of the individual locations.

The eigenvector elements resulting from a PCA of a vector field can be displayed graphically in a manner that is similar to the maps drawn for ordinary, scalar fields. Here, each of the L groups of K eigenvector elements are either overlaid on the same base map, or plotted on separate maps. Figure 11.6, from Kutzbach (1967), illustrates this process for the case of $L = 2$ observations at each location. The two variables are average January surface pressure and average January temperature, measured at $K = 23$ locations in North America. The heavy lines are an analysis of the (first 23) elements of the first eigenvector that pertain to the pressure data, and the dashed lines with shading show an analysis of the

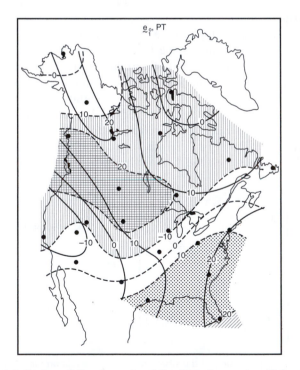

FIGURE 11.6   Spatial display of the elements of the first eigenvector of the (46×46) correlation matrix of average January sea-level pressures and temperatures at 23 locations in North America. The first principal component of this correlation matrix accounts for 28.6% of the joint (standardized) variance of the pressures and temperatures. Heavy lines are a hand analysis of the sea-level pressure elements of the first eigenvector, and dashed lines with shading are a hand analysis of the temperature elements of the same eigenvector. The joint variations of pressure and temperature depicted are physically consistent with temperature advection in response to the pressure anomalies. From Kutzbach (1967).

temperature (second 23) elements of the same eigenvector. The corresponding principal component accounts for 28.6% of the joint variance of the $KL = 23 \times 2 = 46$ standardized variables.

In addition to effectively condensing very much information, the patterns shown in Figure 11.6 are physically consistent with atmospheric processes. In particular, the temperature anomalies are consistent with patterns of thermal advection implied by the pressure anomalies. If the first principal component u_1 is positive for a particular month, the solid contours imply positive pressure anomalies in the north and east, with lower than average pressures in the southwest. On the west coast, this pressure pattern would result in weaker than average westerly surface winds and stronger than average northerly surface winds. The resulting advection of cold air from the north would produce colder temperatures, and this cold advection is reflected by the negative temperature anomalies in this region. Similarly, the pattern of pressure anomalies in the southeast would enhance southerly flow of warm air from the Gulf of Mexico, resulting in positive temperature anomalies as shown. Conversely, if u_1 is negative, reversing the signs of the pressure eigenvector elements implies enhanced westerlies in the west, and northerly wind anomalies in the southeast, which are consistent with positive and negative temperature anomalies, respectively. These temperature anomalies are indicated by Figure 11.6, when the signs on the temperature contours are also reversed.

Figure 11.6 is a simple example involving familiar variables. Its interpretation is easy and obvious if we are conversant with the climatological relationships of pressure and temperature patterns over North America in winter. However, the physical consistency exhibited in this example (where the "right" answer is known ahead of time) is indicative of the power of this kind of PCA to uncover meaningful joint relationships among atmospheric (and other) fields in an exploratory setting, where clues to possibly unknown underlying physical mechanisms may be hidden in the complex relationships among several fields.

11.2.3 Scaling Considerations and Equalization of Variance

A complication arises in PCA of fields in which the geographical distribution of data locations is not uniform (Karl *et al.* 1982). The problem is that the PCA has no information about the spatial distributions of the locations, or even that the elements of the data vector x may pertain to different locations, but nevertheless finds linear combinations that maximize the joint variance. Regions that are overrepresented in x, in the sense that data locations are concentrated in that region, will tend to dominate the analysis, whereas data-sparse regions will be underweighted.

Data available on a regular latitude-longitude grid is a common cause of this problem. In this case the number of gridpoints per unit area increases with increasing latitude because the meridians converge at the poles, so that a PCA for this kind of gridded data will emphasize high-latitude features and deemphasize low-latitude features. One approach to geographically equalizing the variances is to multiply the data by $\sqrt{\cos \phi}$, where ϕ is the latitude. The same effect can be achieved by multiplying each element of the covariance or correlation matrix being analyzed by $\sqrt{\cos \phi_k} \sqrt{\cos \phi_\ell}$, where k and ℓ are the indices for the two locations (or location/variable combinations) corresponding to that element of the matrix. Of course these rescalings must be compensated when recovering the original data from the principal components, as in Equations 11.5 and 11.6. An alternative procedure is to interpolate irregularly or nonuniformly distributed data onto an equal-area grid (Araneo and Compagnucci 2004; Karl*et al.* 1982). This latter approach is also applicable when the data pertain to an irregularly spaced network, such as climatological observing stations, in addition to data on regular latitude-longitude lattices.

A slightly more complicated problem arises when multiple fields with different spatial resolutions or spatial extents are simultaneously analyzed with PCA. Here an additional rescaling is necessary to equalize the sums of the variances in each field. Otherwise fields with more gridpoints will dominate the PCA, even if all the fields pertain to the same geographic area.

11.2.4 Domain Size Effects: Buell Patterns

In addition to providing an efficient data compression, results of a PCA are sometimes interpreted in terms of underlying physical processes. For example, the spatial eigenvector patterns in Figure 11.4 have been interpreted as teleconnected modes of atmospheric variability, and the eigenvector displayed in Figure 11.6 reflects the connection between pressure and temperature fields that is expressed as thermal advection. The possibility that informative or suggestive interpretations may result can be a strong motivation for computing a PCA.

One problem that can occur when making such interpretations of a PCA for field data arises when the spatial scale of the data variations is comparable to or larger than the spatial domain being analyzed. In cases like this the space/time variations in the data are still efficiently represented by the PCA, and PCA is still a valid approach to data compression. But the resulting spatial eigenvector patterns take on characteristic shapes that are nearly independent of the underlying spatial variations in the data. These patterns are called Buell patterns, after the author of the paper that first pointed out their existence (Buell 1979).

Consider, as an artificial but simple example, a 5×5 array of $K = 25$ points representing a square spatial domain. Assume that the correlations among data values observed at these points are functions only of their spatial separation$d$, according to $r(d) = \exp(-d/2)$. The separations of adjacent points in the horizontal and vertical directions are $d = 1$, and so would exhibit correlation r(1) = 0.61; points adjacent diagonally would exhibit correlation $r(\sqrt{2}/2) = 0.49$; and so on. This correlation function is shown in Figure 11.7a. It is unchanging across the domain, and produces no features, or preferred patterns of variability. Its spatial scale is comparable to the domain size, which is 4×4 distance units vertically and horizontally, corresponding to$r(4) = 0.14$.

Even though there are no preferred regions of variability within the 5× 5 domain, the eigenvectors of the resulting(25×25) correlation matrix [R] appear to indicate that there are. The first of these eigenvectors, which accounts for 34.3% of the variance, is shown in Figure 11.7b. It appears to indicate generally in-phase variations throughout the domain, but with larger amplitude (greater magnitudes of variability) near the center. This first characteristic Buell pattern is an artifact of the mathematics behind the eigenvector calculation if all the correlations are positive, and does not merit interpretation beyond its suggestion that the scale of variation of the data is comparable to or larger than the size of the spatial domain.

The dipole patterns in Figures 11.7c and 11.7d are also characteristic Buell patterns, and result from the constraint of mutual orthogonality among the eigenvectors. They do not reflect dipole oscillations or seesaws in the underlying data, whose correlation

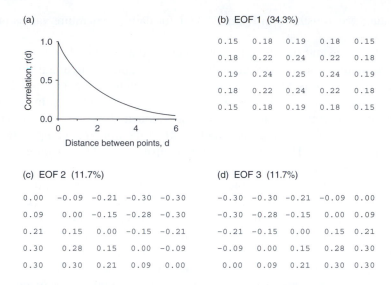

(a)

Correlation, r(d)

1.0

0.5

0.0

0   2   4   6

Distance between points, d

(b) EOF 1 (34.3%)

0.15	0.18	0.19	0.18	0.15
0.18	0.22	0.24	0.22	0.18
0.19	0.24	0.25	0.24	0.19
0.18	0.22	0.24	0.22	0.18
0.15	0.18	0.19	0.18	0.15

(c) EOF 2 (11.7%)

0.00	−0.09	−0.21	−0.30	−0.30
0.09	0.00	−0.15	−0.28	−0.30
0.21	0.15	0.00	−0.15	−0.21
0.30	0.28	0.15	0.00	−0.09
0.30	0.30	0.21	0.09	0.00

(d) EOF 3 (11.7%)

−0.30	−0.30	−0.21	−0.09	0.00
−0.30	−0.28	−0.15	0.00	0.09
−0.21	−0.15	0.00	0.15	0.21
−0.09	0.00	0.15	0.28	0.30
0.00	0.09	0.21	0.30	0.30

FIGURE 11.7   Artificial example of Buell patterns. Data on a 5×5 square grid with unit vertical and horizontal spatial separation exhibit correlations according to the function of their spatial separations shown in (a). Panels (b)–(d) show the first three eigenvectors of the resulting correlation matrix, displayed in the same 5×5 spatial arrangement. The resulting single central hump (b), and pair of orthogonal dipole patterns (c) and (d), are characteristic artifacts of the domain size being comparable to or smaller than the spatial scale of the underlying data.

structure (by virtue of the way this artificial example has been constructed) would be homogeneous and isotropic. Here the patterns are oriented diagonally, because opposite corners of this square domain are further apart than opposite sides, but the characteristic dipole pairs in the second and third eigenvectors might instead have been oriented vertically and horizontally in a differently shaped domain. Notice that the second and third eigenvectors account for equal proportions of the variance, and so are actually oriented arbitrarily within the two-dimensional space that they span (cf. Section 11.4). Additional Buell patterns are sometimes seen in subsequent eigenvectors, the next of which typically suggest tripole patterns of the form $-+-$ or $+-+$.

11.3 Truncation of the Principal Components

11.3.1 Why Truncate the Principal Components?

Mathematically, there are as many eigenvectors of [S] or [R] as there are elements of the data vector x'. However, it is typical of atmospheric data that substantial covariances (or correlations) exist among the original K variables, and as a result there are few or no off-diagonal elements of [S] (or [R]) that are near zero. This situation implies that there is redundant information in x, and that the first few eigenvectors of its dispersion matrix will locate directions in which the joint variability of the data is greater than the variability of any single element, x'_k, of x'. Similarly, the last few eigenvectors will point to directions in the K-dimensional space of x' where the data jointly exhibit very little variation. This feature was illustrated in Example 11.1 for daily temperature values measured at nearby locations.

To the extent that there is redundancy in the original data x', it is possible to capture most of their variance by considering only the most important directions of their joint variations. That is, most of the information content of the data may be represented using some smaller number $M < K$ of the principal components u_m. In effect, the original data set containing the K variables x_k is approximated by the smaller set of new variables u_m. If $M << K$, retaining only the first M of the principal components results in a much smaller data set. This data compression capability of PCA is often a primary motivation for its use.

The truncated representation of the original data can be expressed mathematically by a truncated version of the analysis formula, Equation 11.2, in which the dimension of the truncated u is $(M \times 1)$, and [E] is the (nonsquare, $K \times M$) matrix whose columns consist only of the first M eigenvectors (i.e., those associated with the largest M eigenvalues) of [S]. The corresponding synthesis formula, Equation 11.6, is then only approximately true because the original data cannot be exactly resynthesized without using all K eigenvectors.

Where is the appropriate balance between data compression (choosing M to be as small as possible) and avoiding excessive information loss (truncating only a small number, $K - M$, of the principal components)? There is no clear criterion that can be used to choose the number of principal components that are best retained in a given circumstance. The choice of the truncation level can be aided by one or more of the many available principal component selection rules, but it is ultimately a subjective choice that will depend in part on the data at hand and the purposes of the PCA.

11.3.2 Subjective Truncation Criteria

Some approaches to truncating principal components are subjective, or nearly so. Perhaps the most basic criterion is to retain enough of the principal components to represent a sufficient fraction of the variances of the original x. That is, enough principal components are retained for the total amount of variability represented to be larger than some critical value,

$$\sum_{m=1}^{M} R_m^2 \geq R_{crit}^2, \tag{11.12}$$

where R_m^2 is defined as in Equation 11.4. Of course the difficulty comes in determining how large the fraction R_{crit}^2 must be in order to be considered sufficient. Ultimately this will be a subjective choice, informed by the analyst's knowledge of the data at hand and the uses to which they will be put. Jolliffe (2002) suggests that $70\% \leq R_{crit}^2 \leq 90\%$ may often be a reasonable range.

Another essentially subjective approach to principal component truncation is based on the shape of the graph of the eigenvalues λ_m in decreasing order as a function of their index $m = 1, \ldots, K$, known as the eigenvalue spectrum. Since each eigenvalue measures the variance represented in its corresponding principal component, this graph is analogous to the power spectrum (see Section 8.5.2), extending the parallel between EOF and Fourier analysis.

Plotting the eigenvalue spectrum with a linear vertical scale produces what is known as the scree graph. When using the scree graph qualitatively, the goal is to locate a point separating a steeply sloping portion to the left, and a more shallowly sloping portion to the right. The principal component number at which the separation occurs is then taken as the truncation cutoff, M. There is no guarantee that the eigenvalue spectrum for a given PCA will exhibit a single slope separation, or that it (or they) will be sufficiently abrupt to unambiguously locate a cutoff M. Sometimes this approach to principal component truncation is called the scree test, although this name implies more objectivity and theoretical justification than is warranted: the scree-slope criterion does not involve quantitative statistical inference. Figure 11.8a shows the scree graph (circles) for the PCA summarized in Table 11.1b. This is a relatively well-behaved example, in which the last three eigenvalues are quite small, leading to a fairly distinct bend at $K = 3$, and so a truncation after the first $M = 3$ principal components.

An alternative but similar approach is based on the log-eigenvalue spectrum, or log-eigenvalue (LEV) diagram. Choosing a principal component truncation based on the LEV diagram is motivated by the idea that, if the last $K - M$ principal components represent uncorrelated noise, then the magnitudes of their eigenvalues should decay exponentially with increasing principal component number. This behavior should be identifiable in the LEV diagram as an approximately straight-line portion on its right-hand side. The M retained principal components would then be the ones whose log-eigenvalues lie above the leftward extrapolation of this line. As before, depending on the data set there may no, or more than one, quasi-linear portions, and their limits may not be clearly defined. Figure 11.8b shows the LEV diagram for the PCA summarized in Table 11.1b. Here $M = 3$ would probably be chosen by most viewers of this LEV diagram, although the choice is not unambiguous.

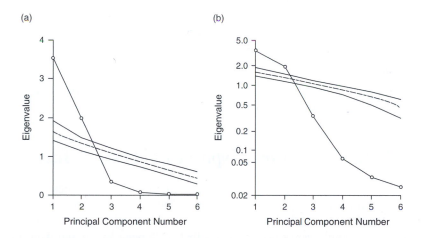

FIGURE 11.8 Graphical displays of eigenvalue spectra; that is, eigenvalue magnitudes as a function of the principal component number (heavier lines connecting circled points), for a $K = 6$ dimensional analysis (see Table 11.1b): (a) Linear scaling, or scree graph, (b) logarithmic scaling, or LEV diagram. Both the scree and LEV criteria would lead to retention of the first three principal components in this analysis. Lighter lines in both panels show results of the resampling tests necessary to apply Rule N of Priesendorfer *et al.* (1981). Dashed line is median of eigenvalues for 1000 (6×6) dispersion matrices of independent Gaussian variables, constructed using the same sample size as the data being analyzed. Solid lines indicate the 5th and 95th percentiles of these simulated eigenvalue distributions. Rule N would indicate retention of only the first two principal components, on the grounds that these are significantly larger than what would be expected from data with no correlation structure.

11.3.3 *Rules Based on the Size of the Last Retained Eigenvalue*

Another class of principal-component selection rules involves focusing on how small an "important" eigenvalue can be. This set of selection rules can be summarized by the criterion

$$\text{Retain}\,\lambda_m \text{ if } \lambda_m > \frac{T}{K}\sum_{k=1}^{K} s_{k,k}, \tag{11.13}$$

where $s_{k,k}$ is the sample variance of the kth element of x, and T is a threshold parameter.

A simple application of this idea, known as Kaiser's rule, involves comparing each eigenvalue (and therefore the variance described by its principal component) to the amount of the joint variance reflected in the average eigenvalue. Principal components whose eigenvalues are above this threshold are retained. That is, Kaiser's rule uses Equation 11.13 with the threshold parameter $T = 1$. Jolliffe (1972, 2002) has argued that Kaiser's rule is too strict (i.e., typically seems to discard too many principal components). He suggests that the alternative $T = 0.7$ often will provide a roughly correct threshold, which allows for the effects of sampling variations.

A third alternative in this class of truncation rules is to use the broken stick model, so called because it is based on the expected length of the mth longest piece of a randomly broken unit line segment. According to this criterion, the threshold parameter in Equation 11.13 is taken to be

$$T(m) = \frac{1}{K}\sum_{j=m}^{K} \frac{1}{j}. \tag{11.14}$$

This rule yields a different threshold for each candidate truncation level—that is, $T = T(m)$, so that the truncation is made at the smallest m for which Equation 11.13 is not satisfied, according to the threshold in Equation 11.14.

All of the three criteria described in this subsection would lead to choosing $M = 2$ for the eigenvalue spectrum in Figure 11.8.

11.3.4 *Rules Based on Hypothesis Testing Ideas*

Faced with a subjective choice among sometimes vague truncation criteria, it is natural to hope for a more objective approach based on the sampling properties of PCA statistics. Section 11.4 describes some large-sample results for the sampling distributions of eigenvalue and eigenvector estimates that have been calculated from multivariate normal samples. Based on these results, Mardia *et al.* (1979) and Jolliffe (2002) describe tests for the null hypothesis that the last $K - M$ eigenvalues are all equal, and so correspond to noise that should be discarded in the principal component truncation. One problem with this approach occurs when the data being analyzed do not have a multivariate normal distribution, and/or are not independent, in which case inferences based on those assumptions may produce serious errors. But a more difficult problem with this approach is that it usually involves examining a sequence of tests that are not independent: Are the last two eigenvalues plausibly equal, and if so, are the last three equal, and if so, are the last four equal . . . ? The true test level for a random number of correlated tests will bear an unknown relationship to the nominal level at which each test in the sequence is conducted. The procedure can be used to choose a truncation level, but it will be as much a rule of thumb as the other possibilities already presented in this section, and not a quantitative choice based on a known small probability for falsely rejecting a null hypothesis.

Resampling counterparts to testing-based truncation rules have been used frequently with atmospheric data, following Preisendorfer *et al.* (1981). The most common of these is known as Rule N. Rule N identifies the largest M principal components to be retained on the basis of a sequence of resampling tests involving the distribution of eigenvalues of randomly generated dispersion matrices. The procedure involves repeatedly generating sets of vectors of independent Gaussian random numbers with the same dimension (K) and sample size (n) as the data x being analyzed, and then computing the eigenvalues of their dispersion matrices. These randomly generated eigenvalues are then scaled in a way that makes them comparable to the eigenvalues λ_m to be tested, for example by requiring that the sum of each set of randomly generated eigenvalues will equal the sum of the eigenvalues computed from the data. Each λ_m from the real data is then compared to the empirical distribution of its synthetic counterparts, and is retained if it is larger than 95% of these.

The light lines in the panels of Figure 11.8 illustrate the use of Rule N to select a principal component truncation level. The dashed lines reflect the medians of 1000 sets of eigenvalues computed from 1000 (6×6) dispersion matrices of independent Gaussian variables, constructed using the same sample size as the data being analyzed. The solid lines show 95th and 5th percentiles of those distributions for each of the six eigenvalues. The first two eigenvalues λ_1 and λ_2 are larger than more than 95% of their synthetic counterparts, and for these the null hypothesis that the corresponding principal components represent only noise would therefore be rejected at the 5% level. Accordingly, Rule N would choose $M = 2$ for this data.

A table of 95% critical values for Rule N, for selected sample sizes n and dimensions K, is presented in Overland and Preisendorfer (1982). Corresponding large-sample tables are given in Preisendorfer *et al.* (1981) and Preisendorfer (1988). Preisendorfer (1988) notes that if there is substantial temporal correlation present in the individual variables x_k, that it may be more appropriate to construct the resampling distributions for Rule N (or to use the tables just mentioned) using the smallest effective sample size (using an equation analogous to Equation 5.12, but appropriate to eigenvalues) among the x_k, rather than using n independent vectors of Gaussian variables to construct each synthetic dispersion matrix. Another potential problem with Rule N, and other similar procedures, is that the data x may not be approximately Gaussian. For example, one or more of the x_ks could be precipitation variables. To the extent that the original data are not Gaussian, the resampling procedure will not simulate accurately the physical process that generated them, and the results of the tests may be misleading. A possible remedy for the problem of non-Gaussian data might be to use a bootstrap version of Rule N, although this approach seems not to have been tried in the literature to date.

Ultimately, Rule N and other similar truncation procedures suffer from the same problem as their parametric counterparts, namely that a sequence of correlated tests must be examined. For example, a sufficiently large first eigenvalue would be reasonable grounds on which to reject a null hypothesis that all the K elements of x are uncorrelated, but subsequently examining the second eigenvalue in the same way would not be an appropriate test for the second null hypothesis, that the last $K - 1$ eigenvalues correspond to uncorrelated noise. Having rejected the proposition that λ_1 is not different from the others, the Monte-Carlo sampling distributions for the remaining eigenvalues are no longer meaningful because they are conditional on all K eigenvalues reflecting noise. That is, these synthetic sampling distributions will imply too much variance if λ_1 has more than a random share, and the sum of the eigenvalues is constrained to equal the total variance. Priesendorfer (1988) notes that Rule N tends to retain too few principal components.

11.3.5 Rules Based on Structure in the Retained Principal Components

The truncation rules presented so far all relate to the magnitudes of the eigenvalues. The possibility that physically important principal components need not have the largest variances (i.e., eigenvalues) has motivated a class of truncation rules based on expected characteristics of physically important principal component series (Preisendorfer *et al.* 1981, Preisendorfer 1988). Since most atmospheric data that are subjected to PCA are time series (e.g., time sequences of spatial fields recorded at K gridpoints), a plausible hypothesis may be that principal components corresponding to physically meaningful processes should exhibit time dependence, because the underlying physical processes are expected to exhibit time dependence. Preisendorfer *et al.* (1981) and Preisendorfer (1988) proposed several such truncation rules, which test null hypotheses that the individual principal component time series are uncorrelated, using either their power spectra or their autocorrelation functions. The truncated principal components are then those for which this null hypothesis is not rejected. This class of truncation rule seems to have been used very little in practice.

11.4 Sampling Properties of the Eigenvalues and Eigenvectors

11.4.1 Asymptotic Sampling Results for Multivariate Normal Data

Principal component analyses are calculated from finite data samples, and are as subject to sampling variations as is any other statistical estimation procedure. That is, we rarely if ever know the true covariance matrix $[\Sigma]$ for the population or underlying generating process, but rather estimate it using the sample counterpart $[S]$. Accordingly the eigenvalues and eigenvectors calculated from $[S]$ are also estimates based on the finite sample, and are thus subject to sampling variations. Understanding the nature of these variations is quite important to correct interpretation of the results of a PCA.

The equations presented in this section must be regarded as approximate, as they are asymptotic (large-n) results, and are based also on the assumption that the underlying $\boldsymbol{x}$ have a multivariate normal distribution. It is also assumed that no pair of the population eigenvalues is equal, implying (in the sense to be explained in Section 11.4.2) that all the population eigenvectors are well defined. The validity of these results is therefore approximate in most circumstances, but they are nevertheless quite useful for understanding the nature of sampling effects on the uncertainty around the estimated eigenvalues and eigenvectors.

The basic result for the sampling properties of estimated eigenvalues is that, in the limit of very large sample size, their sampling distribution is unbiased, and multivariate normal,

$$\sqrt{n}(\hat{\boldsymbol{\lambda}} - \hat{\boldsymbol{\lambda}}) \sim N_K(\mathbf{0}, 2[\Lambda]^2),$$ (11.15a)

or

$$\hat{\boldsymbol{\lambda}} \sim N_K\left(\boldsymbol{\lambda}, \frac{2}{n}[\Lambda]^2\right).$$ (11.15b)

Here $\hat{\boldsymbol{\lambda}}$ is the $(K \times 1)$ vector of estimated eigenvalues, $\boldsymbol{\lambda}$ is its true value; and the $(K \times K)$ matrix $[\Lambda]^2$ is the square of the diagonal, population eigenvalue matrix, having elements λ_k^2. Because $[\Lambda]^2$ is diagonal the sampling distributions for each of the K estimated eigenvalues are (approximately) independent univariate Gaussian distributions,

$$\sqrt{n}(\hat{\lambda}_k - \lambda_k) \sim N(0, 2\lambda_k^2),$$ (11.16a)

or

$$\lambda_k \sim N\left(\lambda_k, \frac{2}{n}\lambda_k^2\right).$$ (11.16b)

Note however that there is a bias in the sample eigenvalues for finite sample size: Equations 11.15 and 11.16 are large-sample approximations. In particular, the largest eigenvalues will be overestimated (will tend to be larger than their population counterparts) and the smallest eigenvalues will tend to be underestimated, and these effects increase with decreasing sample size.

Using Equation 11.16a to construct a standard Gaussian variate provides an expression for the distribution of the relative error of the eigenvalue estimate,

$$z = \frac{\sqrt{n}(\hat{\lambda}_k - \lambda_k) - 0}{\sqrt{2}\lambda_k} = \sqrt{\frac{n}{2}}\left(\frac{\hat{\lambda}_k - \lambda_k}{\lambda_k}\right) \sim N(0, 1).$$ (11.17)

Equation 11.17 implies that

$$\Pr\left\{\left|\sqrt{\frac{n}{2}}\left(\frac{\hat{\lambda}_k - \lambda_k}{\lambda_k}\right)\right| \leq z(1 - \alpha/2)\right\} = 1 - \alpha,$$ (11.18)

which leads to the $(1 - \alpha) \times 100\%$ confidence interval for the kth eigenvalue,

$$\frac{\hat{\lambda}_k}{1 + z(1 - \alpha/2)\sqrt{\frac{2}{n}}} \leq \lambda_k \leq \frac{\hat{\lambda}_k}{1 - z(1 - \alpha/2)\sqrt{\frac{2}{n}}}.$$ (11.19)

The elements of each sample eigenvector are approximately unbiased, and their sampling distributions are approximately multivariate normal. But the variances of the multivariate normal sampling distributions for each of the eigenvectors depend on all the other eigenvalues and eigenvectors in a somewhat complicated way. The sampling distribution for the kth eigenvector is

$$\hat{\mathbf{e}}_k \sim N_K(\mathbf{e}_k, [V_{\mathbf{e}_k}]),$$ (11.20)

where the covariance matrix for this distribution is

$$[V_{\mathbf{e}_k}] = \frac{\lambda_k}{n} \sum_{i \neq k}^{K} \frac{\lambda_i}{(\lambda_i - \lambda_k)^2} \mathbf{e}_i \mathbf{e}_i^{\text{T}}.$$ (11.21)

The summation in Equation 11.21 involves all K eigenvalue-eigenvector pairs, indexed here by i, *except* the kth pair, for which the covariance matrix is being calculated. It is a sum of weighted outer products of these eigenvectors, and so resembles the spectral decomposition of the true covariance matrix $[\Sigma]$ (cf. Equation 9.51). But rather than being weighted only by the corresponding eigenvalues, as in Equation 9.51, they are weighted also by the reciprocals of the squares of the differences between those eigenvalues, and the eigenvalue belonging to the eigenvector whose covariance matrix is being calculated. That is, the elements of the matrices in the summation of Equation 11.21 will be quite small in magnitude, except those that are paired with eigenvalues λ_i that are close in magnitude to the eigenvalue λ_k, belonging to the eigenvector whose sampling distribution is being calculated.

11.4.2 Effective Multiplets

Equation 11.21, for the sampling uncertainty of the eigenvectors of a covariance matrix, has two important implications. First, the pattern of uncertainty in the estimated eigenvectors resembles a linear combination, or weighted sum, of all the *other* eigenvectors. Second, because the magnitudes of the weights in this weighted sum are inversely proportional to the squares of the differences between the corresponding eigenvalues, an eigenvector will be relatively precisely estimated (the sampling variances will be relatively small) if its eigenvalue is well separated from the other $K - 1$ eigenvalues. Conversely, eigenvectors whose eigenvalues are similar in magnitude to one or more of the other eigenvalues will exhibit large sampling variations, and those variations will be larger for the eigenvector elements that are large in eigenvectors with nearby eigenvalues.

The joint effect of these two considerations is that the sampling distributions of a pair (or more) of eigenvectors having similar eigenvalues will be closely entangled. Their sampling variances will be large, and their patterns of sampling error will resemble the patterns of the eigenvector(s) with which they are entangled. The net effect will be that a realization of the corresponding sample eigenvectors will be a nearly arbitrary mixture of the true population counterparts. They will jointly represent the same amount of variance (within the sampling bounds approximated by Equation 11.16), but this joint variance will be arbitrarily mixed between (or among) them. Sets of such eigenvalue-eigenvector pairs are called effectively degenerate multiplets, or *effective multiplets*. Attempts at physical interpretation of their sample eigenvectors will be frustrating if not hopeless.

The source of this problem can be appreciated in the context of a three-dimensional multivariate normal distribution, in which one of the eigenvectors is relatively large, and the two smaller ones are nearly equal. The resulting distribution has ellipsoidal probability contours resembling the cucumbers in Figure 10.2. The eigenvector associated with the single large eigenvalue will be aligned with the long axis of the ellipsoid. But this multivariate normal distribution has (essentially) no preferred direction in the plane perpendicular to the long axis (exposed face on the left-hand cucumber in Figure 10.2b). Any pair of perpendicular vectors that are also perpendicular to the long axis could jointly represent variations in this plane. The leading eigenvector calculated from a sample covariance matrix from this distribution would be closely aligned with the true eigenvector (long axis of the cucumber) because its sampling variations will be small. In terms of Equation 11.21, both of the two terms in the summation would be small because $\lambda_1 >> \lambda_2 \approx \lambda_3$. On the other hand, each of the other two eigenvectors would be subject to large sampling variations: the term in Equation 11.21 corresponding to the other of them will be large, because $(\lambda_2 - \lambda_3)^{-2}$ will be large. The pattern of sampling error for e_2 will resemble e_3, and vice versa. That is, the orientation of the two sample eigenvectors in this plane will be arbitrary, beyond the constraints that they will be perpendicular to each other, and to e_1. The variations represented by each of these two sample eigenvectors will accordingly be an arbitrary mixture of the variations represented by their two population counterparts.

11.4.3 *The North et al. Rule of Thumb*

Equations 11.15 and 11.20, for the sampling distributions of the eigenvalues and eigenvectors, depend on the values of their true but unknown counterparts. Nevertheless, the sample estimates approximate the true values, so that large sampling errors are expected for those eigenvectors whose sample eigenvalues are close to other sample eigenvalues. The idea that it is possible to diagnose instances where sampling variations are expected to cause problems with eigenvector interpretation in PCA was expressed as a rule of thumb by North et al. (1982): "The rule is simply that if the sampling error of a particular eigenvalue λ [$\delta\lambda \sim \lambda(2/n)^{1/2}$] is comparable to or larger than the spacing between λ and a neighboring eigenvalue, then the sampling errors for the EOF associated with λ will be comparable to the size of the neighboring EOF. The interpretation is that if a group of true eigenvalues lie within one or two $\delta\lambda$ of each other, then they form an 'effectively degenerate multiplet,' and sample eigenvectors are a random mixture of the true eigenvectors."

North et al. (1982) illustrated this concept with an instructive example. They constructed synthetic data from a set of known EOF patterns, the first four of which are shown in Figure 11.9a, together with their respective eigenvalues. Using a full set of such patterns, the covariance matrix $[\Sigma]$ from which they could be extracted was assembled using the spectral decomposition (Equation 9.51). Using $[\Sigma]^{1/2}$ (see Section 9.3.4), realizations of data vectors x from a distribution with covariance $[\Sigma]$ were generated as in Section 10.4. Figure 11.9b shows the first four eigenvalue-eigenvector pairs calculated from a sample of $n = 300$ synthetic data vectors, and Figure 11.9c shows the leading eigenvalue-eigenvector pairs for $n = 1000$.

The first four true eigenvector patterns in Figure 11.9a are visually distinct, but their eigenvalues are relatively close. Using Equation 11.16b and $n = 300$, 95% sampling intervals for the four eigenvalues are 14.02 ± 2.24, 12.61 ± 2.02, 10.67 ± 1.71, and 10.43 ± 1.67 (because $z(0.975) = 1.96$), all of which include the adjacent eigenvalues. Therefore it is expected that the sample eigenvectors will be random mixtures of their population counterparts for this sample size, and Figure 11.9b bears out this expectation: the patterns in those four panels appear to be random mixtures of the four panels in Figure 11.9a. Even if the true eigenvectors were unknown, this conclusion would be expected from the North *et al.* rule of thumb, because adjacent sample eigenvectors in Figure 11.9b are within two estimated standard errors, or $2\,\delta\lambda = 2\lambda(2/n)^{1/2}$ of each other.

The situation is somewhat different for the larger sample size (Figure 11.9c). Again using Equation 11.16b but with $n = 1000$, the 95% sampling intervals for the four eigenvalues are 14.02 ± 1.22, 12.61 ± 1.10, 10.67 ± 0.93, and 10.43 ± 0.91. These intervals indicate that the first two sample EOFs should be reasonably distinct from each other and from the other EOFs, but that the third and fourth eigenvectors will probably still be

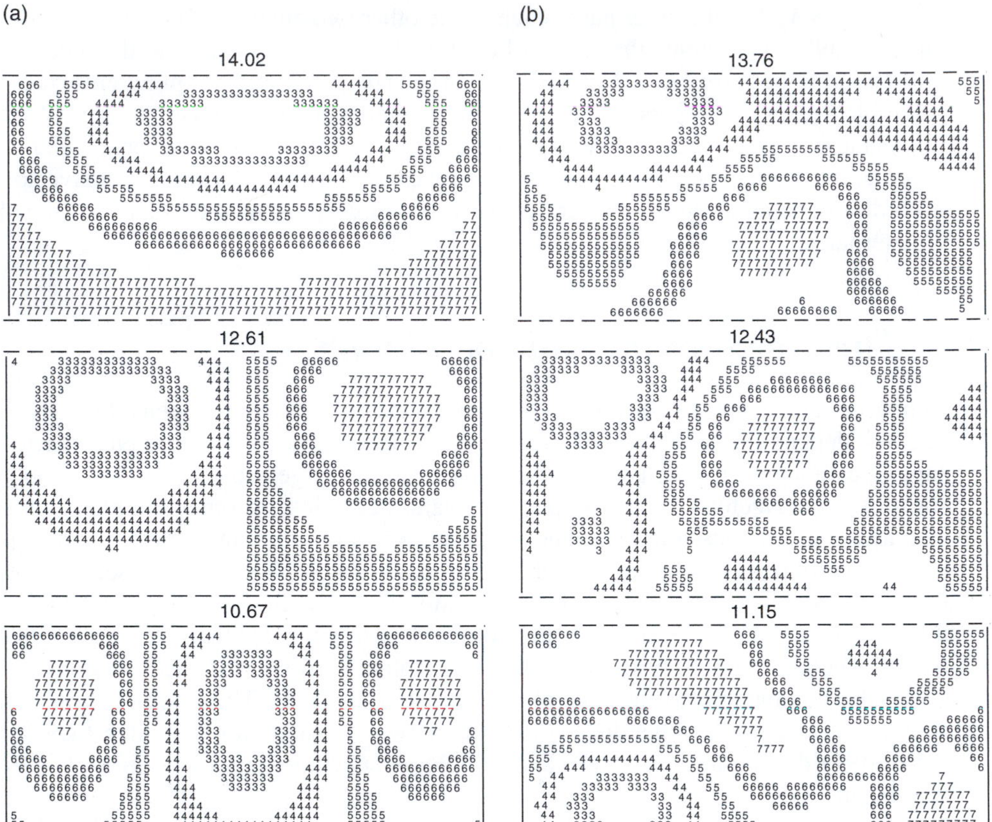

FIGURE 11.9   The North *et al.* (1982) example for effective degeneracy. (a) First four eigenvectors for the population from which synthetic data were drawn, with corresponding eigenvalues. (b) The first four eigenvectors calculated from a sample of $n = 300$, and the corresponding sample eigenvalues. (c) The first four eigenvectors calculated from a sample of $n = 1000$, and the corresponding sample eigenvalues.

(c)

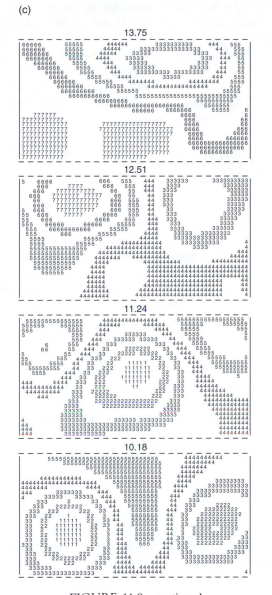

FIGURE 11.9 continued

entangled. Applying the rule of thumb to the sample eigenvalues in Figure 11.9c indicates that the separation between all adjacent pairs is close to 2 $\delta\lambda$. The additional sampling precision provided by the larger sample size allows an approximation to the true EOF patterns to emerge, although an even larger sample still would be required before the sample eigenvectors would correspond well to their population counterparts.

The synthetic data realizations x in this artificial example were chosen independently of each other. If the data being analyzed are serially correlated, the unadjusted rule of thumb will imply better eigenvalue separation than is actually the case, because the variance of the sampling distribution of the sample eigenvalues will be larger than 2 λ_k^2/n (as given in Equation 11.16). The cause of this discrepancy is that the sample eigenvalues are less consistent from batch to batch when calculated from autocorrelated data, so the qualitative effect is the same as was described for the sampling distribution of sample means, in Section 5.2.4. However, the effective sample size adjustment in Equation 5.12 is *not* appropriate for the sampling distribution of the eigenvalues, because they are variances. An appropriate modification appropriate to the effective sample size adjustment for eigenvalue estimation appears not to have been published, but based on the result offered by Livezey (1995), a reasonable guess for an approximate counterpart to Equation 5.12 (assuming AR(1) time dependence) might be $n' \approx n[(1-\rho_1^2)/(1+\rho_1^2)]^2$.

11.4.4 Bootstrap Approximations to the Sampling Distributions

The conditions specified in Section 11.4.1, of large sample size and/or underlying multivariate normal data, may be too unrealistic to be practical in some situations. In such cases it is possible to build good approximations to the sampling distributions of sample statistics using the bootstrap (see Section 5.3.4). Beran and Srivastava (1985) Efron and Tibshirani (1993) specifically describe bootstrapping sample covariance matrices to produce sampling distributions for their eigenvalues and eigenvectors. The basic procedure is to repeatedly resample the underlying data vectors x with replacement; to produce some large number, n_B, of bootstrap samples, each of size n. Each of the n_B bootstrap samples yields a bootstrap realization of [S], whose eigenvalues and eigenvectors can be computed. Jointly these bootstrap realizations of eigenvalues and eigenvectors form reasonable approximations to the respective sampling distributions, which will reflect properties of the underlying data that may not conform to those assumed in Section 11.4.1.

Be careful in interpreting these bootstrap distributions. A (correctable) difficulty arises from the fact that the eigenvectors are determined up to sign only, so that in some bootstrap samples the resampled counterpart of e_k may very well be $-e_k$. Failure to rectify such arbitrary sign switchings will lead to large and unwarranted inflation of the sampling distributions for the eigenvector elements. Difficulties can also arise when resampling effective multiplets, because the random distribution of variance with a multiplet may be different from resample to resample, so the resampled eigenvectors may not bear one-to-one correspondences with their original sample counterparts. Finally, the bootstrap procedure destroys any serial correlation that may be present in the underlying data, which would lead to unrealistically narrow bootstrap sampling distributions. The moving-blocks bootstrap can be used for serially correlated data vectors (Wilks 1997) as well as scalars.

11.5 Rotation of the Eigenvectors

11.5.1 Why Rotate the Eigenvectors?

When PCA eigenvector elements are plotted geographically, there is a strong tendency to try to ascribe physical interpretations to the corresponding principal components. The results shown in Figures 11.4 and 11.6 indicate that it can be both appropriate and informative to do so. However, the orthogonality constraint on the eigenvectors (Equation 9.48) can lead to problems with these interpretations, especially for the second and subsequent principal components. Although the orientation of the first eigenvector is determined solely by the direction of the maximum variation in the data, subsequent vectors must be orthogonal to previously determined eigenvectors, regardless of the nature of the physical processes that may have given rise to the data. To the extent that those underlying physical processes are not independent, interpretation of the corresponding principal components as being independent modes of variability will not be justified (North 1984). The first principal component may represent an important mode of variability or physical process, but it may well also include aspects of other correlated modes or processes. Thus, the orthogonality constraint on the eigenvectors can result in the influences of several distinct physical processes being jumbled together in a single principal component.

When physical interpretation rather than data compression is a primary goal of PCA, it is often desirable to rotate a subset of the initial eigenvectors to a second set of new coordinate vectors. Usually it is some number M of the leading eigenvectors (i.e., eigenvectors with largest corresponding eigenvalues) of the original PCA that are rotated, with M chosen using a truncation criterion such as Equation 11.13. Rotated eigenvectors are less prone to the artificial features resulting from the orthogonality constraint on the unrotated eigenvectors, such as Buell patterns (Richman 1986). They also appear to exhibit better sampling properties (Richman 1986, Cheng et al. 1995) than their unrotated counterparts.

A number of procedures for rotating the original eigenvectors exist, but all seek to produce what is known as simple structure in the resulting analysis. Roughly speaking, simple structure generally is understood to have been achieved if a large number of the elements of the resulting rotated vectors are near zero, and few of the remaining elements correspond to (have the same index k as) elements that are not near zero in the other rotated vectors. The desired result is that each rotated vector represents mainly the few original variables corresponding to the elements not near zero, and that the representation of the original variables is split between as few of the rotated principal components as possible. Simple structure aids interpretation of a rotated PCA by allowing association of rotated eigenvectors with the small number of the original K variables whose corresponding elements are not near zero.

Following rotation of the eigenvectors, a second set of new variables is defined, called rotated principal components. The rotated principal components are obtained from the original data analogously to Equation 11.1 and 11.2, as the dot products of data vectors and the rotated eigenvectors. They can be interpreted as single-number summaries of the similarity between their corresponding rotated eigenvector and a data vector x. Depending on the method used to rotate the eigenvectors, the resulting rotated principal components may or may not be mutually uncorrelated.

A price is paid for the improved interpretability and better sampling stability of the rotated eigenvectors. One cost is that the dominant-variance property of PCA is lost. The first rotated principal component is no longer that linear combination of the original data with the largest variance. The variance represented by the original unrotated eigenvectors is spread more uniformly among the rotated eigenvectors, so that the corresponding eigenvalue spectrum is flatter. Also lost is either the orthogonality of the eigenvectors, or the uncorrelatedness of the resulting principal components, or both.

11.5.2 *Rotation Mechanics*

Rotated eigenvectors are produced as a linear transformation of a subset of M of the original K eigenvectors,

$$\underset{(K \times M)}{[\tilde{E}]} = \underset{(K \times M)}{[E]} \underset{(M \times M)}{[T]} , \tag{11.22}$$

where [T] is the rotation matrix, and the matrix of rotated eigenvectors is denoted by the tilde. If [T] is orthogonal, that is, if $[T][T]^T = [I]$, then the transformation Equation 11.22 is called an orthogonal rotation. Otherwise the rotation is called oblique.

Richman (1986) lists 19 approaches to defining the rotation matrix [T] in order to achieve simple structure, although his list is not exhaustive. However, by far the most commonly used approach is the orthogonal rotation called the varimax (Kaiser 1958). A varimax rotation is determined by choosing the elements of [T] to maximize

$$\sum_{m=1}^{M} \left[\sum_{k=1}^{K} e_{k,m}^{*4} - \frac{1}{K} \left(\sum_{k=1}^{K} e_{k,m}^{*2} \right)^2 \right], \qquad (11.23a)$$

where

$$e_{k,m}^* = \frac{\tilde{e}_{k,m}}{\left(\sum_{m=1}^{M} \tilde{e}_{k,m}^2 \right)^{1/2}} \qquad (11.23b)$$

are scaled versions of the rotated eigenvector elements. Together Equations 11.23a and 11.23b define the normal varimax, whereas Equation 11.23 alone, using the unscaled eigenvector elements $\tilde{e}_{k,m}$, is known as the raw varimax. In either case the transformation is sought that maximizes the sum of the variances of the (either scaled or raw) squared rotated eigenvector elements, which tends to move them toward either their maximum or minimum (absolute) values (which are 0 and 1), and thus tends toward simple structure. The solution is iterative, and is a standard feature of many statistical software packages.

The results of eigenvector rotation can depend on how many of the original eigenvectors are selected for rotation. That is, some or all of the leading rotated eigenvectors may be different if, say, $M + 1$ rather than M eigenvectors are rotated (e.g., O'Lenic and Livezey, 1988). Unfortunately there is often not a clear answer to the question of what the best choice for M might be, and typically an essentially subjective choice is made. Some guidance is available from the various truncation criteria in Section 11.3, although these may not yield a single answer. Sometimes a trial-and-error procedure is used, where M is increased slowly until the leading rotated eigenvectors are stable; that is, insensitive to further increases in M. In any case, however, it makes sense to include either all, or none, of the eigenvectors making up an effective multiplet, since jointly they carry information that has been arbitrarily mixed. Jolliffe (1987, 1989) suggests that it may be helpful to separately rotate groups of eigenvectors within effective multiplets in order to more easily interpret the information that they jointly represent.

Figure 11.10, from Horel (1981), shows spatial displays of the first two rotated eigenvectors of monthly-averaged hemispheric winter 500 mb heights. Using the truncation criterion of Equation 11.13 with $T = 1$, the first 19 eigenvectors of the correlation matrix for these data were rotated. The two patterns in Figure 11.10 are similar to the first two unrotated eigenvectors derived from the same data (see Figure 11.4a and b), although the signs have been (arbitrarily) reversed. However, the rotated vectors conform more to the idea of simple structure in that more of the hemispheric fields are fairly flat (near zero) in Figure 11.10, and each panel emphasizes more uniquely a particular feature of the variability of the 500 mb heights corresponding to the teleconnection patterns in Figure 3.28. The rotated vector in Figure 11.10a focuses primarily on height differences in the northwestern and western tropical Pacific, called the western Pacific teleconnection pattern. It thus represents variations in the 500 mb jet at these longitudes, with positive

(a)   (b)

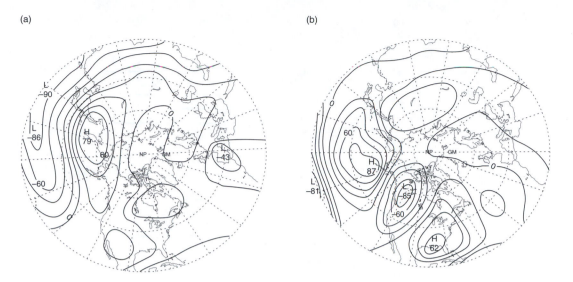

FIGURE 11.10   Spatial displays of the first two rotated eigenvectors of monthly-averaged hemispheric winter 500 mb heights. The data are the same as those underlying Figure 11.4, but the rotation has better isolated the patterns of variability, allowing a clearer interpretation in terms of the teleconnection patterns in Figure 3.28. From Horel (1981).

values of the corresponding rotated principal component indicating weaker than average westerlies, and negative values indicating the reverse. Similarly, the PNA pattern stands out exceptionally clearly in Figure 11.10b, where the rotation has separated it from the eastern hemisphere pattern evident in Figure 11.4b.

Figure 11.11 shows schematic representations of eigenvector rotation in two dimensions. The left-hand diagrams in each section represent the eigenvectors in the two-dimensional plane defined by the underlying variables x_1 and x_2, and the right-hand diagrams represent "maps" of the eigenvector elements plotted at the two "locations" x_1 and x_2, (corresponding to such real-world maps as those shown in Figures 11.4 and 11.10). Figure 11.11a illustrates the case of the original unrotated eigenvectors. The leading eigenvector e_1 is defined as the direction onto which a projection of the data points (i.e., the principal components) has the largest variance, which locates a compromise between the two clusters of points (modes). That is, it locates much of the variance of both groups, without really characterizing either. The leading eigenvector e_1 points in the positive direction for both x_1 and x_2, but is more strongly aligned toward x_2, so the corresponding e_1 map to the right shows a large positive + for x_2, and a smaller + for x_1. The second eigenvector is constrained to be orthogonal to the first, and so corresponds to large negative x_1, and mildly positive x_2, as indicated in the corresponding "map" to the right.

Figure 11.11b represents orthogonally rotated eigenvectors. Within the constraint of orthogonality they approximately locate the two point clusters, although the variance of the first rotated principal component is no longer maximum since the projections onto $\tilde{e}_1$ of the three points with $x_1 < 0$ are quite small. However, the interpretation of the two features is enhanced in the maps of the two eigenvectors on the right, with $\tilde{e}_1$ indicating large positive x_1 together with modest but positive x_2, whereas $\tilde{e}_2$ shows large positive x_2 together with modestly negative x_1. The idealizations in Figures 11.11a and 11.11b are meant to correspond to the real-world maps in Figures 11.4 and 11.10, respectively.

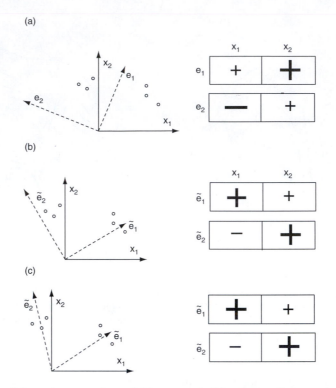

FIGURE 11.11   Schematic comparison of (a) unrotated, (b) orthogonally rotated, and (c) obliquely rotated unit-length eigenvectors in $K = 2$ dimensions. Left panels show eigenvectors in relation to scatterplots of the data, which exhibit two groups or modes. Right panels show schematic two-point maps of the two eigenvectors in each case. After Karl and Koscielny (1982).

Finally, Figure 11.11c illustrates an oblique rotation, where the resulting rotated eigenvectors are no longer constrained to be orthogonal. Accordingly they have more flexibility in their orientations, and can better accommodate features in the data that are not orthogonal.

11.5.3 Sensitivity of Orthogonal Rotation to Initial Eigenvector Scaling

An underappreciated aspect of orthogonal eigenvector rotation is that the orthogonality of the result may depend strongly on the scaling of the original eigenvectors before rotation (Jolliffe 1995, 2002; Mestas-Nuñez, 2000). This dependence is usually surprising because of the name orthogonal rotation, which derives from the orthogonality of the transformation matrix [T] in Equation 11.22; that is, $[T]^T[T] = [T][T]^T = [I]$. The confusion is multiplied because of the incorrect assertion in a number of papers that an orthogonal rotation produces both orthogonal rotated eigenvectors and uncorrelated rotated principal components. At most one of these two results are obtained by an orthogonal rotation, but neither will occur unless the eigenvectors are scaled correctly before the rotation matrix is calculated. Because of the confusion about the issue, an explicit analysis of this counterintuitive phenomenon is worthwhile.

Denote as [E] the possibly truncated $(K \times M)$ matrix of eigenvectors of [S]. Because these eigenvectors are orthogonal (Equation 9.48) and are originally scaled to unit length, the matrix [E] is orthogonal, and so satisfies Equation 9.42b. The resulting principal components can be arranged in the matrix

$$\underset{(n \times M)}{[U]} = \underset{(n \times K)(K \times M)}{[X] \ [E]} \ , \tag{11.24}$$

each of the n rows of which contain values for the M retained principal components, u_m^T. As before, [X] is the original data matrix whose K columns correspond to the n observations on each of the original K variables. The uncorrelatedness of the unrotated principal components can be diagnosed by calculating their covariance matrix,

$$\begin{aligned}
\underset{(M \times M)}{(n-1)^{-1}[U]^T[U]} &= (n-1)^{-1}([X][E])^T[X][E] \\
&= (n-1)^{-1}[E]^T[X]^T[X][E] \\
&= [E]^T([E][\Lambda][E]^T)[E] = [I][\Lambda][I] \\
&= [\Lambda].
\end{aligned} \tag{11.25}$$

The u_m are uncorrelated because their covariance matrix $[\Lambda]$ is diagonal, and the variance for each u_m is λ_m. The steps on the third line of Equation 11.25 follow from the diagonalization of $[S] = (n-1)^{-1}[X]^T[X]$ (Equation 9.50a), and the orthogonality of the matrix [E].

Consider now the effects of the three eigenvector scalings listed in Table 11.3 on the results of an orthogonal rotation. In the first case, the original eigenvectors are not rescaled from unit length, so the matrix of rotated eigenvectors is simply

$$\underset{(K \times M)}{[\tilde{E}]} = \underset{(K \times M)(M \times M)}{[E] \ [T]} \ . \tag{11.26}$$

That these rotated eigenvectors are still orthogonal, as expected, can be diagnosed by calculating

$$\begin{aligned}
[\tilde{E}]^T[\tilde{E}] &= ([E][T])^T[E][T] = [T]^T[E]^T[E][T] \\
&= [T]^T[I][T] = [T]^T[T] = [I].
\end{aligned} \tag{11.27}$$

That is, the resulting rotated eigenvectors are still mutually perpendicular and of unit length. The corresponding rotated principal components are

$$[\tilde{U}] = [X][\tilde{E}] = [X][E][T], \tag{11.28}$$

and their covariance matrix is

$$\begin{aligned}
(n-1)^{-1}[\tilde{U}]^T[\tilde{U}] &= (n-1)^{-1}([X][E][T])^T[X][E][T] \\
&= (n-1)^{-1}[T]^T[E]^T[X]^T[X][E][T] \\
&= [T]^T[E]^T([E][\Lambda][E]^T)[E][T] \\
&= [T]^T[I][\Lambda][I][T] \\
&= [T]^T[\Lambda][T].
\end{aligned} \tag{11.29}$$

This matrix is not diagonal, reflecting the fact that the rotated principal components are no longer uncorrelated. This result is easy to appreciate geometrically, by looking at scatterplots such as Figure 11.1 or Figure 11.3. In each of these cases the point cloud is inclined relative to the original (x_1, x_2) axes, and the angle of inclination of the long axis of the cloud is located by the first eigenvector. The point cloud is not inclined in the (e_1, e_2) coordinate system defined by the two eigenvectors, reflecting the uncorrelatedness of the unrotated principal components (Equation 11.25). But relative to any other pair of mutually orthogonal axes in the plane, the points would exhibit some inclination, and therefore the projections of the data onto these axes would exhibit some nonzero correlation.

The second eigenvector scaling in Table 11.3, $||e_m|| = (\lambda_m)^{1/2}$, is commonly employed, and indeed is the default scaling in many statistical software packages for rotated principal components. In the notation of this section, employing this scaling is equivalent to rotating the scaled eigenvector matrix $[E][\Lambda]^{1/2}$, yielding the matrix of rotated eigenvalues

$$[\tilde{E}] = ([E][\Lambda]^{1/2})[T]. \tag{11.30}$$

The orthogonality of the rotated eigenvectors in this matrix can be checked by calculating

$$
\begin{aligned}
[\tilde{E}]^T[\tilde{E}] &= ([E][\Lambda]^{1/2}[T])^T[E][\Lambda]^{1/2}[T] \\
&= [T]^T[\Lambda]^{1/2}[E]^T[E][\Lambda]^{1/2}[T] \\
&= [T]^T[\Lambda]^{1/2}[I][\Lambda]^{1/2}[T] = [T]^T[\Lambda][T].
\end{aligned} \tag{11.31}
$$

Here the equality in the second line is valid because the diagonal matrix $[\Lambda]^{1/2}$ is symmetric, so that $[\Lambda]^{1/2} = ([\Lambda]^{1/2})^T$. The rotated eigenvectors corresponding to the second, and frequently used, scaling in Table 11.3 are *not* orthogonal, because the result of Equation 11.31 is not a diagonal matrix. Neither are the corresponding rotated principal components independent. This can be seen by calculating their covariance matrix, which is also not diagonal; that is,

$$
\begin{aligned}
(n-1)^{-1}[\tilde{U}]^T[\tilde{U}] &= (n-1)^{-1}([X][E][\Lambda]^{1/2}[T])^T[X][E][\Lambda]^{1/2}[T] \\
&= (n-1)^{-1}[T]^T[\Lambda]^{1/2}[E]^T[X]^T[X][E][\Lambda]^{1/2}[T] \\
&= [T]^T[\Lambda]^{1/2}[E]^T([E][\Lambda][E]^T)[E][\Lambda]^{1/2}[T] \\
&= [T]^T[\Lambda]^{1/2}[I][\Lambda][I][\Lambda]^{1/2}[T] \\
&= [T]^T[\Lambda]^{1/2}[\Lambda][\Lambda]^{1/2}[T] \\
&= [T]^T[\Lambda]^2[T].
\end{aligned} \tag{11.32}
$$

The third eigenvector scaling in Table 11.3, $||e_m|| = (\lambda_m)^{-1/2}$, is used relatively rarely, although it can be convenient in that it yields unit variance for all the principal components u_m. The resulting rotated eigenvectors are not orthogonal, so that the matrix product

$$
\begin{aligned}
[\tilde{E}]^T[\tilde{E}] &= ([E][\Lambda]^{-1/2}[T])^T[E][\Lambda]^{-1/2}[T] \\
&= [T]^T[\Lambda]^{-1/2}[E]^T[E][\Lambda]^{-1/2}[T] \\
&= [T]^T[\Lambda]^{-1/2}[I][\Lambda]^{-1/2}[T] = [T]^T[\Lambda][T],
\end{aligned} \tag{11.33}
$$

is not diagonal. However, the resulting rotated principal components are uncorrelated, so that their covariance matrix,

$$
\begin{aligned}
(n-1)^{-1}[\tilde{U}]^{T}[\tilde{U}] &= (n-1)^{-1}([X][E][\Lambda]^{-1/2}[T])^{T}[X][E][\Lambda]^{-1/2}[T] \\
&= (n-1)^{-1}[T]^{T}[\Lambda]^{-1/2}[E]^{T}[X]^{T}[X][E][\Lambda]^{-1/2}[T] \\
&= [T]^{T}[\Lambda]^{-1/2}[E]^{T}([E][\Lambda][E]^{T})[E][\Lambda]^{-1/2}[T] \\
&= [T]^{T}[\Lambda]^{-1/2}[I][\Lambda][I][\Lambda]^{-1/2}[T] \\
&= [T]^{T}[\Lambda]^{-1/2}[\Lambda]^{1/2}[\Lambda]^{1/2}[\Lambda]^{-1/2}[T] \\
&= [T]^{T}[I][I][T] = [T]^{T}[T] = [I],
\end{aligned}
\tag{11.34}
$$

is diagonal, and also reflects unit variances for all the rotated principal components.

Most frequently in meteorology and climatology, the eigenvectors in a PCA describe spatial patterns, and the principal components are time series reflecting the importance of the corresponding spatial patterns in the original data. When calculating orthogonally rotated principal components in this context, we can choose to have either orthogonal rotated spatial patterns but correlated rotated principal component time series (by using $||\mathbf{e}_m|| = 1$), or nonorthogonal rotated spatial patterns whose time sequences are mutually uncorrelated (by using $||\mathbf{e}_m|| = (\lambda_m)^{-1/2}$), but not both. It is not clear what the advantage of having neither property (using $||\mathbf{e}_m|| = (\lambda_m)^{1/2}$, as is most often done) might be. Differences in the results for the different scalings will be small if sets of effective multiplets are rotated separately, because their eigenvalues will necessarily be similar in magnitude, resulting in similar lengths for the scaled eigenvectors.

11.6 Computational Considerations

11.6.1 *Direct Extraction of Eigenvalues and Eigenvectors from [S]*

The sample covariance matrix [S] is real and symmetric, and so will always have real-valued and nonnegative eigenvalues. Standard and stable algorithms are available to extract the eigenvalues and eigenvectors for real, symmetric matrices (e.g., Press *et al.* 1986), and this approach can be a very good one for computing a PCA. As noted earlier, it is sometimes preferable to calculate the PCA using the correlation matrix [R], which is also the covariance matrix for the standardized variables. The computational considerations presented in this section are equally appropriate to PCA based on the correlation matrix.

One practical difficulty that can arise is that the required computational time increases very quickly as the dimension of the covariance matrix increases. A typical application of PCA in meteorology or climatology involves a field observed at K grid- or other space-points, at a sequence of n times, where $K \gg n$. The typical conceptualization is in terms of the $(K \times K)$ covariance matrix, which is very large—it is not unusual for K to include thousands of gridpoints. Using currently (2004) available fast workstations, the computer time required to extract this many eigenvalue-eigenvector pairs can be hours or even days. Yet since $K > n$ the covariance matrix is singular, implying that the last $K - n$ of its eigenvalues are zero. It is pointless to calculate numerical approximations to these zero eigenvalues, and their associated arbitrary eigenvectors.

In this situation fortunately it is possible to focus the computational effort on the n nonzero eigenvalues and their associated eigenvectors, using a computational trick (von Storch and Hannoschöck, 1984). Recall that the $(K \times K)$ covariance matrix [S] can be computed from the centered data matrix [X'] using Equation 9.30. Reversing the roles of the time and space points, we also can compute the $(n \times n)$ covariance matrix

$$\underset{(n \times n)}{[S^*]} = \frac{1}{n-1} \underset{(n \times K)\,(K \times n)}{[X'][X']^T}. \tag{11.35}$$

Both [S] and [S*] have the same $\min(n, K)$ nonzero eigenvalues, $\lambda_k = \lambda_k^*$, so the required computational time may be much shorter if they are extracted from the smaller matrix [S*].

The eigenvectors of [S] and [S*] are different, but the leading n (i.e., the meaningful) eigenvectors of [S] can be computed from the eigenvectors $\mathbf{e}_k^*$ of [S*] using

$$\mathbf{e}_k = \frac{[X']^T \mathbf{e}_k^*}{\|[X']^T \mathbf{e}_k^*\|}, \quad k = 1, \ldots, n. \tag{11.36}$$

The dimension of the multiplications in both numerator and denominator are $(K \times n)$ $(n \times 1) = (K \times 1)$, and the role of the denominator is to ensure that the calculated $\mathbf{e}_k$ have unit length.

11.6.2 PCA *via* SVD

The eigenvalues and eigenvectors in a PCA can also be computed using the SVD (singular value decomposition) algorithm (see Section 9.3.5), in two ways. First, as illustrated in Example 9.5, the eigenvalues and eigenvectors of a covariance matrix [S] can be computed through SVD of the matrix $(n-1)^{-1}[X']$, where the centered $(n \times K)$ data matrix [X'] is related to the covariance matrix [S] through Equation 9.30. In this case, the eigenvalues of [S] are the squares of the singular values of $(n-1)^{-1}[X']$—that is, $\lambda_k = \omega_k^2$—and the eigenvectors of [S] are the same as the right singular vectors of $(n-1)^{-1}[X']$—that is, [E] = [R], or $\mathbf{e}_k = \mathbf{r}_k$.

An advantage of using SVD to compute a PCA in this way is that left singular vectors (the columns of the $(n \times K)$ matrix [L] in Equation 9.68) are proportional to the principal components (i.e., to the projections of the centered data vectors $\mathbf{x}_i'$ onto the eigenvectors $\mathbf{e}_k$). In particular,

$$u_{i,k} = \mathbf{e}_k^T \mathbf{x}_i' = \sqrt{n-1}\, \ell_{i,k} \sqrt{\lambda_k}, \quad i = 1, \ldots, n, \quad k = 1, \ldots, K; \tag{11.37a}$$

or

$$\underset{(n \times K)}{[U]} = \sqrt{n-1}\, \underset{(n \times K)\,(K \times K)}{[L]\,[\Lambda]^{1/2}}. \tag{11.37b}$$

Here the matrix [U] is used in the same sense as in Section 11.5.3, that is, each of its K columns contains the principal component series u_k corresponding to the sequence of n data values x_i, $i = 1, \ldots, n$.

The SVD algorithm can also be used to compute a PCA by operating on the covariance matrix directly. Comparing the spectral decomposition of a square, symmetric matrix (Equation 9.50a) with its SVD (Equation 9.68), it is clear that these unique decompositions are one in the same. In particular, since a covariance matrix [S] is square and symmetric, both the left and right matrices of its SVD are equal, and contain the eigenvectors; that is, [E] = [L] = [R]. In addition, the diagonal matrix of singular values is exactly the diagonal matrix of eigenvalues, $[\Lambda] = [\Omega]$.

11.7 Some Additional Uses of PCA

11.7.1 Singular Spectrum Analysis (SSA): Time-Series PCA

Principal component analysis can also be applied to scalar or multivariate time series. This approach to time-series analysis is known both as singular spectrum analysis and singular systems analysis (SSA, in either case). Fuller developments of SSA than is presented here can be found in Broomhead and King (1986), Elsner and Tsonis (1996), Golyandina *et al.* (2001), Vautard *et al.* (1992), and Vautard (1995).

SSA is easiest to see in terms of a scalar time series $x_t, t = 1, \ldots, n$; although the generalization to multivariate time series of a vector x_t is reasonably straightforward. As a variant of PCA, SSA involves extraction of eigenvalues and eigenvectors from a covariance matrix. This covariance matrix is calculated from a scalar time series by passing a delay window, or imposing an embedding dimension, of length M on the time series. The process is illustrated in Figure 11.12. For $M = 3$, the first M-dimensional data vector, $x_{(1)}$ is composed of the first three members of the scalar time series, $x_{(2)}$ is composed of the second three members of the scalar time series, and so on, yielding a total of $n - M + 1$ overlapping lagged data vectors.

If the time series x_t is covariance stationary, that is, if its mean, variance, and lagged correlations do not change through time, the $(M \times M)$ population covariance matrix of the lagged time-series vectors $x_{(t)}$ takes on a special banded structure known as Toeplitz, in which the elements $\sigma_{i,j} = \gamma_{|i-j|} = E[x_t' \, x_{t+|i-j|}']$ are arranged in diagonal parallel bands. That is, the elements of the resulting covariance matrix are taken from the autocovariance function (Equation 3.33), with lags arranged in increasing order away from the main diagonal. All the elements of the main diagonal are $\sigma_{i,i} = \gamma_0$; that is, the variance. The elements adjacent to the main diagonal are all equal to γ_1, reflecting the fact that, for example, the covariance between the first and second elements of the vectors $x_{(t)}$ in Figure 11.12 is the same as the covariance between the second and third elements. The elements separated from the main diagonal by one position are all equal to γ_2, and so on. Because of edge effects at the beginnings and ends of sample time series, the sample covariance matrix may be only approximately Toeplitz, although the diagonally banded Toeplitz structure is sometimes enforced before calculation of the SSA (Allen and Smith 1996; Elsner and Tsonis 1996).

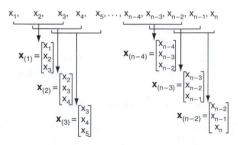

FIGURE 11.12   Illustration of the construction of the vector time series $x_{(t)}, t = 1, \ldots, n - M + 1$, by passing a delay window of embedding dimension $M = 3$ over consecutive members of the scalar time series x_t.

Since SSA is a PCA, the same mathematical considerations apply. In particular, the principal components are linear combinations of the data according to the eigenvectors (Equations 11.1 and 11.2). The analysis operation can be reversed to synthesize, or approximate, the data from all (Equation 11.15) or some (Equation 11.16) of the principal components. What makes SSA different follows from the different nature of the data, and the implications of that different nature on interpretation of the eigenvectors and principal components. In particular, the data vectors are fragments of time series rather than the more usual spatial distribution of values at a single time, so that the eigenvectors in SSA represent characteristic time patterns exhibited by the data, rather than characteristic spatial patterns. Accordingly, the eigenvectors in SSA are sometimes called T-EOFs. Since the overlapping time series fragments x_t themselves occur in a time sequence, the principal components also have a time ordering, as in Equation 11.11. These temporal principal components u_m, or T-PCs, index the degree to which the corresponding time-series fragment x_t resembles the corresponding T-EOF, e_m. Because the data are consecutive fragments of the original time series, the principal components are weighted averages of these time-series segments, with the weights given by the T-EOF elements. The T-PCs are mutually uncorrelated, but in general will exhibit temporal correlations.

The analogy between SSA and Fourier analysis of time series is especially strong, with the T-EOFs corresponding to the sine and cosine functions, and the T-PCs corresponding to the amplitudes. However, there are two major differences. First, the orthogonal basis functions in a Fourier decomposition are the fixed sinusoidal functions, whereas the basis functions in SSA are the data-adaptive T-EOFs. Similarly, the Fourier amplitudes are time-independent constants, but their counterparts, the T-PCs, are themselves functions of time. Therefore SSA can represent time variations that may be localized in time, and so not necessarily recurring throughout the time series.

In common with Fourier analysis, SSA can detect and represent oscillatory or quasi-oscillatory features in the underlying time series. A periodic or quasi-periodic feature in a time series is represented in SSA by pairs of T-PCs and their corresponding eigenvectors. These pairs have eigenvalues that are equal or nearly equal. The characteristic time patterns represented by these pairs of eigenvectors have the same (or very similar) shape, but are offset in time by a quarter cycle (as are a pair of sine and cosine functions). But unlike the sine and cosine functions these pairs of T-EOFs take on shapes that are determined by the time patterns in the underlying data. A common motivation for using SSA is to search, on an exploratory basis, for possible periodicities in time series, which periodicities may be intermittent and/or nonsinusoidal in form. Features of this kind are indeed identified by a SSA, but false periodicities arising only from sampling variations may also easily occur in the analysis (Allen and Smith 1996).

An important consideration in SSA is choice of the window length or embedding dimension, M. Obviously the analysis cannot represent variations longer than this length, although choosing too large a value results in a small sample size, $n - M + 1$, from which to estimate the covariance matrix. Also, the computational effort increases quickly as M increases. Usual rules of thumb are that an adequate sample size may be achieved for $M < n/3$, and that the analysis will be successful in representing time variations with periods between $M/5$ and M.

EXAMPLE 11.3   SSA for an AR(2) Series

Figure 11.13 shows an $n = 100$-point realization from the AR(2) process (Equation 8.27) with parameters $\phi_1 = 0.9$, $\phi_2 = -0.6$, $\mu = 0$, and $\sigma_\varepsilon = 1$. This is a purely random series, but the parameters ϕ_1 and ϕ_2 have been chosen in a way that allows the process to exhibit pseudoperiodicities. That is, there is a tendency for the series to oscillate, although the oscillations are irregular with respect to their frequency and phase. The spectral density

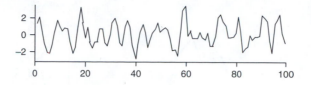

FIGURE 11.13   A realization from an AR(2) process with $\phi_1 = 0.9$ and $\phi_2 = -0.6$.

function for this AR(2) process, included in Figure 8.21, shows a maximum centered near $f = 0.15$, corresponding to a typical period near $\tau = 1/f \approx 6.7$ time steps.

Analyzing the series using SSA requires choosing a delay window length, M, that should be long enough to capture the feature of interest yet short enough for reasonably stable covariance estimates to be calculated. Combining the rules of thumb for the window length, $M/5 < \tau < M < n/3$, a plausible choice is $M = 10$. This choice yields $n - M + 1 = 91$ overlapping time series fragments x_t of length $M = 10$.

Calculating the covariances for this sample of 91 data vectors x_t in the conventional way yields the (10×10) matrix

$$[S] = \begin{bmatrix}
1.792 \\
0.955 & 1.813 \\
-0.184 & 0.958 & 1.795 \\
-0.819 & -0.207 & 0.935 & 1.800 \\
-0.716 & -0.851 & -0.222 & 0.959 & 1.843 \\
-0.149 & -0.657 & -0.780 & -0.222 & 0.903 & 1.805 \\
0.079 & -0.079 & -0.575 & -0.783 & -0.291 & 0.867 & 1.773 \\
0.008 & 0.146 & -0.011 & -0.588 & -0.854 & -0.293 & 0.873 & 1.809 \\
-0.199 & 0.010 & 0.146 & -0.013 & -0.590 & -0.850 & -0.289 & 0.877 & 1.809 \\
-0.149 & -0.245 & -0.044 & 0.148 & 0.033 & -0.566 & -0.828 & -0.292 & 0.874 & 1.794
\end{bmatrix}$$

$$(11.38)$$

For clarity, only the elements in the lower triangle of this symmetric matrix have been printed. Because of edge effects in the finite sample, this covariance matrix is approximately, but not exactly, Toeplitz. The 10 elements on the main diagonal are only approximately equal, and each are estimating the lag-0 autocovariance $\gamma_0 = \sigma_x^2 \approx 1.80$. Similarly, the nine elements on the second diagonal are approximately equal, with each estimating the lag-1 autocovariance $\gamma_0 \approx 0.91$, the eight elements on the third diagonal estimate the lag-2 autocovariance $\gamma_2 \approx -0.25$, and so on. The pseudoperiodicity in the data is reflected in the large negative autocovariance at three lags, and the subsequent damped oscillation in the autocovariance function.

Figure 11.14 shows the leading four eigenvectors of the covariance matrix in Equation 11.38, and their associated eigenvalues. The first two of these eigenvectors (see Figure 11.14a), which are associated with nearly equal eigenvalues, are very similar in shape and are separated by approximately a quarter of the period τ corresponding to the middle of the spectral peak in Figure 8.21. Jointly they represent the dominant feature of the data series in Figure 11.13, namely the pseudoperiodic behavior, with successive peaks and crests tending to be separated by six or seven time units.

The third and fourth T-EOFs in Figure 11.14b represent other, nonperiodic aspects of the time series in Figure 11.13. Unlike the leading T-EOFs in Figure 11.14a, they are not offset images of each other, and do not have nearly equal eigenvalues. Jointly the

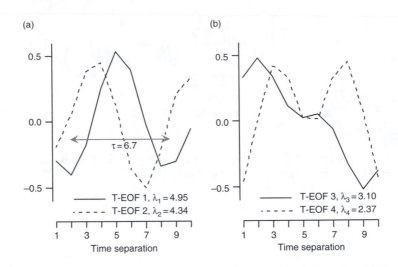

FIGURE 11.14   (a) First two eigenvectors of the covariance matrix in Equation 11.38, and (b) the third and fourth eigenvectors.

four patterns in Figure 11.14 represent 83.5% of the variance within the 10-element time series fragments (but not including variance associated with longer time scales). ◊

It is conceptually straightforward to extend SSA to simultaneous analysis of multiple (i.e., vector) time series, which is called multichannel SSA, or MSSA (Plaut and Vautard 1994; Vautard 1995). The relationship between SSA and MSSA parallels that between an ordinary PCA for a single field and simultaneous PCA for multiple fields as described in Section 11.2.2. The multiple channels in a MSSA might be the K gridpoints representing a spatial field at time t, in which case the time series fragments corresponding to the delay window length M would be coded into a $(KM \times 1)$ vector x_t, yielding a $(KM \times KM)$ covariance matrix from which to extract space-time eigenvalues and eigenvectors (ST-EOFs). The dimension of such a matrix may become unmanageable, and one solution (Plaut and Vautard 1994) can be to first calculate an ordinary PCA for the spatial fields, and then subject the first few principal components to the MSSA. In this case each channel corresponds to one of the spatial principal components calculated in the initial data compression step. Vautard (1995), and Vautard *et al.* (1996, 1999) describe MSSA-based forecasts of fields constructed by forecasting the space-time principal components, and then reconstituting the forecast fields through a truncated synthesis.

11.7.2 *Principal-Component Regression*

A pathology that can occur in multiple linear regression (see Section 6.2.8) is that a set of predictor variables having strong mutual correlations can result in the calculation of an unstable regression relationship, in the sense that the sampling distributions of the estimated regression parameters may have very high variances. The problem can be appreciated in the context of Equation 9.40, for the covariance matrix of the joint sampling distribution of the estimated regression parameters. This equation depends on the inverse of the matrix $[X]^T[X]$, which is proportional to the covariance matrix $[S_x]$ of the predictors. Very strong intercorrelations among the predictors leads to their covariance matrix (and thus also $[X]^T[X]$) being nearly singular, or small in the sense that its determinant is near zero. The inverse, $([X]^T[X])^{-1}$ is then large, and inflates the covariance matrix $[S_b]$ in Equation 9.40. The result is that the estimated regression parameters may be very far from their correct values as a consequence of sampling variations, leading the fitted regression equation to perform poorly on independent data: the prediction intervals (based upon Equation 9.41) are also inflated.

An approach to remedying this problem is first to transform the predictors to their principal components, the correlations among which are zero. The resulting principal-component regression is convenient to work with, because the uncorrelated predictors can be added to or taken out of a tentative regression equation at will without affecting the contributions of the other principal-component predictors. If all the principal components are retained in a principal-component regression, then nothing is gained over the conventional least-squares fit to the full predictor set, but Jolliffe (2002) shows that multi-colinearities, if present, are associated with the principal components having the smallest eigenvalues. As a consequence, the effects of the multicolinearities, and in particular the inflated covariance matrix for the estimated parameters, can be removed by truncating the last principal components associated with the very small eigenvalues. However, in practice the appropriate principal-component truncation may not be so straightforward, in common with conventional least-squares regression.

There are problems that may be associated with principal-component regression. Unless the principal components that are retained as predictors are interpretable in the context of the problem being analyzed, the insight to be gained from the regression may be limited. It is possible to reexpress the principal-component regression in terms of the original predictors using the synthesis equation (Equation 11.6), but the result will in general involve all the original predictor variables even if only one or a few principal component predictors have been used. (This reconstituted regression will be biased, although often the variance is much smaller, resulting in a smaller MSE overall.) Finally, there is no guarantee that the leading principal components provide the best predictive information. If it is the small-variance principal components that are most strongly related to the predictand, then computational instability cannot be removed in this way without also removing the corresponding contributions to the predictability.

11.7.3 The Biplot

It was noted in Section 3.6 that graphical EDA for high-dimensional data is especially difficult. Since principal component analysis excels at data compression using the minimum number of dimensions, it is natural to think about applying PCA to EDA. The biplot, originated by Gabriel (1971), is such a tool. The "bi-" in biplot refers to the simultaneous representation of the n rows (the observations) and the K columns (the variables) of a data matrix, $[X]$.

The biplot is a two-dimensional graph, whose axes are the first two eigenvectors of $[S_x]$. The biplot represents the n observations as their projections onto the plane defined by these two eigenvectors; that is, as the scatterplot of the first two principal components. To the extent that $(\lambda_1 + \lambda_2)/\Sigma_k \lambda_k \approx 1$, this scatterplot will be a close approximation to their relationships, in a graphable two-dimensional space. Exploratory inspection of the data plotted in this way may reveal such aspects of the data as the points clustering into natural groups, or time sequences of points that are organized into coherent trajectories in the plane of the plot.

The other element of the biplot is the simultaneous representation of the K variables. Each of the coordinate axes of the K-dimensional data space defined by the variables can be thought of as a unit basis vector indicating the direction of the corresponding variable; that is, $\boldsymbol{b}_1^T = [1, 0, 0, \ldots, 0], \boldsymbol{b}_2^T = [0, 1, 0, \ldots, 0], \ldots, \boldsymbol{b}_K^T = [0, 0, 0, \ldots, 1]$. These basis vectors can also be projected onto the two leading eigenvectors defining the plane of the biplot; that is,

$$\mathbf{e}_1^T \mathbf{b}_k = \sum_{k=1}^K e_{1,k} b_k \tag{11.39a}$$

and

$$\mathbf{e}_2^T \mathbf{b}_k = \sum_{k=1}^K e_{2,k} b_k. \tag{11.39b}$$

Since each of the elements of each of the basis vectors $\boldsymbol{b}_k$ are zero except for the kth, these dot products are simply the kth elements of the two eigenvectors. Therefore, each of the K basis vectors $\boldsymbol{b}_k$ is located on the biplot by coordinates given by the corresponding eigenvector elements. Because both the data values and their original coordinate axes are both projected in the same way, the biplot amounts to a projection of the full K-dimensional scatterplot of the data onto the plane defined by the two leading eigenvectors.

Figure 11.15 shows a biplot for the $K = 6$ dimensional January 1987 data in Table A.1, after standardization to zero mean and unit variance. The PCA for these data is given in Table 11.1b. The projections of the six original basis vectors (plotted longer than the actual projections in Equation 11.39 for clarity, but with the correct relative magnitudes) are indicated by the line segments diverging from the origin. P, N, and X indicate precipitation, minimum temperature, and maximum temperature, respectively; and the subscripts I and C indicate Ithaca and Canandaigua. It is immediately evident that the pairs of lines corresponding to like variables at the two locations are oriented nearly in the same directions, and that the temperature variables are oriented nearly perpendicularly to the precipitation variables. Approximately (because the variance described is 92% rather than 100%), the correlations among these six variables are equal to the cosines of the angles between the corresponding lines in the biplot (cf. Table 3.5), so the variables oriented in very similar directions form natural groupings.

The scatter of the n data points not only portrays their K-dimensional behavior in a potentially understandable way, their interpretation is informed further by their relationship to the orientations of the variables. In Figure 11.15 most of the points are oriented nearly horizontally, with a slight inclination that is about midway between the angles of the minimum and maximum temperature variables, and perpendicular to the precipitation variables. These are the days corresponding to small or zero precipitation, whose main variability characteristics relate to temperature differences. They are mainly located below the origin, because the mean precipitation is a bit above zero, and the precipitation variables are oriented nearly vertically (i.e., correspond closely to the second principal component). Points toward the right of the diagram, that are oriented similarly to the temperature variables, represent relatively warm days (with little or no precipitation), whereas points to the left are the cold days. Focusing on the dates for the coldest days, we can see that these occurred in a single run, toward the end of the month. Finally, the scatter of data points indicates that the few values in the upper portion of the biplot are different from the remaining observations, but it is the simultaneous display of the variables that allows us to see that these result from large positive values for precipitation.

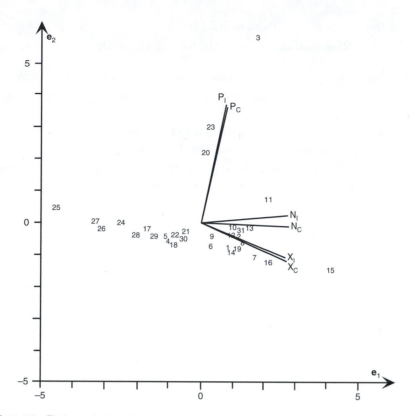

FIGURE 11.15 Biplot of the January 1987 data in Table A.1, after standardization. P = precipitation, X = maximum temperature, and N = minimum temperature. Numbered points refer to the corresponding calendar dates. The plot is a projection of the full six-dimensional scatterplot onto the plane defined by the first two principal components.

11.8 Exercises

11.1. Using information from Exercise 9.6,

a. Calculate the values of the first principal components for 1 January and for 2 January.

b. Estimate the variance of all 31 values of the first principal component.

c. What proportion of the total variability of the maximum temperature data is represented by the first principal component?

11.2. A principal component analysis of the data in Table A.3 yields the three eigenvectors $e_1^T = [.593, .552, -.587]$, $e_2^T = [.332, -.831, -.446]$, and $e_3^T = [.734, -.069, .676]$, where the three elements in each vector pertain to the temperature, precipitation, and pressure data, respectively. The corresponding three eigenvalues are $\lambda_1 = 2.476$, $\lambda_2 = 0.356$, and $\lambda_3 = 0.169$.

a. Was this analysis done using the covariance matrix or the correlation matrix? How can you tell?

b. How many principal components should be retained according to Kaiser's rule, Jolliffe's modification, and the broken stick model?

c. Reconstruct the data for 1951, using a synthesis truncated after the first two principal components.

11.3. Use the information in Exercise 11.2 to

   a. Compute 95% confidence intervals for the eigenvalues, assuming large samples and multinormal data.
   b. Examine the eigenvalue separation using the North *et al*. rule of thumb.

11.4. Using the information in Exercise 11.2, calculate the eigenvector matrix [E] to be orthogonally rotated if

   a. The resulting rotated eigenvectors are to be orthogonal.
   b. The resulting principal components are to be uncorrelated.

11.5. Use the SVD in Equation 9.70 to find the first three values of the first principal component of the minimum temperature data in Table A.1.

11.6. Construct a biplot for the data in Table A.3, using the information in Exercise 11.2.

CHAPTER · 12

Canonical Correlation Analysis (CCA)

12.1 Basics of CCA

12.1.1 Overview

Canonical correlation analysis (CCA) is a statistical technique that identifies a sequence of pairs of patterns in two multivariate data sets, and constructs sets of transformed variables by projecting the original data onto these patterns. The approach thus bears some similarity to PCA, which searches for patterns within a single multivariate data set that represent maximum amounts of the variation in the data. In CCA, the patterns are chosen such that the new variables defined by projection of the two data sets onto them exhibit maximum correlation, while being uncorrelated with the projections of the data onto any of the other identified patterns. That is, CCA identifies new variables that maximize the interrelationships between two data sets, in contrast to the patterns describing the internal variability within a single data set identified in PCA. It is this sense that CCA has been referred to as a double-barreled PCA.

Canonical correlation analysis can also be viewed as an extension of multiple regression to the case of a vector-valued predictand variable y (Glahn, 1968). Ordinary multiple regression finds a weighted average, or pattern, of the vector of predictor variables x such that the correlation between the dot product $b^T x$ and the scalar predictand y is maximized. Here the elements of the vector b are the ordinary least-squares regression coefficients computed using the methods described in Section 6.1, and $b^T x$ is a new variable called the predicted value of y, or $\hat{y}$. Canonical correlation analysis looks for pairs of sets of weights analogous to the regression coefficients, such that the correlations between the new variables defined by the respective dot products with x and (the vector) y are maximized.

As is also the case with PCA, CCA has been most widely applied to geophysical data in the form of fields. In this setting the vector x contains observations of one variable at a collection of gridpoints or locations, and the vector y contains observations of a different variable at a set of locations that may or may not be the same as those represented in x. Typically the data consists of time series of observations of the two fields. When individual observations of the fields x and y are made simultaneously, a CCA can be useful in diagnosing aspects of the coupled variability of the two fields (e.g., Nicholls 1987). When observations of x precede observations of y in time, the CCA may lead to statistical forecasts of the y field using the x field as a predictor (e.g., Barnston and Ropelewski 1992). A more comprehensive comparison between CCA and PCA in the context of atmospheric data analysis is included in Bretherton *et al.* (1992).

12.1.2 *Canonical Variates, Canonical Vectors, and Canonical Correlations*

CCA extracts relationships between pairs of data vectors x and y that are contained in their joint covariance matrix. To compute this matrix, the two centered data vectors are concatenated into a single vector $c'^T = [x'^T, y'^T]$. This partitioned vector contains $I + J$ elements, the first I of which are the elements of x', and the last J of which are the elements of y'. The $((I + J) \times (I + J))$ covariance matrix of c', $[S_C]$, is then partitioned into four blocks, in a manner similar to that done for the correlation matrix in Figure 11.5. That is,

$$[S_C] = \frac{1}{n-1}[C]^T[C] = \begin{bmatrix} [S_{xx}] & [S_{xy}] \\ [S_{yx}] & [S_{yy}] \end{bmatrix}. \tag{12.1}$$

Each of the n rows of the $(n \times (I + J))$ matrix $[C]$ contains one observation of the vector x' and one observation of the vector y', with the primes indicating centering of the data by subtraction of each of the respective sample means. The $(I \times I)$ matrix $[S_{xx}]$ is the variance-covariance matrix of the I variables in x. The $(J \times J)$ matrix $[S_{yy}]$ is the variance-covariance matrix of the J variables in y. The matrices $[S_{xy}]$ and $[S_{yx}]$ contain the covariances between all combinations of the elements of x and the elements of y, and are related according to $[S_{xy}] = [S_{yx}]^T$.

A CCA transforms pairs of original centered data vectors x' and y' into sets of new variables, called canonical variates, v_m and w_m, defined by the dot products

$$v_m = \mathbf{a}_m^T x' = \sum_{i=1}^{I} a_{m,i} x_i', \quad m = 1, \ldots, \min(I, J); \tag{12.2a}$$

and

$$w_m = \mathbf{b}_m^T y' = \sum_{j=1}^{J} b_{m,j} y_j', \quad m = 1, \ldots, \min(I, J). \tag{12.2b}$$

This construction of the canonical variates is similar to that of the principal components u_m (Equation 11.1), in that each is a linear combination (a sort of weighted average) of elements of the respective data vectors x' and y'. These vectors of weights, $\mathbf{a}_m$ and $\mathbf{b}_m$, are called the canonical vectors. One data- and canonical-vector pair need not have the same dimension as the other. The vectors x' and $\mathbf{a}_m$ each have I elements, and the vectors y' and $\mathbf{b}_m$ each have J elements. The number of pairs, M, of canonical variates that can be extracted from the two data sets is equal to the smaller of the dimensions of x and y; that is, $M = \min(I, J)$.

The canonical vectors $\mathbf{a}_m$ and $\mathbf{b}_m$ are the unique choices that result in the canonical variates having the properties

$$\text{Corr}[v_1, w_1] \geq \text{Corr}[v_2, w_2] \geq \cdots \geq \text{Corr}[v_M, w_M] \geq 0; \tag{12.3a}$$

$$\text{Corr}[v_k, w_m] = \begin{cases} r_{C_m}, & k = m \\ 0, & k \neq m \end{cases}; \tag{12.3b}$$

and

$$\text{Var}[v_m] = \mathbf{a}_m^T[S_{x,x}]\mathbf{a}_m = \text{Var}[w_m] = \mathbf{b}_m^T[S_{y,y}]\mathbf{b}_m = 1, \quad m = 1, \dots, M. \quad (12.3c)$$

Equation 12.3a states that each of the M successive pairs of canonical variates exhibits no greater correlation than the previous pair. These (Pearson product-moment) correlations between the pairs of canonical variates are called the canonical correlations, r_C. The canonical correlations can always be expressed as positive numbers, since either $\mathbf{a}_m$ or $\mathbf{b}_m$, can be multiplied by −1 if necessary. Equation 12.3b states that each canonical variate is uncorrelated with all the other canonical variates except its specific counterpart in the mth pair. Finally, Equation 12.3c states that each of the canonical variates has unit variance. Some restriction on the lengths of $\mathbf{a}_m$ and $\mathbf{b}_m$ is required for definiteness, and choosing these lengths to yield unit variances for the canonical variates turns out to be convenient for some applications. Accordingly, the joint $(2M \times 2M)$ covariance matrix for the resulting canonical variates then takes on the simple and interesting form

$$\text{Var}\left(\begin{bmatrix} v \\ w \end{bmatrix}\right) = \begin{bmatrix} [S_v] & [S_{vw}] \\ [S_{wv}] & [S_v] \end{bmatrix} = \begin{bmatrix} [I] & [R_C] \\ [R_C] & [I] \end{bmatrix}, \quad (12.4a)$$

where $[R_C]$ is the diagonal matrix of the canonical correlations,

$$[R_C] = \begin{bmatrix} r_{C_1} & 0 & 0 & \cdots & 0 \\ 0 & r_{C_2} & 0 & \cdots & 0 \\ 0 & 0 & r_{C_3} & \cdots & 0 \\ \vdots & \vdots & \vdots & & \vdots \\ 0 & 0 & 0 & \cdots & r_{C_M} \end{bmatrix}. \quad (12.4b)$$

The definition of the canonical vectors is reminiscent of PCA, which finds a new orthonormal basis for a single multivariate data set (the eigenvectors of its covariance matrix), subject to a variance maximizing constraint. In CCA, two new bases are defined by the canonical vectors $\mathbf{a}_m$ and $\mathbf{b}_m$. However, these basis vectors are neither orthogonal nor of unit length. The canonical variates are the projections of the centered data vectors $\mathbf{x}'$ and $\mathbf{y}'$ onto the canonical vectors, and can be expressed in matrix form through the analysis formulae

$$\underset{(M \times 1)}{\mathbf{v}} = \underset{(M \times I)}{[A]} \quad \underset{(I \times 1)}{\mathbf{x}'} \quad (12.5a)$$

and

$$\underset{(M \times 1)}{\mathbf{w}} = \underset{(M \times J)}{[B]} \quad \underset{(J \times 1)}{\mathbf{y}'}. \quad (12.5b)$$

Here the rows of the matrices $[A]$ and $[B]$ are the transposes of the $M = \min(I, J)$ canonical vectors, $\mathbf{a}_m$ and $\mathbf{b}_m$, respectively. Exposition of how the canonical vectors are calculated from the joint covariance matrix (Equation 12.1) will be deferred to Section 12.3.

12.1.3 Some Additional Properties of CCA

Unlike the case of PCA, calculating a CCA on the basis of standardized (unit variance) variables yields results that are simple functions of the results from an unstandardized analysis. In particular, because the variables x_i' and y_j' in Equation 12.2 would be divided by their respective standard deviations, the corresponding elements of the canonical vectors would be larger by factors of those standard deviations. In particular, if $\boldsymbol{a}_m$ is the mth canonical ($I \times 1$) vector for the $\boldsymbol{x}$ variables, its counterpart $\boldsymbol{a}_m{}^*$ in a CCA of the standardized variables would be

$$\mathbf{a}_m^* = \mathbf{a}_m[D_x], \tag{12.6}$$

where the ($I \times I$) diagonal matrix $[D_x]$ (Equation 9.31) contains the standard deviations of the $\boldsymbol{x}$ variables, and a similar equation would hold for the canonical vectors $\boldsymbol{b}_m$ and the ($J \times J$) diagonal matrix $[D_y]$ containing the standard deviations of the y variables. Regardless of whether a CCA is computed using standardized or unstandardized variables, the resulting canonical correlations are the same.

Correlations between the original and canonical variables can be calculated easily. The correlations between corresponding original and canonical variables, sometimes called homogeneous correlations, are given by

$$\underset{(1 \times I)}{\text{corr}[v_m, \mathbf{x}^T]} = \underset{(1 \times I)}{\mathbf{a}_m^T} \underset{(I \times I)}{[S_{x,x}]} \underset{(I \times I)}{[D_x]^{-1}} \tag{12.7a}$$

and

$$\underset{(1 \times J)}{\text{corr}[w_m, \mathbf{y}^T]} = \underset{(1 \times J)}{\mathbf{b}_m^T} \underset{(J \times J)}{[S_{y,y}]} \underset{(J \times J)}{[D_y]^{-1}}. \tag{12.7b}$$

These equations specify vectors of correlations, between the mth canonical variable v_m and each of the I original variables x_i, and between the canonical variable w_m and each of the J original variables y_k. Similarly, the vectors of heterogeneous correlations, between the canonical variables and the other original variables are

$$\underset{(1 \times J)}{\text{corr}[v_m, \mathbf{y}^T]} = \underset{(1 \times I)}{\mathbf{a}_m^T} \underset{(I \times J)}{[S_{x,y}]} \underset{(J \times J)}{[D_y]^{-1}} \tag{12.8a}$$

and

$$\underset{(1 \times J)}{\text{corr}[w_m, \mathbf{x}^T]} = \underset{(1 \times J)}{\mathbf{b}_m^T} \underset{(J \times I)}{[S_{y,x}]} \underset{(I \times I)}{[D_x]^{-1}}. \tag{12.8b}$$

The canonical vectors $\boldsymbol{a}_m$ and $\boldsymbol{b}_m$ are chosen to maximize correlations between the resulting canonical variates $\boldsymbol{v}$ and $\boldsymbol{w}$, but (unlike PCA) may or may not be particularly effective at summarizing the variances of the original variables $\boldsymbol{x}$ and $\boldsymbol{y}$. If canonical pairs with high correlations turn out to represent small fractions of the underlying variability, their physical significance may be limited. Therefore, it is often worthwhile to calculate the variance proportions R_m^2 captured by each of the leading canonical variables for its underlying original variable.

How well the canonical variables represent the underlying variability is related to how accurately the underlying variables can be synthesized from the canonical variables. Solving the analysis equations (Equation 12.5) yields the CCA synthesis equations

$$\underset{(I \times 1)}{\mathbf{x}'} = \underset{(I \times I)}{[\tilde{A}]^{-1}} \underset{(I \times 1)}{\mathbf{v}} \tag{12.9a}$$

and

$$\mathbf{y}' = [\tilde{B}]^{-1} \mathbf{w} \ . \tag{12.9b}$$
$$\underset{(J \times 1)}{} \underset{(J \times J)}{} \underset{(J \times 1)}{}$$

If $I = J$ (i.e., if the dimensions of the data vectors $\mathbf{x}$ and $\mathbf{y}$ are equal), then the matrices [A] and [B], whose rows are the corresponding M canonical vectors, are both square. In this case $[\tilde{A}] = [A]$ and $[\tilde{B}] = [B]$ in Equation 12.9, and the indicated matrix inversions can be calculated. If $I \neq J$ then one of the matrices [A] or [B] is nonsquare, and so not invertable. In that case, the last $M - J$ rows of [A] (if $I > J$), or the last $M - I$ rows of [B] (if $I < J$), are filled out with the "phantom" canonical vectors corresponding to the zero eigenvalues, as described in Section 12.3.

Equation 12.9 describes the synthesis of individual observations of $\mathbf{x}$ and $\mathbf{y}$ on the basis of their corresponding canonical variables. In matrix form (i.e., for the full set of n observations), these become

$$[X']^{T} = [\tilde{A}]^{-1} [V]^{T} \tag{12.10a}$$
$$\underset{(I \times n)}{} \underset{(I \times I)}{} \underset{(I \times n)}{}$$

and

$$[Y']^{T} = [\tilde{B}]^{-1} [W]^{T}. \tag{12.10b}$$
$$\underset{(J \times n)}{} \underset{(J \times J)}{} \underset{(J \times n)}{}$$

Because the covariance matrices of the canonical variates are $(n-1)^{-1} [V]^{T}[V] = [I]$ and $(n-1)^{-1} [W]^{T}[W] = [I]$ (cf. Equation 12.4a), substituting Equation 12.10 into Equation 9.30 yields

$$[S_{x,x}] = \frac{1}{n-1}[X']^{T}[X] = [\tilde{A}]^{-1}([\tilde{A}]^{-1})^{T} = \sum_{m=1}^{I} \tilde{\mathbf{a}}_{m} \tilde{\mathbf{a}}_{m}^{T} \tag{12.11a}$$

and

$$[S_{y,y}] = \frac{1}{n-1}[Y']^{T}[Y] = [\tilde{B}]^{-1}([\tilde{B}]^{-1})^{T} = \sum_{m=1}^{J} \tilde{\mathbf{b}}_{m} \tilde{\mathbf{b}}_{m}^{T}, \tag{12.11b}$$

where the canonical vectors with tilde accents indicate *columns* of the *inverses* of the corresponding matrices. These decompositions are akin to the spectral decompositions (Equation 9.51a) of the two covariance matrices. Accordingly, the proportions of the variance of $\mathbf{x}$ and $\mathbf{y}$ represented by their m^{th} canonical variables are

$$R_m^2(\mathbf{x}) = \frac{\text{tr}(\tilde{\mathbf{a}}_m \tilde{\mathbf{a}}_m^T)}{\text{tr}([S_{x,x}])} \tag{12.12a}$$

and

$$R_m^2(\mathbf{y}) = \frac{\text{tr}(\tilde{\mathbf{b}}_m \tilde{\mathbf{b}}_m^T)}{\text{tr}([S_{y,y}])}. \tag{12.12b}$$

EXAMPLE 12.1 CCA of the January 1987 Temperature Data

A simple illustration of the mechanics of a small CCA can be provided by again analyzing the January 1987 temperature data for Ithaca and Canandaigua, New York, given in

Table A.1. Let the $I = 2$ Ithaca temperature variables be $x = [T_{max}, T_{min}]^T$, and similarly let the $J = 2$ Canandaigua temperature variables be y. The joint covariance matrix $[S_C]$ of these quantities is then the (4×4) matrix

$$[S_C] = \begin{bmatrix} 59.516 & 75.433 & 58.070 & 51.697 \\ 75.433 & 185.467 & 81.633 & 110.800 \\ 58.070 & 81.633 & 61.847 & 56.119 \\ 51.697 & 110.800 & 56.119 & 77.581 \end{bmatrix}. \tag{12.13}$$

This symmetric matrix contains the sample variances of the four variables on the diagonal, and the covariances between the variables in the other positions. It is related to the corresponding elements of the correlation matrix involving the same variables (see Table 3.5) through the square roots of the diagonal elements: dividing each element by the square root of the diagonal elements in its row and column produces the corresponding correlation matrix. This operation is shown in matrix notation in Equation 9.31.

Since $I = J = 2$, there are $M = 2$ canonical vectors for each of the two data sets being correlated. These are shown in Table 12.1, although the details of their computation will be left until Example 12.3. The first element of each pertains to the respective maximum temperature variable, and the second elements pertain to the minimum temperature variables. The correlation between the first pair of projections of the data onto these vectors, v_1 and w_1, is $r_{C_1} = 0.969$; and the second canonical correlation, between v_2 and w_2, is $r_{C_2} = 0.770$.

Each of the canonical vectors defines a direction in the two-dimensional data space, but their absolute magnitudes are meaningful only in that they produce unit variances for their corresponding canonical variates. However, the relative magnitudes of the canonical vector elements can be interpreted in terms of which linear combinations of one underlying data vector are is most correlated with which linear combination of the other. All the elements of a_1 and b_1 are positive, reflecting positive correlations among all four temperature variables; although the elements corresponding to the maximum temperatures are larger, reflecting the larger correlation between them than between the minima (cf. Table 3.5). The pairs of elements in a_2 and b_2 are comparable in magnitude but opposite in sign, suggesting that the next most important pair of linear combinations with respect to correlation relate to the diurnal ranges at the two locations (recall that the signs of the canonical vectors are arbitrary, and chosen to produce positive canonical correlations; reversing the signs on the second canonical vectors would put positive weights on the maxima and negative weights of comparable magnitudes on the minima).

TABLE 12.1   The canonical vectors a_m (corresponding to Ithaca temperatures) and b_m (corresponding to Canandaigua temperatures) for the partition of the covariance matrix in Equation 12.13 with $I = J = 2$. Also shown are the eigenvalues λ_m (cf. Example 12.3) and the canonical correlations, which are their square roots.

	a_1 (Ithaca)	b_1 (Canandaigua)	a_2 (Ithaca)	b_2 (Canandaigua)
T_{max}	.0923	.0946	−.1618	−.1952
T_{min}	.0263	.0338	.1022	.1907
λ_m	0.938		0.593	
$r_{C_m} = \sqrt{\lambda_m}$	0.969		0.770	

The time series of the first pair of canonical variables is given by the dot products of a_1 and b_1 with the pairs of centered temperature values for Ithaca and Canandaigua, respectively, from Table A.1. The value of v_1 for 1 January would be constructed as $(33 - 29.87)(.0923) + (19 - 13.00)(.0263) = .447$. The time series of v_1 (pertaining to the Ithaca temperatures) would consist of the 31 values (one for each day): .447, .512, .249, −.449, −.686, . . . , −.041, .644. Similarly, the time series for w_1 (pertaining to the Canandaigua temperatures) is .474, .663, .028, −.304, −.310, . . . , −.283, .683. Each of these first two canonical variables are scalar indices of the general warmth at its respective location, with more emphasis on the maximum temperatures. Both series have unit sample variance. The first canonical correlation coefficient, $r_{C_1} = 0.969$, is the correlation between this first pair of canonical variables, v_1 and w_1, and is the largest possible correlation between pairs of linear combinations of these two data sets.

Similarly, the time series of v_2 is .107, .882, .899, −1.290, −.132, . . . , −.225, .354; and the time series of w_2 is 1.046, .656, 1.446, .306, −.461, . . . , −1.038, −.688. Both of these series also have unit sample variance, and their correlation is $r_{C_2} = 0.767$. On each of the $n = 31$ days, (the negatives of) these second canonical variates provide an approximate index of the diurnal temperature ranges at the corresponding locations.

The homogeneous correlations (Equation 12.7) for the leading canonical variates, v_1 and w_1, are

$$\text{corr}[v_1, \mathbf{x}^T] = [.0923, .0263] \begin{bmatrix} 59.516 & 75.433 \\ 75.433 & 185.467 \end{bmatrix} \begin{bmatrix} .1296 & 0 \\ 0 & .0734 \end{bmatrix} = [.969, .869] \quad (12.14a)$$

and

$$\text{corr}[w_1, \mathbf{y}^T] = [.0946, .0338] \begin{bmatrix} 61.847 & 56.119 \\ 56.119 & 77.581 \end{bmatrix} \begin{bmatrix} .1272 & 0 \\ 0 & .1135 \end{bmatrix} = [.985, .900]. \quad (12.14b)$$

All the four homogeneous correlations are strongly positive, reflecting the strong positive correlations among all four of the variables (see Table 3.5), and the fact that the two leading canonical variables have been constructed with positive weights on all four. The homogeneous correlations for the second canonical variates v_2 and w_2 are calculated in the same way, except that the second canonical vectors a_2^T and b_2^T are used in Equations 12.14a and 12.14b, respectively, yielding $\text{corr}[v_2, \mathbf{x}^T] = [−.249, .495]$, and $\text{corr}[w_2, \mathbf{y}^T] = [−.174, .436]$. The second canonical variables are less strongly correlated with the underlying temperature variables, because the magnitude of the diurnal temperature range is only weakly correlated with the overall temperatures: wide or narrow diurnal ranges can occur on both relatively warm and cool days. However, the diurnal ranges are evidently more strongly correlated with the minimum temperatures, with cooler minima tending to be associated with large diurnal ranges.

Similarly, the heterogeneous correlations (Equation 12.8) for the leading canonical variates are

$$\text{corr}[v_1, \mathbf{y}^T] = [.0923, .0263] \begin{bmatrix} 58.070 & 51.697 \\ 81.633 & 110.800 \end{bmatrix} \begin{bmatrix} .1272 & 0 \\ 0 & .1135 \end{bmatrix} = [.955, .872] \quad (12.15a)$$

and

$$\text{corr}[w_1, \mathbf{x}^T] = [.0946, .0338] \begin{bmatrix} 58.070 & 81.633 \\ 51.697 & 110.800 \end{bmatrix} \begin{bmatrix} .1296 & 0 \\ 0 & .0734 \end{bmatrix} = [.938, .842]. \quad (12.15b)$$

Because of the symmetry of these data (like variables at similar locations), these correlations are very close to the homogeneous correlations in Equation 12.14. Similarly, the heterogeneous correlations for the second canonical vectors are also close to their homogeneous counterparts: $\text{corr}[v_2, y^T] = [-.132, .333]$, and $\text{corr}[w_2, x^T] = [-.191, .381]$.

Finally the variance fractions for the temperature data at each of the two locations that are described by the canonical variates depend, through the synthesis equations (Equation 12.9), on the matrices [A] and [B], whose rows are the canonical vectors. Because $I = J$,

$$[\tilde{A}] = [A] = \begin{bmatrix} .0923 & .0263 \\ -.1618 & .1022 \end{bmatrix}, \quad \text{and} \quad [\tilde{B}] = [B] = \begin{bmatrix} .0946 & .0338 \\ -.1952 & .1907 \end{bmatrix}; \tag{12.16a}$$

so that

$$[\tilde{A}]^{-1} = \begin{bmatrix} 7.466 & -1.921 \\ 11.820 & 6.743 \end{bmatrix}, \quad \text{and} \quad [\tilde{B}]^{-1} = \begin{bmatrix} 7.740 & -1.372 \\ 7.923 & 3.840 \end{bmatrix}. \tag{12.16b}$$

Contributions made by the canonical variates to the respective covariance matrices for the underlying data depend on the outer products of the columns of these matrices (terms in the summations of Equations 12.11); that is,

$$\tilde{a}_1 \tilde{a}_1^T = \begin{bmatrix} 7.466 \\ 11.820 \end{bmatrix} [7.466, 11.820] = \begin{bmatrix} 55.74 & 88.25 \\ 88.25 & 139.71 \end{bmatrix}, \tag{12.17a}$$

$$\tilde{a}_2 \tilde{a}_2^T = \begin{bmatrix} -1.921 \\ 6.743 \end{bmatrix} [-1.921, 6.743] = \begin{bmatrix} 3.690 & -12.95 \\ -12.95 & 45.47 \end{bmatrix}, \tag{12.17b}$$

$$\bar{b}_1 \bar{b}_1^T = \begin{bmatrix} 7.740 \\ 7.923 \end{bmatrix} [7.740, 7.923] = \begin{bmatrix} 59.91 & 61.36 \\ 61.36 & 62.77 \end{bmatrix}, \tag{12.17c}$$

$$\tilde{b}_2 \tilde{b}_2^T = \begin{bmatrix} -1.372 \\ 3.840 \end{bmatrix} [-1.372, 3.840] = \begin{bmatrix} 1.882 & 5.279 \\ 5.279 & 14.75 \end{bmatrix}. \tag{12.17d}$$

Therefore the proportions of the Ithaca temperature variance described by its two canonical variates (Equation 12.12a) are

$$R_1^2(\mathbf{x}) = \frac{55.74 + 139.71}{59.52 + 185.47} = 0.798 \tag{12.18a}$$

and

$$R_2^2(\mathbf{x}) = \frac{3.690 + 45.47}{59.52 + 185.47} = 0.202, \tag{12.18b}$$

and the corresponding variance fractions for Canandaigua are

$$R_1^2(\mathbf{y}) = \frac{59.91 + 62.77}{61.85 + 77.58} = 0.880, \tag{12.19a}$$

and

$$R_2^2(\mathbf{y}) = \frac{1.882 + 14.75}{61.85 + 77.58} = 0.120. \tag{12.19b}$$

◇

12.2 CCA Applied to Fields

12.2.1 Translating Canonical Vectors to Maps

Canonical correlation analysis is usually most interesting for atmospheric data when applied to fields. Here the spatially distributed observations (either at gridpoints or observing locations) are encoded into the vectors x and y in the same way as for PCA. That is, even though the data may pertain to a two- or three-dimensional field, each location is numbered sequentially and pertains to one element of the corresponding data vector. It is not necessary for the spatial domains encoded into x and y to be the same, and indeed in the applications of CCA that have appeared in the literature they are usually different.

As is the case with the use of PCA with spatial data, it is often informative to plot maps of the canonical vectors by associating the magnitudes of their elements and the geographic locations to which they pertain. In this context the canonical vectors are sometimes called canonical patterns, since the resulting maps show spatial patterns of the ways in which the original variables contribute to the canonical variables. Examining the pairs of maps formed by corresponding vectors a_m and b_m can be informative about the nature of the relationship between variations in the data over the two domains encoded in x and y, respectively. Figures 12.2 and 12.3 show examples of maps of canonical vectors.

It can also be informative to plot pairs of maps of the homogeneous (Equation 12.7) or heterogeneous correlations (Equation 12.8). Each of these vectors contain correlations between an underlying data field and one of the canonical variables, and these correlations can also be plotted at the corresponding locations. Figure 12.1, from Wallace *et al.* (1992), shows one such pair of homogeneous correlation patterns. Figure 12.1a shows the spatial distribution of correlations between a canonical variable v, and the values of the corresponding data x that contains values of average December-February sea-surface temperatures (SSTs) in the north Pacific Ocean. This canonical variable accounts for 18% of the total variance of the SSTs in the data set analyzed (Equation 12.12). Figure 12.1b shows the spatial distribution of the correlations for the corresponding canonical variable w, that pertains to average hemispheric 500 mb heights y during the same winters included in the SST data in x. This canonical variable accounts for 23% of the total variance of the winter hemispheric height variations. The correlation pattern in Figure 12.1a corresponds to either cold water in the central north Pacific and warm water along the west coast of North America, or warm water in the central north Pacific and cold water along the west coast of North America. The pattern of 500 mb height correlations in Figure 12.1b is remarkably similar to the PNA pattern (cf. Figures 11.10b and 3.28).

The correlation between the two time series v and w is the canonical correlation $r_C = 0.79$. Because v and w are well correlated, these figures indicate that cold SSTs in the central Pacific simultaneously with warm SSTs in the northeast Pacific (relatively large positive v) tend to coincide with a 500 mb ridge over northwestern North America and a 500 mb trough over southeastern North America (relatively large positive w). Similarly, warm water in the central north Pacific and cold water in the northwestern Pacific (relatively large negative v) are associated with the more zonal PNA flow (relatively large negative w).

12.2.2 Combining CCA with PCA

The sampling properties of CCA can be poor when the available data are few relative to the dimensionality of the data vectors. The result can be that sample estimates for

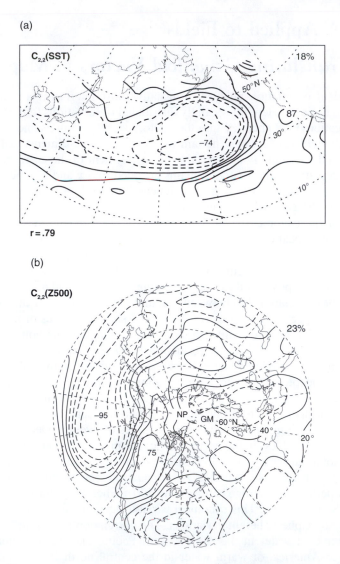

FIGURE 12.1   Homogeneous correlation maps for a pair of canonical variables pertaining to (a) average winter sea-surface temperatures (SSTs) in the northern Pacific Ocean, and (b) hemispheric winter
500 mb heights. The pattern of SST correlation in the left-hand panel (and its negative) are associated with the PNA pattern of 500 mb height correlations shown in the right-hand panel. The canonical
correlation for this pair of canonical variables is 0.79. From Wallace *et al.* (1992).

CCA parameters may be unstable (i.e., exhibit large variations from batch to batch)
for small samples (e.g., Bretherton *et al.* 1992; Cherry 1996; Friedrerichs and Hense
2003). Friedrerichs and Hense (2003) describe, in the context of atmospheric data, both
conventional parametric tests and resampling tests to help assess whether sample canonical
correlations may be spurious sampling artifacts. These tests examine the null hypothesis
that all the underlying population canonical correlations are zero.

Relatively small sample sizes are common when analyzing time series of atmospheric
fields. In CCA, it is not uncommon for there to be fewer observations n than the
dimensions I and J of the data vectors, in which case the necessary matrix inversions

cannot be computed (see Section 12.3). However, even if the sample sizes are large enough to carry through the calculations, sample CCA statistics are erratic unless $n \gg M$. Barnett and Preisendorfer (1987) suggested that a remedy for this problem is to prefilter the two fields of raw data using separate PCAs before subjecting them to a CCA, and this has become a conventional procedure. Rather than directly correlating linear combinations of the fields x' and y', the CCA operates on the vectors u_x and u_y, which consist of the leading principal components of x' and y'. The truncations for these two PCAs (i.e., the dimensions of the vectors u_x and u_y) need not be the same, but should be severe enough for the larger of the two to be substantially smaller than the sample size n. Livezey and Smith (1999) provide some guidance for the subjective choices that need to be made in this approach.

This combined PCA/CCA approach is not always best, and can be inferior if important information is discarded when truncating the PCA. In particular, there is no guarantee that the most strongly correlated linear combinations of x and y will be well related to the leading principal components of one field or the other.

12.2.3 Forecasting with CCA

When one of the fields, say x, is observed prior to y, and some of the canonical correlations between the two are large, it is natural to use CCA as a purely statistical forecasting method. In this case the entire $(I \times 1)$ field $x(t)$ is used to forecast the $(J \times 1)$ field $y(t + \tau)$, where τ is the time lag between the two fields in the training data, which becomes the forecast lead time. In applications with atmospheric data it is typical that there are too few observations n relative to the dimensions I and J of the fields for stable sample estimates (which are especially important for out-of-sample forecasting) to be calculated, even if $n > \max(I, J)$ so that the calculations can be performed. It is therefore usual for both the x and y fields to be represented by series of separately calculated truncated principal components, as described in the previous section. However, in order not to clutter the notation in this section, the mathematical development will be expressed in terms of the original variables x and y, rather than their principal components u_x and u_y.

The basic idea behind forecasting with CCA is straightforward: simple linear regressions are constructed that relate the predictand canonical variates w_m to the predictor canonical variates v_m,

$$w_m = \hat{\beta}_{0,m} + \hat{\beta}_{1,m} v_m, \quad m = 1, \dots, M. \tag{12.20}$$

Here the estimated regression coefficients are indicated by the β's in order to distinguish clearly from the canonical vectors b, and the number of canonical pairs considered can be any number up to the smaller of the numbers of principal components retained for the x and y fields. These regressions are all simple linear regressions, because canonical variables from different canonical pairs are uncorrelated (Equation 12.3b).

Parameter estimation for the regressions in Equation 12.20 is straightforward also. Using Equation 6.7a for the regression slopes,

$$\hat{\beta}_{1,m} = \frac{n \, \text{cov}(v_m, w_m)}{n \, \text{var}(v_m)} = \frac{n \, s_v s_w r_{v,w}}{n \, s_v^2} = r_{v,w} = r_{C_m}, \quad m = 1, \dots, M. \tag{12.21}$$

That is, because the canonical variates are scaled to have unit variance (Equation 12.3c), the regression slopes are simply equal to the corresponding canonical correlations. Similarly, Equation 6.7b yields the regression intercepts

$$\hat{\beta}_0 = \bar{w}_m - \hat{\beta}_1 \bar{v}_m = b_m^T E(y') + \hat{\beta}_1 a_m^T E(x') = 0, \quad m = 1, \dots, M. \tag{12.22}$$

That is, because the CCA is calculated from the centered data x' and y' whose mean vectors are both $\mathbf{0}$, the averages of the canonical variables v_m and w_m are both zero, so that all the intercepts in Equation 12.20 are also zero. Equation 12.22 also holds when the CCA has been calculated from a principal component truncation of the original (centered) variables, because $E(u_x) = E(u_y) = \mathbf{0}$.

Once the CCA has been fit, the basic forecast procedure is as follows. First, centered values for the predictor field x' (or its first few principal components, u_x) are used in Equation 12.5a to calculate the M canonical variates v_m to be used as regression predictors. Combining Equations 12.20 through 12.22, the $(M \times 1)$ vector of predictand canonical variates is forecast to be

$$\hat{\mathbf{w}} = [R_C]\mathbf{v}, \tag{12.23}$$

where $[R_C]$ is the diagonal $(M \times M)$ matrix of the canonical correlations. In general, the forecast map $\hat{y}$ will need to be synthesized from its predicted canonical variates using Equation 12.9b, in order to see the forecast in a physically meaningful way. However, in order to be invertable, the matrix $[B]$, whose rows are the predictand canonical vectors b_m^T, must be square. This condition implies that the number of regressions M in Equation 12.20 needs to be equal to the dimensionality of y (or, more usually, to the number of predictand principal components that have been retained), although the dimension of x (or the number of predictor principal components retained) is not constrained in this way. If the CCA has been calculated using predictand principal components u_y, the centered predicted values $\hat{y}'$ are next recovered with the PCA synthesis, Equation 11.6. Finally, the full predicted field is produced by adding back its mean vector. If the CCA has been computed using standardized variables, so that Equation 12.1 is a correlation matrix, the dimensional values of the predicted variables need to be reconstructed by multiplying by the appropriate standard deviation before adding the appropriate mean (i.e., by reversing Equation 3.21 or Equation 4.26 to yield Equation 4.28).

EXAMPLE 12.2   An Operational CCA Forecast System

Canonical correlation is used as one of the elements of the seasonal forecasts produced operationally at the U.S. Climate Prediction Center (Barnston *et al.* 1999). The predictands are seasonal (three-month) average temperature and total precipitation over the United States, made at lead times of 0.5 through 12.5 months.

The CCA forecasts contributing to this system are modified from the procedure described in Barnston (1994), whose temperature forecast procedure will be outlined in this example. The (59×1) predictand vector y represents temperature forecasts jointly at 59 locations in the conterminous United States. The predictors x consist of global sea-surface temperatures (SSTs) discretized to a 235-point grid, northern hemisphere 700 mb heights discretized to a 358-point grid, and previously observed temperatures at the 59 prediction locations. The predictors are three-month averages also, but in each of the four nonoverlapping three-month seasons for which data would be available preceding the season to be predicted. For example, the predictors for the January-February-March (JFM) forecast, to be made in mid-December, are seasonal averages of SSTs, 700 mb heights, and U.S. surface temperatures during the preceding September-October-November (SON), June-July-August (JJA), March-April-May (MAM), and December-January-February (DJF) seasons, so that the predictor vector x has $4(235 + 358 + 59) = 2608$ elements. Using sequences of four consecutive predictor seasons allows the forecast procedure to incorporate information about the time evolution of the predictor fields.

Since only $n = 37$ years of training data were available when this system was developed, drastic reductions in the dimensionality of both the predictors and predictands was

(a)

(b)

(c)

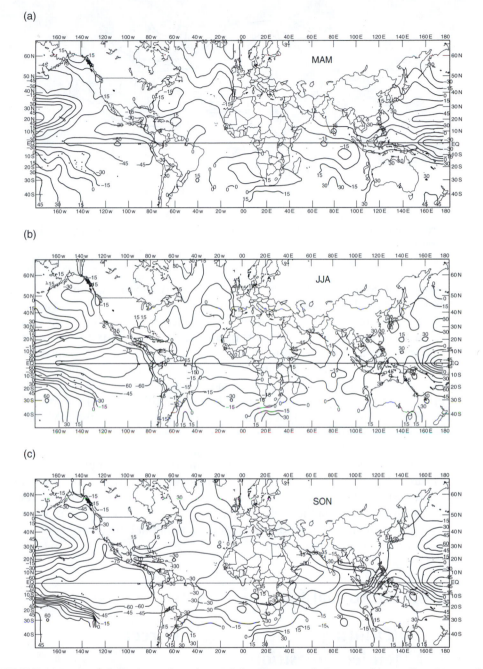

FIGURE 12.2   Spatial displays of portions of the first canonical vector for predictor sea-surface temperatures, in the three seasons preceding the JFM for which U.S. surface temperatures are forecast. The corresponding canonical vector for this predictand is shown in Figure 12.3. From Barnston (1994).

necessary. Separate PCAs were calculated for the predictor and predictand vectors, which retained the leading six predictor principal components $\boldsymbol{u}_x$ and (depending on the forecast season) either five or six predictand principal components $\boldsymbol{u}_y$. The CCAs for these pairs of principal component vectors yield either $M = 5$ or $M = 6$ canonical pairs. Figure 12.2

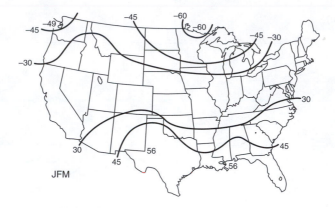

FIGURE 12.3   Spatial display of the first canonical vector for predicted U.S. JFM surface temperatures. A portion of the corresponding canonical vector for the predictors is shown in Figure 12.2. From
Barnston (1994).

shows that portion of the first predictor canonical vector a_1 pertaining to the three seasons
MAM, JJA, and SON, relating to the forecast for the following JFM. That is, each of these
three maps expresses the six elements of a_1 in terms of the original 235 spatial locations,
through the corresponding elements of the eigenvector matrix [E] for the predictor PCA.
The most prominent feature in Figure 12.2 is the progressive evolution of increasingly
negative values in the eastern tropical Pacific, which clearly represents an intensifying
El Niño (warm) event when $v_1 < 0$, and development of a La Niña (cold) event when
$v_1 > 0$, in the spring, summer, and fall before the JFM season to be forecast.

Figure 12.3 shows the first canonical predictand vector for the JFM forecast, b_1, again
projected back to the 59 forecast locations. Because the CCA is constructed to have
positive canonical correlations, a developing El Niño yielding $v_1 < 0$ results in a forecast
$\hat{w}_1 < 0$ (Equation 12.23). The result is that the first canonical pair contributes a tendency
toward relative warmth in the northern United States and relative coolness in the southern
United States during El Niño winters. Conversely, this canonical pair forecasts cold in
the north and warm in the south for La Niña winters. Evolving SSTs not resembling the
patterns in Figure 12.2 would yield $v_1 \approx 0$, resulting in little contribution from the pattern
in Figure 12.3 to the forecast. ◊

12.3 Computational Considerations

12.3.1 Calculating CCA through Direct Eigendecomposition

Finding canonical vectors and canonical correlations requires calculating pairs of eigenvectors e_m, corresponding to the x variables, and eigenvectors f_m, corresponding to the y
variables; together with their corresponding eigenvalues λ_m, which are the same for each
pair e_m and f_m. There are several computational approaches available to find these e_m, f_m,
and $\lambda_m, m = 1, \ldots, M$.

One approach is to find the eigenvectors e_m and f_m of the matrices

$$[S_{xx}]^{-1}[S_{xy}][S_{yy}]^{-1}[S_{yx}]$$

(12.24a)

and

$$[S_{yy}]^{-1}[S_{yx}][S_{xx}]^{-1}[S_{xy}], \qquad (12.24b)$$

respectively. The factors in these equations correspond to the definitions in Equation 12.1. Equation 12.24a is dimensioned $(I \times I)$, and Equation 12.24b is dimensioned $(J \times J)$. The first $M = \min(I, J)$ eigenvalues of these two matrices are equal, and if $I \neq J$, the remaining eigenvalues of the larger matrix are zero. The corresponding "phantom" eigenvectors would fill the extra rows of one of the matrices in Equation 12.9. Equation 12.24 can be difficult computationally because in general these matrices are not symmetric, and algorithms to find eigenvalues and eigenvectors for general matrices are less stable numerically than routines designed specifically for real and symmetric matrices.

The eigenvalue-eigenvector computations are easier and more stable, and the same results are achieved, if the eigenvectors e_m and f_m are calculated from the symmetric matrices

$$[S_{xx}]^{-1/2}[S_{xy}][S_{yy}]^{-1}[S_{yx}][S_{xx}]^{-1/2} \qquad (12.25a)$$

and

$$[S_{yy}]^{-1/2}[S_{yx}][S_{xx}]^{-1}[S_{xy}][S_{yy}]^{-1/2}, \qquad (12.25b)$$

respectively. Equation 12.25a is dimensioned $(I \times I)$, and Equation 12.25b is dimensioned $(J \times J)$. Here the reciprocal square-root matrices must be symmetric (Equation 9.64), and not derived from Cholesky decompositions of the corresponding inverses or obtained by other means. The eigenvalue-eigenvector pairs for the symmetric matrices in Equation 12.25 can be computed using an algorithm specialized to the task, or through the singular value decomposition (Equation 9.68) operating on these matrices. In the latter case, the results are $[E][\Lambda][E]^T$ and $[F][\Lambda][F]^T$, respectively (compare Equations 9.68 and 9.50a), where the columns of $[E]$ are the e_m and the columns of $[F]$ are the f_m.

Regardless of how the eigenvectors e_m and f_m, and their common eigenvalues λ_m, are arrived at, the canonical correlations and canonical vectors are calculated from them. The canonical correlations are simply the positive square roots of the M nonzero eigenvalues,

$$r_{C_m} = \sqrt{\lambda_m}, \quad m = 1, \dots, M. \qquad (12.26)$$

The pairs of canonical vectors are calculated from the corresponding pairs of eigenvectors, using

$$\left. \begin{aligned} \mathbf{a}_m &= [S_{xx}]^{-1/2}\mathbf{e}_m \\[6pt] \mathbf{b}_m &= [S_{yy}]^{-1/2}\mathbf{f}_m \end{aligned} \right\} \quad m = 1, \dots, M. \qquad \begin{aligned} &(12.27a) \\[12pt] &(12.27b) \end{aligned}$$

Since $\|e_m\| = \|f_m\| = 1$, this transformation ensures unit variances for the canonical variates; that is,

$$\operatorname{var}(v_m) = \mathbf{a}_m^T[S_{xx}]\mathbf{a}_m = \mathbf{e}_m^T[S_{xx}]^{-1/2}[S_{xx}][S_{xx}]^{-1/2}\mathbf{e}_m = \mathbf{e}_m^T\mathbf{e}_m = 1, \qquad (12.28)$$

because $[S_{xx}]^{-1/2}$ is symmetric and the eigenvectors e_m are mutually orthogonal. An obvious analogous equation can be written for the variances $\operatorname{var}(w_m)$.

Extraction of eigenvalue-eigenvector pairs from large matrices can require large amounts of computing. However, the eigenvector pairs e_m and f_m are related in a way that makes it unnecessary to compute the eigendecompositions of both Equations 12.25a and 12.25b (or, both Equations 12.24a and 12.24b). For example, each f_m can be computed from the corresponding e_m using

$$\mathbf{f}_m = \frac{[S_{yy}]^{-1/2}[S_{yx}][S_{xx}]^{-1/2}\mathbf{e}_m}{\|[S_{yy}]^{-1/2}[S_{yx}][S_{xx}]^{-1/2}\mathbf{e}_m\|}, \quad m = 1, \ldots, M. \tag{12.29}$$

Here the Euclidean norm in the denominator ensures $\|f_m\| = 1$. The eigenvectors e_m can be calculated from the corresponding f_m by reversing the matrix subscripts in this equation.

12.3.2 Calculating CCA through SVD

The special properties of the singular value decomposition (Equation 9.68) can be used to find both sets of the e_m and f_m pairs, together with the corresponding canonical correlations. This is achieved by computing the SVD

$$\underset{(I\times I)}{[S_{xx}]^{-1/2}}\underset{(I\times J)}{[S_{xy}]}\underset{(J\times J)}{[S_{yy}]^{-1/2}} = \underset{(I\times J)}{[E]}\underset{(J\times J)}{[R_C]}\underset{(J\times J)}{[F]}^{\mathrm{T}}. \tag{12.30}$$

As before the columns of $[E]$ are the e_m, the columns of $[F]$ (not $[F]^{\mathrm{T}}$) are the f_m, and the diagonal matrix $[R_C]$ contains the canonical correlations. Here it has been assumed that $I \geq J$, but if $I < J$ the roles of x and y can be reversed in Equation 12.30. The canonical vectors are calculated as before, using Equation 12.27.

EXAMPLE 12.3   The Computations behind Example 12.1

In Example 12.1 the canonical correlations and canonical vectors were given, with their computations deferred. Since $I = J$ in this example, the matrices required for these calculations are obtained by quartering $[S_C]$ (Equation 12.13) to yield

$$[S_{xx}] = \begin{bmatrix} 59.516 & 75.433 \\ 75.433 & 185.467 \end{bmatrix}, \tag{12.31a}$$

$$[S_{yy}] = \begin{bmatrix} 61.847 & 56.119 \\ 56.119 & 77.581 \end{bmatrix}, \tag{12.31b}$$

and

$$[S_{yx}] = [S_{xy}]^{\mathrm{T}} = \begin{bmatrix} 58.070 & 81.633 \\ 51.697 & 110.800 \end{bmatrix}. \tag{12.31c}$$

The eigenvectors e_m and f_m, respectively, can be computed either from the pair of asymmetric matrices (Equation 12.24)

$$[S_{xx}]^{-1}[S_{xy}][S_{yy}]^{-1}[S_{yx}] = \begin{bmatrix} .830 & .377 \\ .068 & .700 \end{bmatrix} \tag{12.32a}$$

and

$$[S_{yy}]^{-1}[S_{yx}][S_{xx}]^{-1}[S_{xy}] = \begin{bmatrix} .845 & .259 \\ .091 & .686 \end{bmatrix}; \tag{12.32b}$$

or the symmetric matrices (Equation 12.25)

$$[S_{xx}]^{-1/2}[S_{xy}][S_{yy}]^{-1}[S_{yx}][S_{xx}]^{-1/2} = \begin{bmatrix} .768 & .172 \\ .172 & .757 \end{bmatrix} \tag{12.33a}$$

and

$$[S_{yy}]^{-1/2}[S_{yx}][S_{xx}]^{-1}[S_{xy}][S_{yy}]^{-1/2} = \begin{bmatrix} .800 & .168 \\ .168 & .726 \end{bmatrix}. \tag{12.33b}$$

The numerical stability of the computations is better if Equations 12.33a and 12.33b are used, but in either case the eigenvectors of Equations 12.32a and 12.33a are

$$\mathbf{e}_1 = \begin{bmatrix} .719 \\ .695 \end{bmatrix} \quad \text{and} \quad \mathbf{e}_2 = \begin{bmatrix} -.695 \\ .719 \end{bmatrix}, \tag{12.34}$$

with corresponding eigenvalues $\lambda_1 = 0.938$ and $\lambda_2 = 0.593$. The eigenvectors of Equations 12.32b and 12.33b are

$$\mathbf{f}_1 = \begin{bmatrix} .780 \\ .626 \end{bmatrix} \quad \text{and} \quad \mathbf{f}_2 = \begin{bmatrix} -.626 \\ .780 \end{bmatrix}, \tag{12.35}$$

again with eigenvalues $\lambda_1 = 0.938$ and $\lambda_2 = 0.593$. However, once the eigenvectors $\mathbf{e}_1$ and $\mathbf{e}_2$ have been computed it is not necessary to compute the eigendecomposition for either Equation 12.32b or Equation 12.33b, because their eigenvectors can also be obtained through Equation 12.29:

$$\mathbf{f}_1 = \begin{bmatrix} .8781 & .0185 \\ .1788 & .8531 \end{bmatrix}\begin{bmatrix} .719 \\ .695 \end{bmatrix} \bigg/ \left\| \begin{bmatrix} .8781 & .0185 \\ .1788 & .8531 \end{bmatrix}\begin{bmatrix} .719 \\ .695 \end{bmatrix} \right\| = \begin{bmatrix} .780 \\ .626 \end{bmatrix} \tag{12.36a}$$

and

$$\mathbf{f}_2 = \begin{bmatrix} .8781 & .0185 \\ .1788 & .8531 \end{bmatrix}\begin{bmatrix} -.695 \\ .719 \end{bmatrix} \bigg/ \left\| \begin{bmatrix} .8781 & .0185 \\ .1788 & .8531 \end{bmatrix}\begin{bmatrix} -.695 \\ .719 \end{bmatrix} \right\| = \begin{bmatrix} -.626 \\ .780 \end{bmatrix}, \tag{12.36b}$$

since

$$[S_{xx}]^{-1/2}[S_{xy}][S_{yy}]^{-1/2} = \begin{bmatrix} .1788 & -.0522 \\ -.0522 & .0917 \end{bmatrix}\begin{bmatrix} 58.070 & 51.697 \\ 81.633 & 110.800 \end{bmatrix}\begin{bmatrix} .1959 & -.0930 \\ -.0930 & .1699 \end{bmatrix}$$

$$= \begin{bmatrix} .8781 & .0185 \\ .1788 & .8531 \end{bmatrix}. \tag{12.36c}$$

The two canonical correlations are $r_{C_1} = \sqrt{\lambda_1} = 0.969$ and $r_{C_2} = \sqrt{\lambda_2} = 0.770$. The four canonical vectors are

$$\mathbf{a}_1 = [S_{xx}]^{-1/2}\mathbf{e}_1 = \begin{bmatrix} .1788 & -.0522 \\ -.0522 & .0917 \end{bmatrix}\begin{bmatrix} .719 \\ .695 \end{bmatrix} = \begin{bmatrix} .0923 \\ .0263 \end{bmatrix} \tag{12.37a}$$

$$\mathbf{a}_2 = [S_{xx}]^{-1/2}\mathbf{e}_2 = \begin{bmatrix} .1788 & -.0522 \\ -.0522 & .0917 \end{bmatrix}\begin{bmatrix} -.695 \\ .719 \end{bmatrix} = \begin{bmatrix} -.1618 \\ .1022 \end{bmatrix} \tag{12.37b}$$

$$\mathbf{b}_1 = [S_{yy}]^{-1/2}\mathbf{f}_1 = \begin{bmatrix} .1960 & -.0930 \\ -.0930 & .1699 \end{bmatrix}\begin{bmatrix} .780 \\ .626 \end{bmatrix} = \begin{bmatrix} .0946 \\ .0338 \end{bmatrix} \tag{12.37c}$$

and

$$\mathbf{b}_2 = [S_{yy}]^{-1/2}\mathbf{f}_2 = \begin{bmatrix} .1960 & -.0930 \\ -.0930 & .1699 \end{bmatrix} \begin{bmatrix} -.626 \\ .780 \end{bmatrix} = \begin{bmatrix} -.1952 \\ .1907 \end{bmatrix}. \tag{12.37d}$$

Alternatively, the eigenvectors $\mathbf{e}_m$ and $\mathbf{f}_m$ can be obtained through the SVD (Equation 12.30) of the matrix in Equation 12.36c (compare the left-hand sides of these two equations). The result is

$$\begin{bmatrix} .8781 & .0185 \\ .1788 & .8531 \end{bmatrix} = \begin{bmatrix} .719 & -.695 \\ .695 & .719 \end{bmatrix} \begin{bmatrix} .969 & 0 \\ 0 & .770 \end{bmatrix} \begin{bmatrix} .780 & .626 \\ -.626 & .780 \end{bmatrix}. \tag{12.38}$$

The canonical correlations are in the diagonal matrix $[R_C]$ in the middle of Equation 12.38. The eigenvectors are in the matrices [E] and [F] on either side of it, and can be used to compute the corresponding canonical vectors, as in Equation 12.37.◊

12.4 Maximum Covariance Analysis

Maximum covariance analysis is a similar technique to CCA, in that it finds pairs of linear combinations of two sets of vector data $\mathbf{x}$ and $\mathbf{y}$,

$$\left. \begin{aligned} v_m &= \boldsymbol{\ell}_m^T \mathbf{x} \\[4pt] w_m &= \mathbf{r}_m^T \mathbf{y} \end{aligned} \right\} \quad m = 1, \ldots, M; \tag{12.39a} \tag{12.39b}$$

such that their covariances

$$\mathrm{cov}(v_m, w_m) = \boldsymbol{\ell}_m^T [S_{xy}] \mathbf{r}_m \tag{12.40}$$

(rather than their correlations, as in CCA) are maximized, subject to the constraint that the vectors $\boldsymbol{\ell}_m$ and $\mathbf{r}_m$ are orthonormal. As in CCA, the number of such pairs $M = \min(I, J)$ is equal to the smaller of the dimensions of the data vectors $\mathbf{x}$ and $\mathbf{y}$, and each succeeding pair of projection vectors are chosen to maximize covariance, subject to the orthonormality constraint. In a typical application to atmospheric data, $\mathbf{x}(t)$ and $\mathbf{y}(t)$ are both time series of spatial fields, and so their projections in Equation 12.39 form time series also.

Computationally, the vectors $\boldsymbol{\ell}_m$ and $\mathbf{r}_m$ are found through a singular value decomposition (Equation 9.68) of the matrix $[S_{xy}]$ in Equation 12.1, containing the cross-covariances between the elements of $\mathbf{x}$ and $\mathbf{y}$,

$$\underset{(I \times J)}{[S_{xy}]} = \underset{(I \times J)(J \times J)(J \times J)}{[L][\Omega][R]^T}. \tag{12.41}$$

The left singular vectors $\boldsymbol{\ell}_m$ are the columns of the matrix [L] and the right singular vectors $\mathbf{r}_m$ are the columns of the matrix [R] (i.e., the rows of $[R]^T$). The elements ω_m of the diagonal matrix $[\Omega]$ of singular values are the maximized covariances (Equation 12.40) between the pairs of linear combinations in Equation 12.39. Because the machinery of the singular value decomposition is used to find the vectors $\boldsymbol{\ell}_m$ and $\mathbf{r}_m$, and the associated covariances ω_m, maximum covariance analysis sometimes unfortunately is known as SVD analysis; although as illustrated earlier in this chapter and elsewhere in this book, the singular value decomposition has a rather broader range of uses. In recognition of the parallels with CCA, the technique is also sometimes called canonical covariance analysis and the ω_m are sometimes called the canonical covariances.

There are two main distinctions between CCA and maximum covariance analysis. The first is that CCA maximizes correlation, whereas maximum covariance analysis maximizes covariance. The leading CCA modes may capture relatively little of the corresponding variances (and thus yield small covariances even if the canonical correlations are high). On the other hand, maximum covariance analysis will find linear combinations with large covariances, which may result more from large variances than a large correlation. The second difference is that, the vectors ℓ_m and r_m in maximum covariance analysis are orthogonal, and the projections v_m and w_m of the data onto them are in general correlated, whereas the canonical variates in CCA are uncorrelated but the corresponding canonical vectors are not generally orthogonal. Bretherton *et al.* (1992), Cherry (1996), and Wallace *et al.* (1992) compare the two methods in greater detail.

EXAMPLE 12.4   Maximum Covariance Analysis of the January 1987 Temperature Data

Singular value decomposition of the cross-covariance submatrix $[S_{xy}]$ in Equation 12.31c yields

$$\begin{bmatrix} 58.07 & 51.70 \\ 81.63 & 110.8 \end{bmatrix} = \begin{bmatrix} .4876 & .8731 \\ .8731 & -.4876 \end{bmatrix} \begin{bmatrix} 157.4 & 0 \\ 0 & 14.06 \end{bmatrix} \begin{bmatrix} .6325 & .7745 \\ .7745 & -.6325 \end{bmatrix}. \tag{12.42}$$

The results are qualitatively similar to the CCA of the same data in Example 12.1. The first left and right vectors, $\ell_1 = [.4876, .8731]^T$ and $r_1 = [.6325, .7745]^T$, respectively, resemble the first pair of canonical vectors a_1 and b_1 in Example 12.1; in that both put positive weights on both variables in both data sets, but here the weights are closer in magnitude, and emphasize the minimum temperatures rather than the maximum temperatures. The covariance between the linear combinations defined by these vectors is 157.4, which is larger than the covariance between any other pair of linear combinations for these data, subject to $\|\ell_1\| = \|r_1\| = 1$. The corresponding correlation is

$$\begin{aligned} \text{corr}(v_1, w_1) &= \frac{\omega_1}{(\text{var}[v_1]\text{var}[w_1])^{1/2}} = \frac{\omega_1}{(\ell_1^T[S_{xx}]\ell_1)^{1/2}(r_1^T[S_{yy}]r_1)^{1/2}} \\ &= \frac{157.44}{(219.8)^{1/2}(126.3)^{1/2}} = 0.945, \end{aligned} \tag{12.43}$$

which is large, but necessarily smaller than $r_{C_1} = 0.969$ for the CCA of the same data.

The second pair of vectors, $\ell_2 = [.8731, -.4876]^T$ and $r_2 = [.7745, -.6325]^T$, are also similar to the second pair of canonical vectors for the CCA in Example 12.1, in that they also describe a contrast between the maximum and minimum temperatures that can be interpreted as being related to the diurnal temperature ranges. The covariance of the second pair of linear combinations is ω_2, corresponding to a correlation of 0.772. This correlation is slightly larger than the second canonical correlation in Example 12.1, but has not been limited by the CCA constraint that the correlations between v_1 and v_2, and w_1 and w_2 must be zero. ◊

The papers of Bretherton *et al.* (1992) and Wallace *et al.* (1992) have been influential advocates for the use of maximum covariance analysis. One advantage over CCA that sometimes is cited is that no matrix inversions are required, so that a maximum covariance analysis can be computed even if $n < \max(I, J)$. However, both techniques are subject to similar sampling problems in limited-data situations, so it is not clear that this advantage is of practical importance. Some cautions regarding maximum covariance analysis have been offered by Cherry (1997), Hu (1997), and Newman and Sardeshmukh (1995).

12.5 Exercises

12.1. Using the information in Table 12.1 and the data in Table A.1, calculate the values of the canonical variables v_1 and w_1 for 6 January and 7 January.

12.2. The Ithaca maximum and minimum temperatures for 1 January 1988 were $x = [38°F, 16°F]^T$. Use the CCA in Example 12.1 to "forecast" the Canandaigua temperatures for that day.

12.3. Separate PCAs of the correlation matrices for the Ithaca and Canandaigua data in Table A.1 (after square-root transformation of the precipitation data) yields

$$[E_{Ith}] = \begin{bmatrix} .599 & .524 & .606 \\ .691 & .044 & -.721 \\ .404 & -.851 & .336 \end{bmatrix} \quad \text{and} \quad [E_{Can}] = \begin{bmatrix} .702 & .161 & .694 \\ .709 & -.068 & -.702 \\ .066 & -.985 & .161 \end{bmatrix}, \quad (12.44)$$

with corresponding eigenvalues $\lambda_{Ith} = [1.883, 0.927, 0.190]^T$ and $\lambda_{Can} = [1.814, 1.019, 0.168]^T$. Given also the cross-correlations for these data

$$[R_{Can\text{-}Ith}] = \begin{bmatrix} .957 & .762 & .076 \\ .761 & .924 & .358 \\ -.021 & .162 & .742 \end{bmatrix}, \quad (12.45)$$

compute the CCA after truncation to the two leading principal components for each of the locations (and notice that computational simplifications follow from using the principal components), by

a. Computing $[S_C]$, where c is the (4×1) vector $[u_{Ith}, u_{Can}]^T$, and then
b. Finding the canonical vectors and canonical correlations.

CHAPTER ♦ 13

Discrimination and Classification

13.1 Discrimination vs. Classification

This chapter deals with the problem of discerning membership among some number of groups, on the basis of a K-dimensional vector x of attributes that is observed for each member of each group. It is assumed that the number of groups G is known in advance; that this collection of groups constitutes a MECE partition of the sample space; that each data vector belongs to one and only one group; and that a set of training data is available, in which the group membership of each of the data vectors $x_i, i = 1, \ldots, n$, is known with certainty. The related problem, in which we know neither the group membership of the data nor the number of groups overall, is treated in Chapter 14.

The term *discrimination* refers to the process of estimating functions of the training data x_i that best describe the features separating the known group membership of each x_i. In cases where this can be achieved well with three or fewer functions, it may be possible to express the discrimination graphically. The statistical basis of discrimination is the notion that each of the G groups corresponds to a different multivariate PDF for the data, $f_g(x), g = 1, \ldots, G$. It is not necessary to assume multinormality for these distributions, but when this assumption is supported by the data, informative connections can be made with the material presented in Chapter 10.

Classification refers to use of the discrimination rule(s) to assign data that were not part of the original training sample to one of the G groups; or to the estimation of probabilities $p_g(x), g = 1, \ldots, G$, that the observation x belongs to group g. If the groupings of x pertain to a time after x itself has been observed, then classification is a natural tool to use for forecasting discrete events. That is, the forecast is made by classifying the current observation x as belonging to the group that is forecast to occur, or by computing the probabilities $p_g(x)$ for the probabilities of occurrence of each of the G events.

13.2 Separating Two Populations

13.2.1 *Equal Covariance Structure: Fisher's Linear Discriminant*

The simplest form of discriminant analysis involves distinguishing between $G = 2$ groups on the basis of a K-dimensional vector of observations x. A training sample must exist, consisting of n_1 observations of x known to have come from Group 1, and n_2 observations of x known to have come from Group 2. That is, the basic data are the two matrices $[X_1]$, dimensioned $(n_1 \times K)$, and $[X_2]$, dimensioned $(n_2 \times K)$. The goal is to find a linear function of the K elements of the observation vector, that is, the linear combination $a^T x$, called the discriminant function, that will best allow a future K-dimensional vector of observations to be classified as belonging to either Group 1 or Group 2.

Assuming that the two populations corresponding to the groups have the same covariance structure, the approach to this problem taken by the statistician R.A. Fisher was to find the vector a as that direction in the K-dimensional space of the data that maximizes the separation of the two means, in standard deviation units, when the data are projected onto a. This criterion is equivalent to choosing a in order to maximize

$$\frac{(a^T \bar{x}_1 - a^T \bar{x}_2)^2}{a^T [S_{pool}] a}. \tag{13.1}$$

Here the two mean vectors are calculated separately for each group, as would be expected, according to

$$\bar{x}_g = \frac{1}{n_g} [X_g]^T \mathbf{1} = \begin{bmatrix} \frac{1}{n_g} \sum_{i=1}^{n_g} x_{i,1} \\ \frac{1}{n_g} \sum_{i=1}^{n_g} x_{i,2} \\ \vdots \\ \frac{1}{n_g} \sum_{i=1}^{n_g} x_{i,K} \end{bmatrix}, \quad g = 1, 2; \tag{13.2}$$

where $\mathbf{1}$ is a $(n \times 1)$ vector containing only 1's, and n_g is the number of training-data vectors x in each of the two groups. The estimated common covariance matrix for the two groups, $[S_{pool}]$ is calculated using Equation 10.39b. If $n_1 = n_2$, the result is that each element of $[S_{pool}]$ is the simple average of the corresponding elements of $[S_1]$ and $[S_2]$. Note that multivariate normality has not been assumed for either of the groups. Rather, regardless of their distributions and whether or not those distributions are of the same form, all that has been assumed is that their underlying population covariance matrices $[\Sigma_1]$ and $[\Sigma_2]$ are equal.

Finding the direction a maximizing Equation 13.1 reduces the discrimination problem from one of sifting through and comparing relationships among the K elements of the data vectors, to looking at a single number. That is, the data vector x is transformed to a new scalar variable, $\delta_1 = a^T x$, known as Fisher's linear discriminant function. The groups of K-dimensional multivariate data are essentially reduced to groups of univariate data with different means (but equal variances), distributed along the $\boldsymbol{a}$ axis. The discriminant vector locating this direction of maximum separation is given by

$$\mathbf{a} = [\mathbf{S}_{\text{pool}}]^{-1}(\bar{\mathbf{x}}_1 - \bar{\mathbf{x}}_2), \tag{13.3}$$

so that Fisher's linear discriminant function is

$$\delta_1 = \mathbf{a}^{\mathsf{T}}\mathbf{x} = (\bar{\mathbf{x}}_1 - \bar{\mathbf{x}}_2)^{\mathsf{T}}[\mathbf{S}_{\text{pool}}]^{-1}\mathbf{x}. \tag{13.4}$$

As indicated in Equation 13.1, this transformation to Fisher's linear discriminant function maximizes the scaled distance between the two sample means in the training sample, which is

$$\mathbf{a}^{\mathsf{T}}(\bar{\mathbf{x}}_1 - \bar{\mathbf{x}}_2) = (\bar{\mathbf{x}}_1 - \bar{\mathbf{x}}_2)^{\mathsf{T}}[\mathbf{S}_{\text{pool}}]^{-1}(\bar{\mathbf{x}}_1 - \bar{\mathbf{x}}_2) = D^2. \tag{13.5}$$

That is, this maximum distance between the projections of the two sample means is exactly the Mahalanobis distance between them, according to $[\mathbf{S}_{\text{pool}}]$.

A decision to classify a future observation $\boldsymbol{x}$ as belonging to either Group 1 or Group 2 can now be made according to the value of the scalar $\delta_1 = \boldsymbol{a}^{\mathsf{T}}\boldsymbol{x}$. This dot product is a one-dimensional (i.e., scalar) projection of the vector $\boldsymbol{x}$ onto the direction of maximum separation, $\boldsymbol{a}$. The discriminant function δ_1 is essentially a new variable, analogous to the new variable u in PCA and the new variables v and w in CCA, produced as a linear combination of the elements of a data vector $\boldsymbol{x}$. The simplest way to classify an observation $\boldsymbol{x}$ is to assign it to Group 1 if the projection $\boldsymbol{a}^{\mathsf{T}}\boldsymbol{x}$ is closer to the projection of the Group 1 mean onto the direction $\boldsymbol{a}$, and assign it to Group 2 if $\boldsymbol{a}^{\mathsf{T}}\boldsymbol{x}$ is closer to the projection of the mean of Group 2. Along the $\boldsymbol{a}$ axis, the midpoint between the means of the two groups is given by the projection of the average of these two mean vectors onto the vector $\boldsymbol{a}$,

$$\hat{m} = \frac{1}{2}(\mathbf{a}^{\mathsf{T}}\bar{\mathbf{x}}_1 + \mathbf{a}^{\mathsf{T}}\bar{\mathbf{x}}_2) = \frac{1}{2}\mathbf{a}^{\mathsf{T}}(\bar{\mathbf{x}}_1 + \bar{\mathbf{x}}_2) = \frac{1}{2}(\bar{\mathbf{x}}_1 - \bar{\mathbf{x}}_2)^{\mathsf{T}}[\mathbf{S}_{\text{pool}}]^{-1}(\bar{\mathbf{x}}_1 + \bar{\mathbf{x}}_2). \tag{13.6}$$

Given an observation $\boldsymbol{x}_0$ whose group membership is unknown, this simple midpoint criterion classifies it according to the rule

$$\text{Assign } \mathbf{x}_0 \text{ to Group 1 if } \quad \mathbf{a}^{\mathsf{T}}\mathbf{x}_0 \geq \hat{m}, \tag{13.7a}$$

or

$$\text{Assign } \mathbf{x}_0 \text{ to Group 2 if } \quad \mathbf{a}^{\mathsf{T}}\mathbf{x}_0 < \hat{m}. \tag{13.7b}$$

This classification rule divides the K-dimensional space of $\boldsymbol{x}$ into two regions, according to the (hyper-) plane perpendicular to $\boldsymbol{a}$ at the midpoint given by Equation 13.6. In two dimensions, the plane is divided into two regions according to the line perpendicular to $\boldsymbol{a}$ at this point. The volume in three dimensions is divided into two regions according to the plane perpendicular to $\boldsymbol{a}$ at this point, and so on for higher dimensions.

EXAMPLE 13.1 Linear Discrimination in $K = 2$ Dimensions

Table 13.1 shows average July temperature and precipitation for stations in three regions of the United States. The data vectors include $K = 2$ elements each: one temperature element and one precipitation element. Consider the problem of distinguishing between

TABLE 13.1   Average July temperature (°F) and precipitation (inches) for locations in three regions of the United States. Averages are for the period 1951–1980, from Quayle and Presnell (1991).

Group 1: Southeast U.S. (O)			Group 2: Central U.S. (X)			Group 3: Northeast U.S. (+)		
Station	Temp.	Ppt.	Station	Temp.	Ppt.	Station	Temp.	Ppt.
Athens, GA	79.2	5.18	Concordia, KS	79.0	3.37	Albany, NY	71.4	3.00
Atlanta, GA	78.6	4.73	Des Moines, IA	76.3	3.22	Binghamton, NY	68.9	3.48
Augusta, GA	80.6	4.4	Dodge City, KS	80.0	3.08	Boston, MA	73.5	2.68
Gainesville, FL	80.8	6.99	Kansas City, MO	78.5	4.35	Bridgeport, CT	74.0	3.46
Huntsville, AL	79.3	5.05	Lincoln, NE	77.6	3.2	Burlington, VT	69.6	3.43
Jacksonville, FL	81.3	6.54	Springfield, MO	78.8	3.58	Hartford, CT	73.4	3.09
Macon, GA	81.4	4.46	St. Louis, MO	78.9	3.63	Portland, ME	68.1	2.83
Montgomery, AL	81.7	4.78	Topeka, KS	78.6	4.04	Providence, RI	72.5	3.01
Pensacola, FL	82.3	7.18	Wichita, KS	81.4	3.62	Worcester, MA	69.9	3.58
Savannah, GA	81.2	7.37						
Averages:	80.6	5.67		78.7	3.57		71.3	3.17

membership in Group 1 vs. Group 2. This problem might arise if the stations in Table 13.1 represented the core portions of their respective climatic regions, and on the basis of these data we wanted to classify stations not listed in this table as belonging to one or the other of these two groups.

The mean vectors for the $n_1 = 10$ and $n_2 = 9$ data vectors in Groups 1 and 2 are

$$\bar{\mathbf{x}}_1 = \begin{bmatrix} 80.6°F \\ 5.67\,in. \end{bmatrix} \quad \text{and} \quad \bar{\mathbf{x}}_2 = \begin{bmatrix} 78.7°F \\ 3.57\,in. \end{bmatrix}, \tag{13.8a}$$

and the two sample covariance matrices are

$$[S_1] = \begin{bmatrix} 1.47 & 0.65 \\ 0.65 & 1.45 \end{bmatrix} \quad \text{and} \quad [S_2] = \begin{bmatrix} 2.08 & 0.06 \\ 0.06 & 0.17 \end{bmatrix}. \tag{13.8b}$$

Since $n_1 \neq n_2$ the pooled estimate for the common variance-covariance matrix is obtained by the weighted average specified by Equation 10.39b. The vector $\mathbf{a}$ pointing in the direction of maximum separation of the two sample mean vectors is then computed using Equation 13.3 as

$$\mathbf{a} = \begin{bmatrix} 1.76 & 0.37 \\ 0.37 & 0.85 \end{bmatrix}^{-1} \left(\begin{bmatrix} 80.6 \\ 5.67 \end{bmatrix} - \begin{bmatrix} 78.7 \\ 3.57 \end{bmatrix} \right)$$

$$= \begin{bmatrix} 0.625 & -.272 \\ -.272 & 1.295 \end{bmatrix} \begin{bmatrix} 1.9 \\ 2.10 \end{bmatrix} = \begin{bmatrix} 0.62 \\ 2.20 \end{bmatrix}. \tag{13.9}$$

Figure 13.1 illustrates the geometry of this problem. Here the data for the warmer and wetter southeastern stations of Group 1 are plotted as circles, and the central U.S. stations of Group 2 are plotted as Xs. The vector means for the two groups are indicated by the heavy symbols. The projections of these two means onto the direction $\mathbf{a}$, are indicated by the lighter dashed lines. The midpoint between these two projections locates the dividing point between the two groups in the one-dimensional discriminant space defined by $\mathbf{a}$.

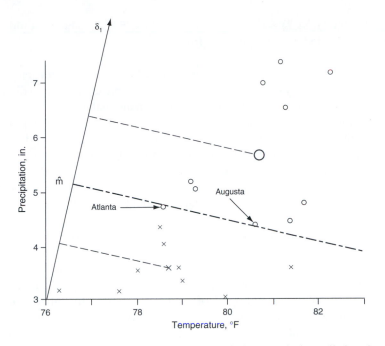

FIGURE 13.1   Illustration of the geometry of linear discriminant analysis applied to the southeastern (circles) and central (Xs) U.S. data in Table 13.1. The (vector) means of the two groups of data are indicated by the heavy symbols, and their projections onto the discriminant function are indicated by the light dashed lines. The midpoint between these two projections, $\hat{m}$, defines the dividing line (heavier dashed line) used to assign future (temperature, precipitation) pairs to the groups. Of this training data, only the data point for Atlanta has been misclassified. Note that the discriminant function has been shifted to the right (i.e., does not pass through the origin, but is parallel to the vector a in Equation 13.9) in order to improve the clarity of the plot, which does not affect the relative positions of the projections of the data points onto it.

The heavy dashed line perpendicular to the discriminant function δ_1 at this point divides the (temperature, precipitation) plane into two regions. Future points of unknown group membership falling above and to the right of this heavy line would be classified as belonging to Group 1, and points falling below and to the left would be classified as belonging to Group 2.

Since the average of the mean vectors for Groups 1 and 2 is $[79.65, 4.62]^T$, the value of the dividing point is $\hat{m} = (0.62)(79.65) + (2.20)(4.62) = 59.55$. Of the 19 points in this training data, only that for Atlanta has been misclassified. For this station, $\delta_1 = a^T x = (0.62)(78.6) + (2.20)(4.73) = 59.14$. Since this value of δ_1 is slightly less than the midpoint value, Atlanta would be falsely classified as belonging to Group 2 (Equation 13.7). By contrast, the point for Augusta lies just to the Group 1 side of the heavy dashed line. For Augusta, $\delta_1 = a^T x = (0.62)(80.6) + (2.20)(4.40) = 59.65$, which is slightly greater than the cutoff value.

Consider now the assignment to either Group 1 or Group 2 of two stations not listed in Table 13.1. For New Orleans, Louisiana, the average July temperature is 82.1°F, and the average July precipitation is 6.73 in. Applying Equation 13.7, we find $a^T x = (0.62)(82.1) + (2.20)(6.73) = 65.71 > 59.55$. Therefore, New Orleans would be classified as belonging to Group 1. Similarly the average July temperature and precipitation for Columbus, Ohio, are 74.7°F and 3.37 in., respectively. For this station,

$a^{\mathrm{T}}x = (0.62)(74.7) + (2.20)(3.37) = 53.73 < 59.55$, which would result in Columbus being classified as belonging to Group 2. ◊

Example 13.1 was constructed with $K = 2$ observations in each data vector in order to allow the geometry of the problem to be easily represented in two dimensions. However, it is not necessary to restrict the use of discriminant analysis to situations with only bivariate observations. In fact, discriminant analysis is potentially most powerful when allowed to operate on higher-dimensional data. For example, it would be possible to extend Example 13.1 to classifying stations according to average temperature and precipitation for all 12 months. If this were done, each data vector x would consist of $K = 24$ values. The discriminant vector a would also consist of $K = 24$ elements, but the dot product $\delta_1 = a^{\mathrm{T}}x$ would still be a single scalar that could be used to classify the group membership of x.

Usually high-dimensional data vectors of atmospheric data exhibit substantial correlation among the K elements, and thus carry some redundant information. For example, the 12 monthly mean temperatures and 12 monthly mean precipitation values are not mutually independent. If only for computational economy, it can be a good idea to reduce the dimensionality of this kind of data before subjecting it to a discriminant analysis. This reduction in dimension is most commonly achieved through a principal component analysis (see Chapter 11). When the groups in discriminant analysis are assumed to have the same covariance structure, it is natural to perform this PCA on the estimate of their common variance-covariance matrix, $[S_{\mathrm{pool}}]$. However, if the dispersion of the group means (as measured by Equation 13.18) is substantially different from $[S_{\mathrm{pool}}]$, its leading principal components may not be good discriminators, and better results might be obtained from a discriminant analysis based on the overall covariance, $[S]$ (Jolliffe 2002). If the data vectors are not of consistent units (some temperatures and some precipitation amounts, for example), it will make more sense to perform the PCA on the corresponding (i.e., pooled) correlation matrix. The discriminant analysis can then be carried out using M-dimensional data vectors composed of elements that are the first M principal components, rather than the original K-dimensional raw data vectors. The resulting discriminant function will then pertain to the principal components in the $(M \times 1)$ vector u, rather than to the original $(K \times 1)$ data, x. In addition, if the first two principal components account for a large fraction of the total variance, the data can effectively be visualized in a plot like Figure 13.1, where the horizontal and vertical axes are the first two principal components.

13.2.2 Fisher's Linear Discriminant for Multivariate Normal Data

Use of Fisher's linear discriminant requires no assumptions about the specific nature of the distributions for the two groups, $f_1(x)$ and $f_2(x)$, except that they have equal covariance matrices. If in addition these are two multivariate normal distributions, or they are sufficiently close to multivariate normal for the sampling distributions of their means to be essentially multivariate normal according to the Central Limit Theorem, there are connections to the Hotelling T^2 test (see Section 10.5) regarding differences between the two means.

In particular, Fisher's linear discriminant vector (Equation 13.3) identifies a direction that is identical to the linear combination of the data that is most strongly significant

(Equation 10.54b), under the null hypothesis that the two population mean vectors are equal. That is, the vector a defines the direction maximizing the separation of the two means for both a discriminant analysis and the T^2 test. Furthermore, the distance between the two means in this direction (Equation 13.5) is their Mahalanobis distance, with respect to the pooled estimate $[S_{pool}]$ of the common covariance $[\Sigma_1] = [\Sigma_2]$, which is proportional (through the factor $n_1^{-1} + n_2^{-1}$, in Equation 10.39a) to the 2-sample T^2 statistic itself (Equation 10.37).

In light of these observations, one way to look at Fisher's linear discriminant, when applied to multivariate normal data, is as an implied test relating to the null hypothesis that $\mu_1 = \mu_2$. Even if this null hypothesis is true, the corresponding sample means in general will be different, and the result of the T^2 test is an informative necessary condition regarding the reasonableness of conducting the discriminant analysis. A multivariate normal distribution is fully defined by its mean vector and covariance matrix. Since $[\Sigma_1] = [\Sigma_2]$ already has been assumed, if in addition the two multivariate normal data groups are consistent with $\mu_1 = \mu_2$, then there is no basis on which to discriminate between them. Note, however, that rejecting the null hypothesis of equal means in the corresponding T^2 test is not a sufficient condition for good discrimination: arbitrarily small mean differences can be detected by this test for increasing sample size, even though the scatter of the two data groups may overlap to such a degree that discrimination is completely pointless.

13.2.3 Minimizing Expected Cost of Misclassification

The point $\hat{m}$ on Fisher's discriminant function halfway between the projections of the two sample means is not always the best point at which to make a separation between groups. One might have prior information that the probability of membership in Group 1 is higher than that for Group 2, perhaps because Group 2 members are rather rare overall. If this is so, it would usually be desirable to move the classification boundary toward the Group 2 mean, with the result that more future observations x would be classified as belonging to Group 1. Similarly, if misclassifying a Group 1 data value as belonging to Group 2 were to be a more serious error than misclassifying a Group 2 data value as belonging to Group 1, again we would want to move the boundary toward the Group 2 mean.

One rational way to accommodate these considerations is to define the classification boundary based on the expected cost of misclassification of a future data vector. Let p_1 be the prior probability (the unconditional probability according to previous information) that a future observation x_0 belongs to Group 1, and let p_2 be the prior probability that the observation x_0 belongs to Group 2. Define $P(2|1)$ to be the conditional probability that a Group 1 object is misclassified as belonging to Group 2, and $P(1|2)$ as the conditional probability that a Group 2 object is classified as belonging to Group 1. These conditional probabilities will depend on the two PDFs $f_1(x)$ and $f_2(x)$, respectively; and on the placement of the classification criterion, because these conditional probabilities will be given by the integrals of their respective PDFs over the regions in which classifications would be made to the other group. That is,

$$P(2|1) = \int_{R_2} f_1(\mathbf{x}) \, d\mathbf{x} \qquad (13.10a)$$

and

$$P(1|2) = \int_{R_1} f_2(\mathbf{x}) \, d\mathbf{x}, \qquad (13.10b)$$

where R_1 and R_2 denote the regions of the K-dimensional space of x in which classifications into Group 1 and Group 2, respectively, would be made. Unconditional probabilities of misclassification are given by the products of these conditional probabilities with the corresponding prior probabilities; that is, $P(2|1)p_1$ and $P(1|2)p_2$.

If $C(1|2)$ is the cost, or penalty, incurred when a Group 2 member is incorrectly classified as part of Group 1, and $C(2|1)$ is the cost incurred when a Group 1 member is incorrectly classified as part of Group 2, then the expected cost of misclassification will be

$$\text{ECM} = C(2|1)\,P(2|1)p_1 + C(1|2)\,P(1|2)p_2. \tag{13.11}$$

The classification boundary can be adjusted to minimize this expected cost of misclassification, through the effect of the boundary on the misclassification probabilities (Equations 13.10). The resulting classification rule is

$$\text{Assign } \mathbf{x}_0 \text{ to Group 1 if } \frac{f_1(\mathbf{x}_0)}{f_2(\mathbf{x}_0)} \geq \frac{C(1|2)p_2}{C(2|1)p_1}, \tag{13.12a}$$

or

$$\text{Assign } \mathbf{x}_0 \text{ to Group 2 if } \frac{f_1(\mathbf{x}_0)}{f_2(\mathbf{x}_0)} < \frac{C(1|2)p_2}{C(2|1)p_1}. \tag{13.12b}$$

That is, classification of $\mathbf{x}_0$ depends on the ratio of its likelihood according to the PDFs for the two groups, in relation to the ratios of the products of the misclassification costs and prior probabilities. Accordingly, it is not actually necessary to know the two misclassification costs specifically, but only their ratio, and likewise it is necessary only to know the ratio of the prior probabilities. If $C(1|2) \gg C(2|1)$—that is, if misclassifying a Group 2 member as belonging to Group 1 is especially undesirable—then the ratio of likelihoods on the left-hand side of Equation 13.12 must be quite large [$\mathbf{x}_0$ must be substantially more plausible according to $f_1(\mathbf{x})$] in order to assign $\mathbf{x}_0$ to Group 1. Similarly, if Group 1 members are intrinsically rare, so that $p_1 \ll p_2$, a higher level of evidence must be met in order to classify $\mathbf{x}_0$ as a member of Group 1. If both misclassification costs and prior probabilities are equal, then classification is made according to the larger of $f_1(\mathbf{x}_0)$ or $f_2(\mathbf{x}_0)$.

Minimizing the ECM (Equation 13.11) does not require assuming that the distributions $f_1(x)$ or $f_2(x)$ have specific forms, or even that they are of the same parametric family. But it is necessary to know or assume a functional form for each of them in order to numerically evaluate the left-hand side of Equation 13.12. Often it is assumed that both $f_1(x)$ and $f_2(x)$ are multivariate normal (possibly after data transformations for some or all of the elements of x), with equal covariance matrices that are estimated using $[S_{\text{pool}}]$. In this case, Equation 13.12a, for the conditions under which $\mathbf{x}_0$ would be assigned to Group 1, becomes

$$\frac{2\pi^{-K/2}|[S_{\text{pool}}]|^{-1/2}\exp\left[-\frac{1}{2}(\mathbf{x}_0 - \bar{\mathbf{x}}_1)^{\text{T}}[S_{\text{pool}}]^{-1}(\mathbf{x}_0 - \bar{\mathbf{x}}_1)\right]}{2\pi^{-K/2}|[S_{\text{pool}}]|^{-1/2}\exp\left[-\frac{1}{2}(\mathbf{x}_0 - \bar{\mathbf{x}}_2)^{\text{T}}[S_{\text{pool}}]^{-1}(\mathbf{x}_0 - \bar{\mathbf{x}}_2)\right]} \geq \frac{C(1|2)}{C(2|1)}\frac{p_2}{p_1}, \tag{13.13a}$$

which, after some rearrangement, is equivalent to

$$(\bar{\mathbf{x}}_1 - \bar{\mathbf{x}}_2)^{\text{T}}[S_{\text{pool}}]^{-1}\mathbf{x}_0 - \frac{1}{2}(\bar{\mathbf{x}}_1 - \bar{\mathbf{x}}_2)^{\text{T}}[S_{\text{pool}}]^{-1}(\bar{\mathbf{x}}_1 + \bar{\mathbf{x}}_2) \geq \ln\left[\frac{C(1|2)}{C(2|1)}\frac{p_2}{p_1}\right]. \tag{13.13b}$$

The left-hand side of Equation 13.13b looks elaborate, but its elements are familiar. In particular, its first term is exactly the linear combination $a^{\text{T}}x_0$ in Equation 13.7. The second term is the midpoint $\hat{m}$ between the two means when projected onto $\boldsymbol{a}$, defined in Equation 13.6. Therefore, if $C(1|2) = C(2|1)$ and $p_1 = p_2$ (or if other combinations of these quantities yield $\ln[1]$ on the right-hand side of Equation 13.13b), the minimum ECM classification criterion for two multivariate normal populations with equal covariance is exactly the same as Fisher's linear discriminant. To the extent that the costs and/or prior probabilities are not equal, Equation 13.13 results in movement of the classification boundary away from the midpoint defined in Equation 13.6, and toward the projection of one of the two means onto $\boldsymbol{a}$.

13.2.4 Unequal Covariances: Quadratic Discrimination

Discrimination and classification are much more straightforward, both conceptually and mathematically, if equality of covariance for the G populations can be assumed. For example, it is the equality-of-covariance assumption that allows the simplification from Equation 13.13a to Equation 13.13b for two multivariate normal populations. If it cannot be assumed that $[\Sigma_1] = [\Sigma_2]$, and instead these two covariance matrices are estimated separately by $[S_1]$ and $[S_2]$, respectively, minimum ECM classification for two multivariate populations leads to classification of $\boldsymbol{x}_0$ as belonging to Group 1 if

$$\frac{1}{2}\boldsymbol{x}_0^T([S_1]^{-1} - [S_2]^{-1})\boldsymbol{x}_0 + (\bar{\boldsymbol{x}}_1^T[S_1]^{-1} - \bar{\boldsymbol{x}}_2^T[S_2]^{-1})\,\boldsymbol{x}_0 - \mathrm{const} \geq \ln\left[\frac{C(1|2)}{C(2|1)}\frac{p_2}{p_1}\right], \quad (13.14a)$$

where

$$\mathrm{const} = \frac{1}{2}\left(\ln\frac{|[S_1]|}{|[S_2]|} + \bar{\boldsymbol{x}}_1^T[S_1]^{-1}\bar{\boldsymbol{x}}_1 - \bar{\boldsymbol{x}}_2^T[S_2]^{-1}\bar{\boldsymbol{x}}_2\right) \qquad (13.14b)$$

contains scaling constants not involving $\boldsymbol{x}_0$.

The mathematical differences between Equations 13.13b and 13.14 result because cancellations and recombinations that are possible when the covariance matrices are equal, result in additional terms in Equation 13.14. Classification and discrimination using Equation 13.14 are more difficult conceptually because the regions R_1 and R_2 are no longer necessarily contiguous. Equation 13.14, for classification with unequal covariances, is also less robust to non-Gaussian data than classification with Equation 13.13, when equality of covariance structure can reasonably be assumed.

Figure 13.2 illustrates quadratic discrimination and classification with a simple, one-dimensional example. Here it has been assumed for simplicity that the right-hand side

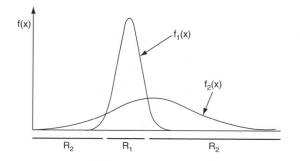

FIGURE 13.2 Discontinuous classification regions resulting from unequal variances for the populations described by two Gaussian PDFs $f_1(x)$ and $f_2(x)$.

of Equation 13.14a is $\ln[1] = 0$, so the classification criterion reduces to assigning x_0 to whichever group yields the larger likelihood, $f_g(x_0)$. Because the variance for Group 1 is so much smaller, both very large and very small x_0 will be assigned to Group 2. Mathematically, this discontinuity for the region R_2 results from the first (i.e., the quadratic) term in Equation 13.14a, which in $K = 1$ dimension is equal to $x_0^2(1/s_1^2 - 1/s_2^2)/2$. In higher dimensions the shapes of quadratic classification regions will be more complicated.

13.3 Multiple Discriminant Analysis (MDA)

13.3.1 Fisher's Procedure for More Than Two Groups

Fisher's linear discriminant, described in Section 13.2.1, can be generalized for discrimination among $G = 3$ or more groups. This generalization is called multiple discriminant analysis (MDA). Here the basic problem is to allocate a K-dimensional data vector x to one of $G > 2$ groups on the basis $J = \min(G - 1, K)$ discriminant vectors, $a_j, j = 1, \ldots, J$. The projection of the data onto these vectors yield the J discriminant functions

$$\delta_j = a_j^T x, \quad j = 1, \ldots, J. \tag{13.15}$$

The discriminant functions are computed on the basis of a training set of G data matrices $[X_1], [X_2], [X_3], \ldots, [X_G]$, dimensioned, respectively, $(n_g \times K)$. A sample variance-covariance matrix can be computed from each of the G sets of data, $[S_1], [S_2], [S_3], \ldots, [S_G]$, according to Equation 9.30. Assuming that the G groups represent populations having the same covariance matrix, the pooled estimate of this common covariance matrix is estimated by the weighted average

$$[S_{pool}] = \frac{1}{n - G} \sum_{g=1}^{G} (n_g - 1)[S_g], \tag{13.16}$$

where there are n_g observations in each group, and the total sample size is

$$n = \sum_{g=1}^{G} n_g. \tag{13.17}$$

The estimated pooled covariance matrix in Equation 13.16 is sometimes called the within-groups covariance matrix. Equation 10.39b is a special case of Equation 13.16, with $G = 2$.

Computation of multiple discriminant functions also requires calculation of the between-groups covariance matrix

$$[S_B] = \frac{1}{G - 1} \sum_{g=1}^{G} (\bar{x}_g - \bar{x}_\bullet)(\bar{x}_g - \bar{x}_\bullet)^T, \tag{13.18}$$

where the individual group means are calculated as in Equation 13.2, and

$$\bar{x}_\bullet = \frac{1}{n} \sum_{g=1}^{G} n_g \bar{x}_g \tag{13.19}$$

is the grand, or overall vector mean of all n observations. The matrix $[S_B]$ is essentially a covariance matrix describing the dispersion of the G sample means around the overall mean (compare Equation 9.35).

The number J of discriminant functions that can be computed is the smaller of $G-1$ and K. Thus for the two-group case discussed in Section 13.2, there is only $G-1=1$ discriminant function, regardless of the dimensionality K of the data vectors. In the more general case, the discriminant functions are derived from the first J eigenvectors (corresponding to the nonzero eigenvalues) of the matrix

$$[S_{pool}]^{-1}[S_B].\tag{13.20}$$

This $(K \times K)$ matrix in general is not symmetric. The discriminant vectors a_j are aligned with these eigenvectors, but are often scaled to yield unit variances for the data projected onto them; that is,

$$\mathbf{a}_j^T[S_{pool}]\mathbf{a}_j = 1, \quad j = 1, \ldots, J.\tag{13.21}$$

Usually computer routines for calculating eigenvectors will scale eigenvectors to unit length, that is, $\|e_j\| = 1$, but the condition in Equation 13.21 can be achieved by calculating

$$\mathbf{a}_j = \frac{\mathbf{e}_j}{(\mathbf{e}_j^T[S_{pool}]\mathbf{e}_j)^{1/2}}.\tag{13.22}$$

The first discriminant vector a_1, which is associated with the largest eigenvalue of the matrix in Equation 13.20, makes the largest contribution to separating the G group means, in aggregate; and a_J, which is associated with the smallest nonzero eigenvalue, makes the least contribution overall.

The J discriminant vectors a_j define a J-dimensional discriminant space, in which the G groups exhibit maximum separation. The projections δ_j (Equation 13.15) of the data onto these vectors are sometimes called the discriminant coordinates or canonical variates. This second name is unfortunate and a cause for confusion, since they do not pertain to canonical correlation analysis. As was also the case when distinguishing between $G=2$ groups, observations x can be assigned to groups according to which of the G group means is closest in discriminant space. For the $G=2$ case the discriminant space is one-dimensional, consisting only of a line. The group assignment rule (Equation 13.7) is then particularly simple. More generally, it is necessary to evaluate the Euclidean distances between the candidate vector x_0 and each of the G group means in order to find which is closest. It is actually easier to evaluate these in terms of squared distances, yielding the classification rule, assign x_0 to group g if:

$$\sum_{j=1}^{J}[\mathbf{a}_j(\mathbf{x}_0 - \bar{\mathbf{x}}_g)]^2 \leq \sum_{j=1}^{J}[\mathbf{a}_j(\mathbf{x}_0 - \bar{\mathbf{x}}_h)]^2, \quad \text{for all } h \neq g.\tag{13.23}$$

That is, the sum of the squared distances between x_0 and each of the group means, along the directions defined by the vectors a_j, are compared in order to find the closest group mean.

EXAMPLE 13.2   Multiple Discriminant Analysis with $G = 3$ Groups

Consider discriminating among all three groups of data in Table 13.1. Using Equation 13.16 the pooled estimate of the common covariance matrix is

$$[S_{pool}] = \frac{1}{28-3} \left(9 \begin{bmatrix} 1.47 & 0.65 \\ 0.65 & 1.45 \end{bmatrix} + 8 \begin{bmatrix} 2.08 & 0.06 \\ 0.06 & 0.17 \end{bmatrix} + 8 \begin{bmatrix} 4.85 & -0.17 \\ -0.17 & 0.10 \end{bmatrix} \right) = \begin{bmatrix} 2.75 & 0.20 \\ 0.20 & 0.61 \end{bmatrix},$$

(13.24)

and using Equation 13.18 the between-groups covariance matrix is

$$[S_B] = \frac{1}{2} \left(\begin{bmatrix} 12.96 & 5.33 \\ 5.33 & 2.19 \end{bmatrix} + \begin{bmatrix} 2.89 & -1.05 \\ -1.05 & 0.38 \end{bmatrix} + \begin{bmatrix} 32.49 & 5.81 \\ 5.81 & 1.04 \end{bmatrix} \right) = \begin{bmatrix} 24.17 & 5.04 \\ 5.04 & 1.81 \end{bmatrix}. \quad (13.25)$$

The directions of the two discriminant functions are specified by the eigenvectors of the matrix

$$[S_{pool}]^{-1}[S_B] = \begin{bmatrix} 0.373 & -0.122 \\ -0.122 & 1.685 \end{bmatrix} \begin{bmatrix} 24.17 & 5.04 \\ 5.04 & 1.81 \end{bmatrix} = \begin{bmatrix} 8.40 & 1.65 \\ 5.54 & 2.43 \end{bmatrix}, \quad (13.26a)$$

which, when scaled according to Equation 13.22 are

$$\mathbf{a}_1 = \begin{bmatrix} 0.542 \\ 0.415 \end{bmatrix} \quad \text{and} \quad \mathbf{a}_2 = \begin{bmatrix} -0.282 \\ 1.230 \end{bmatrix}. \quad (13.26b)$$

The discriminant vectors $\mathbf{a}_1$ and $\mathbf{a}_2$ define the directions of the first discriminant function $\delta_1 = \mathbf{a}_1^T \mathbf{x}$ and the second discriminant function $\delta_2 = \mathbf{a}_2^T \mathbf{x}$. Figure 13.3 shows the data for all three groups in Table 13.1 plotted in the discriminant space defined by these two functions. Points for Groups 1 and 2 are shown by circles and Xs, as in Figure 13.1, and points for Group 3 are shown by +s. The heavy symbols locate the respective vector means for the three groups. Note that the point clouds for Groups 1 and 2 appear to

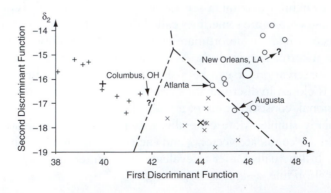

FIGURE 13.3   Illustration of the geometry of multiple discriminant analysis applied to the $G = 3$ groups of data in Table 13.1. Group 1 stations are plotted as circles, Group 2 stations are plotted as Xs, and Group 3 stations are plotted as +s. The three vector means are indicated by the corresponding heavy symbols. The two axes are the first and second discriminant functions, and the heavy dashed lines divide sections of this discriminant space allocated to each group. The data for Atlanta and Augusta are misclassified as belonging to Group 2. The two stations Columbus and New Orleans, which are not part of the training data in Table 13.1, are shown as question marks, and are allocated to Groups 3 and 1, respectively.

be stretched and distorted relative to their arrangement in Figure 13.1. This is because the matrix in Equation 13.26a is not symmetric, so that the two discriminant vectors in Equation 13.26b are not orthogonal.

The heavy dashed lines in Figure 13.3 divide the portions of the discriminant space that are assigned to each of the three groups by the classification criterion in Equation 13.23. These are the regions closest to each of the group means. Here the data for Atlanta and Augusta have both been misclassified as belonging to Group 2 rather than Group 1. For Atlanta, for example, the squared distance to the Group 1 mean is $[.542(78.6 - 80.6) + .415(4.73 - 5.67)]^2 + [-.282(78.6 - 80.6) + 1.230(4.73 - 5.67)]^2 = 2.52$, and the squared distance to the Group 2 mean is $[.542(78.6 - 78.7) + .415(4.73 - 3.57)]^2 + [-.282(78.6 - 78.7) + 1.230(4.73 - 3.57)]^2 = 2.31$. A line in this discriminant space could be drawn by eye that would include these two stations in the Group 1 region. That the discriminant analysis has not specified this line is probably a consequence of the assumption of equal covariance matrices not being well satisfied. In particular, the points in Group 1 appear to be more positively correlated in this discriminant space than the members of the other two groups.

The data points for the two stations Columbus and New Orleans, which are not part of the training data in Table 13.1, are shown by the question marks in Figure 13.3. The location in the discriminant space of the point for New Orleans is $\delta_1 = (.542)(82.1) + (.415)(6.73) = 47.3$ and $\delta_2 = (-.282)(82.1) + (1.230)(6.73) = -14.9$, which is within the region assigned to Group 1. The coordinates in discriminant space for the Columbus data are $\delta_1 = (.542)(74.7) + (.415)(3.37) = 41.9$ and $\delta_2 = (-.282)(74.7) + (1.230)(3.37) = -16.9$, which is within the region assigned to Group 3. $\lozenge$

Graphical displays of the discriminant space such as that in Figure 13.3 can be quite useful for visualizing the separation of data groups. If $J = \min(G - 1, K) > 2$, we cannot plot the full discriminant space in only two dimensions, but it is still possible to calculate and plot its first two components, δ_1 and δ_2. The relationships among the data groups rendered in this reduced discriminant space will be a good approximation to those in the full J-dimensional discriminant space, if the corresponding eigenvalues of Equation 13.20 are large relative to the eigenvalues of the omitted dimensions. Similarly to the idea expressed in Equation 11.4 for PCA, the reduced discriminant space will be a good approximation to the full discriminant space, to the extent that $(\lambda_1 + \lambda_2)/\Sigma_j \lambda_j \approx 1$.

13.3.2 *Minimizing Expected Cost of Misclassification*

The procedure described in Section 13.2.3, accounting for misclassification costs and prior probabilities of group memberships, generalizes easily for MDA. Again, if equal covariances for each of the G populations can be assumed, there are no other restrictions on the PDFs $f_g(x)$ for each of the populations, except that that these PDFs can be evaluated explicitly. The main additional complication is to specify cost functions for all possible $G(G - 1)$ misclassifications of Group g members into Group h,

$$C(h|g); \quad g = 1, \ldots, G; \quad h = 1, \ldots, G; \quad g \neq h. \tag{13.27}$$

The resulting classification rule is to assign an observation x_0 to the group g for which

$$\sum_{\substack{h=1 \\ h \neq g}}^{G} C(g|h) p_h f_h(x_0) \tag{13.28}$$

is minimized. That is, the candidate Group g is selected for which the probability-weighted sum of misclassification costs, considering each of the other $G - 1$ groups h as the potential true home of x_0, is smallest. Equation 13.28 is the generalization of Equation 13.12 to $G \geq 3$ groups.

If all the misclassification costs are equal, minimizing Equation 13.28 simplifies to classifying x_0 as belonging to that group g for which

$$p_g f_g(\mathbf{x}_0) \geq p_h f_h(\mathbf{x}_0), \quad \text{for all } h \neq g. \tag{13.29}$$

If in addition the PDFs $f_g(x)$ are all multivariate normal distributions, with possibly different covariance matrices $[\Sigma_g]$, (the logs of) the terms in Equation 13.29 take on the form

$$\ln(p_g) - \frac{1}{2} \ln |[S_g]| - \frac{1}{2}(\mathbf{x}_0 - \bar{\mathbf{x}}_g)^T [S_g](\mathbf{x}_0 - \bar{\mathbf{x}}_g). \tag{13.30}$$

The observation x_0 would be allocated to the group whose multinormal PDF $f_g(x)$ maximizes Equation 13.30. The unequal covariances $[S_g]$ result in this classification rule being quadratic. If, in addition, all the covariance matrices $[\Sigma_g]$ are assumed equal and are estimated by $[S_{pool}]$, the classification rule in Equation 13.30 simplifies to choosing that Group g maximizing the linear discriminant score

$$\ln(p_g) + \bar{\mathbf{x}}_g^T [S_{pool}]^{-1} \mathbf{x}_0 - \frac{1}{2} \bar{\mathbf{x}}_g^T [S_{pool}]^{-1} \bar{\mathbf{x}}_g. \tag{13.31}$$

This rule minimizes the total probability of misclassification.

13.3.3 Probabilistic Classification

The classification rules presented so far choose only one of the G groups in which to place a new observation x_0. Except in very easy cases, in which group means are well separated relative to the data scatter, these rules rarely will yield perfect results. Accordingly, probability information describing classification uncertainties is often useful.

Probabilistic classification, that is, specification of probabilities for x_0 belonging to each of the G groups, can be achieved through an application of Bayes' Theorem:

$$\Pr\{\text{Group } g | \mathbf{x}_0\} = \frac{p_g f_g(\mathbf{x}_0)}{\sum\limits_{h=1}^{G} p_h f_h(\mathbf{x}_0)}. \tag{13.32}$$

Here the p_g are the prior probabilities for group membership, which often will be the relative frequencies with which each of the groups are represented in the training data. The PDFs $f_g(x)$ for each of the groups can be of any form, so long as they can be evaluated explicitly for particular values of x_0.

Often it is assumed that each of the $f_g(x)$ are multivariate normal distributions. In this case, Equation 13.32 becomes

$$\Pr\{\text{Group } g | \mathbf{x}_0\} = \frac{p_g \left(|[S_g]|^{-1/2} \exp\left[-\frac{1}{2}(\mathbf{x}_0 - \bar{\mathbf{x}}_g)^T [S_g]^{-1}(\mathbf{x}_0 - \bar{\mathbf{x}}_g) \right] \right)}{\sum\limits_{h=1}^{G} p_h \left(|[S_h]|^{-1/2} \exp\left[-\frac{1}{2}(\mathbf{x}_0 - \bar{\mathbf{x}}_h)^T [S_h]^{-1}(\mathbf{x}_0 - \bar{\mathbf{x}}_h) \right] \right)}. \tag{13.33}$$

Equation 13.33 simplifies if all G of the covariance matrices are assumed to be equal, because the factors involving determinants cancel. This equation also simplifies if all the prior probabilities are equal (i.e., $p_g = 1/G$, $g = 1, \ldots, G$), because these probabilities then cancel.

EXAMPLE 13.3   Probabilistic Classification with $G = 3$ Groups

Consider probabilistic classification of Columbus, Ohio, into the three climate-region groups of Example 13.2. The July mean vector for Columbus is $x_0 = [74.7°\text{F}, 3.37 \text{ in.}]^T$. Figure 13.3 shows that this point is near the boundary between the (nonprobabilistic) classification regions for Groups 2 (Central United States) and 3 (Northeastern United States) in the two-dimensional discriminant space, but the calculations in Example 13.2 do not quantify the certainty with which Columbus has been placed in Group 3.

For simplicity, it will be assumed that the three prior probabilities are equal, and that the three groups are all samples from multivariate normal distributions with a common covariance matrix. The pooled estimate for the common covariance is given in Equation 13.24, and its inverse is indicated in the middle equality of Equation 13.26a. The groups are then distinguished by their mean vectors, indicated in Table 13.1.

The differences between x_0 and the three group means are

$$x_0 - \bar{x}_1 = \begin{bmatrix} -5.90 \\ -2.30 \end{bmatrix}, \quad x_0 - \bar{x}_2 = \begin{bmatrix} -4.00 \\ -0.20 \end{bmatrix}, \quad \text{and } x_0 - \bar{x}_3 = \begin{bmatrix} 3.40 \\ 0.20 \end{bmatrix}; \tag{13.34a}$$

yielding the likelihoods (cf. Equation 13.33)

$$f_1(x_0) \propto \exp\left(-\frac{1}{2}[-5.90, -2.30]\begin{bmatrix} .373 & -.122 \\ -.122 & 1.679 \end{bmatrix}\begin{bmatrix} -5.90 \\ -2.30 \end{bmatrix}\right) = 0.000094, \tag{13.34b}$$

$$f_2(x_0) \propto \exp\left(-\frac{1}{2}[-4.00, -0.20]\begin{bmatrix} .373 & -.122 \\ -.122 & 1.679 \end{bmatrix}\begin{bmatrix} -4.00 \\ -0.20 \end{bmatrix}\right) = 0.054, \tag{13.34c}$$

and

$$f_3(x_0) \propto \exp\left(-\frac{1}{2}[3.40, 0.20]\begin{bmatrix} .373 & -.122 \\ -.122 & 1.679 \end{bmatrix}\begin{bmatrix} 3.40 \\ 0.20 \end{bmatrix}\right) = 0.122. \tag{13.34d}$$

Substituting these likelihoods into Equation 13.33 yields the three classification probabilities

$$\Pr(\text{Group 1}|x_0) = .000094/(.000094 + .054 + .122) = 0.0005, \tag{13.35a}$$

$$\Pr(\text{Group 2}|x_0) = .054/(.000094 + .054 + .122) = 0.31, \tag{13.35b}$$

and

$$\Pr(\text{Group 3}|x_0) = .122/(.000094 + .054 + .122) = 0.69. \tag{13.35c}$$

Even though the group into which Columbus was classified in Example 13.2 is most probable, there is still a substantial probability that it might belong to Group 2 instead. The possibility that Columbus is really a Group 1 station appears to be remote.◊

13.4 Forecasting with Discriminant Analysis

Discriminant analysis is a natural tool to use in forecasting when the predictand consists of a finite set of discrete categories (groups), and vectors of predictors x are known sufficiently far in advance of the discrete observation to be predicted. Apparently the first use of discriminant analysis for forecasting in meteorology was described by Miller (1962), who forecast airfield ceiling in five MECE categories at a lead time of 0–2 hours, and also made five-group forecasts of precipitation type (none, rain/freezing rain, snow/sleet) and amount (≤ 0.05 in., and > 0.05 in., if nonzero). Both of these applications today would be called *nowcasting*. Some other examples of the use of discriminant analysis for forecasting can be found in Lawson and Cerveny (1985), and Ward and Folland (1991).

An informative case study in the use of discriminant analysis for forecasting is provided by Lehmiller *et al.* (1997). They consider the problem of forecasting hurricane occurrence (i.e., whether or not at least one hurricane will occur) during summer and autumn, within subbasins of the northwestern Atlantic Ocean, so that $G = 2$. They began with a quite large list of potential predictors and so needed to protect against overfitting in their $n = 43$-year training sample, 1950–1992. Their approach to predictor selection was computationally intensive, but statistically sound: different discriminant analyses were calculated for all possible subsets of predictors, and for each of these subsets the calculations were repeated 43 times, in order to produce leave-one-out cross-validations. The chosen predictor sets were those with the smallest number of predictors that minimized the number of cross-validated incorrect classifications.

Figure 13.4 shows one of the resulting discriminant analyses, for occurrence or nonoccurrence of hurricanes in the Caribbean Sea, using standardized African rainfall predictors that would be known as of 1 December in the preceding year. Because this is a binary forecast (two groups), there is only a single linear discriminant function, which would be perpendicular to the dividing line labeled discriminant partition line in Figure 13.4. This line compares to the long-short dashed dividing line in Figure 13.1. (The discriminant

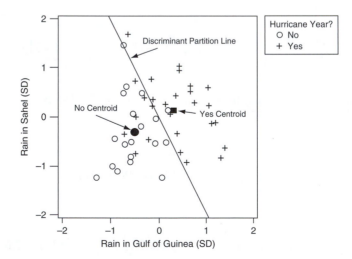

FIGURE 13.4   Binary (yes/no) forecasts for occurrence of at least one hurricane in the Caribbean Sea during summer and autumn, using two standardized predictors observed as of 1 December of the previous year to define a single linear discriminant function. Circles and plusses show the training data, and the two solid symbols locate the two group means (centroids). From Lehmiller *et al.*, 1997.

vector a would be perpendicular to this line, and pass through the origin.) The $n = 43$-year training sample is indicated by the open circles and plusses. Seven of the 18 hurricane years have been misclassified as no years, and only two of 25 nonhurricane years have been misclassified as yes years. Since there are more yes years, accounting for unequal prior probabilities would have moved the dividing line down and to the left, toward the no group mean (solid circle). Similarly, for some purposes it might be reasonable to assume that the cost of an incorrect no forecast would be greater than that of an incorrect yes forecast, and incorporating this asymmetry would also move the partition down and to the left, producing more yes forecasts.

13.5 Alternatives to Classical Discriminant Analysis

Traditional discriminant analysis, as described in the first sections of this chapter, continues to be widely employed and extremely useful. Newer alternative approaches to discrimination and classification are also available. Two of these, relating to topics treated in earlier chapters, are described in this section. Additional alternatives are also presented in Hand (1997) and Hastie *et al.* (2001).

13.5.1 Discrimination and Classification Using Logistic Regression

Section 6.3.1 described logistic regression, in which the nonlinear logistic function (Equation 6.27) is used to relate a linear combination of predictors, x, to the probability of one of the elements of a dichotomous outcome. Figure 6.12 shows a simple example of logistic regression, in which the probability of occurrence of precipitation at Ithaca has been specified as a logistic function of the minimum temperature on the same day.

Figure 6.12 could also be interpreted as portraying classification into $G = 2$ groups, with $g = 1$ indicating precipitation days, and $g = 2$ indicating dry days. The densities (points per unit length) of the dots along the top and bottom of the figure suggest the magnitudes of the two underlying PDFs, $f_1(x)$ and $f_2(x)$, respectively, as functions of the minimum temperature, x. The medians of these two conditional distributions for minimum temperature are near 23°F and 3°F, respectively. However, the classification function in this case is the logistic curve (solid), the equation for which is also given in the figure. Simply evaluating the function using the minimum temperature for a particular day provides an estimate of the probability that that day belonged to Group 1 (nonzero precipitation). A nonprobabilistic classifier could be constructed at the point of equal probability for the two groups, by setting the classification probability ($= y$ in Figure 6.12) to 1/2. This probability is achieved when the argument of the exponential is zero, implying a nonprobabilistic classification boundary of 15°F: days are classified as belonging to Group 1(wet) if the minimum temperature is warmer, and are classified as belonging to Group 2 (dry) if the minimum temperature is colder. Seven days (the five warmest dry days, and the two coolest wet days) in the training data are misclassified by this rule. In this example the relative frequencies of the two groups are nearly equal, but logistic regression automatically accounts for relative frequencies of group memberships in the training sample (which estimate the prior probabilities) in the fitting process.

Figure 13.5 shows a forecasting example of two-group discrimination using logistic regression, with a (2×1) predictor vector x. The two groups are years with (solid dots)

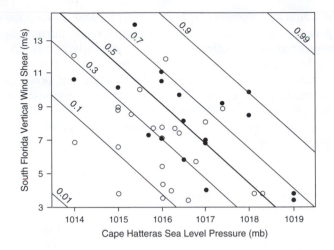

FIGURE 13.5   Two-dimensional logistic regression surface, estimating the probability of at least one landfalling hurricane on the southeastern U.S. coastline from August onward, on the basis of July sea-level pressure at Cape Hatteras and 200–700 mb wind shear over south Florida. Solid dots indicate hurricane years, and open dots indicate non-hurricane years, in the training data. Adapted from Lehmiller *et al.*, 1997.

and without (open circles) landfalling hurricanes on the southeastern U.S. coast from August onward, and the two elements of x are July average values of sea-level pressure at Cape Hatteras, and 200–700 mb wind shear over southern Florida. The contour lines indicate the shape of the logistic function, which in this case is a surface deformed into an S shape, analogously to the logistic function in Figure 6.12 being a line deformed in the same way. High surface pressures and wind shears simultaneously result in large probabilities for hurricane landfalls, whereas low values for both predictors yield small probabilities. This surface could be calculated as indicated in Equation 6.31, except that the vectors would be dimensioned (3×1) and the matrix of second derivatives would be dimensioned (3×3).

13.5.2 *Discrimination and Classification Using Kernel Density Estimates*

It was pointed out in Sections 13.2 and 13.3 that the G PDFs $f_g(x)$ need not be of particular parametric forms in order to implement Equations 13.12, 13.29, and 13.32, but rather it is necessary only that they can be evaluated explicitly. Gaussian or multivariate normal distributions often are assumed, but these and other parametric distributions may be poor approximations to the data. Viable alternatives are provided by kernel density estimates (see Section 3.3.6), which are nonparametric PDF estimates. Indeed, nonparametric discrimination and classification motivated much of the early work on kernel density estimation (Silverman, 1986).

Nonparametric discrimination and classification are straightforward conceptually, but may be computationally demanding. The basic idea is to separately estimate the PDFs $f_g(x)$ for each of the G groups, using the methods described in Section 3.3.6. Somewhat subjective choices for appropriate kernel form and (especially) bandwidth are necessary. But having estimated these PDFs, they can be evaluated for any candidate x_0, and thus lead to specific classification results.

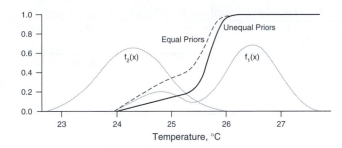

FIGURE 13.6   Separate kernel density estimates (quartic kernel, bandwidth = 0.92) for Guayaquil June temperatures during El Nino $f_1(x)$ and non-El Niño years $f_2(x)$, 1951–1970 (gray PDFs); and posterior probabilities for an El Niño year according to Equation 13.32, assuming equal prior probabilities (dashed), and prior probabilities estimated by the training-sample relative frequencies (solid).

Figure 13.6 illustrates the discrimination procedure for the same June Guayaquil temperature data (see Table A.3) used in Figures 3.6 and 3.8. The distribution of these data is bimodal, as a consequence of four of the five El Niño years being warmer than 26°C, whereas the warmest of the 15 non-El Niño years is 25.2°C. Discriminant analysis could be used to diagnose the presence or absence of El Niño, based on the June Guayaquil temperature, by specifying the two PDFs $f_1(x)$ for El Niño years and $f_2(x)$ for non-El Niño years. Parametric assumptions about the mathematical forms for these PDFs can be avoided through the use of kernel density estimates. The gray curves in Figure 13.6 show these two estimated PDFs. They exhibit fairly good separation, although $f_1(x)$, for El Niño years, is bimodal because the fifth El Niño year in the data set has a temperature of 24.8°C.

The posterior probability of an El Niño year as a function of the June temperature is calculated using Equation 13.32. The dashed curve is the result when equal prior probabilities $p_1 = p_2 = 1/2$ are assumed. Of course, El Niño occurs in fewer than half of all years, so it would be more reasonable to estimate the two prior probabilities as $p_1 = 1/4$ and $p_2 = 3/4$, which are the relative frequencies in the training sample. The resulting posterior probabilities are shown by the solid black curve in Figure 13.6.

Nonprobabilistic classification regions could be constructed using either Equation 13.12 or Equation 13.29, which would be equivalent if the two misclassification costs in Equation 13.12 were equal. If the two prior probabilities were also equal, the boundary between the two classification region would occur at the point where $f_1(x) = f_2(x)$, or $x \approx 25.45°C$. This temperature corresponds to a posterior probability of 1/2, according to the dashed curve. For unequal prior probabilities the classification boundary would shifted toward the less likely group (i.e., requiring a warmer temperature to classify as an El Niño year), occurring at the point where $f_1(x) = (p_2/p_1)f_2(x) = 3f_2(x)$, or $x \approx 25.65$. Not coincidentally, this temperature corresponds to a posterior probability of 1/2 according to the solid black curve.

13.6 Exercises

13.1. Use Fisher's linear discriminant to classify years in Table A.3 as either El Niño or non-El Niño, on the basis of the corresponding temperature and pressure data.

  a. What is the discriminant vector, scaled to have unit length?
  b. Which, if any, of the El Niño years have been misclassified?
  c. Assuming bivariate normal distributions, repeat part (b) accounting for unequal prior probabilities.

TABLE 13.2   Likelihoods calculated from the forecast verification data for subjective 12-24h projection probability-of-precipitation forecasts for the United States during October 1980–March 1981, in Table 7.2.

y_i	0.00	0.05	0.10	0.20	0.30	0.40	0.50	0.60	0.70	0.80	0.90	1.00
$p(y_i|o_1)$	.0152	.0079	.0668	.0913	.1054	.0852	.0956	.0997	.1094	.1086	.0980	.1169
$p(y_i|o_2)$	.4877	.0786	.2058	.1000	.0531	.0272	.0177	.0136	.0081	.0053	.0013	.0016

13.2. Average July temperature and precipitation at Ithaca, New York, are 68.6°F and 3.54 in.
   a. Classify Ithaca as belonging to one of the three groups in Example 13.2.
   b. Calculate probabilities that Ithaca is a member of each of the three groups.

13.3. Using the forecast verification data in Table 7.2, we can calculate the likelihoods (i.e., conditional probabilities for each of the 12 possible forecasts, given either precipitation or no precipitation) in Table 13.2. The unconditional probability of precipitation is $p(o_1) = 0.162$. Considering the two precipitation outcomes as two groups to be discriminated between, calculate the posterior probabilities of precipitation if the forecast, y_i, is

   a. 0.00
   b. 0.10
   c. 1.00

CHAPTER • 14

Cluster Analysis

14.1 Background

14.1.1 Cluster Analysis vs. Discriminant Analysis

Cluster analysis deals with separating data into groups whose identities are not known in advance. This more limited state of knowledge is in contrast to the situation of discrimination methods, which require a training data set for which group membership is known. In general, in cluster analysis even the correct number of groups into which the data should be sorted is not known ahead of time. Rather, it is the degree of similarity and difference between individual observations x that is used to define the groups, and to assign group membership. Examples of use of cluster analysis in the climatological literature include grouping daily weather observations into synoptic types (Kalkstein *et al.* 1987), defining weather regimes from upper-air flow patterns (Mo and Ghil 1988; Molteni *et al.* 1990), grouping members of forecast ensembles (Legg *et al.* 2002; Molteni *et al.* 1996; Tracton and Kalnay 1993), grouping regions of the tropical oceans on the basis of ship observations (Wolter 1987), and defining climatic regions based on surface climate variables (DeGaetano and Shulman 1990; Fovell and Fovell 1993; Galliani and Filippini 1985; Guttman 1993). Gong and Richman (1995) have compared various clustering approaches in a climatological context, and catalog the literature with applications of clustering to atmospheric data through 1993. Romesburg (1984) contains a general-purpose overview.

Cluster analysis is primarily an exploratory data analysis tool, rather than an inferential tool. Given a sample of data vectors x defining the rows of a $(n \times K)$ data matrix [X], the procedure will define groups and assign group memberships at varying levels of aggregation. Unlike discriminant analysis, the procedure does not contain rules for assigning membership to future observations. However, a cluster analysis can bring out groupings in the data that might otherwise be overlooked, possibly leading to an empirically useful stratification of the data, or helping to suggest physical bases for observed structure in the data. For example, cluster analyses have been applied to geopotential height data in order to try to identify distinct atmospheric flow regimes (e.g., Cheng and Wallace 1993; Mo and Ghil 1988).

14.1.2 *Distance Measures and the Distance Matrix*

Central to the idea of the clustering of data points is the idea of distance. Clusters should be composed of points separated by small distances, relative to the distances between clusters. The most intuitive and commonly used distance measure in cluster analysis is the Euclidean distance (Equation 9.6) in the K-dimensional space of the data vectors. Euclidean distance is by no means the only available choice for measuring distance between points or clusters, and in some instances may be a poor choice. In particular, if the elements of the data vectors are unlike variables with inconsistent measurement units, the variable with the largest values will tend to dominate the Euclidean distance. A more general alternative is the weighted Euclidean distance between two vectors x_i and x_j,

$$d_{i,j} = \left[\sum_{k=1}^{K} w_k (x_{i,k} - x_{j,k})^2 \right]^{1/2}. \tag{14.1}$$

For $w_k = 1$ for each $k = 1, \ldots, K$, Equation 14.1 reduces to the ordinary Euclidean distance. If the weights are the reciprocals of the corresponding variances, that is $w_k = 1/s_{k,k}$, the resulting function of the standardized variables is called the Karl Pearson distance. Other choices for the weights are also possible. For example, if one or more of the K variables in x contains large outliers, it might be better to use weights that are reciprocals of the ranges of each of the variables.

Euclidean distance and Karl Pearson distance are the most frequent choices in cluster analysis, but other alternatives are also possible. One alternative is to use the Mahalanobis distance (Equation 9.85), although deciding on an appropriate dispersion matrix [S] may be difficult, since group membership is not known in advance. A yet more general form of Equation 14.1 is the Minkowski metric,

$$d_{i,j} = \left[\sum_{k=1}^{K} w_k \left| x_{i,k} - x_{j,k} \right|^{\lambda} \right]^{1/\lambda}, \quad \lambda \geq 1. \tag{14.2}$$

Again, the weights w_k can equalize the influence of variables with incommensurate units. For $\lambda = 2$, Equation 14.2 reduced to the weighted Euclidean distance in Equation 14.1. For $\lambda = 1$, Equation 14.2 is known as the city-block distance.

The angle between pairs of vectors (Equation 9.15) is another possible choice for a distance measure, as are the many alternatives presented in Mardia *et al.* (1979) or Romesburg (1984). Tracton and Kalnay (1993) have used the anomaly correlation (Equation 7.59) to group members of forecast ensembles, and the ordinary Pearson correlation sometimes is used as a clustering criterion as well. These latter two criteria are inverse distance measures, which should be maximized within groups, and minimized between groups.

Having chosen a distance measure to quantify dissimilarity or similarity between pairs of vectors x_i and x_j, the next step in cluster analysis is to calculate the distances between all $n(n-1)/2$ possible pairs of the n observations. It can be convenient, either organizationally or conceptually, to arrange these into a $(n \times n)$ matrix of distances, $[\Delta]$. This symmetric matrix has zeros along the main diagonal, indicating zero distance between each x and itself.

14.2 Hierarchical Clustering

14.2.1 Agglomerative Methods Using the Distance Matrix

Most commonly implemented cluster analysis procedures are hierarchical and agglomerative. That is, they construct a hierarchy of sets of groups, each of which is formed by merging one pair from the collection of previously defined groups. These procedures begin by considering that the n observations of x have no group structure or, equivalently, that the data set consists of n groups containing one observation each. The first step is to find the two groups (i.e., data vectors) that are closest in their K-dimensional space, and to combine them into a new group. There are then $n - 1$ groups, one of which has two members. On each subsequent step, the two groups that are closest are merged to form a larger group. Once a data vector x has been assigned to a group, it is not removed. Its group membership changes only when the group to which it has been assigned is merged with another group. This process continues until, at the final $(n - 1)^{st}$, step all n observations have been aggregated into a single group.

The n-group clustering at the beginning of this process and the one-group clustering at the end of this process are neither useful nor enlightening. Hopefully, however, a natural clustering of the data into a workable number of informative groups will emerge at some intermediate stage. That is, we hope that the n data vectors cluster or clump together in their K-dimensional space into some number $G, 1 < G < n$, groups that reflect similar data generating processes. The ideal result is a division of the data that both minimizes differences between members of a given cluster, and maximizes differences between members of different clusters.

Distances between pairs of points can be unambiguously defined and stored in a distance matrix. However, even after calculating a distance matrix there are alternative definitions for distances between groups of points if the groups contain more than a single member. The choice made for the distance measure together with the criterion used to define cluster-to-cluster distances essentially define the method of clustering. A few of the most common definitions for intergroup distances based on the distance matrix are:

- *Single-linkage*, or minimum-distance clustering. Here the distance between clusters G_1 and G_2 is the smallest distance between one member of G_1 and one member of G_2. That is,

$$d_{G_1, G_2} = \min_{i \in G_1, j \in G_2} (d_{i,j}).$$
(14.3)

- *Complete-linkage*, or maximum-distance clustering groups data points on the basis of the largest distance between points in the two groups G_1 and G_2,

$$d_{G_1, G_2} = \max_{i \in G_1, j \in G_2} (d_{i,j}).$$
(14.4)

- *Average-linkage* clustering defines cluster-to-cluster distance as the average distance between all possible pairs of points in the two groups being compared. If G_1 contains n_1 points and G_2 contains n_2 points, this measure for the distance between the two groups is

$$d_{G_1, G_2} = \frac{1}{n_1 n_2} \sum_{i=1}^{n_1} \sum_{j=1}^{n_2} d_{i,j}.$$
(14.5)

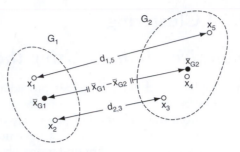

FIGURE 14.1   Illustration of three measures of the distance in $K = 2$ dimensional space, between a cluster G_1 containing the two elements x_1 and x_2 , and a cluster G_2 containing the elements x_3, x_4, and x_5. The data points are indicated by open circles, and centroids of the two groups are indicated by the solid circles. According to the maximum-distance, or complete-linkage criterion, the distance between the two groups is $d_{1,5}$, or the greatest distance between all of the six possible pairs of points in the two groups. The minimum-distance, or single-linkage criterion computes the distance between the groups as equal to the distance between the nearest pair of points, or $d_{2,3}$. According to the centroid method, the distance between the two clusters is the distance between the sample means of the points contained in each.

- *Centroid* clustering compares distances between the centroids, or vector averages, of pairs of clusters. According to this measure the distance between G_1 and G_2 is

$$d_{G_1,G_2} = \|\bar{\mathbf{x}}_{G_1} - \bar{\mathbf{x}}_{G_2}\|, \tag{14.6}$$

where the vector means are taken over all members of each of the groups separately, and the notation $\|\bullet\|$ indicates distance according to whichever point-to-point distance measure has been adopted.

Figure 14.1 illustrates single-linkage, complete-linkage, and centroid clustering for two hypothetical groups G_1 and G_2 in a $K = 2$-dimensional space. The open circles denote data points, of which there are $n_1 = 2$ in G_1 and $n_2 = 3$ in G_2. The centroids of the two groups are indicated by the solid circles. The single-linkage distance between G_1 and G_2 is the distance $d_{2,3}$ between the closest pair of points in the two groups. The complete-linkage distance is that between the most distant pair, $d_{1,5}$. The centroid distance is the distance between the two vector means $\|\bar{\mathbf{x}}_{G1} - \bar{\mathbf{x}}_{G2}\|$. The average-linkage distance can also be visualized in Figure 14.1, as the average of the six possible distances between individual members of G_1 and G_2; that is, $(d_{1,5} + d_{1,4} + d_{1,3} + d_{2,5} + d_{2,4} + d_{2,3})/6$.

The results of a cluster analysis can depend strongly on which definition is chosen for the distances between clusters. Single-linkage clustering rarely is used, because it is susceptible to chaining, or the production of a few large clusters, which are formed by virtue of nearness of points to be merged at different steps to opposite edges of a cluster. At the other extreme, complete-linkage clusters tend to be more numerous, as the criterion for merging clusters is more stringent. Average-distance clustering is usually intermediate between these two extremes, and appears to be the most commonly used approach to hierarchical clustering based on the distance matrix.

14.2.2 *Ward's Minimum Variance Method*

Ward's minimum variance method, or simply Ward's method, is a popular hierarchical clustering method that does not operate on the distance matrix. As a hierarchical method, it begins with n single-member groups, and merges two groups at each step, until all the data are in a single group after $n-1$ steps. However, the criterion for choosing which pair of groups to merge at each step is that, among all possible ways of merging two groups, the pair to be merged is chosen that minimizes the sum of squared distances between the points and the centroids of their respective groups, summed over the resulting groups. That is, among all possible ways of merging two of $G+1$ groups to make G groups, that merger is made that minimizes

$$W = \sum_{g=1}^{G} \sum_{i=1}^{n_g} \|\mathbf{x}_i - \bar{\mathbf{x}}_g\|^2 = \sum_{g=1}^{G} \sum_{i=1}^{n_g} \sum_{k=1}^{K} (x_{i,k} - \bar{x}_{g,k})^2. \tag{14.7}$$

In order to implement Ward's method to choose the best pair from $G+1$ groups to merge, Equation 14.7 must be calculated for all of the $G(G+1)/2$ possible pairs of existing groups. For each trial pair, the centroid, or group mean, for the trial merged group is recomputed using the data for both of the previously separate groups, before the squared distances are calculated. In effect, Ward's method minimizes the sum, over the K dimensions of $\mathbf{x}$, of within-groups variances. At the first (n-group) stage this variance is zero, and at the last (1-group) stage this variance is $\text{tr}[S_x]$, so that $W = n\,\text{tr}[S_x]$. For data vectors whose elements have incommensurate units, operating on nondimensionalized values (dividing by standard deviations) will prevent artificial domination of the procedure by one or a few of the K variables.

14.2.3 The Dendrogram, or Tree Diagram

The progress and intermediate results of a hierarchical cluster analysis are conventionally illustrated using a *dendrogram*, or tree diagram. Beginning with the "twigs" at the beginning of the analysis, when each of the n observations $\mathbf{x}$ constitutes its own cluster, one pair of "branches" is joined at each step as the closest two clusters are merged. The distances between these clusters before they are merged are also indicated in the diagram by the distance of the points of merger from the initial n-cluster stage of the twigs.

Figure 14.2 illustrates a simple dendrogram, reflecting the clustering of the five points plotted as open circles in Figure 14.1. The analysis begins at the left of Figure 14.2, when all five points constitute separate clusters. At the first stage, the closest two points, $\mathbf{x}_3$ and $\mathbf{x}_4$, are merged into a new cluster. Their distance $d_{3,4}$ is proportional to the distance between the vertical bar joining these two points and the left edge of the figure. At the

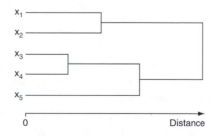

FIGURE 14.2 Illustration of a dendrogram, or tree diagram, for a clustering of the five points plotted as open circles in Figure 14.1. The results of the four clustering steps are indicated as the original five lines are progressively joined from left to right, with the distances between joined clusters indicated by the positions of the vertical lines.

next stage, the points x_1 and x_2 are merged into a single cluster because the distance between them is smallest of the six distances between the four clusters that existed at the previous stage. The distance $d_{1,2}$ is necessarily larger than the distance $d_{3,4}$, since x_1 and x_2 were not chosen for merger on the first step, and the vertical line indicating the distance between them is plotted further to the right in Figure 14.2 than the distance between x_3 and x_4. The third step merges x_5 and the pair (x_3, x_4), to yield the two-group stage indicated by the dashed lines in Figure 14.1.

14.2.4 How Many Clusters?

A hierarchical cluster analysis will produce a different grouping of n observations at each of the $n-1$ steps. At the first step each observation is in a separate group, and after the last step all the observations are in a single group. An important practical problem in cluster analysis is the choice of which intermediate stage will be chosen as the final solution. That is, we need to choose the level of aggregation in the tree diagram at which to stop further merging of clusters. The principal goal guiding this choice is to find that level of clustering that maximizes similarity within clusters and minimizes similarity between clusters, but in practice the best number of clusters for a given problem is usually not obvious. Generally the stopping point will require a subjective choice that will depend to some degree on the goals of the analysis.

One approach to the problem of choosing the best number of clusters is through summary statistics that relate to concepts in discrimination presented in Chapter 13. Several such criteria are based on the within-groups covariance matrix (Equation 13.16), either alone or in relation to the "between-groups" covariance matrix (Equation 13.18). Some of these objective stopping criteria are discussed in Jolliffe *et al.* (1986) and Fovell and Fovell (1993), who also provide references to the broader literature on such methods.

A traditional subjective approach to determination of the stopping level is to inspect a plot of the distances between merged clusters as a function of the stage of the analysis. When similar clusters are being merged early in the process, these distances are small and they increase relatively little from step to step. Late in the process there may be only a few clusters, separated by large distances. If a point can be discerned where the distances between merged clusters jumps markedly, the process can be stopped just before these distances become large.

Wolter (1987) suggests a Monte-Carlo approach, where sets of random numbers simulating the real data are subjected to cluster analysis. The distributions of clustering distances for the random numbers can be compared to the actual clustering distances for the data of interest. The idea here is that genuine clusters in the real data should be closer than clusters in the random data, and that the clustering algorithm should be stopped at the point where clustering distances are greater than for the analysis of the random data.

EXAMPLE 14.1   A Cluster Analysis in Two Dimensions

The mechanics of cluster analysis are easiest to see when the data vectors have only $K = 2$ dimensions. Consider the data in Table 13.1, where these two dimensions are average July temperature and average July precipitation. These data were collected into three groups for use in the discriminant analysis worked out in Example 13.2. However, the point of a cluster analysis is to try to discern group structure within a data set, without prior knowledge or information about the nature of that structure. Therefore, for purposes of a cluster analysis, the data in Table 13.1 should be regarded as consisting of $n = 28$ observations of two-dimensional vectors x, whose natural groupings we would like to discern.

Because the temperature and precipitation values have different physical units, it is well to divide by the respective standard deviations before subjecting them to a clustering algorithm. That is, the temperature and precipitation values are divided by 4.42°F and 1.36 in., respectively. The result is that the analysis is done using the Karl-Pearson distance, and the weights in Equation 14.1 are $w_1 = 4.42^{-2}$ and $w_2 = 1.36^{-2}$ The reason for this treatment of the data is to avoid the same kind of problem that can occur when conducting a principal component analysis using unlike data, where a variable with a much higher variance than the others will dominate the analysis even if that high variance is an artifact of the units of measurement. For example, if the precipitation had been reported in millimeters there would be apparently more distance between points in the direction of the precipitation axis, and a clustering algorithm would focus on precipitation differences to define groups. If the precipitation were reported in meters there would be essentially no distance between points in the direction of the precipitation axis, and a clustering algorithm would separate points almost entirely on the basis of the temperatures.

Figure 14.3 shows the results of clustering the data in Table 13.1, using the complete-linkage clustering criterion in Equation 14.4. On the left is a tree diagram for the process, with the individual stations listed at the bottom as the leaves. There are 27 horizontal lines in this tree diagram, each of which represents the merger of the two clusters it connects. At the first stage of the analysis the two closest points (Springfield and St. Louis) are merged into the same cluster, because their Karl-Pearson distance $d = [4.42^{-2}(78.8 - 78.9)^2 + 1.36^{-2}(3.58 - 3.63)^2]^{1/2} = 0.043$ is the smallest of the

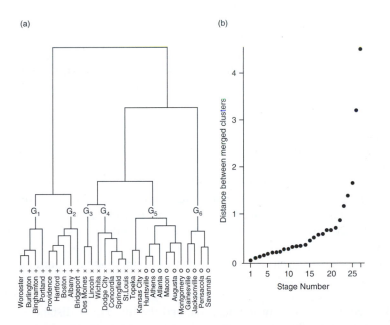

FIGURE 14.3 Dendrogram (a) and the corresponding plot of the distances between merged clusters as a function of the stage of the cluster analysis (b) for the data in Table 13.1. Standardized data (i.e., Karl-Pearson distances) have been clustered according to the complete-linkage criterion. The distances between merged groups appear to increase markedly at stage 22 or 23, indicating that the analysis should stop after 21 or 22 stages, which for these data would yield seven or six clusters, respectively. The six numbered clusters correspond to the grouping of the data shown in Figure 14.4. The seven-cluster solution would split Topeka and Kansas City from the Alabama and Georgia stations in G_5. The five-cluster solution would merge G_3 and G_4.

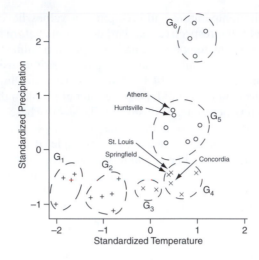

FIGURE 14.4   Scatterplot of the data in Table 13.1 expressed as standardized anomalies, with dashed lines showing the six groups defined in the cluster analysis tree diagram in Figure 14.3a. The five-group clustering would merge the central U.S. stations in Groups 3 and 4. The seven-group clustering would split the two central U.S. stations in Group 5 from six southeastern U.S. stations.

351 distances between the possible pairs. This separation distance can be seen graphically in Figure 14.4: the distance $d = 0.043$ is the height of the first dot in Figure 14.3b. At the second stage Huntsville and Athens are merged, because their Karl-Pearson distance $d = [4.42^{-2}(79.3 \div 79.2)^2 + 1.36^{-2}(5.05 - 5.18)^2]^{1/2} = 0.098$ is the second-smallest separation of the points (cf. Figure 14.4), and this distance corresponds to the height of the second dot in Figure 14.3b. At the third stage, Worcester and Binghamton ($d = 0.130$) are merged, and at the fourth stage Macon and Augusta ($d = 0.186$) are merged. At the fifth stage, Concordia is merged with the cluster consisting of Springfield and St. Louis. Since the Karl-Pearson distance between Concordia and St. Louis is larger than the distance between Concordia and Springfield (but smaller than the distances between Concordia and the other 25 points), the complete-linkage criterion merges these three points at the larger distance $d = [4.42^{-2}(79.0 - 78.9)^2 + 1.36^{-2}(3.37 - 3.63)^2]^{1/2} = 0.193$ (height of the fifth dot in Figure 14.3b).

The heights of the horizontal lines in Figure 14.3a, indicating group mergers, also correspond to the distances between the merged clusters. Since the merger at each stage is between the two closest clusters, these distances become greater at later stages. Figure 14.3b shows the distance between merged clusters as a function of the stage in the analysis. Subjectively, these distances climb gradually until perhaps stage 22 or stage 23, where the distances between combined clusters begin to become noticeably larger. A reasonable interpretation of this change in slope is that natural clusters have been defined at this point in the analysis, and that the larger distances at later stages indicate mergers of unlike clusters that should be distinct groups. Note, however, that a single change in slope does not occur in every cluster analysis, so that the choice of where to stop group mergers may not always be so clear cut. It is possible, for example, for there to be two or more relatively flat regions in the plot of distance versus stage, separated by segments of larger slope. Different clustering criteria may also produce different breakpoints. In such cases the choice of where to stop the analysis is more ambiguous.

If Figure 14.3b is interpreted as exhibiting its first major slope increase between stages 22 and 23, a plausible point at which to stop the analysis would be after stage 22. This stopping point would result in the definition of the six clusters labeled $G_1 - G_6$ on the tree diagram in Figure 14.3a. This level of clustering assigns the nine northeastern stations (+ symbols) into two groups, assigns seven of the nine central stations (× symbols) into two groups, allocates the central stations Topeka and Kansas City to Group 5 with six of the southeastern stations (o symbols), and assigns the remaining four southeastern stations to a separate cluster.

Figure 14.4 indicates these six groups in the $K = 2$-dimensional space of the standardized data, by separating points in each cluster with dashed lines. If this solution seemed too highly aggregated on the basis of the prior knowledge and information available to the analyst, we could choose the seven-cluster solution produced after stage 21, separating the central U.S. cities Topeka and Kansas City (xs) from the six southeastern cities in Group 5. If the six-cluster solution seemed too finely split, the five-cluster solution produced after stage 23 would merge the central U.S. stations in Groups 3 and 4. None of the groupings indicated in Figure 14.3a corresponds exactly to the group labels in Table 13.1, and we should not necessarily expect them to. It could be that limitations of the complete-linkage clustering algorithm operating on Karl-Pearson distances has produced some misclassifications, or that the groups in Table 13.1 have been imperfectly defined, or both.

Finally, Figure 14.5 illustrates the fact that different clustering algorithms will usually yield somewhat different results. Figure 14.5a shows distances at which groups are merged for the data in Table 13.1, according to single linkage operating on Karl-Pearson distances. There is a large jump after stage 21, suggesting a possible natural stopping point with seven groups. These seven groups are indicated in Figure 14.5b, which can be compared with the complete-linkage result in Figure 14.4. The clusters denoted G_2 and G_6 in Figure 14.4 occur also in Figure 14.5b. However, one long and thin group has developed in Figure 14.5b, composed of stations from G_3, G_4, and G_5. This result illustrates the chaining phenomenon to which single-linkage clusters are prone, as additional stations or groups are accumulated that are close to a point at one edge or another of a group, even though the added points may be quite far from other points in the same group. ◊

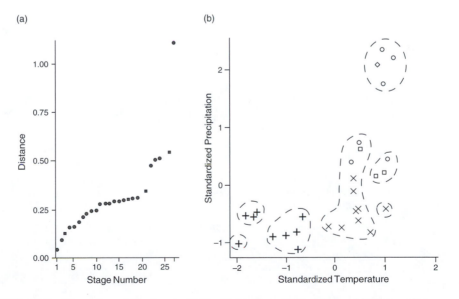

FIGURE 14.5   Clustering of the data in Table 13.1, using single linkage. (a) Merger distances as a function of stage, showing a large jump after 22 stages. (b) The seven clusters existing after stage 22, illustrating the chaining phenomenon.

Section 6.6 describes ensemble forecasting, in which the effects of uncertainty about the initial state of the atmosphere on the evolution of a forecast is addressed by calculating multiple forecasts beginning at an ensemble of similar initial conditions. The method has proved to be an extremely useful advance in forecasting technology, but requires extra effort to absorb the large amount of information produced. One way to summarize the information in a large collection of maps from a forecast ensemble is to group them according to a cluster analysis. If the smooth contours on each map have been interpolated from K gridpoint values, then each $(K \times 1)$ vector $\boldsymbol{x}$ included in the cluster analysis corresponds to one of the forecast maps. Figure 14.6 shows the result of one such cluster analysis, for $n = 14$ ensemble members forecasting hemispheric 500 mb heights at a

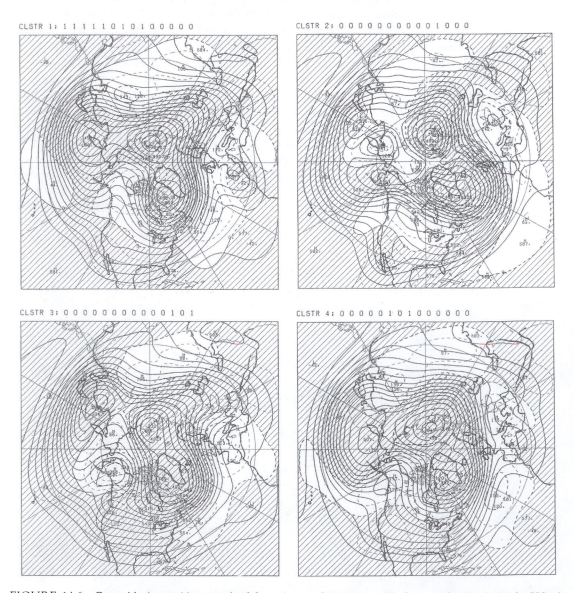

FIGURE 14.6   Centroids (ensemble means) of four clusters for an ensemble forecast for hemispheric 500 mb height at a projection of eight days. Solid contours show forecast heights, and dashed contours and shading show corresponding anomaly fields. From Tracton and Kalnay (1993).

projection of eight days. Here the clustering has been calculated on the basis of anomaly correlation, a similarity measure, rather than using a more conventional distance measure.

Molteni *et al.* (1996) illustrate the use of Ward's method to group $n = 33$ ensemble members forecasting 500 mb heights over Europe, at lead times of five to seven days. An innovation in their analysis is that it was conducted in a way that brings out the time-trajectories of the forecasts, by simultaneously clustering maps for the five-, six-, and seven-day forecasts. That is, if each forecast map consists of K gridpoints, the x vectors being clustered would be dimensioned $(3K \times 1)$, with the first K elements pertaining to day 5, the second K elements pertaining to day 6, and the last K elements pertaining to day 7. Because there are a large number of gridpoints underlying each map, the analysis actually was conducted using the first $K = 10$ principal components of the height fields, which was sufficient to capture 80% of the variance, so the clustered x vectors had dimension (30×1).

Another interesting aspect of the example of Molteni *et al.* (1996) is that the use of Ward's method provided an apparently natural stopping criterion for the clustering, that is related to forecast accuracy. Ward's method (Equation 14.7) is based on the sum of squared differences between the xs being clustered, and their respective group means. Regarding the group means as forecasts, these squared differences would be contributions to the overall expected mean squared error if the ensemble members x were different realizations of plausible observed maps. Molteni *et al.* (1996) stopped their clustering at the point where Equation 14.7 yields squared errors comparable to (the typically modest) 500 mb forecast errors obtained at the three-day lead time, so that their medium-range ensemble forecasts were grouped together if their differences were comparable to or smaller than typical short-range forecast errors.

14.2.5 Divisive Methods

In principle, a hierarchical clustering could be achieved by reversing the agglomerative clustering process. That is, beginning with a singe cluster containing all n observations, we could split this cluster into the two most similar possible groups; at the third stage one of these groups could be split into the three most similar groups possible; and so on. The procedure would proceed, in principle, to the point of n clusters each populated by a single data vector, with an appropriate intermediate solution determined by a stopping criterion. This approach to clustering, which is opposite to agglomeration, is called divisive clustering.

Divisive clustering is almost never used, because it is computationally impractical for all except the smallest sample sizes. Agglomerative hierarchical clustering requires examination of all $G(G-1)/2$ possible pairs of G groups, in order to choose the most similar two for merger. In contrast, divisive clustering requires examination, for each group of size n_g members, all $2^{n_g-1} - 1$ possible ways to make a split. This number of potential splits is 511 for $n_g = 10$, and rises to 524,287 for $n_g = 20$, and 5.4×10^8 for $n_g = 30$.

14.3 Nonhierarchical Clustering

14.3.1 The K-Means Method

A potential drawback of hierarchical clustering methods is that once a data vector x has been assigned to a group it will remain in that group, and groups with which it is merged. That is, hierarchical methods have no provision for reallocating points that may have been misgrouped at an early stage. Clustering methods that allow reassignment of observations as the analysis proceeds are called nonhierarchical. Like hierarchical methods, nonhierarchical clustering algorithms also group observations according to some distance measure in the K-dimensional space of x.

The most widely used nonhierarchical clustering approach is called the K-means method. The K in K-means refers to the number of groups, called G in this text, and not to the dimension of the data vector. The K-means method is named for the number of clusters into which the data will be grouped, because this number must be specified in advance of the analysis, together with an initial guess for the group membership of each of the x_i, $i = 1, \ldots, n$.

The K-means algorithm can begin either from a random partition of the n data vectors into the prespecified number G of groups, or from an initial selection of G seed points. The seed points might be defined by a random selection of G of the n data vectors; or by some other approach that is unlikely to bias the results. Another possibility is to define the initial groups as the result of a hierarchical clustering that has been stopped at G groups, allowing reclassification of xs from their initial placement in the hierarchical clustering.

Having defined the initial membership of the G groups in some way, the K-means algorithm proceeds as follows:

1) Compute the centroids (i.e., vector means) $\bar{x}_g$, $g = 1, \ldots, G$; for each cluster.
2) Calculate the distances between the current data vector x_i and each of the G $\bar{x}_g$s. Usually Euclidean or Karl-Pearson distances are used, but distance can be defined by any measure that might be appropriate to the particular problem.
3) If x_i is already a member of the group whose mean is closest, repeat step 2 for x_{i+1} (or for x_1, if $i = n$). Otherwise, reassign x_i to the group whose mean is closest, and return to step 1.

The algorithm is iterated until each x_i is closest to its group mean; that is, until a full cycle through all n data vectors produces no reassignments.

The need to prespecify the number of groups and their initial membership can be a disadvantage of the K-means method, that may or may not compensate its ability to reassign potentially misclassified observations. Unless there is prior knowledge of the correct number of groups, and/or the clustering is a precursor to subsequent analyses requiring a particular number of groups, it is probably wise to repeat K-means clustering for a range of initial group numbers G, and for different initial assignments of observations for each of the trial values of G.

14.3.2 Nucleated Agglomerative Clustering

Elements of agglomerative clustering and K-means clustering can be combined in an iterative procedure called nucleated agglomerative clustering. This method reduces somewhat the effects of arbitrary initial choices for group seeds in the K-means method, and automatically produces a sequence of K-means clusters through a range of group sizes G.

The nucleated agglomerative method begins by specifying a number of groups G_{init} that is larger than the number of groups G_{final} that will exist at the end of the procedure.

A K-means clustering into G_{init} groups is calculated, as described in Section 14.3.1. The following steps are then iterated:

1) The two closest groups are merged according to Ward's method. That is, the two groups are merged that minimize the increase in Equation 14.7.
2) K-means clustering is performed for the reduced number of groups, using the result of step 1 as the initial point. If the result is G_{final} groups, the algorithm stops. Otherwise, step 1 is repeated.

This algorithm produces a hierarchy of clustering solutions for the range of group sizes $G_{\text{init}} \geq G \geq G_{\text{final}}$, while allowing reassignment of observations to different groups at each stage in the hierarchy.

14.3.3 Clustering Using Mixture Distributions

Another approach to nonhierarchical clustering is to fit mixture distributions (see Section 4.4.6) (e.g., Everitt and Hand 1981; Titterington *et al.* 1985). In the statistical literature, this approach to clustering is called model-based, referring to the statistical model embodied in the mixture distribution (Banfield and Raftery 1993). For multivariate data the most usual approach is to fit mixtures of multivariate normal distributions, for which maximum likelihood estimation using the EM algorithm (see Section 4.6.3) is straightforward (the algorithm is outlined in Hannachi and O'Neill 2001, and Smyth *et al.* 1999). This approach to clustering has been applied to atmospheric data to identify large-scale flow regimes by Haines and Hannachi (1995), Hannachi (1997), and Smyth *et al.* (1999).

The basic idea in this approach to clustering is that each of the component PDFs $f_g(x), g = 1, \ldots, G$, represents one of the G groups from which the data have been drawn. As illustrated in Example 4.13, using the EM algorithm to estimate a mixture distribution produces (in addition to the distribution parameters) posterior probabilities (Equation 4.74) for membership in each of the component PDFs given each of the observed data values x_i. Using these posterior probabilities, a hard (i.e., nonprobabilistic) classification can be achieved by assigning each data vector x_i to that PDF $f_g(x)$ having the largest probability. However, in many applications retention of these probability estimates regarding group membership may be informative.

As is the case for other nonhierarchical clustering approaches, the number of groups G (in this case, the number of component PDFs $f_g(x)$) typically is specified in advance. However, Banfield and Raftery (1993) and Smyth *et al.* (1999) describe nonsubjective algorithms for choosing the number of groups, using a cross-validation approach.

14.4 Exercises

14.1. Compute the distance matrix $[\Delta]$ for the Guayaquil temperature and pressure data in Table A.3 for the six years 1965–1970, using Karl-Pearson distance.

14.2. From the distance matrix computed in Exercise 14.1, cluster the six years using
   a. Single linkage.
   b. Complete linkage.
   c. Average linkage.

14.3. Cluster the Guayaquil pressure data (Table A.3) for the six years 1965–1970, using

   a. The centroid method and Euclidean distance.

   b. Ward's method operating on the raw data.

14.4. Cluster the Guayaquil temperature data (Table A.3) for the six years 1965–1970 into two groups using the K-means method, beginning with $G_1 = \{1965, 1966, 1967\}$ and $G_2 = \{1968, 1969, 1970\}$.

Appendixes

APPENDIX · A

Example Data Sets

In real applications of climatological data analysis we would hope to use much more data (e.g., all available January daily data, rather than data for just a single year), and would have a computer perform the calculations. This small data set is used in a number of examples in this book so that the calculations can be performed by hand, and a clearer understanding of procedures can be achieved.

TABLE A.1   Daily precipitation (inches) and temperature (°F) observations at Ithaca and Canandaigua, New York, for January 1987.

Date	Ithaca			Canandaigua		
	Precipitation	Max Temp.	Min Temp.	Precipitation	Max Temp.	Min Temp.
1	0.00	33	19	0.00	34	28
2	0.07	32	25	0.04	36	28
3	1.11	30	22	0.84	30	26
4	0.00	29	−1	0.00	29	19
5	0.00	25	4	0.00	30	16
6	0.00	30	14	0.00	35	24
7	0.00	37	21	0.02	44	26
8	0.04	37	22	0.05	38	24
9	0.02	29	23	0.01	31	24
10	0.05	30	27	0.09	33	29
11	0.34	36	29	0.18	39	29
12	0.06	32	25	0.04	33	27
13	0.18	33	29	0.04	34	31
14	0.02	34	15	0.00	39	26
15	0.02	53	29	0.06	51	38
16	0.00	45	24	0.03	44	23
17	0.00	25	0	0.04	25	13
18	0.00	28	2	0.00	34	14
19	0.00	32	26	0.00	36	28
20	0.45	27	17	0.35	29	19
21	0.00	26	19	0.02	27	19
22	0.00	28	9	0.01	29	17
23	0.70	24	20	0.35	27	22
24	0.00	26	−6	0.08	24	2
25	0.00	9	−13	0.00	11	4
26	0.00	22	−13	0.00	21	5
27	0.00	17	−11	0.00	19	7
28	0.00	26	−4	0.00	26	8
29	0.01	27	−4	0.01	28	14
30	0.03	30	11	0.01	31	14
31	0.05	34	23	0.13	38	23
sum/avg.	3.15	29.87	13.00	2.40	31.77	20.23
std. dev.	0.243	7.71	13.62	0.168	7.86	8.81

TABLE A.2   January precipitation at Ithaca, New York, 1933–1982, inches.

1933	0.44	1945	2.74	1958	4.90	1970	1.03
1934	1.18	1946	1.13	1959	2.94	1971	1.11
1935	2.69	1947	2.50	1960	1.75	1972	1.35
1936	2.08	1948	1.72	1961	1.69	1973	1.44
1937	3.66	1949	2.27	1962	1.88	1974	1.84
1938	1.72	1950	2.82	1963	1.31	1975	1.69
1939	2.82	1951	1.98	1964	1.76	1976	3.00
1940	0.72	1952	2.44	1965	2.17	1977	1.36
1941	1.46	1953	2.53	1966	2.38	1978	6.37
1942	1.30	1954	2.00	1967	1.16	1979	4.55
1943	1.35	1955	1.12	1968	1.39	1980	0.52
1944	0.54	1956	2.13	1969	1.36	1981	0.87
		1957	1.36			1982	1.51

TABLE A.3   June climate data for Guayaquil, Ecuador, 1951–1970. Asterisks indicate El Niño years.

Year	Temperature, °C	Precipitation, mm	Pressure, mb
1951*	26.1	43	1009.5
1952	24.5	10	1010.9
1953*	24.8	4	1010.7
1954	24.5	0	1011.2
1955	24.1	2	1011.9
1956	24.3	Missing	1011.2
1957*	26.4	31	1009.3
1958	24.9	0	1011.1
1959	23.7	0	1012.0
1960	23.5	0	1011.4
1961	24.0	2	1010.9
1962	24.1	3	1011.5
1963	23.7	0	1011.0
1964	24.3	4	1011.2
1965*	26.6	15	1009.9
1966	24.6	2	1012.5
1967	24.8	0	1011.1
1968	24.4	1	1011.8
1969*	26.8	127	1009.3
1970	25.2	2	1010.6

APPENDIX • B

Probability Tables

Following are probability tables for selected common probability distributions, for which closed-form expressions for the cumulative distribution functions do not exist.

TABLE B.1   Left-tail cumulative probabilities for the standard Gaussian distribution, $\Phi(z) = \Pr\{Z \le z\}$. Values of the standardized Gaussian variable, z, are listed to tenths in the rightmost and leftmost columns. Remaining column headings index the hundredth place of z. Right-tail probabilities are obtained using $\Pr\{Z > z\} = 1 - \Pr\{Z \le z\}$. Probabilities for $Z > 0$ are obtained using the symmetry of the Gaussian distribution, $\Pr\{Z \le z\} = 1 - \Pr\{Z \le -z\}$.

Z	.09	.08	.07	.06	.05	.04	.03	.02	.01	.00	Z
−4.0	.00002	.00002	.00002	.00002	.00003	.00003	.00003	.00003	.00003	.00003	−4.0
−3.9	.00003	.00003	.00004	.00004	.00004	.00004	.00004	.00004	.00005	.00005	−3.9
−3.8	.00005	.00005	.00005	.00006	.00006	.00006	.00006	.00007	.00007	.00007	−3.8
−3.7	.00008	.00008	.00008	.00008	.00009	.00009	.00010	.00010	.00010	.00011	−3.7
−3.6	.00011	.00012	.00012	.00013	.00013	.00014	.00014	.00015	.00015	.00016	−3.6
−3.5	.00017	.00017	.00018	.00019	.00019	.00020	.00021	.00022	.00022	.00023	−3.5
−3.4	.00024	.00025	.00026	.00027	.00028	.00029	.00030	.00031	.00032	.00034	−3.4
−3.3	.00035	.00036	.00038	.00039	.00040	.00042	.00043	.00045	.00047	.00048	−3.3
−3.2	.00050	.00052	.00054	.00056	.00058	.00060	.00062	.00064	.00066	.00069	−3.2
−3.1	.00071	.00074	.00076	.00079	.00082	.00084	.00087	.00090	.00094	.00097	−3.1
−3.0	.00100	.00104	.00107	.00111	.00114	.00118	.00122	.00126	.00131	.00135	−3.0
−2.9	.00139	.00144	.00149	.00154	.00159	.00164	.00169	.00175	.00181	.00187	−2.9
−2.8	.00193	.00199	.00205	.00212	.00219	.00226	.00233	.00240	.00248	.00256	−2.8
−2.7	.00264	.00272	.00280	.00289	.00298	.00307	.00317	.00326	.00336	.00347	−2.7
−2.6	.00357	.00368	.00379	.00391	.00402	.00415	.00427	.00440	.00453	.00466	−2.6
−2.5	.00480	.00494	.00508	.00523	.00539	.00554	.00570	.00587	.00604	.00621	−2.5
−2.4	.00639	.00657	.00676	.00695	.00714	.00734	.00755	.00776	.00798	.00820	−2.4
−2.3	.00842	.00866	.00889	.00914	.00939	.00964	.00990	.01017	.01044	.01072	−2.3
−2.2	.01101	.01130	.01160	.01191	.01222	.01255	.01287	.01321	.01355	.01390	−2.2
−2.1	.01426	.01463	.01500	.01539	.01578	.01618	.01659	.01700	.01743	.01786	−2.1
−2.0	.01831	.01876	.01923	.01970	.02018	.02068	.02118	.02169	.02222	.02275	−2.0
−1.9	.02330	.02385	.02442	.02500	.02559	.02619	.02680	.02743	.02807	.02872	−1.9
−1.8	.02938	.03005	.03074	.03144	.03216	.03288	.03362	.03438	.03515	.03593	−1.8
−1.7	.03673	.03754	.03836	.03920	.04006	.04093	.04182	.04272	.04363	.04457	−1.7
−1.6	.04551	.04648	.04746	.04846	.04947	.05050	.05155	.05262	.05370	.05480	−1.6
−1.5	.05592	.05705	.05821	.05938	.06057	.06178	.06301	.06426	.06552	.06681	−1.5
−1.4	.06811	.06944	.07078	.07215	.07353	.07493	.07636	.07780	.07927	.08076	−1.4
−1.3	.08226	.08379	.08534	.08692	.08851	.09012	.09176	.09342	.09510	.09680	−1.3
−1.2	.09853	.10027	.10204	.10383	.10565	.10749	.10935	.11123	.11314	.11507	−1.2
−1.1	.11702	.11900	.12100	.12302	.12507	.12714	.12924	.13136	.13350	.13567	−1.1
−1.0	.13786	.14007	.14231	.14457	.14686	.14917	.15151	.15386	.15625	.15866	−1.0
−0.9	.16109	.16354	.16602	.16853	.17106	.17361	.17619	.17879	.18141	.18406	−0.9
−0.8	.18673	.18943	.19215	.19489	.19766	.20045	.20327	.20611	.20897	.21186	−0.8
−0.7	.21476	.21770	.22065	.22363	.22663	.22965	.23270	.23576	.23885	.24196	−0.7
−0.6	.24510	.24825	.25143	.25463	.25785	.26109	.26435	.26763	.27093	.27425	−0.6
−0.5	.27760	.28096	.28434	.28774	.29116	.29460	.29806	.30153	.30503	.30854	−0.5
−0.4	.31207	.31561	.31918	.32276	.32636	.32997	.33360	.33724	.34090	.34458	−0.4
−0.3	.34827	.35197	.35569	.35942	.36317	.36693	.37070	.37448	.37828	.38209	−0.3
−0.2	.38591	.38974	.39358	.39743	.40129	.40517	.40905	.41294	.41683	.42074	−0.2
−0.1	.42465	.42858	.43251	.43644	.44038	.44433	.44828	.45224	.45620	.46017	−0.1
−0.0	.46414	.46812	.47210	.47608	.48006	.48405	.48803	.49202	.49601	.50000	0.0

TABLE B.2   Quantiles of the standard ($\beta = 1$) Gamma distribution. Tabulated elements are values of the standardized random variable ξ corresponding to the cumulative probabilities $F(\xi)$ given in the column headings, for values of the shape parameter (α) given in the first column. To find quantiles for distributions with other scale parameters, enter the table at the appropriate row, read the standardized value in the appropriate column, and multiply the tabulated value by the scale parameter. To extract cumulative probabilities corresponding to a given value of the random variable, divide the value by the scale parameter, enter the table at the row appropriate to the shape parameter, and interpolate the result from the column headings.

α	.001	.01	.05	.10	.20	.30	.40	.50	.60	.70	.80	.90	.95	.99	.999
								Cumulative Probability							
0.05	0.0000	0.0000	0.0000	0.000	0.000	0.000	0.000	0.000	0.000	0.000	0.007	0.077	0.262	1.057	2.423
0.10	0.0000	0.0000	0.0000	0.000	0.000	0.000	0.000	0.001	0.004	0.018	0.070	0.264	0.575	1.554	3.035
0.15	0.0000	0.0000	0.0000	0.000	0.000	0.000	0.001	0.006	0.021	0.062	0.164	0.442	0.820	1.894	3.439
0.20	0.0000	0.0000	0.0000	0.000	0.000	0.002	0.007	0.021	0.053	0.122	0.265	0.602	1.024	2.164	3.756
0.25	0.0000	0.0000	0.0000	0.000	0.001	0.006	0.018	0.044	0.095	0.188	0.364	0.747	1.203	2.395	4.024
0.30	0.0000	0.0000	0.0000	0.000	0.003	0.013	0.034	0.073	0.142	0.257	0.461	0.882	1.365	2.599	4.262
0.35	0.0000	0.0000	0.0001	0.001	0.007	0.024	0.055	0.108	0.192	0.328	0.556	1.007	1.515	2.785	4.477
0.40	0.0000	0.0000	0.0004	0.002	0.013	0.038	0.080	0.145	0.245	0.398	0.644	1.126	1.654	2.958	4.677
0.45	0.0000	0.0000	0.0010	0.005	0.022	0.055	0.107	0.186	0.300	0.468	0.733	1.240	1.786	3.121	4.863
0.50	0.0000	0.0001	0.0020	0.008	0.032	0.074	0.138	0.228	0.355	0.538	0.819	1.349	1.913	3.274	5.040
0.55	0.0000	0.0002	0.0035	0.012	0.045	0.096	0.170	0.272	0.411	0.607	0.904	1.454	2.034	3.421	5.208
0.60	0.0000	0.0004	0.0057	0.018	0.059	0.120	0.204	0.316	0.467	0.676	0.987	1.556	2.150	3.562	5.370
0.65	0.0000	0.0008	0.0086	0.025	0.075	0.146	0.240	0.362	0.523	0.744	1.068	1.656	2.264	3.698	5.526
0.70	0.0001	0.0013	0.0123	0.033	0.093	0.173	0.276	0.408	0.579	0.811	1.149	1.753	2.374	3.830	5.676
0.75	0.0001	0.0020	0.0168	0.043	0.112	0.201	0.314	0.455	0.636	0.878	1.227	1.848	2.481	3.958	5.822
0.80	0.0003	0.0030	0.0221	0.053	0.132	0.231	0.352	0.502	0.692	0.945	1.305	1.941	2.586	4.083	5.964
0.85	0.0004	0.0044	0.0283	0.065	0.153	0.261	0.391	0.550	0.749	1.010	1.382	2.032	2.689	4.205	6.103
0.90	0.0007	0.0060	0.0353	0.078	0.176	0.292	0.431	0.598	0.805	1.076	1.458	2.122	2.790	4.325	6.239
0.95	0.0010	0.0080	0.0432	0.091	0.199	0.324	0.471	0.646	0.861	1.141	1.533	2.211	2.888	4.441	6.373

continued

TABLE B.2 continued

							Cumulative Probability									
α	.001	.01	.05	.10	.20	.30	.40	.50	.60	.70	.80	.90	.95	.99	.999
1.00	0.0014	0.0105	0.0517	0.106	0.224	0.357	0.512	0.694	0.918	1.206	1.607	2.298	2.986	4.556	6.503
1.05	0.0019	0.0133	0.0612	0.121	0.249	0.391	0.553	0.742	0.974	1.270	1.681	2.384	3.082	4.669	6.631
1.10	0.0022	0.0166	0.0713	0.138	0.275	0.425	0.594	0.791	1.030	1.334	1.759	2.469	3.177	4.781	6.757
1.15	0.0023	0.0202	0.0823	0.155	0.301	0.459	0.636	0.840	1.086	1.397	1.831	2.553	3.270	4.890	6.881
1.20	0.0024	0.0240	0.0938	0.173	0.329	0.494	0.678	0.889	1.141	1.460	1.903	2.636	3.362	4.998	7.003
1.25	0.0031	0.0271	0.1062	0.191	0.357	0.530	0.720	0.938	1.197	1.523	1.974	2.719	3.453	5.105	7.124
1.30	0.0037	0.0321	0.1192	0.210	0.385	0.566	0.763	0.987	1.253	1.586	2.045	2.800	3.544	5.211	7.242
1.35	0.0044	0.0371	0.1328	0.230	0.414	0.602	0.806	1.036	1.308	1.649	2.115	2.881	3.633	5.314	7.360
1.40	0.0054	0.0432	0.1451	0.250	0.443	0.639	0.849	1.085	1.364	1.711	2.185	2.961	3.722	5.418	7.476
1.45	0.0066	0.0493	0.1598	0.272	0.473	0.676	0.892	1.135	1.419	1.773	2.255	3.041	3.809	5.519	7.590
1.50	0.0083	0.0560	0.1747	0.293	0.504	0.713	0.935	1.184	1.474	1.834	2.324	3.120	3.897	5.620	7.704
1.55	0.0106	0.0632	0.1908	0.313	0.534	0.750	0.979	1.234	1.530	1.896	2.392	3.199	3.983	5.720	7.816
1.60	0.0136	0.0708	0.2070	0.336	0.565	0.788	1.023	1.283	1.585	1.957	2.461	3.276	4.068	5.818	7.928
1.65	0.0177	0.0780	0.2238	0.359	0.597	0.826	1.067	1.333	1.640	2.018	2.529	3.354	4.153	5.917	8.038
1.70	0.0232	0.0867	0.2411	0.382	0.628	0.865	1.111	1.382	1.695	2.079	2.597	3.431	4.237	6.014	8.147
1.75	0.0306	0.0958	0.2588	0.406	0.661	0.903	1.155	1.432	1.750	2.140	2.664	3.507	4.321	6.110	8.255
1.80	0.0360	0.1041	0.2771	0.430	0.693	0.942	1.199	1.481	1.805	2.200	2.731	3.584	4.405	6.207	8.362
1.85	0.0406	0.1145	0.2958	0.454	0.726	0.980	1.244	1.531	1.860	2.261	2.798	3.659	4.487	6.301	8.469
1.90	0.0447	0.1243	0.3142	0.479	0.759	1.020	1.288	1.580	1.915	2.321	2.865	3.735	4.569	6.396	8.575
1.95	0.0486	0.1361	0.3338	0.505	0.790	1.059	1.333	1.630	1.969	2.381	2.931	3.809	4.651	6.490	8.679
2.00	0.0525	0.1514	0.3537	0.530	0.823	1.099	1.378	1.680	2.024	2.442	2.997	3.883	4.732	6.582	8.783
2.05	0.0565	0.1637	0.3741	0.556	0.857	1.138	1.422	1.729	2.079	2.501	3.063	3.958	4.813	6.675	8.887
2.10	0.0657	0.1751	0.3949	0.583	0.891	1.178	1.467	1.779	2.133	2.561	3.129	4.032	4.894	6.767	8.989

2.15	0.0697	0.1864	0.4149	0.610	0.925	1.218	1.512	1.829	2.188	2.620	3.195	4.105	4.973	6.858	9.091
2.20	0.0740	0.2002	0.4365	0.637	0.959	1.258	1.557	1.879	2.242	2.680	3.260	4.179	5.053	6.949	9.193
2.25	0.0854	0.2116	0.4584	0.664	0.994	1.298	1.603	1.928	2.297	2.739	3.325	4.252	5.132	7.039	9.294
2.30	0.0898	0.2259	0.4807	0.691	1.029	1.338	1.648	1.978	2.351	2.799	3.390	4.324	5.211	7.129	9.394
2.35	0.0945	0.2378	0.5023	0.718	1.064	1.379	1.693	2.028	2.405	2.858	3.455	4.396	5.289	7.219	9.493
2.40	0.0996	0.2526	0.5244	0.747	1.099	1.420	1.738	2.078	2.459	2.917	3.519	4.468	5.367	7.308	9.592
2.45	0.1134	0.2680	0.5481	0.775	1.134	1.460	1.784	2.127	2.514	2.976	3.584	4.540	5.445	7.397	9.691
2.50	0.1184	0.2803	0.5754	0.804	1.170	1.500	1.829	2.178	2.568	3.035	3.648	4.612	5.522	7.484	9.789
2.55	0.1239	0.2962	0.5978	0.833	1.205	1.539	1.875	2.227	2.622	3.093	3.712	4.683	5.600	7.572	9.886
2.60	0.1297	0.3129	0.6211	0.862	1.241	1.581	1.920	2.277	2.676	3.152	3.776	4.754	5.677	7.660	9.983
2.65	0.1468	0.3255	0.6456	0.890	1.277	1.622	1.966	2.327	2.730	3.210	3.840	4.825	5.753	7.746	10.079
2.70	0.1523	0.3426	0.6705	0.920	1.314	1.663	2.011	2.376	2.784	3.269	3.903	4.896	5.830	7.833	10.176
2.75	0.1583	0.3561	0.6938	0.950	1.350	1.704	2.058	2.427	2.838	3.328	3.967	4.966	5.906	7.919	10.272
2.80	0.1647	0.3735	0.7188	0.980	1.386	1.746	2.103	2.476	2.892	3.386	4.030	5.040	5.982	8.004	10.367
2.85	0.1861	0.3919	0.7441	1.009	1.423	1.787	2.149	2.526	2.946	3.444	4.093	5.120	6.058	8.090	10.461
2.90	0.1919	0.4056	0.7697	1.040	1.460	1.829	2.195	2.576	2.999	3.502	4.156	5.190	6.133	8.175	10.556
2.95	0.1982	0.4242	0.7936	1.070	1.497	1.871	2.241	2.626	3.054	3.560	4.220	5.260	6.208	8.260	10.649
3.00	0.2050	0.4388	0.8193	1.101	1.534	1.913	2.287	2.676	3.108	3.618	4.283	5.329	6.283	8.345	10.743
3.05	0.2123	0.4577	0.8454	1.134	1.571	1.954	2.333	2.726	3.161	3.676	4.346	5.398	6.357	8.429	10.837
3.10	0.2385	0.4778	0.8717	1.165	1.607	1.996	2.378	2.776	3.215	3.734	4.408	5.468	6.432	8.513	10.930
3.15	0.2447	0.4922	0.8982	1.197	1.645	2.038	2.425	2.825	3.268	3.792	4.471	5.537	6.506	8.596	11.023
3.20	0.2514	0.5125	0.9251	1.227	1.682	2.080	2.471	2.875	3.322	3.850	4.533	5.605	6.580	8.680	11.113
3.25	0.2588	0.5278	0.9498	1.259	1.720	2.123	2.517	2.925	3.376	3.907	4.595	5.675	6.654	8.763	11.205
3.30	0.2667	0.5483	0.9767	1.291	1.758	2.165	2.563	2.975	3.430	3.965	4.658	5.743	6.727	8.845	11.298
3.35	0.2995	0.5704	1.0039	1.323	1.796	2.207	2.610	3.025	3.483	4.022	4.720	5.811	6.801	8.928	11.389

continued

TABLE B.2   continued

α	.001	.01	.05	.10	.20	.30	.40	.50	.60	.70	.80	.90	.95	.99	.999
									Cumulative Probability						
3.40	0.3057	0.5850	1.0313	1.354	1.834	2.250	2.656	3.075	3.537	4.079	4.782	5.879	6.874	9.010	11.480
3.45	0.3126	0.6072	1.0590	1.386	1.872	2.292	2.702	3.125	3.590	4.137	4.843	5.948	6.947	9.093	11.570
3.50	0.3201	0.6228	1.0870	1.418	1.910	2.334	2.748	3.175	3.644	4.194	4.905	6.015	7.020	9.174	11.660
3.55	0.3282	0.6450	1.1152	1.451	1.948	2.377	2.795	3.225	3.697	4.252	4.967	6.084	7.092	9.255	11.749
3.60	0.3370	0.6614	1.1405	1.483	1.985	2.420	2.841	3.274	3.750	4.309	5.028	6.152	7.165	9.337	11.840
3.65	0.3767	0.6837	1.1687	1.516	2.024	2.462	2.887	3.324	3.804	4.366	5.091	6.219	7.237	9.418	11.929
3.70	0.3830	0.7084	1.1972	1.549	2.062	2.505	2.934	3.374	3.858	4.423	5.152	6.286	7.310	9.499	12.017
3.75	0.3900	0.7233	1.2259	1.582	2.101	2.547	2.980	3.425	3.911	4.480	5.214	6.354	7.381	9.579	12.107
3.80	0.3978	0.7480	1.2549	1.613	2.140	2.590	3.027	3.474	3.964	4.537	5.275	6.420	7.454	9.659	12.195
3.85	0.4064	0.7637	1.2843	1.646	2.179	2.633	3.073	3.524	4.018	4.594	5.336	6.488	7.525	9.740	12.284
3.90	0.4157	0.7883	1.3101	1.680	2.218	2.676	3.120	3.574	4.071	4.651	5.397	6.555	7.596	9.820	12.371
3.95	0.4259	0.8049	1.3393	1.713	2.257	2.719	3.163	3.624	4.124	4.708	5.458	6.622	7.668	9.900	12.459
4.00	0.4712	0.8294	1.3687	1.746	2.295	2.762	3.209	3.674	4.177	4.765	5.519	6.689	7.739	9.980	12.546
4.05	0.4779	0.8469	1.3984	1.780	2.334	2.805	3.256	3.724	4.231	4.822	5.580	6.755	7.811	10.059	12.634
4.10	0.4853	0.8714	1.4285	1.814	2.373	2.848	3.302	3.774	4.284	4.879	5.641	6.821	7.882	10.137	12.721
4.15	0.4937	0.8999	1.4551	1.848	2.413	2.891	3.350	3.823	4.337	4.936	5.701	6.888	7.952	10.216	12.807
4.20	0.5030	0.9141	1.4850	1.882	2.451	2.935	3.396	3.874	4.390	4.992	5.762	6.954	8.023	10.295	12.894
4.25	0.5133	0.9424	1.5150	1.916	2.491	2.978	3.443	3.924	4.444	5.049	5.823	7.020	8.093	10.374	12.981
4.30	0.5244	0.9575	1.5454	1.950	2.531	3.021	3.489	3.974	4.497	5.105	5.883	7.086	8.170	10.453	13.066
4.35	0.5779	0.9856	1.5762	1.985	2.572	3.065	3.537	4.024	4.550	5.162	5.944	7.153	8.264	10.531	13.152
4.40	0.5842	1.0016	1.6034	2.017	2.612	3.108	3.584	4.074	4.603	5.218	6.005	7.219	8.334	10.609	13.238
4.45	0.5916	1.0294	1.6339	2.051	2.653	3.152	3.630	4.123	4.656	5.274	6.065	7.284	8.405	10.687	13.324

4.50	0.6001	1.0463	1.6646	2.085	2.691	3.195	3.677	4.173	4.709	5.331	6.126	7.350	8.475	10.765	13.410
4.55	0.6096	1.0739	1.6956	2.120	2.731	3.239	3.724	4.223	4.762	5.387	6.186	7.415	8.544	10.843	13.495
4.60	0.6202	1.0917	1.7271	2.155	2.771	3.283	3.771	4.273	4.815	5.443	6.246	7.480	8.615	10.920	13.578
4.65	0.6319	1.1191	1.7547	2.190	2.812	3.326	3.817	4.323	4.868	5.501	6.306	7.546	8.684	10.998	13.663
4.70	0.6978	1.1378	1.7857	2.225	2.852	3.369	3.864	4.373	4.921	5.557	6.366	7.611	8.754	11.075	13.748
4.75	0.7031	1.1649	1.8170	2.260	2.890	3.412	3.911	4.423	4.974	5.613	6.426	7.676	8.823	11.152	13.832
4.80	0.7095	1.1844	1.8487	2.295	2.930	3.456	3.958	4.474	5.027	5.669	6.486	7.742	8.892	11.229	13.916
4.85	0.7172	1.2113	1.8809	2.330	2.970	3.500	4.005	4.524	5.081	5.725	6.546	7.807	8.962	11.306	14.000
4.90	0.7262	1.2465	1.9088	2.366	3.011	3.544	4.052	4.573	5.134	5.781	6.606	7.872	9.031	11.382	14.084
4.95	0.7365	1.2582	1.9403	2.398	3.051	3.588	4.099	4.623	5.186	5.837	6.665	7.937	9.100	11.457	14.168
5.00	0.7482	1.2931	1.9722	2.434	3.091	3.632	4.146	4.673	5.239	5.893	6.725	8.002	9.169	11.534	14.251

TABLE B.3   Right-tail quantiles of the Chi-square distribution. For large ν, the Chi-square distribution is approximately Gaussian, with mean ν and variance 2ν.

ν	Cumulative Probability					
	0.50	0.90	0.95	0.99	0.999	0.9999
1	0.455	2.706	3.841	6.635	10.828	15.137
2	1.386	4.605	5.991	9.210	13.816	18.421
3	2.366	6.251	7.815	11.345	16.266	21.108
4	3.357	7.779	9.488	13.277	18.467	23.512
5	4.351	9.236	11.070	15.086	20.515	25.745
6	5.348	10.645	12.592	16.812	22.458	27.855
7	6.346	12.017	14.067	18.475	24.322	29.878
8	7.344	13.362	15.507	20.090	26.124	31.827
9	8.343	14.684	16.919	21.666	27.877	33.719
10	9.342	15.987	18.307	23.209	29.588	35.563
11	10.341	17.275	19.675	24.725	31.264	37.366
12	11.340	18.549	21.026	26.217	32.910	39.134
13	12.340	19.812	22.362	27.688	34.528	40.871
14	13.339	21.064	23.685	29.141	36.123	42.578
15	14.339	22.307	24.996	30.578	37.697	44.262
16	15.338	23.542	26.296	32.000	39.252	45.925
17	16.338	24.769	27.587	33.409	40.790	47.566
18	17.338	25.989	28.869	34.805	42.312	49.190
19	18.338	27.204	30.144	36.191	43.820	50.794
20	19.337	28.412	31.410	37.566	45.315	52.385
21	20.337	29.615	32.671	38.932	46.797	53.961
22	21.337	30.813	33.924	40.289	48.268	55.523
23	22.337	32.007	35.172	41.638	49.728	57.074
24	23.337	33.196	36.415	42.980	51.179	58.613
25	24.337	34.382	37.652	44.314	52.620	60.140
26	25.336	35.563	38.885	45.642	54.052	61.656
27	26.336	36.741	40.113	46.963	55.476	63.164
28	27.336	37.916	41.337	48.278	56.892	64.661
29	28.336	39.087	42.557	49.588	58.301	66.152
30	29.336	40.256	43.773	50.892	59.703	67.632
31	30.336	41.422	44.985	52.191	61.098	69.104
32	31.336	42.585	46.194	53.486	62.487	70.570
33	32.336	43.745	47.400	54.776	63.870	72.030
34	33.336	44.903	48.602	56.061	65.247	73.481
35	34.336	46.059	49.802	57.342	66.619	74.926
36	35.336	47.212	50.998	58.619	67.985	76.365
37	36.336	48.363	52.192	59.892	69.347	77.798
38	37.335	49.513	53.384	61.162	70.703	79.224

TABLE B.3   continued

ν	Cumulative Probability					
	0.50	0.90	0.95	0.99	0.999	0.9999
39	38.335	50.660	54.572	62.428	72.055	80.645
40	39.335	51.805	55.758	63.691	73.402	82.061
41	40.335	52.949	56.942	64.950	74.745	83.474
42	41.335	54.090	58.124	66.206	76.084	84.880
43	42.335	55.230	59.304	67.459	77.419	86.280
44	43.335	56.369	60.481	68.710	78.750	87.678
45	44.335	57.505	61.656	69.957	80.077	89.070
46	45.335	58.641	62.830	71.201	81.400	90.456
47	46.335	59.774	64.001	72.443	82.721	91.842
48	47.335	60.907	65.171	73.683	84.037	93.221
49	48.335	62.038	66.339	74.919	85.351	94.597
50	49.335	63.167	67.505	76.154	86.661	95.968
55	54.335	68.796	73.311	82.292	93.168	102.776
60	59.335	74.397	79.082	88.379	99.607	109.501
65	64.335	79.973	84.821	94.422	105.988	116.160
70	69.334	85.527	90.531	100.425	112.317	122.754
75	74.334	91.061	96.217	106.393	118.599	129.294
80	79.334	96.578	101.879	112.329	124.839	135.783
85	84.334	102.079	107.522	118.236	131.041	142.226
90	89.334	107.565	113.145	124.116	137.208	148.626
95	94.334	113.038	118.752	129.973	143.344	154.989
100	99.334	118.498	124.342	135.807	149.449	161.318

APPENDIX ◆ C

Answers to Exercises

Chapter 2

2.1. b. $\Pr\{A \cup B\} = 0.7$
 c. $\Pr\{A \cup B^C\} = 0.1$
 d. $\Pr\{A^C \cup B^C\} = 0.3$
2.2. b. $\Pr\{A\} = 9/31, \Pr\{B\} = 15/31, \Pr\{A, B\} = 9/31$
 c. $\Pr\{A|B\} = 9/15$
 d. No: $\Pr\{A\} \neq \Pr\{A|B\}$
 2.3 a. 18/22
 b. 22/31
2.4. b. $\Pr\{E_1, E_2, E_3\} = .000125$
 c. $\Pr\{E_1^C, E_2^C, E_3^C\} = .857$
2.5. 0.20

Chapter 3

3.1. median $= 2$ mm, trimean $= 2.75$ mm, mean $= 12.95$ mm
3.2. MAD $= 0.4$ mb, IQ $= 0.8$ mb, s $= 1.26$ mb
3.3. $\gamma_{YK} = 0.273$, $\gamma = 0.877$
3.7. $\lambda = 0$
3.9. z $= 1.36$
3.10. $r_0 = 1.000$, $r_1 = 0.652$, $r_2 = 0.388$, $r_3 = 0.281$
3.12. Pearson: 1.000   Spearman   1.000
             0.703   1.000   0.606   1.000
           -0.830 -0.678 1.000 -0.688 -0.632 1.000

Chapter 4

4.1. 0.168
4.2. a. 0.037
 b. 0.331

4.3. a. $\mu_{drought} = 0.056, \mu_{wet} = 0.565$
      b. 0.054
      c. 0.432
4.4. \$280 million, \$2.825 billion
4.5. a. $\mu = 24.8°C, \sigma = 0.98°C$
      b. $\mu = 76.6°F, \sigma = 1.76°F$
4.6. a. 0.00939
      b. 22.9°C
4.7. a. $\alpha = 3.785, \beta = 0.934''$
      b. $\alpha = 3.785, \beta = 23.7\,mm$
4.8. a. $q_{30} = 2.41'' = 61.2\,mm; q_{70} = 4.22'' = 107.2\,mm$
      b. $0.30''$, or 7.7 mm
      c. $\cong 0.05$
4.9. a. $q_{30} = 2.30'' = 58.3\,mm; q_{70} = 4.13'' = 104.9\,mm$
      b. $0.46''$, or 11.6 mm
      c. $\cong 0.07$
4.10. a. $\beta = 35.1\,cm, \xi = 59.7\,cm$
       b. $x = \xi - \beta\ln[-\ln(F)]; Pr\{X \le 221\,cm\} = 0.99$
4.11. a. $\mu_{max} = 31.8°F, \sigma_{max} = 7.86°F, \mu_{min} = 20.2°F, \sigma_{min} = 8.81°F, \rho = 0.810$
       b. 0.728
4.13. a. $\beta = \Sigma x/n$
       b. $-I^{-1}(\hat{\beta}) = \hat{\beta}^2/n$
4.14. $x(u) = \beta[-\ln(1-u)]^{1/\alpha}$

Chapter 5

5.1. a. $z = 4.88$, reject H_0
      b. $[1.26°C, 2.40°C]$
5.2. 6.53 days (Ithaca), 6.08 days (Canandaigua)
5.3. $z = -3.94$
      a. $p = 0.00008$
      b. $p = 0.00004$
5.4. $|r| \ge 0.366$
5.5. a. $D_n = 0.152$ (reject at 10%, not at 5% level)
      b. For classes: $[<2, 2–3, 3–4, 4–5, \ge 5], \chi^2 = 0.522$ (do not reject)
      c. r = 0.971 (do not reject)
5.6. $\Lambda = 21.86$, reject $(p < .001)$
5.7. a. $U_1 = 1$, reject $(p < .005)$
      b. $z = -1.88$, reject $(p = .03)$
5.8. $\approx [1.02, 3.59]$
5.9. a. Observed $(s_{E-N}^2/s_{non-E-N}^2) = 329.5$; permutation distribution critical value (1%, 2-tailed) ≈ 141, reject $H_0(p < 0.01)$
      b. 15/10000 members of bootstrap sampling distribution for $s_{E-N}^2/s_{non-E-N}^2 \le 1$; 2-tailed $p = 0.003$
5.10. Not significant, even assuming spatial independence $(p = 0.19,$ one-tailed)

Chapter 6

6.1. a. $a = 959.8°C$, $b = -0.925°C/mb$
     c. $z = -6.33$
     d. 0.690
     e. 0.876
     f. 0.925
6.2. a. 3
     b. 117.9
     c. 0.974
     d. 0.715
6.3. $\ln[\bar{y}/(1 - \bar{y})]$
6.4. a. 1.74 mm
     b. [0 mm, 13.1 mm]
6.5. Range of slopes, -0.850 to -1.095; MSE $= 0.369$
6.6. a. -59 n.m.
     b. -66 n.m.
6.7. a. 65.8°F
     b. 52.5°F
     c. 21.7°F
     d. 44.5°F
6.8. a. 0.65
     b. 0.49
     c. 0.72
     d. 0.56
6.9. $f_{MOS} = 30.8°F + (0)(Th)$
6.10. 0.20
6.11. a. 12 mm
     b. [5 mm, 32 mm], [1 mm, 55 mm]
     c. 0.625

Chapter 7

7.1. a. .0025 .0013 .0108 .0148 .0171 .0138 .0155 .0161 .0177 .0176 .0159 .0189
         .4087 .0658 .1725 .0838 .0445 .0228 .0148 .0114 .0068 .0044 .0011 .0014
     b. 0.162
7.2. 1644   1330
      364   9064
7.3. a. 0.863
     b. 0.493
     c. 0.578
     d. 0.691
     e. 0.456
7.4. a. 0.074
     b. 0.097
     c. 0.761
     d. 0.406

7.5. a. .9597   .0127   .0021   .0007
         .0075   .0043   .0014   .0005
         .0013   .0013   .0009   .0003
         .0007   .0006   .0049   .0009
     b. 0.966
     c. 0.369
     d. 0.334
7.6. a. 5.37°F
     b. 7.54°F
     c. −0.03°F
     d. 1.95%
7.7. a. 0.1215
     b. 0.1699
     c. 28.5%
7.8. a. .0415   .0968   .1567   .1428   .1152   .0829   .1060   .0829   .0783   .0553   .0415
         .3627   .2759   .1635   .0856   .0498   .0230   .0204   .0102   .0051   .0026   .0013
     c. $H = $ .958, .862, .705, .562, .447, .364, .258, .175, .097, .042
         $F = $ .637, .361, .198, .112, .062, .039, .019, .009, .004, .001
     d. $A = 0.831, z = -14.9$
7.9. a. 0.298
     b. 16.4%
7.10. a. 30.3
      b. 5.31 dam^2
      c. 46.9%
      d. 0.726
      e. 0.714
7.11. a. 5 rank 1, 2 rank 2, 3 rank 3, 2 rank 4, 2 rank 5, 6 rank 6
      b. underdispersed
7.12 .352, .509, .673, .598, .504, .426, .343, .275, .195, .128, −.048

Chapter 8

8.1. a. $p_{01} = 0.45, p_{11} = 0.79$
     b. $\chi^2 = 3.51, p \approx 0.064$
     c. $\pi_1 = 0.682, n_{\bullet 1}/n = 0.667$
     d. $r_0 = 1.00, r_1 = 0.34, r_2 = 0.12, r_3 = 0.04$
     e. 0.624
8.2. a. $r_0 = 1.00, r_1 = 0.40, r_2 = 0.16, r_3 = 0.06, r_4 = 0.03, r_5 = 0.01$
     b. $r_0 = 1.00, r_1 = 0.41, r_2 = -0.41, r_3 = -0.58, r_4 = -0.12, r_5 = 0.32$
8.3. a. $AR(1) : \phi = 0.80; s_\varepsilon^2 = 36.0$
         $AR(2) : \phi_1 = 0.89, \phi_2 = -0.11; s_\varepsilon^2 = 35.5$
         $AR(3) : \phi_1 = 0.91, \phi_2 = -0.25, \phi_3 = 0.16; s_\varepsilon^2 = 34.7$
     b. $AR(1) : BIC = 369.6$
     c. $AR(1) : AIC = 364.4$
8.4. $x_1 = 71.5, x_2 = 66.3, x_3 = 62.1$
8.5. a. 28.6
     b. 19.8
     c. 4.5
8.6. a. $C_1 = 16.92°F, \phi_1 = 199°; C_2 = 4.16°F, \phi_2 = 256°$

8.7. a. 82.0°F
 b. 74.8°F
8.8. b. 0.990
8.9. 56
8.10. a. e.g., $f_A = 1 - .0508\ mo^{-1} = .9492\ mo^{-1}$
 b. $\approx$twice monthly
8.12. a. [0.11, 16.3]
 b. $C_{11}^2 < 0.921$, do not reject

Chapter 9

9.1. 216.0 −4.32
 135.1 7.04
9.2. $([X]^T y)^T = [627, 11475], [X^T X]^{-1} = \begin{matrix} 0.06263 & -0.002336 \\ -0.002336 & 0.0001797 \end{matrix}, b^T = [12.46, 0.60]$

9.3. 90°

9.6. a. 59.5 58.1
 58.1 61.8
 b. .205 −.193
 −.193 .197
 c. .205 −.193
 −.193 .197
 d. 6.16 4.64
 4.64 6.35
 e. 1.765

9.7. a. 59.52
 75.43 185.47
 58.07 81.63 61.85
 51.70 110.80 56.12 77.58
 b. $\mu_y^T = [21.4, 26.0]$
 $[S_y] = $ 98.96 75.55
 75.55 62.92

Chapter 10

10.2. a. $\mu = [29.87, 13.00]^T, [S] = $ 5.09 −0.41
 −0.41 26.23
 b. $N_2(\mu[\Sigma]);\quad \mu = [-1.90, 5.33]^T\quad [\Sigma] = $ 5.23 7.01
 7.01 50.24
10.3. $r = 0.974 > r_{crit}(10\%) = 0.970$; do not reject
10.4. a. $T^2 = 68.5 >> 18.421 = \chi_2^2(.9999)$; reject
 b. $a \propto [-.6217, .1929]^T$
10.5. a. $T^2 = 7.80$, reject @ 5%
 b. $a \propto [-.0120, .0429]^T$

Chapter 11

11.1. a. 3.78, 4.51
    b. 118.8
    c. 0.979

11.2. a. Correlation matrix: $\Sigma \lambda_k = 3$
    b. 1, 1, 3
    c. $\boldsymbol{x}_1^{\mathrm{T}} \approx [26.2, 42.6, 1009.6]$

11.3 a. [1.51, 6.80], [0.22, 0.98], [0.10, 0.46]
    b. λ_1 and λ_2 may be entangled

11.4 a.

$$
\begin{array}{rrr}
.593 & .332 & .734 \\
.552 & -.831 & -.069 \\
-.587 & -.446 & .676
\end{array}
$$

    b.

$$
\begin{array}{rrr}
.377 & .556 & 1.785 \\
.351 & -1.39 & -.168 \\
-.373 & -.747 & 1.644
\end{array}
$$

11.5 9.18, 14.34, 10.67

Chapter 12

12.1 6 Jan: $v_1 = .038$, $w_1 = .433$; 7 Jan: $v_1 = .868$, $w_1 = 1.35$

12.2 39.0°F, 23.6°F

12.3 a.

$$
\begin{array}{rrrr}
1.883 & 0 & 1.698 & -0.295 \\
0 & 0.927 & .0384 & 0.692 \\
1.698 & 0.38 & 1.814 & 0 \\
-0.295 & 0.692 & 0 & 1.019
\end{array}
$$

    b. $\boldsymbol{a}_1 = [.6862, .3496]^{\mathrm{T}}$, $\boldsymbol{b}_1 = [.7419, .0400]^{\mathrm{T}}$, $r_{C_1} = 0.965$
        $\boldsymbol{a}_2 = [-.2452, .9784]^{\mathrm{T}}$, $\boldsymbol{b}_2 = [-.0300, .9898]^{\mathrm{T}}$, $r_{C_2} = 0.746$

Chapter 13

13.1. a. $\boldsymbol{a}_1^{\mathrm{T}} = [0.83, -0.56]$
    b. 1953
    c. 1953

13.2 a. $\delta_1 = 38.65$, $\delta_2 = -14.99$; Group 3
    b. 5.2×10^{-12}, 2.8×10^{-9}, 0.99999997

13.3 a. 0.006
    b. 0.059
    c. 0.934

Chapter 14

14.1

$$
\begin{array}{llllll}
0 \\
3.63 & 0 \\
2.30 & 1.61 & 0 \\
3.14 & 0.82 & 0.90 & 0 \\
0.73 & 4.33 & 2.93 & 3.80 & 0 \\
1.64 & 2.28 & 0.72 & 1.62 & 2.22 & 0
\end{array}
$$

14.2 a. $1967 + 1970, d = 0.72$; $1965 + 1969, d = 0.73$; $1966 + 1968$,
         $d = 0.82$; $(1967 + 1970) + (1966 + 1968), d = 1.61$; all, $d = 1.64$.
      b. $1967 + 1970, d = 0.72$; $1965 + 1969, d = 0.73$; $1966 + 1968$,
         $d = 0.82$; $(1967 + 1970) + (1966 + 1968), d = 2.28$; all, $d = 4.33$.
      c. $1967 + 1970, d = 0.72$; $1965 + 1969, d = 0.73$; $1966 + 1968$,
         $d = 0.82$; $(1967 + 1970) + (1966 + 1968), d = 1.60$; all, $d = 3.00$.
14.3 a. $1967 + 1970, d = 0.50$; $1965 + 1969, d = 0.60$; $1966 + 1968$,
         $d = 0.70$; $(1967 + 1970) + (1965 + 1969), d = 1.25$; all, $d = 1.925$.
      b. $1967 + 1970, d = 0.125$; $1965 + 1969, d = 0.180$; $1966 + 1968$,
         $d = .245$; $(1967 + 1970) + (1965 + 1969), d = 1.868$; all, $d = 7.053$.
14.4 {1966, 1967}, {1965, 1968, 1969, 1970}; {1966, 1967, 1968}, {1965, 1969, 1970};
      {1966, 1967, 1968, 1970}, {1965, 1969}.

References

Abramowitz, M., and I.A. Stegun, eds., 1984. *Pocketbook of Mathematical Functions.* Frankfurt, Verlag Harri Deutsch, 468 pp.

Agresti, A., 1996. *An Introduction to Categorical Data Analysis.* Wiley, 290 pp.

Agresti, A., and B.A. Coull, 1998. Approximate is better than "exact" for interval estimation of binomial proportions. *American Statistician*, **52**, 119–126.

Akaike, H., 1974. A new look at the statistical model identification. *IEEE Transactions on Automatic Control*, **19**, 716–723.

Allen, M.R., and L.A. Smith, 1996. Monte Carlo SSA: Detecting irregular oscillations in the presence of colored noise. *Journal of Climate*, **9**, 3373–3404.

Anderson, J.L., 1996. A method for producing and evaluating probabilistic forecasts from ensemble model integrations. *Journal of Climate*, **9**, 1518–1530.

Anderson, J.L., 1997. The impact of dynamical constraints on the selection of initial conditions for ensemble predictions: low-order perfect model results. *Monthly Weather Review*, **125**, 2969–2983.

Anderson, J.L., 2003. A local least squares framework for ensemble filtering. *Monthly Weather Review*, **131**, 634–642.

Anderson, J., H. van den Dool, A. Barnston, W. Chen, W. Stern, and J. Ploshay, 1999. Present-day capabilities of numerical and statistical models for atmospheric extratropical seasonal simulation and prediction. *Bulletin of the American Meteorological Society*, **80**, 1349–1361.

Andrews, D.F., P.J. Bickel, F.R. Hampel, P.J. Huber, W.N. Rogers, and J.W. Tukey, 1972. *Robust Estimates of Location—Survey and Advances.* Princeton University Press.

Andrews, D.F., R. Gnanadesikan, and J.L. Warner, 1971. Transformations of multivariate data. *Biometrics*, **27**, 825–840.

Applequist, S., G.E. Gahrs, and R.L. Pfeffer, 2002. Comparison of methodologies for probabilistic quantitative precipitation forecasting. *Weather and Forecasting*, **17**, 783–799.

Araneo, D.C., and R.H. Compagnucci, 2004. Removal of systematic biases in S-mode principal components arising from unequal grid spacing. *Journal of Climate*, **17**, 394–400.

Arkin, P.A., 1989. The global climate for December 1988–February 1989: Cold episode in the tropical pacific continues. *Journal of Climate*, **2**, 737–757.

Atger, F., 1999. The skill of ensemble prediction systems. *Monthly Weather Review*, **127**, 1941–1953.

Azcarraga, R., and A.J. Ballester G., 1991. Statistical system for forecasting in Spain. In: H.R. Glahn, A.H. Murphy, L.J. Wilson, and J.S. Jensenius, Jr., eds., *Programme on Short- and Medium-Range Weather Prediction Research*. World Meteorological Organization WM/TD No. 421. XX23–25.

Baker, D.G., 1981. Verification of fixed-width, credible interval temperature forecasts. *Bulletin of the American Meteorological Society*, **62**, 616–619.

Banfield, J.D., and A.E. Raftery, 1993. Model-based Gaussian and non-Gaussian clustering. *Biometrics*, **49**, 803–821.

Barker, T.W., 1991. The relationship between spread and forecast error in extended-range forecasts. *Journal of Climate*, **4**, 733–742.

Barnett, T.P., and R.W. Preisendorfer, 1987. Origins and levels of monthly and seasonal forecast skill for United States surface air temperatures determined by canonical correlation analysis. *Monthly Weather Review*, **115**, 1825–1850.

Barnston, A.G., 1994. Linear statistical short-term climate predictive skill in the northern hemisphere. *Journal of Climate*, **7**, 1513–1564.

Barnston, A.G., M.H. Glantz, and Y. He, 1999. Predictive skill of statistical and dynamical climate models in SST forecasts during the 1997–1998 El Niño episode and the 1998 La Niña onset. *Bulletin of the American Meteorological Society*, **80**, 217–243.

Barnston, A.G., A. Leetmaa, V.E. Kousky, R.E. Livezey, E.A. O'Lenic, H. van den Dool, A.J. Wagner, and D.A. Unger, 1999. NCEP forecasts of the El Niño of 1997–98 and its U.S. impacts. *Bulletin of the American Meteorological Society*, **80**, 1829–1852.

Barnston, A.G., S.J. Mason, L. Goddard, D.G. DeWitt, and S.E. Zebiak, 2003. Multi-model ensembling in seasonal climate forecasting at IRI. *Bulletin of the American Meteorological Society*, **84**, 1783–1796.

Barnston, A.G., and C.F. Ropelewski, 1992. Prediction of ENSO episodes using canonical correlation analysis. *Journal of Climate*, **5**, 1316–1345.

Barnston, A.G., and T.M. Smith, 1996. Specification and prediction of global surface temperature and precipitation from global SST using CCA. *Journal of Climate*, **9**, 2660–2697.

Barnston, A.G., and H.M. van den Dool, 1993. A degeneracy in cross-validated skill in regression-based forecasts. *Journal of Climate*, **6**, 963–977.

Baughman, R.G., D.M. Fuquay, and P.W. Mielke, Jr., 1976. Statistical analysis of a randomized lightning modification experiment. *Journal of Applied Meteorology*, **15**, 790–794.

Beran, R., and M.S. Srivastava, 1985. Bootstrap tests and confidence regions for functions of a covariance matrix. *Annals of Statistics*, **13**, 95–115.

Beyth-Marom, R., 1982. How probable is probable? A numerical translation of verbal probability expressions. *Journal of Forecasting*, **1**, 257–269.

Bjerknes, J., 1969. Atmospheric teleconnections from the equatorial Pacific. *Monthly Weather Review*, **97**, 163–172.

Blackmon, M.L., 1976. A climatological spectral study of the 500 mb geopotential height of the northern hemisphere. *Journal of the Atmospheric Sciences*, **33**, 1607–1623.

Bloomfield, P., and D. Nychka, 1992. Climate spectra and detecting climate change. *Climatic Change*, **21**, 275–287.

Boswell, M.T., S.D. Gore, G.P. Patil, and C. Taillie, 1993. The art of computer generation of random variables. In: C.R. Rao, ed., *Handbook of Statistics, Vol. 9*, Elsevier, 661–721.

Box, G.E.P, and D.R. Cox, 1964. An analysis of transformations. *Journal of the Royal Statistical Society*, **B26**, 211–243.

Box, G.E.P., and G.M. Jenkins, 1994. *Time Series Analysis: Forecasting and Control*. Prentice-Hall, 598 pp.

Bradley, A.A., T. Hashino, and S.S. Schwartz, 2003. Distributions-oriented verification of probability forecasts for small data samples. *Weather and Forecasting*, **18**, 903–917.

Bras, R.L., and I. Rodríguez-Iturbe, 1985. *Random Functions and Hydrology*. Addison-Wesley, 559 pp.

Bratley, P., B.L. Fox, and L.E. Schrage, 1987. *A Guide to Simulation*, Springer, 397 pp.

Bremnes, J.B., 2004. Probabilistic forecasts of precipitation in terms of quantiles using NWP model output. *Monthly Weather Review*, **132**, 338–347.

Bretherton, C.S., C. Smith, and J.M. Wallace, 1992. An intercomparison of methods for finding coupled patterns in climate data. *Journal of Climate*, **5**, 541–560.

Brier, G.W., 1950. Verification of forecasts expressed in terms of probabilities. *Monthly Weather Review*, **78**, 1–3.

Brier, G.W., and R.A. Allen, 1951. Verification of weather forecasts. In: T.F. Malone, ed., *Compendium of Meteorology*, American Meteorological Society, 841–848.

Briggs, W.M., and R.A. Levine, 1997. Wavelets and field forecast verification. *Monthly Weather Review*, **125**, 1329–1341.

Brooks, C.E.P, and N. Carruthers, 1953. *Handbook of Statistical Methods in Meteorology*. London, Her Majesty's Stationery Office, 412 pp.

Brooks, H.E., C.A. Doswell III, and M.P. Kay, 2003. Climatological estimates of local daily tornado probability for the United States *Weather and Forecasting*, **18**, 626–640.

Broomhead, D.S., and G. King, 1986. Extracting qualitative dynamics from experimental data. *Physica D*, **20**, 217–236.

Brown, B.G., and R.W. Katz, 1991. Use of statistical methods in the search for teleconnections: past, present, and future. In: M. Glantz, R.W. Katz, and N. Nicholls, eds., *Teleconnections Linking Worldwide Climate Anomalies*, Cambridge University Press.

Brunet, N., R. Verret, and N. Yacowar, 1988. An objective comparison of model output statistics and "perfect prog" systems in producing numerical weather element forecasts. *Weather and Forecasting*, **3**, 273–283.

Buell, C.E., 1979. On the physical interpretation of empirical orthogonal functions. Preprints, 6th Conference on Probability and Statistics in the Atmospheric Sciences. American Meteorological Society, 112–117.

Buishand, T.A., M.V. Shabalova, and T. Brandsma, 2004. On the choice of the temporal aggregation level for statistical downscaling of precipitation. *Journal of Climate*, **17**, 1816–1827.

Buizza, R., 1997. Potential forecast skill of ensemble prediction and ensemble spread and skill distributions of the ECMWF Ensemble Prediction System. *Monthly Weather Review*, **125**, 99–119.

Buizza, R., A. Hollingsworth, F. Lalaurette, and A. Ghelli, 1999a. Probabilistic predictions of precipitation using the ECMWF ensemble prediction system. *Weather and Forecasting*, **14**, 168–189.

Buizza, R., M. Miller, and T.N. Palmer, 1999b. Stochastic representation of model uncertainties in the ECMWF Ensemble Prediction System. *Quarterly Journal of the Royal Meteorological Society*, **125**, 2887–2908.

Burman, P., E. Chow, and D. Nolan, 1994. A cross-validatory method for dependent data. *Biometrika*, **81**, 351–358.

Carter, G.M., J.P. Dallavalle, and H.R. Glahn, 1989. Statistical forecasts based on the National Meteorological Center's numerical weather prediction system. *Weather and Forecasting*, **4**, 401–412.

Casati, B., G. Ross, and D.B. Stephenson, 2004. A new intensity-scale approach for the verification of spatial precipitation forecasts. *Meteorological Applications*, **11**, 141–154.

Chen, W.Y., 1982a. Assessment of southern oscillation sea-level pressure indices. *Monthly Weather Review*, **110**, 800–807.

Chen, W.Y., 1982b. Fluctuations in northern hemisphere 700 mb height field associated with the southern oscillation. *Monthly Weather Review*, **110**, 808–823.

Cheng, X., G. Nitsche, and J.M. Wallace, 1995. Robustness of low-frequency circulation patterns derived from EOF and rotated EOF analyses. *Journal of Climate*, **8**, 1709–1713.

Cheng, X., and J.M. Wallace, 1993. Cluster analysis of the northern hemisphere wintertime 500-hPa height field: spatial patterns. *Journal of the Atmospheric Sciences*, **50**, 2674–2696.

Cherry, S., 1996. Singular value decomposition and canonical correlation analysis. *Journal of Climate*, **9**, 2003–2009.

Cherry, S., 1997. Some comments on singular value decomposition. *Journal of Climate*, **10**, 1759–1761.

Cheung, K.K.W., 2001. A review of ensemble forecasting techniques with a focus on tropical cyclone forecasting. *Meteorological Applications*, **8**, 315–332.

Chowdhury, J.U., J.R. Stedinger, and L.-H. Lu, 1991. Goodness-of-fit tests for regional GEV flood distributions. *Water Resources Research*, **27**, 1765–1776.

Chu, P.-S., and R.W. Katz, 1989. Spectral estimation from time series models with relevance to the southern oscillation. *Journal of Climate*, **2**, 86–90.

Clayton, H.H., 1927. A method of verifying weather forecasts. *Bulletin of the American Meteorological Society*, **8**, 144–146.

Clayton, H.H., 1934. Rating weather forecasts. *Bulletin of the American Meteorological Society*, **15**, 279–283.

Clemen, R.T., 1996. *Making Hard Decisions: an Introduction to Decision Analysis*. Duxbury, 664 pp.

Coles, S., 2001. *An Introduction to Statistical Modeling of Extreme Values*. Springer, 208 pp.

Conover, W.J., 1999. *Practical Nonparametric Statistics*, Wiley, 584 pp.

Conte, M., C. DeSimone, and C. Finizio, 1980. Post-processing of numerical models: forecasting the maximum temperature at Milano Linate. *Rev. Meteor. Aeronautica*, **40**, 247–265.

Cooke, W.E., 1906a. Forecasts and verifications in western Australia. *Monthly Weather Review*, **34**, 23–24.

Cooke, W.E., 1906b. Weighting forecasts. *Monthly Weather Review*, **34**, 274–275.

Crochet, P., 2004. Adaptive Kalman filtering of 2-metre temperature and 10-metre wind-speed forecasts in Iceland. *Meteorological Applications*, **11**, 173–187.

Crutcher, H.L., 1975. A note on the possible misuse of the Kolmogorov-Smirnov test. *Journal of Applied Meteorology*, **14**, 1600–1603.

Cunnane, C., 1978. Unbiased plotting positions—a review. *Journal of Hydrology*, **37**, 205–222.

Daan, H., 1985. Sensitivity of verification scores to the classification of the predictand. *Monthly Weather Review*, **113**, 1384–1392.

D'Agostino, R.B., 1986. Tests for the normal distribution. In: D'Agostino, R.B., and M.A. Stephens, 1986. *Goodness-of-Fit Techniques*, Marcel Dekker, 367–419.

D'Agostino, R.B., and M.A. Stephens, 1986. *Goodness-of-Fit Techniques*. Marcel Dekker, 560 pp.

Dagpunar, J., 1988. *Principles of Random Variate Generation*. Oxford, 228 pp.

Dallavalle, J.P., J.S. Jensenius Jr., and S.A. Gilbert, 1992. NGM-based MOS guidance—the FOUS14/FWC message. Technical Procedures Bulletin 408. NOAA/National Weather Service, Washington, D.C., 9 pp.

Daniel, W.W., 1990. *Applied Nonparametric Statistics*. Kent, 635 pp.

Davis, R.E., 1976. Predictability of sea level pressure anomalies over the north Pacific Ocean. *Journal of Physical Oceanography*, **6**, 249–266.

de Elia, R., R. Laprise, and B. Denis, 2002. Forecasting skill limits of nested, limited-area models: A perfect-model approach. *Monthly Weather Review*, **130**, 2006–2023.

DeGaetano, A.T., and M.D. Shulman, 1990. A climatic classification of plant hardiness in the United States and Canada. *Agricultural and Forest Meteorology*, **51**, 333–351.

Dempster, A.P., N.M. Laird, and D.B. Rubin, 1977. Maximum likelihood from incomplete data via the EM algorithm. *Journal of the Royal Statistical Society*, **B39**, 1–38.

Denis, B., J. Côté, and R. Laprise, 2002. Spectral decomposition of two-dimensional atmospheric fields on limited-area domains using the discrete cosine transform (DCT). *Monthly Weather Review*, **130**, 1812–1829.

Déqué, M., 2003. Continuous variables. In: I.T. Jolliffe and D.B. Stephenson, *Forecast Verification*. Wiley, 97–119.

Devroye, L., 1986. *Non-Uniform Random Variate Generation*. Springer, 843 pp.

Doolittle, M.H., 1888. Association ratios. *Bulletin of the Philosophical Society, Washington*, **7**, 122–127.

Doswell, C.A., 2004. Weather forecasting by humans—heuristics and decision making. *Weather and Forecasting*, **19**, 1115–1126.

Doswell, C.A., R. Davies-Jones, and D.L. Keller, 1990. On summary measures of skill in rare event forecasting based on contingency tables. *Weather and Forecasting*, **5**, 576–585.

Downton, M.W., and R.W. Katz, 1993. A test for inhomogeneous variance in time-averaged temperature data. *Journal of Climate*, **6**, 2448–2464.

Draper, N.R., and H. Smith, 1998. *Applied Regression Analysis*. Wiley, 706 pp.

Drosdowsky, W., and H. Zhang, 2003. Verification of spatial fields. In: I.T. Jolliffe and D.B. Stephenson, eds., *Forecast Verification*. Wiley, 121–136.

Du, J., S.L. Mullen, and F. Sanders, 2000. Removal of distortion error from an ensemble forecast. *Journal of Applied Meteorology*, **35**, 1177–1188.

Durban, J., and G.S. Watson, 1971. Testing for serial correlation in least squares regression. III. *Biometrika*, **58**, 1–19.

Ebert, E.E., and J.L. McBride, 2000. Verification of precipitation in weather systems: determination of systematic errors. *Journal of Hydrology*, **239**, 179–202.

Efron, B., 1979. Bootstrap methods: another look at the jackknife. *Annals of Statistics*, **7**, 1–26.

Efron, B., 1982. *The Jackknife, the Bootstrap and Other Resampling Plans*. Society for Industrial and Applied Mathematics, 92 pp.

Efron, B., 1987. Better bootstrap confidence intervals. *Journal of the American Statistical Association*, **82**, 171–185.

Efron, B., and G. Gong, 1983. A leisurely look at the bootstrap, the jackknife, and cross-validation. *The American Statistician*, **37**, 36–48.

Efron, B., and R.J. Tibshirani, 1993. *An Introduction to the Bootstrap*. Chapman and Hall, 436 pp.

Ehrendorfer, M., 1994. The Liouville equation and its potential usefulness for the prediction of forecast skill. Part I: Theory. *Monthly Weather Review*, **122**, 703–713.

Ehrendorfer, M., 1997. Predicting the uncertainty of numerical weather forecasts: a review. *Meteorol. Zeitschrift*, **6**, 147–183.

Ehrendorfer, M., and A.H. Murphy, 1988. Comparative evaluation of weather forecasting systems: sufficiency, quality, and accuracy. *Monthly Weather Review* , **116**, 1757–1770.

Ehrendorfer, M., and J.J. Tribbia, 1997. Optimal prediction of forecast error covariances through singular vectors. *Journal of the Atmospheric Sciences* , **54**, 286–313.

Elsner, J.B., and C.P. Schmertmann, 1993. Improving extended-range seasonal predictions of intense Atlantic hurricane activity. *Weather and Forecasting* , **8**, 345–351.

Elsner, J.B., and C.P. Schmertmann, 1994. Assessing forecast skill through cross validation. *Journal of Climate* , **9**, 619–624.

Elsner, J.B., and A.A. Tsonis, 1996. *Singular Spectrum Analysis, A New Tool in Time Series Analysis* . Plenum, 164 pp.

Epstein, E.S., 1969a. The role of initial uncertainties in prediction. *Journal of Applied Meteorology* , **8**, 190–198.

Epstein, E.S., 1969b. A scoring system for probability forecasts of ranked categories. *Journal of Applied Meteorology* , **8**, 985–987.

Epstein, E.S., 1969c. Stochastic dynamic prediction. *Tellus* , **21**, 739–759.

Epstein, E.S., 1985. *Statistical Inference and Prediction in Climatology: A Bayesian Approach*. Meteorological Monograph 20(42). American Meteorological Society, 199 pp.

Epstein, E.S., 1991. On obtaining daily climatological values from monthly means. *Journal of Climate* , **4**, 365–368.

Epstein, E.S., and A.G. Barnston, 1988. *A Precipitation Climatology of Five-Day Periods* . NOAA Tech. Report NWS 41, Climate Analysis Center, National Weather Service, Camp Springs MD, 162 pp.

Epstein, E.S., and R.J. Fleming, 1971. Depicting stochastic dynamic forecasts. *Journal of the Atmospheric Sciences* , **28**, 500–511.

Erickson, M.C., J.B. Bower, V.J. Dagostaro, J.P. Dallavalle, E. Jacks, J.S. Jensenius, Jr., and J.C. Su, 1991. Evaluating the impact of RAFS changes on the NGM-based MOS guidance. *Weather and Forecasting* , **6**, 142–147.

Errico, R.M., and R. Langland, 1999a. Notes on the appropriateness of "bred modes" for generating initial perturbations used in ensemble prediction. *Tellus* , **51A**, 431–441.

Errico, R.M., and R. Langland, 1999b. Reply to: Comments on "Notes on the appropriateness of 'bred modes' for generating initial perturbations." *Tellus* , **51A**, 450–451.

Everitt, B.S., and D.J. Hand, 1981. *Finite Mixture Distributions*. Chapman and Hall, 143 pp.

Feller, W., 1970. *An Introduction to Probability Theory and its Applications* . Wiley, 509 pp.

Filliben, J.J., 1975. The probability plot correlation coefficient test for normality. *Technometrics*, **17**, 111–117.

Finley, J.P., 1884. Tornado prediction. *American Meteorological Journal* , **1**, 85–88.

Fisher, R.A., 1929. Tests of significance in harmonic analysis. *Proceedings of the Royal Society, London*, **A125**, 54–59.

Flueck, J.A., 1987. A study of some measures of forecast verification. Preprints, Tenth Conference on Probability and Statistics in Atmospheric Sciences. American Meteorological Society, 69–73.

Folland, C., and C. Anderson, 2002. Estimating changing extremes using empirical ranking methods. *Journal of Climate* , **15**, 2954–2960.

Foufoula-Georgiou, E., and D.P. Lettenmaier, 1987. A Markov renewal model for rainfall occurrences. *Water Resources Research* , **23**, 875–884.

Fovell, R.G., and M.-Y. Fovell, 1993. Climate zones of the conterminous United States defined using cluster analysis. *Journal of Climate* , **6**, 2103–2135.

Francis, P.E., A.P. Day, and G.P. Davis, 1982. Automated temperature forecasting, an application of Model Output Statistics to the Meteorological Office numerical weather prediction model. *Meteorological Magazine*, **111**, 73–87.

Friederichs, P., and A. Hense, 2003. Statistical inference in canonical correlation analyses exemplified by the influence of North Atlantic SST on European climate *Journal of Climate*, **16**, 522–534.

Friedman, R.M., 1989. *Appropriating the Weather : Vilhelm Bjerknes and the Construction of a Modern Meteorology*. Cornell University Press, 251 pp.

Fuller, W.A., 1996. *Introduction to Statistical Time Series* . Wiley, 698 pp.

Gabriel, R.K., 1971. The biplot—graphic display of matrices with application to principal component analysis. *Biometrika*, **58**, 453–467.

Galanis, G., and M. Anadranistakis, 2002. A one-dimensional Kalman filter for the correction of near surface temperature forecasts. *Meteorological Applications* , **9**, 437–441.

Galliani, G., and F. Filippini, 1985. Climatic clusters in a small area. *Journal of Climatology*, **5**, 487–501.

Gandin, L.S., and A.H. Murphy, 1992. Equitable skill scores for categorical forecasts. *Monthly Weather Review*, **120**, 361–370.

Garratt, J.R., R.A. Pielke, Sr., W.F. Miller, and T.J. Lee, 1990. Mesoscale model response to random, surface-based perturbations—a sea-breeze experiment. *Boundary-Layer Meteorology*, **52**, 313–334.

Gerrity, J.P., Jr., 1992. A note on Gandin and Murphy's equitable skill score. *Monthly Weather Review*, **120**, 2709–2712.

Gershunov, A., and D.R. Cayan, 2003. Heavy daily precipitation frequency over the contiguous United States: Sources of climatic variability and seasonal predictability. *Journal of Climate*, **16**, 2752–2765.

Gilbert, G.K., 1884. Finley's tornado predictions. *American Meteorological Journal* , **1**, 166–172.

Gillies, D., 2000. *Philosophical Theories of Probability* . Routledge, 223 pp.

Gilman, D.L., F.J. Fuglister, and J.M. Mitchell, Jr., 1963. On the power spectrum of "red noise." *Journal of the Atmospheric Sciences*, **20**, 182–184.

Glahn, H.R., 1968. Canonical correlation analysis and its relationship to discriminant analysis and multiple regression. *Journal of the Atmospheric Sciences*, **25**, 23–31.

Glahn, H.R., 1985. Statistical weather forecasting. In: A.H. Murphy and R.W. Katz, eds., *Probability, Statistics, and Decision Making in the Atmospheric Sciences* . Boulder, Westview, 289–335.

Glahn, H.R., and D.L. Jorgensen, 1970. Climatological aspects of the Brier p-score. *Monthly Weather Review*, **98**, 136–141.

Glahn, H.R., and D.A. Lowry, 1972. The use of Model Output Statistics (MOS) in objective weather forecasting. *Journal of Applied Meteorology*, **11**, 1203–1211.

Gleeson, T.A., 1961. A statistical theory of meteorological measurements and predictions. *Journal of Meteorology*, **18**, 192–198.

Gleeson, T.A., 1967. Probability predictions of geostrophic winds. *Journal of Applied Meteorology*, **6**, 355–359.

Gleeson, T.A., 1970. Statistical-dynamical predictions. *Journal of Applied Meteorology*, **9**, 333–344.

Goldsmith, B.S., 1990. NWS verification of precipitation type and snow amount forecasts during the AFOS era. NOAA Technical Memorandum NWS FCST 33. National Weather Service, 28 pp.

Golub, G.H., and C.F. van Loan, 1996. *Matrix Computations* . Johns Hopkins Press, 694 pp.

Golyandina, N., V. Nekrutkin, and A. Zhigljavsky, 2001. *Analysis of Time Series Structure, SSA and Related Techniques*. Chapman & Hall, 305 pp.

Gong, X., and M.B. Richman, 1995. On the application of cluster analysis to growing season precipitation data in North America east of the Rockies. *Journal of Climate*, **8**, 897–931.

Good, P., 2000. *Permutation Tests*. Springer, 270 pp.

Goodall, C., 1983. M-Estimators of location: an outline of the theory. In: D.C. Hoaglin, F. Mosteller, and J.W. Tukey, eds., *Understanding Robust and Exploratory Data Analysis*. New York, Wiley, 339–403.

Gordon, N.D., 1982. Comments on "verification of fixed-width credible interval temperature forecasts." *Bulletin of the American Meteorological Society*, **63**, 325.

Graedel, T.E., and B. Kleiner, 1985. Exploratory analysis of atmospheric data. In: A.H. Murphy and R.W. Katz, eds., *Probability, Statistics, and Decision Making in the Atmospheric Sciences*. Boulder, Westview, 1–43.

Gray, W.M., 1990. Strong association between West African rainfall and U.S. landfall of intense hurricanes. *Science*, **249**, 1251–1256.

Gray, W.M., C.W. Landsea, P.W. Mielke, Jr., and K.J. Berry, 1992. Predicting seasonal hurricane activity 6–11 months in advance. *Weather and Forecasting*, **7**, 440–455.

Greenwood, J.A., and D. Durand, 1960. Aids for fitting the gamma distribution by maximum likelihood. *Technometrics*, **2**, 55–65.

Grimit, E.P., and C.F. Mass, 2002. Initial results of a mesoscale short-range ensemble forecasting system over the Pacific Northwest. *Weather and Forecasting*, **17**, 192–205.

Gringorten, I.I., 1967. Verification to determine and measure forecasting skill. *Journal of Applied Meteorology*, **6**, 742–747.

Guttman, N.B., 1993. The use of L-moments in the determination of regional precipitation climates. *Journal of Climate*, **6**, 2309–2325.

Haines, K., and A. Hannachi, 1995. Weather regimes in the Pacific from a GCM. *Journal of the Atmospheric Sciences*, **52**, 2444–2462.

Hall, P., and S.R. Wilson, 1991. Two guidelines for bootstrap hypothesis testing. *Biometrics*, **47**, 757–762.

Hamill, T.M., 1999. Hypothesis tests for evaluating numerical precipitation forecasts. *Weather and Forecasting*, **14**, 155–167.

Hamill, T.M., 2001. Interpretation of rank histograms for verifying ensemble forecasts. *Monthly Weather Review*, **129**, 550–560.

Hamill, T.M., 2006. Ensemble-based atmospheric data assimilation: a tutorial. In: Palmer, T.N., and R. Hagedorn, eds., *Predictability of Weather and Climate*. Cambridge, in press.

Hamill, T.M., and S.J. Colucci, 1997. Verification of Eta–RSM short-range ensemble forecasts. *Monthly Weather Review*, **125**, 1312–1327.

Hamill, T.M., and S.J. Colucci, 1998. Evaluation of Eta–RSM ensemble probabilistic precipitation forecasts. *Monthly Weather Review*, **126**, 711–724.

Hamill, T. M., J. S. Whitaker, and X. Wei, 2004. Ensemble re-forecasting: improving medium-range forecast skill using retrospective forecasts, *Monthly Weather Review*, **132**, 1434–1447.

Hand, D.J., 1997. *Construction and Assessment of Classification Rules*. Wiley, 214 pp.

Hannachi, A., 1997. Low-frequency variability in a GCM: three dimensional flow regimes and their dynamics. *Journal of Climate*, **10**, 1357–1379.

Hannachi, A., and A. O'Neill, 2001. Atmospheric multiple equilibria and non-Gaussian behavior in model simulations. *Quarterly Journal of the Royal Meteorological Society*, **127**, 939–958.

Hansen, J.A. 2002. Accounting for model error in ensemble-based state estimation and forecasting. *Monthly Weather Review*, **130**, 2373–2391.

Hanssen, A.W., and W.J.A. Kuipers, 1965. On the relationship between the frequency of rain and various meteorological parameters. *Mededeelingen en Verhandelingen*, **81**, 2–15.

Harrison, M.S.J., T.N. Palmer, D.S. Richardson, and R. Buizza, 1999. Analysis and model dependencies in medium-range ensembles: two transplant case-studies. *Quarterly Journal of the Royal Meteorological Society*, **125**, 2487–2515.

Harter, H.L., 1984. Another look at plotting positions. *Communications in Statistics. Theory and Methods*, **13**, 1613–1633.

Hasselmann, K., 1976. Stochastic climate models. Part I: Theory. *Tellus*, **28**, 474–485.

Hastie, T., R. Tibshirani, and J. Friedman, 2001. *The Elements of Statistical Learning*. Springer, 533 pp.

Healy, M.J.R., 1988. *Glim: An Introduction*. Oxford, 130 pp.

Heidke, P., 1926. Berechnung des Erfolges und der Güte der Windstärkevorhersagen im Sturmwarnungsdienst. *Geografika Annaler*, **8**, 301–349.

Hersbach, H., 2000. Decomposition of the continuous ranked probability score for ensemble prediction systems. *Weather and Forecasting*, **15**, 559–570.

Hilliker, J.L., and J.M. Fritsch, 1999. An observations-based statistical system for warm-season hourly probabilistic precipitation forecasts of low ceiling at the San Francisco international airport. *Journal of Applied Meteorology*, **38**, 1692–1705.

Hinkley, D., 1977. On quick choice of power transformation. *Applied Statistics*, **26**, 67–69.

Hoffman, M.S., ed., 1988. *The World Almanac Book of Facts*. Pharos Books, 928 pp.

Hoffman, R.N., Z. Liu, J.-F. Louis, and C. Grassotti, 1995: Distortion representation of forecast errors. *Monthly Weather Review*, **123**, 2758–2770.

Hoke, J.E., N.A. Phillips, G.J. DiMego, J.J. Tuccillo, and J.G. Sela, 1989. The regional analysis and forecast system of the National Meteorological Center. *Weather and Forecasting*, **4**, 323–334.

Hollingsworth, A., K. Arpe, M. Tiedtke, M. Capaldo, and H. Savijärvi, 1980. The performance of a medium range forecast model in winter—impact of physical parameterizations. *Monthly Weather Review*, **108**, 1736–1773.

Homleid, M., 1995. Diurnal corrections of short-term surface temperature forecasts using the Kalman filter. *Weather and Forecasting*, **10**, 689–707.

Horel, J.D., 1981. A rotated principal component analysis of the interannual variability of the Northern Hemisphere 500 mb height field. *Monthly Weather Review*, **109**, 2080–2902.

Hosking, J.R.M., 1990. L-moments: analysis and estimation of distributions using linear combinations of order statistics. *Journal of the Royal Statistical Society*, **B52**, 105–124.

Houtekamer, P.L., L. Lefaivre, J. Derome, H. Ritchie, and H.L. Mitchell, 1996. A system simulation approach to ensemble prediction. *Monthly Weather Review*, **124**, 1225–1242.

Houtekamer, P.L., and H.L. Mitchell, 1998. Data assimilation using an ensemble Kalman filtering technique. *Monthly Weather Review*, **126**, 796–811.

Hsu, W.-R., and A.H. Murphy, 1986. The attributes diagram: a geometrical framework for assessing the quality of probability forecasts. *International Journal of Forecasting*, **2**, 285–293.

Hu, Q., 1997. On the uniqueness of the singular value decomposition in meteorological applications. *Journal of Climate*, **10**, 1762–1766.

Iglewicz, B., 1983. Robust scale estimators and confidence intervals for location. In: D.C. Hoaglin, F. Mosteller, and J.W. Tukey, eds., *Understanding Robust and Exploratory Data Analysis*. New York, Wiley, 404–431.

Ivarsson, K.-I., R. Joelsson, E. Liljas, and A.H. Murphy, 1986. Probability forecasting in Sweden: some results of experimental and operational programs at the Swedish Meteorological and Hydrological Institute. *Weather and Forecasting*, **1**, 136–154.

Jacks, E., J.B. Bower, V.J. Dagostaro, J.P. Dallavalle, M.C. Erickson, and J. Su, 1990. New NGM-based MOS guidance for maximum/minimum temperature, probability of precipitation, cloud amount, and surface wind. *Weather and Forecasting*, **5**, 128–138.

Jenkins, G.M., and D.G. Watts, 1968. *Spectral Analysis and its Applications*. Holden-Day, 523 pp.

Johnson, M.E., 1987. *Multivariate Statistical Simulation*. Wiley, 230 pp.

Johnson, N.L,. and S. Kotz, 1972. *Distributions in Statistics—4. Continuous Multivariate Distributions*. New York, Wiley, 333 pp.

Johnson, N.L., S. Kotz, and N. Balakrishnan, 1994. *Continuous Univariate Distributions, Volume 1*. Wiley, 756 pp.

Johnson, N.L., S. Kotz, and N. Balakrishnan, 1995. *Continuous Univariate Distributions, Volume 2*. Wiley, 719 pp.

Johnson, N.L., S. Kotz, and A.W. Kemp, 1992. *Univariate Discrete Distributions*. Wiley, 565 pp.

Johnson, S.R., and M.T. Holt, 1997. The value of weather information. In: R.W. Katz and A.H. Murphy, eds., *Economic Value of Weather and Climate Forecasts*. Cambridge, 75–107.

Jolliffe, I.T., 1972. Discarding variables in a principal component analysis, I: Artificial data. *Applied Statistics*, **21**, 160–173.

Jolliffe, I.T., 1987. Rotation of principal components: some comments. *Journal of Climatology*, **7**, 507–510.

Jolliffe, I.T., 1989. Rotation of ill-defined principal components. *Applied Statistics*, **38**, 139–147.

Jolliffe, I.T., 1995. Rotation of principal components: choice of normalization constraints. *Journal of Applied Statistics*, **22**, 29–35.

Jolliffe, I.T., 2002. *Principal Component Analysis*, 2nd Ed. Springer, 487 pp.

Jolliffe, I.T., B. Jones, and B.J.T. Morgan, 1986. Comparison of cluster analyses of the English personal social services authorities. *Journal of the Royal Statistical Society*, **A149**, 254–270.

Jolliffe, I.T., and D.B. Stephenson, 2003. *Forecast Verification*. Wiley, 240 pp.

Jones, R.H., 1975. Estimating the variance of time averages. *Journal of Applied Meteorology*, **14**, 159–163.

Juras, J., 2000. Comments on "Probabilistic predictions of precipitation using the ECMWF ensemble prediction system." *Weather and Forecasting*, **15**, 365–366.

Kaiser, H.F., 1958. The varimax criterion for analytic rotation in factor analysis. *Psychometrika*, **23**, 187–200.

Kalkstein, L.S., G. Tan, and J.A. Skindlov, 1987. An evaluation of three clustering procedures for use in synoptic climatological classification. *Journal of Climate and Applied Meteorology*, **26**, 717–730.

Kalnay, E., 2003. *Atmospheric Modeling, Data Assimilation and Predictability*. Cambridge, 341 pp.

Kalnay, E., and A. Dalcher, 1987. Forecasting the forecast skill. *Monthly Weather Review*, **115**, 349–356.

Kalnay, E., M. Kanamitsu, and W.E. Baker, 1990. Global numerical weather prediction at the National Meteorological Center. *Bulletin of the American Meteorological Society* , **71**, 1410–1428.

Karl, T.R., and A.J. Koscielny, 1982. Drought in the United States, 1895–1981. *Journal of Climatology* , **2**, 313–329.

Karl, T.R., A.J. Koscielny, and H.F. Diaz, 1982. Potential errors in the application of principal component (eigenvector) analysis to geophysical data. *Journal of Applied Meteorology* , **21**, 1183–1186.

Karl, T.R., M.E. Schlesinger, and W.C. Wang, 1989. A method of relating general circulation model simulated climate to the observed local climate. Part I: Central tendencies and dispersion. Preprints, Sixth Conference on Applied Climatology, American Meteorological Society, 188–196.

Karlin, S., and H.M Taylor, 1975. *A First Course in Stochastic Processes* . Academic Press, 557 pp.

Katz, R.W., 1977. Precipitation as a chain-dependent process. *Journal of Applied Meteorology* , **16**, 671–676.

Katz, R.W., 1981. On some criteria for estimating the order of a Markov chain. *Technometrics* , **23**, 243–249.

Katz, R.W., 1982. Statistical evaluation of climate experiments with general circulation models: a parametric time series modeling approach. *Journal of the Atmospheric Sciences* , **39**, 1446–1455.

Katz, R.W., 1985. Probabilistic models. In: A.H. Murphy and R.W. Katz, eds., *Probability, Statistics, and Decision Making in the Atmospheric Sciences* . Westview, 261–288.

Katz, R.W., and A.H. Murphy, 1997a. *Economic Value of Weather and Climate Forecasts* . Cambridge, 222 pp.

Katz, R.W., and A.H. Murphy, 1997b. Forecast value: prototype decision-making models. In: R.W. Katz and A.H. Murphy, eds., *Economic Value of Weather and Climate Forecasts* . Cambridge, 183–217.

Katz, R.W., A.H. Murphy, and R.L. Winkler, 1982. Assessing the value of frost forecasts to orchardists: a dynamic decision-making approach. *Journal of Applied Meteorology* , **21**, 518–531.

Katz, R.W., and M.B. Parlange, 1993. Effects of an index of atmospheric circulation on stochastic properties of precipitation. *Water Resources Research* , **29**, 2335–2344.

Katz, R.W., M.B. Parlange, and P. Naveau, 2002. Statistics of extremes in hydrology. *Advances in Water Resources* , **25**, 1287–1304.

Katz, R.W., and X. Zheng, 1999. Mixture model for overdispersion of precipitation. *Journal of Climate* , **12**, 2528–2537.

Kendall, M., and J.K. Ord, 1990. *Time Series* . Edward Arnold, 296.

Kharin, V.V., and F.W. Zwiers, 2003a. Improved seasonal probability forecasts. *Journal of Climate* , **16**, 1684–1701.

Kharin, V.V., and F.W. Zwiers, 2003b. On the ROC score of probability forecasts. *Journal of Climate* , **16**, 4145–4150.

Klein, W.H., B.M. Lewis, and I. Enger, 1959. Objective prediction of five-day mean temperature during winter. *Journal of Meteorology* , **16**, 672–682.

Knaff, J.A., and C.W. Landsea, 1997. An El Niño-southern oscillation climatology and persistence (CLIPER) forecasting scheme. *Weather and Forecasting* , **12**, 633–647.

Krzysztofowicz, R., 1983. Why should a forecaster and a decision maker use Bayes' theorem? *Water Resources Research* , **19**, 327–336.

Krzysztofowicz, R., 2001. The case for probabilistic forecasting in hydrology. *Journal of Hydrology* , **249**, 2–9.

Krzysztofowicz, R., W.J. Drzal, T.R. Drake, J.C. Weyman, and L.A. Giordano, 1993. Probabilistic quantitative precipitation forecasts for river basins. *Weather and Forecasting*, **8**, 424–439.

Krzysztofowicz, R., and D. Long, 1990. Fusion of detection probabilities and comparison of multisensor systems. *IEEE Transactions on Systems, Man, and Cybernetics*, **20**, 665–677.

Krzysztofowicz, R., and D. Long, 1991. Beta probability models of probabilistic forecasts. *International Journal of Forecasting*, **7**, 47–55.

Kutzbach, J.E., 1967. Empirical eigenvectors of sea-level pressure, surface temperature and precipitation complexes over North America. *Journal of Applied Meteorology*, **6**, 791–802.

Lahiri, S.N., 2003. *Resampling Methods for Dependent Data*. Springer, 374 pp.

Lall, U., and A. Sharma, 1996. A nearest neighbor bootstrap for resampling hydrologic time series. *Water Resources Research*, **32**, 679–693.

Landsea, C.W., and J.A. Knaff, 2000. How much skill was there in forecasting the very strong 1997-1998 El Niño? *Bulletin of the American Meteorological Society*, **81**, 2107–2119.

Lawson, M.P., and R.S. Cerveny, 1985. Seasonal temperature forecasts as products of antecedent linear and spatial temperature arrays. *Journal of Climate and Applied Meteorology*, **24**, 848–859.

Leadbetter, M.R., G. Lindgren, and H. Rootzen, 1983. *Extremes and Related Properties of Random Sequences and Processes*. Springer, 336 pp.

Leger, C., D.N. Politis, and J.P. Romano, 1992. Bootstrap technology and applications. *Technometrics*, **34**, 378–398.

Legg, T.P., K.R. Mylne, and C. Woodcock, 2002. Use of medium-range ensembles at the Met Office I: PREVIN—a system for the production of probabilistic forecast information from the ECMWF EPS. *Meteorological Applications*, **9**, 255–271.

Lehmiller, G.S., T.B. Kimberlain, and J.B. Elsner, 1997. Seasonal prediction models for North Atlantic basin hurricane location. *Monthly Weather Review*, **125**, 1780–1791.

Leith, C.E., 1973. The standard error of time-average estimates of climatic means. *Journal of Applied Meteorology*, **12**, 1066–1069.

Leith, C.E., 1974. Theoretical skill of Monte-Carlo forecasts. *Monthly Weather Review*, **102**, 409–418.

Lemcke, C., and S. Kruizinga, 1988. Model output statistics forecasts: three years of operational experience in the Netherlands. *Monthly Weather Review*, **116**, 1077–1090.

Liljas, E., and A.H. Murphy, 1994. Anders Angstrom and his early papers on probability forecasting and the use/value of weather forecasts. *Bulletin of the American Meteorological Society*, **75**, 1227–1236.

Lilliefors, H.W., 1967. On the Kolmogorov-Smirnov test for normality with mean and variance unknown. *Journal of the American Statistical Association*, **62**, 399–402.

Lin, J. W.-B., and J.D. Neelin, 2000. Influence of a stochastic moist convective parameterization on tropical climate variability. *Geophysical Research Letters*, **27**, 3691–3694.

Lin, J. W.-B., and J.D. Neelin, 2002. Considerations for stochastic convective parameterization. *Journal of the Atmospheric Sciences*, **59**, 959–975.

Lindgren, B.W., 1976. *Statistical Theory*. MacMillan, 614 pp.

Lipschutz, S., 1968. *Schaum's Outline of Theory and Problems of Linear Algebra*. McGraw-Hill, 334 pp.

Livezey, R.E., 1985. Statistical analysis of general circulation model climate simulation: sensitivity and prediction experiments. *Journal of the Atmospheric Sciences*, **42**, 1139–1149.

Livezey, R.E., 1995a. Field intercomparison. In: H. von Storch and A. Navarra, eds., *Analysis of Climate Variability* . Springer, 159–176.

Livezey, R.E., 1995b. The evaluation of forecasts. In: H. von Storch and A. Navarra, eds., *Analysis of Climate Variability* . Springer, 177–196.

Livezey, R.E., 2003. Categorical events. In: I.T. Jolliffe and D.B. Stephenson, *Forecast Verification*. Wiley, 77–96.

Livezey, R.E., and W.Y. Chen, 1983. Statistical field significance and its determination by Monte Carlo techniques. *Monthly Weather Review* , **111**, 46–59.

Livezey, R.E., J.D. Hoopingarner, and J. Huang, 1995. Verification of official monthly mean 700-hPa height forecasts: an update. *Weather and Forecasting* , **10**, 512–527.

Livezey, R.E., and T.M. Smith, 1999. Considerations for use of the Barnett and Preisendorfer (1987) algorithm for canonical correlation analysis of climate variations. *Journal of Climate* , **12**, 303–305.

Lorenz, E.N., 1956. Empirical orthogonal functions and statistical weather prediction. Science Report 1, Statistical Forecasting Project, Department of Meteorology, MIT (NTIS AD 110268), 49 pp.

Lorenz, E.N., 1963. Deterministic nonperiodic flow. *Journal of the Atmospheric Sciences* , **20**, 130–141.

Lorenz, E.N., 1975. Climate predictability. In: *The Physical Basis of Climate and Climate Modelling* , GARP Publication Series, vol. 16, WMO 132–136.

Lorenz, E.N., 1996. Predictability—A problem partly solved. *Proceedings, Seminar on Predictability*, Vol. 1., 1–18. ECMWF, Reading, UK.

Loucks, D.P., J.R. Stedinger, and D.A. Haith, 1981. *Water Resource Systems Planning and Analysis*. Prentice-Hall, 559 pp.

Lu, R., 1991. The application of NWP products and progress of interpretation techniques in China. In: H.R. Glahn, A.H. Murphy, L.J. Wilson, and J.S. Jensenius, Jr., eds., *Programme on Short-and Medium-Range Weather Prediction Research* . World Meteorological Organization WM/TD No. 421. XX19–22.

Madden, R.A., 1979. A simple approximation for the variance of meteorological time averages. *Journal of Applied Meteorology* , **18**, 703–706.

Madden, R.A., and R.H. Jones, 2001. A quantitative estimate of the effect of aliasing in climatological time series. *Journal of Climate* , **14**, 3987–3993.

Madden, R.A., and D.J. Shea, 1978. Estimates of the natural variability of time-averaged temperatures over the United States. *Monthly Weather Review* , **106**, 1695–1703.

Madsen, H., P.F. Rasmussen, and D. Rosbjerg, 1997. Comparison of annual maximum series and partial duration series methods for modeling extreme hydrologic events. 1. At-site modeling. *Water Resources Research* , **33**, 747–757.

Mao, Q., R.T. McNider, S.F. Mueller, and H.-M.H. Juang, 1999. An optimal model output calibration algorithm suitable for objective temperature forecasting. *Weather and Forecasting*, **14**, 190–202.

Mardia, K.V., 1970. Measures of multivariate skewness and kurtosis with applications. *Biometrika*, **57**, 519–530.

Mardia, K.V., J.T. Kent, and J.M. Bibby, 1979. *Multivariate Analysis* . Academic, 518 pp.

Marzban, C., 2004. The ROC curve and the area under it as performance measures. *Weather and Forecasting* , **19**, 1106–1114.

Marzban, C., and V. Lakshmanan, 1999. On the uniqueness of Gandin and Murphy's equitable performance measures. *Monthly Weather Review* , **127**, 1134–1136.

Mason, I.B., 1979. On reducing probability forecasts to yes/no forecasts. *Monthly Weather Review*, **107**, 207–211.

Mason., I.B., 1982. A model for assessment of weather forecasts. *Australian Meteorological Magazine* , **30**, 291–303.

Mason, I.B., 2003. Binary events. In: I.T. Jolliffe and D.B. Stephenson, eds., *Forecast Verification*. Wiley, 37–76.

Mason, S.J., L. Goddard, N.E. Graham, E. Yulaleva, L. Sun, and P.A. Arkin, 1999. The IRI seasonal climate prediction system and the 1997/98 El Niño event. *Bulletin of the American Meteorological Society*, **80**, 1853–1873.

Mason, S.J., and N.E. Graham, 2002. Areas beneath the relative operating characteristics (ROC) and relative operating levels (ROL) curves: statistical significance and interpretation. *Quarterly Journal of the Royal Meteorological Society*, **128**, 2145–2166.

Mason, S.J., and G.M. Mimmack, 1992. The use of bootstrap confidence intervals for the correlation coefficient in climatology. *Theoretical and Applied Climatology*, **45**, 229–233.

Mason, S.J., and G.M. Mimmack, 2002. Comparison of some statistical methods of probabilistic forecasting of ENSO. *Journal of Climate*, **15**, 8–29.

Matalas, N.C., 1967. Mathematical assessment of synthetic hydrology. *Water Resources Research*, **3**, 937–945.

Matalas, N.C., and W.B. Langbein, 1962. Information content of the mean. *Journal of Geophysical Research*, **67**, 3441–3448.

Matalas, N.C., and A. Sankarasubramanian, 2003. Effect of persistence on trend detection via regression. *Water Resources Research*, **39**, 1342–1348.

Matheson, J.E., and R.L. Winkler, 1976. Scoring rules for continuous probability distributions. *Management Science*, **22**, 1087–1096.

Matsumoto, M., and T. Nishimura, 1998. Mersenne twister: a 623-dimensionally equidistributed uniform pseudorandom number generator. *ACM* (Association for Computing Machinery) *Transactions on Modeling and Computer Simulation*, **8**, 3–30.

Mazany, R.A., S. Businger, S.I. Gutman, and W. Roeder, 2002. A lightning prediction index that utilizes GPS integrated precipitable water vapor. *Weather and Forecasting*, **17**, 1034–1047.

McCullagh, P., and J.A. Nelder, 1989. *Generalized Linear Models*. Chapman and Hall, 511 pp.

McDonnell, K.A., and N.J. Holbrook, 2004. A Poisson regression model of tropical cyclogenesis for the Australian-southwest Pacific Ocean region. *Weather and Forecasting*, **19**, 440–455.

McGill, R., J.W. Tukey, and W.A. Larsen, 1978. Variations of boxplots. *The American Statistician*, **32**, 12–16.

McLachlan, G.J., and T. Krishnan, 1997. *The EM Algorithm and Extensions*. Wiley, 274 pp.

McLachlan, G.J. and D. Peel 2000. *Finite Mixture Models*. Wiley 419 pp.

Mestas-Nuñez, A.M., 2000. Orthogonality properties of rotated empirical modes. *International Journal of Climatology*, **20**, 1509–1516.

Michaelson, J., 1987. Cross-validation in statistical climate forecast models. *Journal of Climate and Applied Meteorology*, **26**, 1589–1600.

Mielke, P.W., Jr., K.J. Berry, and G.W. Brier, 1981. Application of multi-response permutation procedures for examining seasonal changes in monthly mean sea-level pressure patterns. *Monthly Weather Review*, **109**, 120–126.

Mielke, P.W., Jr., K.J. Berry, C.W. Landsea, and W.M. Gray, 1996. Artificial skill and validation in meteorological forecasting. *Weather and Forecasting*, **11**, 153–169.

Miller, B.I., E.C. Hill, and P.P. Chase, 1968. A revised technique for forecasting hurricane movement by statistical methods. *Monthly Weather Review*, **96**, 540–548.

Miller, R.G., 1962. Statistical prediction by discriminant analysis. *Meteorological Monographs*, **4**, No. 25. American Meteorological Society, 53 pp.

Miyakoda, K., G.D. Hembree, R.F. Strikler, and I. Shulman, 1972. Cumulative results of extended forecast experiments. I: Model performance for winter cases. *Monthly Weather Review*, **100**, 836–855.

Mo, K.C., and M. Ghil, 1987. Statistics and dynamics of persistent anomalies. *Journal of the Atmospheric Sciences*, **44**, 877–901.

Mo., K.C., and M. Ghil, 1988. Cluster analysis of multiple planetary flow regimes. *Journal of Geophysical Research*, **D93**, 10927–10952.

Molteni, F., R. Buizza, T.N. Palmer, and T. Petroliagis, 1996. The new ECMWF Ensemble Prediction System: methodology and validation. *Quarterly Journal of the Royal Meteorological Society*, **122**, 73–119.

Molteni, F., S. Tibaldi, and T.N. Palmer, 1990. Regimes in wintertime circulation over northern extratropics. I: Observational evidence. *Quarterly Journal of the Royal Meteorological Society*, **116**, 31–67.

Moore, A.M., and R. Kleeman, 1998. Skill assessment for ENSO using ensemble prediction. *Quarterly Journal of the Royal Meteorological Society*, **124**, 557–584.

Moritz, R.E., and A. Sutera, 1981. The predictability problem: effects of stochastic perturbations in multiequilibrium systems. *Reviews of Geophysics*, **23**, 345–383.

Moura, A.D., and S. Hastenrath, 2004. Climate prediction for Brazil's Nordeste: performance of empirical and numerical modeling methods. *Journal of Climate*, **17**, 2667–2672.

Mullen, S.L, and R. Buizza, 2001. Quantitative precipitation forecasts over the United States by the ECMWF Ensemble Prediction System. *Monthly Weather Review*, **129**, 638–663.

Mullen, S.L., J. Du, and F. Sanders, 1999. The dependence of ensemble dispersion on analysis-forecast systems: implications to short-range ensemble forecasting of precipitation. *Monthly Weather Review*, **127**, 1674–1686.

Muller, R.H., 1944. Verification of short-range weather forecasts (a survey of the literature). *Bulletin of the American Meteorological Society*, **25**, 18–27, 47–53, 88–95.

Murphy, A.H., 1966. A note on the utility of probabilistic predictions and the probability score in the cost-loss ratio situation. *Journal of Applied Meteorology*, **5**, 534–537.

Murphy, A.H., 1971. A note on the ranked probability score. *Journal of Applied Meteorology*, **10**, 155–156.

Murphy, A.H., 1973a. Hedging and skill scores for probability forecasts. *Journal of Applied Meteorology*, **12**, 215–223.

Murphy, A.H., 1973b. A new vector partition of the probability score. *Journal of Applied Meteorology*, **12**, 595–600.

Murphy, A.H., 1977. The value of climatological, categorical, and probabilistic forecasts in the cost-loss ratio situation. *Monthly Weather Review*, **105**, 803–816.

Murphy, A.H., 1985. Probabilistic weather forecasting. In: A.H. Murphy and R.W. Katz, eds., *Probability, Statistics, and Decision Making in the Atmospheric Sciences*. Boulder, Westview, 337–377.

Murphy, A.H., 1988. Skill scores based on the mean square error and their relationships to the correlation coefficient. *Monthly Weather Review*, **116**, 2417–2424.

Murphy, A.H., 1991. Forecast verification: its complexity and dimensionality. *Monthly Weather Review*, **119**, 1590–1601.

Murphy, A.H., 1992. Climatology, persistence, and their linear combination as standards of reference in skill scores. *Weather and Forecasting*, **7**, 692–698.

Murphy, A.H., 1993. What is a good forecast? An essay on the nature of goodness in weather forecasting. *Weather and Forecasting*, **8**, 281–293.

Murphy, A.H., 1995. The coefficients of correlation and determination as measures of performance in forecast verification. *Weather and Forecasting* **10**, 681–688.

Murphy, A.H., 1996. The Finley affair: a signal event in the history of forecast verification. *Weather and Forecasting*, **11**, 3–20.

Murphy, A.H., 1997. Forecast verification. In: R.W. Katz and A.H. Murphy, Eds., *Economic Value of Weather and Climate Forecasts*. Cambridge, 19–74.

Murphy, A.H., 1998. The early history of probability forecasts: some extensions and clarifications. *Weather and Forecasting*, **13**, 5–15.

Murphy, A.H., and B.G. Brown, 1983. Forecast terminology: composition and interpretation of public weather forecasts. *Bulletin of the American Meteorological Society*, **64**, 13–22.

Murphy, A.H., B.G. Brown, and Y.-S. Chen, 1989. Diagnostic verification of temperature forecasts. *Weather and Forecasting*, **4**, 485–501.

Murphy, A.H., and H. Daan, 1985. Forecast evaluation. In: A.H. Murphy and R.W. Katz, eds., *Probability, Statistics, and Decision Making in the Atmospheric Sciences*. Boulder, Westview, 379–437.

Murphy, A.H., and E.S. Epstein, 1989. Skill scores and correlation coefficients in model verification. *Monthly Weather Review*, **117**, 572–581.

Murphy, A.H., and D.S. Wilks, 1998. A case study in the use of statistical models in forecast verification: precipitation probability forecasts. *Weather and Forecasting*, **13**, 795–810.

Murphy, A.H., and R.L. Winkler, 1974. Credible interval temperature forecasting: some experimental results. *Monthly Weather Review*, **102**, 784–794.

Murphy, A.H., and R.L. Winkler, 1979. Probabilistic temperature forecasts: the case for an operational program. *Bulletin of the American Meteorological Society*, **60**, 12–19.

Murphy, A.H., and R.L. Winkler, 1984. Probability forecasting in meteorology. *Journal of the American Statistical Association*, **79**, 489–500.

Murphy, A.H., and R.L. Winkler, 1987. A general framework for forecast verification. *Monthly Weather Review*, **115**, 1330–1338.

Murphy, A.H., and R.L. Winkler, 1992. Diagnostic verification of probability forecasts. *International Journal of Forecasting*, **7**, 435–455.

Murphy, A.H., and Q. Ye, 1990. Comparison of objective and subjective precipitation probability forecasts: the sufficiency relation. *Monthly Weather Review*, **118**, 1783–1792.

Mylne, K.R., 2002. Decision-making from probability forecasts based on forecast value. *Meteorological Applications*, **9**, 307–315.

Mylne, K.R., R.E. Evans, and R.T. Clark, 2002a. Multi-model multi-analysis ensembles in quasi-operational medium-range forecasting. *Quarterly Journal of the Royal Meteorological Society*, **128**, 361–384.

Mylne, K.R., C. Woolcock, J.C.W. Denholm-Price, and R.J. Darvell, 2002b. Operational calibrated probability forecasts from the ECMWF ensemble prediction system: implementation and verification. *Preprints, Symposium on Observations, Data Analysis, and Probabilistic Prediction*, (Orlando, Florida), American Meteorological Society, 113–118.

Namias, J., 1952. The annual course of month-to-month persistence in climatic anomalies. *Bulletin of the American Meteorological Society*, **33**, 279–285.

National Bureau of Standards, 1959. *Tables of the Bivariate Normal Distribution Function and Related Functions*. Applied Mathematics Series, **50**, U.S. Government Printing Office, 258 pp.

Nehrkorn, T., R.N. Hoffman, C. Grassotti, and J.-F. Louis, 2003: Feature calibration and alignment to represent model forecast errors: Empirical regularization. *Quarterly Journal of the Royal Meteorological Society*, **129**, 195–218.

Neilley, P.P., W. Myers, and G. Young, 2002. Ensemble dynamic MOS. *Preprints, 16th Conference on Probability and Statistics in the Atmospheric Sciences* (Orlando, Florida), American Meteorological Society, 102–106.

Neter, J., W. Wasserman, and M.H. Kutner, 1996. *Applied Linear Statistical Models*. McGraw-Hill, 1408 pp.

Neumann, C.J., M.B. Lawrence, and E.L. Caso, 1977. Monte Carlo significance testing as applied to statistical tropical cyclone prediction models. *Journal of Applied Meteorology*, **16**, 1165–1174.

Newman, M., and P. Sardeshmukh, 1995. A caveat concerning singular value decomposition. *Journal of Climate*, **8**, 352–360.

Nicholls, N., 1987. The use of canonical correlation to study teleconnections. *Monthly Weather Review*, **115**, 393–399.

North, G.R., 1984. Empirical orthogonal functions and normal modes. *Journal of the Atmospheric Sciences*, **41**, 879–887.

North, G.R., T.L. Bell, R.F. Cahalan, and F.J. Moeng, 1982. Sampling errors in the estimation of empirical orthogonal functions. *Monthly Weather Review*, **110**, 699–706.

O'Lenic, E.A., and R.E. Livezey, 1988. Practical considerations in the use of rotated principal component analysis (RPCA) in diagnostic studies of upper-air height fields. *Monthly Weather Review*, **116**, 1682–1689.

Overland, J.E., and R.W. Preisendorfer, 1982. A significance test for principal components applied to a cyclone climatology. *Monthly Weather Review*, **110**, 1–4.

Paciorek, C.J., J.S. Risbey, V. Ventura, and R.D. Rosen, 2002. Multiple indices of Northern Hemisphere cyclone activity, winters 1949–99. *Journal of Climate*, **15**, 1573–1590.

Palmer, T.N., 1993. Extended-range atmospheric prediction and the Lorenz model. *Bulletin of the American Meteorological Society*, **74**, 49–65.

Palmer, T.N., 2001. A nonlinear dynamical perspective on model error: A proposal for non-local stochastic-dynamic parameterization in weather and climate prediction models. *Quarterly Journal of the Royal Meteorological Society*, **127**, 279–304.

Palmer, T.N., 2002. The economic value of ensemble forecasts as a tool for risk assessment: from days to decades. *Quarterly Journal of the Royal Meteorological Society*, **128**, 747–774.

Palmer, T.N., R. Mureau, and F. Molteni, 1990. The Monte Carlo forecast. *Weather*, **45**, 198–207.

Palmer, T.N., and S. Tibaldi, 1988. On the prediction of forecast skill. *Monthly Weather Review*, **116**, 2453–2480.

Panofsky, H.A., and G.W. Brier, 1958. *Some Applications of Statistics to Meteorology*. Pennsylvania State University, 224 pp.

Peirce, C.S., 1884. The numerical measure of the success of predictions. *Science*, **4**, 453–454.

Penland, C., and P.D. Sardeshmukh, 1995. The optimal growth of tropical sea surface temperatures anomalies. *Journal of Climate*, **8**, 1999–2024.

Peterson, C.R., K.J. Snapper, and A.H. Murphy 1972. Credible interval temperature forecasts. *Bulletin of the American Meteorological Society*, **53**, 966–970.

Pitcher, E.J., 1977. Application of stochastic dynamic prediction to real data. *Journal of the Atmospheric Sciences*, **34**, 3–21.

Pitman, E.J.G., 1937. Significance tests which may be applied to samples from any populations. *Journal of the Royal Statistical Society*, **B4**, 119–130.

Plaut, G., and R. Vautard, 1994. Spells of low-frequency oscillations and weather regimes in the Northern Hemisphere. *Journal of the Atmospheric Sciences*, **51**, 210–236.

Preisendorfer, R.W., 1988. *Principal Component Analysis in Meteorology and Oceanography*. C.D. Mobley, ed. Elsevier, 425 pp.

Preisendorfer, R.W., and T.P. Barnett, 1983. Numerical-reality intercomparison tests using small-sample statistics. *Journal of the Atmospheric Sciences*, **40**, 1884–1896.

Preisendorfer, R.W., and C.D. Mobley, 1984. Climate forecast verifications, United States Mainland, 1974–83. *Monthly Weather Review*, **112**, 809–825.

Preisendorfer, R.W., F.W. Zwiers, and T.P. Barnett, 1981. *Foundations of Principal Component Selection Rules*. SIO Reference Series 81-4, Scripps Institution of Oceanography, 192 pp.

Press, W.H., B.P. Flannery, S.A. Teukolsky, and W.T. Vetterling, 1986. *Numerical Recipes: The Art of Scientific Computing*. Cambridge University Press, 818 pp.

Quayle, R. and W. Presnell, 1991. *Climatic Averages and Extremes for U.S. Cities*. Historical Climatology Series 6-3, National Climatic Data Center, Asheville, NC. 270 pp.

Rajagopalan, B., U. Lall, and D.G. Tarboton, 1997. Evaluation of kernel density estimation methods for daily precipitation resampling. *Stochastic Hydrology and Hydraulics*, **11**, 523–547.

Richardson, C.W., 1981. Stochastic simulation of daily precipitation, temperature, and solar radiation. *Water Resources Research*, **17**, 182–190.

Richardson, D.S., 2000. Skill and economic value of the ECMWF ensemble prediction system. *Quarterly Journal of the Royal Meteorological Society*, **126**, 649–667.

Richardson, D.S., 2001. Measures of skill and value of ensemble predictions systems, their interrelationship and the effect of ensemble size. *Quarterly Journal of the Royal Meteorological Society*, **127**, 2473–2489.

Richardson, D.S., 2003. Economic value and skill. In: I.T. Jolliffe and D.B. Stephenson, eds., *Forecast Verification*. Wiley, 165–187.

Richman, M.B., 1986. Rotation of principal components. *Journal of Climatology*, **6**, 293–335.

Roebber, P.J., and L.F. Bosart, 1996. The complex relationship between forecast skill and forecast value: a real-world analysis. *Weather and Forecasting*, **11**, 544–559.

Romesburg, H.C., 1984. *Cluster Analysis for Researchers*. Wadsworth/Lifetime Learning Publications, 334 pp.

Ropelewski, C.F., and P.D. Jones, 1987. An extension of the Tahiti-Darwin Southern Oscillation index. *Monthly Weather Review*, **115**, 2161–2165.

Rosenberger, J.L., and M. Gasko, 1983. Comparing location estimators: trimmed means, medians, and trimean. In: D.C. Hoaglin, F. Mosteller, and J.W. Tukey, eds., *Understanding Robust and Exploratory Data Analysis*. New York, Wiley, 297–338.

Roulston, M.S., and L.A. Smith, 2003. Combining dynamical and statistical ensembles. *Tellus*, **55A**, 16–30.

Sansom, J. and P.J. Thomson, 1992. Rainfall classification using breakpoint pluviograph data. *Journal of Climate*, **5**, 755–764.

Santer, B.D., T.M.L. Wigley, J.S. Boyle, D.J. Gaffen, J.J. Hnilo, D. Nychka, D.E. Parker, and K.E. Taylor, 2000. Statistical significance of trends and trend differences in layer-average atmospheric temperature series. *Journal of Geophysical Research*, **105**, 7337–7356.

Saravanan, R., and J.C. McWilliams, 1998. Advective ocean-atmosphere interaction: an analytical stochastic model with implications for decadal variability. *Journal of Climate*, **11**, 165–188.

Sauvageot, H., 1994. Rainfall measurement by radar: a review. *Atmospheric Research*, **35**, 27–54.

Schaefer, J.T., 1990. The critical success index as an indicator of warning skill. *Weather and Forecasting*, **5**, 570–575.

Schwarz, G., 1978. Estimating the dimension of a model. *Annals of Statistics*, **6**, 461–464.

Scott, D.W., 1992. *Multivariate Density Estimation*. Wiley, 317 pp.

Shapiro, S.S., and M.B. Wilk, 1965. An analysis of variance test for normality (complete samples). *Biometrika*, **52**, 591–610.

Sharma, A., U. Lall, and D.G. Tarboton, 1998. Kernel bandwidth selection for a first order nonparametric streamflow simulation model. *Stochastic Hydrology and Hydraulics*, **12**, 33–52.

Sheets, R.C., 1990. The National Hurricane Center—past, present and future. *Weather and Forecasting*, **5**, 185–232.

Silverman, B.W., 1986. *Density Estimation for Statistics and Data Analysis*, Chapman and Hall, 175 pp.

Smith, L.A., 2001. Disentangling uncertainty and error: on the predictability of nonlinear systems. In: A.I. Mees, ed., *Nonlinear Dynamics and Statistics*. Birkhauser, 31–64.

Smith, L.A., and J.A. Hansen, 2004. Extending the limits of ensemble forecast verification with the minimum spanning tree. *Monthly Weather Review*, **132**, 1522–1528.

Smith, R.L., 1989. Extreme value analysis of environmental time series: An application to trend detection in ground-level ozone. *Statistical Science*, **4**, 367–393.

Smyth, P., K. Ide, and M. Ghil, 1999. Multiple regimes in Northern Hemisphere height fields via mixture model clustering. *Journal of the Atmospheric Sciences*, **56**, 3704–3723.

Solow, A.R., 1985. Bootstrapping correlated data. *Mathematical Geology*, **17**, 769–775.

Solow, A.R., and L. Moore, 2000. Testing for a trend in a partially incomplete hurricane record. *Journal of Climate*, **13**, 3696–3710.

Spetzler, C.S., and C.-A. S. Staël von Holstein, 1975. Probability encoding in decision analysis. *Management Science*, **22**, 340–358.

Sprent, P., and N.C. Smeeton, 2001. *Applied Nonparametric Statistical Methods*. Chapman and Hall, 461 pp.

Staël von Holstein, C-.A. S., and A.H. Murphy, 1978. The family of quadratic scoring rules. *Monthly Weather Review*, **106**, 917–924.

Stanski, H.R., L.J. Wilson, and William R. Burrows, 1989. *Survey of Common Verification Methods in Meteorology*. World Weather Watch Technical Report No. 8, World Meteorological Organization, TD No. 358, 114 pp.

Stedinger, J.R., R.M. Vogel, and E. Foufoula-Georgiou, 1993. Frequency analysis of extreme events. In: D.R. Maidment, ed., *Handbook of Hydrology*. McGraw-Hill, 66 pp.

Stensrud, D.J., J.-W. Bao, and T.T. Warner, 2000. Using initial conditions and model physics perturbations in short-range ensemble simulations of mesoscale convective systems. *Monthly Weather Review*, **128**, 2077–2107.

Stensrud, D.J., H.E. Brooks, J. Du, M.S. Tracton, and E. Rogers, 1999. Using ensembles for short-range forecasting. *Monthly Weather Review*, **127**, 433–446.

Stensrud, D.J., and M.S. Wandishin, 2000. The correspondence ratio in forecast evaluation. *Weather and Forecasting*, **15**, 593–602.

Stephens, M., 1974. E.D.F. statistics for goodness of fit. *Journal of the American Statistical Association*, **69**, 730–737.

Stephenson, D.B., 2000. Use of the "odds ratio" for diagnosing forecast skill. *Weather and Forecasting*, **15**, 221–232.

Stephenson, D.B., and F.J. Doblas-Reyes, 2000. Statistical methods for interpreting Monte-Carlo ensemble forecasts. *Tellus*, **52A**, 300–322.

Stephenson, D.B., and I.T. Jolliffe, 2003. Forecast verification: past, present, and future. In: I.T. Jolliffe and D.B. Stephenson, eds., *Forecast Verification*. Wiley, 189–201.

Stern, R.D., and R. Coe, 1984. A model fitting analysis of daily rainfall data. *Journal of the Royal Statistical Society*, **A147**, 1–34.

Strang, G., 1988. *Linear Algebra and its Applications*. Harcourt, 505 pp.

Stull, R.B., 1988. *An Introduction to Boundary Layer Meteorology*. Kluwer, 666 pp.

Swets, J.A., 1973. The relative operating characteristic in psychology. *Science*, **182**, 990–1000.

Swets, J.A., 1979. ROC analysis applied to the evaluation of medical imaging techniques. *Investigative Radiology*, **14**, 109–121.

Talagrand, O., R. Vautard, and B. Strauss, 1997. Evaluation of probabilistic prediction systems. *Proceedings, ECMWF Workshop on Predictability*. ECMWF, 1–25.

Tezuka, S., 1995. *Uniform Random Numbers: Theory and Practice*. Kluwer, 209 pp.

Thiébaux, H.J., and M.A. Pedder, 1987. *Spatial Objective Analysis: with Applications in Atmospheric Science*. London, Academic Press, 299 pp.

Thiébaux, H.J., and F.W. Zwiers, 1984. The interpretation and estimation of effective sample size. *Journal of Climate and Applied Meteorology*, **23**, 800–811.

Thom, H.C.S., 1958. A note on the gamma distribution. *Monthly Weather Review*, **86**, 117–122.

Thompson, C.J., and D.S. Battisti, 2001. A linear stochastic dynamical model of ENSO. Part II: Analysis. *Journal of Climate*, **14**, 445–466.

Thompson, J.C., 1962. Economic gains from scientific advances and operational improvements in meteorological prediction. *Journal of Applied Meteorology*, **1**, 13–17.

Thornes, J.E., and D.B. Stephenson, 2001. How to judge the quality and value of weather forecast products. *Meteorological Applications*, **8**, 307–314.

Titterington, D.M., A.F.M. Smith, and U.E. Makov, 1985. *Statistical Analysis of Finite Mixture Distributions*. Wiley, 243 pp.

Todorovic, P., and D.A. Woolhiser, 1975. A stochastic model of n-day precipitation. *Journal of Applied Meteorology*, **14**, 17–24.

Tong, H., 1975. Determination of the order of a Markov chain by Akaike's Information Criterion. *Journal of Applied Probability*, **12**, 488–497.

Toth, Z., and E. Kalnay, 1993. Ensemble forecasting at NMC: the generation of perturbations. *Bulletin of the American Meteorological Society*, **74**, 2317–2330.

Toth, Z., and E. Kalnay, 1997. Ensemble forecasting at NCEP the breeding method. *Monthly Weather Review*, **125**, 3297–3318.

Toth, Z., E. Kalnay, S.M. Tracton, R.Wobus, and J. Irwin, 1997. A synoptic evaluation of the NCEP ensemble. *Weather and Forecasting*, **12**, 140–153.

Toth, Z., I. Szunyogh, E. Kalnay, and G. Iyengar, 1999. Comments on: "Notes on the appropriateness of 'bred modes' for generating initial perturbations." *Tellus*, **51A**, 442–449.

Toth, Z., O. Talagrand, G. Candille and Y. Zhu, 2003. Probability and Ensemble Forecasts. In: I.T. Jolliffe and D.B. Stephenson, eds., *Forecast Verification*. Wiley, 137–163.

Toth, Z., Y. Zhu, and T. Marchok, 2001. The use of ensembles to identify forecasts with small and large uncertainty. *Weather and Forecasting*, **16**, 463–477.

Tracton, M.S., and E. Kalnay, 1993. Operational ensemble prediction at the National Meteorological Center: practical aspects. *Weather and Forecasting*, **8**, 379–398.

Tracton, M.S., K. Mo, W. Chen, E. Kalnay, R. Kistler, and G. White, 1989. Dynamical extended range forecasting (DERF) at the National Meteorological Center. *Monthly Weather Review*, **117**, 1604–1635.

Tukey, J.W., 1977. *Exploratory Data Analysis*. Reading, Mass., Addison-Wesley, 688 pp.

Unger, D.A., 1985. A method to estimate the continuous ranked probability score. *Preprints, 9th Conference on Probability and Statistics in the Atmospheric Sciences*. American Meteorological Society, 206–213.

Valée, M., L.J. Wilson, and P. Bourgouin, 1996. New statistical methods for the interpretation of NWP output and the Canadian meteorological center*Preprints*, 13*th* *Conference on Probability and Statistics in the Atmospheric Sciences* (San Francisco, California). American Meteorological Society, 37–44.

Vautard, R., 1995. Patterns in Time: SSA and MSSA. In: H. von Storch and A. Navarra eds., *Analysis of Climate Variability*. Springer, 259–279.

Vautard, R., C. Pires, and G. Plaut, 1996. Long-range atmospheric predictability using space-time principal components. *Monthly Weather Review*, **124**, 288–307.

Vautard, R., G. Plaut, R. Wang, and G. Brunet, 1999. Seasonal prediction of North American surface air temperatures using space-time principal components. *Journal of Climate*, **12**, 380–394.

Vautard, R., P. Yiou, and M. Ghil, 1992. Singular spectrum analysis: a toolkit for short, noisy and chaotic series. *Physica D*, **58**, 95–126.

Velleman, P.F., 1988. *Data Desk*. Ithaca, NY, Data Description, Inc.

Velleman, P.F., and D.C. Hoaglin, 1981. *Applications, Basics, and Computing of Exploratory Data Analysis*. Boston, Duxbury Press, 354 pp.

Vislocky, R.L., and J.M. Fritsch, 1997. An automated, observations-based system for short-term prediction of ceiling and visibility. *Weather and Forecasting*, **12**, 31–43.

Vogel, R.M., 1986. The probability plot correlation coefficient test for normal, lognormal, and Gumbel distributional hypotheses. *Water Resources Research*, **22**, 587–590.

Vogel, R.M., and C.N. Kroll, 1989. Low-flow frequency analysis using probability-plot correlation coefficients. *Journal of Water Resource Planning and Management*, **115**, 338–357.

Vogel, R.M., and D.E. McMartin, 1991. Probability-plot goodness-of-fit and skewness estimation procedures for the Pearson type III distribution. *Water Resources Research*, **27**, 3149–3158.

von Storch, H., 1982. A remark on Chervin-Schneider's algorithm to test significance of climate experiments with GCMs. *Journal of the Atmospheric Sciences*, **39**, 187–189.

von Storch, H., 1995. Misuses of statistical analysis in climate research. In: H. von Storch and A. Navarra, eds., *Analysis of Climate Variability*. Springer, 11–26.

von Storch, H., and G. Hannoschöck, 1984. Comments on "empirical orthogonal function-analysis of wind vectors over the tropical Pacific region." *Bulletin of the American Meteorological Society*, **65**, 162.

von Storch, H., and F.W. Zwiers, 1999. *Statistical Analysis in Climate Research*. Cambridge, 484 pp.

Wallace, J.M., and M.L. Blackmon, 1983. Observations of low-frequency atmospheric variability. In: B.J. Hoskins and R.P. Pearce, *Large-Scale Dynamical Processes in the Atmosphere*. Academic Press, 55–94.

Wallace, J.M., and D.S. Gutzler, 1981. Teleconnections in the geopotential height field during the northern hemisphere winter*Monthly Weather Review*, **109**, 784–812.

Wallace, J.M., C. Smith, and C.S. Bretherton, 1992. Singular value decomposition of wintertime sea surface temperature and 500-mb height anomalies. *Journal of Climate*, **5**, 561–576.

Walshaw, D., 2000. Modeling extreme wind speeds in regions prone to hurricanes. *Applied Statistics*, **49**, 51–62.

Wandishin, M.S., and H.E. Brooks, 2002. On the relationship between Clayton's skill score and expected value for forecasts of binary events. *Meteorological Applications*, **9**, 455–459.

Wang, X., and C.H. Bishop, 2005. Improvement of ensemble reliability with a new dressing kernel. *Quarterly Journal of the Royal Meteorological Society*, **131**, 965–986.

Ward, M.N., and C.K. Folland, 1991. Prediction of seasonal rainfall in the north Nordeste of Brazil using eigenvectors of sea-surface temperature. *International Journal of Climatology*, **11**, 711–743.

Watson, J.S., and S.J. Colucci, 2002. Evaluation of ensemble predictions of blocking in the NCEP global spectral model. *Monthly Weather Review*, **130**, 3008–3021.

Waymire, E., and V.K. Gupta, 1981. The mathematical structure of rainfall representations. 1. A review of stochastic rainfall models. *Water Resources Research*, **17**, 1261–1272.

Whitaker, J.S., and A.F. Loughe, 1998. The relationship between ensemble spread and ensemble mean skill. *Monthly Weather Review*, **126**, 3292–3302.

Wigley, T.M.L, and B.D. Santer, 1990. Statistical comparison of spatial fields in model validation, perturbation, and predictability experiments *Journal of Geophysical Research*, **D95**, 851–865.

Wilks, D.S., 1989. Conditioning stochastic daily precipitation models on total monthly precipitation. *Water Resources Research*, **25**, 1429–1439.

Wilks, D.S., 1990. Maximum likelihood estimation for the gamma distribution using data containing zeros. *Journal of Climate*, **3**, 1495–1501.

Wilks, D.S., 1992. Adapting stochastic weather generation algorithms for climate change studies. *Climatic Change*, **22**, 67–84.

Wilks, D.S., 1993. Comparison of three-parameter probability distributions for representing annual extreme and partial duration precipitation series. *Water Resources Research*, **29**, 3543–3549.

Wilks, D.S., 1997a. Forecast value: prescriptive decision studies. In: R.W. Katz and A.H. Murphy, eds., *Economic Value of Weather and Climate Forecasts*. Cambridge, 109–145.

Wilks, D.S., 1997b. Resampling hypothesis tests for autocorrelated fields. *Journal of Climate*, **10**, 65–82.

Wilks, D.S., 1998. Multisite generalization of a daily stochastic precipitation generalization model. *Journal of Hydrology*, **210**, 178–191.

Wilks, D.S., 1999a. Interannual variability and extreme-value characteristics of several stochastic daily precipitation models. *Agricultural and Forest Meteorology*, **93**, 153–169.

Wilks, D.S., 1999b. Multisite downscaling of daily precipitation with a stochastic weather generator. *Climate Research*, **11**, 125–136.

Wilks, D.S., 2001. A skill score based on economic value for probability forecasts. *Meteorological Applications*, **8**, 209–219.

Wilks, D.S., 2002a. Realizations of daily weather in forecast seasonal climate. *Journal of Hydrometeorology*, **3**, 195–207.

Wilks, D.S., 2002b. Smoothing forecast ensembles with fitted probability distributions. *Quarterly Journal of the Royal Meteorological Society*, **128**, 2821–2836.

Wilks, D.S., 2004. The minimum spanning tree histogram as a verification tool for multidimensional ensemble forecasts. *Monthly Weather Review*, **132**, 1329–1340.

Wilks, D.S., 2005. Effects of stochastic parametrizations in the Lorenz '96 system. *Quarterly Journal of the Royal Meteorological Society*, **131**, 389–407.

Wilks, D.S., and K.L. Eggleson, 1992. Estimating monthly and seasonal precipitation distributions using the 30- and 90-day outlooks. *Journal of Climate*, **5**, 252–259.

Wilks, D.S., and C.M. Godfrey, 2002. Diagnostic verification of the IRI new assessment forecasts, 1997–2000. *Journal of Climate*, **15**, 1369–1377.

Wilks, D.S., and R.L. Wilby, 1999. The weather generation game: a review of stochastic weather models. *Progress in Physical Geography*, **23**, 329–357.

Williams, P.D., P.L. Read, and T.W.N. Haine, 2003. Spontaneous generation and impact of intertia-gravity waves in a stratified, two-layer shear flow. *Geophysical Research Letters*, **30**, 2255–2258.

Wilmott, C.J., S.G. Ackleson, R.E. Davis, J.J. Feddema, K.M. Klink, D.R. Legates, J.O'Donnell, and C.M. Rowe, 1985. Statistics for the evaluation and comparison of models. *Journal of Geophysical Research*, **90**, 8995–9005.

Wilson, L.J., W.R. Burrows, and A. Lanzinger, 1999. A strategy for verification of weather element forecasts from an ensemble prediction system. *Monthly Weather Review*, **127**, 956–970.

Wilson, L.J., and M. Vallée, 2002. The Canadian updateable model output statistics (UMOS) system: design and development tests. *Weather and Forecasting*, **17**, 206–222.

Wilson, L.J., and M. Vallée, 2003. The Canadian updateable model output statistics (UMOS) system: validation against perfect prog. *Weather and Forecasting*, **18**, 288–302.

Winkler, R.L., 1972a. A decision-theoretic approach to interval estimation. *Journal of the American Statistical Association*, **67**, 187–191.

Winkler, R.L., 1972b. *Introduction to Bayesian Inference and Decision*. New York, Holt, Rinehart and Winston, 563 pp.

Winkler, R.L., 1994. Evaluating probabilities: asymmetric scoring rules. *Management Science*, **40**, 1395–1405.

Winkler, R.L., 1996. Scoring rules and the evaluation of probabilities. *Test*, **5**, 1–60.

Winkler, R.L., and A.H. Murphy, 1979. The use of probabilities in forecasts of maximum and minimum temperatures. *Meteorological Magazine*, **108**, 317–329.

Winkler, R.L., and A.H. Murphy, 1985. Decision analysis. In: A.H. Murphy and R.W. Katz, eds., *Probability, Statistics and Decision Making in the Atmospheric Sciences*. Westview, 493–524.

Wolter, K., 1987. The southern oscillation in surface circulation and climate over the tropical Atlantic, eastern Pacific, and Indian Oceans as captured by cluster analysis. *Journal of Climate and Applied Meteorology*, **26**, 540–558.

Woodcock, F., 1976. The evaluation of yes/no forecasts for scientific and administrative purposes. *Monthly Weather Review*, **104**, 1209–1214.

Woolhiser, D.A., and J. Roldan, 1982. Stochastic daily precipitation models, 2. A comparison of distributions of amounts. *Water Resources Research*, **18**, 1461–1468.

Yeo, I.-K., and R.A. Johnson, 2000. A new family of power transformations to improve normality or symmetry. *Biometrika*, **87**, 954–959.

Young, M.V., and E.B. Carroll, 2002. Use of medium-range ensembles at the Met Office 2: applications for medium-range forecasting. *Meteorological Applications*, **9**, 273–288.

Yue, S., and C.-Y. Wang, 2002. The influence of serial correlation in the Mann-Whitney test for detecting a shift in median. *Advances in Water Research*, **25**, 325–333.

Yule, G.U., 1900. On the association of attributes in statistics. *Philosophical Transactions of the Royal Society, London*, **194A**, 257–319.

Yuval, and W.W. Hsieh, 2003. An adaptive nonlinear MOS scheme for precipitation forecasts using neural networks. *Weather and Forecasting*, **18**, 303–310.

Zeng, X., R.A. Pielke and R. Eykhold, 1993. Chaos theory and its application to the atmosphere. *Bulletin of the American Meteorological Society*, **74**, 631–644.

Zepeda-Arce, J., E. Foufoula-Georgiou, and K.K. Droegemeier, 2000. Space-time rainfall organization and its role in validating quantitative precipitation forecasts. *Journal of Geophysical Research*, **D105**, 10129–10146.

Zhang, P., 1993. Model selection via multifold cross validation. *Annals of Statistics*, **21**, 299–313.

Zheng, X., R.E. Basher, and C.S. Thomson, 1997. Trend detection in regional-mean temperature series: maximum, minimum, mean, diurnal range, and SST. *Journal of Climate*, **10**, 317–326.

Zhang, X., F.W. Zwiers, and G. Li, 2004. Monte Carlo experiments on the detection of trends in extreme values. *Journal of Climate*, **17**, 1945–1952.

Ziehmann, C., 2001. Skill prediction of local weather forecasts based on the ECMWF ensemble. *Nonlinear Processes in Geophysics*, **8**, 419–428.

Zwiers, F.W., 1987. Statistical considerations for climate experiments. Part II: Multivariate tests. *Journal of Climate and Applied Meteorology*, **26**, 477–487.

Zwiers, F.W., 1990. The effect of serial correlation on statistical inferences made with resampling procedures. *Journal of Climate*, **3**, 1452–1461.

Zwiers, F.W., and H.J. Thiébaux, 1987. Statistical considerations for climate experiments. Part I: scalar tests. *Journal of Climate and Applied Meteorology*, **26**, 465–476.

Zwiers, F.W., and H. von Storch, 1995. Taking serial correlation into account in tests of the mean. *Journal of Climate*, **8**, 336–351.

Index

International Geophysics Series

EDITED BY

RENATA DMOWSKA

Division of Applied Science
Harvard University
Cambridge, Massachusetts

DENNIS HARTMANN

Department of Atmospheric Sciences
University of Washington
Seattle, Washington

H. THOMAS ROSSBY

Graduate School of Oceanography
University of Rhode Island
Narragansett, Rhode Island

* Out of print